Principles of
Physical Chemistry
Fourth Edition

Principles of
Physical Chemistry

Fourth Edition

Samuel H Maron
Professor of Physical Chemistry and
Polymer Science, Case Institute of Technology

Carl F Prutton
Former Professor of Chemistry and
Chemical Engineering, Case Institute of Technology

Oxford & IBH Publishing Co. Pvt. Ltd.
New Delhi
(A Unit of CBS Publishers & Distributors Pvt Ltd)

CBS Publishers & Distributors Pvt Ltd

New Delhi • Bengaluru • Chennai • Kochi • Kolkata • Mumbai
Bhopal • Bhubaneswar • Hyderabad • Jharkhand • Nagpur • Patna
• Pune • Uttarakhand • Dhaka (Bangladesh) • Kathmandu (Nepal)

Principles of
Physical Chemistry
Fourth Edition

ISBN-13: 978-81-204-1758-8
ISBN-10: 81-204-1758-5

CBS Reprint 2017, 2020

OXFORD & IBH
New Delhi
(A Unit of CBS Publishers & Distributors Pvt Ltd)

Published by Satish Kumar Jain and produced by Varun Jain for
CBS Publishers & Distributors Pvt Ltd
4819/XI Prahlad Street, 24 Ansari Road, Daryaganj, New Delhi 110 002, India.
Ph: 23289259, 23266861, 23266867   Fax: 011-23243014   Website: www.cbspd.com
e-mail: delhi@cbspd.com; cbspubs@airtelmail.in.

Corporate Office: 204 FIE, Industrial Area, Patparganj, Delhi 110 092
Ph: 011-4934 4934   Fax: 011-4934 4935   e-mail:publishing@cbspd.com;
                                                                  publicity@cbspd.com

Branches
- **Bengaluru:** Seema House 2975, 17th Cross, K.R. Road, Banasankari 2nd Stage, Bengaluru 560 070, Karnataka, India
  Ph: +91-80-26771678/79   Fax: +91-80-26771680   e-mail: bangalore@cbspd.com
- **Chennai:** 7, Subbaraya Street, Shenoy Nagar, Chennai 600 030, Tamil Nadu, India
  Ph: +91-44-26260666, 26208620   Fax: +91-44-42032115   e-mail: chennai@cbspd.com
- **Kochi:** 42/1325, 1326, Power House Road, Opp KSEB, Kochi 682 018, Kerala
  Ph: +91-484-4059061-65   Fax: +91-484-4059065   e-mail: kochi@cbspd.com
- **Kolkata:** 6/B, Ground Floor, Rameswar Shaw Road, Kolkata-700014 (West Bengal), India
  Ph: +91-33-2289-1126, 2289-1127, 2289-1128   e-mail: kolkata@cbspd.com
- **Mumbai:** 83-C, Dr E Moses Road, Worli, Mumbai-400018, Maharashtra, India
  Ph: +91-22-24902340/41   Fax: +91-22-24902342   e-mail: mumbai@cbspd.com

Representatives

Bhopal	0-8319310552	Bhubaneswar	0-9911037372	Hyderabad	0-9885175004
Jharkhand	0-9811541605	Nagpur	0-9421945513	Patna	0-9334159340
Pune	0-9623451994	Uttarakhand	0-9716462459	Dhaka	01912-003485
Kathmandu (Nepal)	977-9818742655			(Bangladesh)	

Printed at Chaman Enterprises, Daryaganj, New Delhi, India

Preface

I N THE PREFACE to the first edition, the authors stated that their aim in writing a text on physical chemistry was to place in the hands of teachers and students a book covering the fundamental principles of the subject in a thorough, sound, up-to-date, and clear manner. To preserve this aim in times of rapid expansion of knowledge and modified philosophy of course content, extensive changes had to be made in the subject matter and in the order of presentation. Specifically, the changes were directed toward the following major ends: (1) more rigorous, extensive, and consecutive treatment of thermodynamics; (2) more complete discussion of atomic and molecular structure in quantum-mechanical terms; and (3) introduction of basic ideas of statistical mechanics. How these ends were attained can be judged by perusal of the table of contents and of the topics included in each chapter. Extensive rearrangements, amplification, and rewriting of older material were necessary, as was the preparation of completely new chapters on atomic structure, nature of chemical bonding, investigation of molecular structure, nuclear chemistry, and statistical mechanics. At the same time the number of problems was increased from 634 to over 800, all nonrepetitive and many more challenging than before.

The book as a whole is intended primarily for a solid, full-year course in physical chemistry. However, by judicious selection of portions of the contents, the book can also be adapted to a one-semester course. What is to be omitted and what is to be covered is largely a matter of circumstance and need, and can be safely left to the discretion and decision of the individual teacher.

The authors wish to express their appreciation to the colleagues, friends, and teachers who were kind enough to offer suggestions and criticisms for improvement of the text, and who called attention to the inevitable errors which, despite all efforts, still managed to escape detection.

S. H. M.
C. F. P.

Contents

Introduction

THE AREA of chemistry which concerns itself with the study of the physical properties and structure of matter, with the laws of chemical interaction, and with the theories governing these is called *physical chemistry*. The purpose of physical chemistry is, first, to collect the appropriate data required to define the properties of gases, liquids, solids, solutions, and colloidal dispersions, to systematize them into laws, and to give them a theoretical foundation. Next, physical chemistry is interested in establishing the energy relations obtaining in physical and chemical transformations, in ascertaining the extent and speed with which they take place, and in defining quantitatively the controlling factors. In this connection must be considered not only the more common variables of temperature, pressure, and concentration, but also the effects of the intimate interaction of matter with electricity and light. Finally, matter itself must be examined to determine its nature and structure. This is necessary in order that we may be able to arrive at a basic understanding of physical and chemical behavior in terms of the properties of the fundamental constituents of matter.

To accomplish its purposes physical chemistry must rely to a large degree on experiment. Experimental methods and techniques thus play a very important role. The subject also draws generously on the laws and methods of physics and mathematics. In fact, physical chemistry may be looked upon as the field where physics and mathematics are applied extensively to the study and solution of problems of prime chemical interest. With the appropriate data at hand, physical chemistry then proceeds to its correlational and theoretical goal through two general modes of attack, namely, the *thermodynamic* and the *kinetic*. In the thermodynamic approach the fundamental laws of thermodynamics are utilized to yield deductions based on the energy relations connecting the initial and final stages of a process. By circumventing the steps intervening between the start and end of a process, thermodynamics enables us to arrive at many valuable deductions without our knowing all the intimate details of the intermediate stages. Consequently, although this approach is able to tell us what can happen, and to what extent, it is unable, by its very nature, to give us information on *how*, or *how rapidly*, a change will actually occur. On the other hand, the kinetic approach requires for its operation an intimate and detailed "picture" of the process. From the mechanism postulated then may be deduced the law for the over-all process and its various stages. Evidently the kinetic approach to a problem is more explanatory in character, but unfortunately it is generally more complicated and difficult to apply. These two modes of attack will be illustrated at various stages in the text. From the examples given there the student will be able to differentiate more clearly between them and come to appreciate their respective powers and utilities.

The roots of physical chemistry lie in the fields of both chemistry and physics. At first these two branches of science developed more or less independently. However, in the nineteenth century it was found that the discoveries in physics had important bearing on and application to chemistry, and hence need arose for a field dealing primarily with the application of physical laws to chemical phenomena. This need finally impelled Wilhelm Ostwald, van't Hoff, and Arrhenius to organize and systematize the subject matter generally included now under the head of physical chemistry, and led them in 1881 to found the *Zeitschrift für physikalische Chemie*. The inception of physical chemistry as a formal branch of chemical science may be dated from the appearance of this journal.

Stimulated by this publication, and fostered by the contributions of the men mentioned, physical chemistry entered a period of very rapid growth. Aiding this progress were not only advances in chemistry, but also the remarkable series of discoveries in physics which started with the discovery of the electron, and which include the discovery of X rays and radioactivity, the establishment of the quantum theory, and the unfolding of our understanding of subatomic phenomena. Thanks to these contributions, physical chemistry has developed to a position of importance and utility not only to chemistry but to other sciences as well.

Since physical chemistry deals with the principles and theories of chemistry, it goes without saying that any student or practitioner of this science must be familiar with physical chemistry in order to understand his own subject. The same applies also to the chemical engineer. The essential difference between a chemist and a chemical engineer is that whereas the former conducts his reactions and operations on a small scale, the chemical engineer carries them out in large commercial units. To transfer an operation from the laboratory to a plant the chemical engineer must of course be able to apply engineering and economic principles. However, at the same time he must understand also the fundamentally chemical nature of the processes he is dealing with, and for that he needs physical chemistry. As a matter of fact, chemical engineering has frequently been described as applied physical chemistry. Viewed in this light, many of the aspects of chemical engineering fall within the realm of physical chemistry and can be handled in terms of well-established and familiar physicochemical laws. On the other hand, any attempt to consider chemical engineering as a purely empirical pursuit robs it of the attributes of a science and translates it into an art.

What has been said about the importance of physical chemistry to the chemist and chemical engineer applies equally well to the metallurgist and metallurgical engineer. The latter two perform essentially the same functions as the two former, except that their attention is confined primarily to metals. From this point of view the prominent position of physical chemistry, whether under this or other titles, in these subjects becomes clear, and accounts for the valuable contributions made to these fields by the application of physicochemical principles.

Finally, physical chemistry finds application also in physics, geology, and in the various ramifications of the biological sciences. To appreciate the extent of its utility it is only necessary to compare a book on chemical physics, geology, or biochemistry with one on physical chemistry. From such a comparison it becomes quite evident why physical chemistry is often included in curricula in these subjects, and why it can be applied with effect in these sciences.

1
Gases and Liquids

ALL MATTER exists in one of three states of aggregation, solid, liquid, or gaseous. A solid may be defined as a body possessing both definite volume and definite shape at a given temperature and pressure. Further, to be classed as a solid the body must be crystalline, i.e., the atoms, molecules, or ions composing the body must be arranged in a definite geometric configuration characteristic of the substance in question. A liquid in bulk, on the other hand, has a definite volume but no definite shape, while a gas has neither definite shape nor volume. Liquids and gases are both termed *fluids*. A liquid, insofar as it fills the container, will always adopt the shape of the container in which it is placed, but will retain its definite volume, while a gas will always fill completely any container in which it may be confined.

The distinctions among the three states of matter are not always as clear cut as the above definitions would imply. For example, a liquid at the critical point is indistinguishable from its vapor. Again, such substances as glass or asphalt, although exhibiting many of the properties of a solid, will, under certain conditions of temperature, become plastic and exhibit properties not ascribed to pure solids. For this reason such substances are usually considered to be supercooled liquids with very high viscosity.

The particular state of aggregation of a substance is determined by the temperature and pressure under which it exists. However, within certain limits of temperature and pressure a substance may exist in more than one state at the same time. In fact, under special conditions a substance may exist in all three states simultaneously. Thus at 4.57 mm Hg pressure and at 0.010°C, ice, water, and water vapor may all be present simultaneously, and all be stable. This subject of simultaneous existence in more than one state will be discussed more completely in various places in the book.

IDEAL AND REAL GASES

For purposes of discussion, it is convenient to classify all gases into two types, namely, (a) *ideal* gases, and (b) *nonideal* or *real* gases. An ideal gas is one that obeys certain laws which will be presented shortly, while a real gas is one that obeys these laws only at low pressures.

In ideal gases the volume occupied by the molecules themselves is negligible compared with the total volume at all pressures and temperatures, and the intermolecular attraction is extremely small under all conditions. For nonideal or real gases both of these factors are appreciable, the magnitude of each depending on the nature, the temperature, and the pressure of the gas. We can easily see that an ideal gas must be a hypothetical gas, as all actual gases must contain molecules which occupy a definite volume and exert attractions between each other. However, very often the influence of these factors becomes negligible, and the gas may then be considered to be ideal. We shall find that the latter condition will obtain in particular at low pressures and relatively high temperatures, conditions under which the "free" space within the gas is large and the attractive forces between molecules small.

GENERALIZATIONS OF IDEAL GAS BEHAVIOR

Through the study of gases there have been evolved certain laws or generalizations which are the starting point in any discussion of gas behavior. These are: (a) Boyle's law, (b) Charles's or Gay-Lussac's law, (c) Dalton's law of partial pressures, and (d) Graham's law of diffusion. Another generalization is Avogadro's principle, but this will be considered later.

BOYLE'S LAW

In 1662 Robert Boyle reported that the volume of a gas at constant temperature decreased with increasing pressure, and that, within the limits of his experimental accuracy, *the volume of any definite quantity of gas at constant temperature varied inversely as the pressure on the gas.* This highly important generalization is known as *Boyle's law.* Expressed mathematically, this law states that at *constant temperature* $V \propto 1/P$, or that

$$V = \frac{K_1}{P}$$

where V is the volume and P the pressure of the gas, while K_1 is a proportionality factor whose value is dependent on the temperature, the weight

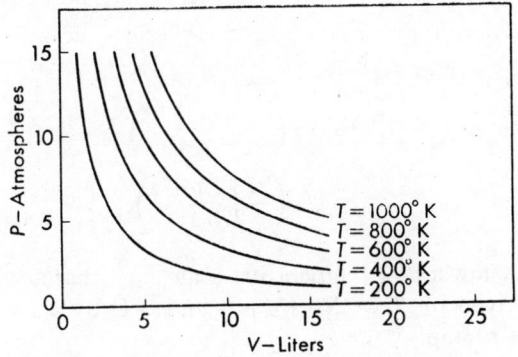

Figure 1-1. Isothermal plot of P vs. V according to Boyle's law (1 mole of gas).

of the gas, its nature, and the units in which P and V are expressed. On rearrangement this equation becomes

$$PV = K_1 \tag{1}$$

from which it follows that if in a certain state the pressure and volume of the gas are P_1 and V_1, while in another state they are P_2 and V_2, then at constant temperature

$$P_1 V_1 = K_1 = P_2 V_2$$

and

$$\frac{P_1}{P_2} = \frac{V_2}{V_1} \tag{2}$$

When the pressure of a gas is plotted against the volume in accordance with Eq. (1), we obtain a family of curves such as that shown in Figure 1-1. Each curve is a hyperbola with a different value of K_1. Since for a given weight of gas K_1 varies only with temperature, each curve corresponds to a different fixed temperature and is known as an *isotherm* (constant temperature plot). The higher curves correspond to the higher temperatures.

THE CHARLES OR GAY - LUSSAC LAW

Charles in 1787 observed that the gases hydrogen, air, carbon dioxide, and oxygen expanded an equal amount upon being heated from 0 to 80°C at constant pressure. However, it was Gay-Lussac in 1802 who first found that for *all* gases the increase in volume for each degree centigrade rise in temperature was equal approximately to $\frac{1}{273}$ of the volume of the gas at 0°C. A more precise value of this fraction is $\frac{1}{273.15}$. If we designate by V_0 the volume of a gas at 0°C and by V its volume at any temperature t°C, then in terms of Gay-Lussac's finding V may be written as

$$V = V_0 + \frac{t}{273.15}\ V_0$$

$$= V_0 \left(1 + \frac{t}{273.15} \right)$$

$$= V_0 \left(\frac{273.15 + t}{273.15} \right) \tag{3}$$

We may define now a new temperature scale such that any temperature t on it will be given by $T = 273.15 + t$, and 0°C by $T_0 = 273.15$. Then Eq. (3) becomes simply

$$\frac{V}{V_0} = \frac{T}{T_0}$$

or generally
$$\frac{V_2}{V_1} = \frac{T_2}{T_1} \tag{4}$$

This new temperature scale, designated as the absolute or Kelvin scale of temperature, is of fundamental importance in all science. In terms of this temperature scale, Eq. (4) tells us that *the volume of a definite quantity of gas at constant pressure is directly proportional to the absolute temperature,* or that

$$V = K_2T \tag{5}$$

where K_2 is a proportionality factor determined by the pressure, the nature and amount of gas, and the units of V. The above statement and Eq. (5) are expressions of *Charles's or Gay-Lussac's law of volumes.*

According to Eq. (5) the volume of a gas should be a straight line function of the absolute temperature at any constant pressure. Such a plot of V vs. T at selected pressures is shown in Figure 1–2. Since for a given amount of gas K_2 will have different values at different pressures, we obtain a series of straight lines, one for each constant pressure. Each constant pressure line is called an *isobar.* For every isobar the slope is the greater the lower the pressure.

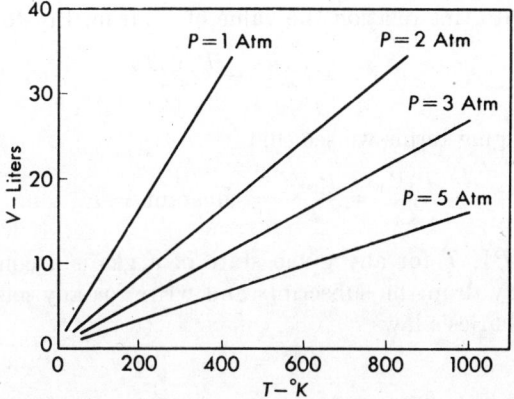

Figure 1-2. Isobaric plots of V vs. T according to Charles's law (1 mole of gas).

Equation (5) suggests also that if we were to cool a gas to 0°K (-273°C), its volume would become zero. However, no such phenomenon is ever encountered, for usually long before 0°K is approached a gas liquefies or solidifies. Again, as will be shown in the following, under such drastic conditions the equation itself cannot be considered to hold.

THE COMBINED GAS LAW

The two laws discussed give the separate variation of the volume of a gas with pressure and with temperature. To obtain the simultaneous variation of the volume with temperature and pressure, we proceed as follows. Consider a quantity of gas at P_1, V_1, and T_1, and suppose that it is desired to obtain the volume of the gas, V_2, at P_2 and T_2. First, let us compress (or expand) the gas from P_1 to P_2 at constant temperature T_1. The resulting volume V_x then will be, according to Boyle's law,

$$\frac{V_x}{V_1} = \frac{P_1}{P_2}$$

$$V_x = \frac{V_1 P_1}{P_2} \tag{6}$$

If the gas at V_x, P_2, and T_1 is heated now at constant pressure P_2 from T_1 to T_2, the final state at P_2 and T_2 will have the volume V_2 given by Charles's law, namely,

$$\frac{V_2}{V_x} = \frac{T_2}{T_1}$$

$$V_2 = \frac{V_x T_2}{T_1}$$

Substituting into this relation the value of V_z from Eq. (6), V_2 becomes

$$V_2 = \frac{V_z T_2}{T_1} = \frac{P_1 V_1 T_2}{P_2 T_1}$$

and on rearranging terms we see that

$$\frac{P_1 V_1}{T_1} = \frac{P_2 V_2}{T_2} = \text{constant} = K \tag{7}$$

i.e., the ratio PV/T for any given state of a gas is a constant. Consequently we may drop the subscripts and write for any gas which obeys Boyle's and Charles's laws

$$PV = KT \tag{8}$$

Equation (8) is known as the *combined gas law*. It gives the complete relationship between the pressure, volume, and temperature of any gas as soon as the constant K is evaluated. That Boyle's and Charles's laws are merely special cases of Eq. (8) is easily shown. When T is constant, Eq. (8) reduces to $PV =$ constant, or Boyle's law. Again, when P is constant, Eq. (8) becomes

$$V = \frac{K}{P} T = K_2 T$$

or Charles's law.

THE GAS CONSTANT

The numerical value of the constant K in Eq. (8) is determined by the number of moles[1] of gas involved and the units in which P and V are expressed; but it is *totally independent of the nature of the gas*. Equation (8) shows that for any given pressure and temperature an increase in the quantity of gas increases the volume, and thereby also correspondingly the magnitude of K. In other words, K is directly proportional to the number of moles of gas involved. For convenience this constant may be replaced, therefore, by the expression $K = nR$, where n is the number of moles of gas occupying volume V at P and T, while R is the *gas constant per mole*. Thus expressed, R becomes a *universal constant for all gases* and Eq. (8) takes the final form

$$PV = nRT \tag{9}$$

Equation (9) is the *ideal gas equation*, one of the most important relations in physical chemistry. It connects directly the volume, temperature, pressure, and number of moles of a gas, and permits all types of gas

---

[1] A *mole* is the mass of a substance in grams equal numerically to its molecular weight.

calculations as soon as the constant R is known. R may be found from the fact that 1 mole of *any* ideal gas at standard conditions, i.e., at 0°C and 1 atm pressure, occupies a volume of 22.413 liters. If we express then the *volume in liters* and the *pressure in atmospheres*, R follows from Eq. (9) as

$$R = \frac{PV}{nT} = \frac{1 \times 22.413}{1 \times 273.15} = 0.08205 \text{ liter-atm degree}^{-1} \text{ mole}^{-1}$$

This value of R can be used only when volume is taken in liters and pressure in atmospheres. For other combinations of units R will have other values. Thus, if the pressure be expressed in atmospheres while the volume in cubic centimeters, R becomes

$$R = \frac{1 \times 22{,}413}{1 \times 273.15} = 82.05 \text{ cc-atm degree}^{-1} \text{ mole}^{-1}$$

Since pressure is force per unit area and volume is area times length, it immediately follows that the units of PV/nT and hence of R are

$$\frac{PV}{nT} = R = \frac{\dfrac{\text{force}}{\text{area}} \times \text{area} \times \text{length}}{\text{moles} \times \text{degrees}} = \frac{\text{force} \times \text{length}}{\text{moles} \times \text{degrees}} = \frac{\text{work degree}^{-1}}{\text{mole}^{-1}}$$

Consequently R may be expressed in any set of units representing work or energy. Although in gas calculations in the metric system the units given above are the most useful, there is necessity in other types of calculations to employ R in some alternate energy units. These are usually ergs, joules, and calories.

To obtain R in ergs the pressure must be expressed in dynes per square centimeter and the volume in cubic centimeters. For the volume at standard conditions we have $V = 22{,}413$ cc. Again, a pressure of 1 atm is the pressure of a column of mercury 76 cm high and 1 cm² in cross section at 0°C. The total volume of such a column is thus 76 cc, and the mass 76×13.595, where the latter quantity is the density of mercury at 0°C. The pressure in dynes per square centimeter will be then this mass multiplied by the acceleration of gravity, 980.66 cm sec⁻². Inserting these values of V and P into the expression for R, we find that

$$R = \frac{(76 \times 13.595 \times 980.66)(22{,}413)}{1 \times 273.15} = 8.314 \times 10^7 \text{ ergs degree}^{-1} \text{ mole}^{-1}$$

Further, since 1 joule $= 10^7$ ergs, and 1 calorie $= 4.184$ joules, we arrive also at

$$R = 8.314 \text{ joules degree}^{-1} \text{ mole}^{-1}$$
$$= \frac{8.314}{4.184} = 1.987 \text{ cal degree}^{-1} \text{ mole}^{-1}$$

It should be clearly understood that, although R may be expressed in different units, for pressure-volume calculations involving gases *R must always be taken in the same units as those used for pressure and volume.* To facilitate calculations, Table 1–1 gives a summary of the values of *R* in various units.

TABLE 1–1.   VALUES OF *R* IN VARIOUS UNITS

Pressure	Volume	Temperature	*n*	*R*
Atmospheres	liters	°K	gram-moles	0.08205 liter-atm (°K)$^{-1}$ (g-mole)$^{-1}$
Atmospheres	cc	°K	gram-moles	82.05 cc-atm (°K)$^{-1}$ (g-mole)$^{-1}$
Dynes/cm^2	cc	°K	gram-moles	8.314 × 10^7 ergs (°K)$^{-1}$ (g-mole)$^{-1}$
mm Hg	cc	°K	gram-moles	62,360 cc-mm Hg (°K)$^{-1}$ (g-mole)$^{-1}$
R in joules		°K	gram-moles	8.314 joules (°K)$^{-1}$ (g-mole)$^{-1}$
R in calories		°K	gram-moles	1.987 cal (°K)$^{-1}$ (g-mole)$^{-1}$

CALCULATIONS INVOLVING IDEAL GAS LAW

The ideal gas law may be employed to find any one of the variables *P*, *V*, *T*, or *n* from any specified set of three of these. As an illustration, suppose that we want to ascertain the volume occupied by 10.0 g of oxygen at 25.0°C and 650 mm Hg pressure. From the data we know that

$$n = \frac{10.0}{32.0} = 0.312 \text{ mole}$$

$$T = 273.2 + 25.0 = 298.2°\text{K}$$

$$P = \frac{650}{760} = 0.855 \text{ atm}$$

$$R = 0.0821 \text{ liter-atm degree}^{-1} \text{ mole}^{-1}$$

Insertion of these into Eq. (9) yields for the volume

$$V = \frac{nRT}{P} = \frac{0.312 \times 0.0821 \times 298.2}{0.855}$$

$$= 8.94 \text{ liters}$$

Similarly, from appropriately specified data the other quantities involved in the ideal gas equation may be found.

DALTON'S LAW OF PARTIAL PRESSURES

Different gases introduced into the same container interdiffuse or mix rapidly. Dalton's *law of partial pressures* states that *at constant temperature the total pressure exerted by a mixture of gases in a definite volume is equal to the sum of the individual pressures which each gas would exert if it occupied the same total volume alone.* In other words,

$$P_{total} = P_1 + P_2 + P_3 + \cdots \qquad (1?)$$

where the individual pressures, P_1, P_2, P_3, etc., are termed the *partial pressures* of the respective gases. The partial pressure of each constituent may be thought of as the pressure which that constituent would exert if it were isolated in the same volume and at the same temperature as that of the mixture. In terms of the partial pressures, Dalton's law may be restated as follows: *The total pressure of a mixture of gases is equal to the sum of the partial pressures of the individual components of the mixture.*

The significance of Dalton's law and of the concept of partial pressures is best brought out by the following example. If we were to take three *1-liter* flasks filled respectively with hydrogen at 70 mm Hg pressure, carbon monoxide at 500 mm, and nitrogen at 1000 mm, all at the same temperature, and were to force all these gases into a fourth *1-liter* flask, the total pressure within the fourth flask would be

$$P = P_{H_2} + P_{CO} + P_{N_2}$$
$$= 70 + 500 + 1000$$
$$= 1570 \text{ mm Hg}$$

and the pressures of the individual gases within their 1-liter flasks would be the partial pressures of these gases in the mixture.

Consider now a gaseous mixture composed of n_1 moles of one gas, n_2 moles of another gas, and n_3 moles of still a third. Let the total volume be V and the temperature T. If the conditions of pressure and temperature are not too extreme, the ideal gas laws would be valid for each gas in the mixture, and we obtain for the respective partial pressures

$$P_1 = \frac{n_1 RT}{V} \qquad \textbf{(11a)}$$

$$P_2 = \frac{n_2 RT}{V} \qquad \textbf{(11b)}$$

$$P_3 = \frac{n_3 RT}{V} \qquad \textbf{(11c)}$$

According to Dalton's law the total pressure P thus becomes

$$P = \frac{n_1 RT}{V} + \frac{n_2 RT}{V} + \frac{n_3 RT}{V}$$
$$= \frac{(n_1 + n_2 + n_3) RT}{V}$$
$$= \frac{n_t RT}{V} \qquad \textbf{(12)}$$

where $n_t = (n_1 + n_2 + n_3)$ = total number of moles of gas in the mixture. We see from Eq. (12), therefore, that the gas laws may be applied to mixtures as well as to pure gases, and in exactly the same way.

On division of Eqs. (11a)–(11c) by Eq. (12) it is found that

$$P_1 = \frac{n_1}{n_t} P \tag{13a}$$

$$P_2 = \frac{n_2}{n_t} P \tag{13b}$$

and
$$P_3 = \frac{n_3}{n_t} P \tag{13c}$$

Equations such as (13) are very important in chemical and chemical engineering calculations, for they relate the partial pressure of a gas to the total pressure of the mixture. Since the fractions n_1/n_t, n_2/n_t, and n_3/n_t represent the moles of a particular constituent present in the mixture divided by the total number of moles of all gases present, these quantities are called *mol fractions* and are designated by the respective symbols N_1, N_2, N_3, etc. Of necessity the sum of all the mol fractions for a system will have to be unity, namely,

$$N_1 + N_2 + N_3 + \cdots = 1 \tag{14}$$

In terms of these definitions *the partial pressure of any component in a gas mixture is equal to the mol fraction of that component multiplied by the total pressure*. This is true only when the ideal gas law applies to each constituent of the gas mixture.

AMAGAT'S LAW OF PARTIAL VOLUMES

A law similar to Dalton's is *Amagat's law of partial volumes*. This law states that *in any gas mixture the total volume may be considered to be the sum of the partial volumes of the constituents of the mixture*, i.e.,

$$V = V_1 + V_2 + V_3 + \cdots \tag{15}$$

where V is the total volume while V_1, V_2, etc., are the partial volumes. By the partial volume of a constituent is meant the volume which that constituent would occupy if present alone at the given temperature and at the *total pressure* of the mixture. By an argument similar to the one employed for partial pressures it is readily shown that, if the ideal gas laws are again applicable, then

$$V_1 = N_1 V, \qquad V_2 = N_2 V, \text{ etc.} \tag{16}$$

where V_1, V_2, etc., are the partial volumes, N_1, N_2, etc., the mol fractions, and V the total volume at any pressure and temperature.

Dalton's and Amagat's laws are equivalent and hold equally well with gases that approximate ideal behavior, i.e., with gases that are not too close to their condensation temperatures or at too elevated pressures. At high pressures and near their condensation temperatures gases begin to exhibit considerable intermolecular attractions and effects which are no longer general but are specific to the composition and nature of the substances. Under such conditions deviations appear not only from Eqs. (13) and (16), but also from Eqs. (10) and (15). In general, the law of partial volumes holds somewhat better than the law of partial pressures at high pressures and low temperatures.

GRAHAM'S LAW OF DIFFUSION

Different gases diffuse through a tube or escape from a container having a fine opening at different rates dependent on the densities or molecular weights of the gases. The law governing such diffusions was first enunciated by Graham in 1829 and bears his name. This law states that *at constant temperature and pressure the rates of diffusion of various gases vary inversely as the square roots of their densities or molecular weights*. Thus, if we let u_1 and u_2 be the rates of diffusion of two gases, and ρ_1 and ρ_2 be their respective densities, then

$$\frac{u_1}{u_2} = \frac{\sqrt{\rho_2}}{\sqrt{\rho_1}} \tag{17}$$

Again, since at the same pressure and temperature both gases must have the same molar volume V_m, we have also that

$$\frac{u_1}{u_2} = \frac{\sqrt{\rho_2 V_m}}{\sqrt{\rho_1 V_m}} = \frac{\sqrt{M_2}}{\sqrt{M_1}} \tag{18}$$

where M_1 and M_2 are the molecular weights of the two gases.

THE KINETIC THEORY OF IDEAL GASES

All the principles of gas behavior which have been discussed so far have been arrived at by experiment. The *kinetic theory of gases*, on the other hand, attempts to elucidate the behavior of gases by theoretical means in terms of a postulated "picture" of a gas and certain assumptions regarding its behavior. The theory was first proposed by Bernoulli in 1738, and was considerably elaborated and extended by Clausius, Maxwell, Boltzmann, van der Waals, and Jeans.

The kinetic theory is based on the following fundamental postulates:

1. Gases are considered to be composed of minute discrete particles called *molecules*. For any one gas all molecules are thought to be of the same mass and size but to differ in these from gas to gas.
2. The molecules within a container are believed to be in ceaseless chaotic motion during which they collide with each other and with the walls of the container.
3. The bombardment of the container walls by the molecules gives rise to the phenomenon we call *pressure*, i.e., the force exerted on the walls per unit area is the average force per unit area which the molecules exert in their collisions with the walls.
4. Inasmuch as the pressure of a gas within a container does not vary with time at any given pressure and temperature, the molecular collisions must involve no energy loss due to friction. In other words, all molecular collisions are elastic.
5. The absolute temperature is a quantity proportional to the *average* kinetic energy of all the molecules in a system.
6. At relatively low pressures the average distances between molecules are large compared with molecular diameters, and hence the attractive forces between molecules, which depend on the distance of molecular separation, may be considered negligible.
7. Finally, since the molecules are small compared with the distances between them, their volume may be considered to be negligible compared with the total volume of the gas.

Postulates 6 and 7, by ignoring the size of the molecules and the interactions between them, limit the theoretical treatment to ideal gases.

A mathematical analysis of this concept of a gas leads to fundamental conclusions that are directly verifiable by experiment. Consider a cubical container filled with n' molecules of gas, all the same, and all with molecular mass m and velocity u. This velocity u may be resolved into its three components along the x, y, and z axes, as is shown in Figure 1–3. If we

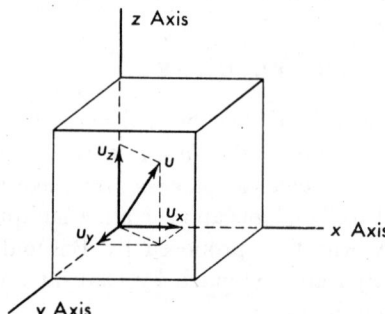

Figure 1–3.   Resolution of velocity along x, y, and z axes.

designate these velocity components by u_x, u_y, u_z, then

$$u^2 = u_x^2 + u_y^2 + u_z^2 \tag{19}$$

where u is called the *root-mean-square velocity*. Each of these components may now be treated as though a single molecule of mass m were to move independently with each of the component velocities in the appropriate directions x, y, or z. The total effect of these independent motions is obtained by combining the velocities according to Eq. (19).

Suppose now that the molecule of mass m is moving in the x direction to the right with velocity u_x. It will strike the yz plane with a momentum mu_x, and, since the collision is elastic, it will rebound with velocity $-u_x$ and momentum $-mu_x$. Consequently the *change in momentum* per molecule per single collision in the x direction is $mu_x - (-mu_x) = 2\,mu_x$. Before the molecule can strike the same wall again it must travel to the opposite wall, collide with it, rebound, and return. To do this it must cover the distance $2\,l$, where l is the length of the cube edge. Hence the number of collisions with the right-hand wall which the molecule will experience per second will be $u_x/2\,l$, and thereby the change in momentum per second for the one molecule on the given wall will be

$$(2\,mu_x)\frac{u_x}{2l} = \frac{mu_x^2}{l} \tag{20}$$

But the same change in momentum will be experienced also by the same molecule at the other yz plane, so that the total change in momentum per molecule per second in the x direction is twice the quantity in Eq. (20), or

$$\text{Change in momentum/second/molecule in } x \text{ direction} = 2\frac{mu_x^2}{l} \tag{21}$$

A moment's reflection will show that analogous changes in momentum take place in the y and z directions, and that these are given by $2\,mu_y^2/l$ and $2\,mu_z^2/l$ per molecule per second. From these the

$$\text{Total change in momentum/molecule/second} = \frac{2\,mu_x^2}{l} + \frac{2\,mu_y^2}{l} + \frac{2\,mu_z^2}{l}$$

$$= \frac{2\,m}{l}(u_x^2 + u_y^2 + u_z^2)$$

$$= \frac{2\,m}{l}u^2 \tag{22}$$

by Eq. (19). As there are n' molecules in the cube, the change in momentum per second for all of them will be Eq. (22) multiplied by n', or

$$\text{Total change in momentum per second} = \frac{2\,n'mu^2}{l} \tag{23}$$

However, the rate of change of momentum is the acting force, f. Again, pressure is the force per unit area. Consequently,

$$P = \frac{f}{A} = \frac{2\,mn'u^2}{lA} \tag{24}$$

where P is the pressure while A is the total area over which the force is applied. For the cube in question $A = 6\,l^2$, and hence

$$P = \frac{mn'u^2}{3\,l^3} \tag{25}$$

But l^3 is the volume V of the cube, and so

$$P = \frac{mn'u^2}{3\,V}$$

or

$$PV = \frac{1}{3}\,mn'u^2 \tag{26}$$

According to Eq. (26) the product PV for any gas should equal one-third the mass of all the molecules (mn') multiplied by the square of the root-mean-square velocity. Although this equation was derived on the assumption of a cubical vessel, it can be shown that the same result is obtained no matter what shape of vessel is considered, and consequently the above deduction must be perfectly general.

DEDUCTIONS FROM KINETIC THEORY OF GASES

Boyle's Law.   We have seen that one of the fundamental postulates of the kinetic theory is the direct proportionality between the kinetic energy of the molecules, i.e., $\frac{1}{2}\,mn'u^2$, and the absolute temperature, namely, that

$$\frac{1}{2}\,mn'u^2 = k_1 T \tag{27}$$

where k_1 is a proportionality constant. If now Eq. (26) is multiplied and divided by 2, we have

$$PV = \frac{2}{3}\left(\frac{1}{2}\,mn'u^2\right)$$

and hence, on insertion of Eq. (27),

$$PV = \frac{2}{3}\,k_1 T \tag{28}$$

At constant temperature Eq. (28) becomes thus $PV = $ constant, which is Boyle's law.

Charles's Law. This law holds at constant pressure. If this condition is imposed on Eq. (28), we get

$$V = \left(\frac{2}{3}\frac{k_1}{P}\right) T$$
$$= K_2 T \tag{29}$$

which is a statement of Charles's law.

Avogadro's Principle. In 1811 Avogadro enunciated the principle that *equal volumes of all gases at the same pressure and temperature contain equal numbers of molecules*. This principle is readily deducible from the kinetic theory of gases. Since the volumes and pressures are equal, $P_1 V_1 = P_2 V_2$ for two different gases, and hence it follows from Eq. (26) that

$$\frac{1}{3} n_1' m_1 u_1^2 = \frac{1}{3} n_2' m_2 u_2^2$$

Again, as the temperature is also constant, the average kinetic energy per molecule must be the same, or

$$\frac{1}{2} m_1 u_1^2 = \frac{1}{2} m_2 u_2^2$$

Inserting the latter relation into the preceding, we see that

$$n_1' = n_2' \tag{30}$$

which is a statement of Avogadro's principle.

The actual number of molecules in a gram-mole of any gas is an important physical constant known as *Avogadro's number*, symbol N. This constant may be arrived at by a number of methods. The best present value for this quantity is 6.0229×10^{23} molecules per gram-mole. Once this constant is available, the mass of any particular molecule can readily be computed by merely dividing the molecular weight of the substance by Avogadro's number. Thus, since the molecular weight of oxygen is 32.00, the mass of an individual molecule must be

$$m_{O_2} = \frac{32.00}{6.023 \times 10^{23}} = 5.31 \times 10^{-23} \text{ g/molecule}$$

Graham's Law of Diffusion. Graham's law also follows readily from the kinetic theory of gases. Since at constant volume and pressure for two different gases

$$\frac{1}{3} n_1' m_1 u_1^2 = \frac{1}{3} n_2' m_2 u_2^2$$

then

$$\frac{u_1^2}{u_2^2} = \frac{m_2 n_2'}{m_1 n_1'}$$

and

$$\frac{u_1}{u_2} = \sqrt{\frac{m_2 n_2'}{m_1 n_1'}} \tag{31}$$

Further, if $n'_2 = n'_1 = N$, then

$$\frac{u_1}{u_2} = \sqrt{\frac{m_2 N}{m_1 N}} = \sqrt{\frac{M_2}{M_1}} \tag{32}$$

Again, since at constant temperature and pressure the molar volumes are identical, we have also

$$\frac{u_1}{u_2} = \sqrt{\frac{\rho_2}{\rho_1}} \tag{33}$$

where ρ_2 and ρ_1 are the densities of the two gases. Equations (32) and (33) are identical with (17) and (18), and are, of course, statements of Graham's law.

All these deductions point to the fact that the theoretical relation $PV = \frac{1}{3} n'mu^2$ is in agreement with the empirical ideal gas law $PV = nRT$. Consequently we may write without further hesitation that

$$PV = \frac{1}{3} n'mu^2 = nRT$$

and, since $n' = nN$,

$$PV = \frac{1}{3} n(Nm)u^2 = nRT$$

$$= \frac{nMu^2}{3} = nRT \tag{34}$$

where $M = Nm$ is the molecular weight of the gas in question, and n is the number of *moles* of gas in the volume V at pressure P and temperature T.

FURTHER DEDUCTIONS FROM THE KINETIC THEORY

The value of any theory lies not only in its ability to account for known experimental facts but also in its suggestiveness of new modes of attack. In this respect the kinetic theory of gases has been very fruitful. We have seen that Eq. (26), a direct consequence and expression of the theory, gives all the laws of ideal gas behavior. At the same time, however, many other highly important relations can be deduced from it, some of which are outlined in the following.

The Velocity of Gas Molecules.   According to the kinetic theory all molecules at the same temperature must have the same average kinetic energy, i.e.,

$$\frac{1}{2} m_1 u_1^2 = \frac{1}{2} m_2 u_2^2 = \frac{1}{2} m_3 u_3^2, \text{ etc.}$$

It follows, therefore, that the higher the mass of a molecule, the more slowly must it be moving. It is of considerable interest to ascertain the actual velocity with which various molecules move. From Eq. (34) we have that

$$\frac{1}{3} nMu^2 = nRT$$

and hence
$$u = \sqrt{\frac{3\,RT}{M}} \tag{35a}$$

Again, since $RT = PV/n$, and $nM/V = \rho$, the density of the gas in question at temperature T and pressure P, Eq. (35a) may be written also as

$$u = \sqrt{\frac{3\,P}{\rho}} \tag{35b}$$

By either of these equations the root-mean-square velocity of a gas may be calculated from directly measurable quantities. In doing this, if R is expressed in ergs per degree per mole, P in dynes per square centimeter, and the density in grams per cubic centimeter, then u will be given in centimeters per second.

To calculate the velocity of hydrogen molecules at 0°C we know that $R = 8.314 \times 10^7$ ergs per mole per degree, $T = 273.15$°K, and $M = 2.016$. Hence Eq. (35a) yields for u

$$\begin{aligned}
u &= \sqrt{\frac{3\,RT}{M}} \\
&= \sqrt{\frac{3 \times 8.314 \times 10^7 \times 273.15}{2.016}} \\
&= 184,000 \text{ cm/sec} \\
&= 68 \text{ miles/min}
\end{aligned}$$

Since hydrogen is the lightest of all elements, this tremendously high velocity represents an upper limit for rates of molecular motion. For all other molecules the speeds will be lower in accordance with Graham's law. Thus for sulfur dioxide, with $M = 64$, the velocity at 0°C would be

$$\frac{u_{SO_2}}{68} = \sqrt{\frac{2}{64}}$$
$$u_{SO_2} = 12 \text{ miles/min}$$

The Kinetic Energy of Translation. The only type of energy we have ascribed thus far to gas molecules is that due to molecular motion along three coordinate axes, i.e., *kinetic energy of translation.* The amount of this energy is again deducible from Eq. (34). Since from this equation $nRT = nMu^2/3$, and since the kinetic energy, E_k, is given by

$E_k = nMu^2/2$, then for n moles of gas

$$E_k = \frac{3}{2}\left(\frac{1}{3}nMu^2\right)$$

$$= \frac{3}{2}nRT \tag{36a}$$

and per mole

$$E_k = \frac{3}{2}RT \tag{36b}$$

Consequently, the translational energy of an ideal gas is completely independent of the nature or pressure of the gas, and depends only on the absolute temperature. At, say, 300°K all ideal gases will thus contain per mole

$$E_k = \frac{3}{2}R(300)$$

$$= 450\ R$$

$$= 895\ \text{cal}$$

or approximately 900 cal of translational kinetic energy.

The average kinetic energy per molecule follows from Eq. (36b) on division by Avogadro's number, N, namely,

$$\frac{E_k}{N} = \frac{3}{2}\frac{RT}{N} = \frac{3}{2}kT \tag{37}$$

where $k = R/N$ is called the *Boltzmann constant*, and is equal to 1.3805×10^{-16} erg per degree.

Distribution of Molecular Velocities. For convenience of treatment all molecules in a given gas and at a given temperature were considered to be composed of molecules moving with a constant root-mean-square velocity u. Actually, however, all molecules do not possess the same velocity, for as a result of collisions a redistribution of both energy and velocity takes place. Maxwell and Boltzmann, utilizing probability considerations, have in fact shown that the actual distribution of molecular velocities depends on the temperature and molecular weight of a gas, and is given by

$$\frac{dn_c}{n'} = 4\pi\left(\frac{M}{2\pi RT}\right)^{3/2} e^{-\frac{Mc^2}{2RT}}c^2 dc \tag{38}$$

In this equation dn_c is the number of molecules out of a total n' having velocities between c and $c + dc$, while M and T are, respectively, the molecular weight and temperature of the gas. Obviously dn_c/n' is the fraction of the total number of molecules having velocities between c and $c + dc$. Equation (38) is known as the *Maxwell–Boltzmann distribution law* for molecular velocities. If Eq. (38) is divided by dc, we get

$$p = \frac{1}{n'}\frac{dn_c}{dc} = 4\pi\left(\frac{M}{2\pi RT}\right)^{3/2} e^{-\frac{Mc^2}{2RT}}c^2 \tag{38a}$$

where p may be taken as the probability of finding molecules with the velocity c.

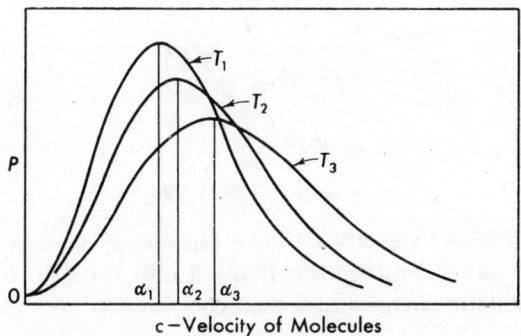

Figure 1-4. Distribution of molecular velocities in a gas.

Figure 1-4 shows schematic plots of p vs. c for several temperatures which increase in the order T_1, T_2, T_3. From these plots it may be seen that the probability of a molecule being motionless at any instant is very small. Further, for incidence of velocities greater than zero the probability increases with c, passes through a maximum, and then falls away more or less rapidly toward zero again for very high rates of motion. It is evident, therefore, that both very low and very high speeds are highly improbable, and that most of the molecules in a gas have velocities grouped about the *most probable velocity* corresponding to the peak of the curve at each temperature. The most probable velocity is in any gas not a constant, but shifts toward higher values of c with increase in temperature; i.e., at higher temperatures higher velocities are more probable than at low.

Mathematical analysis shows that the most probable velocity, α, is not equal either to the root-mean-square velocity u or the average velocity of all the molecules v. If we designate by $c_1, c_2, c_3, \cdots c_n$ the individual velocities of n' molecules in a gas, then the average velocity v is defined as

$$v = \frac{c_1 + c_2 + c_3 + \cdots c_n}{n'} \tag{39}$$

and the root-mean-square velocity as

$$u = \sqrt{\frac{c_1^2 + c_2^2 + c_3^2 + \cdots c_n^2}{n'}} \tag{40}$$

Kinetic theory arguments reveal that these various velocities are related by the equations

$$v = 0.921\ u \tag{41}$$

$$\alpha = \sqrt{\frac{2}{3}}u \tag{42}$$

and hence, on substitution of the value of u from Eq. (35a), we have

$$u = \sqrt{\frac{3\ RT}{M}} \tag{35a}$$

$$v = 0.921\sqrt{\frac{3\ RT}{M}} \tag{43}$$

$$\alpha = \sqrt{\frac{2\ RT}{M}} \tag{44}$$

or $\qquad\qquad \alpha:v:u = 1:1.128:1.224$

Since for any specific velocity c the kinetic energy of a gas per mole is $E = Mc^2/2$, we may substitute this relation into Eq. (38) to obtain the distribution of kinetic energies in a gas. The result is

$$\frac{dn_c}{n'} = \frac{2\ \pi}{(\pi RT)^{3/2}}\ e^{-\frac{E}{RT}}E^{1/2}dE \tag{45}$$

where now dn_c/n' is the fraction of the total number of molecules having kinetic energies between E and $E + dE$. Again, division of Eq. (45) by dE yields

$$p' = \frac{1}{n'}\frac{dn_c}{dE} = \frac{2\pi}{(\pi RT)^{3/2}}\ e^{-\frac{E}{RT}}E^{1/2} \tag{45a}$$

where p' is the probability for incidence of kinetic energies of translation of magnitude E. The plots of p' vs. E obtained from Eq. (45a) are very similar to those shown in Figure 4.

Frequency of Collisions and Mean Free Path.   It can be shown that in a gas containing n^* identical molecules per cubic centimeter, the number of molecules with which a *single* gas molecule will collide per second is

$$\sqrt{2}\ \pi v\sigma^2 n^*$$

where v is the average molecular velocity in centimeters per second and σ the molecular diameter in centimeters. Hence the total number of colliding molecules per cubic centimeter per second, Z, must be n^* times this quantity, or

$$Z = \sqrt{2}\ \pi v\sigma^2(n^*)^2 \tag{46}$$

Further, since each collision involves two molecules, the number of molecular collisions occurring in each cubic centimeter per second, N_c, will be one-half this number, namely,

$$N_c = \frac{Z}{2} = \frac{1}{\sqrt{2}}\,\pi v \sigma^2 (n^*)^2 \tag{47}$$

Another important quantity in kinetic theory considerations is the average distance a molecule traverses before colliding, or the *mean free path*, l. If a molecule has an average velocity v cm per sec, and if within this period it experiences, as we have seen, $\sqrt{2}\,\pi v \sigma^2 n^*$ collisions, then the average distance between collisions, or mean free path, must be

$$l = \frac{v}{\sqrt{2}\,\pi v \sigma^2 n^*}$$

$$= \frac{1}{\sqrt{2}\,\pi \sigma^2 n^*} \tag{48}$$

The quantities N_c, Z, and l are readily calculable as soon as the molecular diameters σ are available. These are usually obtained from gas viscosity measurements in a manner to be discussed toward the end of the chapter.

APPLICABILITY OF THE IDEAL GAS LAWS

The concordance between the empirical generalizations embodied in the expression $PV = nRT$ and the deductions of the kinetic theory of gases lends considerable credence to our conception of the nature of gases and their behavior. However, there still remains the question of how completely and accurately can the expression $PV = nRT$ reproduce the actual P–V–T relations of gases. To test this point we may resort to the fact that at constant temperature the combined gas law reduces to $PV = nRT =$ constant. Hence, as long as T does not vary, the product PV for a given quantity of gas should remain the same at all pressures. A plot of PV vs. P at constant T should yield, therefore, a straight line parallel to the abscissa.

Such a plot of PV vs. P constructed from actual data for several gases at 0°C is shown in Figure 1–5. The fact immediately apparent is that PV is not constant over most of the pressure range shown. The curves are in general of two types. One, including only hydrogen and helium here, starts at the value of PV demanded by $PV = nRT$ for the temperature in question and increases continually with pressure. In every case the product PV is greater than demanded by theory. On the other hand, in the second type the plot starts again at the same point as before, but now the product PV decreases at first with pressure, passes through a minimum characteristic of each gas and the temperature, and then increases to values which may rise appreciably above the theoretical.

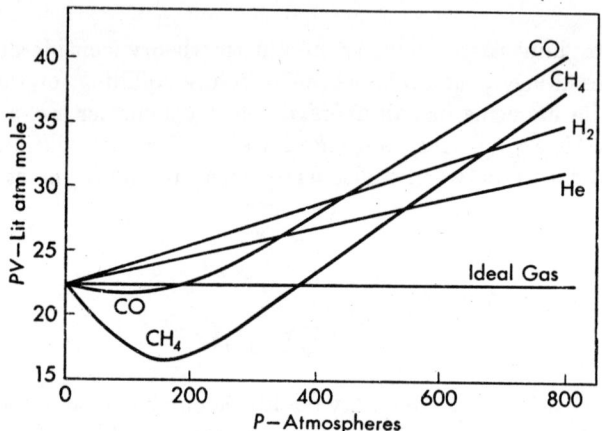

Figure 1-5. PV vs. P plot for several gases at 0°C.

Actually, both types of curves are part of a single pattern of behavior exhibited by all gases. To show this, it is convenient to employ a quantity z, called the *compressibility factor*, which is defined as

$$z = \frac{PV}{nRT} \tag{49}$$

For an ideal gas $z = 1$ at all temperatures and pressures. In the case of actual gases the compressibility factor may vary with both of these variables, and hence its deviation from a value of unity is an index of deviation from ideal behavior.

From the experimental $P-V-T$ data for a gas z may be calculated by means of Eq. (49), and its variation with temperature and pressure observed. Figure 1-6 shows a plot of z vs. P for nitrogen at various constant temperatures. Inspection of this plot reveals that all the isotherms start with $z = 1$ at $P = 0$, and change with pressure in a manner dependent on the temperature. However, there is one temperature, in this case 51°C, at which z remains close to unity over an appreciable pressure range. In fact, between $P = 0$ and $P = 100$ atm z changes only from 1.00 to 1.02. Beyond 100 atm z rises quite rapidly with increasing pressure and attains values considerably above $z = 1$. This temperature at which a real gas obeys the ideal gas law over an appreciable pressure range is called the *Boyle temperature* or *Boyle point*. The Boyle temperature is also a dividing line in the types of isotherms exhibited by the gas. Above the Boyle point the gas shows only positive deviations from ideality, and hence all values of z are greater than unity. Below the Boyle temperature, however, the values of z first decrease with increasing pressure, go through a minimum, and then increase to values which may climb appreciably above $z = 1$. It should also be observed that the lower the temperature the lower the minima, and that they occur at pressures which vary with the temperature.

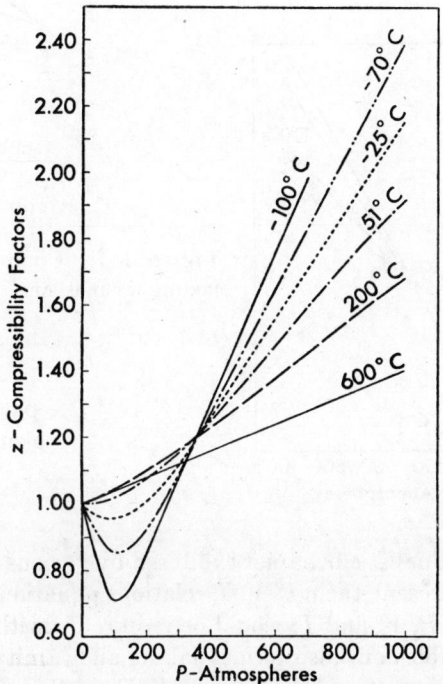

Figure 1-6.   Compressibility factors for nitrogen.

The plots shown in Figure 1–6 are typical of the P–V–T behavior of *all* gases when data covering wide ranges of pressure and temperature are considered. The only differences observed are in the Boyle temperatures and in the positions of the isotherms on the plots, since these are dependent in each instance on the gas in question. Nevertheless, it is always found that above the Boyle temperature only positive deviations from ideality are observed. Below the Boyle temperature, on the other hand, increase of pressure causes the z values first to decrease below $z = 1$, to go through a minimum, and then to increase to values which eventually go appreciably above $z = 1$.

In terms of the above discussion a glance at Figure 1–5 should suffice to indicate that at 0°C hydrogen and helium are already above their Boyle temperatures, whereas carbon monoxide and methane are still below theirs. It is also to be expected that at sufficiently low temperatures hydrogen and helium should exhibit minima in their z vs. P plots, while at higher temperatures carbon monoxide and methane should show plots similar to those of hydrogen and helium at 0°C. Such is actually the case, as may be seen from the compressibility factor plot for methane given in Figure 1-7.

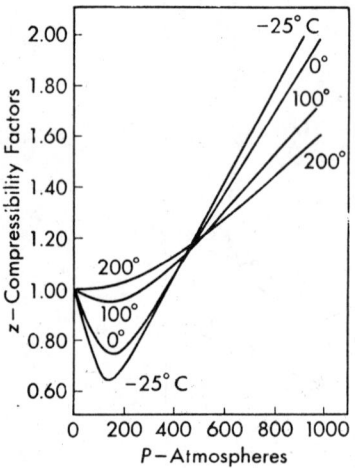

Figure 1-7.   Compressibility factors for methane.

The highly individualistic behavior exhibited by various gases indicates that in order to represent their $P-V-T$ relations equations of state, i.e., equations involving P, V, and T, would be required which would contain not only these variables but also terms making allowance for the specific forces operative in each gas. However, $P-V-T$ studies on gases at low pressures do show that when the pressures are lowered gases begin to approximate more closely the ideal gas law, and, furthermore, the lower the pressure the better is the agreement between the observed PV product and that calculated from the combined gas law. At these low pressures all gases lose their individualistic behavior and merge to obey the simple and general expression obtained from the kinetic theory of gases. For this reason the expression $PV = nRT$ is considered to be a *limiting law* only, a law which gases obey strictly only when they are diluted highly enough so that the volume of the molecules themselves is negligible compared with the total volume, and the intermolecular attractive forces are too feeble to exercise any influence on the pressure of the gas. It may be concluded, therefore, that a gas becomes more ideal as the pressure is lowered and will become completely ideal as the pressure approaches zero. This conclusion is confirmed by the fact that, as P

approaches zero, the compressibility factors at all temperatures go to $z = 1$.

How far this concordance between the ideal gas law and observation will extend into the range of higher pressures depends on the nature of the gas and the temperature. For gases which are permanent at ordinary temperatures, i.e., which are above their critical temperatures,[2] such as hydrogen, nitrogen, oxygen, and helium, this concordance may extend within 5 per cent or so up to pressures as high as 50 atm. On the other hand, with easily condensible gases, such as carbon dioxide, sulfur dioxide, chlorine, and methyl chloride, discrepancies as large as 2 or 3 per cent may appear at 1 atm pressure. The use of the ideal gas law for such gases is considerably limited, therefore, when fairly precise calculations are required. In any case, before using the ideal gas law at any appreciable pressure it is always advisable to consider the nature of the gas in question and how far it is above its critical temperature. The greater this distance, the wider in general will be the pressure range over which calculations can be made within a given accuracy.

USE OF COMPRESSIBILITY FACTORS

When the compressibility factors of a gas are known under various conditions, they may be employed quite readily for making exact gas calculations. For instance, suppose that the volume of 10 moles of methane is required at 100 atm pressure and 0°C. At this pressure and temperature $z = 0.783$, and hence, according to Eq. (49),

$$
\begin{aligned}
V &= \frac{znRT}{P} \\
&= \frac{0.783 \times 10 \times 0.08205 \times 273.2}{100} \\
&= 1.754 \text{ liters}
\end{aligned}
$$

The experimentally observed volume is 1.756 liters. Again, suppose that a certain quantity of methane occupies a volume of 0.138 liter under a pressure of 300 atm at 200°C, and the volume is required at 600 atm at 0°C. For 300 atm at 200°C, $z_2 = 1.067$, while for 600 atm at 0°C, $z_1 = 1.367$. Since for the lower temperature we have $P_1V_1 = z_1 nRT_1$ while for the higher one $P_2V_2 = z_2 nRT_2$, then

$$
\frac{P_1V_1}{P_2V_2} = \frac{z_1 nRT_1}{z_2 nRT_2} = \frac{z_1 T_1}{z_2 T_2} \tag{50}
$$

[2] The critical temperature is the highest temperature at which a gas may be liquefied.

and hence on substitution of the given values we get

$$V_1 = \left(\frac{z_1 T_1}{z_2 T_2}\right)\left(\frac{P_2 V_2}{P_1}\right)$$

$$= \left(\frac{1.367 \times 273.2}{1.067 \times 473.2}\right)\left(\frac{300 \times 0.138}{600}\right)$$

$$= 0.051 \text{ liter}$$

THE VAN DER WAALS EQUATION OF STATE

Because of the deviation of real gases from the ideal gas law, many attempts have been made to set up equations of state which will reproduce more satisfactorily the P–V–T relations of gases. Of these equations one of the earliest and best known is that of van der Waals.

The van der Waals equation differs from the ideal gas law in that it makes allowance both for the volume occupied by the molecules themselves and for the attractive forces between them. To make correction for the finite dimensions of the molecules, let b be the effective volume of the molecules in *one* mole of gas and V be the total volume of n moles of gas. In this total volume, that occupied by the molecules themselves thus will be nb, and hence the volume available for compression will be not V but $(V - nb)$. Since the latter is the "free space," it should be substituted for V in the ideal gas law. It may be anticipated that b will be characteristic and different for each gas.

The second factor for cognizance is the attractive force operative between molecules. Consider the wall of a container which is being bombarded by gaseous molecules. When the gas molecules are not constrained by attractions for each other, they will bombard the wall with the full force due to their outward motion. If, however, under the same conditions a molecule moving outward is subjected by molecular attraction to an inward "pull," it will not strike the wall with as high a force as if it were not "dragged back" by the other molecules within the gas. Consequently, the pressure resulting from the bombardment will be lessened by an amount P'. The observed pressure, P, will thus be less than the ideal pressure, P_i, by the amount P', or

$$P = P_i - P'$$

Since in the expression $P_i V = nRT$ the pressure P_i refers to the ideal pressure, we must substitute for it its value from the expression given above, or $P_i = (P + P')$. If we combine this corrected pressure with the expression for the corrected volume we obtain

$$(P + P')(V - nb) = nRT \tag{51}$$

van der Waals indicated that the magnitude of the pressure correction P' for n moles of gas present in volume V is given by

$$P' = \frac{n^2 a}{V^2}$$

where a is a constant characteristic of each gas and independent of pressure and temperature. It is for each gas a measure of the magnitude of the intermolecular attractive forces within the gas. If this expression for P' is substituted in Eq. (51), we get

$$\left(P + \frac{n^2 a}{V^2} \right)(V - nb) = nRT \tag{52}$$

This is the celebrated equation of state which was first developed by van der Waals in 1873 and which bears his name.

In applying van der Waals' equation care must be exercised in the choice of appropriate units for the constants a and b. Since $n^2 a / V^2$ represents a pressure, the units of a must be pressure $\times$ (volume)^2/(mole)^2, i.e., atm-liter^2 mole^{-2}, or atm-cc^2 mole^{-2}. In any event, the units used must be the same as those of P and V, and this applies also to R. In turn, b is a volume and must correspond to the units of V.

The use of the equation can be illustrated with an example. Suppose it is desired to calculate by van der Waals' equation the pressure at which 2 moles of ammonia will occupy a volume of 5 liters at 27°C. For ammonia, $a = 4.17$ atm-liter^2 mole^{-2}, while $b = 0.0371$ liter per mole. Hence,

$$
\begin{aligned}
P &= \frac{nRT}{V - nb} - \frac{n^2 a}{V^2} \\
&= \frac{2(0.0821)300.2}{5 - 2(0.0371)} - \frac{(2)^2 \times 4.17}{(5)^2} \\
&= 9.33 \text{ atm}
\end{aligned}
$$

The corresponding pressure calculated from the ideal gas law is 9.86 atm.

Table 1–2 lists the van der Waals constants for a number of gases. Such gases as carbon disulfide, ammonia, sulfur dioxide, chloroform, etc., which are easily condensible, have relatively high values of a, indicating strong intermolecular attractions. On the other hand, for the permanent gases such as argon, carbon monoxide, helium, and hydrogen, the a values are considerably lower, and hence in these the intermolecular forces are considerably weaker.

The van der Waals equation is much more accurate than the ideal gas law and is valid over a much wider pressure range, as may be seen from Table 1–3. However, under extreme conditions, such as temperatures near the critical and at very high pressures, its predictions deviate considerably in many instances from experimentally observed values. It is very doubtful whether it is justifiable to consider a and b as constants

TABLE 1-2.   VAN DER WAALS CONSTANTS FOR VARIOUS GASES
(a in atm-liter^2 mole^{-2}; b in liter mole^{-1})

Gas	Formula	a	b
Ammonia	NH_3	4.17	0.0371
Argon	Ar	1.35	0.0322
Carbon dioxide	CO_2	3.59	0.0427
Carbon disulfide	CS_2	11.62	0.0769
Carbon monoxide	CO	1.49	0.0399
Carbon tetrachloride	CCl_4	20.39	0.1383
Chlorine	Cl_2	6.49	0.0562
Chloroform	$CHCl_3$	15.17	0.1022
Ethane	C_2H_6	5.49	0.0638
Ethylene	C_2H_4	4.47	0.0571
Helium	He	0.034	0.0237
Hydrogen	H_2	0.244	0.0266
Hydrogen bromide	HBr	4.45	0.0443
Methane	CH_4	2.25	0.0428
Neon	Ne	0.211	0.0171
Nitric oxide	NO	1.34	0.0279
Nitrogen	N_2	1.39	0.0391
Oxygen	O_2	1.36	0.0318
Sulfur dioxide	SO_2	6.71	0.0564
Water	H_2O	5.46	0.0305

TABLE 1-3.   COMPARISON OF IDEAL GAS LAW AND VAN DER WAALS'
EQUATION AT 100°C

Observed P (atm)	Hydrogen				Carbon Dioxide			
	P Calc. Ideal	% Deviation	P Calc. van der Waals	% Deviation	P Calc. Ideal	% Deviation	P Calc. van der Waals	% Deviation
50	48.7	−2.6	50.2	+0.4	57.0	+14.0	49.5	−1.0
75	72.3	−3.6	75.7	+0.9	92.3	+17.3	73.3	−2.3
100	95.0	−5.0	100.8	+0.8	133.5	+33.5	95.8	−4.2

independent of pressure and temperature. In fact, in order to fit the equation to experimental data with a relatively high order of fidelity, it is necessary to choose different values of a and b over different ranges of pressure and temperature.

OTHER EQUATIONS OF STATE

A large number of other equations of state have been proposed to represent the P–V–T relations of gases. Some of these are based to some extent on theoretical considerations, while other are entirely empirical. We shall consider now several of the more important of these equations.

The Kamerlingh Onnes Equation of State.   This empirical equation expresses PV as a power series of the pressure at any given temperature, namely,

$$PV_m = A + BP + CP^2 + DP^3 + \cdots \tag{53}$$

P is the pressure, generally in atmospheres, and V_m is the *molar* volume in liters or cubic centimeters. The coefficients A, B, C, etc., are known respectively as the first, second, third, etc., *virial coefficients*. At very low pressures only the first of these coefficients is significant, and it is equal to RT. At higher pressures, however, the others as well are important and must be considered. In general the order of significance of the coefficients is their order in the equation. These coefficients, although constant at any given temperature, change in value as the temperature is changed. Of necessity the first virial coefficient A is always positive and increases with temperature. The second coefficient, on the other hand, is negative at low temperatures, passes through zero, and becomes increasingly positive as the temperature is raised. The temperature at which $B = 0$ is the *Boyle temperature*, for at this temperature Boyle's law is valid over a fairly wide pressure range.

By using a sufficient number of terms this equation can be fitted to experimental data with a high order of accuracy. The virial coefficients for several gases are shown in Table 1–4. With these it is possible to calculate PV up to 1000 atm.

TABLE 1–4.   VIRIAL COEFFICIENTS OF SOME GASES
(P in atm, V_m in liters mole^{-1})

$t°C$	A	$B \times 10^2$	$C \times 10^5$	$D \times 10^8$	$E \times 10^{11}$
			Nitrogen		
−50	18.312	−2.8790	14.980	−14.470	4.657
0	22.414	−1.0512	8.626	−6.910	1.704
100	30.619	0.6662	4.411	−3.534	0.9687
200	38.824	1.4763	2.775	−2.379	0.7600
			Carbon Monoxide		
−50	18.312	−3.6878	17.900	−17.911	6.225
0	22.414	−1.4825	9.823	−7.721	1.947
100	30.619	0.4036	4.874	−3.618	0.9235
200	38.824	1.3163	3.052	−2.449	0.7266
			Hydrogen		
−50	18.312	1.2027	1.164	−1.741	1.022
0	22.414	1.3638	0.7851	−1.206	0.7354
500	63.447	1.7974	0.1003	−0.1619	0.1050

The Berthelot Equation.   The high-pressure form of this equation is rather difficult to handle. For low pressures the equation reduces to

$$PV = nRT \left[1 + \frac{9\,PT_c}{128\,P_cT} \left(1 - \frac{6\,T_c^2}{T^2} \right) \right] \tag{54}$$

where P, V, R, T, and n have the same meaning as in the ideal gas law, while P_c and T_c are the critical pressure and critical temperature respectively.[3] For pressures of about an atmosphere and below this equation is very accurate, and it is consequently very useful in calculating the molecular weights of gases from their densities. Its use will be illustrated in that connection.

TABLE 1–5.   BEATTIE-BRIDGEMAN CONSTANTS FOR SOME GASES*
(For P in atm, V in liters mole^{-1})

Gas	A_0	a	B_0	b	c
He	0.0216	0.05984	0.01400	0	0.004×10^4
Ne	0.2125	0.2196	0.02060	0	0.101×10^4
Ar	1.2907	0.02328	0.03931	0	5.99×10^4
H_2	0.1975	−0.00506	0.02096	−0.04359	0.050×10^4
N_2	1.3445	0.02617	0.05046	−0.00691	4.20×10^4
O_2	1.4911	0.02562	0.04624	0.004208	4.80×10^4
Air	1.3012	0.01931	0.04611	−0.01101	4.34×10^4
CO_2	5.0065	0.07132	0.10476	0.07235	66.00×10^4
CH_4	2.2769	0.01855	0.05587	−0.01587	12.83×10^4
$(C_2H_5)_2O$	31.278	0.12426	0.45446	0.11954	33.33×10^4

* *J. Am. Chem. Soc.*, **50**, 3136 (1928). See also Maron and Turnbull, *Ind. Eng. Chem.*, **33**, 408 (1941).

The Beattie-Bridgeman Equation of State.   This equation of state involving five constants may be stated in two forms, one explicit in pressure, the other in molar volume V_m, namely,

$$P = \frac{RT}{V_m} + \frac{\beta}{V_m^2} + \frac{\gamma}{V_m^3} + \frac{\delta}{V_m^4} \tag{55}$$

$$V_m = \frac{RT}{P} + \frac{\beta}{RT} + \frac{\gamma P}{(RT)^2} + \frac{\delta P^2}{(RT)^3} \tag{56}$$

where

$$\beta = RTB_0 - A_0 - \frac{Rc}{T^2} \tag{57a}$$

$$\gamma = -RTB_0 b + A_0 a - \frac{RcB_0}{T^2} \tag{57b}$$

$$\delta = \frac{RB_0 bc}{T^2} \tag{57c}$$

In these relations T is again the absolute temperature and R the gas constant, while A_0, B_0, a, b, and c are constants characteristic of each gas.

[3] See page 47.

Of the two forms, Eqs. (55) and (56), the first is the more accurate, for the second was deduced from it with certain approximations.

This equation is applicable over wide ranges of temperature and pressure with excellent accuracy. Volumes and pressures calculated by it agree with experiment to 0.3 per cent or less up to pressures of 100 atm and temperatures as low as −150°C. With lower accuracy the equation may be extended to considerably higher pressures. Beattie-Bridgeman constants for a number of gases are given in Table 1-5.

MOLECULAR WEIGHTS OF GASES

The molecular weight of a gas is an important quantity essential for all types of calculations. It should be clearly realized that chemical analysis alone is insufficient to establish the molecular weight of a substance. Chemical analysis merely establishes the elements entering into the composition of a molecule and their proportions, but it does not tell us how many atoms of each substance are involved. For instance, chemical analysis of ethane shows that it is composed of carbon and hydrogen in the proportion of three atoms of hydrogen for each atom of carbon. From this alone we should be tempted to write the formula as CH_3. Actually, however, density measurements on the gas show that the formula is not CH_3 but a multiple of it, $(CH_3)_2$ or C_2H_6; i.e., the molecule is composed of two atoms of carbon and six atoms of hydrogen. In brief, chemical analysis can yield only the composition and empirical formula. Physicochemical measurements, on the other hand, can establish the molecular weight, and they can give us then the multiple by which the empirical formula weight must be multiplied in order to arrive at the actual molecular weight of the substance.

Until 1961 all molecular weights were based on the arbitrarily assumed standard of 16.0000 for the chemical atomic weight of oxygen. In that year the International Union of Pure and Applied Chemistry adopted a new atomic weight system based on the most abundant isotope of carbon, namely, C^{12}, as being 12.0000. On this new basis the chemical atomic weight of oxygen is changed to 15.9994. Since it has been proved that the oxygen molecule contains two atoms, it follows immediately that the molecular weight of oxygen must be 31.9988. Knowing the molecular weight of oxygen, the molecular weights of all other gases may be determined by physicochemical methods through application of the Avogadro hypothesis.

The Avogadro hypothesis states that under the same conditions of temperature and pressure equal volumes of all ideal gases contain the same number of molecules. If we were to determine, then, the volume that a mole of oxygen occupies under a specified set of conditions, this would be also the volume that a mole of any other gas would occupy under the same conditions; and the weight of this volume would yield directly the molecular weight of the gas. Since by Avogadro's hypothesis the two volumes would contain the same number of molecules, the molecular weights would be in the same proportion as the actual masses of the individual molecules.

In physical chemistry the unit of mass commonly employed is the gram, and the gram-mole is the weight of a substance in grams corresponding to the molecular weight.[4] For oxygen the gram-mole is 31.9988 grams. Further, by direct measurement it has been found that the density of oxygen at standard conditions, i.e., 1 atm pressure and 273.15°K, is 1.4276 grams per liter when corrected for nonideality of the gas. Since density is mass per unit volume, it follows that the molar volume of oxygen under the conditions specified must be

$$\frac{31.9988}{1.4276} = 22.413 \text{ liters}$$

As this is also the molar volume of any other ideal gas at standard conditions, the problem of determining the exact molecular weight of any other gas reduces itself to the determination of the weight of 22.413 liters of the gas at 1 atm pressure and 273.15°K after correction for nonideal behavior.

Actually, it is not necessary or convenient to make measurements under the above conditions. Measurements may be made under any desired conditions, and the molecular weight may be calculated conveniently from these. The procedure followed for obtaining exact molecular weights will be presented later in the chapter. At present we shall concern ourselves with the determination of approximate molecular weights which, along with the chemical analysis, are generally sufficient to establish the molecular weight of a substance. For the latter purpose we employ the ideal gas law as follows: If we let W be the weight of gas under consideration, then $n = W/M$, and

$$PV = nRT = \frac{W}{M} RT$$

or
$$M = \frac{WRT}{PV} \tag{58}$$

Therefore, to obtain the molecular weight of any gas we need only determine the temperature and pressure at which a weight of gas W occupies the volume V, substitute these quantities in Eq. (58), and solve for M. Equation (58) may be expressed also in terms of the density of the gas ρ.

[4] Similarly, a pound-mole is the weight of a substance in pounds corresponding to the molecular weight.

Since $\rho = W/V$,

$$M = \frac{\rho R T}{P} \tag{59}$$

and the molecular weight follows from the density of the gas at any given temperature and pressure. Most methods for determining the molecular weights of gases are based on these equations.

REGNAULT'S METHOD FOR DETERMINATION OF MOLECULAR WEIGHTS

This method is employed to determine the molecular weights of substances which are gaseous at room temperature. The procedure in outline is as follows: A dry glass bulb of 300- to 500-cc capacity fitted with a stopcock is evacuated and weighed. It is then filled at a definite temperature and pressure with the gas whose molecular weight is to be determined and weighed again. The difference in weights represents the weight of gas W in the flask. The volume of the flask is determined by filling it with water or mercury, whose densities are known, and again weighing. From the data thus obtained the molecular weight may be calculated by Eq. (58).

For more precise work a larger bulb is used to increase the mass of gas and a similar bulb is employed as a counterpoise. The observed weights are also reduced to vacuo.

DUMAS' METHOD FOR DETERMINATION OF VAPOR DENSITIES

This method is used to determine the molecular weights in the vapor phase of readily volatile liquids. A retort-shaped bulb, having a small opening drawn to a capillary, is first weighed full of air. A sample of several cubic centimeters of the liquid in question is drawn into the bulb by cooling it with the tip below the surface of the liquid, and the bulb is then immersed in a bath whose temperature is above the boiling point of the liquid. The boiling is permitted to proceed until the vapors of boiling liquid have expelled all the air from the bulb, and the liquid in the flask has completely vaporized. The flask is then sealed, cooled to room temperature, and weighed. The volume of the bulb is determined as in Regnault's method. The pressure of the vapor when the bulb is sealed is the same as atmospheric, while the temperature is that of the bath. The weight of vapor, after corrections for buoyancy, is obtained from the equation

$$W_{vapor} = W_{(bulb+vapor)} - W_{(bulb+air)} + W_{air} \tag{60}$$

W_{air} is obtained by multiplying the volume of the flask by the density of the air. Knowing P, V, T, and W_{vapor}, the molecular weight of the liquid in the vapor phase may be calculated as before.

THE VICTOR MEYER METHOD FOR VAPOR DENSITIES

This method serves the same purpose as the Dumas method for the determination of vapor densities but is considerably simpler and more flexible. A sketch of the apparatus is shown in Figure 1–8. It consists of

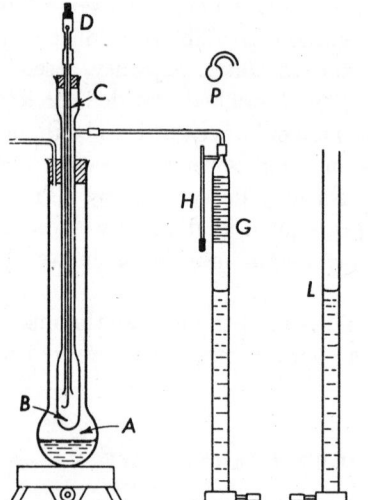

Figure 1–8. Victor Meyer apparatus.

an inner tube B, approximately 50 cm long, which is surrounded by a jacket A, partly filled as indicated with a liquid whose boiling point is at least 30 degrees higher than that of the substance to be studied. The function of the outer jacket is to keep the temperature of the inner tube constant by boiling the liquid in A throughout a run. Inside the inner tube, in turn, is another tube C, open at the bottom, down which passes a metal or glass rod, anchored with rubber tubing at the top in the manner shown and fitted with a hook at the bottom. The outlet from B communicates with a gas burette G, filled either with water, in which case correction for the aqueous pressure must be applied, or preferably mercury. L is a leveling bulb to permit adjustment of gas pressure in G to that of the atmosphere.

The liquid whose molecular weight is to be determined is enclosed in a small glass ampoule with finely drawn tip, P. This ampoule is first weighed empty. Next, enough of the liquid is drawn in to yield 40 to 60 cc of vapor, and the bulb is sealed carefully in a flame and weighed again. The difference between the first and second weighings gives the

weight of the liquid W to be vaporized. This ampoule then is hung o͘
hook projecting from C, and the entire apparatus assembled as shͻ
the figu:e.

To make a measurement, the liquid in A is brought to boiling and kept
there for the entire run. When thermal equilibrium has been established,
the levels in G and L are equalized and the burette reading is taken. Next
the ampoule is smashed by pulling upward on the rod at D so as to bring
the neck of the ampoule up against the bottom of C. With the bulb
broken the liquid vaporizes, and the vapors generated displace air from
the bottom of B into the gas burette G. The volume of air thus displaced
is equal to the volume of the vapors formed at the temperature of the
inner tube. Once in the gas burette the air cools to room temperature,
and its volume can be measured by again reading the burette. Provided
the levels in G and L are equalized, the pressure of this air is the same as
that of the atmosphere outside the burette, while the temperature is that
read on the thermometer H. The volume of displaced air thus obtained,
i.e., final minus initial burette readings, is equal to the volume which the
vapors of the liquid would occupy if they could be cooled to the temperature of the room and atmospheric pressure. Having measured in this
manner the weight of liquid W, and its volume as a *vapor* at room temperature T and barometric pressure P, the density of the vapor and its
molecular weight may readily be calculated from the observed data.

DETERMINATION OF EXACT MOLECULAR WEIGHTS

The molecular weights calculated from the ideal gas law are, even with
good data, only approximate. The reason is that already at atmospheric
pressure the ideal gas law fails to represent accurately the behavior of the
vapors. If an exact molecular weight is desired, this must be obtained from
either a more precise gas equation or by special treatment of the ideal gas
law.

When the constants a and b of a substance are known, use of van der
Waals' equation will give better concordance between observed and
calculated values of the molecular weight. For the purpose at hand, however, the Berthelot equation is more convenient and gives good results.
It can be used, of course, only when the critical temperature and pressure
of the substance are available. Since $n = W/M$, Eq. (54) gives for M

$$M = \left(\frac{W}{V}\right)\left(\frac{RT}{P}\right)\left[1 + \frac{9\,PT_c}{128\,P_cT}\left(1 - \frac{6\,T_c^2}{T^2}\right)\right] \tag{61}$$

Further, since $W/V = \rho$, Eq. (61) may also be written as

$$M = \frac{\rho RT}{P}\left[1 + \frac{9\,PT_c}{128\,P_cT}\left(1 - \frac{6\,T_c^2}{T^2}\right)\right] \tag{62}$$

from which the density follows when M is known or vice versa.

The higher accuracy of the Berthelot equation can be illustrated with the following data on methyl chloride. For methyl chloride, $T_c = 416.2°K$, $P_c = 65.8$ atm, while the density at standard conditions is 2.3076 g per liter. Hence, by Eq. (62),

$$M = \frac{2.3076 \times 0.08205 \times 273.2}{1} \left[1 + \frac{9 \times 1 \times 416.2}{128 \times 65.8 \times 273.2} \left(1 - 6 \frac{(416.2)^2}{(273.2)^2} \right) \right]$$

$$= 50.62 \text{ g mole}^{-1}$$

as against the theoretically calculated 50.49. Using the same data and the ideal gas law, the molecular weight obtained is 51.71.

A means of obtaining exact molecular weights is the method of *limiting densities*. This method, which gives excellent results, is based upon the fact that as zero pressure is approached the ideal gas laws become exact for all gases. The densities of a gas or vapor are determined at a given temperature at atmospheric pressure and at several other pressures below one atmosphere. The ratio ρ/P is then plotted against P. If the vapor or gas were ideal, this ratio would be the same at all pressures, for

$$P = \frac{\rho}{M} RT$$

and
$$\frac{\rho}{P} = \frac{M}{RT} = \text{constant} \tag{63}$$

However, since this is not true for real gases, the ratio ρ/P changes with decreasing pressure. Fortunately the plot is practically linear and can be extrapolated to zero pressure without any difficulty. At zero pressure the limiting ratio ρ/P is that for the ideal gas, and so

$$\left(\frac{\rho}{P} \right)_{P=0} = \frac{M}{RT}$$

and
$$M = RT \left(\frac{\rho}{P} \right)_{P=0} \tag{64}$$

This method can be illustrated with the data on hydrogen bromide given in Table 1–6, while the plot of ρ/P vs. P is shown in Fig. 1–9. The

TABLE 1–6.   DENSITIES OF HBr AT VARIOUS PRESSURES (0°C)

P (atm)	ρ (g/liter)	ρ/P
1	3.6444	3.6444
$\frac{2}{3}$	2.4220	3.6330
$\frac{1}{3}$	1.2074	3.6222
0	—.	3.6108 (extp'd)

extrapolated value of ρ/P is 3.6108 g per liter per atm at 0°C. Hence the molecular weight of hydrogen bromide is

$$M = 3.6108 \times 0.082054 \times 273.15 = 80.93 \text{ g mole}^{-1}$$

The value calculated from atomic weights is 80.92.

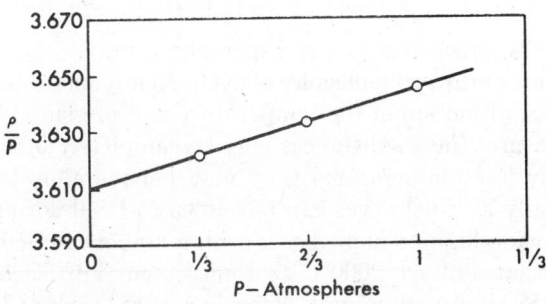

Figure 1-9.   Plot of ρ/P vs. P for HBr at 0°C.

RESULTS OF VAPOR DENSITY MEASUREMENTS

The measurement of the vapor densities of a large number of substances shows that the molecular weight of these substances in the gas phase over a certain temperature interval is what would be expected from their simple formula. Among these may be mentioned ammonia, carbon dioxide, hydrogen, nitrogen, carbon monoxide, methyl chloride, methyl fluoride, ethyl ether, methyl ether, carbon tetrachloride, chloroform, carbon disulfide, acetone. There are other substances, however, which exhibit a highly anomalous behavior. These may be segregated into two groups: (a) those which exhibit vapor densities, and consequently molecular weights, higher than would be expected on the basis of their simple formulas, and (b) those which exhibit vapor densities lower than expected from their simple formulas. All these abnormalities are very much greater than can be accounted for by either experimental uncertainty or deviation from ideal behavior.

The substances exhibiting abnormally high vapor densities are considered to be associated in the vapor phase, i.e., the molecules are considered to be composed of more than a single structural unit. In line with this view is the fact that the calculated molecular weight is usually a whole-number multiple of the simple formula. Thus aluminum chloride is shown in the vapor phase to be $(AlCl_3)_2$ or Al_2Cl_6, ferric chloride Fe_2Cl_6, beryllium chloride Be_2Cl_4, and gallium chloride Ga_2Cl_6. Sulfur is another substance which shows different stages of association in the gas phase at different temperatures.

Substances exhibiting abnormally low vapor densities break down or dissociate in the vapor phase under the influence of heat into simpler molecules, leading thereby to a greater number of particles and a lower density for any given pressure. Thus the vapor of ammonium chloride contains ammonia and hydrogen chloride as a result of the reaction

$$NH_4Cl = NH_3 + HCl$$

Similarly, PCl_5 dissociates in the vapor phase into PCl_3 and Cl_2, while N_2O_4 dissociates into two molecules of NO_2. In any instance the extent of dissociation is a function of the temperature and pressure. At sufficiently high temperatures these substances may be completely dissociated, while at sufficiently low temperatures they may behave almost normally. In fact, practically all substances can be shown to be abnormal if the temperature is made high enough. Even such a stable compound as carbon dioxide dissociates above 2000°C to some extent into carbon monoxide and oxygen. Similarly, aluminum chloride at 400°C is Al_2Cl_6, at 500°C it is a mixture of Al_2Cl_6 and $AlCl_3$, while at 1100°C it is all $AlCl_3$. If heated further, $AlCl_3$ will actually dissociate into aluminum and chlorine. Hence when we speak of the molecular weight of a substance in the gas phase, it is very important to keep in mind the temperature to which reference is made.

HEAT CAPACITY OF GASES

The specific heat of any substance is defined as the quantity of heat required to raise the temperature of unit weight of the substance 1 degree of temperature. In terms of calories and degrees centigrade, the specific heat is the number of calories of heat required to raise the temperature of 1 g of a substance 1°C. Chemical calculations are most frequently made on a molar basis, and for that reason it is more convenient to deal with the *heat capacity* per mole. The heat capacity per mole is the amount of heat required to raise the temperature of 1 mole of a substance 1°C. It is equal, of necessity, to the specific heat per gram multiplied by the molecular weight of the substance.

Two types of heat capacities are recognized, depending on whether the substance is heated at constant volume or at constant pressure. When a substance is heated at constant volume, all of the energy supplied goes to increase the internal energy of the substance, and we speak then of the *heat capacity at constant volume*, C_v. On the other hand, when a substance is heated at constant pressure, energy must be supplied not only to increase its internal energy, but also to make possible expansion of th substance against the confining atmospheric pressure. The *heat capacity at constant pressure*, C_p, must therefore be larger than that at constan' volume by the amount of work which must be performed in the expansion accompanying 1 degree rise in temperature. In liquids and solids, where volume changes on heating are small, this difference between C_p and C_v is usually slight. With gases, however, where the volume changes with temperature are always large, the difference $C_p - C_v$ is always significant and cannot be disregarded.

Some important deductions concerning the specific heats of gases can be made from the kinetic theory of gas behavior. According to equation (36b), the kinetic energy of translation of an ideal gas per mole is

$$E_k = \frac{3}{2} RT$$

If this is the only type of energy the gas possesses (monatomic gas), the energy difference of the gas $(E_{k_2} - E_{k_1})$ between two temperatures T_2 and T_1 is

$$\Delta E = E_{k_2} - E_{k_i} = \frac{3}{2} R(T_2 - T_1)$$

When the temperature difference $T_2 - T_1 = 1$, ΔE becomes the energy required to raise the translational energy of 1 mole of gas 1 degree without involving any external work, or, in other words, the heat capacity per mole at constant volume C_v. Hence we may write

$$C_v = \frac{3}{2} R = \frac{3}{2} \times 1.987$$
$$= 2.98 \text{ cal degree}^{-1} \text{ mole}^{-1} \qquad (65)$$

The kinetic theory predicts, therefore, that C_v for any ideal gas *containing only translational energy* should be approximately 3 cal per mole, and, further, that this heat capacity should be constant and independent of temperature.

A similar prediction can be arrived at with respect to the heat capacity at constant pressure, C_p. In view of the preceding considerations it follows that

$$C_p = C_v + w \qquad \text{cal degree}^{-1} \text{ mole}^{-1} \qquad (66)$$

where w is the work which must be performed against a confining pressure P when 1 mole of an ideal gas is expanded from a volume V_1 at T_1 to a volume V_2 at $T_2 = T_1 + 1$. The value of w can be obtained from the relation

$$w = \int_{V_1}^{V_2} P dV \qquad (67)$$

which will be discussed in greater detail in Chapter 3. If we now differentiate $PV = RT$ at constant pressure, we have $PdV = RdT$, and on substitution of RdT for PdV in Eq. (67), we see that

$$w = \int_{V_1}^{V_2} PdV = \int_{T_1}^{T_2} RdT$$
$$= R(T_2 - T_1)$$

For $T_2 - T_1 = 1$ this reduces to $w = R$ per mole, and hence for an ideal gas

$$C_p = C_v + R \quad \text{cal degree}^{-1} \text{ mole}^{-1} \tag{68}$$

Equation (68) is valid for all ideal gases, and permits the simple conversion of C_p to C_v or vice versa. Inserting the values of C_v from Eq. (65) and of R, we see that for any ideal gas *involving only translational energy* C_p should be

$$C_p = \frac{5}{2} R$$
$$= 4.97 \text{ cal degree}^{-1} \text{ mole}^{-1} \tag{69}$$

Consequently, like C_v, C_p should be constant and independent of temperature for all gases. Again, the ratio C_p/C_v, commonly designated by γ, should also be a constant equal to

$$\gamma = \frac{C_p}{C_v} = \frac{5/2\ R}{3/2\ R}$$
$$= 1.67 \tag{70}$$

In Table 1–7 are listed values of C_p, C_v, $C_p - C_v$, and γ for various gases at 15°C. It will be observed, first of all, that the requirement $C_p - C_v = R = 1.99$ cal per mole is met fairly well by practically all the gases in the table. Second, the predictions of the kinetic theory that $C_p = 4.97$ and $C_v = 2.98$ cal per mole are borne out by the heat capacities of a group of gases which includes, besides argon and helium, also krypton, xenon, and a number of metallic vapors. However, for all the other gases in the table the prediction is not valid. Inspection of the table reveals that the various gases can be divided into classes based upon their values of γ. The first group, comprising gases that obey the kinetic theory, has the expected $\gamma = 1.67$. The others, in turn, may be grouped as those with γ equal approximately to 1.4, 1.3, and lower. Further, the decrease in γ is always associated with an increase in the complexity of the molecules involved. Thus argon and helium with $\gamma = 1.67$ are monatomic, i.e., the molecules contain a single atom of the element. Again, the substances with γ equal to about 1.4, such as oxygen, nitrogen, and chlorine, are diatomic, those with γ equal to about 1.3 triatomic, while all others with

γ still lower are more complex. Finally, all substances exhibiting γ values lower than 1.67 also have values of C_p and C_v considerably greater than the predicted $C_p = \frac{5}{2} R$ and $C_v = \frac{3}{2} R$.

TABLE 1-7.   HEAT CAPACITIES OF GASES AT 15°C
(Cal mole^{-1} degree^{-1})

Gas	Formula	C_p	C_v	$C_p - C_v$	γ
Argon	Ar	5.00	3.01	1.99	1.66
Helium	He	4.99	3.00	1.99	1.66
Carbon monoxide	CO	6.94	4.95	1.99	1.40
Chlorine	Cl$_2$	8.15	6.02	2.13	1.35
Hydrogen	H$_2$	6.83	4.84	1.99	1.41
Hydrogen chloride	HCl	7.07	5.01	2.06	1.41
Nitrogen	N$_2$	6.94	4.94	2.00	1.40
Oxygen	O$_2$	6.96	4.97	1.99	1.40
Carbon dioxide	CO$_2$	8.75	6.71	2.04	1.30
Hydrogen sulfide	H$_2$S	8.63	6.54	2.09	1.32
Nitrous oxide	N$_2$O	8.82	6.77	2.05	1.30
Sulfur dioxide	SO$_2$	9.71	7.53	2.18	1.29
Acetylene	C$_2$H$_2$	9.97	7.91	2.06	1.26
Ethylene	C$_2$H$_4$	10.07	8.01	2.06	1.26
Ethane	C$_2$H$_6$	11.60	9.51	2.09	1.22

These high heat capacities suggest that the fundamental assumption made, that the only energy present in a gas is kinetic energy of translation, is not always correct. A monatomic molecule can execute only translational motion along the coordinate axes, and for such a gas the deductions of the kinetic theory are valid. A more complex molecule, however, may be subject not only to translational motion as a unit, but to rotation and vibration as well. If we think of a diatomic molecule simply as a "dumbbell" held together by an elastic spring, then the two atoms may execute vibrations with respect to each other along their line of centers. Further, the molecule as a whole may undergo rotation about axes perpendicular to the line joining the centers of mass of these molecules. These extra motions involve additional terms for the energy of the gas; and if these motions are subject to temperature variation, as they are, additional terms will appear in the heat capacity equation for the gas. A more detailed discussion of rotational and vibrational energies of gas molecules will be given in Chapters 16 and 18.

THEORY OF NONIDEAL GASES

The theory of nonideal gases reduces itself essentially to the theory of intermolecular or van der Waals forces. Although we are not prepared at present to discuss the nature of these forces,[5] it is nevertheless of importance to summarize briefly the results obtained. Theoretical arguments show that the interaction energy, E', between a pair of molecules is given by

$$E' = -\frac{A}{r^6} + \frac{B}{r^n} \tag{71}$$

and the force of interaction by

$$f' = \frac{A}{r^7} - \frac{B}{r^{(n+1)}} \tag{72}$$

where A and B are constants characteristic of the molecules involved, r is their distance of separation, and n is a constant whose value may range from 9 to 12. In these equations the first term on the right represents attraction, while the second represents repulsion between the molecules. From these equations it is evident that the forces between molecules are short-range in character, and that they increase very rapidly as the distance between molecules is made small.

The consideration of molecular interactions has made possible the theoretical explanation of the second virial coefficient and, less successfully, of the third virial. However, we do not have as yet a theory which can account completely for the P–V–T behavior of gases over very wide ranges of temperature and pressure.

LIQUIDS

From the standpoint of kinetic theory, a liquid may be considered as a continuation of the gas phase into the region of small volumes and very high molecular attractions. The cohesive forces in a liquid must be stronger than those in a gas even at high pressures, for they are high enough to keep the molecules confined to a definite volume. Still, the molecules within the liquid must not be thought of as rigidly fixed. They have some freedom of motion, but this motion is considerably restricted, and hence the mean free path is much shorter than in the gas phase.

Our knowledge of the nature of the liquid state is still very incomplete.

[5] See Chapter 16.

Because of the proximity of molecules to each other, effects frequently manifest themselves in liquids which, if present, are of only secondary significance in gases. Thus we encounter clustering, association, and in general orientation of the molecules into some, even though not very pronounced, order. At best, the situation within a liquid is very complex, and the progress made in unraveling the multitudinous effects has been rather slow.

CRITICAL PHENOMENA IN LIQUIDS

If a liquid, such as water, is sealed in an evacuated tube, a certain amount will evaporate to form vapor. This vapor will exert a pressure just as any gas does, and, provided the temperature is maintained constant, an equilibrium will be established between the liquid and vapor phases. The vapor pressure established is characteristic for each liquid and is a constant at any given temperature; it is known as the *saturated vapor pressure* of the liquid. The saturated vapor pressure increases continuously with temperature. Thus, at 25°C the vapor pressure of water is 23.76 mm Hg, while at 100°C it is 760 mm Hg. As the water in the sealed tube is heated further, more and more water evaporates and the pressure continues to increase. At all times there is a definite line of demarcation, or meniscus, between the liquid and vapor phases. When we reach the temperature of 374°C, however, the meniscus becomes indefinite, fades into the vapor, and disappears. At this temperature the physical properties of liquid and vapor become identical, and no distinction can be observed between the two. A liquid in this condition is said to be at the *critical point*. The temperature, saturated vapor pressure, and molar volume corresponding to this point are designated the *critical temperature, critical pressure,* and *critical volume* respectively. Their values, which are constant and characteristic for each substance, are known as the *critical constants.* For water the critical constants are: $t_c = 374.4$°C, $P_c = 219.5$ atm, and $V_c = 58.7$ cc per mole.

On heating the sealed tube even slightly above the critical temperature, no evidence can be found of the presence of liquid. The whole mass is gaseous and remains in that state no matter how high it is heated, or how large an external pressure is applied. Since the phenomena described for water are exhibited by all liquids, it must be concluded that *no liquid can exist as such at temperatures above the critical under any applied pressure.*

The critical phenomena are reversible. When the gas in the sealed tube is cooled below the critical temperature, if the pressure is sufficiently high the meniscus reappears, and again we have the two phases, liquid and vapor.

THE *P–V–T* RELATIONS OF GASES AND LIQUIDS

The first complete data on the *P–V–T* relations of a substance in both gaseous and liquid states were obtained by Andrews[6] on carbon dioxide.

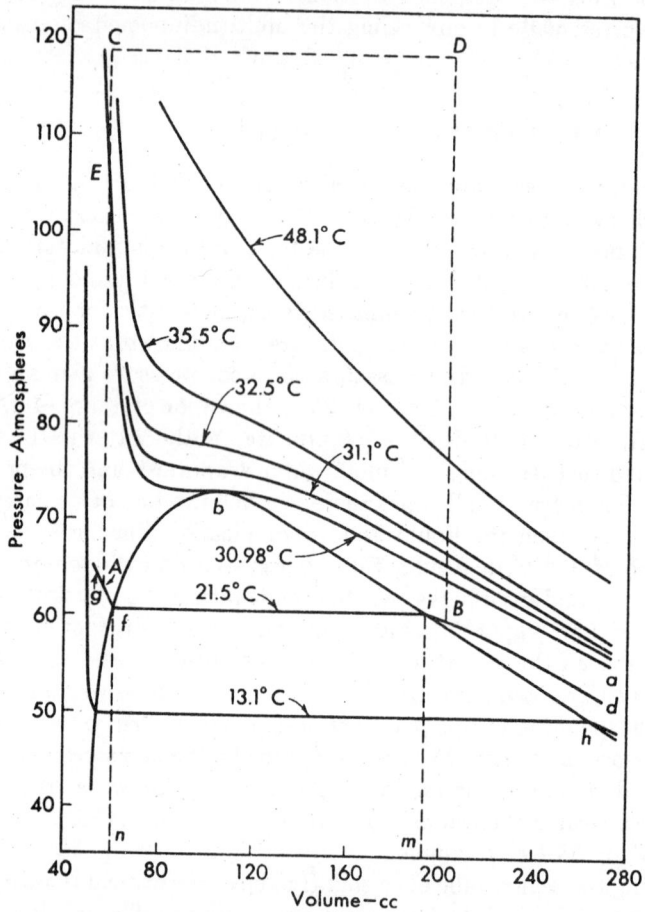

Figure 1–10.   Isothermals of CO_2.

Andrews measured the variation of the volume of carbon dioxide with pressure at various constant temperatures, and he was able to show that the critical temperature of carbon dioxide is 31°C at a critical pressure of 73 atm.

Figure 1–10 shows the plot of pressure vs. volume for carbon dioxide at various constant temperatures. Each *P–V* plot is called an *isothermal*.

[6] Andrews, *Trans. Roy. Soc.*, **159**, 583 (1869).

The data on which the plot is based are not due to Andrews but are the composite results of several subsequent investigators. The 48.1°C isothermal is very similar to the hyperbolic plot demanded by Boyle's law and shows no presence of liquid carbon dioxide even at the highest pressures attained. The same conditions obtain at 35.5°C, 32.5°C, and 31.10°C, except that now the data indicate that Boyle's law when applied to carbon dioxide is considerably in error, since the gas does not behave ideally. At 30.98°C, however, the carbon dioxide remains gaseous only up to a pressure of 73 atm (line ab). At 73 atm (point b) liquid first appears, and since this is the highest temperature at which liquid is observed, 30.98°C must be the critical temperature of carbon dioxide. Further increase in pressure at this temperature (line bE) shows only the presence of liquid, and consequently this line must represent the compressibility of liquid carbon dioxide at this temperature. Below 30.98°C, the behavior of the gas on compression is quite different, as may be judged from the 21.5°C and 13.1°C isotherms. At 21.5°C, for instance, only gas exists on compression along line di. At i liquid, of specific volume n, first appears, and the pressure of the system remains constant thereafter as long as both gas and liquid are present. At this stage further application of pressure results merely in further condensation of gas until point f is reached. At f all the gas has been condensed, and further application of pressure results merely in compression of the liquid, as is shown by the steep line fg. At lower temperatures the behavior is similar to that at 21.5°C, except that the horizontal portions, corresponding to the range of coexistence of liquid and vapor, become longer the lower the temperature.

It may be concluded from this explanation that, in the area to the left of the dome-shaped area and below the line bE, only liquid carbon dioxide will exist; to the right of the line bE and to the right of the dome-shaped area, only gaseous carbon dioxide will exist; while within the dome-shaped area is the range of coexistence of liquid and vapor carbon dioxide.

All gases upon isothermal compression behave similarly to carbon dioxide. For each, of course, the curves will be displaced in line with the characteristics and critical temperature of the gas in question. Thus, for example, the critical temperature of helium is −268°C and the dome-shaped area is moved downward, while for chlorine the critical temperature is 144°C and the dome-shaped area is moved above that for carbon dioxide.

THE PRINCIPLE OF CONTINUITY OF STATES

For further theoretical considerations it is essential to show that the liquid state does not represent a sharp and discontinuous transition from the gaseous state, but is rather a continuation of the gaseous phase into the region of very strong intermolecular attractions and small volumes. This can be shown from the following considerations. Suppose we wish to convert liquid carbon dioxide at 21.5°C and the pressure given by point A in Figure 1–10 to gaseous carbon dioxide at the same temperature and the pressure given by point B. The most obvious way to accomplish this transformation is to follow the 21.5°C isotherm and reduce the pressure along AfiB. In doing this gas appears suddenly and discontinuously, and coexists with liquid along fi until finally all liquid disappears at i. The same transformation may, however, be accomplished in another way. If the liquid at A is heated at constant volume, increase of temperature will lead to increased pressure, and the mass will move along the line AEC. As long as the carbon dioxide is below the critical isotherm, point E, the carbon dioxide is liquid; as soon as the carbon dioxide passes the critical isotherm, however, it becomes gaseous. At the critical temperature, as we have seen, the liquid passes to gas imperceptibly and continuously, and hence in heating the liquid from A to C we convert it without discontinuity to gas. The gas at C may now be expanded to D at constant pressure by heating, and then cooled at constant volume from D to B. By this series of operations we can convert liquid to gaseous carbon dioxide at 21.5°C without introducing any discontinuity between the phases.

The implication involved in this principle of the continuity of the gaseous and liquid states is highly important. It suggests that if we have an equation of state which is satisfactory in the region of high pressures and low temperatures, that equation should be applicable also to the conditions prevailing at the critical point and to the liquid itself. We shall see now how the van der Waals equation meets these requirements.

APPLICATION OF VAN DER WAALS' EQUATION TO THE ISOTHERMALS OF CARBON DIOXIDE

By substituting $n = 1$ and the values of the constants a and b for carbon dioxide in van der Waals' equation, namely,

$$\left(P + \frac{a}{V^2}\right)(V - b) = RT$$

we can calculate for any given temperature the P–V relationships above, at, and below the critical temperature. The results of such a calculation are summarized in Figure 1–11. The plot is, in general, similar to the one obtained experimentally. At t_1, for instance, which is above the critical temperature, the P–V relationship corresponds closely to that of the 48.1°C isotherm in Figure 1–10. At t_c, which is the critical temperature, a slight break is observed at *a*, the critical point, which is again in accord with observation. However, below the critical temperature, the range determining the coexistence of liquid and gas is indicated by a continuous S-shaped portion as *bcd* at t_3, rather than by the horizontal constant pressure range actually observed. In this respect, therefore, and in point of strict quantitative agreement with observed data, the van der Waals

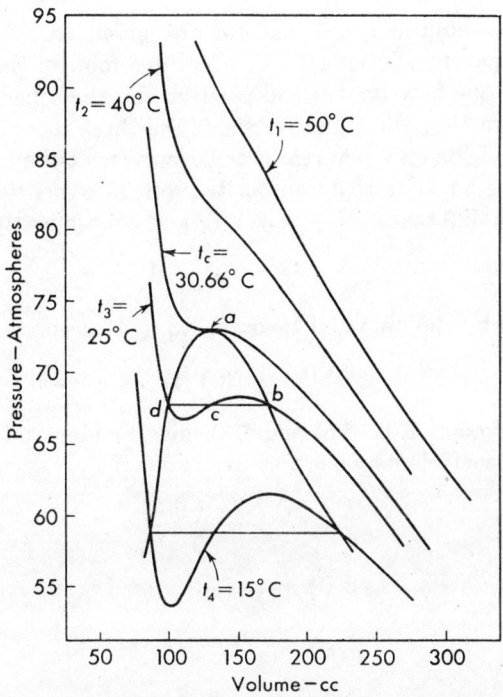

Figure 1–11. Isothermals of CO_2 according to van der Waals' equation.

equation leaves something to be desired. Nevertheless, some investigators have found that by compressing the gas very carefully part of the curve *bc* may be realized, though only in an unstable condition. Similarly, if the pressure on a liquid be released slowly, part of curve *cd* can be obtained, but again the condition is unstable.

DETERMINATION OF VAN DER WAALS CONSTANTS

If it be assumed that van der Waals' equation is applicable at the critical point, then the van der Waals constants for any gas can be calculated from the critical constants of the gas in the following manner. On expanding and rearranging the equation we have

$$\left(P + \frac{a}{V^2}\right)(V - b) = RT$$

$$PV^3 - V^2(RT + Pb) + aV - ab = 0$$

and
$$V^3 - \left(\frac{RT + Pb}{P}\right)V^2 + \left(\frac{a}{P}\right)V - \left(\frac{ab}{P}\right) = 0 \qquad (73)$$

This is a cubic equation in V and for any given value of P and T will yield three separate solutions for V. The three roots of this equation may all be real, or one may be real and positive and the other two imaginary. Thus in Figure 1–11 the equation yields the three roots d, c, and b at t_2, while at t_1 it yields only one real root. However, at the critical point the three roots are not only real and positive but also identical and equal to V_c. Hence the difference $(V - V_c) = 0$, and consequently,

$$(V - V_c)^3 = 0 \qquad (74)$$

On expansion by the binomial theorem Eq. (74) becomes

$$V^3 - (3 V_c)V^2 + (3 V_c^2)V - V_c^3 = 0 \qquad (75)$$

At the critical point Eqs. (75) and (73) must be identical. On comparing and equating coefficients we get

$$3 V_c = \frac{RT_c + bP_c}{P_c} \qquad (76)$$

$$3 V_c^2 = \frac{a}{P_c} \qquad (77)$$

$$V_c^3 = \frac{ab}{P_c} \qquad (78)$$

From Eq. (77) a follows as

$$a = 3 V_c^2 P_c \qquad (79)$$

while from Eqs. (77) and (78) b is given by

$$b = \frac{V_c}{3} \qquad (80)$$

Thus a and b may be calculated from known values of P_c and V_c, or vice versa.

Usually V_c is the critical constant known least accurately, and it is therefore preferable to calculate a and b from T_c and P_c only. This can readily be done. On eliminating V_c between Eqs. (76) and (80) we get

$$b = \frac{RT_c}{8 P_c} \qquad (81)$$

Again, on combining Eqs. (76), (80), and (77), a follows as

$$a = \frac{27}{64} \frac{R^2 T_c^2}{P_c} \tag{82}$$

Combination of Eqs. (76) and (80) leads also to the value of R in terms of the critical constants, namely,

$$R = \frac{8}{3} \frac{P_c V_c}{T_c} = 2.67 \frac{P_c V_c}{T_c} \tag{83}$$

Although the van der Waals equation predicts the coefficient in Eq. (83) to be 2.67, the values for it calculated from experimental data are generally higher and differ for various gases. Thus for helium this constant comes out to be 3.18, while for water it is 4.97. These differences are due to inaccuracies inherent in the van der Waals equation.

THE CRITICAL CONSTANTS OF GASES

Table 1–8 gives the critical constants of a number of gases. Instead of the critical volume, the critical density is given; this is the weight of substance at the critical point per cubic centimeter. The critical volume is obtained by dividing the molecular weight of the substance by the critical density.

Cailletet and Mathias found that when the mean values of the sum of the densities of liquid and saturated vapor of a substance are plotted

TABLE 1–8.   CRITICAL CONSTANTS OF GASES

Gas	t_c (°C)	P_c (atm)	d_c (g/cc)
Ammonia	132.4	111.5	0.235
Argon	−122	48	0.531
Carbon dioxide	30.98	73.0	0.460
Carbon monoxide	−139	35	0.311
Chlorine	144.0	76.1	0.573
Ethane	32.1	48.8	0.21
Ethyl alcohol	243.1	63.1	0.2755
Ethylene	9.7	50.9	0.22
Helium	−267.9	2.26	0.0693
Hydrogen	−239.9	12.8	0.0310
Neon	−228.7	25.9	0.484
Nitric oxide	−94	65	0.52
Nitrogen	−147.1	33.5	0.3110
Oxygen	−118.8	49.7	0.430
Propane	96.81	42.01	0.226
Toluene	320.6	41.6	0.292
Water	374.4	219.5	0.307

against the temperature, the plot is a straight line. This is shown in Figure 1–12. The equation of the line is

$$t = A + B\left(\frac{d_l + d_v}{2}\right) \tag{84}$$

where d_l is the density of the liquid at any temperature t, d_v the density of the saturated vapor at the same temperature, and A and B constants

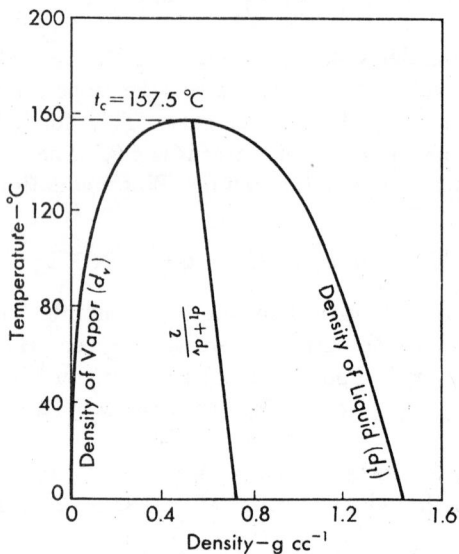

Figure 1–12. Linear variation of mean density of SO_2 with temperature.

evaluated from the plot. Once the equation is determined, the critical density may be calculated with ease, for at the critical temperature $d_v = d_l = d_c$, and the equation reduces to

$$t_c = A + B\left(\frac{2\,d_c}{2}\right) = A + Bd_c \tag{85}$$

Substitution of t_c then yields the critical density. Critical densities can usually be obtained more accurately in this manner than by direct measurement at the critical point.

THE PRINCIPLE OF CORRESPONDING STATES

If we substitute in the van der Waals equation the values of a, b, and R as given by Eqs. (81), (82), and (83), we obtain

$$\left(P + \frac{3\,V_c^2 P_c}{V^2}\right)\left(V - \frac{V_c}{3}\right) = \frac{8}{3}\frac{P_c V_c T}{T_c} \tag{86}$$

Dividing both sides of Eq. (86) by $P_c V_c$, we get

$$\left(\frac{P}{P_c} + \frac{3 \, V_c^2}{V^2}\right)\left(\frac{V}{V_c} - \frac{1}{3}\right) = \frac{8}{3}\frac{T}{T_c}$$

or
$$\left(P_r + \frac{3}{V_r^2}\right)(3 \, V_r - 1) = 8 \, T_r \qquad (87)$$

where $P_r = P/P_c$, $V_r = V/V_c$, and $T_r = T/T_c$. P_r is termed the *reduced pressure*, V_r the *reduced volume*, and T_r the *reduced temperature*. Expressed in terms of P_r, V_r, and T_r, Eq. (87) involves no constants characterizing the individuality of various substances and should therefore be generally applicable to all liquids and gases. It is known as a *reduced equation of state*. Its physical meaning is that at any given value of T_r and P_r, all liquids and gases should have the same corresponding volumes, V_r.

The principle of corresponding states is only approximately correct, but it does suggest that frequently better correlation of experimental data may be obtained when the various substances are in corresponding states, i.e., at equal values of T_r, V_r, or P_r. The principle finds frequent and useful application in thermodynamic and chemical engineering calculations, especially at elevated pressures. For examples see Maron and Turnbull,[7] Dodge,[8] and Gouq-Jen Su.[9]

LIQUEFACTION OF GASES

The particular method employed in the liquefaction of a gas depends on the nature of the gas. Vapors of substances which are liquid at or near room temperature and atmospheric pressure are condensed simply by cooling. Other substances which are liquid at lower temperatures may be condensed either by pressure or by a combination of cooling and compression. Cooling reduces considerably the pressure required for liquefaction, as may be seen from Figure 1–10. With the "permanent" gases, however, such as oxygen, nitrogen, hydrogen, and helium, application of pressure alone will not produce liquefaction, and more involved methods of cooling, compression, and even expansion, are required before the gases will liquefy.

Before liquefaction is possible, a gas must be cooled below its critical temperature. Since their critical temperatures are very low, as may be seen from Table 1–8, liquefaction of the "permanent" gases requires intense cooling as well as considerable compression. To attain these low temperatures, two general principles, or a combination of the two, are

[7] Maron and Turnbull, *Ind. Eng. Chem.*, **34,** 544 (1942).

[8] Dodge, *Chemical Engineering Thermodynamics*, McGraw-Hill Book Company Inc., New York, 1944.

[9] Gouq-Jen Su, *Ind. Eng. Chem.*, **38,** 803 (1946).

employed, namely, (a) adiabatic expansion, in which advantage is taken
of the Joule-Thomson effect[10] to attain cooling; and (b) allowing the gas
to cool itself by performing work in an adiabatic expansion against a
piston. These two methods are exemplified in the Linde and Claude
processes for the liquefaction of air.

The basic principle of the Linde process is the adiabatic Joule-Thomson
expansion and consequent cooling of the air. The steps in the process are
in outline as shown in Figure 1–13. Air is first compressed to approxi-

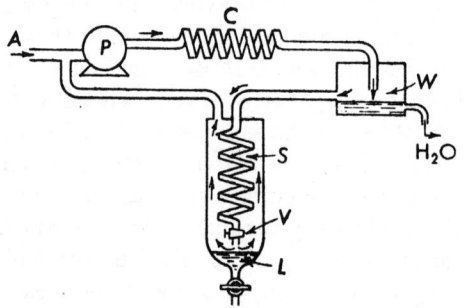

Figure 1–13.   Linde process for liquefaction of air.

mately 100 atm. During the compression most of the water in the air con-
denses and is removed. The heat generated in compression is removed
by passing the gas through coils C, refrigerated by water or ammonia.
The dry gas is passed, then, through a copper spiral coil S, from which
it is expanded to almost atmospheric pressure through a controlled valve
V. The issuing gas, cooled now due to the Joule-Thomson effect, passes
over the copper coil and cools further the incoming compressed gas. On
repeating the cycle several times, the temperature of the expanding gas
finally drops far enough to condense part of the air to liquid, which
collects in the bottom of the chamber L and can be drawn off. Any uncon-
densed air is recirculated.

In the Claude process the gas, instead of being permitted to expand
freely, is forced to do work against a confining piston. Since the gas is
adiabatically insulated, work is achieved at the expense of the internal
energy of the gas, and a cooling results. The work thus gained may be
utilized to operate the compressors.

Easily liquefiable gases, such as sulfur dioxide, ammonia, methyl
chloride, and dichloro-difluoromethane (freon), are used in refrigeration
and air conditioning. In the laboratory other refrigerants frequently
employed are ice, liquid air, liquid hydrogen and mixtures of "dry ice"
(solid carbon dioxide) and alcohol, ether, or acetone. With one of the

---

[10] See Chapter 3.

last-named mixtures temperatures of -80 to $-90°C$ can be obtained. Liquid air will give a temperature of $-180°C$, while, if needed, liquid hydrogen can give a temperature of $-250°C$.

VISCOSITY

Gases and liquids possess a property known as *viscosity*, which may be defined as the resistance that one part of a fluid offers to the flow of another part of the fluid. Viscosity is produced by the shearing effect of moving one layer of the fluid past another, and is quite distinct from inter-molecular attraction. It may be thought of as caused by the internal friction of the molecules themselves and it is present in ideal gases as well as in real gases and liquids.

To define viscosity, let us visualize a fluid as being stratified into layers or planes of molecules. Let the area of each plane be A, and the distance between planes be dy. Further, consider each of the planes to be moving to the right with velocities v_1, v_2, etc., where each succeeding velocity is greater than the preceding by an amount dv. Flow occurring in this manner is called laminar flow, as distinct from turbulent flow where parallelism of the planes is not preserved. In laminar flow the force f required to maintain a steady velocity difference dv between any two parallel planes is directly proportional to A and dv, and is inversely pro-portional to dy. Consequently,

$$f = \eta A \left(\frac{dv}{dy} \right) \tag{88}$$

where η is a proportionality constant called the *viscosity coefficient* of the fluid. The quantity dv/dy in Eq. (88) is referred to as the rate of shear G, while f/A, the force per unit area, is called the shear stress F. In terms of F and G Eq. (88) becomes

$$\eta = \frac{F}{G} \tag{89}$$

Either Eq. (88) or Eq. (89) may be taken as the defining expression for η.

The viscosity coefficient may be thought of as the force per unit area required to move a layer of fluid with a velocity difference of 1 cm per second past another parallel layer 1 cm away. Although the force f may vary with experimental conditions, the viscosity coefficient η is a physical quantity characteristic of each fluid. In the cgs system of units the vis-cosity coefficient of a fluid is expressed in *poises*, a poise being the viscosity coefficient requiring a force of 1 dyne when A, dv, and dy are all unity in Eq. (88). Since this unit is rather large, the viscosities of gases are usually given in *micropoises*, or 10^{-6} poise, while those of liquids in poises or *centipoises*, i.e., 10^{-2} poise.

THE VISCOSITY OF GASES

The viscosity of gases can be measured by various methods, some of which will be described in the next section. Results show that the viscosity coefficients of gases *increase* with increase in temperature. Thus chlorine at 1 atm pressure has an η of 132.7 micropoises at 20°C, 167.9 at 100°C, and 208.5 at 200°C. Again, although η is almost independent of pressure at low pressures, such is not the case at higher pressures. For instance, for carbon dioxide at 35°C and 1 atm pressure $\eta = 156$ micropoises, but at 80 atm and the same temperature $\eta = 361$ micropoises.

The kinetic theory of gases ascribes viscosity to a transfer of momentum from one moving plane to another. Considerations of this momentum transfer between flow planes show that for ideal gases η is related to the density of the gas ρ, the mean free path l, and the average velocity of the gas molecules v by the equation

$$\eta = \frac{1}{3} v l \rho \tag{90}$$

Since the mean free path varies inversely as the density of the gas, it may be concluded that the viscosity of an ideal gas should be independent of density, and hence also the pressure. This deduction has been confirmed at relatively low pressures.

Equation (90) may be employed to calculate the mean free path directly from the viscosity coefficients. To do this we need only substitute the value of v from Eq. (43), in which case l becomes

$$l = \frac{3\,\eta}{v\rho} = \frac{3\,\eta}{0.921\,\rho\,\sqrt{3\,RT/M}}$$
$$= \frac{1.88\,\eta}{\rho\,\sqrt{RT/M}} \tag{91}$$

Once l is thus found, it may be inserted into Eq. (48) to obtain the molecular diameter σ of the gas molecules involved.

THE VISCOSITY OF LIQUIDS

Liquids exhibit much greater resistance to flow than gases, and consequently they have much higher viscosity coefficients. The viscosity coefficients of gases increase with temperature, while those of most liquids decrease. Again, we have seen that the viscosity coefficients for gases at moderate pressures are essentially independent of pressure, whereas with liquids increase of pressure leads to an increase in viscosity.

Most methods employed for the measurement of the viscosity of liquids are based on either the Poiseuille or Stokes equations. The Poiseuille equation for the coefficient of viscosity of a fluid is

$$\eta = \frac{\pi P r^4 t}{8 L V} \qquad (92)$$

where V is the volume of liquid of viscosity η which flows in time t through a capillary tube of radius r and length L under a pressure head of P dynes per square centimeter. This equation has been verified repeatedly. To determine the viscosity of a liquid by this equation it is not always necessary to measure all the quantities indicated when once the viscosity of some reference liquid, usually water, is known with accuracy. If we measure the time of flow of the same volume of two different liquids through the same capillary, then according to the Poiseuille equation the ratio of the viscosity coefficients of the two liquids is given by

$$\frac{\eta_1}{\eta_2} = \frac{\pi P_1 r^4 t_1}{8 L V} \cdot \frac{8 L V}{\pi P_2 r^4 t_2} = \frac{P_1 t_1}{P_2 t_2}$$

Since the pressures P_1 and P_2 are proportional to the densities of the two liquids ρ_1 and ρ_2, we may write also

$$\frac{\eta_1}{\eta_2} = \frac{P_1 t_1}{P_2 t_2} = \frac{\rho_1 t_1}{\rho_2 t_2} \qquad (93)$$

Consequently, once ρ_1, ρ_2, and η_2 are known, determination of t_1 and t_2 permits the calculation of η_1, the viscosity coefficient of the liquid under consideration.

The quantities t_1 and t_2 are most conveniently measured with an Ostwald viscometer, Figure 1–14. A definite quantity of liquid is introduced into the viscometer immersed in a thermostat and is then drawn up

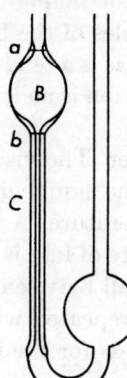

Figure 1–14.   Ostwald viscometer.

by suction into bulb B until the liquid level is above the mark a. The liquid is then allowed to drain, and the time necessary for the liquid level to fall from a to b is measured with a stopwatch. The viscometer is now cleaned, the reference liquid added, and the whole operation repeated. In this simple manner t_1 and t_2 are obtained, and the viscosity of the liquid is calculated by Eq. (93).

Stokes's law is concerned with the fall of bodies through fluid media. A *spherical* body of radius r and density ρ, falling under gravity through a fluid of density ρ_m, is acted on by the gravitational force f_1,

$$f_1 = \frac{4}{3}\pi r^3(\rho - \rho_m)g \tag{94}$$

where g is the acceleration of gravity. This force, which tends to accelerate the body falling through the fluid medium, is opposed by frictional forces within the medium which increase with increase in velocity of the falling body. Eventually a uniform rate of fall is reached at which the frictional forces become equal to the gravitational force, and thereafter the body will continue to fall with a *constant velocity* v. Sir George G. Stokes showed that, for a spherical body falling under the conditions of constant uniform velocity, the force of friction, f_2, is given by

$$f_2 = 6\,\pi r \eta v \tag{95}$$

Equating the gravitational and frictional forces, we see that

$$\frac{4}{3}\pi r^3(\rho - \rho_m)g = 6\,\pi r \eta v$$

$$\eta = \frac{2\,r^2(\rho - \rho_m)g}{9\,v} \tag{96}$$

This equation, known as *Stokes's law*, is applicable to the fall of spherical bodies in all types of fluid media provided the radius of the falling body r is large compared with the distance between the molecules of the fluid. When r is smaller than the distance between molecules there is a tendency for the falling body to "drop" or "channel," and the equation is no longer applicable.

Stokes's law is the basis of the falling sphere viscometer. The viscometer consists of a vertical cylindrical tube filled with the liquid under test and immersed in a thermostat at the desired temperature. A steel ball, of density ρ and a diameter suitable to give a slow rate of fall, is now dropped through the neck of the tube, and the time of fall between two marks is determined with a stopwatch. If the process is repeated with a liquid of known density and viscosity, then Eq. (96) yields for the ratio of the two viscosities

$$\frac{\eta_1}{\eta_2} = \frac{(\rho - \rho_{m_1})t_1}{(\rho - \rho_{m_2})t_2} \qquad (97)$$

Therefore, knowing one of the viscosities, the density of the ball, and the densities of the two liquids, the viscosity of the liquid under study can be calculated by means of Eq. (97) from the observed values of t_1 and t_2.

A term frequently employed in connection with viscosity is *fluidity*. The fluidity, ϕ, of a substance is merely the reciprocal of the viscosity coefficient, namely, $\phi = 1/\eta$.

Table 1-9 gives the viscosity coefficients in centipoises of several liquids at various temperatures. With very rare exceptions (liquid carbon

TABLE 1-9.   VISCOSITY COEFFICIENTS OF LIQUIDS
(Centipoises)

Liquid	0°C	20°C	40°C	60°C	80°C
Benzene	0.912	0.652	0.503	0.392	0.329
Carbon tetrachloride	1.329	0.969	0.739	0.585	0.468
Ethyl alcohol	1.773	1.200	0.834	0.592	—
Ethyl ether	0.284	0.233	0.197	0.140	0.118
Mercury	1.685	1.554	1.450	1.367	1.298
Water	1.792	1.002	0.656	0.469	0.357

dioxide at low temperatures), the viscosity of a liquid decreases with increase in temperature. Various equations have been proposed to represent η as a function of T, of which the simplest is

$$\log \eta = \frac{A}{T} + B \qquad (98)$$

A and B are constants, and T is the absolute temperature. This equation holds quite well for a large number of pure liquids.

REFERENCES

1. B. F. Dodge, *Chemical Engineering Thermodynamics*, McGraw-Hill Book Company, Inc., New York, 1944.
2. H. S. Green, *The Molecular Theory of Fluids*, Interscience Publishers, Inc., New York, 1952.
3. J. O. Hirshfelder, C. F. Curtis, and R. B. Bird, *The Molecular Theory of Gases and Liquids*, John Wiley & Sons, Inc., New York, 1954.
4. E. Kennard, *Kinetic Theory of Gases*, McGraw-Hill Book Company, Inc., New York, 1938.
5. C. J. Pings and B. H. Sage, "Equations of State," *Ind. Eng. Chem.*, **49**, 1315 (1957).

6. H. S. Taylor and S. Glasstone, *A Treatise on Physical Chemistry*, D. Van Nostrand Company, Inc., New York, 1951, Vol. II.
7. A. Weissberger, *Physical Methods of Organic Chemistry*, Interscience Publishers, Inc., New York, 1959, Vol. I, Chap. 12.

PROBLEMS

Note: Unless otherwise indicated, assume all gases in the following problems to be ideal.

1. Four grams of CH_4 at 27.0°C and a pressure of 2.50 atm occupy a volume of 2.46 liters. Calculate the value of the gas constant R in cc-atm degree^{-1} mole^{-1}.

2. Two grams of O_2 are confined in a 2-liter vessel by a pressure of 1.21 atm. What is the temperature of the gas in °C?   *Ans.* 200°C.

3. A certain gas occupies a volume of 6 liters under a pressure of 720 mm Hg at 25°C. What volume will this gas occupy under standard conditions of temperature and pressure?

4. At 0°C and under a pressure of 1000 mm Hg, a given weight of N_2 occupies a volume of 1 liter. At −100°C the same weight of gas under the same pressure occupies a volume of 0.6313 liter. Calculate the absolute zero in degrees centigrade, and give reasons for the observed difference from the accepted value.
Ans. −271.2°C.

5. Find the density of ammonia gas at 100°C when confined by a pressure of 1600 mm Hg.

6. Assuming that dry air contains 79% N_2 and 21% O_2 by volume, calculate the density of moist air at 25°C and 1 atm pressure when the relative humidity is 60%. The vapor pressure of water at 25°C is 23.76 mm Hg.
Ans. 1.171 g/liter.

7. The composition of a mixture of gases in percentage by volume is 30% N_2, 50% CO, 15% H_2, and 5% O_2. Calculate the percentage by weight of each gas in the mixture.

8. (a) Find the weight of helium gas necessary to fill a balloon whose capacity is 1,000,000 liters at 1 atm pressure and 25°C. (b) What will be the lifting power of this balloon in grams per liter in the air described in problem 6?   (c) What will be its total lifting power in kilograms?

9. At 27°C, 500 cc of H_2, measured under a pressure 400 mm Hg, and 1000 cc of N_2, measured under a pressure of 600 mm Hg, are introduced into an evacuated 2-liter flask. Calculate the resulting pressure.   *Ans.* 400 mm Hg.

10. Find the total pressure exerted by 2 g of ethane and 3 g of CO_2 contained in a 5-liter vessel at 50°C.

11. The time required for a given volume of N_2 to diffuse through an orifice is 35 sec. Calculate the molecular weight of a gas which requires 50 sec to diffuse through the same orifice under identical conditions.   *Ans.* 57.15 g/mole.

12. Compare the times of diffusion through a given orifice, and under the same conditions of temperature and pressure, of the gases H_2, NH_3, and CO_2 relative to that of N_2.

13. By means of a mercury vapor pump a vacuum of 10^{-7} mm Hg is obtained within a certain apparatus. Calculate the number of molecules which still remain in 1 cc of the apparatus at 27°C. *Ans.* 3.24×10^9.

14. What is the total kinetic energy of translation in ergs of 2 moles of a perfect gas at 27°C? In calories?

15. Calculate the root-mean-square velocity in centimeters per second of N_2 molecules at 27°C. Repeat the calculation at 127°C.

16. Calculate the root-mean-square, average, and most probable velocities in centimeters per second of H_2 molecules at 0°C.

17. The molecular diameter of CO is 3.19×10^{-8} cm. At 300°K and a pressure of 100 mm Hg what will be (a) the number of molecules colliding per cubic centimeter per second; (b) the number of bimolecular collisions; and (c) the mean free path of the gas? *Ans.* (a) 2.23×10^{27}; (b) 1.12×10^{27}; (c) 6.87×10^5 cm.

18. Repeat the calculations called for in problem 17 for the same temperature but a pressure of 200 mm Hg. How pronounced is the effect of pressure on the quantities sought?

19. Repeat the calculations called for in problem 17 for a pressure of 100 mm Hg and a temperature of 600°K. How pronounced is the effect of temperature on the quantities calculated?

20. By use of the van der Waals equation, find the temperature at which 3 moles of SO_2 will occupy a volume of 10 liters at a pressure of 15 atm.
Ans. 350°C.

21. (a) Using the van der Waals equation, calculate the pressure developed by 100 g of CO_2 contained in a volume of 5 liters at 40°C. (b) Compare this value with that calculated using the ideal gas law.

22. At 0°C and under a pressure of 100 atm the compressibility factor of O_2 is 0.927. Calculate the weight of O_2 necessary to fill a gas cylinder of 100-liter capacity under the given conditions.

23. Using the Beattie-Bridgeman equation explicit in volume, calculate the density in grams per cubic centimeter of N_2 at 0°C and 100 atm pressure.
Ans. 0.127 g/cc.

24. Utilizing the virial coefficients listed in Table 1–4, determine analytically the pressure at which the PV vs. P plot for N_2 at −50°C exhibits a minimum.

25. Employing the Kamerlingh Onnes equation of state, find the compressibility factors of CO at −50°C and pressures of (a) 10, (b) 100, and (c) 1000 atm pressure. *Ans.* (a) $z = 0.981$.

26. The following data were taken in measuring the molecular weight of a certain gas by the Regnault method:

$$
\begin{aligned}
\text{Wt. of evacuated bulb} &= 42.5050 \text{ g} \\
\text{Wt. of bulb} + \text{gas} &= 43.3412 \text{ g} \\
\text{Wt. of bulb} + H_2O &= 365.31 \text{ g} \\
\text{Temperature} &= 25°C \\
\text{Pressure (corrected)} &= 745 \text{ mm}
\end{aligned}
$$

Find the molecular weight of the gas.

27. In a Victor Meyer experiment involving the determination of the molecular weight of ethyl alcohol the data obtained were

$$
\begin{aligned}
\text{Wt. of liquid taken} &= 0.1211 \text{ g} \\
\text{Volume of air measured over water} &= 67.30 \text{ cc} \\
\text{Temperature} &= 28.0°C \\
\text{Atmospheric pressure} &= 755.2 \text{ mm Hg (corrected)} \\
\text{Vapor pressure of water at } 28°C \text{ (from tables)} &= 28.3 \text{ mm Hg}
\end{aligned}
$$

From these data (a) calculate the molecular weight of the alcohol, and (b) compare the result with that calculated from the atomic weights.

Ans. (a) 46.5 g/mole.

28. The elementary analysis of a compound yielded the following results: C, 39.98%; H, 6.72%; and O, 53.30%. In a Victor Meyer determination 0.1510 g of the vaporized compound displaced 33.8 cc of air measured at 25°C over H_2O and at a barometric pressure of 745 mm. Calculate (a) the empirical formula, (b) the approximate molecular weight, and (c) the molecular formula of the compound.

29. A sample of vapor weighing 0.180 g occupies a volume of 53.1 cc at 27°C and 760 mm pressure (corrected). The critical pressure of the vapor is 47.7 atm, while the critical temperature is 288.5°C. By use of the Berthelot equation calculate the molecular weight of the vapor, and compare the result with that calculated by the ideal gas law.

30. The densities of CH_4 at 0°C were measured at several pressures with the following results:

Pressure (atm)	Density (g/liter)
¼	0.17893
½	0.35808
¾	0.53745
1	0.71707

Find the exact molecular weight of CH_4.   Ans. 16.03 g/mole.

31. How much heat will be required to raise the temperature of 3 moles of helium from 0°C to 100°C at (a) constant volume and (b) constant pressure?

32. Utilizing the data given in Table 1–4, find the Boyle temperature of carbon monoxide.

33. (a) Calculate the van der Waals constants for C_2H_6 from the critical temperatures and pressures listed in Table 1–8. (b) Using the constants thus calculated find the pressure exerted by 10 g of C_2H_6 when contained in a liter flask at 13°C. *Ans.* (b) 7.39 atm.

34. The van der Waals constants for HCl are $a = 3.67$ atm-liter^2 mole^{-2}, and $b = 40.8$ cc mole^{-1}. Find the critical constants of this substance.

35. A modified form of the van der Waals equation (Berthelot) is

$$\left(P + \frac{n^2\alpha}{TV^2}\right)(V - n\beta) = nRT$$

where all the terms have their usual significance, and α and β are constants. Deduce the expressions for α, β, and R in terms of the critical constants.

36. Calculate the critical density of methyl alcohol from the following data:

t°C	P (atm)	$d_{liq.}$ (g/cc)	$d_{vap.}$ (g/cc)
150	13.57	0.6495	0.01562
225	61.25	0.4675	0.1003

The critical temperature is 240.0°C.

37. Compare the reduced pressures of N_2 and NH_3 when each exerts a pressure of 100 atm. *Ans.* N_2: 2.99; NH_3: 0.90.

38. Compare the reduced temperatures of ethylene and H_2 at 27°C.

39. Set up the reduced equation of state for the modified van der Waals equation given in problem 35.

40. The equation of state of a liquid gives the volume as a function of the temperature and pressure. Further, the thermal coefficient of expansion, α, is defined as

$$\alpha = \frac{1}{V}\left(\frac{\partial V}{\partial T}\right)_P$$

while the compressibility coefficient, β, is defined as

$$\beta = -\frac{1}{V}\left(\frac{\partial V}{\partial P}\right)_T$$

Assuming α to be independent of temperature and β to be independent of pressure, deduce the expression for V as a function of T and P.

41. (a) For liquid benzene $\alpha = 1.24 \times 10^{-3}$ degree^{-1} at 20°C and 1 atm pressure. Utilizing the equation derived in problem 40 and assuming α to be independent of temperature, find the percentage change in volume of a sample of benzene on being heated at 1 atm pressure from 20 to 50°C. (b) What would be the percentage change in volume of an ideal gas heated over the same temperature interval at constant pressure? *Ans.* (a) 3.8%; (b) 10.2%.

42. (a) For liquid benzene $\beta = 9.30 \times 10^{-5}$ atm^{-1} at 20°C and 1 atm pressure. Utilizing the equation derived in problem 40 and assuming β to be independent of pressure, find the percentage change in volume of a sample of benzene on being compressed at constant temperature from a pressure of 1 atm to a pressure of

11 atm. (b) What would be the percentage change in volume of an ideal gas compressed over the same pressure interval at constant temperature?

43. (a) Suppose that a sample of benzene, initially at 20°C and 1 atm pressure, is subjected to a pressure of 11 atm at 50°C. Assuming α and β to be constant, find the percentage change in volume of the benzene. (b) What would be the percentage change in volume exhibited by an ideal gas on being subjected to the same change in pressure and temperature?

44. The viscosity coefficient of gaseous Cl_2 at 1 atm pressure and 20°C is 147.0 micropoises. Find the molecular diameter of the chlorine molecule.

Ans. 4.30×10^{-8} cm.

45. Consider two parallel layers of NH_3 gas, one of large area and stationary, and the other 10 cm^2 in area and moving at a fixed distance of 1×10^{-6} cm above the first. What force in dynes will be required to keep the upper layer moving with a velocity of 5 cm per second when the pressure of the gas is 10 mm Hg and the temperature is 300°K? The molecular diameter of the NH_3 molecule is 3.0×10^{-8} cm.

46. A gas whose viscosity is 200 micropoises flows through a capillary tube 2 mm in diameter and 2 meters long. If 5 liters of gas pass through the tube every 10 seconds, what must be the pressure head under which the gas is flowing?

Ans. 5.09×10^5 dynes/cm^2.

47. The time of efflux of H_2O through an Ostwald viscometer is 1.52 minutes. For the same volume of an organic liquid of density 0.800 g/cc the time is 2.25 minutes. Find the viscosity of the liquid relative to that of water and its absolute value in millipoises. The temperature is 20°C.

48. A steel ball of density 7.90 g/cc and 4 mm diameter requires 55 seconds to fall a distance of 1 meter through a liquid of density 1.10 g/cc. Calculate the viscosity of the liquid in poises.

49. A sphere of radius 5×10^{-2} cm and density of 1.10 g/cc falls at constant velocity through a liquid of density 1.00 g/cc and viscosity of 1.00 poise. What is the velocity of the falling sphere? *Ans.* 5.45×10^{-2} cm/sec.

50. Suppose that all the conditions given in problem 49 are the same except that the density of the sphere is 0.90 g/cc. What is the velocity of the sphere in this case? Explain the significance of the result.

51. Using the data for the viscosity coefficients of C_2H_5OH as a function of temperature given in Table 1-9, find for this substance the constants A and B in Eq. (98).

2
The Solid
State

$\mathcal{S}$OLIDS differ from liquids and gases in possessing both definite volume and definite shape. The geometric stability of a solid is not due to any difference in compactness between the solid and liquid states, for the density of a substance in the solid state may actually be less than that of the corresponding liquid, as in the case of ice and water. The definite shape of a solid is to be ascribed rather to the fact that the structural units, instead of being in random motion like the molecules of a liquid or gas, are confined to definite positions of equilibrium within the crystal of the solid, positions about which the particles may vibrate but which they cannot readily leave.

Solid substances are frequently classed as either *crystalline* or *amorphous*. A crystalline solid is one in which the constituent structural units are arranged in a definite geometrical configuration characteristic of the substance. Amorphous substances, on the other hand, although possessing many of the attributes of a solid, such as definite shape, a certain rigidity, and hardness, do not show under test a definite configurational arrangement. For that reason they are not considered to be true solids but rather highly supercooled liquids of very high viscosity. Further, crystalline substances such as ice, sodium chloride, or naphthalene melt sharply at a constant and definite temperature, while amorphous substances like glass or asphalt melt gradually and over a temperature interval. However, under certain conditions an amorphous substance may acquire crystalline characteristics. Thus glass may crystallize on long standing or heating. Again, natural rubber when stretched exhibits a definite pattern on examination with X rays, an indication of the production of a definite configurational arrangement.

CRYSTALLIZATION AND FUSION

A pure liquid on being cooled at constant pressure suffers a decrease in the average translational energy of its molecules, and hence its temperature drops until the freezing point is reached. At this temperature the attractive forces of the molecules are sufficient to overcome the translational energy, and the molecules are forced to arrange themselves in a geometric pattern which is characteristic for each substance. When crystallization starts, heat is evolved. This heat evolution arrests further temperature drop, and the temperature of the mixture of solid and liquid remains constant as long as both phases are present. Further removal of heat results merely in the crystallization of more liquid, until finally the whole mass solidifies; only then does the temperature begin to fall again on cooling. The amount of heat evolved per mole of substance is called the *heat of crystallization* of the substance.

The reverse of crystallization is the fusion or melting of the solid. As the pure solid is heated, its average vibrational energy increases, until at the melting point some particles are vibrating with sufficient energy to overcome the confining forces. The solid then begins to fuse. For a given pressure the temperature at which this occurs is the *same* as the crystallization temperature. To accomplish further fusion, heat must be supplied to compensate for the loss of the particles with high energy. The amount of heat which must be *absorbed* to accomplish the transition of one mole of solid to one mole of liquid is known as the *heat of fusion*. This amount of heat is equal in magnitude, but is opposite in sign to the heat of crystallization of the substance.

CRYSTALLOGRAPHY

Crystallography is the branch of science which deals with the geometry, properties, and structure of crystals and crystalline substances. Geometric crystallography is concerned with the outward spatial arrangement of crystal planes and the geometric shape of crystals, and is based on three fundamental laws, namely: (a) the law of constancy of interfacial angles; (b) the law of rationality of indices; and (c) the law of symmetry. The law of constancy of interfacial angles states that for a given substance the corresponding faces or planes forming the external surface of a crystal always intersect at a definite angle and that this angle remains constant no matter how the faces develop. Commonly it is observed that the crystal planes are unequally developed so as to produce faces of variable size and shape; but the angle of intersection of any two corresponding faces is always found to be the same for any crystal of the same substance.

For any crystal a set of three coordinate axes can be so chosen that all the faces of the crystal will either intercept these axes at definite distances from the origin, or be parallel to some of the axes, in which case the intercepts are at infinity. The law of rationality of indices or intercepts, proposed in 1784 by Haüy, states that it is possible to choose along the three coordinate axes unit distances (a, b, c), not necessarily the same length, such that the ratio of the three intercepts of any plane in the crystal is given by $(ma:nb:pc)$, where m, n, and p are either integral whole numbers, including infinity, or fractions of whole numbers. The law may be illustrated with data on crystals of the mineral topaz, $Al_2(FOH)_2SiO_4$, for which four different planes have the parameters,

$$
\begin{aligned}
&1. \quad m = 1 \qquad n = 1 \qquad p = 1 \\
&2. \quad m = 1 \qquad n = 1 \qquad p = \infty \\
&3. \quad m = 1 \qquad n = 1 \qquad p = \tfrac{2}{3} \\
&4. \quad m = 2 \qquad n = 1 \qquad p = \infty
\end{aligned}
$$

and hence the intercept ratios are

$$
\begin{aligned}
&1. \quad a:b:c \\
&2. \quad a:b:\infty c \\
&3. \quad a:b:\tfrac{2}{3}\,c \\
&4. \quad 2\,a:b:\infty c
\end{aligned}
$$

For any particular plane these ratios characterize the plane and may consequently be used to represent it. The coefficients of a, b, and c are known as the *Weiss indices* of a plane. However, Weiss indices are rather awkward in use and have consequently been replaced by Miller indices. The Miller indices of a plane are obtained by taking the reciprocals of the Weiss coefficients and multiplying through by the smallest number that will express all the reciprocals as integers. Thus a plane which in the Weiss notation is given by $a:b:\infty c$ becomes in the Miller notation $a:b:0\,c$, or simply (110), since the order a, b, c is understood. Similarly, a face $a:\infty b:\tfrac{1}{4}\,c$ becomes (104). As an exercise the student may verify the statement that the four planes mentioned above for topaz are respectively (111), (110), (223), and (120) in the Miller system of crystal face notation.

The third law of crystallography states simply that all crystals of the same substance possess the same elements of symmetry. There are three possible types of symmetry. First, if a crystal can be divided by an imaginary plane passed through its center into two equal portions each of which is a mirror image of the other, the crystal is said to possess a *plane of symmetry*. Second, a crystal is said to possess *line symmetry* if it is possible to draw an imaginary line through the center of the crystal and then revolve the crystal about this line through 360° in such a way as to cause the crystal to appear unchanged two, three, four, or six times. Depending on the number of times the crystal appears unchanged on revolution, the crystal is said to possess two-, three-, four-, or sixfold symmetry. Finally, a crystal is said to possess a center of symmetry if every face has an identical face at an equal distance on the opposite side of this center. The total number of plane, line, and center symmetries possessed by a crystal is termed the *elements of symmetry* of the crystal.

THE CRYSTAL SYSTEMS

There are 230 crystal forms possible, and practically all have been observed. On the basis of their symmetry these 230 crystal forms may be grouped into 32 classes, and these in turn may be referred to six crystal systems. All the crystals belonging to a particular system are characterized by the fact that, although they may not all have the same elements of symmetry, they can all be referred to a particular set of crystallographic axes which differ from system to system in length of the various axes and the angles of inclination between axes. Table 2–1 lists the six crystal systems, their axial characteristics, the *maximum* symmetry which may be expected in each system, and some examples of substances crystallizing in the various systems. For a discussion of the various geometric forms which correspond to each of these systems, the student must be referred to treatises on the subject.[1]

Such elementary forms as the cube, the octahedron, and the dodecahedron, a figure possessing 12 sides each of which is a rhombus, are all forms corresponding to the regular system, and all possess the maximum symmetry. A form of lower symmetry, but still in the same system, is the tetrahedron.

PROPERTIES OF CRYSTALS

For gases, liquids, and unstrained amorphous solids, such properties as index of refraction, coefficient of thermal expansion, thermal and electrical conductivity, and rate of solubility are all independent of direction. The same is true of substances crystallizing in the regular system. Such substances exhibiting the same properties in all directions are said to be *isotropic*. However, for substances crystallizing in the other crystal sys-

[1] See Kraus, Hunt, and Ramsdell, *Mineralogy*, McGraw-Hill Book Company, Inc., New York, 1951.

TABLE 2–1. CRYSTAL SYSTEMS AND THEIR CHARACTERISTICS

System	Axial Characteristics	Maximum Symmetry	Examples
1. Regular (cubic or isometric)	Three axes at right angles. Unit distances: $a = b = c$	Nine planes Thirteen axes	NaCl KCl Alums Diamond CaF_2 (Fluorspar)
2. Tetragonal	Three axes at right angles, only two of equal length. Unit distances: $a = b \neq c$	Five planes Five axes	TiO_2 (Rutile) $ZrSiO_4$ (Zircon) SnO_2 (Cassiterite)
3. Hexagonal	Two axes of equal length in one plane making an angle of 120° with each other, and a third axis at right angles to these and of unequal length. Unit distances: $a = b \neq c$	Seven planes Seven axes	PbI_2 Mg Beryl CdS (Greenockite) ZnO (Zincite)
4. Orthorhombic (rhombic)	Three axes at right angles, but all of different length. Unit distances: $a \neq b \neq c$	Three planes Three axes	KNO_3 Rhombic sulfur K_2SO_4 $BaSO_4$ (Baryte) $PbCO_3$ (Cerrusite)
5. Monoclinic	Three axes, all unequal. Two axes at right angles, the third inclined to these at an angle other than 90°. Unit distances: $a \neq b \neq c$	One plane One axis	$Na_2SO_4 \cdot 10\ H_2O$ $Na_2B_4O_7 \cdot 10\ H_2O$ $CaSO_4 \cdot 2\ H_2O$ Monoclinic sulfur
6. Triclinic	Three axes of unequal length, all inclined at angles other than 90°. Unit distances: $a \neq b \neq c$	No planes No axes	$CuSO_4 \cdot 5\ H_2O$ $K_2Cr_2O_7$ H_3BO_3

tems the properties enumerated above may vary according to the axis along which observation is made. Substances exhibiting directional differences in properties are said to be *anisotropic*. For such substances the coefficients of thermal expansion may not only differ in different directions along the crystal but may actually be positive in one direction and negative in another, as is the case with silver iodide. Again, unlike isotropic solids, anisotropic solids exhibit more than one index of refraction for the same crystal. From the standpoint of optics, anisotropic substances are subdivided into *uniaxial* and *biaxial* crystals. Uniaxial crystals, embracing the tetragonal and hexagonal systems, possess two indices of refraction, according to the axis along which observation is made. Thus ice, which is hexagonal, has the refractive indices 1.3090 and 1.3104, both for sodium D-line. Biaxial crystals, on the other hand, embracing the orthorhombic, monoclinic, and triclinic systems, have three indices of refraction, as for instance the triclinic potassium dichromate, for which the indices of refraction, depending on the axis of observation, are 1.7202, 1.7380, and 1.8197, all for the sodium D-line.

The anisotropic character of certain crystals is also responsible for directional differences in solubility observed with many solids. In such solids certain faces of the crystal will dissolve faster than others to produce characteristic patterns on the surface called *etch figures*. These etch figures are quite typical and may be employed to characterize the nature and even composition of the substance. This method of identification and estimation is employed very extensively in metallographic analysis.

POLYMORPHISM

Many substances exist only in a single solid crystalline form. Quite frequently it is found, however, that certain substances occur in more than one solid modification or undergo a change of crystalline form on heating or under pressure. The existence of a substance in more than one modification is known as *polymorphism*. Thus carbon exists in crystalline form as either diamond or graphite, calcium carbonate occurs as calcite or aragonite, while sulfur has been found to exist in a variety of solid modifications. Polymorphism occurring in elements, as in the last illustration, is more commonly referred to as *allotropy*. Some polymorphs exist as modifications in the same crystal system, as in ammonium chloride where both forms are cubic, or they may crystallize in different crystal systems, as in rhombic and monoclinic sulfur, or hexagonal and cubic silver iodide.

Each polymorphic form of a substance is thermodynamically stable within a particular temperature and pressure range, and the transformation from one form to another takes place at any given pressure at a fixed temperature known as the *transition temperature* or *transition point*. The rhombic variety of sulfur is stable at atmospheric pressure up to 95.6°C, at which temperature it is transformed into the monoclinic form. Conversely, on cooling slowly monoclinic sulfur, rhombic sulfur will not appear until the temperature of 95.6°C has been reached. At this temperature monoclinic and rhombic sulfur are in equilibrium, and the temperature will not change until *all* the monoclinic sulfur has been transformed to the rhombic variety. In this respect the transition temperatures bear a very striking similarity to fusion points and are sometimes employed as fixed points in thermometry.

In instances like the preceding the change is entirely reversible, i.e., it proceeds in both directions, and the temperature at which both forms are in equilibrium is definite and constant. However, the change from one form to the other often is not reversible but proceeds in one direction only. The fact is that if more than one form of a substance is found to exist within the same temperature and pressure range, only one of these

TABLE 2-2. TRANSITION POINTS OF POLYMORPHIC SUBSTANCES
AT ATMOSPHERIC PRESSURE

Substance	Transition	Transition Temp.
Sulfur	Rhombic $\rightleftharpoons$ Monoclinic	95.6°
Tin	Gray $\rightleftharpoons$ White (tetragonal)	18°
	White $\rightleftharpoons$ Rhombic	170°
Ammonium nitrate	Tetragonal $\rightleftharpoons$ (Rhombic)$_1$	−17°
	(Rhombic)$_1$ $\rightleftharpoons$ (Rhombic)$_2$	32.1°
	(Rhombic)$_2$ $\rightleftharpoons$ Rhombohedral	84.2°
	Rhombohedral $\rightleftharpoons$ Cubic	125.2°
Potassium nitrate	Rhombic $\rightleftharpoons$ Rhombohedral	128.5°
Silver iodide	Hexagonal $\rightleftharpoons$ Cubic	146.5°
Silver nitrate	Rhombic $\rightleftharpoons$ Rhombohedral	159.5°

is stable. All others are unstable and *tend* to change continuously and irreversibly into the stable form. Nevertheless, the rate of change is often extremely slow, and the unstable form or forms may exhibit to all appearances all the attributes of stability. An instance of such a metastable substance is aragonite ($CaCO_3$), which, although apparently stable, may be transformed into the more stable calcite. However, the reverse transformation is impossible at ordinary pressures, a fact which speaks for the thermodynamic instability of aragonite.

Table 2-2 lists several polymorphic substances, the transformations which they undergo, and the transition temperatures at one atmosphere pressure.

THE STRUCTURE OF CRYSTALS

We have seen above that on purely geometrical grounds crystallographers have found it possible to classify all crystals into 32 classes of symmetry and into 6 crystal systems. However, such a classification tells nothing about the *internal* structure of the crystals. To gain insight on the latter point, crystallography postulates that any macroscopic crystal of a substance is built up by repetition and extension in all directions of a fundamental structural unit known as a *unit crystal lattice* or *space group*. Each unit lattice, in turn, must be constituted of atoms, molecules, or ions, as the case may be, arranged to give the particular geometrical configuration of the lattice. Further, the geometric shape of the unit lattice must be the same as that of the macroscopic crystal; i.e., if the crystal is a cube, the unit lattice will also have its constituents arranged so as to give a tiny cube.

TABLE 2–3.   DISTRIBUTION OF SPACE LATTICES
AND GROUPS

System	Number of Space Lattices	Number of Space Groups
Cubic	3	36
Tetragonal	2	68
Hexagonal	2	52
Orthorhombic	4	59
Monoclinic	2	13
Triclinic	1	2
	14	230

A mathematical analysis of these ideas, combined with the fundamental laws of crystallography, has shown that there are only 14 basic arrangements, known as *space lattices*, in terms of which the internal structure of crystals can be described. These space lattices are shown in Figure 2–1, where the black circles indicate the locations of the atoms, molecules, or ions composing the structure. The number of these space lattices falling within a given system is listed in the second column of Table 2–3. Starting with the space lattices in a particular system, it is possible to combine these into a definite number of more intricate groupings, the *space groups*, whose number is shown in the last column of the table. Thus the three space lattices occurring in the cubic system may be elaborated into 36 space groups without violating the symmetry requirements, while the two basic designs of the tetragonal system may be combined to give 68 possible patterns of arrangement. Table 2–3 shows that the 14 space ttices can yield a total of 230 space groups, a number practically iden- al with that of the kinds of crystals actually observed.

Although crystallography has been able to delimit the number of space ps occurring in a given system, it could not define the specific space

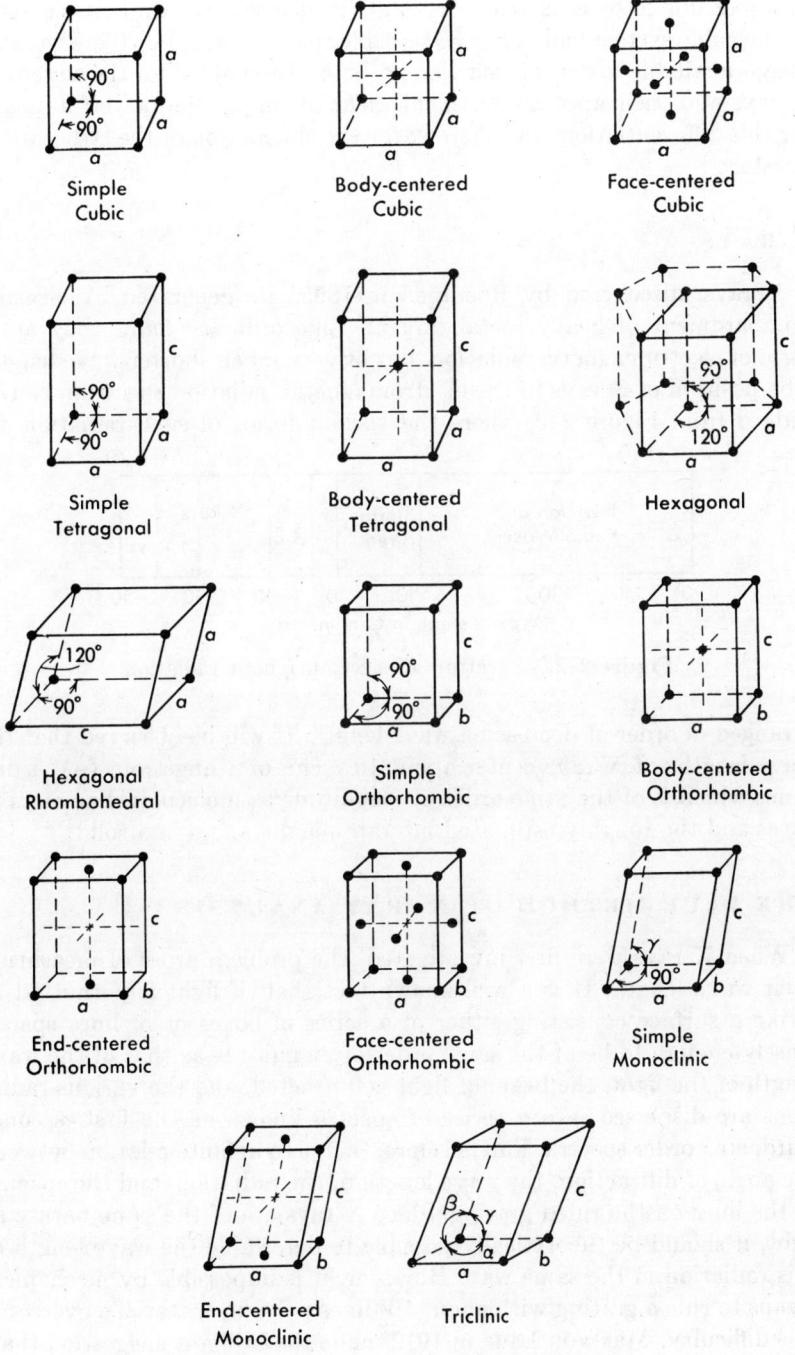

Figure 2–1.   The fourteen space lattices.

group exhibited by a particular crystal. Thus it was known that a crystal in the cubic system had one of 36 possible space groups, but there were no means available to distinguish among these. It remained for the advent of X rays and their application to this field to supply the powerful means for this differentiation, and thereby for the elucidation of the structure of crystals.

X RAYS

X rays, discovered by Roentgen in 1895, are generated by electron bombardment of heavy metal targets. Like ordinary light, they are a form of electromagnetic radiation, but of very much shorter wave length. The position of X rays in the electromagnetic radiation spectrum can be judged from Figure 2–2, where the various forms of such radiation are

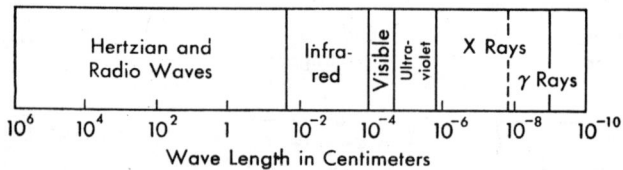

Figure 2–2.   Spectrum of electromagnetic radiation.

arranged in order of decreasing wave length. It will be observed that the wave lengths of X rays center about 10^{-8} cm, or 1 angstrom (Å), a distance which is of the same order of magnitude as molecular diameters in gases and the roughly estimated interatomic distances in a solid.

THE LAUE METHOD OF X-RAY ANALYSIS

When X rays were first investigated, the problem arose of measuring their wave length. It is a well-known fact that, if light is permitted to strike a surface consisting either of a series of edges or of lines spaced closely enough to be of the same order of magnitude as that of the wave length of the light, the beam of light is diffracted, and the various radiations are dispersed into a series of spectra known as the first, second, third, etc., order spectra. Furthermore, there is a definite relation between the angle of diffraction, the wave length of the radiation, and the spacing of the lines on the ruled grating. Since X rays are of the same nature as light, it should be theoretically possible to determine the wave length of this radiation in the same way. However, it is impossible by mechanical means to rule a grating with about 10^8 lines per centimeter. To overcome this difficulty, Max von Laue in 1912 made the brilliant suggestion that, if a crystal consists actually of an orderly arrangement of atoms, then the atomic planes in the crystal should be spaced at intervals of about 10^{-8} cm, and the crystal should act then as a natural and very fine three-dimensional diffraction grating for X rays. He further predicted that, if a beam of inhomogeneous X rays were directed against a crystal and a photographic plate placed behind it, the image obtained on the latter would show a series of spots arranged in a geometrical fashion about the center of the beam.

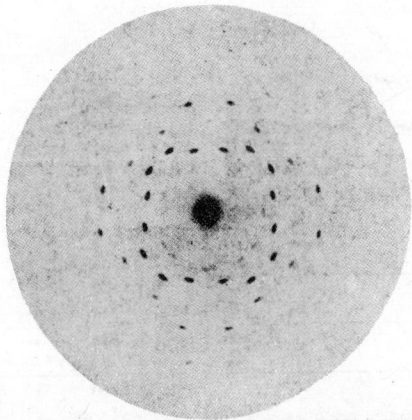

Figure 2–3.   Laue diffraction pattern of zinc blende. (From A. S. Shankland, *Atomic and Nuclear Physics, 2/E,* The Macmillan Company, New York.)

Experimental investigation of the idea in 1913 verified von Laue's predictions in every respect. A series of diffraction patterns were obtained for various substances which showed differences characteristic of the materials examined. One of these Laue diagrams, that of zinc blende (ZnS), is shown in Figure 2–3. These diagrams speak for a definite arrangement of the atoms in a crystal, and actually permit a reconstruction of the crystal arrangement which would account for the particular distribution of each Laue pattern. However, the method of reconstruction is highly complicated and difficult. A much simpler method of crystal analysis is that suggested by W. H. and W. L. Bragg, and this will be discussed now in some detail.

THE BRAGG METHOD OF CRYSTAL ANALYSIS

The Braggs first called attention to the fact that, since a crystal is composed of a series of equally spaced atomic planes, it may be employed not only as a transmission grating, as in the Laue method, but also as a reflection grating. A beam of X rays striking the atoms which constitute these planes will be diffracted then in such a manner as to cause either interference with or reinforcement of the beam diffracted from the first, or outer, plane, and the whole beam will behave as if it had been *reflected* from the surface of the crystal.

To understand better the theory of this method consider, as shown in Figure 2–4, a wave front $GG'G''$ of X rays approaching at an angle θ a series of parallel, equidistant planes, W, X, Y, Z, etc., which constitute the atomic planes of the crystal. Part of the beam HGO will be reflected at

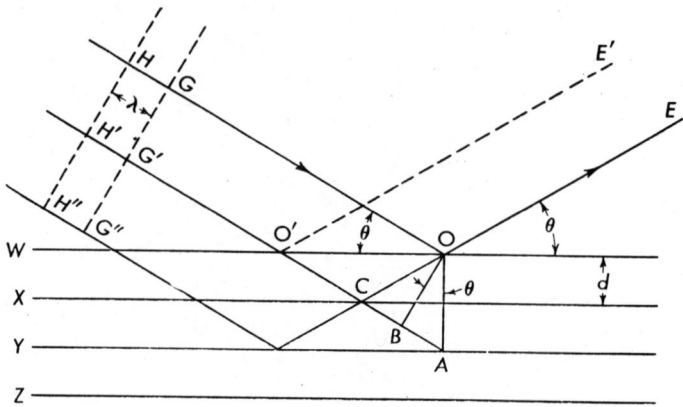

Figure 2–4.   Reflection of X rays by parallel planes.

O along OE at the angle of reflection θ, which is the same as the angle of incidence. Similarly, the beam $H'G'O'$ will be reflected partly at O' along $O'E'$, and then again at C on the second plane along COE. In order to emerge along OE the second beam has to travel a longer distance than the first, namely, the distance $H'G'O'CO$ as against the distance HGO. If the difference in distance of the two paths is exactly equal to an integral multiple of the wave length of the radiation, the two beams will be in phase at O, will reinforce each other, and the intensity of the reflected rays will be at a maximum. When the two beams are out of phase, however, interference will result, and the intensity of the reflected beam will be less than the maximum. The condition for maximum reflection intensity is then that the distance

$$G'CO - GO = n\lambda$$

where λ is the wave length of the X rays employed, while n is an integer taking on the values 1, 2, 3, $\cdots$ and known as the *order of reflection*.

If a perpendicular is drawn now from O to the extension of the line $G'O'C$, while another is dropped from O to A perpendicular to W, X, and

Y, it follows that

$$G'CO = G'A - CA + CO$$
$$= G'A$$

since $CO = CA$ from the construction of the figure. But

$$GO = G'B$$

Therefore, $\qquad G'CO - GO = G'A - G'B$
$$= BA$$

and hence $\qquad BA = n\lambda$

It can readily be shown that the angle BOA is also θ. Then, since OB was constructed perpendicular to $G'A$,

$$\sin \theta = \frac{BA}{OA}$$

and $\qquad BA = OA \sin \theta$
$$= 2\,d \sin \theta$$

where d is the distance between any two atomic planes in the crystal.

Therefore, $\qquad n\lambda = 2\,d \sin \theta \qquad\qquad$ (1)

This simple equation connects directly the wave length and order of reflection of the X rays with the interplanar distance d and the angle of maximum reflection θ. Without any further information the ratio λ/d is deducible by measuring n and θ. On the other hand, if λ is known, d may be calculated; or vice versa, if d is known, the crystal may be employed to determine the wave length of the X rays.

The reflection angles θ and the intensities of the reflected beams corresponding to these angles can be determined with the Bragg X-ray spectrometer, a diagram of which is shown in Figure 2–5. The X rays generated in tube A by bombardment of a suitable target B are passed through a series of slits and screen (C, D, E) to give a sharp and monochromatic beam, and are then directed to strike the face of a crystal which is suitably mounted on a turntable F. The graduated turntable may be rotated to give any angle of incidence desired. Coaxially with the table and crystal is mounted an ionization chamber H, into which the reflected beam is passed. The ionization of the gas filling the chamber, usually sulfur dioxide, is proportional to the intensity of the X rays passing through the chamber. Since the current passing through the ionization chamber is proportional to the ionization of the gas, the intensity of such a current as measured by an electrometer gives directly a measure of the intensities of the X rays that are reflected from the crystal. By determining the intensities of the reflected beam at various angles of reflection, the angles at which maximum reflection occurs may be readily found.

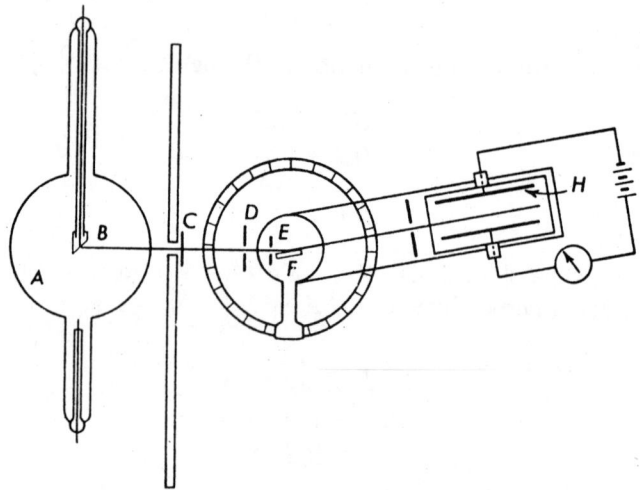

Figure 2–5.   Bragg X-ray spectrometer.

THE X‑RAY ANALYSIS OF SODIUM CHLORIDE

The procedure employed to elucidate the structure of sodium chloride will now be described in some detail, and this example may be taken as a general indication of the methods involved in studying other types of crystals. Sodium chloride has been chosen because it belongs to the cubic system, exhibits the highest type of symmetry in crystals, and is simple to study.

The unit lattice of sodium chloride, like the macroscopic crystal, must be a cube, and the atoms of sodium and chlorine must be arranged, as we have seen, in some combination of only three possible space lattices. These are known respectively as the *simple cubic*, the *face-centered cubic*, and the *body-centered cubic* arrangements. In the simple cubic lattice an atom is located at each of the corners of the cube. The face-centered cubic lattice in turn involves a simple cubic arrangement modified by the location of an atom in the center of each of the six faces of the cube. Finally, the body-centered cubic lattice is again a simple cubic arrangement but modified this time by the presence of a single atom in the center of the cube.

In the cubic system the planes that can be passed through the atoms have the Miller indices (100), (110), or (111), as may be seen from Figure 2–6. However, the ratios of the distances among the (100), (110), and (111) planes in the three types of cubic lattices are not the same. If we arbitrarily designate by a the distance between 100 planes in the simple cubic lattice, then the perpendicular distance between (110) planes will be $a/\sqrt{2}$, and the distance between (111) planes will be $a/\sqrt{3}$. In the face-centered cubic lattice, on the other hand, parallel planes can be interposed halfway between the (100) and the (110) planes in the simple cubic lattice, and hence the interplanar distances in this case are respectively

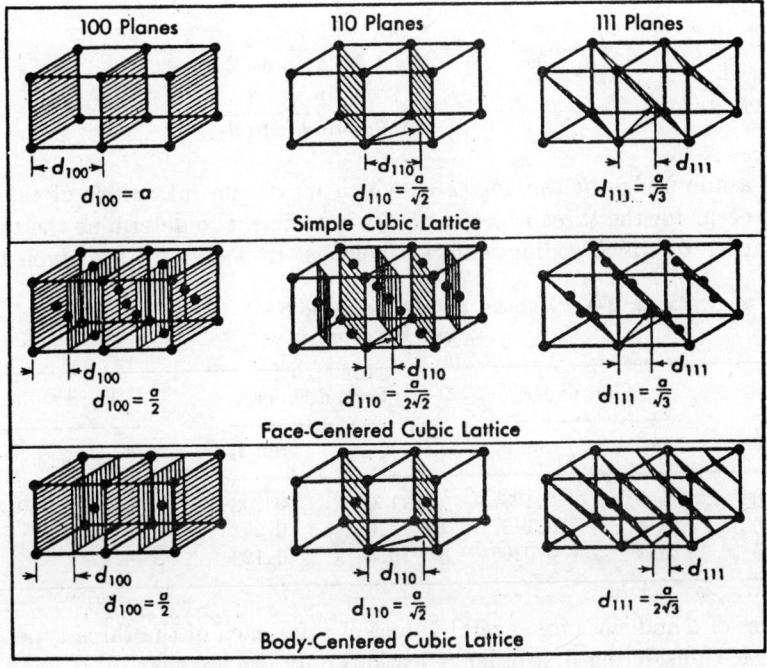

Figure 2–6.   Planes in cubic lattices.

$a/2$, $a/(2\sqrt{2})$, and $a/\sqrt{3}$. Finally, in the body-centered cubic lattice parallel planes can be interposed halfway between any (100) or (111) planes in the simple cubic lattices, so that in terms of a the respective distances become $a/2$, $a/\sqrt{2}$, and $a/(2\sqrt{3})$. These distances for the various planes in the three types of cubic lattices are summarized in Figure 2–6. The ratios $d_{100}:d_{110}:d_{111}$ for the simple cubic, face-centered cubic, and body-centered cubic lattices are then:

$$\text{Simple cubic:} \quad d_{100}:d_{110}:d_{111} = a:\frac{a}{\sqrt{2}}:\frac{a}{\sqrt{3}} = 1:0.707:0.577$$

$$\text{Face-centered cubic:} \quad d_{100}:d_{110}:d_{111} = \frac{a}{2}:\frac{a}{2\sqrt{2}}:\frac{a}{\sqrt{3}} = 1:0.707:1.154$$

$$\text{Body-centered cubic:} \quad d_{100}:d_{110}:d_{111} = \frac{a}{2}:\frac{a}{\sqrt{2}}:\frac{a}{2\sqrt{3}} = 1:1.414:0.577$$

It will be observed that the ratios of the distances in the three cases are all different, and hence a determination of the interplanar distance ratios in sodium chloride should permit a decision as to the type of lattice to which this substance belongs. Since for any given reflection order and wave length of X rays $d = (n\lambda)/(2 \sin \theta)$, then for nth order reflection maxima from the (100), (110), and (111) planes we have

$$d_{100}:d_{110}:d_{111} = \frac{n\lambda}{2 \sin \theta_1}:\frac{n\lambda}{2 \sin \theta_2}:\frac{n\lambda}{2 \sin \theta_3}$$

$$= \frac{1}{\sin \theta_1}:\frac{1}{\sin \theta_2}:\frac{1}{\sin \theta_3} \tag{2}$$

and a knowledge of the angles at which maximum intensities of reflection occur for the three types of planes is sufficient to determine the type of lattice to which sodium chloride belongs. In Table 2–4 are given the

TABLE 2–4.   ANGLES AT MAXIMA FOR NaCl USING K LINE
FROM PALLADIUM

Planes	First Order		Second Order		Third Order	
	θ_1	Sin θ_1	θ_2	Sin θ_2	θ_3	Sin θ_3
(100)	5.9	0.103	11.9	0.208	18.2	0.312
(110)	8.4	0.146	17.0	0.292	—	—
(111)	5.2	0.0906	10.5	0.182	—	—

values of θ and sin θ for several values of n for each of the three types of planes. Considering first order reflections only, we see that

$$d_{100}:d_{110}:d_{111} = \frac{1}{\sin \theta_1}:\frac{1}{\sin \theta_2}:\frac{1}{\sin \theta_3} = \frac{1}{0.103}:\frac{1}{0.146}:\frac{1}{0.0906}$$

$$= 1:0.705:1.14$$

Comparison of these ratios with those previously established for the three possible types of cubic lattices shows that in sodium chloride the atoms must be arranged in a face-centered cubic lattice. The same conclusion may be arrived at by considering the results for second order reflection maxima.

POSITIONS OF SODIUM AND CHLORINE ATOMS IN LATTICE

Determining the type of lattice along which the atoms in sodium chloride are arranged does not solve completely the problem of the structure of sodium chloride, for the question of the relative arrangement of the atoms of sodium and chlorine in the lattice still remains. This question can be resolved by considering the relative intensities of the reflection maxima for the different orders and planes. In Table 2–5 are given the intensities for various orders of reflection for the (100), (110), and (111) planes, the intensity of the first order maximum for the (100) plane being taken as 100.

The intensity of a diffracted beam depends, in the first place, on the mass of the particle responsible for the diffraction, the larger the mass the greater being the intensity; and, second, on the order of diffraction, the

TABLE 2–5.   RELATIVE INTENSITIES OF REFLECTION
MAXIMA IN NaCl

Order	Intensities for (100) Planes	Intensities for (110) Planes	Intensities for (111) Planes
First	100.00	50.4	9.00
Second	19.90	6.10	33.1
Third	4.87	0.71	0.58
Fourth	0.79		2.82
Fifth	0.12		0.14

intensity decreasing in a definite manner with increase in order. These two facts are employed to explain the results of Table 2–5. It will be observed that in the (100) and (110) planes the intensities decrease progressively with order. This systematic decrease can be accounted for quantitatively by assuming that such planes contain equal numbers of sodium and chlorine atoms. In the (111) planes, however, an alternation of intensities is observed, the first order being weaker than the second and the third weaker than both first and second, while the fourth is weaker than the second but stronger than the third. It is possible to account for this by postulating that the (111) planes are composed alternately of sodium atoms and chlorine atoms, and that the planes containing sodium atoms only are interposed halfway between the (111) planes containing only chlorine atoms.

Study has shown that the only possible arrangement of sodium and chlorine atoms which will satisfy the facts elicited from the study of the intensities is that shown in Figure 2–7. In this arrangement atoms of sodium, shown by black circles, are located in the corners of the cube and in the center of each of the six faces. Halfway between every two sodium atoms is located a chlorine atom, shown by open circles. A moment's reflection will show that the indicated structure meets the required conditions and that it consists essentially of two interpenetrating face-centered cubic lattices, one composed entirely of sodium atoms, the other of chlorine atoms. The chlorine lattice is merely displaced the distance a along any edge of the cube.

Although for purposes of discussion we have considered the structural units of sodium chloride to be sodium and chlorine atoms, general consensus at present is that the sodium chloride in the crystal is ionized and that the units are rather sodium and chloride *ions*. Again, in terms of the X-ray interpretation of the structure as outlined here, the designation "molecule of sodium chloride" loses a great deal of its definite meaning. It can hardly be said that any particular chloride ion belongs to any

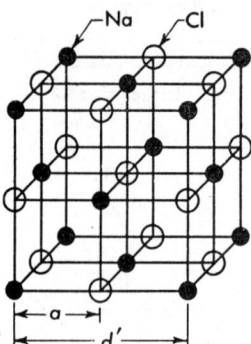

Figure 2–7.   The sodium chloride lattice.

definite sodium ion; rather, each sodium is shared equally by six chloride ions, and each chloride is held equally by six sodium ions. All that can be said is that to each sodium corresponds *one-sixth* of *six* chloride ions, so that each sodium has the equivalent of a chloride, but not any one ion exclusively.

CALCULATION OF d AND λ

The cube indicated in Figure 2–7 is considered to be the unit lattice of sodium chloride, and the edge of the cube is then the distance between any two sodium or any two chloride ions which lie on the edge. This is *twice* the distance between (100) planes. The macroscopic crystal is built up by the extension of unit lattices in all directions.

Once the nature of the unit lattice is established, it is a simple matter to calculate the length of the edge of the unit cube and the distance between (100) planes from the molar volume and Avogadro's number. Each of the sodium ions at the corners is shared by eight cubes, hence to each cube may be ascribed $\frac{1}{8} \times 8$, or one sodium ion. Further, each of the sodium ions in the centers of the faces is shared by two cubes. This adds $\frac{1}{2} \times 6$ or three sodium ions more, making a total of four sodium ions. Similarly, each of the chloride ions along the edges is shared by four cubes, and since there are 12 of these, $\frac{1}{4} \times 12$ or three chloride ions are part of the cube. Adding to these the one in the center of the cube, we obtain four chloride ions as the average chloride content of the cube. It may be said, therefore, that each unit lattice of sodium chloride contains on the average four sodium ions and four chloride ions, or a *total equivalent to four molecules of sodium chloride.*

The molar volume of sodium chloride is the molecular weight, 58.443, divided by the density, 2.165 g/cc, or

$$V_m = \frac{58.443}{2.165} = 26.99 \text{ cc}$$

This is the volume occupied by $N = 6.023 \times 10^{23}$ molecules. The volume occupied by four molecules, V, which is also the volume of the unit cube, is, then,

$$V = \frac{V_m \times 4}{6.023 \times 10^{23}} = \frac{26.99 \times 4}{6.023 \times 10^{23}} = 179.2 \times 10^{-24} \text{ cc}$$

From this volume the edge of the unit cube follows as

$$d' = \sqrt[3]{179.2 \times 10^{-24}}$$
$$= 5.64 \times 10^{-8} \text{ cm}$$

Since the edge of the unit cube is twice the distance between (100) planes, the distance d to be employed in the Bragg equation is $d'/2$, or,

$$d = \frac{5.64 \times 10^{-8}}{2} = 2.82 \times 10^{-8} \text{ cm}$$

The best value accepted at present for this distance is 2.81384×10^{-8} cm at 18°C.

Once the interplanar distance for sodium chloride is known, the crystal may be employed to determine the wave length of any X rays which may be directed against it. Thus in the Bragg experiments $\sin \theta$ was found to be 0.103 for $n = 1$. Consequently the X rays employed in these experiments had a wave length of

$$\lambda = \frac{2 \, d \sin \theta}{n} = \frac{2(2.814 \times 10^{-8})0.103}{1}$$
$$= 0.58 \times 10^{-8} \text{ cm} = 0.58 \text{ Å}$$

THE POWDER METHOD OF CRYSTAL ANALYSIS

Debye and Scherrer in 1916, and Hull independently in 1917, devised a method of X-ray crystal analysis which permits the use of a substance in powder form. In this method a monochromatic beam of X rays is focused upon a small tube containing the finely ground substance to be examined. Since in the powder the crystal planes are oriented at all possible angles to the beam, there will be always some crystals with just the proper orientation to give reinforced diffraction images, and from all the planes simultaneously. These diffraction maxima are photographed on a film fixed behind the sample in the form of a circular arc.

The diffracted rays obtained in this manner form concentric cones originating from the powder under examination. Photographed on a narrow strip of film, these images appear as nearly vertical lines arranged on each side of a bright center spot due to the undiffracted beam, although actually these lines are portions of arcs of circles whose center

Figure 2–8.   Powder diffraction diagram of ZnO

is the point of focus of the X-ray beam. Each pair of lines equidistant to the right and left of the center spot corresponds to a single order of diffraction for a family of planes, a pair being obtained for each order of diffraction for each type of plane present. A powder diffraction pattern of zinc oxide is shown in Figure 2–8.

The diffraction patterns obtained from various solids are characteristic of the substances involved. For this reason the Debye-Scherrer-Hull method of X-ray diffraction is frequently employed in qualitative and quantitative chemical analysis to identify and estimate both pure substances and mixtures. When used for analytical purposes, a comparison plate is prepared showing the position of various lines for the various substances which may be expected in the unknown sample. Then several milligrams of the unknown sample are also exposed, and the diffraction plate thus obtained is compared with the reference plate. Occurrence of identical lines in both plates testifies to the presence of the particular substance exhibiting those lines in both samples. The identity of each line is determined by comparison of plates with diagrams obtained from the pure constituents.

For quantitative estimation the intensity of each line must be determined and compared with the intensities obtained from definite amounts of the pure constituents under identical operating conditions. Since the intensity of a line is proportional to the amount of substance present, such a comparison gives directly an estimate of amount.

In one respect X-ray analysis supplies more information than a chemical analysis. From a qualitative chemical analysis of a mixture of, say, calcium chloride and sodium bromide, it is impossible to tell whether the constituents are calcium chloride and sodium bromide or calcium bromide and sodium chloride. The X-ray method answers this question directly, for it will show the lines of calcium chloride and sodium bromide and no lines for the reciprocal salts.

RESULTS OF X - RAY STUDY OF CRYSTALS

Studies similar to that described for sodium chloride have been made upon a great many solid substances to determine their structure. Some of the results of these studies will now be summarized.

Many metals crystallize in the cubic system, with the atoms arranged in face-centered or body-centered lattices. As examples of the first may

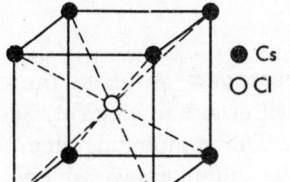

● Cs
○ Cl

Figure 2–9.   The cesium chloride lattice.

be cited aluminum, calcium, nickel, cobalt, copper, silver, platinum, gold, and lead; and of the second, lithium, chromium, sodium, potassium, iron, and tungsten. The simple cubic arrangement does not appear to be overly favored. On the other hand, the sodium chloride structure of two interpenetrating face-centered lattices is very common among compounds in the regular system. It is exhibited by oxides such as those of magnesium, calcium, strontium, barium, nickel, and cobalt, sulfides such as those of magnesium, barium, manganese, and lead, and all of the halides of the alkali metals except those of cesium. The cesium halides occur as two simple cubic lattices, one of cesium and another of the halide, interlocking to form a resultant body-centered lattice with the equivalent of one molecule of cesium halide per unit cube. This structure is shown in Figure 2–9.

Another type of cubic arrangement is found in zinc sulfide, diamond, silicon, germanium, and gray tin. Here we have a face-centered lattice which also contains an atom in the center of each alternate small cube within the larger lattice. On this basis each atom is equidistant from four other atoms. A modification of this scheme is the arrangement found in calcium, strontium, and barium fluorides. In these salts the metal ions are located on a face-centered lattice, while the nonmetal ions are situated in the centers of each of the eight small cubes composing the metallic lattice. The nonmetallic ions thus form a simple cubic lattice inside the face-centered lattice of the metal ions.

A still different cubic structure is that shown by cuprous oxide and silver oxide, Figure 2–10. In crystals of these the oxygens lie on a body-centered cubic lattice, while the metal atoms interpenetrate this lattice to fall at the center of each alternate small cube within the oxygen framework.

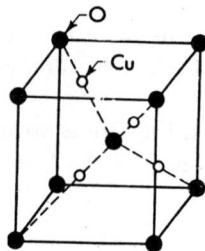

Figure 2–10.   The cuprous oxide lattice.

As an example of a noncubic arrangement may be given the close-packed hexagonal lattice, Figure 2–11, which occurs in elements such as magnesium, zinc, cadmium, and titanium. The compounds zinc oxide, beryllium oxide, cobalt sulfide, and stannous sulfide appear as two such interpenetrating hexagonal lattices, one composed of the metal, the other of the nonmetal. On the other hand, calcium, magnesium, manganese, and iron carbonates and sodium nitrate have a rhombohedral lattice in

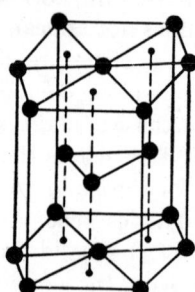

Figure 2–11.   Close-packed hexagonal arrangement.

which the axes are all equal but inclined at equal angles other than 90°. If we imagine this lattice to be a distorted cube, then the metal ions and the anions are on two interpenetrating distorted face-centered lattices comparable to the undistorted sodium chloride lattice. It is of interest to point out that the anions CO_2^{-2} and NO_3^- have been shown to be present in the crystal as a unit, with the oxygens distributed about the central atom. The same is true of other inorganic anions like SO_4^{-2}, PO_4^{-3}, ClO_3^-, and MnO_4^-. These findings are excellent evidence for the dissociation of salts in the solid state and for the existence of these radicals as individual entities.

LATTICE UNITS AND FORCES

Depending on the nature of the substance in question, the units entering into the construction of a crystal lattice may be ions, atoms, or molecules. Inorganic compounds generally have ionic lattices. In such lattices the binding forces are the electrostatic attractions between the oppositely charged ions. As has already been pointed out, for such substances the ordinary definition of a molecule is meaningless, for no particular negative ion, say, may be said to belong exclusively to any particular positive ion. Rather, because of the requirements of electroneutrality, each positive ion has associated with it on the average a sufficient number of negative charges to balance the positive charge. Ionic lattices are generally very stable, and crystals constituted of these melt at relatively high temperatures.

In diamond and graphite the lattice-building units are atoms. Within the diamond lattice each atom of carbon is surrounded by four others in the form of a tetrahedron, an arrangement which confirms the valency of carbon and the directionality of the valence bonds as postulated by the organic chemists. The atoms are held together by covalent forces, and, as these are very strong, the crystals are hard, strong, and high melting. The graphite structure, on the other hand, consists of planar hexagonal rings in which the carbon atoms are tetrahedrally joined by covalent bonds. These planes, in turn, are held together by van der Waals forces. Since the latter are much weaker than covalent forces, the planes separate readily. This fact accounts for the flakiness of graphite.

Metals also have atoms as structural units. Since electrons in metals are loosely held, metals are good conductors of electricity. Again, most metallic crystals are strong, and generally malleable and ductile

In substances like carbon dioxide, hydrogen chloride, and stannic iodide the complete molecule acts as a structural unit within the lattice. X-ray evidence indicates that such molecules occupy the key spatial positions, while the atoms within the molecules are arranged in a definite configuration about the mean position of equilibrium for each molecule. The molecules are held in their geometrical distribution within the lattice by van der Waals forces. These forces are much weaker than the electrostatic attractions in ionic lattices, and consequently such crystals are less strong and melt at considerably lower temperatures. The lower rigidity and higher vapor pressure of these substances as compared with ionic crystals may also be ascribed to the relative weakness of the van der Waals forces.

Another force present in some crystals is hydrogen bonding. Hydrogen bonding forces are somewhat stronger than van der Waals forces, but they are considerably weaker than either covalent or electrovalent bonds. An example of a crystal where hydrogen bonding plays an important role is ice, which crystallizes in the hexagonal system.

THE PACKING OF UNIFORM SPHERES

In the preceding discussion we considered the occupants of lattice sites to be spheres of unspecified dimensions. Further, from the manner in which they were indicated, the student may have gathered the impression that they were small compared with the distances between them. Actually, this is not the case. It is fruitful to consider a lattice as being composed of rigid spheres of definite dimensions packed tightly enough to touch each other. On this basis it is possible to understand more fully the packing differences involved in the various types of lattices, and also to arrive at the radii of the spheres involved.

For what follows, consider a cubic unit lattice of edge length a composed of uniform spheres of radius r. A lattice which will meet these conditions is one for an element, where all the atoms are identical. Suppose, first, that the lattice involved is simple cubic. In a lattice of this type the spheres are packed by laying down a base of spheres, and then piling upon the base other layers in such a way that each sphere is immediately above the one below it. With such an arrangement the front view of the unit lattice will be that shown within the square in Figure 2–12(a). From this figure it may be seen that the distance between the centers of the spheres is a, the radius of each sphere is $a/2$, and the volume of the sphere is $(4\pi r^3)/3 = (4\pi)(a/2)^3/3$. Since in simple cubic packing there is only one sphere present in the unit lattice, and since the volume of the lattice is a^3, then the fraction of the total volume occupied by the sphere is

$$\frac{\frac{4\pi}{3}\left(\frac{a}{2}\right)^3}{a^3} = \frac{\pi}{6} = 0.5236$$

while the remainder, i.e., 0.4764 of the total volume, is empty space or *void volume*.

In the body-centered cubic lattice the situation is different. Here the packing consists of a base of spheres, followed by a second layer where each sphere rests in the hollow at the junction of four spheres below it. The third layer then rests on these in an arrangement which corresponds exactly to that in the first layer. A front view of this arrangement in a body-centered unit lattice is given in Figure 2–12(b). Here the radius of the spheres can be calculated to be $(\sqrt{3}\,a)/4$; and, as there are two spheres in each unit lattice, we obtain for the fraction of the total volume occupied by the spheres

$$\frac{2\left[\dfrac{4\,\pi}{3}\left(\dfrac{\sqrt{3}\,a}{4}\right)^{3}\right]}{a^{3}} = \frac{\sqrt{3}\,\pi}{8} = 0.6802$$

A front view of the packing in a face-centered cubic unit lattice is depicted in Figure 2–12(c). Considerations similar to those used above

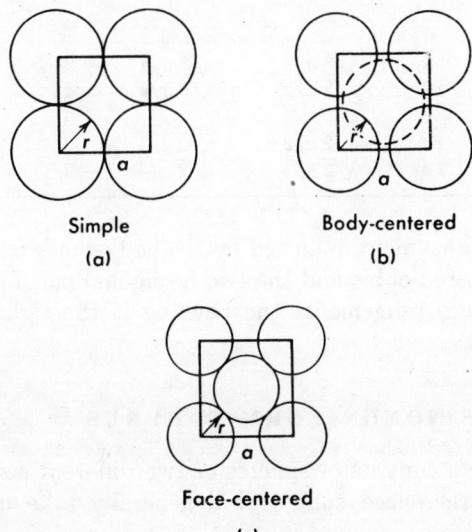

Simple

(a)

Body-centered

(b)

Face-centered

(c)

Figure 2–12.   Packing of spheres in cubic lattices.

show that the sphere radius in such packing is given by $(\sqrt{2}\,a)/4$. Since a face-centered cubic unit lattice contains four spheres, the volume fraction occupied by these is

$$\frac{4\left[\dfrac{4\,\pi}{3}\left(\dfrac{\sqrt{2}\,a}{4}\right)^{3}\right]}{a^{3}} = \frac{\sqrt{2}\,\pi}{6} = 0.7404$$

Table 2–6 gives a summary of the packing in cubic and hexagonal lattices. Given in the table for each situation are the number of spheres nearest to and surrounding a particular central sphere, the distance between sphere centers, the radius of the spheres, the number of spheres per unit lattice, the volume of the lattice, and the packing fraction for each arrangement. In the case of hexagonal lattices a is the same distance as that shown in Figure 2–1. Inspection of the table reveals that the packing fraction is independent of the radius of the spheres, and depends only on the nature of the packing. This packing is least compact for the

TABLE 2–6. PACKING IN CUBIC AND HEXAGONAL LATTICES

Lattice	No. of Nearest Neighbors	Distance Between Neighbors	Radius of Spheres	No. of Spheres per Unit Lattice	Volume of Unit Lattice	Packing Fraction
Cubic						
Simple	6	a	$a/2$	1	a^3	0.5236
Body-centered	8	$\sqrt{3}\,a/2$	$\sqrt{3}\,a/4$	2	a^3	0.6802
Face-centered	12	$\sqrt{2}\,a/2$	$\sqrt{2}\,a/4$	4	a^3	0.7404
Hexagonal						
Close packed	12	$\sqrt{2}\,a/2$	$\sqrt{2}\,a/4$	2	$a^3/2$	0.7404
Rhombohedral	12	$\sqrt{2}\,a/2$	$\sqrt{2}\,a/4$	2	$a^3/2$	0.7404

simple cubic arrangement, followed by the body-centered cubic, and then by the face-centered cubic and the two hexagonal packings. For the last-mentioned three arrangements the packing is the tightest possible for uniform spheres.

PACKING OF NONUNIFORM SPHERES

When a system consists of spheres of two different sizes, as in sodium chloride, the larger-sized spheres will generally take up their definite

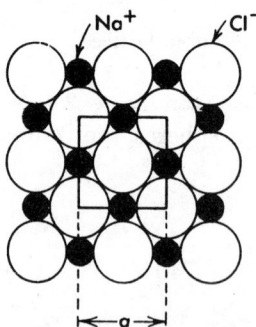

Figure 2–13. Packing of spheres in sodium chloride.

arrangement, and the smaller spheres will have to accommodate them-selves in the voids between the larger spheres. The situation which results in the case of sodium chloride is illustrated for one face of the lattice in Figure 2–13. The face-centered cubic arrangement of the crystal is apparent. Further, the edge of the unit lattice, a, is related to the radii of the two ions, r_+ and r_-, by the equation

$$2\,r_+ + 2\,r_- = a$$

or $$r_+ + r_- = \frac{a}{2} \tag{3}$$

This sum is obviously the interionic distance between the sodium and chloride ions. For body-centered lattices of the CsCl type, the interionic distance is given by

$$r_+ + r_- = \frac{\sqrt{3}\,a}{2} \tag{4}$$

It is possible to set up a set of values for *ionic* radii, Table 2–7, such that their sum gives the interionic distance $r_+ + r_-$. The results obtained

TABLE 2–7. IONIC RADII FOR CRYSTALS
(In Å)

Li⁺	0.60	Be⁺²	0.31	F⁻	1.36
Na⁺	0.95	Mg⁺²	0.65	Cl⁻	1.81
K⁺	1.33	Ca⁺²	0.99	Br⁻	1.95
Rb⁺	1.48	Sr⁺²	1.13	I⁻	2.16
Cs⁺	1.69	Ba⁺²	1.35	O⁻²	1.40
Ag⁺	1.26	Zn⁺²	0.74	S⁻²	1.84

by this means are generally within several per cent of the values obtained by X-ray or electron diffraction. Thus the interionic distance predicted by the radii in Table 2–7 for SrS is $r_+ + r_- = 1.13 + 1.84 = 2.97$ Å, as against the observed value of 3.00 Å.

ELECTRON AND NEUTRON DIFFRACTION

DeBroglie in 1924 first called attention to the fact that moving electrons, besides exhibiting their corpuscular properties, should also possess wave properties characteristic of light. He showed that an electron moving with velocity v should have associated with it a wave of length λ given by

$$\lambda = \frac{h}{mv} \tag{5}$$

where m is the mass of the electron, 9.108×10^{-28} g, while h is Planck's constant, namely, 6.625×10^{-27} erg-sec. The velocity of an electron depends on the potential difference through which it falls. For the potential drop ε in volts it is given as

$$v = 5.93 \times 10^7 \sqrt{\varepsilon} \text{ cm sec}^{-1} \tag{6}$$

Inserting Eq. (6) into Eq. (5), we obtain for λ

$$
\begin{aligned}
\lambda &= \frac{h}{m \times 5.93 \times 10^7 \sqrt{\varepsilon}} \\
&= \frac{6.625 \times 10^{-27}}{9.108 \times 10^{-28} \times 5.93 \times 10^7 \sqrt{\varepsilon}} \\
&= \frac{12.3 \times 10^{-8}}{\sqrt{\varepsilon}} \text{ cm}
\end{aligned}
\tag{7}
$$

For potentials between 10 and 10,000 volts, λ should vary, then, between 3.89 and 0.12 Å, and hence such electrons should behave like X rays toward crystals.

This prediction of DeBroglie's was confirmed in a beautiful manner by Davisson and Germer in 1925. Davisson and Germer investigated the diffraction of electrons by a nickel surface and found that the electron diffraction pattern thus obtained was very similar to the one given with X rays. Further, the wave length of the electrons calculated from the diffraction pattern agreed remarkably well with that calculated from the DeBroglie equation.

These experiments and others not only provide excellent confirmation of DeBroglie's theory but also make available a new tool for investigation of solids. Compared with X rays, electron beams are much less penetrating. Consequently, whereas X rays are diffracted by atomic planes deep within a crystal, electrons are diffracted primarily by planes lying near the surface. Electron diffraction offers, then, a convenient means of investigating the nature of solid surfaces and surface films, and a great deal of work in this field is directed thus toward an elucidation of the nature of oxide and other surface films on solids.

Solids bombarded with high speed neutrons also give diffraction patterns. The wave length of the neutrons is again given by Eq. (5), but m is now 1.675×10^{-24} g. Neutrons are strongly scattered by hydrogen atoms, and hence neutron diffraction is very effective in locating the position of such atoms in crystals.

THE HEAT CAPACITY OF SOLIDS

The variation of the heat capacity with temperature for several solid elements is shown in Figure 2–14. It will be observed that the heat capacities are zero at 0°K, and for elements like aluminum, copper, and silver, they increase very rapidly with temperature, approaching a value of $3 R = 5.97$ cal mole^{-1} at or near room temperature. Carbon and silicon, on the other hand, show a much more gradual increase in heat capacity with temperature and do not attain the $3 R$ value until very much higher temperatures. In fact, the heat capacity of carbon does not become $3\,R$ until above 1300°C.

The first satisfactory approach to the theory of heat capacities of crystalline monatomic solids was made by Albert Einstein in 1907. Einstein pointed out that the atoms constituting a crystal may be considered to be oscillators executing simple harmonic motion about their mean positions of equilibrium in the crystal. He ascribed to each substance a constant and characteristic frequency of vibration, ν, and postulated that

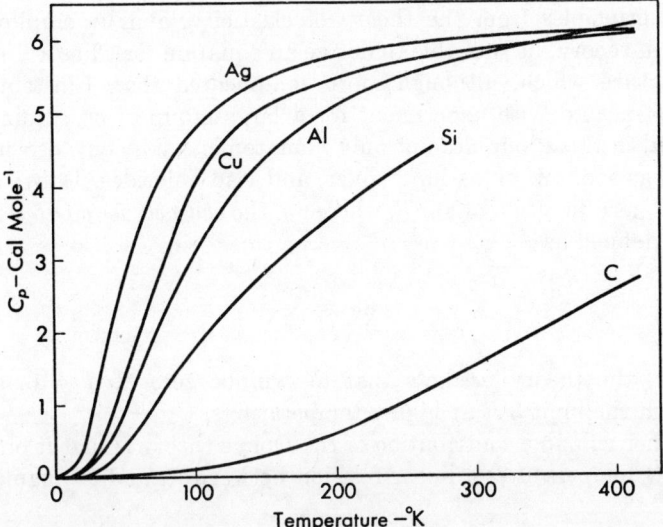

Figure 2–14.   Variation of heat capacity with temperature.

absorption of energy by the oscillators does not take place continuously, but discontinuously in line with Planck's quantum theory.[2] With these assumptions Einstein was able to show that the heat capacity per mole at constant volume C_v, at any temperature T, should be given for a monatomic crystalline solid by the equation

$$C_v = 3\,kN \left(\frac{h\nu}{kT}\right)^2 \frac{e^{h\nu/kT}}{(e^{h\nu/kT} - 1)^2} \tag{8}$$

Here N is Avogadro's number, k the Boltzmann constant R/N, e the base of natural logarithms, h Planck's constant, while θ_E, *the characteristic Einstein temperature*, is defined as

$$\theta_E = \frac{h\nu}{k} \tag{9}$$

[2] See Chapter 14.

Equation (8) indicates that C_v is the same function of θ_E/T for all monatomic solids. Further, this equation predicts that C_v will approach zero at $T = 0$, and that at high temperatures C_v will approach asymptotically the value 3 R. In both these respects the equation is in general agreement with the facts. In the intermediate range, however, the equation gives C_v values which are lower than those actually observed.

A more successful theory of heat capacities of solids is that due to Peter Debye (1912). Debye assumed that a solid may vibrate with any frequency ranging from zero up to a limiting frequency ν_m. By introducing certain principles from the theory of elasticity, and by employing the quantum theory, he was able to derive an equation for C_v as a function of temperature which, although more complicated than Einstein's, is in excellent accord with experiment for a large group of crystalline solids. Included in this group are not only monatomic solids, but also such substances as sodium, potassium, silver, and lead chlorides. Debye's theory yields C_v as a function of θ_D/T, where θ_D, the *characteristic Debye temperature*, is defined as

$$\theta_D = \frac{h\nu_m}{k} \tag{10}$$

Further, the theory predicts that C_v will be zero at $T = 0$, and will approach the limit 3 R at higher temperatures.

Another valuable contribution of the Debye theory is that it predicts at very low temperatures a linear relation between C_v and T^3, namely,

$$C_v = \left(\frac{12\,R\pi^4}{5\,\theta_D^3}\right) T^3 \tag{11a}$$

$$= aT^3 \tag{11b}$$

where a is a constant given by

$$a = \frac{12\,R\pi^4}{5\,\theta_D^3} \tag{12}$$

This equation, known as the *Debye third power law*, has been repeatedly verified. Since experimental determinations of specific heats cannot be carried conveniently below 15 to 20°K, the Debye third power law is employed to estimate the heat capacities of solids between 0 and say 20°K.

A shortcoming of the Debye theory is that it accounts only for heat capacities up to 3 R. Yet certain elements, particularly the alkali metals, reach values of C_v considerably above this limit at high temperatures. The excess absorption of energy is usually ascribed to electrons, of whose displacement by thermal means the Debye theory takes no account.

REFERENCES

1. M. J. Buerger, *Elementary Crystallography*, John Wiley & Sons, Inc., New York, 1956.
2. M. J. Buerger, *Crystal Structure Analysis*, John Wiley & Sons, Inc., New York, 1960.
3. C. W. Bunn, *Chemical Crystallography*, Oxford University Press, New York, 1946.
4. Kraus, Hunt, and Ramsdell, *Mineralogy*, McGraw-Hill Book Company, Inc., New York, 1951.
5. K. Lonsdale, *Crystals and X-Rays*, D. Van Nostrand Company, Inc.. New York, 1949.
6. L. Pauling, *The Nature of the Chemical Bond*, Cornell University Press, Ithaca, N.Y., 1960.
7. R. W. G. Wyckoff, *Crystal Structures*, Interscience Publishers, Inc., New York, 1948–1960.

PROBLEMS

1. A crystal plane intercepts the three crystallographic axes at the multiples of the unit distances $\frac{3}{2}$, 2, and 1. What will be the Miller indices of the plane?
Ans. (436).

2. The Miller indices of a plane in a crystal are (210). (a) What are the Weiss indices of the plane? (b) What do these indices mean?

3. The first order reflection of a beam of X rays from a given crystal occurs at $5° 15'$. At what angle will be the third order reflection?   *Ans.* $15° 56'$.

4. The first order reflection of a beam of X rays from the (100) face of NaCl occurs at an angle of $6° 30'$. (a) What is the wave length of the X rays used? (b) What would be the angle of reflection with X rays of $\lambda = 1.50$ Å?

5. If X rays of $\lambda = 1.540$ Å are used, find the angle at which will be obtained the second order reflection maxima from the (111) planes of NaCl.

6. The first order reflections from the (100), (110), and (111) planes of a given cubic crystal occur at angles of $7° 10'$, $10° 12'$, and $12° 30'$ respectively. To what type of cubic lattice does the crystal belong?

7. CsBr crystallizes in a body-centered cubic unit lattice with an edge length of 4.287 Å. (a) Calculate the angles at which may be expected the second order reflection maxima for the (100), (110), and (111) planes when X rays of $\lambda = 0.500$ Å are used. (b) How many molecules of CsBr will be present in the unit lattice?
Ans. (a) $13° 29'$, $9° 30'$, $23° 49'$; (b) One.

8. The density of CaF_2 is 3.180 g/cc at 20°C. Calculate the dimensions of a unit cube of the substance containing four Ca^{++} and eight F^- ions.   *Ans.* 5.46 Å.

9. Aluminum crystallizes in a face-centered cubic lattice. Its density is 2.70 g/cc at 20°C. Calculate the distance between successive (100) planes and the distance of closest approach of Al atoms in the crystal.

10. Cesium chloride, whose density is 3.97 g/cc, crystallizes in a body-centered cubic lattice 4.12 Å on edge, and the equivalent of one CsCl molecule per unit cube. From these data calculate the value of Avogadro's number.

11. Tungsten crystallizes in a body-centered cubic lattice with a unit cube side length of 3.16 Å. (a) How many atoms of tungsten are present in a unit lattice? (b) What is the density of the metal?   *Ans.* (b) 19.3 g/cc.

12. A certain compound, whose density is 4.56 g/cc, crystallizes in the tetragonal system with unit lattice distances of $a = b = 6.58$ Å and $c = 5.93$ Å. If the unit lattice contains four molecules, calculate the molecular weight of the compound.

13. Show that in face-centered cubic packing of uniform spheres $D = a/\sqrt{2}$, where D is the sphere diameter and a is the length of the cube edge.

14. Show that in body-centered cubic packing of uniform spheres $r = \sqrt{3}\, a/4$, where r is the sphere radius and a the length of the cube edge.

15. At 1425°C Fe(δ) crystallizes in a body-centered cubic lattice whose edge length is 2.93 Å. Assuming the atoms to be packed spheres, calculate (a) the radius of the spheres, (b) the distance between centers of neighboring spheres, (c) the number of atoms of Fe per unit lattice, and (d) the total volume occupied by an atom of Fe.   *Ans.* (a) 1.27 Å; (b) 2.54 Å; (c) 2; (d) 12.6 Å³.

16. At 1100°C Fe (γ) crystallizes in a face-centered cubic lattice with an edge length of 3.63 Å. Assuming the atoms to be packed spheres, calculate (a) the radius of the spheres, (b) the distance between centers of neighboring spheres, (c) the number of atoms of Fe per unit lattice, and (d) the total volume occupied per atom of Fe.

17. Suppose that Fe (γ) were to crystallize in the hexagonal system. Assuming the radius of the Fe atoms to be that found in problem 16, calculate (a) the distance between neighboring atoms, (b) the number of atoms per unit lattice, and (c) the total volume occupied by an atom of Fe. Compare these results with those obtained in the preceding problem.

18. CaO crystallizes in a face-centered cubic lattice with a unit distance of 4.80 Å. (a) Calculate the distance between the two ions in the lattice. (b) How does this distance compare with that predicted by the ionic radii of Ca^{+2} and O^{-2}?

19. NH_4Cl crystallizes in a body-centered cubic lattice with a unit distance of 3.87 Å. (a) Calculate the distance between the two ions in the lattice. (b) Using the radius for the Cl^- ion given in Table 2-7, find the radius for the NH_4^+ ion.
          Ans. (b) 1.54 Å.

20. To accelerate an electron in an electric field to a velocity of v cm/sec, work equal to Ve has to be done. V is the potential drop and $e = 4.803 \times 10^{-10}$ is the electronic charge, both in electrostatic (esu) units. Further, $300\, V = \varepsilon$, where ε is the potential in volts. From these facts deduce Eq. (6) of this chapter.

21. Electrons emitted from a hot filament are accelerated in an electric field until their velocity is 5×10^9 cm/sec. Find the wave length of the electrons, and the potential drop required to obtain this wave length.
          Ans. 0.146 Å; 7110 volts.

22. What will have to be the velocity of neutrons in order for them to have a wave length of 0.60 Å?

23. At 20°K the heat capacity per gram atom of silver is 0.39 cal/°K. Assuming the validity of the Debye third power law for this substance, calculate the heat capacity per gram atom at 1°K.

24. Using the data given in problem 23, calculate the characteristic Debye temperature and the limiting vibration frequency for silver.

3

The First
Law of
Thermodynamics

ONE of the most fundamental manifestations in nature is the energy that accompanies all changes and transformations. Such diversified phenomena as the drop of a stone, the motion of a billiard ball, the impinging of light, the burning of coal, and the growth and reactions of the complex mechanism known as a living being all involve absorption, emission, and redistribution of energy. The most common form in which this energy appears, and the form to which all others tend, is heat. Besides this, there is mechanical energy involved in the motion of all machinery; electrical energy, exhibited by a current in heating a conductor and in doing chemical and mechanical work; radiant energy, inherent in visible light and in radiation in general; and finally, chemical energy, the energy stored in all substances, and which appears when the substances undergo transformation. As diversified and distinct as these various forms may at first glance appear, they are, nevertheless, related to one another, and under certain conditions may be transformed from one into the other. A study of this interrelation of the various forms of energy in a system constitutes the subject of *thermodynamics.*

Since thermodynamic laws deal with energy, they are applicable to all phenomena in nature. They hold rigidly because they are based on the behavior of macroscopic systems, i.e., systems comparatively large and involving many molecules, rather than on the behavior of microscopic systems in which comparatively few molecules are involved. Moreover, thermodynamics does not consider the time element in transformations. It is interested merely in the initial and final states of a system without any curiosity about the speed with which a change is accomplished.

Within any system the energy may be kinetic or potential in nature, or both. Kinetic energy is the energy a system possesses by virtue of its motion, be it molecular or motion of the body as a whole. Potential energy, on the other hand, is the energy a system possesses by virtue of its position, i.e., energy due to the structure of the body or due to its configuration with respect to other bodies. The total energy content of any system is the sum of the potential and kinetic energies.

Although the absolute value of the total energy contained in a system can be calculated from the famous Einstein relation $E = mc^2$, where E is the energy, m the mass, and c the velocity of light, this fact is of little help in ordinary thermodynamic considerations. The reason is that the energies involved are so large that any changes in them as a result of the usual chemical or physical processes would be negligible compared with the totals. Further, the changes in the masses resulting from the energy transfers would be so small as to be beyond detection by our available means of weighing. Consequently thermodynamics prefers to deal with the energy differences which accompany changes in systems since these can be measured. These differences are expressed in the units used in connection with the various forms of energy. Thus the cgs unit of mechanical energy is the *erg;* of electrical energy, the *joule;* of thermal energy, the *calorie.* The relation of the unit of mechanical work to the thermal unit is known as the *mechanical equivalent of heat.* The first determinations of this equivalent by Joule laid one of the foundation stones upon which the first law of thermodynamics was reared.

THE THERMODYNAMIC SYSTEM

In dealing with thermodynamic problems, the term "system" is employed very frequently. A *system* is defined as any portion of the universe isolated in an inert container, which may be real or imaginary, for purposes of study of the effect of various variables upon the contents of the system. In turn, the portion of the universe excluded from the system is called its *surroundings.* The system's contents may range all the way from a small quantity of, say, water, up to the entire universe.

The interactions between a system and its surroundings are very important in thermodynamics. A system which can exchange both matter and energy with its surroundings is said to be *open,* while one which cannot is said to be *isolated.* Again, a *closed* system is one in which no transfer of matter to or from the surroundings is possible, but that of energy is.

A *homogeneous* system contains only one phase, while in a *heterogeneous* system more than a single phase may be present. A *phase* is defined as a homogeneous, physically distinct, and mechanically separable portion of a system. If in the system "water" ice, liquid water, and water vapor coexist, then each form constitutes a separate phase, and each phase is separated from every other by a phase boundary. Further, each phase may be continuous, as the gas and liquid phases, or it may be broken up into smaller portions, as a mass of ice crystals. The term "mechanically separable" in the definition means that each phase can be separated from every other phase by such operations as filtration, sedimentation, decantation, or any other mechanical means of separation, as, say, hand picking of crystals. It does not include, however, such separation methods as evaporation, distillation, adsorption, or extraction. Since all gases are completely miscible, only one gas phase is possible in a system. With liquids and solids there appears to be no theoretical limit to the number of phases possible, but eight[1] is the largest number of liquid phases ever observed in any one system.

The phases present in a system may consist of pure substances, such as N_2, H_2O, NaCl, or C_2H_6, or they may be solutions. A true solution is defined as a *physically homogeneous mixture of two or more substances*. This definition of a solution places no restriction on either the state of aggregation or the relative amounts of the constituents. Consequently a solution may be gaseous, liquid, or solid, and it may vary in composition within wide limits. It is this latter fact which excludes pure compounds from the classification of solutions, for a fixed and definite ratio persists among the constituents of a compound.

PROPERTIES AND VARIABLES OF A SYSTEM

The properties of a system may be divided into two types, namely, extensive and intensive. An *extensive* property of a system is any property whose magnitude depends on the amount of substance present. Examples of such properties are total mass, volume, and energy. On the other hand, *intensive* properties are those whose value is independent of the total amount, but depends instead on the concentration of the substance or substances in a system. Examples of intensive properties are pressure, density, refractive index, and mass, volume, or energy per mole.

In absence of special forces, such as electric or magnetic fields, the extensive properties of a pure substance will depend upon any *two* of the three variables P, V, and T, and the number of moles, n, of substance present. Only two of the state variables are independent because the

---

[1] Kittsley and Goeden, *J. Am. Chem. Soc.*, **72**, 4841 (1950).

system possesses in principle an equation of state which relates P, V, and T, and hence makes one of the variables dependent. In the present discussion we shall take P and T as the two independent variables. Consequently, when n is constant the extensive properties of the system depend only on P and T. The same conclusion applies also to the intensive properties. By definition the latter properties are independent of n, and hence P and T are the only independent variables involved.

The situation is different with solutions. For these the extensive properties depend on P, T, and the numbers of moles of the various constituents present, i.e., n_1, n_2, n_3, etc. The intensive properties, in turn, depend on P, T, and the concentrations of the various species. The number of the latter variables is one less than the total number of solution constituents for the following reason. If, say, we express the concentration in mol fractions, then specification of all but one of these yields the last by subtraction from unity, which is the sum of the mol fractions.

Once the values of the independent variables of a system are specified, the state of the system is determined. If we take now the system in a given state and change the magnitude of the variables, the system will change to a new state. To return the system to its initial conditions the variables will have to be adjusted back to their initial values. This fact, that the state of a system can be reproduced by reproducing the values of the variables, is known as the *principle of reproducibility of states*.

METHODS OF EXPRESSING SOLUTION CONCENTRATION

Depending on the purpose at hand, there are many ways in which the concentration of a solution can be expressed. The more common of these are:

Weight basis

1. Per cent or fraction of dissolved substance by weight.
2. Weight of dissolved substance per definite weight of one of the constituents.
3. Weight of dissolved substance per definite total weight of solution.
4. Molality—number of moles of dissolved substance per 1000 grams of solvent.
5. Mol fraction.

Volume basis

1. Per cent or fraction of dissolved substance by volume.
2. Weight of dissolved substance per given volume of solution.

3. Molarity—number of moles of dissolved substance per liter of solution.
4. Normality—number of equivalents of dissolved substance per liter of solution.

All these schemes are either self-explanatory or familiar to the student. Of the various methods listed above, those expressed on a weight basis are temperature-independent, i.e., the concentrations will be the same at all temperatures. On the other hand, those based on volume will vary with temperature in a manner dependent upon the thermal expansion of the solution.

PARTIAL MOLAL QUANTITIES

Consider a solution composed of j constituents, and let n_1, n_2, . . . n_j be the numbers of moles of the various constituents present. Further, let G be *any* extensive property of the system. Then, in view of what has been said above, G must be a function of P, T, and the numbers of moles of the various constituents present, or

$$G = \phi(P, T, n_1, n_2, \ldots n_j) \tag{1}$$

Partial differentiation of Eq. (1) yields for dG

$$dG = \left(\frac{\partial G}{\partial T}\right)_{P,n_i} dT + \left(\frac{\partial G}{\partial P}\right)_{T,n_i} dP + \left(\frac{\partial G}{\partial n_1}\right)_{P,T,n_.} dn_1 + \left(\frac{\partial G}{\partial n_2}\right)_{P,T,n_i} dn_2$$
$$+ \cdots \left(\frac{\partial G}{\partial n_j}\right)_{P,T,n_i} dn_j \tag{2}$$

where n_i represents in each instance all the n's except the one with respect to which G is differentiated. If we write now

$$\bar{G}_1 = \left(\frac{\partial G}{\partial n_1}\right)_{P,T,n_i} \tag{3}$$

$$\bar{G}_2 = \left(\frac{\partial G}{\partial n_2}\right)_{P,T,n_i} \tag{4}$$

$$\bar{G}_j = \left(\frac{\partial G}{\partial n_j}\right)_{P,T,n_i} \tag{5}$$

Eq. (2) becomes

$$dG = \left(\frac{\partial G}{\partial T}\right)_{P,n_i} dT + \left(\frac{\partial G}{\partial P}\right)_{T,n_i} dP + \bar{G}_1 dn_1 + \bar{G}_2 dn_2 + \cdots \bar{G}_j dn_j \tag{6}$$

Further, at constant temperature and pressure the first two terms in Eq. (6) are zero, and we get thus

$$dG = \bar{G}_1 dn_1 + \bar{G}_2 dn_2 + \cdots \bar{G}_j dn_j \tag{7}$$

By use of Euler's theorem for homogeneous functions, Eq. (7) can be integrated to yield

$$G = \bar{G}_1 n_1 + \bar{G}_2 n_2 + \cdots \bar{G}_j n_j \tag{8}$$

Equation (8) shows that any extensive property of a solution can, at constant T and P, be expressed as a sum of $\bar{G}n$ products for the individual solution components. Further, since in each product the n represents a capacity factor, the $\bar{G}$, called a *partial molal quantity*, must represent an intensity factor. Consequently the partial molal quantities must be intensive properties of a solution.

When Eq. (8) is differentiated completely, we obtain

$$
\begin{aligned}
dG &= \bar{G}_1 dn_1 + n_1 d\bar{G}_1 + \bar{G}_2 dn_2 + n_2 d\bar{G}_2 + \cdots \bar{G}_j dn_j + n_j d\bar{G}_j \\
&= (\bar{G}_1 dn_1 + \bar{G}_2 dn_2 + \cdots \bar{G}_j dn_j) \\
&\qquad\qquad + (n_1 d\bar{G}_1 + n_2 d\bar{G}_2 + \cdots n_j d\bar{G}_j) \tag{9}
\end{aligned}
$$

But, according to Eq. (7) dG is given by the first term in parentheses on the right of Eq. (9), and consequently the second term must be zero. We get thus, in general,

$$n_1 d\bar{G}_1 + n_2 d\bar{G}_2 + \cdots n_j d\bar{G}_j = 0 \tag{10}$$

and for only two constituents, i.e., a binary solution,

$$n_1 d\bar{G}_1 + n_2 d\bar{G}_2 = 0 \tag{11a}$$

or

$$d\bar{G}_1 = -\frac{n_2}{n_1} d\bar{G}_2 \tag{11b}$$

Equations (10) and (11) are two forms of the *Gibbs-Duhem equation*. They show that the partial molal quantities are not independent of each other, and that variation of one partial molal quantity affects the others in the manner given by the equations. For instance, Eq. (11b) indicates that if $d\bar{G}_1$ is positive, i.e., $\bar{G}_1$ increases, then $d\bar{G}_2$ must be negative and $\bar{G}_2$ must decrease at the same time, and vice versa. The significance of this behavior will become more apparent when we discuss solutions in greater detail in Chapter 8.

DETERMINATION OF PARTIAL MOLAL QUANTITIES

Partial molal quantities can be determined by a variety of methods, both graphical and analytical, of which we shall discuss only three. Specifically, suppose that it is desired to ascertain the partial molal volumes of a binary solution, and suppose that for this purpose we measure the total volume V of the solution as a function of the molality m of solute. Since by definition the molality represents the number of moles of solute per 1000 grams of solvent, then the number of moles of solvent present in each instance is fixed and given by $n_1 = 1000/M_1$, where M_1 is the molecular weight of the solvent. Again, m must be n_2. Therefore, if we plot V vs. m, Figure 3–1(a), and determine the slopes of the curve at various concentrations such as A and B, those slopes will be the values of $\bar{V}_2$ at the selected concentrations. Having thus obtained $\bar{V}_2$, the corresponding values of $\bar{V}_1$ can be calculated by substitution of V, n_1, and n_2

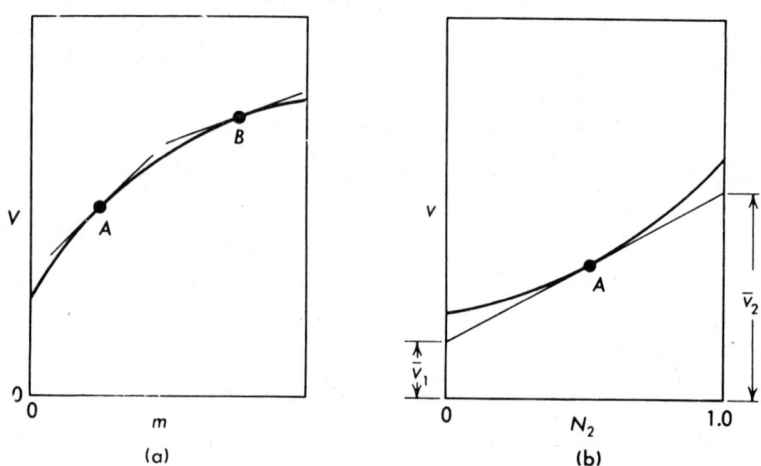

(a)   (b)

Figure 3–1.   Graphical determination of partial molal quantities.

for each concentration into Eq. (8), which in this instance takes the form

$$V = \bar{V}_1 n_1 + \bar{V}_2 n_2 \tag{12}$$

An alternate method of handling the same data is as follows. We calculate from V, n_1, and n_2 the quantity

$$v = \frac{V}{n_1 + n_2}$$

and plot v against the mol fraction of solute N_2, Figure 3–1(b). If we now draw a tangent to the curve at any point such as A, it can be shown that the intercept of the tangent on the $N_2 = 0$ axis gives $\bar{V}_1$, and the intercept on the $N_2 = 1.0$ axis gives $\bar{V}_2$. By repeating this procedure at various points along the curve, $\bar{V}_1$ and $\bar{V}_2$ can be evaluated for various solution concentrations.

Still a third way of handling the data involves the analytic expression of V as a function of m. Suppose that V as a function of m can be expressed by the relation

$$V = a + bm + cm^2 \tag{13}$$

where a, b, and c are constants at a given temperature and pressure. Differentiation of Eq. (13) with respect to m yields then $\bar{V}_2$, namely,

$$\bar{V}_2 = \frac{\partial V}{\partial m} = b + 2\,cm \qquad (14)$$

and substitution of Eqs. (13) and (14) into Eq. (12) gives for $\bar{V}_1$

$$a + bm + cm^2 = \bar{V}_1 n_1 + \bar{V}_2 m$$
$$= \bar{V}_1 n_1 + (b + 2\,cm)m$$
$$\bar{V}_1 = \frac{a - cm^2}{n_1} \qquad (15)$$

Although the above methods for determining the partial molal quantities were described here in terms of solution volumes, they are equally applicable, and in the same manner, to any other extensive properties of a solution.

THE FIRST LAW OF THERMODYNAMICS

After these preliminaries, we are prepared to consider the first law of thermodynamics. This law is merely the law of conservation of energy, namely, that *energy can be neither created nor destroyed*. Worded differently, the law says that for any quantity of a form of energy that disappears, another or other forms of energy will have to appear in total quantity exactly equal to the amount that disappeared. To be more specific, consider the fate of a quantity of heat q added to a system. This heat will go to raise the internal energy of the system and also to do any outside work the system may perform as a result of the absorption of heat. If we let ΔE be the increase in the internal energy of the system and w be the work done by the system on its surroundings, then by the first law

$$\Delta E + w = q$$

and
$$\Delta E = q - w \qquad (16)$$

Equation (16) is a mathematical statement of the first law. Since the internal energy depends only on the state of the system, then the change in energy, ΔE, involved in going from a state where the energy is E_1 to another state where the energy is E_2 must be given by

$$\Delta E = E_2 - E_1 \qquad (17)$$

ΔE thus depends only on the initial and final states of the system, and not at all on the manner in which the change is accomplished.

These considerations do *not* apply to w and q, because the magnitude of these depends on the manner in which the work is performed in passing from the initial to the final state. The symbol w represents the total work done by a system. In a galvanic cell, for instance, w may include the electrical energy supplied, plus, if there is a change in volume, any energy utilized to effect expansion or contraction against an opposing pressure p. At present we are interested only in the pressure-volume or mechanical work involved in a process; and this energy term is readily deducible as follows.

Consider a cylinder, Figure 3–2, of cross-sectional area A fitted with a frictionless and weightless piston. Let the pressure on the piston be p.

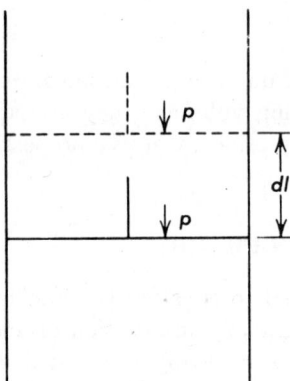

Figure 3–2.   Pressure-volume work.

Then, since pressure is force per unit area, the total force acting on the piston is $f = pA$. If the piston is now moved through a distance dl, the work dw done is

$$dw = fdl = pA\,dl$$

But $A\,dl$ is the element of volume, dV, swept out by the piston in its motion. Hence

$$dw = p\,dV \qquad (18)$$

and, on integration between the limits V_1 and V_2,

$$w = \int_{V_1}^{V_2} p\,dV \qquad (19)$$

If the only work done by the system is of this type, then substitution of Eq. (19) into Eq. (16) yields for the first law expression

$$\Delta E = q - \int_{V_1}^{V_2} p\,dV \qquad (20)$$

Equations (19) and (20) are perfectly general and apply to the calculation of w, q, and ΔE in any expansion or contraction of a system. However, under special conditions these equations may take on special forms as follows:

1. *Constant volume*—When the volume does not change $dV = 0$, $dw = 0$, and Eq. (20) becomes

$$\Delta E = q \tag{21}$$

2. *Opposing pressure zero*—A process of this type is called a *free expansion*. Here $p = 0$, $dw = 0$, and hence again $\Delta E = q$.
3. *Opposing pressure constant*—If $p = $ constant, then Eq. (19) integrates to

$$w = p(V_2 - V_1) \tag{22}$$

and Eq. (20) becomes

$$\Delta E = q - p(V_2 - V_1) \tag{23}$$

4. *Opposing pressure variable*—When p is variable, then p must be known as a function of V for the given situation before Eq. (19) can be integrated. If no analytic function is available, the integration can be performed graphically by plotting p vs. V and determining the area under the curve. Once w is thus obtained, it can be substituted into Eq. (20) to find either ΔE or q, as the case may be.

The work involved in the expansion of solids and liquids is small, and frequently negligible. However, such is not the case with gases, where the volume changes may be large. In all calculations of w involving gas expansions it must be clearly understood that, except under the conditions to be pointed out below, *the pressure which determines the amount of work done is not the pressure of the gas P, but the pressure against which the gas is working, namely, p.* Once this fact is realized, no difficulty will be encountered with such calculations. Further, when $V_2 > V_1$, the process is an expansion, w is positive, and work is done by the gas on its surroundings. If, however, $V_2 < V_1$, then we have a contraction, w is negative, and work is done by the surroundings on the gas.

From the above discussion it should be evident that for a given change the magnitude of w depends on the manner in which the work is performed, and, therefore, w cannot be a state function. Again, since for a given change ΔE is fixed, q must vary with w, and be thus dependent upon the path followed by the process.

The quantities q, w, and ΔE are experimentally measurable, but the absolute magnitudes of E are not. The latter fact is no hindrance in thermodynamics, as we are primarily interested in changes in E and not in the absolute values.

REVERSIBILITY AND MAXIMUM WORK

Any process, which is so conducted that at every stage the driving force is only infinitesimally greater than the opposing force, and which can be reversed by increasing the opposing force by an infinitesimal amount, is called a *reversible* process. On the other hand, any change which does not meet these requirements is said to be *irreversible*.

Strictly speaking, all reversible processes are impossible in nature, as they would require an infinite time for their accomplishment. As a consequence all naturally occurring processes must be irreversible. Nevertheless, the concept of reversibility is so valuable, both theoretically and practically, that its use is justified. Further, reversibility can be approached quite closely. Thus in the potentiometric method for measuring the potentials of galvanic cells, the voltage of the cell is opposed by another voltage until practically no current flows. By making the opposing voltage only very slightly smaller than the voltage of the cell, the cell can be made to discharge while, if the opposing voltage is increased slightly above that of the cell, the cell can be made to charge. In this manner, any current flowing through the circuit has to do work against a potential which is at all times only slightly less than its own. This arrangement is the closest approximation that can at present be made to a truly reversible process.

We will now proceed to show that the work obtainable from a given process under reversible conditions is the maximum possible, and that this work is the minimum required to reverse the process.

From the arguments given above it is evident that the amount of work a system has to perform to bring about a certain change depends on the opposition the system experiences to the change. The greater that resistance is, the more work must be done by the system to overcome it. To be specific, let us consider the expansion of a gas against an external pressure p through an infinitesimal volume change dV. The work done is pdV. When p is zero, i.e., when the system expands into a vacuum, the piston confining the gas experiences no restraining force, and if it is frictionless and weightless, no work is involved, and $pdV = 0$. However, as p is increased from zero, more and more work has to be done as the pressure approaches that of the gas itself. When the latter point is reached, the two forces become balanced and no further change in volume is possible. If we continue to increase the pressure, the pressure on the gas becomes greater than the pressure of the gas, the volume begins to decrease, and work is done *on* rather than *by* the system. From this description it is evident that the work which may be performed by a system is a maximum when the opposing pressure p differs only infinitesimally in magnitude from the internal pressure of the gas itself. But these are exactly the conditions defined for the reversibility of a process. Hence it may be concluded that *maximum work is obtainable from a system when any change taking place in it is entirely reversible.* Further, under reversible conditions the external pressure p differs only by an infinitesimal from the pressure of the gas itself, P. We can replace, therefore, p by P in all the expressions given above for w, and thereby obtain w_m, the maximum work performed in the reversible expansion of a gas.

Consider now the reversible compression of the gas under discussion. Since the pressure of the gas is P, the minimum pressure which will be required to compress it will have to be $P + dP$. Consequently, since the opposing pressure is P, the minimum work required for a volume change dV will be PdV. This work is exactly equal and opposite in sign to the work obtained in the reversible expansion of the gas. If the driving pressure is made greater than $P + dP$, the process becomes irreversible, and more work than PdV will be required to accomplish the compression.

THE ENTHALPY OF A SYSTEM

Thermal changes at constant pressure are most conveniently expressed in terms of another function, H, called the *enthalpy* or *heat content* of a system. This function is defined by the relation

$$H = E + PV \tag{24}$$

where P and V are the pressure and volume *of the system*. Since E and PV are completely characterized by the state of the system, H is also a function of the state of the system and is completely independent of the manner in which the state is achieved. Consequently the change in enthalpy, ΔH, may be written as

$$\Delta H = H_2 - H_1 \tag{25}$$

where H_2 is the enthalpy of the system in the final state and H_1 the enthalpy in the initial state. Substituting for H_2 and H_1 their equivalents from Eq. (24), we obtain for ΔH

$$
\begin{aligned}
\Delta H &= H_2 - H_1 \\
&= (E_2 + P_2 V_2) - (E_1 + P_1 V_1) \\
&= (E_2 - E_1) + (P_2 V_2 - P_1 V_1) \\
&= \Delta E + (P_2 V_2 - P_1 V_1)
\end{aligned}
\tag{26}
$$

Equation (26) is the most general definition of ΔH. When the pressure remains constant throughout the process, then

$$
\begin{aligned}
\Delta H &= \Delta E + P(V_2 - V_1) \\
&= \Delta E + P\Delta V
\end{aligned}
\tag{27}
$$

i.e., the change in enthalpy at constant pressure is equal to the increase in internal energy plus any pressure volume work done. Hence at constant pressure ΔH represents the heat absorbed by a system in going from an initial to a final state, *provided the only work done is P-V work*. When the initial and final pressures are not the same, ΔH is calculated not by Eq. (27) but by Eq. (26).

In this book we shall employ the convention that any gain in a quantity on the part of a system is considered to be positive and any loss negative. On this basis values of q, ΔE, and ΔH which are greater than zero correspond to absorption of heat by the system and to increases in E and H of the system. Conversely, when these quantities are less than zero, they denote loss of heat by the system and decreases in E and H. Finally, when $w > 0$ work is done *by* the system, while when $w < 0$ work is done *on* the system.

HEAT CAPACITY

Consider a very small quantity of heat dq added to a system, and suppose that as a result of the heat absorption the temperature rise produced is dT. Then the amount of heat required to raise the temperature of the system 1 degree is

$$C = \frac{dq}{dT}$$

and C is thus the heat capacity of the system. Now, from Eq. (20) we get $dq = dE + pdV$, and hence

$$C = \frac{dE + pdV}{dT} \tag{28}$$

When the volume is held constant $dV = 0$, and Eq. (28) reduces to

$$C_v = \left(\frac{\partial E}{\partial T}\right)_V \tag{29}$$

This equation is the thermodynamic relation defining C_v and tells us that C_v is the rate of change of the internal energy with temperature at constant volume.

However, when the heat absorption occurs reversibly at constant pressure, then $p = P$, and Eq. (28) becomes

$$C_p = \left(\frac{\partial E}{\partial T}\right)_P + P\left(\frac{\partial V}{\partial T}\right)_P$$

But, if Eq. (24) be differentiated with respect to T at constant P we get

$$\left(\frac{\partial H}{\partial T}\right)_P = \left(\frac{\partial E}{\partial T}\right)_P + P\left(\frac{\partial V}{\partial T}\right)_P$$

Consequently,

$$C_p = \left(\frac{\partial H}{\partial T}\right)_P \tag{30}$$

which is the thermodynamic definition of C_p, i.e., C_p is the rate of change of the enthalpy with temperature at constant pressure.

The difference between the two heat capacities is readily deducible by a thermodynamic argument. From Eqs. (29) and (30) we have

$$C_p - C_v = \left(\frac{\partial H}{\partial T}\right)_P - \left(\frac{\partial E}{\partial T}\right)_V \tag{31}$$

But, $H = E + PV$. Differentiating this quantity with respect to temperature at constant pressure we obtain

$$\left(\frac{\partial H}{\partial T}\right)_P = \left(\frac{\partial E}{\partial T}\right)_P + P\left(\frac{\partial V}{\partial T}\right)_P \tag{32}$$

and on substitution of Eq. (32) into Eq. (31), the latter becomes

$$C_p - C_v = \left(\frac{\partial E}{\partial T}\right)_P + P\left(\frac{\partial V}{\partial T}\right)_P - \left(\frac{\partial E}{\partial T}\right)_V \tag{33}$$

The problem now is to relate the first and third terms on the right of Eq. (33). To do this we proceed as follows. The internal energy E will be, in general, a function of any two of the three variables P, V, T. If we take as our independent variables T and V, then

$$E = f(T,V)$$

and

$$dE = \left(\frac{\partial E}{\partial T}\right)_V dT + \left(\frac{\partial E}{\partial V}\right)_T dV \tag{34}$$

Dividing both sides of the equation by dT and imposing the condition of constant pressure, we get

$$\left(\frac{\partial E}{\partial T}\right)_P = \left(\frac{\partial E}{\partial T}\right)_V + \left(\frac{\partial E}{\partial V}\right)_T \left(\frac{\partial V}{\partial T}\right)_P \tag{35}$$

and Eq. (35) substituted in Eq. (33) yields finally

$$C_p - C_v = \left(\frac{\partial E}{\partial T}\right)_V + \left(\frac{\partial E}{\partial V}\right)_T \left(\frac{\partial V}{\partial T}\right)_P + P\left(\frac{\partial V}{\partial T}\right)_P - \left(\frac{\partial E}{\partial T}\right)_V$$

$$= \left(\frac{\partial E}{\partial V}\right)_T \left(\frac{\partial V}{\partial T}\right)_P + P\left(\frac{\partial V}{\partial T}\right)_P \tag{36}$$

Equation (36) is perfectly general. We shall show later how this equation can be used to obtain the actual difference between the two heat capacities.

DEPENDENCE OF STATE FUNCTIONS ON VARIABLES

We have defined thus far two quantities which are functions of the state of the system, namely, E and H, and also C_v and C_p. If we confine ourselves to pure substances, then these quantities are functions of any two of the three variables P, V, and T. In dealing with these variables it is found that E and C_v are most conveniently expressed in terms of T and V, while for H and C_p the best choice is T and P.

If we start now with the fact that $E = f(T, V)$, then

$$dE = \left(\frac{\partial E}{\partial T}\right)_V dT + \left(\frac{\partial E}{\partial V}\right)_T dV \tag{37}$$

According to Eq. (29), $(\partial E/\partial T)_V = C_v$. A corresponding expression for $(\partial E/\partial V)_T$ cannot be obtained without the second law of thermodynamics. However, for pedagogic reasons we shall assume its value here subject to proof in Chapter 5. On this basis we can state that

$$\left(\frac{\partial E}{\partial V}\right)_T = T \left(\frac{\partial P}{\partial T}\right)_V - P \tag{38}$$

Substituting these values of the partial derivatives into Eq. (37), we obtain

$$dE = C_v dT + \left[T \left(\frac{\partial P}{\partial T}\right)_V - P \right] dV \tag{39}$$

The first term on the right in Eq. (39) gives the effect on E of temperature change at constant volume, while the second gives the effect of volume change at constant temperature.

Similar consideration of H as a function of T and P gives

$$dH = \left(\frac{\partial H}{\partial T}\right)_P dT + \left(\frac{\partial H}{\partial P}\right)_T dP \tag{40}$$

But $(\partial H/\partial T)_P = C_p$. Again, subject to proof in Chapter 5,

$$\left(\frac{\partial H}{\partial P}\right)_T = V - T \left(\frac{\partial V}{\partial T}\right)_P \tag{41}$$

Consequently,

$$dH = C_p dT + \left[V - T \left(\frac{\partial V}{\partial T}\right)_P \right] dP \tag{42}$$

The first term in Eq. (42) gives the variation of H with temperature at constant pressure, while the second gives the effect of pressure at constant temperature.

The effect of volume change on C_v can be deduced by differentiation of Eq. (29) with respect to volume at constant temperature and utilization of Eq. (38). The result is

$$\left(\frac{\partial C_v}{\partial V}\right)_T = T \left(\frac{\partial^2 P}{\partial T^2}\right)_V \tag{43}$$

Similarly, the effect of pressure on C_p follows from differentiation of Eq. (30) with respect to P at constant T, and use of Eq. (41). We get thus

$$\left(\frac{\partial C_p}{\partial P}\right)_T = -T\left(\frac{\partial^2 V}{\partial T^2}\right)_P \tag{44}$$

THE THERMODYNAMIC BEHAVIOR OF IDEAL GASES

To illustrate the use of the above equations and also to deduce some important conclusions, we shall consider now the thermodynamic behavior of ideal gases. Equation (38) can be used to find the dependence of E on V at constant T as follows: For an ideal gas $PV = nRT$. Differentiation of this expression with respect to T at constant V yields

$$\left(\frac{\partial P}{\partial T}\right)_V = \frac{nR}{V}$$

which, on substitution into Eq. (38), gives

$$\left(\frac{\partial E}{\partial V}\right)_T = \frac{nRT}{V} - P$$
$$= 0 \tag{45}$$

Consequently *the internal energy of an ideal gas is independent of volume* and depends only on the temperature. This conclusion is identical with that reached in Chapter 1 on the basis of kinetic theory. Again, if we differentiate the ideal gas law with respect to temperature at constant pressure, we obtain

$$\left(\frac{\partial V}{\partial T}\right)_P = \frac{nR}{P}$$

which, when inserted into Eq. (41), shows that

$$\left(\frac{\partial H}{\partial P}\right)_T = V - \frac{nRT}{P}$$
$$= 0 \tag{46}$$

Hence *the enthalpy of an ideal gas is independent of pressure* and is dependent only on the temperature.

From these deductions it must follow also that C_v and C_p are functions of T only, and are then independent of volume and pressure. Therefore,

$$\left(\frac{\partial C_v}{\partial V}\right)_T = 0 \tag{47}$$

and

$$\left(\frac{\partial C_p}{\partial P}\right)_T = 0 \tag{48}$$

Furthermore, since for an ideal gas $(\partial E/\partial V)_T = 0$, Eq. (36) reduces to

$$C_p - C_v = P\left(\frac{\partial V}{\partial T}\right)_P$$

But, for an ideal gas

$$P\left(\frac{\partial V}{\partial T}\right)_P = nR$$

and so we find that

$$C_p - C_v = nR$$

for n moles, or per mole

$$C_p - C_v = R \tag{49}$$

This conclusion is again the same as obtained from the kinetic theory of gases in Chapter 1.

ISOTHERMAL AND ADIABATIC PROCESSES

The energy relations and the state of any system during a process depend not only on the manner in which work is performed, but also on certain experimental conditions imposed upon the system as a whole. Two such constraints of especial importance involve processes conducted under (a) isothermal and (b) adiabatic conditions. An *isothermal* process is any process conducted in a manner such that the temperature remains constant during the entire operation. In turn, an *adiabatic* process is one in which no heat is absorbed or evolved by the system.

ISOTHERMAL PROCESSES IN IDEAL GASES

Since the internal energy of ideal gases is a function of temperature only, imposition of a constant temperature means also constancy of E and hence $\Delta E = 0$. Insertion of this criterion for an isothermal change in *ideal* gases into Eq. (20) leads to

$$q = w = \int_{V_1}^{V_2} p\,dV \tag{50}$$

We see, therefore, that in such a process all the work performed will have to be done at the expense of absorbed heat, or any heat obtained from a process will have to come from work done on the system. The magnitude of q will obviously depend on the manner in which the work is performed, and may range from zero for a free expansion to a maximum when reversibility obtains.

The expression for the maximum work, w_m, obtained when an ideal gas expands *isothermally and reversibly* may be deduced as follows: Under reversible conditions $p = P = nRT/V$. Substituting this value of p into Eq. (50) we get

$$w_m = \int_{V_1}^{V_2} \frac{nRTdV}{V}$$

$$= nRT \int_{V_1}^{V_2} \frac{dV}{V}$$

and on integration

$$w_m = nRT \ln \frac{V_2}{V_1} \tag{51}$$

Equation (51) gives the *maximum work* obtainable from an isothermal reversible expansion of n moles of an ideal gas from volume V_1 to V_2 at temperature T. Again, since the temperature is constant, $V_2/V_1 = P_1/P_2$ by Boyle's law, and hence Eq. (51) may also be written in the alternate and equivalent form

$$w_m = nRT \ln \frac{P_1}{P_2} \tag{52}$$

The application of Eq. (51), and the distinction between isothermal reversible work and isothermal work against a constant pressure, may best be understood from the following examples.

Example (a): Find the work done when 2 moles of hydrogen expand isothermally from 15 to 50 liters against a *constant pressure* of 1 atm at 25°C. By Eq. (22),

$$w = p(V_2 - V_1) = 1(50 - 15) = 35 \text{ liter-atm}$$
$$= 847 \text{ cal}$$

Example (b): Calculate the work performed when 2 moles of hydrogen expand *isothermally* and *reversibly* at 25°C from 15 to 50 liters. Using Eq. (51),

$$w_m = nRT \ln \frac{V_2}{V_1} = 2.303 \, nRT \log \frac{V_2}{V_1}$$

$$= (2.303 \times 2 \times 1.987 \times 298.2) \log \frac{50}{15}$$

$$= 1427 \text{ cal}$$

The heats absorbed during the expansions are equal to w in both cases and are therefore $q = 847$ cal in (a), and $q = 1427$ cal in (b). The internal energy change is zero in both instances, since there is no change in temperature.

For real gases $\Delta E \neq 0$ under isothermal conditions because, even with T constant, there is a change in internal energy due to change in volume or pressure. The same conclusion applies to ΔH.

ADIABATIC PROCESSES IN IDEAL GASES

Since in an adiabatic process there is no exchange of heat between a system and its surroundings, we must have $q = 0$. Inserting this criterion for an adiabatic process into Eq. (16) we get

$$\Delta E = -w \tag{53}$$

From Eq. (53) it follows that any work in an adiabatic process is done at the expense of the internal energy. As work is performed, the internal energy of the system decreases, and consequently the temperature drops. Conversely, if work is done upon the system, all the work goes to increase the internal energy of the system and the temperature must, therefore, increase.

An equation that every ideal gas must obey at every stage of an adiabatic reversible expansion can be derived readily from Eq. (53). Consider n moles of an ideal gas at a pressure P and a volume V. For an infinitesimal increase in volume dV at the pressure P, the work done by the gas is PdV. Since this work is accomplished at the expense of the internal energy of the gas, the internal energy must *decrease* by an amount dE. According to Eq. (53), therefore,

$$PdV = -dE$$

However, from Eq. (29) $dE = nC_v dT$. Consequently,

$$PdV = -dE = -nC_v dT \tag{54}$$

Further, since $P = nRT/V$, we can transform Eq. (54) to

$$\frac{nRTdV}{V} = -nC_v dT$$

or

$$\frac{dV}{V} = -\frac{C_v}{R}\left(\frac{dT}{T}\right)$$

Considering now C_v to be constant, and integrating between the limits V_1 at T_1 and V_2 at T_2, we have

$$\int_{V_1}^{V_2} \frac{dV}{V} = -\frac{C_v}{R}\int_{T_1}^{T_2} \frac{dT}{T}$$

and, therefore,

$$\ln\left(\frac{V_2}{V_1}\right) = -\frac{C_v}{R}\ln\left(\frac{T_2}{T_1}\right)$$

On rearranging and taking antilogarithms we obtain

$$V_1 T_1^{C_v/R} = V_2 T_2^{C_v/R} = C_1 \tag{55}$$

where C_1 is a constant.

Other forms of Eq. (55) may be easily derived by eliminating dT or dV instead of P from Eq. (54). A very common form is one involving P and V, namely,

$$P_1 V_1^{\gamma} = P_2 V_2^{\gamma} = C_2 \tag{56}$$

where C_2 is a constant and $\gamma = C_p/C_v$, the ratio of heat capacities. It should be realized that Eqs. (55) and (56) do not displace the ideal gas law $PV = nRT$, but merely *supplement* it. The ideal gas law is applicable under *all* conditions of an ideal gas, while Eqs. (55) and (56) apply only under adiabatic reversible conditions.

The constants C_1 and C_2 depend on the amounts of gas present and differ from each other numerically. The constants may be eliminated in calculations by taking the ratio of initial to final conditions. Thus from Eq. (56)

$$P_1 V_1^{\gamma} = P_2 V_2^{\gamma}$$

and

$$\frac{P_1}{P_2} = \left(\frac{V_2}{V_1}\right)^{\gamma} \tag{57}$$

Equation (57), like Boyle's law, permits a recalculation of volumes from pressures alone or vice versa. During the adiabatic expansion, however, the temperature of the gas does not remain constant. The initial and final temperatures may be obtained in any instance by substituting the initial and final values of P and V, along with n, in the expression $PV = nRT$.

The use of these equations can be illustrated by an example. Two moles of hydrogen at standard conditions are compressed adiabatically to a volume of 10 liters. For hydrogen, $\gamma = 1.41$. From these data it is desired to find the final pressure and temperature of the gas. The known and unknown quantities are

Initial	Final
$P_1 = 1$ atm	$P_2 = ?$ ·
$V_1 = 2(22.4) = 44.8$ liters	$V_2 = 10$ liters
$T_1 = 273.2°K$	$T_2 = ?$
$n = 2$	$n = 2$
$\gamma = 1.41$	$\gamma = 1.41$

Applying Eq. (57), we have for P_2,

$$P_2 = P_1\left(\frac{V_1}{V_2}\right)^\gamma = 1\left(\frac{44.8}{10}\right)^{1.41}$$
$$= 8.30 \text{ atm}$$

Had the expansion taken place *isothermally*, the new pressure would have been 4.5 atm. Knowing now P_2 and V_2, T_2 follows as

$$T_2 = \frac{P_2 V_2}{nR} = \frac{8.30 \times 10}{2 \times 0.0821}$$
$$= 505°\text{K} \quad \text{or} \quad 232°\text{C}$$

An instance of maximum work obtainable under nonisothermal conditions is the work performed in the *adiabatic reversible* expansion of an ideal gas. The expression in question is arrived at as follows: Differentiation of $PV^\gamma = $ constant yields

$$\gamma PV^{\gamma-1}dV + V^\gamma dP = 0$$
$$\gamma PdV + VdP = 0$$
or
$$VdP = -\gamma PdV$$

Again, complete differentiation of $PV = nRT$ gives

$$PdV + VdP = nRdT$$

which on substitution of the expression for VdP becomes

$$PdV - \gamma PdV = nRdT$$
and
$$PdV = \frac{nRdT}{(1-\gamma)}$$

On inserting this identity for PdV into Eq. (19) and changing the limits to the temperatures corresponding to the volumes, we obtain

$$w_m = \int_{V_1}^{V_2} PdV = \int_{T_1}^{T_2} \frac{nRdT}{(1-\gamma)}$$
$$= \frac{nR(T_2 - T_1)}{(1-\gamma)} \tag{58}$$

In Eq. (58) T_1 is the initial temperature of n moles of gas, T_2 is the final temperature resulting from the adiabatic reversible expansion, while γ is the ratio C_p/C_v for the gas. Whenever $T_2 > T_1$, w_m is negative and work is done on the gas. On the other hand, when work is done by the gas, $T_2 < T_1$, and w_m is positive.

THE JOULE - THOMSON EFFECT

An ideal gas exhibits no intermolecular attraction, and for it the product PV is a constant at any given temperature at all pressures. Hence when such a gas expands under adiabatic conditions into a vacuum, no heat is absorbed or evolved, no external work or work to separate the molecules has to be performed, and so

$$q = 0, \quad w = 0, \quad \text{and} \quad \Delta E = 0$$

Thus the internal energy of the gas remains constant as well as PV, and consequently the temperature is the same before and after expansion. The abiabatic condition was imposed merely to prevent any interchange of energy between the surroundings and the gas, and thus to avoid an increase or decrease in the internal energy by heat absorption or evolution.

The situation with *real* gases is different and was first investigated by Joule and Thomson (Lord Kelvin). Their experimental setup is illustrated schematically in Figure 3-3. A tube, thoroughly insulated to

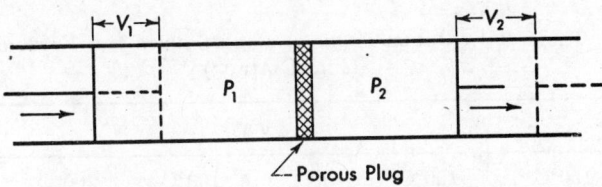

Figure 3–3.   Joule-Thomson experiment.

approximate adiabatic conditions, was fitted with a porous plug, as indicated, to allow gas to be kept on either side of it at the different pressures P_1 and P_2 $(P_1 > P_2)$. By applying pressure on the piston on the left slowly enough so as not to change the pressure P_1, a volume of gas V_1 was forced slowly through the porous plug and allowed then to expand to the pressure P_2 and volume V_2 by moving the piston on the right outward. While the expansion was taking place, accurate temperature readings were taken on the gas in the two chambers to ascertain whether the expansion was accompanied by a temperature change.

The work done *on* the system at the left piston is $-P_1V_1$, the work done *by* the system at the right piston is P_2V_2, and hence the net work done *by* the system is

$$w = P_2V_2 - P_1V_1$$

Since the process was conducted adiabatically, $q = 0$, and, therefore,

$$\Delta E = E_2 - E_1 = -w = -(P_2V_2 - P_1V_1)$$
$$E_2 + P_2V_2 = E_1 + P_1V_1$$
$$H_2 = H_1$$
$$\Delta H = 0$$

The process was conducted, then, at *constant enthalpy*. Under these conditions Joule and Thomson observed near room temperatures that all gases, with the exception of hydrogen, experienced a cooling on expansion, while hydrogen actually became warmer. The extent of the temperature change was found to depend on the initial temperature and pressure of the gas. Later, when helium was discovered, it was shown that this gas as well undergoes a heating effect on expansion. The *Joule-Thomson coefficient*, μ, is defined as

$$\mu = \left(\frac{\partial T}{\partial P}\right)_H \tag{59}$$

It may be thought of as the number of degrees temperature change produced per atmosphere drop in pressure under conditions of constant enthalpy. For a cooling μ is positive, while for an observed heating μ is negative. Table 3–1 shows Joule-Thomson coefficients for nitrogen at

TABLE 3–1.   JOULE-THOMSON COEFFICIENTS FOR NITROGEN
(μ in °C Atm^{-1})

P Atm	μ at					
	−150°C	−100°C	0°C	100°C	200°C	300°C
1	1.266	0.6490	0.2656	0.1292	0.0558	0.0140
20	1.125	0.5958	0.2494	0.1173	0.0472	0.0096
33.5	0.1704	0.5494	0.2377	0.1100	0.0430	0.0050
60	0.0601	0.4506	0.2088	0.0975	0.0372	−0.0013
100	0.0202	0.2754	0.1679	0.0768	0.0262	−0.0075
140	−0.0056	0.1373	0.1316	0.0582	0.0168	−0.0129
180	−0.0211	0.0765	0.1015	0.0462	0.0094	−0.0160
200	−0.0284	0.0087	0.0891	0.0419	0.0070	−0.0171

various pressures and temperatures. Inspection of the table reveals that at any given temperature μ decreases with increase in pressure and actually becomes negative as the pressure is made high enough. This behavior means that at the lower pressures nitrogen cools on adiabatic expansion, whereas at higher pressures it undergoes a heating effect. The pressure at which the gas neither heats nor cools on expansion, i.e., where $\mu = 0$, is called an *inversion point*, and this point varies with the temperature. Again, at a pressure such as 200 atmospheres nitrogen undergoes a heating effect at −150°C, cooling at −100 to 200°C, and heating again at 300°C. At constant pressure we encounter, therefore, two *inversion temperatures*, an upper and a lower one, between which a gas undergoes cooling, and outside of which the gas experiences heating effects. Most gases at room temperature are already below their upper inversion temperatures and hence they undergo cooling on adiabatic expansion. Hydrogen and helium, on the other hand, are still above theirs, and hence they show heating. It may be expected that these gases would also exhibit cooling if their temperature were lowered below the upper inversion point at a given pressure, and such is actually the case.

The Joule-Thomson effect is of great practical importance in the liquefaction of gases. The manner in which it is utilized has already been discussed in Chapter 1.

The Joule-Thomson coefficient can readily be related to other thermodynamic quantities. If we impose upon Eq. (40) the condition for the Joule-Thomson effect, namely that $dH = 0$, we get

$$\left(\frac{\partial H}{\partial T}\right)_P dT = -\left(\frac{\partial H}{\partial P}\right)_T dP$$

On division of both sides of the equation by dP and imposition of the condition of constant H, we obtain

$$\left(\frac{\partial H}{\partial T}\right)_P \left(\frac{\partial T}{\partial P}\right)_H = -\left(\frac{\partial H}{\partial P}\right)_T$$

But, $(\partial H/\partial T)_P = C_p$, and $(\partial T/\partial P)_H = \mu$. Hence

$$\mu C_p = -\left(\frac{\partial H}{\partial P}\right)_T \tag{60}$$

and μ can be calculated from $(\partial H/\partial P)_T$, or vice versa. Again, if we substitute for $(\partial H/\partial P)_T$ its value from Eq. (41), then

$$\mu C_p = V - T\left(\frac{\partial V}{\partial T}\right)_P \tag{61}$$

and Eq. (61) can be used to evaluate μ at various temperatures and pressures from the P-V-T data for a gas.

THE CARNOT CYCLE

Experience has shown that periodically operating heat engines which absorb a quantity of heat at some temperature T_2 and reject the waste heat at a lower temperature T_1 can convert only a fraction of the absorbed heat into work. It can be deduced by theoretical considerations that even an ideal engine, operating under ideal conditions, is able to convert only a fraction of the absorbed heat into work, and that this fraction is determined by the operating temperatures T_2 and T_1 and is totally independent of the nature of the engine. In other words, there is a natural limitation to the convertibility of heat to work above and beyond any imperfections which may be present in any contrivances employed.

To establish the above deductions, let us consider the sequence of operations called a *Carnot cycle*. A cycle is any series of operations so carried out that at the end the system is back to its initial state. Any process so conducted is called a cyclic process. The one we are interested in consists of four steps, two isothermal and two adiabatic, and yields the maximum work that can be derived from a quantity of heat absorbed at one temperature and given out at another, lower, temperature.

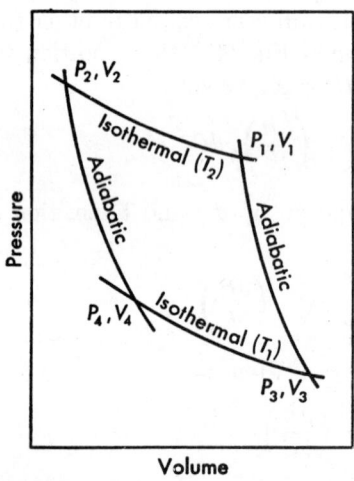

Figure 3–4.   The Carnot cycle.

Imagine a cylinder fitted with a weightless and frictionless piston, and containing a given weight of any substance. Let the pressure, volume, and temperature of the substance be P_2, V_2, and θ_2, where θ_2 is a temperature whose nature we shall not specify at present. Allow this substance to expand isothermally and reversibly at θ_2 to a pressure P_1 and volume V_1. In doing so the substance absorbs a quantity of heat q_2 which goes to increase its internal energy by ΔE_1 and also to do the work w_1. This expansion is represented by the line $P_2V_2P_1V_1$ in Figure 3–4. According to the first law of thermodynamics we have then

$$\Delta E_1 = q_2 - w_1$$

Let this substance now expand adiabatically and reversibly from P_1, V_1 at θ_2 to a new state P_3, V_3. Since the change is adiabatic, $q = 0$, all the work done, w_2, must be at the expense of a change in internal energy, ΔE_2, and hence the temperature must drop from θ_2 to θ_1. For this situation we have thus

$$\Delta E_2 = -w_2$$

In the third stage of the cycle the substance is compressed isothermally and reversibly at θ_1 from point P_3, V_3 to point P_4, V_4. During this step work w_3 is done on the system, a quantity q_1 of heat is given up to the surroundings, and hence

$$\Delta E_3 = -q_1 - w_3$$

Finally, the substance is compressed adiabatically and reversibly from P_4, V_4 at θ_1 to its initial state P_2, V_2 at θ_2. Here again $q = 0$, and so

$$\Delta E_4 = -w_4$$

The total change in internal energy for the complete cycle must be

$$\begin{aligned}
\Delta E &= \Delta E_1 + \Delta E_2 + \Delta E_3 + \Delta E_4 \\
&= (q_2 - w_1) - w_2 - (q_1 + w_3) - w_4 \\
&= (q_2 - q_1) - (w_1 + w_2 + w_3 + w_4) \\
&= (q_2 - q_1) - w_m
\end{aligned}$$

where w_m is the total work done in the cycle. It must represent maximum work since all the steps were reversible. Further, since the system is back to its initial state, we must have $\Delta E = 0$ by application of the principle of reproducibility of states. Hence

$$w_m = q_2 - q_1 \tag{62}$$

and on division of both sides of the equation by q_2

$$\epsilon = \frac{w_m}{q_2} = \frac{q_2 - q_1}{q_2} \tag{63}$$

In Eq. (63) ϵ is the *thermodynamic efficiency* of the process, as it gives the fraction of the total heat q_2 which was absorbed at θ_2 and converted to maximum work w_m. The equation shows further that ϵ is independent of the nature of the working substance used in the cycle, and that it depends entirely upon q_2 and q_1.

Lord Kelvin first showed how Eq. (63) can be used to define a *thermodynamic temperature scale.* Suppose we define the temperatures θ_2 and θ_1 mentioned above by the relation

$$\frac{q_2 - q_1}{q_2} = \frac{\theta_2 - \theta_1}{\theta_2} \tag{64}$$

We may take now a substance and subject it to a Carnot cycle operating between the boiling and freezing points of water. Since in terms of degrees centigrade the difference between these two points is 100°C, we may set $(\theta_2 - \theta_1) = 100$. Further, if we were to determine the thermodynamic efficiency of the process we would be then in a position to find from Eq. (64) θ_2, and arrive thus at the thermodynamic temperature corresponding to the boiling point of water. Knowledge of θ_2 also would fix θ_1, the freezing point of water, and we thus would obtain the ice point on the thermodynamic scale.

Actually it is not necessary to evaluate the θ's in the manner described, for it is possible to arrive at the relation between the thermodynamic temperature θ and the ideal gas temperature T through use of an ideal gas in the Carnot cycle. Since Eq. (64) is valid for any substance, it should be valid also for ideal gases, and the result obtained with the latter should be identical with the general case.

CARNOT CYCLE FOR IDEAL GASES

If we repeat the Carnot cycle between the temperatures T_2 and T_1 corresponding to θ_2 and θ_1, but use this time n moles of an ideal gas, then we have in line with Figure 3–4 the following:

Step 1: Isothermal reversible expansion at T_2

$$\Delta E_1 = 0 \qquad w_1 = q_2 = nRT'_2 \ln\left(\frac{V_1}{V_2}\right)$$

Step 2: Adiabatic reversible expansion from T_2 to T_1

$$q'_2 = 0 \qquad \Delta E_2 = -w_2 = -n \int_{T_2}^{T_1} C_v dT$$

Step 3: Isothermal reversible compression at T_1

$$\Delta E_3 = 0 \qquad w_3 = -q_1 = nRT_1 \ln\left(\frac{V_4}{V_3}\right)$$

Step 4: Adiabatic reversible compression from T_1 to T_2

$$q_4 = 0 \qquad \Delta E_4 = -w_4 = -n \int_{T_1}^{T_2} C_v dT$$

For the complete cycle $\Delta E = 0$. Hence the total maximum work done, w_m, must be given by $w_m = q_2 - q_1$. Further,

$$w_m = w_1 + w_2 + w_3 + w_4$$
$$= nRT_2 \ln\left(\frac{V_1}{V_2}\right) + n \int_{T_2}^{T_1} C_v dT + nRT_1 \ln\left(\frac{V_4}{V_3}\right) + n \int_{T_1}^{T_2} C_v dT$$
$$= nRT_2 \ln\left(\frac{V_1}{V_2}\right) + nRT_1 \ln\left(\frac{V_4}{V_3}\right)$$

because the two integrals cancel each other. Consequently,

$$w_m = q_2 - q_1 = nRT_2 \ln\left(\frac{V_1}{V_2}\right) + nRT_1 \ln\left(\frac{V_4}{V_3}\right) \qquad \text{(A}`$$

Again, since $q_2 = nRT_2 \ln (V_1/V_2)$, division of Eq. (A) by q_2 yields

$$\frac{w_m}{q_2} = \frac{(q_2 - q_1)}{q_2} = \frac{RT_2 \ln (V_1/V_2) + RT_1 \ln (V_4/V_3)}{RT_2 \ln (V_1/V_2)} \qquad \text{(B)}$$

Equation (B) can be simplified considerably. Since the points (P_1, V_1) and (P_3, V_3) lie on the same adiabatic, then by Eq. (55),

$$T_2^{C_v/R} V_1 = T_1^{C_v/R} V_3$$

Similarly, we have for the points (P_2, V_2) and (P_4, V_4), which lie on the same adiabatic,

$$T_2^{C_v/R} V_2 = T_1^{C_v/R} V_4$$

On dividing the first of these equations by the second, we obtain

$$\frac{V_1}{V_2} = \frac{V_3}{V_4}$$

and, consequently,

$$\ln \left(\frac{V_4}{V_3} \right) = - \ln \left(\frac{V_1}{V_2} \right)$$

Substituting this value for $\ln (V_4/V_3)$ in Eq. (B) we get, finally,

$$\frac{w_m}{q_2} = \frac{q_2 - q_1}{q_2} = \frac{RT_2 \ln (V_1/V_2) - RT_1 \ln (V_1/V_2)}{RT_2 \ln (V_1/V_2)}$$
$$= \frac{T_2 - T_1}{T_2} \qquad \text{(65)}$$

Comparison of Eqs. (64) and (65) shows that

$$\frac{\theta_2 - \theta_1}{\theta_2} = \frac{T_2 - T_1}{T_2} \qquad \text{(66)}$$

Further, if we compare any substance and an ideal gas operating in a Carnot cycle between 0 and 100°C, then $(\theta_2 - \theta_1) = (T_2 - T_1) = 100°C$, and, therefore, $\theta_2 = T_2$. Likewise, we must have $\theta_1 = T_1$. From this argument we see that the thermodynamic and ideal gas temperature scales are identical, and we may write thus for all substances

$$\frac{w_m}{q_2} = \frac{q_2 - q_1}{q_2} \qquad \text{(67a)}$$

$$= \frac{T_2 - T_1}{T_2} \qquad \text{(67b)}$$

THE THERMODYNAMIC EFFICIENCY

Equation (67b) states that when a system during a reversible cyclical process absorbs a quantity of heat q_2 at temperature T_2 and then undergoes a temperature drop $(T_2 - T_1)$, the maximum work that may be recovered from the process is equal to the heat absorbed at T_2 multiplied by the ratio $(T_2 - T_1)/T_2$. It is important to observe that the absorption takes place at the higher temperature and that the heat passes from the higher temperature to the lower. The magnitude of the work involved is given by the enclosed area in Figure 3-4. Furthermore, *the thermodynamic efficiency must be the same for all processes operating under given temperature conditions.* The necessity for this deduction was pointed out by Carnot, who argued that if any machine were more efficient than one executing a Carnot cycle, the two could be so coupled together as to obtain during a complete cycle a net quantity of work at the higher temperature at the expense of the heat at the lower temperature. But such a situation is contrary to our experience with the convertibility of heat into work, and hence there can be no engine more efficient than a Carnot engine.

From Eq. (67b) it may be seen that 100% thermodynamic efficiency could be obtained only by making $T_1 = 0$. To reach this temperature, we would require surroundings at least dT below absolute zero in order to absorb the heat rejected by the system at $T_1 = 0$. However, since the latter temperature is the lowest possible, we are forced to the conclusion that absolute zero is approachable but not attainable. Actually temperatures several thousandths of a degree above absolute zero have been obtained experimentally, but not $0°K$. Consequently 100% conversion of heat into work is impossible in any cyclic process.

REFERENCES

1. B. F. Dodge, *Chemical Engineering Thermodynamics*, McGraw-Hill Book Company, Inc., New York, 1944.
2. S. Glasstone, *Thermodynamics for Chemists*, D. Van Nostrand Company, Inc., Princeton, N.J., 1947.
3. I. M. Klotz, *Chemical Thermodynamics*, Prentice-Hall, Inc., Englewood Cliffs, N.J., 1950.
4. G. N. Lewis, M. Randall, K. S. Pitzer, and L. Brewer, *Thermodynamics*, McGraw-Hill Book Company, Inc., New York, 1961.
5. F. D. Rossini, *Chemical Thermodynamics*, John Wiley & Sons, Inc., New York, 1950.
6. L. Steiner, *Introduction to Chemical Thermodynamics*, McGraw-Hill Book Company, Inc., New York, 1948.

7. F. T. Wall, *Chemical Thermodynamics*, W. H. Freeman & Company, San Francisco, 1958.

PROBLEMS

1. A solution contains 50% of water, 35% of ethyl alcohol, and 15% of acetic acid by weight. Calculate the mol fraction of each component in the mixture.
Ans. $N_{H_2O} = 0.733$; $N_{HAc} = 0.066$; $N_{C_2H_5OH} = 0.201$.

2. A Na_2CO_3 solution is made up by dissolving 22.5 g of $Na_2CO_3 \cdot 10\ H_2O$ in H_2O and adding H_2O until the total volume is 200 cc. The density of the resulting solution is 1.040 g/cc. Calculate the molarity, normality, and mol fraction of Na_2CO_3 in the solution. *Ans.* 0.393 m; 0.786 N; 0.00704 mol fraction.

3. A solution containing 10% of NaCl by weight has a density of 1.071 g/cc Calculate the molality and molarity of NaCl in the solution.

4. A gaseous solution was analyzed and found to contain 15% of H_2, 10% of CO, and 75% of N_2 by volume. What is the mol fraction and percentage by weight of each gas in the mixture?

5. Consider a solution composed of three constituents. (a) How many independent variables will be required to express the total energy of the solution? (b) How many independent variables will be required to express the density of the solution?

6. In Eqs. (13), (14), and (15) of this chapter, what is the significance of the parameters a, b, and a/n_1?

7. Starting with Eq. (14) for $\bar{V}_2$, deduce the expression for $\bar{V}_1$ without employing Eq. (13). How does the result obtained compare with Eq. (15)?

8. For a certain binary solution at constant T and P

$$\bar{V}_2 = a_2 + 2\ a_3 m + 3\ a_4 m^2$$

where m is the molality and a_2, a_3, and a_4 are constants. Deduce the expression for the total volume of solution, V, as a function of m.

9. Transform the equations involved in problem 8 into expressions of $\bar{V}_2$ and V as functions of the mol fractions of the two constituents.

10. A weight of 1000 g falls freely to a platform from a height of 10 meters. What amount of heat will be evolved when the weight strikes the platform?
Ans. 23.4 cal.

11. A piston whose area is 60 cm² moves through a distance of 20 cm against a pressure of 3 atm. Calculate the work done in (a) joules and (b) calories.

12. A gas in expanding against a constant pressure of 2 atm from 10 to 20 liters absorbs 300 cal of heat. What is the change in the internal energy of the gas?

13. A gas expands against a variable opposing pressure given by $p = \dfrac{10}{V}$ atm, where V is the volume of the gas at each stage of the expansion. Further, in

expanding from 10 to 100 liters, the gas undergoes a change in internal energy of $\Delta E = 100$ cal. How much heat is absorbed by the gas during the process?

$Ans.\ q = 658$ cal.

14. For a certain ideal gas $C_v = 6.76$ cal mole^{-1} degree^{-1}. If 10 moles of the gas are heated from 0 to 100°C, what will be ΔE and ΔH for the process?

15. Two liters of N_2 at 0°C and 5 atm pressure are expanded isothermally against a constant pressure of 1 atm until the pressure of the gas is also 1 atm. Assuming the gas to be ideal, what are the values of w, ΔE, ΔH, and q for the process? $Ans.\ w = q = 194$ cal; $\Delta E = \Delta H = 0$.

16. Calculate the work done by 5 moles of an ideal gas during expansion from 5 atm at 25°C to 2 atm at 50°C against a constant pressure of 0.5 atm. If for the gas $C_p = 5.0$ cal mole^{-1} degree^{-1}, find also ΔE, ΔH, and q for the process.

17. Three moles of an ideal gas at 1 atm pressure and 20°C are heated at constant pressure until the final temperature is 80°C. For the gas $C_v = 7.50 + 3.2 \times 10^{-3}\ T$ cal mole^{-1} degree^{-1}. Calculate w, ΔE, ΔH, and q for the process.

18. Assuming CO_2 to be an ideal gas, calculate the work done by 10 g of CO_2 in expanding isothermally and reversibly from a volume of 5 liters to 10 liters at 27°C. What are q, ΔE, and ΔH for the process?

$Ans.\ w = q = 93.9$ cal; $\Delta E = \Delta H = 0$.

19. Calculate the minimum work necessary to compress 20 g of O_2 from 10 to 5 liters at 0°C. How much heat is evolved in the process?

20. Two liters of N_2 at 0°C and 5 atm pressure are expanded isothermally and reversibly until the confining pressure is 1 atm. Assuming the gas to be ideal, calculate w, q, ΔE, and ΔH for the expansion.

21. Using the van der Waals equation of state, derive the expression for the isothermal reversible expansion of n moles of gas from volume V_1 to volume V_2 at temperature T.

22. For a certain gas the van der Waals constants are $a = 6.69$ atm-liter^2/mole^2 and $b = 0.057$ liter. What will be the maximum work performed in the expansion of 2 moles of the gas from 4 to 40 liters at 300°K? $Ans.\ w_m = 2640$ cal.

23. For CO_2 the van der Waals constants are $a = 3.59$ atm-liter^2/mole^2 and $b = 0.0427$ liter. (a) Find the minimum work required to compress 1 mole of CO_2 from a volume of 10 liters to 1 liter at 27°C. (b) How does this work compare with that obtained on the assumption that CO_2 is an ideal gas?

24. Employing the Beattie-Bridgeman equation of state explicit in volume, Eq. (56) of Chapter 1, deduce the expression for the maximum work performed in the isothermal expansion of n moles of a gas from pressure P_1 to P_2 at temperature T.

25. Eight grams of O_2 at 27°C and under a pressure of 10 atm are permitted to expand adiabatically and reversibly until the final pressure is 1 atm. Find the final temperature and the work done in the process. Assume that $C_p = \frac{7}{2}\ R$ for O_2.

$Ans.\ t = -117.7$°C; $w = 179.6$ cal.

26. Ten grams of N_2 at 17°C are compressed adiabatically and reversibly from

8 to 5 liters. Calculate the final temperature and the work done on the gas. What are ΔE and ΔH for the process? Assume $C_p = \frac{7}{2} R$.

27. For a certain ideal gas $C_p = 8.58$ cal mole^{-1} degree^{-1}. What will be the final volume and temperature when 2 moles of the gas at 20°C and 15 atm are allowed to expand adiabatically and reversibly to 5 atm pressure?

Ans. $V = 7.45$ liters; $t = -46$°C.

28. Find w, q, ΔE, and ΔH for the process given in problem 27.

29. Consider again the gas in problem 27, but suppose that now the expansion takes place adiabatically against a constant pressure of 5 atm. What will be the final volume and temperature of the gas? *Ans.* $V = 8.15$ liters; $t = -25.0$°C.

30. Find w, q, ΔE, and ΔH for the process in problem 29.

31. Deduce the expressions for ΔE and ΔC_v accompanying the expansion of n moles of a van der Waals gas from volume V_1 to V_2 at temperature T.

32. (a) Using the gas described in problem 22, calculate ΔE and ΔC_v involved in the given process. (b) What would be the values of ΔE and ΔC_v for the process with an ideal gas?

33. A gas obeys the van der Waals equation with $a = 6.69$ atm-liter^2/mole^2 and $b = 0.057$ liter For the gas $C_v = 7.00$ cal mole^{-1} degree^{-1}. What will be ΔE for a process involving the compression of 5 moles of the gas from a volume of 100 liters at 300°K to a volume of 10 liters at 400°K. *Ans.* $\Delta E = 3140$ cal.

34. What will be the value of ΔH for the process described in problem 33?

35. Suppose that a gas obeys at low pressures the equation

$$PV_m = RT + \left(A + \frac{B}{T^2} \right) P$$

where A and B are constants independent of pressure and temperature, and V_m is the molar volume. Derive the expression for the change in enthalpy which will accompany the expansion of n moles of gas from a pressure P_2 to a pressure P_1 at temperature T.

36. What will be the expression for the change in C_p accompanying the process described in problem 36?

37. (a) What expressions are predicted by the low-pressure form of the Berthelot equation for the constants A and B mentioned in problem 35? (b) Using these expressions, find ΔH accompanying the expansion of 1 mole of CO_2 from $P = 1$ to $P = 0$ at 300°K. For CO_2, $t_c = 31.0$°C and $P_c = 73.0$ atm.

Ans. $\Delta H = -10.2$ cal.

38. For CO_2(g) at 300°K and 1 atm pressure, $C_p = 8.919$ cal mole^{-1} degree^{-1}. Using the information given and obtained in the preceding problem, find C_p for 1 mole of the gas at zero pressure.

39. For CO_2(g) at 300°K and 1 atm pressure $(\partial H/\partial P)_T = -10.2$ cal mole^{-1} atm^{-1} and $C_p = 8.919$ cal mole^{-1} degree^{-1}. Calculate the Joule-Thomson coefficient of the gas for the given temperature and pressure conditions.

40. Ascertain from the data given in Table 3–1 the pressure at which $N_2(g)$ undergoes a Joule-Thomson inversion at 300°C.

41. From the data given in Table 3–1 determine the temperature range over which $N_2(g)$ undergoes cooling on adiabatic Joule-Thomson expansion at 200 atm.

42. At 300°C and at pressures of 0–60 atm, the Joule-Thomson coefficient of $N_2(g)$ can be represented by the equation

$$\mu = 0.0142 - 2.60 \times 10^{-4}\, P$$

Assuming this equation to be temperature-independent near 300°C, find the temperature drop which may be expected on Joule-Thomson expansion of the gas from 60 to 20 atm. *Ans.* $\Delta t = -0.152°C$.

43. A certain fuel furnishes 7000 cal of heat per gram. (a) Calculate the maximum work which can be obtained from this heat in an engine which operates with mercury between its boiling point, 356.6°C, and 40°C. (b) How much maximum work can be obtained from this heat in an engine which operates with water between its boiling point and 40°C?

44. What are the thermodynamic efficiencies of the processes described in problem 43?

45. Compare the thermodynamic efficiencies to be expected: (a) When an engine is allowed to operate between temperatures of 1000°K and 300°K. (b) When an engine is allowed to operate between 1000°K and 600°K, and then the waste heat is passed on to another engine which operates between 600°K and 300°K.

46. What will be the minimum amount of work required to operate a refrigerating machine which removes 1000 cal of heat at 0°C and rejects it at 50°C?

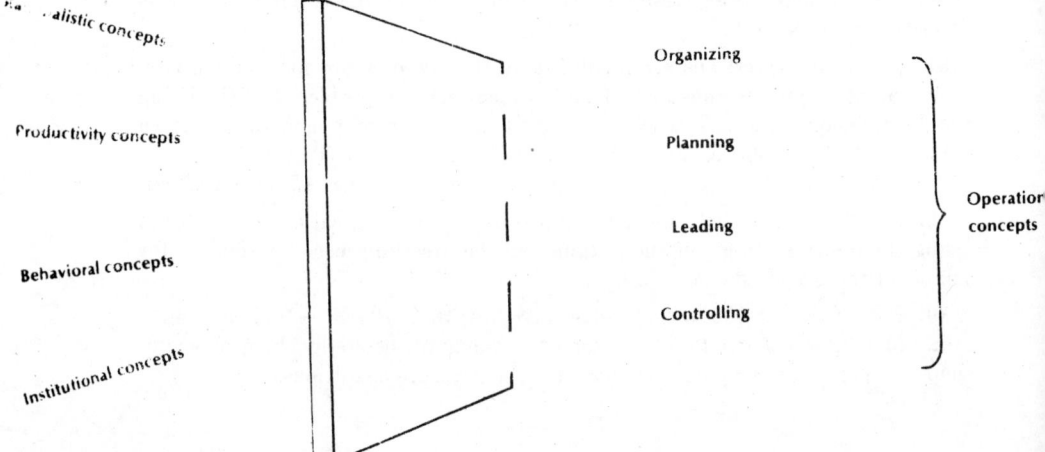

4

Thermochemistry

T HERMOCHEMISTRY is the branch of physical chemistry which deals with the thermal changes accompanying chemical and physical transformations. Its aims are the determination of the quantities of energy emitted or absorbed as heat in various processes, as well as the development of methods for calculating these thermal readjustments without recourse to experiment.

From a practical point of view it is essential to know whether heat is evolved or absorbed in a particular reaction, and how much, for in one case provision must be made for removing that heat in order to effect the reaction, while in the second case provision must be made for supplying the necessary quantity of heat. Again, the heats of various metamorphoses are required for many types of physicochemical calculations. For these reasons it is imperative to consider how heats of reaction are determined experimentally, as well as some of the methods and principles that have been deduced from thermodynamics for evaluation of such thermal changes without appeal to experiment in each case.

The energy units in which thermal changes are usually expressed are the *calorie,* the *joule,* and the *kilocalorie.* A kilocalorie is equal to 1000 calories.

MEASUREMENT OF THERMAL CHANGES

To determine directly the heat change involved in a reaction, calorimeters are employed. A calorimeter consists essentially of an insulated container filled with water in which is immersed the reaction chamber. In an exothermic reaction the heat generated is transferred to the water, and the consequent temperature rise of the water is read from an accurate thermometer immersed in it. Knowing the quantity of water present, its specific heat, and the change in temperature, the amount of heat evolved in the reaction may be calculated. Special corrections must be applied for radiation, rate of cooling of the calorimeter, temperature rise of the vessels, stirrer, etc. To avoid the latter corrections it is best to determine the heat capacity of the calorimeter by burning a definite amount of substance whose heat of combustion has been accurately measured. For this purpose specially prepared samples of benzoic acid, naphthalene, or sugar may be used. An alternate means of avoiding these corrections is to reproduce by electrical heating the temperature change produced in the calorimeter by the process being studied. The amount of electrical energy required to do this is equal to the heat evolved in the process.

A similar apparatus may serve also with endothermic reactions, except that the temperature drop instead of temperature increase is measured.

HEAT OF REACTION AT CONSTANT VOLUME OR PRESSURE

There are two general conditions under which thermochemical measurements are made, namely, (a) constant volume and (b) constant pressure. Under the first of these conditions the volume of the system whose thermal change is sought is kept constant during the whole course of the measurement. In experiments at constant pressure, however, the system is kept either open to the atmosphere or is confined within a vessel upon which a constant external pressure is exerted. Under these conditions any volume change which accompanies the transformation or reaction can take place, and the system is able to adjust itself to the constant external pressure.

The magnitudes of the thermal changes obtained under these two conditions are, in general, different. At constant volume any thermal change taking place must be due only to the difference in the sum of the internal energies of the products and the sum of the internal energies of the reactants. At constant pressure, however, not only does the change in the internal energy take place, but work is also involved. This work must modify the amount of heat observed in the calorimeter at constant volume.

The exact significance of both types of measurement can best be obtained by applying the first law of thermodynamics to the thermal change occurring within the calorimeter. According to the first law, any heat q added to a system will in general go to increase the internal energy of the system and to perform external work, namely,

$$q = \Delta E + w = \Delta E + \int_{V_1}^{V_2} p dV \tag{1}$$

Since at constant volume $dV = 0$, no work can be performed. Hence

$$(q)_V = \Delta E \tag{2}$$

i.e., any thermal change occurring in the calorimeter at constant volume must be the change in the internal energy due to the chemical reaction or physical transformation. On the other hand, when the pressure is kept constant, and the same inside and outside the system, then $p = P$, and

$$w = \int_{V_1}^{V_2} P dV = P(V_2 - V_1) = P\Delta V$$

Consequently,

$$(q)_P = \Delta E + P\Delta V \tag{3}$$

According to Eq. (3) the thermal change observed in the calorimeter at constant pressure involves not only the change in the internal energy, but also the work performed in any expansion or contraction of the system.

Equation (3) may be rewritten to

$$\begin{aligned} (q)_P &= (E_p - E_r) + P(V_p - V_r) \\ &= (E_p + PV_p) - (E_r + PV_r) \end{aligned} \tag{4}$$

where the subscripts p and r refer to products and reactants respectively. But, since H, the enthalpy or heat content, is defined by $H = E + PV$, $H_p = E_p + PV_p$, $H_r = E_r + PV_r$, and Eq. (4) becomes

$$\begin{aligned} (q)_P &= H_p - H_r \\ &= \Delta H \end{aligned} \tag{5}$$

We see, therefore, that the thermal effect which at constant volume measures the change in the internal energy, gives at constant pressure the change in enthalpy of products and reactants. Equation (2) defines the *heat of reaction at constant volume* and shows that it is equal to the thermodynamic quantity ΔE. Similarly, Eq. (5) defines the *heat of reaction at constant pressure* and shows that it is equal to ΔH. The relation between the two is, of course,

$$\Delta H = \Delta E + P\Delta V \tag{6}$$

In line with the conventions established in Chapter 3 a *positive value* *of ΔH or ΔE shows that heat is absorbed* during the process, while a *negative value of ΔH or ΔE shows that heat is evolved.*

THERMOCHEMICAL EQUATIONS

The heat associated with a process depends not only on whether the change occurs at constant volume or pressure, but also on the amounts of substances considered, their physical state, the temperature, and the pressure. At present we shall consider the effects of the first two of these variables, while the influences of temperature and pressure will be discussed toward the end of the chapter.

The quantity of heat obtained in a reaction depends on the amount of substance reacted and is directly proportional to it. Thus, when 2 g of hydrogen are burned in oxygen to form liquid water, 68,320 cal of heat are evolved. For 4 g the amount of heat evolved would be twice 68,320 cal, and so on. Instead of recording the amount of heat evolved per gram of substance reacted, it is customary, rather, to give the heat for a particular reaction, namely,

$$H_2 + \frac{1}{2} O_2 = H_2O \qquad \Delta H = -68,320 \text{ cal} \qquad (7a)$$

Equation (7a) is called a *thermochemical equation*. It indicates that when 1 mole of hydrogen reacts with 0.5 mole of oxygen to form 1 mole of water 68,320 cal of heat are evolved at constant pressure. If it is desired to indicate the interaction of 2 moles of hydrogen with oxygen to form 2 moles of water, then Eq. (7a) becomes

$$2 H_2 + O_2 = 2 H_2O \qquad \Delta H = 2(-68,320) = -136,640 \text{ cal} \quad (7b)$$

In writing a thermochemical equation it is essential that the equation be balanced and that the values of ΔE or ΔH indicated in the reaction correspond to the quantities of substance given by the equation.

Equation (7a) is still incomplete in two respects. In the first place, the heat of reaction given corresponds to a definite temperature, in this instance 25°C, and this must be stated. Second, the equation as it stands does not tell us anything about the physical state of reactants and products. There is not much ambiguity about the states of hydrogen and oxygen, for both are definitely gaseous at this temperature, but the water may be liquid or gaseous. For the formation of liquid water from the elements at 25°C, $\Delta H = -68,320$ cal per mole. However, if a mole of gaseous water is formed from the elements at the same temperature, $\Delta H = -57,800$ cal per mole, the two values of ΔH differing from each other by the heat of vaporization of a mole of water at 25°C, namely,

10,520 cal. To make the thermochemical equation complete, therefore, we write, instead of Eq. (7a),

$$H_2(g) + \frac{1}{2} O_2(g) = H_2O(l) \qquad \Delta H_{25°C} = -68,320 \text{ cal} \qquad (8)$$

The symbols (g) and (l) indicate that the hydrogen and oxygen are gaseous, the water is liquid, while the subscript on the ΔH shows that the heat of reaction given is for 25°C. In a similar manner the symbol (s) is used to designate a solid phase, while the symbol (aq) is used to represent a dilute aqueous solution.

It is very important that the student appreciate the physical significance of a thermochemical equation. Since in general $\Delta H = H_{(products)} - H_{(reactants)}$, ΔH for the reaction given in Eq. (8) defines the difference in enthalpy of 1 mole of liquid water and the sum of the enthalpies of 1 mole of gaseous hydrogen and 0.5 mole of gaseous oxygen, namely,

$$\Delta H = H_{H_2O(l)} - \left(H_{H_2(g)} + \frac{1}{2} H_{O_2(g)} \right)$$

As this ΔH is negative, the sum of the heat contents of the reactants must exceed that of the product by 68,320 cal, and consequently when a mole of liquid water is formed from the elements that amount of heat is *evolved*. Conversely, if we were to consider the reverse reaction, namely, the decomposition of a mole of water to form the elements, the sum of the heat contents of products would exceed that of the reactant by the same amount, and 68,320 cal of heat would have to be *absorbed* to accomplish the reaction. It is evident, therefore, that once ΔH for a particular reaction is known, the ΔH for the reverse reaction is also known from the relation

$$\Delta H_{\text{direct reaction}} = -\Delta H_{\text{reverse reaction}} \qquad (9)$$

CALCULATION OF ΔE FROM ΔH AND VICE VERSA

According to Eq. (6), the difference between ΔH and ΔE is given by $P\Delta V$, where P is constant. The significant factor determining the difference is, therefore, ΔV, the change in volume which occurs during a reaction. For reactions involving solids and liquids only, the volume changes are usually very slight and may be disregarded if the pressure is not very high. For such reactions, then, ΔH is equal to ΔE. In reactions involving gases, however, the volume changes may be large and cannot be disregarded. For such reactions the difference $\Delta H - \Delta E$ can be calculated very simply if we assume that the gases are ideal. If in general we have a reaction in which are involved n_r moles of *gaseous* reactants and n_p moles of *gaseous* products, then $PV_r = n_rRT$, $PV_p = n_pRT$, and, since the pressure is constant,

$$
\begin{aligned}
P\Delta V &= P(V_p - V_r) \\
&= (n_p - n_r)RT \\
&= \Delta n_g RT
\end{aligned}
\tag{10}
$$

where $\Delta n_g = (n_p - n_r)$, i.e., *the difference between the number of moles of gaseous products and reactants.* If Eq. (10) is inserted into Eq. (6), we obtain

$$
\Delta H = \Delta E + \Delta n_g RT
\tag{11}
$$

The usefulness of this equation in converting ΔH into ΔE, and vice versa, can best be illustrated by an example. For the combustion of 1 mole of liquid benzene, the heat of reaction at constant pressure is given by

$$
C_6H_6(l) + 7\tfrac{1}{2} O_2(g) = 6\ CO_2(g) + 3\ H_2O(l) \qquad \Delta H_{25^\circ C} = -780{,}980\ \text{cal}
$$

In this reaction there is a contraction in volume from 7.5 to 6 moles of gas, and hence $\Delta n_g = 6 - 7.5 = -1.5$. From Eq. (11) we obtain, then, for ΔE,

$$
\begin{aligned}
\Delta E_{25^\circ C} &= \Delta H_{25^\circ C} - \Delta n_g RT \\
&= -780{,}980 + 1.5(1.99)(298.2) \\
&= -780{,}980 + 894 \\
&= -780{,}090\ \text{cal}
\end{aligned}
$$

HESS'S LAW OF HEAT SUMMATION

We have seen in the last chapter that E and H are functions of the state of the system only, and the same must be true of ΔE and ΔH. Consequently, *the heat evolved or absorbed in a given reaction must be independent of the particular manner in which the reaction takes place.* It depends only on the initial and final states of the system and is not affected by the number of steps which may intervene between the reactants and products. Stated differently, if a reaction proceeds in several steps, the heat of the over-all reaction will be the algebraic sum of the heats of the various stages, and this sum in turn will be identical with the heat the reaction would evolve or absorb were it to proceed in a single step. This generalization is known as *Hess's law.*

This principle makes it possible to calculate the heats of many reactions which either do not lend themselves to direct experimental determination or which it is not desired to measure. In these calculations thermochemical equations are handled as if they were ordinary algebraic equations, being added, subtracted, and multiplied or divided. The manner in which this is done can be illustrated by the following example. Suppose it is desired to find ΔH for the reaction

$$2\ C(s) + 2\ H_2(g) + O_2(g) = CH_3COOH(l)\qquad \Delta H_{25°C} = ?\quad (12)$$

The heat of this reaction cannot be determined directly. However, calorimetric measurements are available for the reactions:

$$CH_3COOH(l) + 2\ O_2(g) = 2\ CO_2(g) + 2\ H_2O(l)$$
$$\Delta H_{25°C} = -208,340\ \text{cal}\quad \textbf{(12a)}$$
$$C(s) + O_2(g) = CO_2(g)\qquad\qquad \Delta H_{25°C} = -94,050\ \text{cal}\quad \textbf{(12b)}$$
$$H_2(g) + \frac{1}{2}\ O_2(g) = H_2O(l)\qquad\qquad \Delta H_{25°C} = -68,320\ \text{cal}\quad \textbf{(12c)}$$

If we now multiply Eqs. (12b) and (12c) by 2 and then add, we have

$$2\ C(s) + 2\ O_2(g) = 2\ CO_2(g)\qquad \Delta H_{25°C} = -188,100\ \text{cal}$$
$$2\ H_2(g) + O_2(g) = 2\ H_2O(l)\qquad \Delta H_{25°C} = -136,640\ \text{cal}$$

$$\overline{2\ C(s) + 2\ H_2(g) + 3\ O_2(g) = 2\ CO_2(g) + 2\ H_2O(l)}$$
$$\Delta H_{25°C} = -324,740\ \text{cal}\quad \textbf{(12d)}$$

and on subtraction of Eq. (12a) from Eq. (12d) we obtain

$$2\ C(s) + 2\ H_2(g) + O_2(g) = CH_3COOH(l)\qquad \Delta H_{25°C} = -116,400\ \text{cal}$$

This result is the heat of reaction at constant pressure for Eq. (12).

HEATS OF FORMATION

A substance at any temperature is said to be in its *standard state* when its activity is equal to one. The activity,[1] symbol a, may be looked upon as a thermodynamically corrected pressure or concentration. For pure solids, liquids, and ideal gases the standard state corresponds to the substances at 1 atm pressure. For real gases the pressure in the standard state is not 1 atm, but the difference from unity is not large. In the case of dissolved substances the standard state is the concentration in each instance at which $a = 1$. The enthalpies of substances in their standard states are designated by the symbol H^0, while the ΔH of a reaction where all reactants and products are at unit activity is represented by the symbol ΔH^0.

The thermal change involved in the formation of *1 mole of a substance from the elements* is called the *heat of formation* of the substance. Again, the *standard heat of formation* is the heat of formation when *all* the

---

[1] The activity will be defined more fully in Chapter 6. Here the mere statement that the activity in the standard state is unity will suffice.

substances involved in the reaction are each at unit activity. Thus the thermochemical equation

$$C(s) + 2\ H_2(g) = CH_4(g) \qquad \Delta H^0_{25°C} = -17,890\ \text{cal} \qquad (13a)$$

gives the standard heat of formation of a mole of methane at 25°C. By definition ΔH^0 in Eq. (13a) is given by

$$\Delta H^0_{25°C} = -17,890 = H^0_{CH_4(g)} - (H^0_{C(s)} + 2\ H^0_{H_2(g)}) \qquad (13b)$$

where the H^0's are the standard enthalpies per mole.

Were the enthalpies of the elements known, the above information would be sufficient to calculate the absolute heat content of the methane. However, such information is not available. To get around this difficulty, the assumption is arbitrarily made that *the enthalpies of all elements in their standard states at 25°C are zero.* When an element can exist in more than one allotropic form under these conditions, the form most stable at 1 atm pressure and 25°C is chosen to have zero enthalpy. Thus for carbon graphite rather than diamond is selected, and for sulfur the rhombic rather than the monoclinic modification, since graphite and rhombic sulfur are the stable forms. On the basis of this assumption the enthalpies of $H_2(g)$ and $C(s)$ are zero at 25°C provided the latter is graphite. Therefore, $\Delta H^0_{25°C} = H^0_{CH_4(g)}$, i.e., the standard heat of formation of the compound may be considered to be its heat content at 25°C.

The heat of formation of a compound may be obtained either by measuring the ΔH of the reaction involving the direct formation of the compound from the elements or by calculating the heat of formation from heats of reactions involving the compound. For instance, to find the standard heat of formation of $Fe_2O_3(s)$, the reactions

$$3\ C(s) + 2\ Fe_2O_3(s) = 4\ Fe(s) + 3\ CO_2(g)$$
$$\Delta H^0_{25°C} = +110,800\ \text{cal} \qquad (14a)$$
$$C(s) + O_2(g) = CO_2(g) \qquad \Delta H^0_{25°C} = -94,050\ \text{cal} \qquad (14b)$$

are available. In terms of the assumption made above, the enthalpies of $C(s)$, $Fe(s)$, and $O_2(g)$ are zero. Therefore,

$$\Delta H^0 = +110,800 = 3\ \Delta H^0_{CO_2} - 2\ \Delta H^0_{Fe_2O_3}$$

and

$$\Delta H^0_{Fe_2O_3} = \frac{3\ \Delta H^0_{CO_2} - \Delta H^0}{2}$$
$$= \frac{3(-94,050) - 110,800}{2}$$
$$= -196,500\ \text{cal/mole of } Fe_2O_3$$

The standard heats of formation for a number of compounds at 25°C are given in Table 4–1. With the aid of such a table heats of reaction may be easily calculated. Thus, for the reaction

$$Na_2CO_3(s) + 2 HCl(g) = 2 NaCl(s) + CO_2(g) + H_2O(l) \quad (15)$$
$$\Delta H^0_{25°C} = (2 \Delta H^0_{NaCl} + \Delta H^0_{CO_2} + \Delta H^0_{H_2O}) - (\Delta H^0_{Na_2CO_3} + 2 \Delta H^0_{HCl})$$

Substituting the appropriate heats of formation from the table, we obtain

$$\Delta H^0_{25°C} = [2(-98,230) + (-94,050) + (-68,320)]$$
$$-[(-270,300) + 2(-22,060)]$$
$$= -44,410 \text{ cal}$$

Compounds formed from their elements with evolution of heat are called exothermic compounds and those formed with absorption of heat endothermic compounds. The first type is much more common.

TABLE 4-1. STANDARD HEATS OF FORMATION OF COMPOUNDS AT 25°C

Substance	ΔH^0 (cal mole^{-1})	Substance	ΔH^0 (cal mole^{-1})
$H_2O(l)$	-68,320	$Ag_2O(s)$	-7,310
$H_2O(g)$	-57,800	$CuO(s)$	-38,500
$HCl(g)$	-22,060	$FeO(s)$	-64,300
$HBr(g)$	-8,660	$Fe_2O_3(s)$	-196,500
$HI(g)$	6,200	$Fe_3O_4(s)$	-267,000
$HNO_3(l)$	-41,400	$NaCl(s)$	-98,230
$H_2SO_4(l)$	-193,910	$KCl(s)$	-104,180
$H_2S(g)$	-4,820	$AgCl(s)$	-30,360
$CO(g)$	-26,420	$NaOH(s)$	-102,000
$CO_2(g)$	-94,050	$KOH(s)$	-102,000
$NH_3(g)$	-11,040	$AgNO_3(s)$	-29,400
$NO(g)$	21,600	$Na_2SO_4(s)$	-330,500
$NO_2(g)$	8,090	$PbSO_4(s)$	-219,500
$SO_2(g)$	-70,960	$Na_2CO_3(s)$	-270,300
$SO_3(g)$	-94,450	$CaCO_3(s)$	-288,450
Methane(g), CH_4	-17,890	Acetylene(g), C_2H_2	54,190
Ethane(g), C_2H_6	-20,240	Benzene(l), C_6H_6	11,720
Propane(g), C_3H_8	-24,820	Naphthalene(s), $C_{10}H_8$	14,400
n-Butane(g), C_4H_{10}	-29,810	Methanol(l), CH_3OH	-57,020
n-Hexane(g), C_6H_{14}	-39,960	Ethanol(g), C_2H_5OH	-56,300
n-Octane(g), C_8H_{18}	-49,820	Ethanol(l), C_2H_5OH	-66,360
Ethylene(g), C_2H_4	12,500	Acetic Acid(l), CH_3COOH	-116,400

HEATS OF COMBUSTION

The heats evolved in the complete combustion of many organic compounds in oxygen have been carefully determined. The method ordinarily used is to burn the substance in a combustion bomb and to measure the heat evolved. Since heats of combustion are obtained at constant volume, the experimentally measured ΔE's are converted and corrected to ΔH^0's. The term *heat of combustion* refers to the amount of heat liberated *per mole* of substance burned.

Heats of combustion may be employed directly to calculate heats of formation of organic compounds. If the organic compounds contain only carbon, hydrogen, and oxygen, the supplementary information required is the heats of formation of carbon dioxide and liquid water, the usual final oxidation products of such compounds. The method of calculation may be illustrated with the data on the combustion of propane. For gaseous propane, the heat of combustion is

$$C_3H_8(g) + 5\ O_2(g) = 3\ CO_2(g) + 4\ H_2O(l)$$
$$\Delta H^0_{25°C} = -530,610 \text{ cal} \quad \textbf{(16)}$$

Therefore,

$$\Delta H^0_{25°C} = -530,610 = (3\ \Delta H^0_{CO_2} + 4\ \Delta H^0_{H_2O(l)}) - \Delta H^0_{propane}$$
$$\Delta H^0_{propane} = 530,610 + 3(-94,050) + 4(-68,320)$$
$$= -24,820 \text{ cal}$$

The standard heat of formation of propane from the elements is thus

$$3\ C(s) + 4\ H_2(g) = C_3H_8(g) \qquad \Delta H^0_{25°C} = -24,820 \text{ cal/mole} \quad \textbf{(17)}$$

Another direction in which heats of combustion have proved of value is in the study of the energy differences of allotropic forms of the elements. Carbon exists in two crystalline forms, diamond and graphite. When these two forms of carbon are burned, the heats evolved are found to be

$$C \text{ (diamond)} + O_2(g) = CO_2(g) \qquad \Delta H^0_{25°C} = -94,500 \text{ cal} \quad \textbf{(18)}$$
$$C \text{ (graphite)} + O_2(g) = CO_2(g) \qquad \Delta H^0_{25°C} = -94,050 \text{ cal} \quad \textbf{(19)}$$

Therefore, carbon in the diamond form has the higher heat content, and for the transition,

$$C \text{ (graphite)} = C \text{ (diamond)} \qquad \Delta H^0_{25°C} = 450 \text{ cal} \quad \textbf{(20)}$$

Further information supplied by heats of combustion pertains to the energy associated with certain atomic groups in the molecule. It has been observed, for instance, that the heat of combustion in a homologous series of compounds varies by more or less constant amount as we pass from one member to the next higher one in the series. Thus for addition of a CH_2 group to a normal saturated paraffin hydrocarbon chain, the increase in ΔH^0 of combustion is approximately 157,000 calories. Similar regularities have been observed for other groupings and linkages, and not only in the heats of combustion, but also in heats of formation. As may be anticipated, the thermal increments vary both with the nature of the group and the character of the bond.

Heats of combustion of some organic compounds are given in Table 4–2.

TABLE 4–2. HEATS OF COMBUSTION OF ORGANIC
COMPOUNDS AT 25°C

Substance	Formula	ΔH^0 (cal/mole)
Methane(g)	CH_4	$-212,800$
Ethane(g)	C_2H_6	$-372,820$
Propane(g)	C_3H_8	$-530,600$
n-Butane(g)	C_4H_{10}	$-687,980$
n-Pentane(g)	C_5H_{12}	$-845,160$
Ethylene(g)	C_2H_4	$-337,230$
Acetylene(g)	C_2H_2	$-310,620$
Benzene(g)	C_6H_6	$-787,200$
Benzene(l)	C_6H_6	$-780,980$
Toluene(l)	C_7H_8	$-934,500$
Naphthalene(s)	$C_{10}H_8$	$-1,228,180$
Sucrose(s)	$C_{12}H_{22}O_{11}$	$-1,348,900$
Methanol(l)	CH_3OH	$-173,670$
Ethanol(l)	C_2H_5OH	$-326,700$
Acetic acid(l)	CH_3COOH	$-208,340$
Benzoic acid(s)	C_6H_5COOH	$-771,200$

HEATS OF SOLUTION AND DILUTION

Solution of one substance in another is accompanied by absorption or evolution of heat, and this thermal effect is termed the *integral heat of solution* of the substance. Per mole of dissolved substance the integral heat of solution at any given temperature and pressure depends upon the amount of solvent in which solution takes place, as may be seen from column 2 of Table 4–3. For this reason it is essential to specify the number of moles of solvent per mole of solute in giving a heat of solution, as

$$H_2SO_4(l) + 10 H_2O(l) = H_2SO_4(10 H_2O) \quad \Delta H_{25°C} = -16,240 \text{ cal} \quad \textbf{(21)}$$

However, when the amount of solvent per mole of substance is large, it is usually found that further dilution will produce no significant thermal effect. Once this state of a dilute solution has been reached, the symbol *aq* is employed to indicate this fact. Thus the limiting value of the integral heats of solution in Table 4–3 would be represented by

$$H_2SO_4(l) + aq = H_2SO_4(aq) \quad \Delta H_{25°C} = -22,990 \text{ cal} \quad \textbf{(22)}$$

TABLE 4–3. INTEGRAL AND DIFFERENTIAL HEATS OF SOLUTION
FOR 1 MOLE OF H_2SO_4 IN WATER AT 25°C

Moles of Water (n_1)	ΔH (cal)	$\overline{\Delta H_1}$ (cal/mole)	$\overline{\Delta H_2}$ (cal/mole)
0	0	−6,750	0
0.50	−3,810	−6,740	−438
1	−6,820	−4,730	−2,090
2	−9,960	−2,320	−5,320
3	−11,890	−1,480	−7,450
4	−13,120	−1,040	−8,960
6	−14,740	−570	−11,320
10	−16,240	−233	−13,910
15	−16,990	−89.8	−15,640
25	−17,470	−24.4	−16,860
50	−17,770	−5.70	−17,480
200	−18,130	−2.16	−17,700
800	−18,990	−0.91	−18,260
3,200	−20,050	−0.26	−19,220
∞	−22,990	0	−22,990

Consider in general a solution process at any given temperature and pressure such as

$$n_2A_2 + n_1A_1 = n_2A_2(n_1A_1) \qquad (23)$$

The integral heat of solution, ΔH, for this process is given by

$$\Delta H = H - (n_1H_1^0 + n_2H_2^0) \qquad (24)$$

where H is the enthalpy of the solution and H_1^0 and H_2^0 are the molar enthalpies of the two pure solution constituents. Since H is an extensive solution property, then by Eq. (8) of the last chapter $H = n_1\bar{H}_1 + n_2\bar{H}_2$, and

$$\begin{aligned} \Delta H &= n_1\bar{H}_1 + n_2\bar{H}_2 - (n_1H_1^0 + n_2H_2^0) \\ &= n_1(\bar{H}_1 - H_1^0) + n_2(\bar{H}_2 - H_2^0) \\ &= n_1\overline{\Delta H_1} + n_2\overline{\Delta H_2} \end{aligned} \qquad (25)$$

where $\overline{\Delta H_1} = (\bar{H}_1 - H_1^0)$ and $\overline{\Delta H_2} = (\bar{H}_2 - H_2^0)$. Equation (25) is of the same form as the expression for H. Hence ΔH is an extensive property of the solution, with $\overline{\Delta H_1}$ and $\overline{\Delta H_2}$ being the *partial or differential molal heats of solution*. By employing methods such as those described in the last chapter, $\overline{\Delta H_1}$ and $\overline{\Delta H_2}$ can be evaluated from measured values of ΔH. The $\overline{\Delta H_1}$'s and $\overline{\Delta H_2}$'s for sulfuric acid solutions at 25°C are shown in the last two columns of Table 4–3. The subscripts 1 and 2 refer, respectively, to water and the acid.

The difference between any two integral heats of solution gives the heat involved in the *dilution* of a substance from the initial state to the final state, and is termed the *integral heat of dilution* of the substance. According to Table 4–3, the heat recoverable on diluting with 8 moles of water a solution containing 1 mole of sulfuric acid in 2 moles of water is

$$H_2SO_4(2\ H_2O) + 8\ H_2O(l) = H_2SO_4(10\ H_2O)$$
$$\Delta H_{25^\circ C} = -16{,}240 - (-9960) = -6280\ cal$$

Similarly, the heat evolved on diluting the same solution with a very large quantity of water is

$$H_2SO_4(2\ H_2O) + aq = H_2SO_4(aq)$$
$$\Delta H_{25^\circ C} = -22{,}990 + 9960 = -13{,}030\ cal$$

The latter value represents the maximum heat obtainable from dilution of the given solution.

The *differential heat of dilution* of the solvent is the difference between the $\overline{\Delta H}_1$ values for two different concentrations, namely,

$$(\overline{\Delta H}_1)_d = (\overline{\Delta H}_1)_2 - (\overline{\Delta H}_1)_1 \tag{26}$$

where the subscripts 1 and 2 outside the parentheses represent, respectively, initial and final concentrations. Similarly, the differential heat of dilution of the solute is

$$(\overline{\Delta H}_2)_d = (\overline{\Delta H}_2)_2 - (\overline{\Delta H}_2)_1 \tag{27}$$

From Eqs. (25), (26), and (27) the student can verify for himself that, if no solute is added to the solution, then

$$\Delta H_d = \Delta H_2 - \Delta H_1$$
$$= n_1(\overline{\Delta H}_1)_d + n_2(\overline{\Delta H}_2)_d + \Delta n_1(\overline{\Delta H}_1)_2 \tag{28}$$

when n_1 and n_2 are the numbers of moles of solvent and solute present in the *initial* solution and Δn_1 is the number of moles of solvent *added* during the dilution. Further, introduction of the definitions of $\overline{\Delta H}_1$ and $\overline{\Delta H}_2$ into Eqs. (26) and (27) shows that

$$(\overline{\Delta H}_1)_d = (\bar{H}_1)_2 - (\bar{H}_1)_1 \tag{29}$$

and

$$(\overline{\Delta H}_2)_d = (\bar{H}_2)_2 - (\bar{H}_2)_1 \tag{30}$$

HEATS OF FORMATION IN SOLUTION

From Table 4–1 the standard heat of formation of $H_2SO_4(l)$ is given as

$$H_2(g) + S(s) + 2\ O_2(g) = H_2SO_4(l) \qquad \Delta H^0_{25^\circ C} = -193{,}910\ cal \tag{31}$$

When this equation is added to Eq. (21) we obtain

$$H_2(g) + S(s) + 2\,O_2(g) + 10\,H_2O(l) = H_2SO_4(10\,H_2O)$$
$$\Delta H_{25°C} = -210,150 \text{ cal} \quad (32)$$

Similarly, when we add Eqs. (22) and (31) we get

$$H_2(g) + S(s) + 2\,O_2(g) + aq = H_2SO_4(aq)$$
$$\Delta H_{25°C} = -216,900 \text{ cal} \quad (33)$$

Equations (32) and (33) both give the heat of formation per mole of H_2SO_4 *in aqueous solution*, the first for 1 mole of H_2SO_4 in presence of 10 moles of water, the second for 1 mole of the acid at infinite dilution. From these examples it is evident that the heat of formation of a substance in solution is a function of the concentration, and cognizance must be taken of this fact. Otherwise, such heats of formation can be used in the same manner as those for pure substances in all types of thermochemical calculations involving solutions.

THERMONEUTRALITY OF SALT SOLUTIONS

Since salts of strong acids and bases are considered to be completely ionized in dilute solutions, it might be expected that, if solutions of such salts were mixed without the incidence of chemical action, the heat effect observed should be essentially zero. Such is actually the case. Thus, when a dilute solution of potassium nitrate is mixed with a dilute solution of sodium bromide, we get

$$KNO_3(aq) + NaBr(aq) = KBr(aq) + NaNO_3(aq) \quad \Delta H = 0 \quad (34a)$$

That the exchange indicated in Eq. (34a) involves no chemical reaction may be seen when the equation is written in ionic form, namely,

$$K^+(aq) + NO_3^-(aq) + Na^+(aq) + Br^-(aq)$$
$$= K^+(aq) + Br^-(aq) + Na^+(aq) + NO_3^-(aq) \quad \Delta H = 0 \quad (34b)$$

Since the products and reactants are identical, no thermal change is to be expected. If a slight thermal effect is observed, it is to be ascribed to dilution on mixing.

The principle that dilute solutions of neutral salts of strong acids and strong bases may be mixed without absorption or evolution of heat is termed the *principle of thermoneutrality of salt solutions*.

When a chemical reaction does occur on mixing, the principle of thermoneutrality is no longer valid. Thus, when a dilute solution of barium chloride is mixed with a dilute solution of sodium sulfate, barium sulfate precipitates, and the heat of reaction instead of being zero is actually

$$BaCl_2(aq) + Na_2SO_4(aq) = BaSO_4(s) + 2\,NaCl(aq)$$
$$\Delta H_{25°C} = -4630\text{ cal} \quad \textbf{(35a)}$$

or, in ionic form,

$$Ba^{++}(aq) + SO_4^{--}(aq) = BaSO_4(s) \qquad \Delta H_{25°C} = -4630\text{ cal} \quad \textbf{(35b)}$$

HEATS OF NEUTRALIZATION OF ACIDS AND BASES

When dilute solutions of strong acids are neutralized with dilute solutions of strong bases at room temperature, the *heat of neutralization* per mole of water formed is essentially constant and independent of the nature of the acid or base. This constancy of the heat of neutralization is readily understood when it is remembered that strong acids, bases, and salts are completely dissociated in their dilute solutions and that consequently the neutralization process involves only the combination of hydrogen and hydroxyl ions to form unionized water. Since this process is the same in all neutralizations, ΔH of neutralization should be essentially constant per mole of water formed. The value of this thermal quantity at 25°C, corrected to the standard state, is

$$H^+(a = 1) + OH^-(a = 1) = H_2O(l) \qquad \Delta H^0_{25°C} = -13{,}360\text{ cal} \quad \textbf{(36)}$$

This constancy of heat of neutralization does not carry over to the neutralization of weak acids by strong bases, weak bases by strong acids, or weak acids by weak bases. The behavior of the reactions mentioned is explicable on the ground that in such neutralizations the combination of hydrogen and hydroxyl ions to form water is not the only reaction taking place. Take the case of hydrocyanic acid and sodium hydroxide as an example. In water solution the hydrocyanic acid is practically unionized. Before the hydrogen ion of the acid can react with the hydroxyl ion of the base, ionization must take place. Since this ionization occurs while the neutralization is proceeding, the thermal change observed is the sum of the heat of ionization of the acid and the heat of neutralization of the ionized hydrogen ion; i.e., after correction to the standard state, the over-all reaction

$$HCN(a = 1) + OH^-(a = 1) = CN^-(a = 1) + H_2O(l)$$
$$\Delta H^0_{25°C} = -2460\text{ cal} \quad \textbf{(37a)}$$

is in reality composed of two reactions, namely,

$$HCN(a = 1) = H^+(a = 1) + CN^-(a = 1) \qquad \Delta H^0_{25°C} = \Delta H^0_i \quad \textbf{(37b)}$$
$$H^+(a = 1) + OH^-(a = 1) = H_2O(l) \qquad \Delta H^0_{25°C} = -13{,}360\text{ cal} \quad \textbf{(37c)}$$

The sum of Eqs. (37b) and (37c) gives Eq. (37a), and, consequently,

$$\Delta H_i^0 + (-13,360) = -2460$$
$$\Delta H_i^0 = +10,900 \text{ cal} \tag{38}$$

ΔH_i^0 is the heat of ionization of the hydrocyanic acid per mole. For the purpose at hand, the slight ionization in water of an acid as weak as hydrocyanic acid may be disregarded.

Similarly may be explained the results on the neutralization of weak bases by strong acids and weak bases by weak acids. In the first instance the heat of ionization of the weak base must be considered and in the latter the heats of ionization of both the weak acid and the weak base.

HEATS OF FORMATION OF IONS

To present complete thermal data on aqueous solutions of electrolytes, it would be necessary to give heats of formation at various concentrations. The task could be simplified by listing only the standard heats of formation of the electrolytes, i.e., when all the species are at unit activity. However, it would be much more desirable to tabulate standard heats of formation of the ions themselves. Then, for any given electrolyte in water the standard heat of formation would be the sum of the heats of formation of the individual ions. Further, these heats could be used to calculate thermal changes for other reactions in which the ions participate.

The approach to this problem can be made through Eq. (36). Since the standard heat of formation of a mole of water from hydrogen and hydroxyl ions at 25°C involves the evolution of 13,360 cal of heat, this amount of heat must be supplied in order to dissociate a mole of water into these two ions, namely,

$$H_2O(l) = H^+(a = 1) + OH^-(a = 1) \qquad \Delta H_{25°C}^0 = 13,360 \text{ cal} \quad \textbf{(39)}$$

When Eq. (39) is combined with the equation for the standard heat of formation of 1 mole of $H_2O(l)$, we obtain the standard heat of formation of the hydrogen and hydroxyl ions,

$$H_2(g) + \frac{1}{2}O_2(g) = H^+(a = 1) + OH^-(a = 1)$$

$$\Delta H_{25°C}^0 = -54,960 \text{ cal} \quad \textbf{(40)}$$

This sum cannot be resolved at present without some assumption with respect to the heat of formation of one of these ions. The convention generally adopted is that *the heat of formation of the hydrogen ion in aqueous solution is zero at 25°C and unit activity*, i.e., that for the reaction

$$\frac{1}{2} H_2(g) = H^+(a = 1) \qquad \Delta H^0_{25°C} = 0 \tag{41}$$

With this convention Eq. (40) gives directly the heat of formation of the hydroxyl ion, or

$$\frac{1}{2} H_2(g) + \frac{1}{2} O_2(g) = OH^-(a = 1) \qquad \Delta H^0_{25°C} = -54,960 \text{ cal} \tag{42}$$

Once the heats of formation of these two ions are known, those for other ions may be readily calculated. Since the standard heat of formation of HCl in water at 25°C is $\Delta H^0 = -40,020$ cal, and since $\Delta H^0 = 0$ for H^+, $\Delta H^0 = -40,020$ cal must represent the heat of formation of the chloride ion. Again in the reaction

$$Na(s) + \frac{1}{2} O_2(g) + \frac{1}{2} H_2(g) = Na^+(a = 1) + OH^-(a = 1)$$

$$\Delta H^0_{25°C} = -112,240 \text{ cal} \tag{43}$$

Combining this equation with Eq. (42), we see that

$$-112,240 = \Delta H^0_{Na^+} + \Delta H^0_{OH^-}$$
$$= \Delta H^0_{Na^+} - 54,960$$
$$\Delta H^0_{Na^+} = -57,280 \text{ cal at } 25°C$$

By similar procedures the heats of formation of many other ions in aqueous solution have been evaluated. Some of these are summarized in Table 4–4. The use of this table in the calculation of heats of reactions

TABLE 4–4. STANDARD HEATS OF FORMATION OF IONS AT 25°C

Ion	ΔH^0 cal (g ion)$^{-1}$	Ion	ΔH^0 cal (g ion)$^{-1}$
H^+	0	Cd^{++}	$-17,300$
Li^+	$-66,550$	Fe^{+++}	$-11,400$
Na^+	$-57,280$	OH^-	$-54,960$
K^+	$-60,040$	Cl^-	$-40,020$
NH_4^+	$-31,740$	Br^-	$-28,900$
Ag^+	$+25,310$	I^-	$-13,370$
Mg^{++}	$-110,410$	HSO_4^-	$-211,700$
Ca^{++}	$-129,770$	NO_3^-	$-49,370$
Sr^{++}	$-130,380$	HCO_3^-	$-165,180$
Ba^{++}	$-128,670$	S^{--}	$+10,000$
Fe^{++}	$-21,000$	SO_3^{--}	$-149,200$
Co^{++}	$-16,100$	SO_4^{--}	$-216,900$
Ni^{++}	$-15,300$	CO_3^{--}	$-161,630$
Zn^{++}	$-36,430$	PO_4^{---}	$-306,900$

involving strong acids, bases, or salts in aqueous solutions can best be shown by an example. Suppose it is desired to calculate ΔH^0 for the reaction

$$\frac{1}{2} H_2(g) + AgCl(s) = Ag(s) + H^+(a = 1) + Cl^-(a = 1)$$

$$\Delta H^0_{25°C} = ? \quad (44)$$

Since the enthalpies of $H_2(g)$, $Ag(s)$, and H^+ are all zero, ΔH^0 is given by

$$\Delta H^0 = \Delta H^0_{Cl^-} - \Delta H^0_{AgCl(s)}$$

On inserting the value of $\Delta H^0_{Cl^-}$ from Table 4–4 and $\Delta H^0_{AgCl(s)}$ from Table 4–1, we get

$$\Delta H^0_{25°C} = -40,020 - (-30,360)$$
$$= -9660 \text{ cal}$$

OTHER HEATS OF REACTION

There are many types of heats of reaction besides those mentioned. Examples are heats of fusion, vaporization, sublimation, dissociation, hydrogenation, polymerization, etc. All of them are subject to the same treatment and thermodynamic requirements as the heats discussed in this chapter.

HEATS OF REACTION FROM BOND ENTHALPIES

Various methods have been proposed for estimating heats of reaction for processes for which no thermal data are available. Of these the most popular is the one based on bond enthalpies. This method, applicable to *gaseous reactions involving substances having only covalent bonds*, is based on the assumptions (a) that all the bonds of a particular type, as C—H in methane, are identical, and (b) that the bond enthalpies are independent of the compound in which they appear. Although neither assumption is strictly valid, nevertheless the method offers a simple and fairly satisfactory means for estimating the enthalpies of many and varied reactions.

In establishing the bond enthalpies to be used in calculations, those for substances like H_2, N_2, Cl_2, O_2, etc., are taken to be the values obtained for dissociation of the molecules by either thermal or spectroscopic means. Again, to find the value for C—H, the heat of formation of methane from C(s) and $H_2(g)$ is taken, and the latter combined with the heat of sublimation of carbon and the heat of dissociation of $H_2(g)$ to obtain the heat of dissociation of methane into the gaseous atoms. This result is divided then by 4 to arrive at the value of the C—H bond. By continuing this process with various types of compounds and reactions, it is possible to establish a set of average and best values of bond enthalpies such as those given in Table 4–5. In using these enthalpies a plus sign is affixed to the bond enthalpies when a bond is broken, since heat is absorbed under such conditions, and a minus sign when a bond is formed and heat is evolved.

TABLE 4–5.   EMPIRICAL BOND ENTHALPIES
AT 25°C
(kcal/mole)

Bond	ΔH	Bond	ΔH
H—H	104	C—Cl	79
H—F	135	C—Br	66
H—Cl	103	C—S	62
H—Br	88	C=S	114
O—O	33	C—N	70
O=O	118	C=N	147
O—H	111	C≡N	210
C—H	99	N—N	38
C—O	84	N=N	100
C=O	170	N≡N	226
C—C	83	N—H	93
C=C	147	F—F	37
C≡C	194	Cl—Cl	58
C—F	105	Br—Br	46
		C(s, graphite) = C(g) 172	

We shall illustrate the use of these bond enthalpies with several examples. Suppose, first, that the enthalpy change is desired at 25°C for the reaction

$$H_2C{=}CH_2(g) + H_2(g) = H_3C{-}CH_3(g) \qquad (45a)$$

In this reaction the four C—H bonds in C_2H_4 are unaffected and they may thus be neglected. However, a double bond is broken in C_2H_4 and an H—H bond in H_2. In turn, we form in C_2H_6 one C—C bond and two C—H bonds. Consequently, we can write for ΔH

$$\Delta H_{25°C} = -(\Delta H_{C-C} + 2\,\Delta H_{C-H}) + (\Delta H_{C=C} + \Delta H_{H-H}) \qquad (45b)$$

Inserting into Eq. (45b) the bond enthalpies from Table 4–5, we get

$$\Delta H_{25°} = -(83 + 198) + (147 + 104)$$
$$= -30 \text{ kcal}$$

The experimentally observed value for the reaction in Eq. (45a) is $\Delta H^0_{25°C} = -33$ kcal.

As a second example, consider the reaction

$$2 \, C(s) + 3 \, H_2(g) + \frac{1}{2} \, O_2(g) = H_3C—O—CH_3(g) \cdot \quad \Delta H = \Delta H_1 \quad \textbf{(46a)}$$

Before bond enthalpies can be used here the C(s) has to be converted to C(g). For this purpose let us break up Eq. (46a) into the two reactions

$$2 \, C(g) + 3 \, H_2(g) + \frac{1}{2} \, O_2(g) = H_3C—O—CH_3(g) \quad \Delta H = \Delta H_2 \quad \textbf{(46b)}$$

and

$$2 \, C(s) = 2 \, C(g) \quad \Delta H_3 = 344 \text{ kcal} \quad \textbf{(46c)}$$

whose sum is Eq. (46a). For Eq. (46b) we have

$$\Delta H_2 = -(2 \, \Delta H_{C—O} + 6 \, \Delta H_{C—H}) + (3 \, \Delta H_{H—H} + \frac{1}{2} \Delta H_{O—O})$$
$$= -(168 + 594) + (312 + 59)$$
$$= -391 \text{ kcal}$$

Combining this result with Eq. (46c) we obtain $\Delta H_1 = -391 + 344 = -47$ kcal. The observed value is $\Delta H^0_{25°C} = -44$ kcal.

VARIATION OF HEAT OF REACTION
WITH TEMPERATURE

In general the ΔH of a reaction is a function of temperature and pressure, and ΔE is a function of temperature and volume. However, since the evaluation of the pressure and volume dependencies of heats of reaction can become rather involved, we shall consider here only the effects of temperature on ΔH and ΔE.

The heat of a reaction obtained calorimetrically or by calculation corresponds to some one definite temperature. At other temperatures the heat of the reaction will usually not be the same. As reactions are carried out at various temperatures, it is very frequently necessary to know heats of reaction at temperatures other than those at which they were determined. A method of calculating the heat of reaction at one temperature from that at another is, therefore, highly desirable.

In such calculations heat capacities play a very important part. We have seen that there are two types of heat capacities, C_p and C_v. For gases $C_p - C_v = R$ cal per mole. On the other hand, since the volume changes accompanying the heating of solids and liquids are small, these two heat capacities may be taken as essentially equal for these. Furthermore, heat capacities vary with temperature. This variation is generally expressed by empirical formulas of the type

$$C_p = a + bT + cT^2 + dT^3 \quad \text{cal mole}^{-1} \text{ degree}^{-1} \qquad (47)$$

or
$$C_p = a' + b'T + \frac{c'}{T^2} \quad \text{cal mole}^{-1} \text{ degree}^{-1} \qquad (48)$$

where a, b, c, and d, or a', b', and c', are constants for a given substance. Values of these constants for various substances are given in Table 4–6. The temperature range over which the constants hold is also indicated.

TABLE 4–6.   CONSTANTS FOR MOLAR HEAT CAPACITY EQUATIONS
OF VARIOUS SUBSTANCES

Substance	Range ($^\circ$K)	a	$b \times 10^3$	$c \times 10^7$	$d \times 10^9$
$H_2(g)$	300–2500	6.62	0.81		
$N_2(g)$	300–2500	6.76	0.606	1.3	
$O_2(g)$	300–2500	6.76	0.606	1.3	
$CO(g)$	300–2500	6.60	1.2		
$HCl(g)$	300–1500	6.70	0.84		
$H_2O(g)$	300–1500	7.219	2.374	2.67	
$H_2S(g)$	300–1800	6.955	3.675	7.40	−0.585
$NH_3(g)$	300–1000	6.189	7.887	−7.28	
$CO_2(g)$	300–1500	5.166	15.177	−95.78	2.260
$CH_4(g)$	300–1500	3.422	17.845	−41.65	
$C_2H_4(g)$	300–1500	2.706	29.160	−90.59	
$C_2H_6(g)$	300–1000	1.375	41.852	−138.27	
		a'	$b' \times 10^3$	$c' \times 10^{-5}$	
$Cl_2(g)$	300–1500	8.76	0.271	−0.656	
$NO(g)$	300–2500	8.05	0.233	−1.56	
$C(s, \text{graphite})$	300–1400	2.673	2.617	−1.169	
$C(s, \text{diamond})$	300–1300	2.162	3.059	−1.303	

The differential equation for the variation of heat of reaction with temperature may be obtained as follows. Since

$$\Delta H = H_{\text{products}} - H_{\text{reactants}} \qquad (49)$$

differentiation of both sides with respect to absolute temperature at constant pressure yields

$$\left[\frac{\partial(\Delta H)}{\partial T}\right]_P = \left[\frac{\partial H_{\text{products}}}{\partial T}\right]_P - \left[\frac{\partial H_{\text{reactants}}}{\partial T}\right]_P$$

But the second and third terms are, respectively, $C_{p\text{products}}$ and $C_{p\text{reactants}}$. Therefore,

$$\left[\frac{\partial(\Delta H)}{\partial T}\right]_P = C_{p\text{products}} - C_{p\text{reactants}} = \Delta C_p \qquad (50)$$

Equation (50) is known as *Kirchhoff's equation*. In the same manner can be obtained the equation for the variation of ΔE with temperature, namely,

$$\left[\frac{\partial(\Delta E)}{\partial T}\right]_V = C_{v\text{products}} - C_{v\text{reactants}} = \Delta C_v \tag{51}$$

The integration of Eq. (50) depends on whether ΔC_p is constant or temperature-dependent. When ΔC_p is constant, then integration between two temperatures T_1 and T_2 yields

$$\int_{\Delta H_1}^{\Delta H_2} d(\Delta H) = \int_{T_1}^{T_2} \Delta C_p dT$$

$$\Delta H_2 - \Delta H_1 = \Delta C_p(T_2 - T_1)$$

and
$$\Delta H_2 = \Delta H_1 + \Delta C_p(T_2 - T_1) \tag{52}$$

Here ΔH_1 is the heat of reaction at T_1 and ΔH_2 that at T_2. However, when ΔC_p is not constant, then

$$\int d(\Delta H) = \int \Delta C_p dT + \Delta H_0$$

and
$$\Delta H = \int \Delta C_p dT + \Delta H_0 \tag{53}$$

where ΔH_0 is a constant of integration. To evaluate the integral in Eq. (53), ΔC_p must be available as a function of the temperature.

The use of Eqs. (52) and (53) can best be illustrated by examples. Suppose it is desired to calculate ΔH^0 at 348°K for the reaction

$$\frac{1}{2} H_2(g) + \frac{1}{2} Cl_2(g) = HCl(g) \qquad \Delta H^0_{298°K} = -22{,}060 \text{ cal} \tag{54}$$

The mean heat capacities over this temperature interval are

$$H_2(g): \quad C_p = 6.82 \text{ cal mole}^{-1} \text{ degree}^{-1}$$
$$Cl_2(g): \quad C_p = 7.71 \text{ " \qquad " \qquad "}$$
$$HCl(g): \quad C_p = 6.81 \text{ " \qquad " \qquad "}$$

from which we get

$$\Delta C_p = 6.81 - \frac{1}{2}(6.82) - \frac{1}{2}(7.71) = -0.46$$

Therefore,

$$\Delta H^0_{348°K} = -22{,}060 + (-0.46)(348 - 298)$$
$$= -22{,}080 \text{ cal}$$

However, suppose it is desired to find the heat of formation of ammonia at 1000°K from the following data:

$$\frac{1}{2} N_2(g) + \frac{3}{2} H_2(g) = NH_3(g) \qquad \Delta H^0_{298.2°K} = -11,040 \text{ cal} \qquad (55)$$

$N_2(g): C_p = 6.76 + 0.606 \times 10^{-3} T + 1.3 \times 10^{-7} T^2$
$$\text{cal mole}^{-1} \text{ degree}^{-1} \qquad (56a)$$
$H_2(g): C_p = 6.62 + 0.81 \times 10^{-3} T \text{ cal mole}^{-1} \text{ degree}^{-1} \qquad (56b)$
$NH_3(g): C_p = 6.189 + 7.887 \times 10^{-3} T - 7.28 \times 10^{-7} T^2$
$$\text{cal mole}^{-1} \text{ degree}^{-1} \qquad (56c)$$

Now $C_{p_{\text{products}}} = C_{p_{NH_3}}$. Again, $C_{p_{\text{reactants}}}$ is given by

$$C_{p_{\text{reactants}}} = \frac{1}{2} C_{p_{N_2}} + \frac{3}{2} C_{p_{H_2}}$$
$$= 3.38 + 0.303 \times 10^{-3} T + 0.65 \times 10^{-7} T^2 + 9.93$$
$$+ 1.22 \times 10^{-3} T$$
$$= 13.31 + 1.523 \times 10^{-3} T + 0.65 \times 10^{-7} T^2 \qquad (57)$$

Subtracting Eq. (57) from Eq. (56c), we obtain for ΔC_p

$$\Delta C_p = -7.12 + 6.364 \times 10^{-3} T - 7.93 \times 10^{-7} T^2 \qquad (58)$$

On substitution of Eq. (58) for ΔC_p in Eq. (53) and integration, we get

$$\Delta H^0 = \int \Delta C_p dT + \Delta H_0$$
$$= \int (-7.12 + 6.364 \times 10^{-3} T - 7.93 \times 10^{-7} T^2) dT + \Delta H_0$$
$$= -7.12 T + 3.182 \times 10^{-3} T^2 - 2.64 \times 10^{-7} T^3 + \Delta H_0 \qquad (59)$$

In order to evaluate the constant of integration, a known value of ΔH^0 at some definite temperature must be substituted in Eq. (59), and ΔH_0 solved for. Since we have given that at 298.2°K $\Delta H^0 = -11,040$ cal, then

$$-11,040 = -7.12(298.2) + 3.182 \times 10^{-3}(298.2)^2 - 2.64 \times 10^{-7}(298.2)^3 + \Delta H_0$$
and $$\Delta H_0 = -11,040 + 2120 - 280 + 10$$
$$= -9190 \text{ cal}$$

Inserting the found value of ΔH_0 into Eq. (59), the expression for ΔH^0 becomes

$$\Delta H^0 = -9190 - 7.12 T + 3.182 \times 10^{-3} T^2 - 2.64 \times 10^{-7} T^3 \qquad (60)$$

This equation gives ΔH^0 for the formation of ammonia as a function of the temperature. Equations of this type are very useful, for they permit calculation of the heat of reaction merely by inserting the temperature at which ΔH^0 is sought. Thus for the present case we have at 1000°K

$$\Delta H^0 = -9190 - 7.12(1000) + 3.182 \times 10^{-3}(1000)^2 - 2.64$$
$$\times 10^{-7}(1000)^3$$
$$= -9190 - 7120 + 3180 - 260$$
$$= -13,390 \text{ cal at } 1000°K$$

Hence 2350 more calories of heat are evolved at 1000°K than at 298°K.

The accuracy of these equations is determined by the precision with which the specific heats and the heats of reaction have been determined. Discretion should be exercised as to the temperature range over which these equations are used, since they are valid only for the temperature interval over which the specific heats have been determined. ·

REFERENCES

See references listed at end of Chapter 3. Also:

1. Hougen, Watson, and Ragatz, *Chemical Process Principles*, John Wiley & Sons, Inc., New York. Part I, 1954; Part II, 1959.
2. K. K. Kelley, U.S. Bureau of Mines Bulletins No. 324, 371, 383, 384, 393, 406, and 407.
3. L. Pauling, *The Nature of the Chemical Bond*, Cornell University Press, Ithaca, N.Y., 1960.
4. Rossini, Pitzer, Taylor, Ebert, Kilpatrick, Beckett, Williams, and Werner, *Selected Values of Properties of Hydrocarbons*, Circ. Natl. Bur. Standards C 461, U.S. Government Printing Office, Washington, D.C., 1947.
5. Rossini, Wagman, Evans, Levine, and Jaffe, *Selected Values of Chemical Thermodynamic Properties*, Circ. Natl. Bur. Standards 500, U.S. Government Printing Office, Washington, D.C., 1952.
6. R. R. Wenner, *Thermochemical Calculations*, McGraw-Hill Book Company, Inc., New York, 1941.
7. A. Weissberger, *Physical Methods of Organic Chemistry*, Interscience Publishers, Inc., New York, 1959, Chap. 10.

PROBLEMS

1. The molar heat of combustion of naphthalene (M.W. = 128.17) is -1228.2 kcal/mole. If 0.3000 g of naphthalene burned in a calorimeter causes a rise in temperature of 2.050°C, what is the total heat capacity of the calorimeter?
Ans. 1402 cal/°C.

2. If 1.520 g of an organic compound burned in the calorimeter in problem 1 cause the temperature to rise 1.845°C, what is the heat of combustion of the compound in calories per gram?

3. A 0.500 g sample of n-heptane(l) burned in a constant volume calorimeter to $CO_2(g)$ and $H_2O(l)$ causes a temperature rise of 2.934°C. If the heat capacity of the calorimeter and its accessories is 1954 cal/°C, and the mean temperature of the calorimeter is 25°C, calculate:

(a) The heat of combustion per mole of the heptane at constant volume
(b) The heat of combustion of the heptane per mole at constant pressure.

Ans. (b) $\Delta H = -1151$ kcal.

4. For the following reactions state whether ΔH will be significantly different from ΔE, and tell whether ΔH will be greater or less than ΔE in each case. Assume that all reactants and products are in their normal states at 25°C.

(a) The complete combustion of sucrose $(C_{12}H_{22}O_{11})$
(b) The oxidation of solid naphthalene $(C_{10}H_8)$ with gaseous O_2 to solid phthalic acid, $C_6H_4(COOH)_2$
(c) The complete combustion of ethyl alcohol
(d) The oxidation of PbS with O_2 to PbO and SO_2

5. For the reaction

$$NH_3(g) = \frac{1}{2} N_2(g) + \frac{3}{2} H_2(g) \qquad \Delta H^0_{25°C} = 11,040 \text{ cal}$$

Find the value of ΔE^0 of the reaction at 25°C.

6. The heats of the following reactions at 25°C are

$$Na(s) + \frac{1}{2} Cl_2(g) = NaCl(s) \qquad \Delta H^0 = -98,230 \text{ cal}$$
$$H_2(g) + S(s) + 2 O_2(g) = H_2SO_4(l) \qquad \Delta H^0 = -193,910 \text{ cal}$$
$$2 Na(s) + S(s) + 2 O_2(g) = Na_2SO_4(s) \qquad \Delta H^0 = -330,500 \text{ cal}$$
$$\frac{1}{2} H_2(g) + \frac{1}{2} Cl_2(g) = HCl(g) \qquad \Delta H^0 = -22,060 \text{ cal}$$

From these data find the heat of reaction at constant volume at 25°C for the process

$$2 NaCl(s) + H_2SO_4(l) = Na_2SO_4(s) + 2 HCl(g)$$

Ans. $\Delta E^0 = 14,560$ cal.

7. From the following equations and heats of reaction, calculate the standard molar heat of formation of AgCl at 25°C.

$$Ag_2O(s) + 2 HCl(g) = 2 AgCl(s) + H_2O(l) \qquad \Delta H^0 = -77,610 \text{ cal}$$
$$2 Ag(s) + \frac{1}{2} O_2(g) = Ag_2O(s) \qquad \Delta H^0 = -7310 \text{ cal}$$
$$\frac{1}{2} H_2(g) + \frac{1}{2} Cl_2(g) = HCl(g) \qquad \Delta H^0 = -22,060 \text{ cal}$$
$$H_2(g) + \frac{1}{2} O_2(g) = H_2O(l) \qquad \Delta H^0 = -68,320 \text{ cal}$$

8. For the reaction

$$2 NaHCO_3(s) = Na_2CO_3(s) + CO_2(g) + H_2O(g) \qquad \Delta H^0_{25°C} = 30,920 \text{ cal}$$

Find the standard heat of formation at 25°C of $NaHCO_3(s)$ in calories per mole.

9. From the data of Table 4–1 calculate the heats of the following reactions at 25°C:

(a) $Fe_2O_3(s) + CO(g) = CO_2(g) + 2 FeO(s)$
(b) $2 NO_2(g) = 2 NO(g) + O_2(g)$
(c) $3 C_2H_2(g) = C_6H_6(l)$

10. From the heat of combustion of n-butane in Table 4-2 calculate the standard heat of formation of this compound per mole at 25°C.
                                          Ans. $-29,820$ cal/mole.

11. From the data in Table 4–1 calculate the heat of combustion of $C_2H_5OH(g)$ at 25°C.

12. From the data in Table 4–3 determine at 25°C:

(a) The amount of heat which will be liberated when 2 moles of $H_2SO_4(l)$ are dissolved in 30 moles of $H_2O(l)$
(b) The differential heats of solution of the acid and the water for the given solution.   *Ans.* (a) $-33,980$ cal.

13. (a) How much heat will be evolved or absorbed in the solution of 2 moles of $H_2SO_4(l)$ in a very large volume of water at 25°C? (b) What will be the differential heats of solution of the acid and the water under these conditions?

14. A solution of sulfuric acid in water, containing initially 73.13% by weight of H_2SO_4, is diluted with water until the percentage of acid is 35.25 by weight. (a) What will be the integral heat of dilution when 100 g of the initial solution are diluted at 25°C? (b) What will be the differential heats of dilution of the acid and water under these conditions?   *Ans.* (a) -4680 cal.

15. Suppose that the initial solution described in problem 14 is diluted with a very large volume of water. What will be the integral and differential heats of dilution for the process?

16. Using Eq. (28) of this chapter, calculate the integral heats of dilution for the processes described in problems 14 and 15. How do these results compare with those obtained before?

17. $Na_2CO_3(s)$ dissolves in a large excess of H_2O with the evolution of 5500 cal of heat per mole at 25°C. Calculate the heat of formation per mole of Na_2CO_3 in a dilute solution.   *Ans.* $-275,800$ cal/mole.

18. From the following reactions and thermal data at 25°C

$$2 Fe(s) + \frac{3}{2} O_2(g) = Fe_2O_3(s) \qquad \Delta H^0 = -196,500 \text{ cal}$$

$$2 FeO(s) + \frac{1}{2} O_2(g) = Fe_2O_3(s) \qquad \Delta H^0 = -67,900 \text{ cal}$$

$$Fe(s) + 2 H^+ (a = 1) = Fe^{++} (a = 1) + H_2(g) \qquad \Delta H^0 = -21,000 \text{ cal}$$

$$\frac{1}{2} H_2(g) = H^+ (a = 1) \qquad \Delta H^0 = 0 \text{ cal}$$

$$H_2(g) + \frac{1}{2} O_2(g) = H_2O(l) \qquad \Delta H^0 = -68,320 \text{ cal}$$

calculate ΔH^0 of the reaction:

$$FeO(s) + 2 H^+ (a = 1) = H_2O(l) + Fe^{++} (a = 1)$$

19. The following compounds originally in dilute aqueous solution are mixed together as indicated:

(a) $CH_3COONa + HCl \longrightarrow$
(b) $KCl + MgSO_4 \longrightarrow$
(c) $NH_4Cl + Ba(OH)_2 \longrightarrow$
(d) $CaCl_2 + K_2CO_3 \longrightarrow$

In which cases may an appreciable thermal effect be expected? Explain your answer.

20. From the data of Tables 4–1 and 4–4, calculate the heats of the following reactions at 25°C:

(a) $Ca^{++} + CO_3^{--} = CaCO_3(s)$
(b) $CO_3^{--} + 2 H^+ = H_2O(l) + CO_2(g)$

21. A 150.0-cc portion of 0.40 N HCl is neutralized with an excess of NH_4OH in a Dewar vessel with a resulting rise in temperature of 2.36°C. If the heat capacity of the Dewar and its contents after the reaction is 315 cal/degree, calculate the heat of neutralization in calories per mole.

22. At 18°C the heat of solution of anhydrous $CuSO_4$ in a large volume of water is $-15,800$ cal/mole, while that of $CuSO_4 \cdot 5\ H_2O$ is 2750 cal/mole. Find at 18°C the heat of the reaction

$$CuSO_4(s) + 5\ H_2O(l) = CuSO_4 \cdot 5\ H_2O(s)$$

Ans. $\Delta H = -18,550$ cal.

23. The integral heats of solution at 18°C for the various solid modifications of calcium chloride in the indicated quantities of water are given by

$CaCl_2(s) + 400\ H_2O(l) = CaCl_2(400\ H_2O)$ $\Delta H_1 = -17,990$ cal
$CaCl_2 \cdot 2\ H_2O(s) + 398\ H_2O(l) = CaCl_2(400\ H_2O)$ $\Delta H_2 = -10,030$ cal
$CaCl_2 \cdot 4\ H_2O(s) + 396\ H_2O(l) = CaCl_2(400\ H_2O)$ $\Delta H_3 = -1830$ cal
$CaCl_2 \cdot 6\ H_2O(s) + 394\ H_2O(l) = CaCl_2(400\ H_2O)$ $\Delta H_4 = +4560$ cal

From these data find the heats of the following hydration reactions:

(a) $CaCl_2(s) + 2\ H_2O(l) = CaCl_2 \cdot 2\ H_2O(s)$
(b) $CaCl_2 \cdot 2\ H_2O(s) + 2\ H_2O(l) = CaCl_2 \cdot 4\ H_2O(s)$
(c) $CaCl_2 \cdot 4\ H_2O(s) + 2\ H_2O(l) = CaCl_2 \cdot 6\ H_2O(s)$
(d) $CaCl_2(s) + 6\ H_2O(l) = CaCl_2 \cdot 6\ H_2O(s)$

24. Using bond enthalpies, estimate for 25°C the heat of the reaction

$$H_2(g) + \frac{1}{2} O_2(g) = H_2O(g)$$

How does this result compare with the observed value?

Ans. $\Delta H_{calc.} = -59$ kcal.

25. Using bond enthalpies, estimate the heat of the reaction

$$2 \text{ C(s)} + 3 \text{ H}_2(g) + \frac{1}{2} \text{ O}_2(g) = \text{C}_2\text{H}_5\text{OH}(g)$$

at 25°C. How does this result compare with the observed value?

26. Bond enthalpy calculations cannot be applied to molecules such as $CO(g)$ and $CO_2(g)$ on the supposition that the structure of the first is $C≡O$ and of the second $O=C=O$. Calculate the percentage error, based on the observed value, which would result if we were to calculate the ΔH of the reaction

$$\text{CO}(g) + \frac{1}{2} \text{ O}_2(g) = \text{CO}_2(g)$$

on the assumption that the structures of CO and CO_2 are those given above.

27. Calculate from bond enthalpies the heat of combustion of methane to $CO_2(g)$ and $H_2O(l)$ on the supposition that the structure of the CO_2 molecule is $O=C=O$. (b) How does this estimate compare with the observed value?

28. Using the heat capacity constants given in Table 4–6, calculate the amount of heat required to raise the temperature of 200 g of $CO_2(g)$ from 300 to 500°K at (a) constant pressure and (b) constant volume. Assume $C_p - C_v = R$ cal mole^{-1} degree^{-1}.   *Ans.* (a) 8880 cal; (b) 7080 cal.

29. One mole of $N_2(g)$ and 3 moles of $H_2(g)$ at 25°C are heated to 450°C and subjected to a pressure which results in the conversion of 0.1 mole of the N_2 into $NH_3(g)$. The gases are then cooled rapidly back to 25°C. From the thermal data given in this chapter find how much heat is given up or absorbed in the whole process.

30. A mixture of gases contains 40 % of CO_2, 30 % of CO, and 30 % of N_2 by volume. Calculate the amount of heat necessary to raise the temperature of 1000 g of this mixture from 300 to 500°K at constant pressure.

31. Calculate the heat of vaporization of H_2O at 120°C and 1 atm pressure. The heat capacity of $H_2O(l)$ may be taken as 1.0 cal/g-degree, C_p for the vapor as 0.45 cal/g-degree, and the heat of vaporization at 100°C as 540 cal/g.

Ans. 529 cal/g.

32. One mole of steam at its boiling point under a constant pressure of 1 atm is condensed to liquid water under the same conditions. What are the values of ΔH, q, w, and ΔE for the process?

33. For the reaction $ZnO(s) + CO(g) = Zn(g) + CO_2(g)$ $\Delta H^0 = 47,390 - 0.69 \, T - 3.29 \times 10^{-3} \, T^2 + 1.25 \times 10^{-6} \, T^3$, (a) Deduce the expression for ΔE^0 as a function of T. (b) Find the values of ΔH^0 and ΔE^0 at 500°K.

34. Find ΔH^0 as a function of T for the reaction

$$\text{CO}_2(g) + \text{C(s, graphite)} = 2 \text{ CO}(g)$$

given that $\Delta H^0_{293°K} = 41,400$ calories.

Ans. $\Delta H^0 = 40,810 + 5.361 \, T - 7.697 \times 10^{-3} \, T^2 + 31.93 \times 10^{-7} \, T^3$
$$-0.565 \times 10^{-9} \, T^4 - \frac{1.169 \times 10^5}{T}$$

35. Find the expression for ΔH^0 as a function of the temperature for the reaction

$$N_2(g) + O_2(g) = 2\ NO(g)$$

given that $\Delta H^0_{293°K} = 43,000$ cal.

36. The expression for ΔH^0 of formation of CO_2 as a function of temperature is

$$\Delta H^0 = -93,480 - 0.603\ T - 0.675 \times 10^{-4}\ T^2 - \frac{1.091 \times 10^5}{T}$$

Find ΔC_p^0 for this reaction as a function of T.

37. For the reaction $CaO(s) + CO_2(g) = CaCO_3(s)$

$$\Delta H^0 = -42,500 - 0.66\ T + 2.155 \times 10^{-3}\ T^2 + \frac{4.1 \times 10^3}{T}$$

Find ΔE^0, ΔC_p^0, and ΔC_v^0 as functions of T.

38. Suppose that the reaction $2\ A(g) = A_2(g)$ proceeds vary rapidly and completely at 300°K with the liberation of 50,000 cal of heat per mole of A_2 formed. Under such conditions the process can be considered to be essentially adiabatic. If for the system C_p is constant and equal to 8.00 cal mole^{-1} degree^{-1}, and if A_2 is stable with temperature, what is the maximum temperature attained by the system as a result of the formation of 0.500 mole of A_2?

5

The Second
and Third
Laws of
Thermodynamics

ALTHOUGH the first law of thermodynamics establishes the relationship between the heat absorbed and the work performed by a system, it places absolutely no restrictions on the source of this heat or on the direction of its flow. According to the first law there is nothing impossible about a process in which, without any external aid, ice may be used to heat water by extracting heat from the former at a lower temperature and supplying it to the latter at a higher temperature. Yet we know from experience that such a transfer of heat from a lower to a higher temperature will not take place spontaneously. Instead heat is always found to flow of its own accord from the warmer to the colder body, i.e., *the spontaneous flow of heat is always unidirectional from the higher to the lower temperature*. This statement does not preclude the possibility of cooling a body below the temperature of its surroundings. Such a cooling can be accomplished, but in order to bring it about work has to be expended.

A similar unidirectionality of change is observed in all natural phenomena. Thus electricity tends to flow only from a point of higher electric potential to one of lower, water will move by itself only from a higher level to a lower, diffusion will occur only from the point of higher concentration to the lower, and all chemical systems under given conditions will tend to undergo reaction in a direction which will lead to the establishment of equilibrium. In fact, all the above observations can be summarized by the statement that *all naturally occurring processes always tend to change spontaneously in a direction which will lead to equilibrium.*

There is still another respect in which the first law is insufficient. In demanding the conservation of energy in all types of transformations, the first law of thermodynamics does not define the ease or extent of convertibility of one form of energy into another. Still, it is an empirical fact that, whereas various forms of energy can be converted readily and completely into heat, the converse process, the conversion of heat into work, can be accomplished only under severely limited conditions. At constant temperature heat can be transformed into work only at the expense of some permanent change in the system involved. For instance, heat may be converted into work by the isothermal reversible expansion of a gas in a cylinder. But to retain this work the gas must remain expanded. If we attempt to return the gas to its original condition, we find that the work obtained in the expansion must be utilized in the compression, and as a result we wind up with the original quantity of heat and no work. Again, to obtain work from heat by means of a periodically operating machine, such as a heat engine, it is essential that a temperature drop take place and that a flow of heat occur from a higher to a lower temperature. Even under such conditions not all the heat absorbed by the system can be converted to work, but only a fraction of it determined, under ideal conditions, by the two temperatures between which the operation takes place. Further, even though the heat engine remains unaffected in such an operation, the heat unconverted to work is degraded by having its temperature lowered. From these facts it can be seen that *heat cannot be converted to work without leaving permanent changes either in the systems involved or their surroundings.*

ENTROPY

The italicized passages given above are all limited expressions of the second law of thermodynamics, which was promulgated largely through the efforts of Clausius and Lord Kelvin. To arrive at a general statement of the law and to express it in mathematical form, let us define a new thermodynamic quantity S, called the *entropy* of the system. As we shall see later, *the entropy of a system depends only on the initial and final states of the system,* and hence, as in the case of E and H, we may write that the change in entropy of a process, ΔS, is

$$\Delta S = S_2 - S_1 \tag{1}$$

where S_2 and S_1 are, respectively, the entropies of the system in the final and initial states. Further, let us specify that the differential change in

S, dS, is given by

$$dS = \frac{dq_r}{T} \tag{2}$$

where dq_r is the infinitesimal quantity of heat absorbed in a process taking place under *reversible* conditions at temperature T. In the case of a finite reversible change at constant temperature, dS becomes ΔS, dq_r becomes q_r, and Eq. (2) takes then the form

$$\Delta S = \frac{q_r}{T} \tag{3}$$

Therefore, for any isothermal reversible process in which an amount of heat q_r is absorbed at temperature T, the entropy change involved is simply the absorbed heat divided by the absolute temperature. When q_r is positive, i.e., heat is absorbed, ΔS is also positive, indicating an increase in the entropy of the system. On the other hand, when heat is evolved q_r is negative and so is ΔS, and the system experiences a decrease in entropy.

Entropies and entropy changes are expressed in calories per degree per given amount of substance. The quantity *calorie per degree* is called an *entropy unit* (eu).

ENTROPY CHANGE IN ISOLATED SYSTEMS

Consider a cylinder containing any substance and fitted with a frictionless and weightless piston. Let the cylinder be enclosed in a large heat reservoir which is so thoroughly insulated from its surrounding that no heat can enter or leave the reservoir. Such an arrangement of a system and heat reservoir adiabatically insulated from their surroundings is called an *isolated system*. Let now the temperature in the isolated system be constant and equal to T, and suppose that the substance in the cylinder undergoes an isothermal and reversible expansion from volume V_1 to V_2. During this process the substance will absorb from the reservoir a quantity of heat q_r, and so the entropy change suffered by it, ΔS_s, will be, according to Eq. (3),

$$\Delta S_s = \frac{q_r}{T} \tag{4a}$$

At the same time the reservoir loses a quantity of heat q_r, and hence the entropy change of the reservoir, ΔS_r, is

$$\Delta S_r = \frac{\bar{q}_r}{T} \tag{4b}$$

where the bar over q_r indicates a loss of heat. The total change in entropy for the substance and the reservoir, ΔS_1, is then

$$\Delta S_1 = \Delta S_s + \Delta S_r$$
$$= \frac{q_r}{T} + \frac{\bar{q}_r}{T}$$
$$= 0 \tag{4c}$$

If we compress the substance now isothermally and reversibly from V_2 back to V_1, the heat rejected by the substance will be q_r and so will be the heat gained by the reservoir. The entropy changes involved are

$$\Delta S_s' = \frac{\bar{q}_r}{T} \tag{5a}$$

$$\Delta S_r' = \frac{q_r}{T} \tag{5b}$$

and the total entropy change for the substance and reservoir, ΔS_2, is thus

$$\Delta S_2 = \Delta S_s' + \Delta S_r'$$
$$= \frac{\bar{q}_r}{T} + \frac{q_r}{T}$$
$$= 0 \tag{5c}$$

Further, the total change in entropy for the complete cycle, ΔS, is the sum of ΔS_1 and ΔS_2, or

$$\Delta S = \Delta S_1 + \Delta S_2$$
$$= 0 \tag{6}$$

The above considerations lead to two very important conclusions. First, even though parts of an isolated system may experience a change in entropy, the entropy change for the entire isolated system when a reversible isothermal change occurs in it is zero. Second, the total entropy change for a reversible isothermal cycle is zero, and hence at the end of the cycle the system has the same entropy as it had initially. The entropy behaves, then, as a property of the state of the system only, and this fact justifies Eq. (1). These conclusions are valid for all types of processes and cycles performed under isothermal and reversible conditions in isolated systems.

Consider again the isothermal expansion of the substance from volume V_1 to V_2, but let the change be now irreversible. Since the expansion is irreversible, the heat absorbed by the substance will be q, where $q < q_r$. However, *the entropy change* of the substance must still be the same as it was in the reversible expansion, for it *is determined by the reversible heat and not by the heat actually absorbed*. The actual value of q in an irreversible process depends on the manner of conducting the process and can vary anywhere from $q = 0$ to $q = q_r$ when complete reversibility obtains. Therefore, for an irreversible process ΔS cannot equal q/T, for, if ΔS were equal to q/T in all these cases, we should obtain a series of different values of ΔS between any two given states of the system. As this is impossible for a property characteristic of the states of the system only, it must follow that even in an irreversible process ΔS must be given by q_r/T. On the basis of these arguments the entropy change for the isothermal but irreversible expansion of the substance from V_1 to V_2 is still given by Eq. (4a). However, the loss of q cal of heat by the reservoir can be considered to take place reversibly, and so the entropy change of the reservoir is $\Delta S_r = \bar{q}/T$. The total entropy change of the isolated system is then

$$\Delta S_1 = \frac{q_r}{T} + \frac{\bar{q}}{T} \qquad (7a)$$

But, $q_r > \bar{q}$ and $q_r/T > \bar{q}/T$. Consequently,

$$\Delta S_1 = \frac{q_r}{T} + \frac{\bar{q}}{T} > 0 \qquad (7b)$$

and hence *an irreversible process occurring isothermally in an isolated system leads to an increase in the total entropy of the system.*

If we recompress the substance to its original state isothermally and reversibly at temperature T, then the entropy change of the process will again be that given by Eq. (5c), or $\Delta S_2 = 0$. As a result the total entropy change for the irreversible cycle will be the sum of $\Delta S_2 = 0$ and Eq. (7b), namely,

$$\Delta S = \frac{q_r}{T} + \frac{\bar{q}}{T} > 0 \qquad (8)$$

Therefore, whereas for a complete isothermal and reversible cycle carried out in an isolated system $\Delta S = 0$, for an irreversible cycle $\Delta S > 0$. This conclusion is again valid for all types of irreversible processes.

The increase in entropy which occurs in the irreversible cycle is the result of conversion of work to heat. At the end of the cycle the working substance, by being returned to its initial state, suffers no change of any kind. The reservoir, however, has lost a quantity of heat q and regained q_r. The net heat gained by the reservoir is thus $q_r + \bar{q}$, and hence its entropy gain is $q_r/T + \bar{q}/T$, a quantity greater than zero. At the same time the work performed on the substance was w_m, while that performed by the substance was w. The difference $w_m - w$ was converted into the heat gained by the reservoir and caused the entropy increase.

Any nonisothermal process may be considered to be composed of a successive series of isothermal steps, each occurring at a temperature infinitesimally different from the preceding. If each of these isothermal steps takes place reversibly, the entropy change for each will be given by the heat absorbed in the step, dq_r, divided by T, the temperature at which the heat absorption takes place. The total entropy change for a process occurring between the temperatures T_1 and T_2 will be, then, the sum of the small continuous isothermal entropy changes, or

$$\Delta S = \int_{T_1}^{T_2} \frac{dq_r}{T} \tag{9}$$

From what has been said before it is evident that Eq. (9) is applicable also to irreversible nonisothermal processes provided the q's employed are not the heats actually observed but the heats evaluated for the corresponding reversible processes between the same two states.

When Eq. (9) is applied in a manner similar to the one used above to the changes which take place in an isolated system under nonisothermal conditions, it is found that:

1. For any *reversible* process or cycle $\Delta S = 0$
2. For any *irreversible* process or cycle $\Delta S > 0$

These conclusions are identical to those reached for isothermal processes. Therefore, whether an increase in entropy does or does not occur for processes taking place in isolated systems depends entirely on whether the processes are irreversible or reversible.

The validity of the statement that for any reversible nonisothermal cycle $\Delta S = 0$ can readily be checked with the result for the Carnot cycle. For this cycle we have

$$\frac{q_2 - q_1}{q_2} = \frac{T_2 - T_1}{T_2}$$

$$1 - \frac{q_1}{q_2} = 1 - \frac{T_1}{T_2}$$

and so

$$\frac{q_1}{T_1} = \frac{q_2}{T_2}$$

But $q_1/T_1 = \Delta S_1$ and $q_2/T_2 = \Delta S_2$. Hence

$$\Delta S_1 = \Delta S_2$$

and for the complete cycle $\Delta S = 0$.

THE SECOND LAW OF THERMODYNAMICS

If instead of finite changes we consider infinitesimal ones, then the results of the preceding section can be summarized by the statement that for any process occurring in an isolated system the entropy change dS_i is given by

$$dS_i \gtrless 0 \tag{10}$$

The equality applies to reversible processes and the inequality to irreversible ones. Completely reversible processes, involving as they do a balance of driving and opposing forces, must of necessity occur very slowly. In fact, to carry out any change in a completely reversible manner would require infinite time. Consequently, any process that does take place in a finite time must be irreversible, and it must be attended by an increase in the total entropy of all the bodies involved. The latter conclusion permits a statement of the second law of thermodynamics in its most general form, namely, that *all processes in nature tend to occur only with an increase in entropy and that the direction of change is always such as to lead to the entropy increase.* The several forms of the second law given earlier in the chapter are merely special cases of this general statement.

Suppose we consider the universe as an isolated system, as Clausius did, and apply to it the second law. Since all processes in nature are irreversible, the entropy of the universe must be a unidirectional property which continually increases and tends to reach a maximum. On the other hand, the first law of thermodynamics states that the energy of the universe is constant. These facts led Clausius to the enunciation of the first two laws of thermodynamics in the oft-quoted statement that "the energy of the universe is constant, the entropy of the universe tends to a maximum."

J. W. Gibbs, one of the greatest scientific minds America has produced, referred to entropy as a measure of the "mixtupness" of a system. This term is both descriptive and illuminating. What he meant is this. Energy in useful form, such as electrical, mechanical, or chemical energy, is organized and directed energy which can be utilized for the performance of work. On the other hand, heat is a form of energy due to the random motions of the atoms or molecules in a body and is thus chaotic in character. Therefore, when energy which is organized and which can be utilized for performance of work is converted to heat, the chaos or "mixtupness" of the system is increased. Since entropy is a measure of this "mixtupness," it must also increase. The idea that entropy is a measure of the disorder in a system is the basis of the relation which has been established between entropy and probability, and which is the concern of the field of science known as statistical mechanics.[1]

From the above discussion the essence of the second law of thermodynamics can be summarized in the statement that our stockpile of

[1] See Chapter 18.

available° energy in the universe is continually decreasing and is being converted into the disordered form of energy we call heat.

ENTROPY CHANGE FOR SYSTEMS ONLY

Processes occuring in isolated systems give us an insight into the behavior of the entropy when both the systems and their surroundings are considered. Under such conditions we keep track of all the entropy changes which take place, and we arrive thus at the conclusion that the entropy can either remain constant or increase. However, more commonly we are interested only in the systems themselves and not their surroundings. When we ignore the latter, then the entropy of the systems themselves can be constant, increase, or decrease in various situations, and consequently ΔS may be zero, positive, or negative.

The differential change in entropy experienced by the system itself during a process can be deduced from Eq. (10). In this equation dS_i is the sum of the entropy variations for the system itself, dS, and of the reservoir, dS_r; i.e.,

$$dS_i = dS + dS_r \gtrless 0 \tag{11}$$

In turn, dS_r is the heat lost (or gained) by the reservoir divided by T, and this heat is equal and opposite in sign to the heat gained (or lost) by the system, dq. We can write thus $dS_r = -dq/T$, and

$$dS - \frac{dq}{T} \gtrless 0$$
$$TdS - dq \gtrless 0 \tag{12}$$

However, by the first law $dq = dE + dw$. Again, dw may consist in general of $P\text{-}V$ work, i.e., pdV, as well as of other types of work for which we shall write dw'. Then

$$dq = dE + pdV + dw' \tag{13}$$

and on substitution of Eq. (13) into (12) we get

$$TdS - dE - pdV - dw' \gtrless 0 \tag{14}$$

In Eq. (14) the inequality applies to irreversible processes. Further, when only $P\text{-}V$ work is involved $dw' = 0$, and we get thus

$$TdS - dE - pdV > 0 \tag{15}$$

However, for reversible processes $p = P$, $dw' = dw'_m$, and the equality holds. Now we obtain

$$TdS - dE - PdV - dw'_m = 0 \tag{16}$$

and when $dw'_m = 0$,

$$TdS - dE - PdV = 0 \tag{17}$$

The manner in which these equations can be utilized in ΔS calculations can best be illustrated with the following two examples.

EXAMPLE 1: Calculate the entropy change involved in the isothermal reversible expansion of 5 moles of an ideal gas from a volume of 10 liters to a volume of 100 liters at 300°K. Here Eq. (17) applies with $dE = 0$, since the process is isothermal. We have then

$$dS = \frac{PdV}{T}$$
$$= \frac{nRdV}{V}$$

on making the substitution $P = nRT/V$. Consequently we get on integration

$$\Delta S = S_2 - S_1$$
$$= nR \ln \frac{V_2}{V_1}$$
$$= 5(1.987)(2.303) \log \frac{100}{10}$$
$$= 22.88 \text{ cal degree}^{-1} \text{ or eu}$$

It should be observed that the expression for ΔS is the maximum work divided by T, namely, w_m/T, and this is equal also to q_r/T.

If we wish to treat the system and its surroundings as an isolated system, then the entropy change of the surroundings is $-q_r/T$, or -22.88 eu. The total change in entropy for the combination becomes, therefore, $(22.88 - 22.88) = 0$ in line with Eq. (10).

EXAMPLE 2: Calculate the entropy change involved in the isothermal expansion of 5 moles of an ideal gas against a constant pressure of 1 atm from a volume of 10 liters to a volume of 100 liters at 300°K. The expansion here is irreversible but again isothermal. Therefore, with $dE = 0$ Eq. (15) predicts that $dS > pdV/T$, or

$$\Delta S > \frac{1(V_2 - V_1)}{T} > \frac{1(100 - 10)24.22}{300} \text{ eu}$$

and

$$\Delta S > 7.27 \text{ eu}$$

For the present process $w = q$; however, since $w \neq w_m$, $q \neq q_r$. To find q_r we must imagine the same change conducted reversibly, and this is the one described in the preceding problem. Consequently, we have here also $\Delta S = 22.88$ eu, a value greater than given by pdV/T.

If again we consider the system and its surroundings, then for the latter

$$\Delta S = \frac{q}{T} = -\frac{p(V_2 - V_1)}{T}$$
$$= -7.27 \text{ eu}$$

Therefore, the total entropy change for the isolated system is (22.88 − 7.27) = 15.61 eu. In line with Eq. (10) this entropy change is greater than zero.

DEPENDENCE OF ENTROPY ON VARIABLES OF A SYSTEM

Since the entropy is a function of the state of a system, its value for any *pure substance* will depend on any two of the three variables T, V, and P. Commonly, T is selected as one of the independent variables, and hence the combinations of variables to be dealt with are T and V, or T and P.

Variables T and V. If the entropy of a substance is a function of T and V, then

$$dS = \left(\frac{\partial S}{\partial T}\right)_V dT + \left(\frac{\partial S}{\partial V}\right)_T dV \qquad (18)$$

Again, the first law gives for reversible conditions

$$dE = dq_r - dw_m \qquad (19)$$

But, $dq_r = TdS$, and $dw_m = PdV$. Therefore,

$$dE = TdS - PdV \qquad (20)$$

However, E is also a function of T and V, and hence

$$dE = \left(\frac{\partial E}{\partial T}\right)_V dT + \left(\frac{\partial E}{\partial V}\right)_T dV \qquad (21)$$

On elimination of dE between Eqs. (20) and (21) we get for dS

$$dS = \frac{1}{T}\left(\frac{\partial E}{\partial T}\right)_V dT + \frac{1}{T}\left[\left(\frac{\partial E}{\partial V}\right)_T + P\right] dV \qquad (22)$$

and on comparison of Eqs. (18) and (22) we see that

$$\left(\frac{\partial S}{\partial T}\right)_V = \frac{1}{T}\left(\frac{\partial E}{\partial T}\right)_V \qquad (23a)$$

$$= \frac{C_v}{T} \qquad (23b)$$

and

$$\left(\frac{\partial S}{\partial V}\right)_T = \frac{1}{T}\left[\left(\frac{\partial E}{\partial V}\right)_T + P\right] \qquad (24)$$

Equation (24) can be simplified as follows: Rearrangement of the equation gives

$$\left(\frac{\partial E}{\partial V}\right)_T = T\left(\frac{\partial S}{\partial V}\right)_T - P \tag{25}$$

which on differentiation with respect to T at constant V becomes

$$\frac{\partial^2 E}{\partial T \partial V} = T\left(\frac{\partial^2 S}{\partial T \partial V}\right) + \left(\frac{\partial S}{\partial V}\right)_T - \left(\frac{\partial P}{\partial T}\right)_V \tag{26}$$

Again, rearrangement of Eq. (23a) gives

$$\left(\frac{\partial E}{\partial T}\right)_V = T\left(\frac{\partial S}{\partial T}\right)_V \tag{27}$$

which on differentiation with respect to V at constant T yields

$$\frac{\partial^2 E}{\partial V \partial T} = T\left(\frac{\partial^2 S}{\partial V \partial T}\right) \tag{28}$$

Since the order of differentiation is immaterial, we now can equate Eqs. (26) and (28) and obtain thus

$$\left(\frac{\partial S}{\partial V}\right)_T = \left(\frac{\partial P}{\partial T}\right)_V \tag{29}$$

With Eqs. (23b) and (29) for the partial derivatives, Eq. (18) becomes finally

$$dS = \frac{C_v}{T}\,dT + \left(\frac{\partial P}{\partial T}\right)_V dV \tag{30}$$

Variables T and P. With independent variables T and P, dS is given by

$$dS = \left(\frac{\partial S}{\partial T}\right)_P dT + \left(\frac{\partial S}{\partial P}\right)_T dP \tag{31}$$

Since $H = E + PV$, complete differentiation yields

$$dH = dE + PdV + VdP$$

or $$dE + PdV = dH - VdP \tag{32}$$

Substitution of Eq. (32) in Eq. (20) gives then

$$TdS = dH - VdP \tag{33}$$

However, H is also a function of T and P, and so

$$dH = \left(\frac{\partial H}{\partial T}\right)_P dT + \left(\frac{\partial H}{\partial P}\right)_T dP \tag{34}$$

On elimination of dH between Eqs. (33) and (34) we obtain

$$dS = \frac{1}{T}\left(\frac{\partial H}{\partial T}\right)_P dT + \frac{1}{T}\left[\left(\frac{\partial H}{\partial P}\right)_T - V\right]dP \qquad (35)$$

and hence on comparison of Eqs. (31) and (35) we get

$$\left(\frac{\partial S}{\partial T}\right)_P = \frac{1}{T}\left(\frac{\partial H}{\partial T}\right)_P \qquad (36a)$$

$$= \frac{C_p}{T} \qquad (36b)$$

and

$$\left(\frac{\partial S}{\partial P}\right)_T = \frac{1}{T}\left[\left(\frac{\partial H}{\partial P}\right)_T - V\right] \qquad (37)$$

Equation (37) can again be simplified. Rearrangement of the equation gives

$$\left(\frac{\partial H}{\partial P}\right)_T = T\left(\frac{\partial S}{\partial P}\right)_T + V \qquad (38)$$

and differentiation of Eq. (38) with respect to T at constant P yields

$$\frac{\partial^2 H}{\partial T \partial P} = T\left(\frac{\partial^2 S}{\partial T \partial P}\right) + \left(\frac{\partial S}{\partial P}\right)_T + \left(\frac{\partial V}{\partial T}\right)_P \qquad (39)$$

Further, on rearrangement and differentiation of Eq. (36a) with respect to P at constant T we obtain

$$\frac{\partial^2 H}{\partial P \partial T} = T\left(\frac{\partial^2 S}{\partial P \partial T}\right) \qquad (40)$$

and, therefore, combination of Eqs. (39) and (40) results in

$$\left(\frac{\partial S}{\partial P}\right)_T = -\left(\frac{\partial V}{\partial T}\right)_P \qquad (41)$$

If we now insert Eqs. (36b) and (41) into Eq. (31) we get, finally,

$$dS = \frac{C_p}{T} dT - \left(\frac{\partial V}{\partial T}\right)_P dP \qquad (42)$$

Equations (30) and (42) are perfectly general and they apply equally well to pure solids, liquids, or gases. However, before discussing the use of these equations, we shall digress to show the validity of Eqs. (38) and (41) of Chapter 3, which were assumed as valid subject to later proof. If we substitute Eq. (29) into Eq. (25), we obtain

$$\left(\frac{\partial E}{\partial V}\right)_T = T\left(\frac{\partial P}{\partial T}\right)_V - P \qquad (43)$$

and, similarly, substitution of Eq. (41) into Eq. (38) yields

$$\left(\frac{\partial H}{\partial P}\right)_T = V - T\left(\frac{\partial V}{\partial T}\right)_P \tag{44}$$

Equations (43) and (44) are identical with Eqs. (38) and (41) of Chapter 3.

ENTROPY CHANGE IN IDEAL GASES

In Eq. (30) C_v refers to the heat capacity of whatever quantity of substance is being considered. If we are dealing with n moles of gas, it is more convenient to write for it nC_v, where C_v is now the heat capacity per mole. Again, for an ideal gas, $PV = nRT$, and differentiation yields

$$\left(\frac{\partial P}{\partial T}\right)_V = \frac{nR}{V} \tag{45}$$

Hence for ideal gases Eq. (30) becomes

$$dS = \frac{nC_v dT}{T} + \frac{nR dV}{V} \tag{46}$$

and on integration between the indicated limits we get for ΔS

$$\begin{aligned}
\Delta S &= \int_{T_1}^{T_2} \frac{nC_v dT}{T} + \int_{V_1}^{V_2} \frac{nR dV}{V} \\
&= \int_{T_1}^{T_2} \frac{nC_v dT}{T} + nR \ln \frac{V_2}{V_1}
\end{aligned} \tag{47}$$

When C_v is constant over the temperature interval in question, Eq. (47) reduces to

$$\Delta S = nC_v \ln \frac{T_2}{T_1} + nR \ln \frac{V_2}{V_1} \tag{48}$$

If, however, it is not constant, then C_v has to be substituted as a function of T, and the expression integrated between the limits T_1 and T_2.

The first term on the right in Eqs. (47) or (48) gives the change in entropy of n moles of an ideal gas on change of temperature at constant volume. In turn, the second term gives the ΔS due to change of volume at constant temperature. When T is constant the first term vanishes, and only the second applies. Again, at constant V the second term is zero, and only the first term is used to get ΔS.

In the case of T and P as the independent variables, we may substitute nC_p for C_p in Eq. (42) where C_p is now the heat capacity per mole of gas. Again, from differentiation of $PV = nRT$ we obtain

$$\left(\frac{\partial V}{\partial T}\right)_P = \frac{nR}{P} \tag{49}$$

Substituting these into Eq. (42), dS for n moles of an ideal gas becomes

$$dS = \frac{nC_p}{T}\,dT - \frac{nR}{P}\,dP \tag{50}$$

and hence on integration between the indicated limits

$$\Delta S = \int_{T_1}^{T_2} \frac{nC_p dT}{T} - \int_{P_1}^{P_2} \frac{nR dP}{P}$$
$$= \int_{T_1}^{T_2} \frac{nC_p dT}{T} - nR \ln \frac{P_2}{P_1} \tag{51}$$

If C_p is or may be considered constant, then Eq. (51) yields

$$\Delta S = nC_p \ln \frac{T_2}{T_1} - nR \ln \frac{P_2}{P_1} \tag{52}$$

These relations are handled in the same way as those given for T and V as variables. The first term in Eq. (51) or Eq. (52) gives the entropy change suffered by n moles of an ideal gas due to variation of temperature at constant pressure, while the second term gives the ΔS resulting from changing the pressure at constant temperature.

Equations (30) and (42) may be used to calculate entropy changes for nonideal gases provided equations of state or data are available to yield the necessary expressions for $(\partial P/\partial T)_V$ as a function of volume, or $(\partial V/\partial T)_P$ as a function of pressure. With values of these derivatives available, Eqs. (30) and (42) are then used in the same manner as described above for ideal gases.

ENTROPY OF MIXING FOR IDEAL GASES

Mixing of gases is generally accompanied by a change in entropy. This entropy of mixing, ΔS_m, can readily be calculated for ideal gases as follows: Suppose that we mix n_1 moles of one gas at an initial pressure P_1^0 and n_2 moles of a second gas at pressure P_2^0, and suppose that in the mixture the partial pressures of the two gases are, respectively, P_1 and P_2. For any constant temperature the change in entropy accompanying a change in pressure is given by the second term in Eq. (52). Using this expression, we obtain for the entropy change suffered by the first gas as a result of the mixing

$$\Delta S_1 = -n_1 R \ln \frac{P_1}{P_1^0}$$
$$= n_1 R \ln \frac{P_1^0}{P_1} \tag{53a}$$

Similarly, we have for the second gas

$$\Delta S_2 = n_2 R \ln \frac{P_2^0}{P_2} \tag{53b}$$

and hence the total entropy of mixing is

$$\Delta S_m = \Delta S_1 + \Delta S_2$$
$$= n_1 R \ln \frac{P_1^0}{P_1} + n_2 R \ln \frac{P_2^0}{P_2} \tag{54}$$

But, according to Dalton's law of partial pressures, $P_1 = N_1 P_t$ and $P_2 = N_2 P_t$ where N_1 and N_2 are the mol fractions of the two gases and P_t is the total pressure of the mixture. Introducing these identities into Eq. (54) we get

$$\Delta S_m = n_1 R \ln \frac{P_1^0}{N_1 P_t} + n_2 R \ln \frac{P_2^0}{N_2 P_t} \tag{55}$$

In the special instance of $P_1^0 = P_2^0 = P_t$, Eq. (55) reduces to

$$\Delta S_m = -(n_1 R \ln N_1 + n_2 R \ln N_2) \tag{56}$$

Equation (56) shows that ΔS_m is independent of temperature. Further, since N_1 and N_2 are less than unity, then for the conditions specified ΔS_m is positive, and hence the mixing process is accompanied by an increase in entropy.

ENTROPY CHANGE IN PHYSICAL TRANSFORMATIONS

Changes in entropy accompany not only variations in the temperature, pressure, or volume of a system, but also physical transformations such as fusion, vaporization, or transition from one crystalline form to another. For all such processes the change in entropy is defined as

$$\Delta S = S_2 - S_1 \tag{57}$$

where S_2 is the entropy of the final form and S_1 the entropy of the initial form.

The transitions enumerated take place reversibly at constant temperature T and pressure P and are accompanied by an absorption or evolution of ΔH cal of heat for a given quantity of substance. Therefore, for all such processes

$$\Delta S = \frac{q_r}{T} = \frac{\Delta H}{T} \tag{58}$$

i.e., the change in entropy is given by the heat necessary to accomplish the transition divided by the temperature at which the transition takes place. Of necessity Eq. (58) is valid only when reversible conditions obtain during the transformation, i.e., when equilibrium exists between the two forms.

As an illustration of the use of Eq. (58), consider the problem of finding the entropy difference for the transition

$$H_2O(l, 1 \text{ atm}) = H_2O(g, 1 \text{ atm}) \qquad \Delta H_{373.2°K} = 9717 \text{ cal mole}^{-1}$$

Since at 373.2°K and 1 atm pressure, the normal boiling point of water, $H_2O(l)$ is in equilibrium with $H_2O(g)$, then

$$\Delta S = S_g - S_l = \frac{\Delta H}{T}$$
$$= \frac{9717}{373.2}$$
$$= 26.04 \text{ eu mole}^{-1}$$

Entropy changes may also be calculated for irreversible transitions, but the change in entropy will no longer be given by Eq. (58). Consider first the process

$$H_2O(l, 1 \text{ atm}) = H_2O(g, 0.1 \text{ atm})$$

at 373.2°K. Since the change in entropy does not depend on the manner of accomplishing the change, we may first vaporize the water isothermally and reversibly to steam at 1 atm pressure and then expand the steam reversibly and isothermally from 1 to 0.1 atm. The total change in entropy will then be

$$\Delta S = \Delta S_{\text{vaporization}} + \Delta S_{\text{expansion}}$$
$$= \frac{\Delta H}{T} + R \ln \frac{P_1}{P_2}$$
$$= \frac{9717}{373.2} + 4.58 \log \frac{1}{0.1}$$
$$= 30.62 \text{ eu mole}^{-1}$$

Again, suppose it is desired to calculate the entropy change involved in an irreversible process such as the conversion of a mole of water to ice at $-15°C$ and 1 atm pressure. At this pressure water and ice are in equilibrium only at 0°C, and hence Eq. (58) will not give the entropy change at $-15°C$. However, if we determine the entropy changes involved in the following steps:

1. The heating of a mole of water from -15 to 0°C
2. Reversible conversion of a mole of water to a mole of ice at 0°C
3. Cooling of a mole of ice from 0 to $-15°C$

then the sum of these entropy changes should be the same as the entropy change of the irreversible process $H_2O(l) = H_2O(s)$ at $-15°C$, since the net process is the same in both cases.

ENTROPY CHANGE IN CHEMICAL REACTIONS

The entropy change accompanying a chemical reaction is defined as the difference between the sum of the entropies of all the products and the sum of the entropies of all reactants. For any reaction such as

$$\alpha A + bB + \cdots = cC + dD + \cdots$$

the entropy change is given by

$$\Delta S = (cS_C + dS_D + \cdots) - (\alpha S_A + bS_B + \cdots) \tag{59}$$

where S_A, S_B, etc., are the entropies *per mole* of the various species. When the entropies of the individual substances correspond to a state of unit activity they are called *standard entropies* and they are designated by the symbol S^0. Again, in a reaction where all the substances involved are at unit activity ΔS is written ΔS^0, and the latter is the *standard entropy change* of the reaction.

Entropy changes of chemical reactions are usually evaluated at constant temperature and pressure. For any given reaction at any given temperature and pressure the entropy change is definite and just as characteristic of the reaction as the change in the internal energy or enthalpy. The manner in which the entropy change of a reaction depends on the temperature at constant pressure may readily be deduced from Eqs. (59) and (36b). If Eq. (59) be differentiated with respect to temperature at constant pressure, then

$$\left[\frac{\partial(\Delta S)}{\partial T}\right]_P = \left[c\left(\frac{\partial S_C}{\partial T}\right)_P + d\left(\frac{\partial S_D}{\partial T}\right)_P + \cdots\right]$$
$$- \left[\alpha\left(\frac{\partial S_A}{\partial T}\right)_P + b\left(\frac{\partial S_B}{\partial T}\right)_P + \cdots\right]$$

But, according to Eq. (36b) $(\partial S/\partial T)_P = C_p/T$. Hence,

$$\left[\frac{\partial(\Delta S)}{\partial T}\right]_P = \frac{(cC_{pc} + dC_{pD} + \cdots)}{T} - \frac{(\alpha C_{pA} + bC_{pB} + \cdots)}{T}$$
$$= \frac{\Delta C_p}{T} \tag{60}$$

ΔC_p is the difference between the C_p's of products and reactants. On integration of Eq. (60) between the limits T_1 and T_2, we obtain

$$\int_{\Delta S_1}^{\Delta S_2} d(\Delta S) = \int_{T_1}^{T_2} \frac{\Delta C_p}{T} dT$$

$$\Delta S_2 - \Delta S_1 = \int_{T_1}^{T_2} \frac{\Delta C_p}{T} dT \tag{61}$$

ΔS_2 is the entropy change at T_2, while ΔS_1 is the entropy change of the reaction at T_1. The difference between the two is given by the integral in Eq. (61). When ΔC_p may be assumed constant over the temperature interval in question, then Eq. (61) becomes

$$\Delta S_2 - \Delta S_1 = \Delta C_p \int_{T_1}^{T_2} \frac{dT}{T}$$

$$= \Delta C_p \ln \frac{T_2}{T_1} \tag{62}$$

However, if ΔC_p is not constant, then the expression for ΔC_p as a function of T must be obtained in the same manner as described in the preceding chapter, and the integration must be carried out term by term between the given temperature limits.

The isothermal entropy change of a reaction is calculated by means of Eq. (59), or, more commonly, from ΔH and ΔF of the reaction. ΔF, the free energy change, will be discussed in the next chapter. To obtain ΔS with the aid of Eq. (59), the entropies of the individual substances at the temperature in question must be available. One way of obtaining these is through the *third law of thermodynamics*.

THE THIRD LAW OF THERMODYNAMICS

As a result of the researches of T. W. Richards, Walter Nerst, Max Planck, and others, another fundamental principle of thermodynamics, which deals with the entropy of pure crystalline substances at the absolute zero of temperature, has come into being. This principle, called the third law of thermodynamics, states that the *entropy of all pure crystalline solids may be taken as zero at the absolute zero of temperature*. The statement is confined to pure crystalline solids because theoretical argument and experimental evidence have shown that the entropy of solutions and supercooled liquids is not zero at $0°K$. For pure crystalline solids the law has been verified repeatedly, and at present little doubt remains about the general validity of the above statement of the law.

The importance of the third law lies in the fact that it permits the calculation of absolute values of the entropy of pure substances from thermal data alone. The variation of entropy with temperature at constant pressure is given by Eq. (36b). Further, since the third law states that for any pure crystalline substance $S = 0$ at $T = 0$, then for any such substance the equation may be integrated between this lower limit and any temperature T to yield

$$\int_{S=0}^{S=S} dS = \int_{T=0}^{T=T} \frac{C_p dT}{T}$$

$$S_T = \int_0^T \frac{C_p dT}{T} \qquad \text{(63a)}$$

or

$$S_T = \int_0^T C_p d \ln T \qquad \text{(63b)}$$

S_T, known as the absolute entropy of the solid, is always a positive quantity. All that is necessary for integration of this equation is a knowledge of the heat capacities of the solid from $T = 0$ to any desired temperature T.

EVALUATION OF ABSOLUTE ENTROPIES

The integral in Eq. (63) is evaluated graphically by plotting either C_p vs. $\ln T$ or C_p/T vs. T and determining the area under the curve between $T = 0$ and any temperature T. This area is then the value of the integral and hence of S_T. In practice heat capacities are usually measured from approximately 20°K to temperature T, and extrapolation is resorted to from ca. 20°K to $T = 0$. Such extrapolations are generally made by use of the Debye third power law for the heat capacity of solids at low temperatures, namely,

$$C_p = aT^3 \qquad \text{(64)}$$

This expression substituted into Eq. (36b) yields

$$dS = \frac{C_p}{T} dT$$

$$= aT^2 dT$$

and hence on integration between $T = 0$ and T_1, the lower limit of the heat capacity data, we get

$$S_{T_1} = \int_0^{T_1} dS = \int_0^{T_1} aT^2 dT$$

$$= \frac{aT_1^3}{3} \qquad \text{(65)}$$

Since $C_p = aT^3$, then the value of S_{T_1} is equal to one-third of the value of C_p at temperature T_1.

A plot of C_p vs. $\log_{10} T$ for anhydrous sodium sulfate is shown in Figure 5–1. The area under the curve between $T = 14°$ and $T = 298.15°K$ is 15.488, while the area from $T = 0$ to $T = 14°$ is 0.026. Hence the absolute entropy per mole of sodium sulfate at $298.15°K$ is

$$S^0_{298.15°K} = 2.303(15.488 + 0.026)$$
$$= 35.73 \text{ eu mole}^{-1}$$

Absolute entropies of substances that are liquid or gaseous at room temperatures can also be obtained with the third law. The total absolute

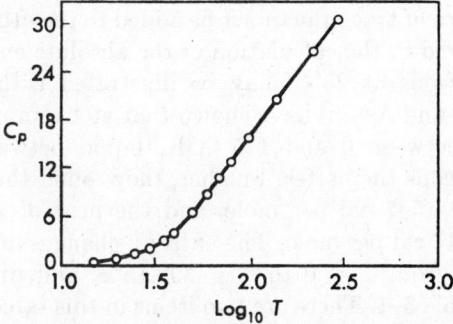

Figure 5–1. Heat capacity of Na_2SO_4 at various temperatures. [From Pitzer and Coulter, *J. Am. Chem. Soc.*, **60**, 1310 (1938).]

entropy of a substance in a particular state at a given temperature will be the sum of all the entropy changes the substance has to undergo in order to reach the particular state from the crystalline solid at absolute zero. Thus, if a substance is gaseous at 1 atm pressure and 25°C, the entropy of the gas must be the sum of the entropies involved in (a) heating the crystalline solid from $T = 0$ to $T = T_f$, the fusion point; (b) the entropy of fusion, $\Delta H_f/T_f$; (c) the entropy of heating the liquid from T_f to T_b, the normal boiling point; (d) the entropy of vaporization, $\Delta H_v/T_b$; and (e) the entropy of heating the gas from T_b to $298°K$. These changes and the attendant entropies are shown in the following scheme:

$$\text{Solid} \xrightarrow[T=0]{\Delta S_1 = \int_0^{T_f} C_{p_s} d\ln T} \text{Solid} \xrightarrow[T=T_f]{\Delta S_f = \frac{\Delta H_f}{T_f}} \text{Liquid} \xrightarrow[T=T_f]{\Delta S_2 = \int_{T_f}^{T_b} C_{p_l} d\ln T} $$

$$\text{Liquid} \xrightarrow[T=T_b]{\Delta S_v = \frac{\Delta H_v}{T_b}} \text{Gas} \xrightarrow[T=T_b]{\Delta S_3 = \int_{T_b}^{298°} C_{p_g} d\ln T} \text{Gas} \quad T=298.15°K$$

It follows, therefore, that the absolute entropy of any substance at a temperature T may be written as

$$S_T = \int_0^{T_f} C_{p_s} d \ln T + \frac{\Delta H_f}{T_f} + \int_{T_f}^{T_b} C_{p_l} d \ln T + \frac{\Delta H_v}{T_b} + \int_{T_b}^{T} C_{p_g} d \ln T \tag{66}$$

If a substance is solid at temperature T, only the first integral applies with $T_f = T$; when it is liquid, the first three terms are used with $T_b = T$ now; when it is gaseous, the full equation must be used. Equation (66) may be complicated further if the solid undergoes any crystalline modifications between $T = 0$ and $T = T_f$. When such transitions occur, the entropy or entropies of transition must be added to the others in Eq. (66).

The steps involved in the calculation of the absolute entropy of a substance that is gaseous at 25°C may be illustrated with the following example. Messerly and Aston[2] have shown that at 1 atm pressure methyl chloride is solid between 0 and 175.44°K, liquid between 175.44 and 248.94°K, and gaseous thereafter. Further, they found the heat of fusion at 175.44°K to be 1537 cal per mole, and the heat of vaporization at 248.94°K to be 5147 cal per mole. The entropy changes involved in heating the substance from $T = 0$ to $T = 298.15$°K and unit activity are summarized in Table 5–1. There are two items in this table which require

TABLE 5–1.   ABSOLUTE ENTROPY OF METHYL CHLORIDE
AT 25°C AND UNIT ACTIVITY

Item	Entropy Change (eu/mole)
Solid: 0–10.00°K, Debye extrapolation	0.075
10.00–175.44°K, graphical	18.402
Fusion: $\Delta S_f = \Delta H_f / T_f = 1537/175.44$	8.760
Liquid: 175.44–248.94°K, graphical	6.239
Vaporization: $\Delta S_r = \Delta H_r / T_b = 5147/248.94$	20.677
Entropy of actual gas at 248.94°K	54.153
Correction for gas imperfection	0.119
Ideal gas at 248.94°K	54.27 ± 0.15
Gas: 248.94–298.15°K, spectroscopic	1.67
S^0 of gas at $a = 1$ and 298.15°K	55.94 ± 0.15

explanation. First, the entropy obtained for the actual gas at 248.94°K does not correspond to $a = 1$ because of presence of nonideality. An activity of unity is, as we shall see in the next chapter, equivalent to a

[2] Messerly and Aston, *J. Am. Chem. Soc.*, **62,** 889 (1940).

pressure of 1 atm in the ideal gas. Consequently a correction has to be applied to the actual gas at 1 atm to convert it to an ideal gas at the same pressure. This correction is generally made by use of the low-pressure form of the Berthelot equation, and it is given by

$$S_{\text{ideal}} = S_{\text{actual}} + \frac{27}{32} \frac{RT_c^3 P}{T^3 P_c} \tag{67}$$

where T_c and P_c are, respectively, the critical temperature and pressure of the substance. For methyl chloride $T_c = 416.4°K$ and $P_c = 65.7$ atm.

The second point of interest in the table is the evaluation of ΔS for the gas between 248.94 and 298.15°K from spectroscopic data instead of from heat capacity measurements. This method of obtaining entropy changes will be discussed in Chapter 18.

USE OF ABSOLUTE ENTROPIES

Absolute entropies are usually calculated and tabulated for 25°C and unit activity. Values of these thermodynamic quantities in the standard state are available at present for most elements and for many compounds. Some of these are listed in Table 5–2. These S^0 values may be used as

TABLE 5–2. STANDARD ABSOLUTE ENTROPIES OF ELEMENTS
AND COMPOUNDS AT 25°C
(Entropy units per gram atom or mole)

Substance	S^0	Substance	S^0
$H_2(g)$	31.21	$H_2O(l)$	16.72
C (diamond)	0.583	$H_2O(g)$	45.11
C (graphite)	1.36	$CO(g)$	47.30
$N_2(g)$	45.77	$CO_2(g)$	51.06
$O_2(g)$	49.00	$HgCl_2(s)$	34.6
Na(s)	12.2	$HgCl(s)$	23.5
Mg(s)	7.77	CuI(s)	23.1
S (rhombic)	7.62	AgCl(s)	22.97
S (monoclinic)	7.78	AgI(s)	27.6
$Cl_2(g)$	53.29	$Fe_2O_3(s)$	21.5
Fe(s)	6.49	MgO(s)	6.55
Cu(s)	7.97	NaCl(s)	17.3
$Br_2(g)$	58.64	$C_2H_6(g)$	54.85
Ag(s)	10.21	$CH_3OH(l)$	30.3
$I_2(s)$	27.9	$C_2H_5OH(l)$	38.4
$I_2(g)$	62.29	$C_6H_6(l)$	41.30
Hg(l)	18.5	$C_6H_5OH(l)$	34.0
Hg(g)	41.80	$CH_3COOH(l)$	38.2

follows to calculate the entropy changes accompanying chemical reactions. Suppose the entropy change is required for the reaction

$$C(s, \text{graphite}) + 2\,H_2(g) + \frac{1}{2}\,O_2(g) = CH_3OH(l)$$

Using the molar entropies given in Table 5–2, we get

$$\Delta S^0_{25°C} = S^0_{CH_3OH} - \left(S^0_C + 2\,S^0_{H_2} + \frac{1}{2}\,S^0_{O_2}\right)$$
$$= 30.3 - 1.36 - 62.42 - 24.50$$
$$= -58.0 \text{ eu}$$

In a similar manner may be calculated ΔS^0 values for other reactions at 25°C provided the necessary absolute entropies are known. To obtain the entropy changes at temperatures other than 25°C, we need only apply Eq. (61). In terms of this equation the entropy change for any reaction at a temperature T in relation to ΔS^0 at 298.15°K is given by

$$\Delta S^0_T - \Delta S^0_{298.15°K} = \int_{298.15°K}^{T} \frac{\Delta C_p dT}{T}$$
$$\Delta S^0_T = \Delta S^0_{298.15°K} + \int_{298.15°K}^{T} \frac{\Delta C_p dT}{T} \tag{68}$$

The evaluation of the integral has already been discussed.

REFERENCES

See the references listed at the end of Chapters 3 and 4.

PROBLEMS

1. A quantity of ideal gas in an isolated system is expanded isothermally and reversibly at 400°K from a volume V_1 to V_2. During the expansion the gas absorbs 200 cal of heat from the reservoir in contact with it. Find (a) the entropy change of the gas, (b) the entropy change of the reservoir, and (c) the entropy change of the whole system.   *Ans.* (a) 0.50 eu; (b) −0.50 eu; (c) zero.

2. If the gas in problem 1 is expanded from V_1 to V_2 isothermally but irreversibly at 400°K with an absorption of 100 cal of heat, what will be the entropy changes for (a) the gas, (b) the reservoir, and (c) the complete system?

3. Suppose the gas in problem 1 expands freely from V_1 to V_2 at 400°K. What will be the entropy changes for (a) the gas, (b) the reservoir, and (c) the entire system?

4. Consider again the processes described in problems 1–3, and determine:

(a) The changes in ΔH and ΔE for the gas involved in these processes
(b) The work performed by the gas in these processes

(c) The qualitative relation between the work done by the gas and the entropy change experienced by the isolated system

(d) The qualitative relation between the degree of irreversibility of the processes and the change in entropy of the isolated system.

5. Starting with Eq. (17) of this chapter, deduce the expression for the change in entropy which will attend the heating of n moles of an ideal gas from volume V_1 at temperature T_1 to volume V_2 at temperature T_2.

$$Ans. \ \Delta S = \int_{T_1}^{T_2} \frac{nC_v}{T} \, dT + nR \ln \frac{V_2}{V_1}.$$

6. Starting with Eq. (17) of this chapter, deduce the expression for the change in entropy which will attend the heating of n moles of an ideal gas from a pressure P_1 at temperature T_1 to a pressure P_2 at temperature T_2.

7. Starting with the fact that the volume of a substance is a function of T and P, show that $(\partial S/\partial V)_T = \dfrac{\alpha}{\beta}$, where α is the thermal coefficient of expansion and β is the compressibility coefficient of the substance. The definitions of α and β are given in problem 40 of Chapter 1.

8. Show that $(\partial S/\partial P)_T = -\alpha V$.

9. For $C_6H_6(l)$ $\alpha = 1.24 \times 10^{-3}$ degree^{-1} and $\beta = 9.30 \times 10^{-5}$ atm^{-1} at 20°C and 1 atm pressure. Assuming α and β to be constant, find the change in molar volume which will be required to produce an entropy change of 0.500 eu per mole at 20°C. *Ans.* 1.55 cc.

10. The density of $C_6H_6(l)$ is 0.8790 g cc^{-1} at 20°C. Assuming α and the molar volume to be independent of pressure, find the change in entropy which will accompany the compression of 1 mole of the substance from 1 to 11 atm.

11. Show the identity of Eqs. (48) and (52) for a given change occurring in an ideal gas.

12. For a certain ideal gas $C_p = \frac{5}{2} R$ cal mole^{-1} degree^{-1}. Calculate the change in entropy suffered by 3 moles of the gas on being heated from 300 to 600°K at (a) constant pressure and (b) constant volume.

Ans. (a) 10.33 eu; (b) 6.20 eu.

13. Assuming C_p for $N_2(g)$ to be $\frac{7}{2} R$ cal mole^{-1} degree^{-1}, find the change in entropy undergone by 10 g of the gas on being cooled from 100 to 0°C at (a) constant pressure and (b) constant volume.

14. The atomic heat capacity of solid Mo is given by the equation

$$C_p = 5.69 + 1.88 \times 10^{-3} T - \frac{0.503 \times 10^5}{T^2}$$

Find the change in entropy which accompanies the heating of 1 atomic weight of Mo from 0°C to its melting point, 2620°C. *Ans.* $\Delta S = 18.6$ eu.

15. Using the equation for C_p as a function of T for $CH_4(g)$ given in the last chapter, calculate the entropy change which results from heating 2 moles of the gas from 300 to 600°K at constant pressure.

16. Assuming that for $CH_4(g)$ $C_p - C_v = R$, find the entropy change resulting from heating 2 moles of the gas from 300 to 600°K at constant volume.

17. Assuming that nitrogen is an ideal gas, calculate ΔS for the compression of 200 g of the gas from a pressure of 1 to 5 atm at 25°C.

Ans. $\Delta S = -22.85$ eu.

18. One mole of a perfect gas contained in a 10-liter vessel at 27°C is permitted to expand freely into an evacuated vessel of 10-liter capacity so that the final volume is 20 liters. What amount of work is done and what amount of heat is absorbed in the process? What is the change in the entropy accompanying the process?

19. Calculate the change in entropy suffered by 2 moles of an ideal gas on being heated from a volume of 100 liters at 50°C to a volume of 150 liters at 150°C. For the gas $C_v = 7.88$ cal mole^{-1} degree^{-1}.   *Ans.* $\Delta S = 5.88$ eu.

20. What is the difference in entropy between 1 mole of N_2 at standard conditions and 1 mole of N_2 at 200°C when the molar volume is 50 liters? Assume that C_p is $\frac{7}{2} R$ and that N_2 behaves ideally.

21. Calculate the change in entropy experienced by 2 moles of an ideal gas on being heated from a pressure of 5 atm at 50°C to a pressure of 10 atm at 100°C. For the gas $C_p = 9.88$ cal mole^{-1} degree^{-1}.   *Ans.* $\Delta S = 0.08$ eu.

22. For a certain gas $C_p = 12.0$ cal mole^{-1} degree^{-1}. What will be the change in entropy of 10 moles of the gas when it is expanded from a volume of 200 liters at 3 atm pressure to a volume of 400 liters at 1 atm pressure?

23. For $CHCl_3(g)$ $C_p = 7.05 + 35.60 \times 10^{-3} T - 216.9 \times 10^{-7} T^2$ cal mole^{-1} degree^{-1}. Assuming this gas to be ideal, calculate the change in entropy involved in heating 2 moles of the gas from a volume of 100 liters at 500°K to a volume of 70 liters at 700°K.

24. A heat engine operates in a reversible cycle consisting of the following steps:

(a) Adiabatic compression from volume V_1 at T_1 to volume V_2 at T_2
(b) Heating at constant V_2 from temperature T_2 to T_3
(c) Adiabatic expansion from volume V_2 at T_3 to volume V_1 at T_4
(d) Cooling at constant V_1 from temperature T_4 to T_1

Taking the working substance to be n moles of an ideal gas for which C_v is a constant, show that $\Delta S = 0$ for the complete cycle.

25. Three moles of $N_2(g)$, originally at 1 atm pressure, are mixed isothermally with 5 moles of $H_2(g)$, also at 1 atm pressure, to yield a mixture whose total pressure is 1 atm. Assuming the gases to be ideal, calculate (a) the total entropy of mixing, and (b) the entropy of mixing per mole of gas.   *Ans.* (a) 10.52 eu.

26. Repeat problem 25 on the basis that the total pressure of the mixture is 10 atm.

27. One mole of an ideal gas at 0°C and 1 atm pressure is mixed adiabatically with one mole of a different gas at 100°C and 1 atm pressure to yield a mixture whose pressure is also 1 atm. If C_p for each gas is $\frac{5}{2} R$ cal mole^{-1} degree^{-1}, what is ΔS for the process?   *Ans.* $\Delta S = 2.87$ eu.

28. One mole of an ideal gas, originally at a volume 8.21 liters at 1000°K, is allowed to expand adiabatically until the final volume is 16.42 liters. For the gas $C_v = \frac{3}{2} R$ cal mole^{-1} degree^{-1}. Calculate values of ΔS for the process when:

(a) The expansion takes place reversibly
(b) The expansion takes place against a constant pressure of 3 atm
(c) The change in volume involves a free expansion

29. Derive the expression for the change in entropy experienced by n moles of a van der Waals gas on expansion from volume V_1 to V_2 at temperature T.

30. For the reversible transformation $AgI(\alpha) = AgI(\beta)$ at 1 atm pressure and 146.5°C, the heat of transition is 1530 cal/mole. What is the entropy change involved in the transformation of 2 moles of the β to the α forms?
                    Ans. $\Delta S_t = -7.29$ eu.

31. At the triple point of water, 0.01°C, ice, liquid water, and water vapor are in equilibrium under a pressure of 4.58 mm Hg. At this temperature the heat of fusion of ice is 1436 cal/mole, while the heat of vaporization is 10,767 cal/mole. From these data find:

(a) The heat of sublimation of ice
(b) The entropies of fusion, vaporization, and sublimation per mole of substance.

32. One g of ice at 0°C is added to 10 g of H_2O at the boiling point. What will be the final temperature, and what is the entropy change accompanying the process? Assume that the heat of fusion of H_2O is 80 cal/g and the specific heat 1 cal/g-degree.   *Ans.* $t = 83.6$°C; $\Delta S = 0.11$ eu.

33. What is the entropy change involved in transforming 1 g of ice at 0°C and 1 atm pressure into vapor at 150°C and 0.1 atm pressure? Assume that the specific heats of liquid and gaseous water are respectively 1.0 and 0.45 cal/g-degree.

34. The heat capacities of the solid and liquid forms of a compound A are respectively 18.3 and 25.2 cal mole^{-1} degree^{-1}. The compound melts under 1 atm pressure at a temperature of 160°C with a heat of fusion of 2460 cal/mole. What will be the entropy change of the process $A(l) \longrightarrow A(s)$ at 150°C?

35. Eastman and McGavock [*J. Am. Chem. Soc.*, **50,** 145 (1937)] list the atomic heat capacities of rhombic S from 15°K to 360°K. From the data of Table II in this article determine by a graphical method the entropy of rhombic S per mole at 298.2°K.

36. Kemp and Egan [*J. Am. Chem. Soc.*, **60,** 1521 (1938)] found that for propane at 1 atm pressure $\Delta H_f = 842.2$ cal/mole at the melting point, 85.45°K, and $\Delta H_v = 4487$ cal/mole at the boiling point, 231.04°K. They found, further, that for heating the solid from 0°K to 85.45°K $\Delta S = 9.92$ eu, while for heating the liquid from 85.45°K to the boiling point $\Delta S = 21.06$ eu. For propane $T_c = 368.8$°K and $P_c = 43$ atm. From these data find the standard entropy of gaseous propane at 231.04°K.   *Ans.* $\Delta S^0 = 60.42$ eu.

37. From the data of Table 5–1 find the standard entropy changes accompanying the following reactions at 25°C:

(a) $CO(g) + 2 H_2(g) = CH_3OH(l)$

(b) $2 HgCl(s) = 2 Hg(l) + Cl_2(g)$

(c) $MgO(s) + H_2(g) = H_2O(l) + Mg(s)$.

38. For a certain reaction $\Delta S^0_{298.2°K} = -59.20$ eu, and

$$\Delta C_p = -7.58 + 17.42 \times 10^{-3}\, T - \frac{3.985 \times 10^5}{T^2}$$

Find from these data ΔS^0 of the reaction of 400°K.

6
Free Energy
and Equilibrium

ALL CHANGES in nature are due to the tendency on the part of systems to reach a condition of maximum stability commensurate with the state of each system, i.e., equilibrium. Once equilibrium has been reached, the propensity toward further change disappears, and we say that the system is stable. As long as a system is away from equilibrium it will experience a tendency to reach that state, and the tendency will be greater the greater the distance from equilibrium.

Work results only when the tendency of systems to attain equilibrium is harnessed in some way. From a system in equilibrium no work can be obtained, but a system on the way to equilibrium may be made to yield useful work. The amount of work that can be recovered from any system undergoing a change depends both on the nature of the change and the manner in which the system is harnessed. However, for each particular process there is a maximum amount of work the system can possibly do, and this maximum work may be taken as a measure of the tendency of the system in question to undergo change.

A system undergoing change can perform maximum work only when the change is carried out reversibly. If the process is not completely reversible, the amount of work obtainable is always below the maximum, the difference appearing as heat. The driving force behind the change is still, however, the maximum work difference between the final and initial states, for this difference still represents the highest possible quantity of energy that *tends* to appear in utilizable form as a result of the process. Whether it does appear or not depends entirely on the manner in which the process is conducted, and in no way affects the conclusion that the maximum work a process *may* perform, if conducted properly, is the true measure of the driving tendency behind the process.

THE HELMHOLTZ FREE ENERGY

Although entropy can be employed to measure the tendency of systems to undergo change, under conditions most frequently encountered it is not as convenient a quantity to use as the so-called free energy functions A and F. The *Helmholtz free energy* of any system, A, is defined as

$$A = E - TS \tag{1}$$

where E and S are, respectively, the internal energy and entropy of a system. Since E, T, and S depend only on the state of the system, A must also be a state function. Consequently, when the system passes from one state to another, the change in A must be given by

$$
\begin{aligned}
\Delta A &= A_2 - A_1 \\
&= (E_2 - T_2 S_2) - (E_1 - T_1 S_1) \\
&= (E_2 - E_1) - (T_2 S_2 - T_1 S_1) \\
&= \Delta E - (T_2 S_2 - T_1 S_1)
\end{aligned}
$$

$$\text{(2a)} \qquad \text{(2b)}$$

Equation (2) gives the most general definition of ΔA. Under isothermal conditions, when $T_2 = T_1 = T$, this equation reduces to

$$\Delta A = \Delta E - T\Delta S \tag{3}$$

Equation (3) allows a physical interpretation of ΔA. Since under isothermal conditions $T\Delta S = q_r$, we have

$$
\begin{aligned}
\Delta A &= \Delta E - q_r \\
&= -(q_r - \Delta E) \\
&= -w_m
\end{aligned} \tag{4}
$$

Hence at constant temperature the maximum work done by a system is accomplished at the expense of a decrease in the Helmholtz free energy of the system. This is why A is sometimes called the maximum work content of a system.

For any pure substance A is most conveniently expressed in terms of T and V as the independent variables. We have thus

$$dA = \left(\frac{\partial A}{\partial T}\right)_V dT + \left(\frac{\partial A}{\partial V}\right)_T dV \tag{5}$$

Now, if we completely differentiate Eq. (1) we get

$$dA = dE - SdT - TdS \tag{6a}$$
$$= dE - SdT - dq_r \tag{6b}$$

since $TdS = dq_r$. But $dq_r = dE + PdV$, and hence

$$dA = dE - SdT - dE - PdV$$
$$= -SdT - PdV \tag{7}$$

On comparing Eqs. (5) and (7) we see that

$$\left(\frac{\partial A}{\partial T}\right)_V = -S \tag{8}$$

and

$$\left(\frac{\partial A}{\partial V}\right)_T = -P \tag{9}$$

An alternate equation for variation of A with T can be obtained as follows: Differentiation of the quantity A/T with respect to T at constant V yields

$$\left[\frac{\partial(A/T)}{\partial T}\right]_V = \frac{T\left(\frac{\partial A}{\partial T}\right)_V - A}{T^2}$$
$$= -\frac{(A + TS)}{T^2}$$

But $A + TS = E$. Therefore,

$$\left[\frac{\partial(A/T)}{\partial T}\right]_V = -\frac{E}{T^2} \tag{10}$$

Equation (7) shows the dependence of A for a pure substance on both the temperature and volume, with the individual effects of these variables being given by Eqs. (8) or (10) for T and by Eq. (9) for V. Again, at any constant temperature the relation between the resulting ΔA for the change and the attending changes in E and S is given by Eq. (3).

ΔA FOR REACTIONS

For any reaction such as

$$aA + bB + \cdots = cC + dD + \cdots \tag{11}$$

ΔA is given by the sum of the A's for the products minus a similar sum for the reactants; i.e.,

$$\Delta A = (cA_C + dA_D + \cdots) - (aA_A + bA_B + \cdots) \tag{12}$$

where A_A, A_B, etc., are the Helmholtz free energies of the various species *per mole* of substance. For such a reaction at any temperature T the relation of ΔA to ΔE and ΔS of the process can be obtained by substituting for each A in Eq. (12) its value from Eq. (1). When this is done, and the E's and S's for the products and reactants are added, we obtain an equation identical in form with Eq. (3), namely,

$$\Delta A = (E_p - E_r) - T(S_p - S_r)$$
$$= \Delta E - T\Delta S \tag{13}$$

where the subscripts p and r refer to products and reactants. Equation (13) allows the calculation of ΔA, ΔE, or ΔS when any two of these quantities are known.

The variation of ΔA with temperature at constant volume follows from Eq. (12) as

$$\left[\frac{\partial(\Delta A)}{\partial T}\right]_V = \left[c\left(\frac{\partial A_C}{\partial T}\right)_V + d\left(\frac{\partial A_D}{\partial T}\right)_V + \cdots\right]$$
$$- \left[\alpha\left(\frac{\partial A_A}{\partial T}\right)_V + b\left(\frac{\partial A_B}{\partial T}\right)_V + \cdots\right]$$
$$= [-cS_C - dS_D - \cdots] - [-\alpha S_A - bS_B - \cdots]$$
$$= -\Delta S \tag{14}$$

If we now insert Eq. (14) into Eq. (13) we get instead of the latter

$$\Delta A = \Delta E + T\left[\frac{\partial(\Delta A)}{\partial T}\right]_V \tag{15}$$

An alternate expression for variation of ΔA with T is obtained by dividing Eq. (12) by T and differentiating with respect to T at constant V. Introduction of Eq. (10) for the individual derivatives yields then

$$\left[\frac{\partial(\Delta A/T)}{\partial T}\right]_V = -\frac{\Delta E}{T^2} \tag{16}$$

THE GIBBS FREE ENERGY

The maximum work a process may yield is not necessarily the amount of energy available for doing useful work, even though the process is conducted reversibly. Of the total amount of work available, a certain amount has to be utilized for the performance of pressure-volume work against the atmosphere due to contraction or expansion of the system during the process. For a process taking place reversibly at constant temperature and pressure, and involving a volume change from V_1 to V_2, the work done against the atmosphere is $P(V_2 - V_1) = P\Delta V$. Since this work is accomplished at the expense of the maximum work yielded by the process, the *net* amount of energy available for work other than pressure-volume against the confining atmosphere must be

$$\text{Net available energy at } T \text{ and } P = w_m - P\Delta V \tag{17}$$

To bring out more precisely the nature of the maximum net energy available from a process, let us define another state function F, called the *Gibbs free energy*, by the relation

$$F = H - TS \tag{18}$$

The change in F between two states of a system will be, therefore,

$$
\begin{aligned}
\Delta F &= F_2 - F_1 \\
&= (H_2 - T_2 S_2) - (H_1 - T_1 S_1) \\
&= (H_2 - H_1) - (T_2 S_2 - T_1 S_1) \\
&= \Delta H - (T_2 S_2 - T_1 S_1)
\end{aligned}
\tag{19}
$$

and when the temperature is constant

$$\Delta F = \Delta H - T\Delta S \tag{20}$$

An alternate, but equivalent, way of defining F is through the relation

$$F = A + PV \tag{21}$$

where A is the Helmholtz free energy, and P and V are the pressure and volume of the system. From Eq. (21) ΔF follows as

$$\Delta F = \Delta A + P_2 V_2 - P_1 V_1 \tag{22}$$

which at constant pressure becomes

$$\Delta F = \Delta A + P\Delta V \tag{23}$$

The equivalence of the two definitions may be shown by inserting Eq. (2b) for ΔA into Eq. (22). We obtain thus

$$
\begin{aligned}
\Delta F &= (\Delta E + P_2 V_2 - P_1 V_1) - (T_2 S_2 - T_1 S_1) \\
&= \Delta H - (T_2 S_2 - T_1 S_1)
\end{aligned}
$$

The latter expression is identical with Eq. (19).

The physical significance of ΔF at constants T and P may be obtained as follows: At constant temperature $T\Delta S = q_r$. Again, when the pressure is also constant, $\Delta H = \Delta E + P\Delta V$. Inserting these quantities into Eq. (20) we get

$$
\begin{aligned}
\Delta F &= \Delta E + P\Delta V - q_r \\
&= -(q_r - \Delta E - P\Delta V)
\end{aligned}
$$

But, by the first law $q_r - \Delta E = w_m$, and, therefore,

$$\Delta F = -(w_m - P\Delta V) \tag{24}$$

Comparison of Eqs. (24) and (17) shows that $-\Delta F$ represents the maximum net energy at constant T and P available for doing useful work; i.e., the net available energy under the specified conditions results from a *decrease* in the free energy content of the system on passing from the initial to the final state.

It is customary to refer to the Gibbs free energy simply as the free energy. We shall follow this procedure, and append the designation Gibbs or Helmholtz only when necessary for clarity.

Like H, F is most conveniently expressed in terms of T and P as independent variables. In terms of these variables we have for dF of any pure substance

$$dF = \left(\frac{\partial F}{\partial T}\right)_P dT + \left(\frac{\partial F}{\partial P}\right)_T dP \tag{25}$$

Again, from complete differentiation of Eq. (18) we get

$$dF = dH - TdS - SdT \tag{26}$$

But, $TdS = dE + PdV$. Further, from $H = E + PV$ we get $dH = dE + PdV + VdP$. Substituting these identities into Eq. (26) we obtain

$$\begin{aligned} dF &= dE + PdV + VdP - dE - PdV - SdT \\ &= -SdT + VdP \end{aligned} \tag{27}$$

Comparison of Eqs. (25) and (27) shows that

$$\left(\frac{\partial F}{\partial T}\right)_P = -S \tag{28}$$

and

$$\left(\frac{\partial F}{\partial P}\right)_T = V \tag{29}$$

An alternate expression for dependence of F on T follows from differentiation of the quantity F/T with respect to T at constant P, namely,

$$\begin{aligned} \left[\frac{\partial(F/T)}{\partial T}\right]_P &= \frac{T\left(\frac{\partial F}{\partial T}\right)_P - F}{T^2} \\ &= -\frac{(F + TS)}{T^2} \end{aligned}$$

But, $F + TS = H$. Consequently,

$$\left[\frac{\partial(F/T)}{\partial T}\right]_P = -\frac{H}{T^2} \tag{30}$$

ΔF FOR REACTIONS

For any reaction $\Delta F = F_p - F_r$, where F_p is the sum total of the free energies of all the products and F_r is the sum total of the free energies of all the reactants. If we insert Eq. (18) for the free energy of each species, then at any temperature T

$$
\begin{aligned}
\Delta F &= (H_p - TS_p) - (H_r - TS_r) \\
&= (H_p - H_r) - T(S_p - S_r) \\
&= \Delta H - T\Delta S
\end{aligned} \tag{31}
$$

where ΔH and ΔS are, respectively, the enthalpy and entropy change of the reaction.

The free energy change is in general a function of T and P, i.e.,

$$
d(\Delta F) = \left[\frac{\partial(\Delta F)}{\partial T}\right]_P dT + \left[\frac{\partial(\Delta F)}{\partial P}\right]_T dP \tag{32}
$$

The first of these derivatives can be obtained by differentiation of Eq. (31) with respect to T at constant P, namely,

$$
\left[\frac{\partial(\Delta F)}{\partial T}\right]_P = \left[\frac{\partial(\Delta H)}{\partial T}\right]_P - T\left[\frac{\partial(\Delta S)}{\partial T}\right]_P - \Delta S
$$

But, the first derivative on the right is ΔC_p, while the second is $\Delta C_p/T$. Consequently,

$$
\begin{aligned}
\left[\frac{\partial(\Delta F)}{\partial T}\right]_P &= \Delta C_p - \frac{T\Delta C_p}{T} - \Delta S \\
&= -\Delta S
\end{aligned} \tag{33}
$$

In turn, the second derivative in Eq. (32) can be obtained from the definition of ΔF as

$$
\left[\frac{\partial(\Delta F)}{\partial P}\right]_T = \left(\frac{\partial F_p}{\partial P}\right)_T - \left(\frac{\partial F_r}{\partial P}\right)_T
$$

However, by use of Eq. (29) $(\partial F_p/\partial P)_T = V_p$ and $(\partial F_r/\partial P)_T = V_r$. Consequently,

$$
\begin{aligned}
\left[\frac{\partial(\Delta F)}{\partial P}\right]_T &= V_p - V_r \\
&= \Delta V
\end{aligned} \tag{34}
$$

where ΔV is the change in volume on reaction. Introducing Eqs. (33) and (34) into Eq. (32), we have

$$
d(\Delta F) = -\Delta S dT + \Delta V dP \tag{35}
$$

In Eq. (35) the first term on the right gives the effect on ΔF of change in temperature at constant pressure, while the second shows the effect of change in pressure at constant temperature. Further, if we substitute Eq. (33) for ΔS in Eq. (31), we obtain the *Gibbs-Helmholtz equation*

$$\Delta F = \Delta H + T\left[\frac{\partial(\Delta F)}{\partial T}\right]_P \tag{36}$$

A more convenient relation for the dependence of ΔF on temperature is obtained by differentiating $\Delta F/T$ with respect to T at constant P. We get thus

$$\left[\frac{\partial(\Delta F/T)}{\partial T}\right]_P = \frac{T[\partial(\Delta F)/\partial T]_P - \Delta F}{T^2}$$

But, according to Eq. (36)

$$T\left[\frac{\partial(\Delta F)}{\partial T}\right]_P = \Delta F - \Delta H$$

Therefore,

$$\left[\frac{\partial(\Delta F/T)}{\partial T}\right]_P = \frac{\Delta F - \Delta H - \Delta F}{T^2}$$

$$= -\frac{\Delta H}{T^2} \tag{37}$$

PROPERTIES AND SIGNIFICANCE OF ΔF

The free energy change for any process, being a function of the initial and final states of the system only, is a definite quantity at any given temperature and pressure and varies as these two variables are changed. As in the case of heat contents and internal energies, the absolute values of the free energies of substances are not known, and hence only differences can be dealt with. The free energy changes of processes are expressed in equations similar to thermochemical ones and can be similarly added and subtracted. Thus, for instance, the free energy changes for the two reactions given below are

$$SO_2(g) + Cl_2(g) = SO_2Cl_2(g) \qquad\qquad \Delta F_{298°K} = -2270 \text{ cal} \quad \text{(38a)}$$
$$S(\text{rhom.}) + O_2(g) + Cl_2(g) = SO_2Cl_2(g) \quad \Delta F_{298°K} = -74,060 \text{ cal} \quad \text{(38b)}$$

When Eq. (38a) is subtracted from (38b), we find that

$$S(\text{rhom.}) + O_2(g) = SO_2(g) \qquad\qquad \Delta F_{298°K} = -71,790 \text{ cal} \quad \text{(38c)}$$

Consequently, the formation of sulfur dioxide from rhombic sulfur and gaseous oxygen proceeds with a free energy decrease of 71,790 cal; i.e., the free energy content of $(S + O_2)$ is greater than that of SO_2 by this amount.

The sign of the free energy change of a process is very important. When the driving tendency of a reaction is from left to right, energy is emitted on reaction, and the sign of ΔF is negative. A minus sign denotes, there-fore, that the reaction tends to proceed spontaneously. When the tendency is from right to left, however, net work equivalent to ΔF has to be absorbed in order for the reaction to proceed in the direction indicated, and ΔF is positive. A positive sign for ΔF signifies, therefore, that the reaction in the given direction is not spontaneous. Finally, when the system is in equilibrium, there is no tendency to proceed in either direction, no work can be done by the system, and hence $\Delta F = 0$. These three possible conditions for the free energy change of a process *at constant temperature and pressure* may be summarized as follows:

$$A + B \longrightarrow C + D \qquad \Delta F = - \qquad \text{(spontaneous)} \qquad \textbf{(39a)}$$
$$A + B \longleftarrow C + D \qquad \Delta F = + \qquad \text{(nonspontaneous)} \qquad \textbf{(39b)}$$
$$A + B \rightleftharpoons C + D \qquad \Delta F = 0 \qquad \text{(equilibrium)} \qquad \textbf{(39c)}$$

The arrows indicate the directions the reaction tends to follow spontaneously for the given sign of the free energy change.

A negative free energy change for a process does not necessarily mean that the process will take place. It is merely an indication that the process *can* occur provided the conditions are right. Thus oxygen and hydrogen can coexist indefinitely at room temperature without combining, although ΔF for the reaction at 25°C is $-56,690$ cal per mole of water. When a catalyst like platinized asbestos is introduced, however, the reaction proceeds with explosive violence. Even with a catalyst the reaction would have been impossible had not the potentiality to react been present. It is the sign of the free energy change which determines whether the potentiality to react exists, and it is the magnitude of the free energy change which tells us how large that potentiality is.

The physical significance of Eq. (31) or (36) is best illustrated by an example. Consider the reaction

$$\text{Zn(s)} + \text{CuSO}_4 \text{ (solution)} = \text{ZnSO}_4 \text{ (solution)} + \text{Cu(s)}$$

When this reaction is permitted to take place in an open beaker by adding zinc to a solution of copper sulfate, heat of reaction equal to ΔH is obtained. If, on the other hand, the same reaction is carried out reversibly by allowing the process to proceed in an electrochemical cell and forcing the voltage established to do work against another voltage only infinitesimally smaller than that of the cell, then, instead of all heat, work equivalent to ΔF will be obtained. The difference between the work so obtained and the heat which would have been liberated had the reaction been carried out completely irreversibly, as in an open beaker, is given by the Gibbs-Helmholtz equation, and is equal to either $-T\Delta S$ or $T[\partial(\Delta F)/\partial T]_P$.

The term $-T\Delta S = T[\partial(\Delta F)/\partial T]_P$ represents the heat interchange between the system and its surroundings when the process is conducted *isothermally and reversibly*, for

$$-T\Delta S = -T\frac{q_r}{T} = -q_r$$

and hence,

$$\Delta H - \Delta F = q_r \tag{40}$$

When ΔH is greater than ΔF, q_r is positive and energy is absorbed as heat from the surroundings. On the other hand, when ΔF is greater than ΔH, q_r is negative and heat is evolved to the surroundings. Finally, in the special case when $\Delta H = \Delta F$, heat is neither absorbed nor evolved by the system, and hence there is no change in entropy.

CALCULATION OF FREE ENERGY CHANGES

Neither Eq. (30) nor Eq. (28) lends itself readily for calculation of the free energy change attending the change in temperature at constant pressure. In the first instance the required absolute values of H of a substance are unknown, while in the second case absolute values of S as a function of T are generally unavailable. However, calculations of the variation of F with pressure at constant temperature can readily be made. From Eq. (29) the change in free energy attending a change in total pressure on a substance from P_1 to P_2 at the constant temperature T follows as

$$\int_{F_1}^{F_2} dF = \int_{P_1}^{P_2} V dP$$

$$\Delta F = F_2 - F_1 = \int_{P_1}^{P_2} V dP \tag{41}$$

To evaluate the integral between the given limits, V must be known as a function of P. Since for n moles of an ideal gas $V = nRT/P$, we obtain for such gases

$$\Delta F = \int_{P_1}^{P_2} V dP$$

$$= \int_{P_1}^{P_2} \frac{nRTdP}{P}$$

$$= nRT \ln \frac{P_2}{P_1} \tag{42}$$

For real gases, either V or dP obtained from suitable equations of state must be substituted in Eq. (41), and the integration performed either analytically or graphically.

As solids and liquids are only slightly compressible, their volume may be considered essentially constant over appreciable pressure ranges. Equation (41) integrates, then, simply to

$$\Delta F = \int_{P_1}^{P_2} V dP$$
$$= V(P_2 - P_1) \tag{43}$$

where ΔF is the free energy change of the solid or liquid due to change in pressure from P_1 to P_2. Such free energy changes with pressure usually are small compared to the free energy changes in gases and may frequently be disregarded; i.e., the free energies of pure solids and liquids may be considered to be constant over a fairly wide pressure range at any given temperature.

Equation (37) gives the variation of the ΔF of a reaction with temperature at constant pressure in terms of ΔH for the process, and it may thus be employed to calculate ΔF at one temperature from that at another. To do this, ΔH as a function of temperature, deduced by the method outlined in Chapter 4, must be available, as well as one value of ΔF at a known temperature. With such information at hand, Eq. (37) can be integrated in the manner illustrated below.

EXAMPLE: Suppose ΔF is sought at 1000°K for the reaction

$$\frac{1}{2} N_2(g) + \frac{3}{2} H_2(g) = NH_3(g) \qquad \Delta F_{298.2°K} = -3980 \text{ cal} \tag{44}$$

For this reaction it was already shown, p. 155, that ΔH is given by

$$\Delta H = -9190 - 7.12\, T + 3.182 \times 10^{-3}\, T^2 - 2.64 \times 10^{-7}\, T^3$$

Substituting this value of ΔH in Eq. (37), we have

$$\left[\frac{\partial(\Delta F/T)}{\partial T} \right]_P = -\frac{\Delta H}{T^2} = \frac{9190}{T'^2} + \frac{7.12}{T} - 3.182 \times 10^{-3} + 2.64 \times 10^{-7}\, T$$

On integration this expression becomes

$$\frac{\Delta F}{T} = -\frac{9190}{T} + 7.12 \ln T - 3.182 \times 10^{-3}\, T + 1.32 \times 10^{-7}\, T^2 + I$$

where I is a constant of integration. Hence,

$$\Delta F = -9190 + 7.12\, T \ln T - 3.182 \times 10^{-3}\, T^2 + 1.32 \times 10^{-7}\, T^3 + IT$$

Inserting now the value of $\Delta F = -3980$ cal at $T = 298.2°K$ from Eq. (44), and solving for I, we find that

$$-3980 = -9190 + 7.12(298.2) \ln (298.2) - 3.182 \times 10^{-3}(298.2)^2$$
$$+ 1.32 \times 10^{-7}(298.2)^3 + I(298.2)$$

$$I = -21.61$$

Consequently ΔF as a function of the temperature follows as

$$\Delta F = -9190 + 7.12\, T \ln T - 3.182 \times 10^{-3}\, T^2$$
$$+ 1.32 \times 10^{-7}\, T^3 - 21.61\, T \quad (45)$$

and hence ΔF at $T = 1000°K$ is

$$\Delta F = -9190 + 7.12(1000) \ln (1000) - 3.182 \times 10^{-3}(1000)^2$$
$$+ 1.32 \times 10^{-7}(1000)^3 - 21.61(1000)$$
$$= +15,340 \text{ cal}$$

We see, therefore, that although the formation of ammonia will proceed spontaneously at 298.2°K, for ΔF is negative, the same process at 1000°K will be nonspontaneous to the extent of 15,340 cals.

The method described for calculating ΔF at one temperature from that at another is thermodynamically exact and is conditioned only by the accuracy of the data employed. As in the case of heats of reaction, great care must be exercised that calculations made are within the range over which the thermal data have been determined. Only with precise data and with due regard for the limits of their validity can great trust be placed in the results of such calculations.

ΔF equations such as Eq. (45) are useful not only for calculating the free energy change at any temperature T, but also for evaluating ΔS and ΔH. ΔS is readily obtained by differentiation with respect to temperature at constant pressure of the ΔF expression in accordance with Eq. (33). In turn, ΔH may be obtained either by differentiating $\Delta F/T$ with respect to T and using Eq. (37), or by evaluating first both ΔF and ΔS and subsequently employing Eq. (31) to find ΔH.

EXAMPLE: Suppose ΔS and ΔH for the ammonia synthesis, Eq. (44), are to be calculated at 1000°K. To obtain ΔS we differentiate first Eq. (45) with respect to T. Then,

$$\left[\frac{\partial(\Delta F)}{\partial T}\right]_P = -\Delta S = 7.12 + 7.12 \ln T - 6.364 \times 10^{-3}\, T + 3.96$$
$$\times 10^{-7}\, T^2 - 21.61$$
$$\Delta S = 14.49 - 7.12 \ln T + 6.364 \times 10^{-3}\, T$$
$$- 3.96 \times 10^{-7}\, T^2$$

On substituting $T = 1000°K$, we find thus $\Delta S_{1000°K} = -28.74$ eu. Having now ΔF and ΔS at 1000°K, Eq. (31) yields for ΔH

$$\Delta H = T\Delta S + \Delta F$$
$$= (1000)(-28.74) + 15,340$$
$$= -13,400 \text{ cal at } 1000°K$$

This result is identical with the one found for the heat of this reaction at 1000°K in Chapter 4.

THE FUGACITY AND ACTIVITY CONCEPTS

When Eq. (42) is applied to real gases, particularly at higher pressures, it is found that the change in free energy is not reproduced by this simple relation. The difficulty is, of course, that in cases of nonideal behavior V is no longer given by nRT/P, but by some more complicated function of the pressure. To obtain the free energy change with pressure for the nonideal substance, the exact dependence of the volume on the pressure must be known before Eq. (41) can be integrated. Since such volume dependence may be highly individual, the result would be that ΔF in each case would be given by an equation different in form, and the simplicity and generality of Eq. (42) would be destroyed.

G. N. Lewis first showed how nonideal systems may be handled without discarding the simple free energy equations deduced for ideal systems. To do this he introduced two new thermodynamic quantities, *fugacity* and *activity*. To understand these quantities, consider first a system composed of liquid water and its vapor. At constant temperature there is a definite pressure of water vapor above the liquid. This vapor comes from the liquid phase and represents a tendency of the liquid to pass into the vapor phase. In turn, the vapor tends to escape the gaseous state by condensing to liquid. When these two *escaping tendencies* become equal, the system reaches equilibrium, i.e., the vapor pressure becomes constant at constant temperature. We may, say, therefore, that a state of equilibrium is the point at which the escaping tendency of a constituent is the same in all parts of the system.

The idea that each substance in a particular state has a definite tendency to escape from that state is perfectly general. Lewis pointed out that this escaping tendency can be measured by a quantity f, called the *fugacity*, which is related to the free energy content of the substance *per mole*, F, by the expression

$$F = RT \ln f + B \qquad (46)$$

where B is a constant dependent only on the temperature and the nature of the substance. Since absolute values of the free energy are not known, B cannot be evaluated. However, we can get around this difficulty by referring all free energy measurements for any given substance to a standard reference point. If we designate by F^0 the free energy and by f^0 the fugacity in this standard state, then F^0 is given by

$$F^0 = RT \ln f^0 + B \qquad (47)$$

and the free energy difference between any state in which the free energy is F and the standard state is given by

$$F - F^0 = RT \ln \frac{f}{f^0} \tag{48}$$

Consequently, the free energy content of a substance in any state in terms of the free energy in the standard state follows as

$$F = F^0 + RT \ln \frac{f}{f^0} \tag{49}$$

If we write now

$$\frac{f}{f^0} = a \tag{50}$$

Eq. (49) becomes

$$F = F^0 + RT \ln a \tag{51}$$

The quantity a is called the *activity*. From Eq. (51) we see that the free energy per mole of any substance at a temperature T may be written as the free energy of the substance in the standard state at temperature T, and the quantity $RT \ln a$. In the standard state $F = F^0$, $RT \ln a = 0$, and hence $a = 1$, i.e., in the standard state the activity must be unity. In any other state the value of the activity will depend on the difference $(F - F^0)$, or in other words, on the distance of the particular state from the standard state. This relation between F and F^0 can best be understood from Figure 6–1. Let the vertical line represent a free energy axis. On

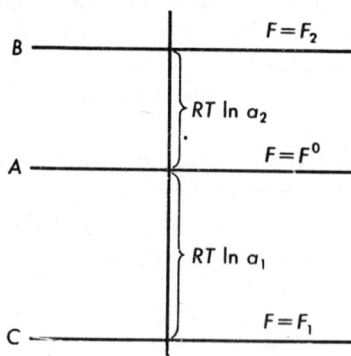

Figure 6–1.

this axis A represents the arbitrary point chosen as standard state, while B represents another point where the free energy is F_2. Then Eq. (51) states that the difference in free energies of a substance between points B and A is $RT \ln a_2$. Similarly, the free energy difference between a point like C and the standard state is given by $RT \ln a_1$.

In terms of Eq. (51) the difference in free energy per mole occasioned on passing from one state where the free energy is F_1 and the activity is a_1, to another state where the free energy is F_2 and the activity is a_2, must be

$$\Delta F = F_2 - F_1 = (F^0 + RT \ln a_2) - (F^0 + RT \ln a_1)$$

$$\Delta F = RT \ln \frac{a_2}{a_1} \tag{52a}$$

or, for n moles,

$$\Delta F = nRT \ln \frac{a_2}{a_1} \tag{52b}$$

The striking similarity of Eq. (52b) to Eq. (42) suggests that we may consider the activity to be the thermodynamic counterpart of gas pressure, and also, as we shall see in subsequent chapters, of the concentration in the case of solution constituents. When activities of gases and solution constituents are substituted for pressures and concentrations, exact free energy calculations become possible. The reason for this exactness is that activities, unlike concentrations or pressures, take into account not only stoichiometric relationships, but also mutual attractions between molecules, interactions between solute and solvent in a solution, and ionization. These effects complicate ideal behavior and are the factors responsible for the breakdown of the thermodynamic equations for ideal systems when applied to real substances.

Before activities can be evaluated, the particular state to be chosen as a standard for each substance must be defined. This definition may be arbitrary, but certain conventions have been established and are in general use at present. We shall discuss here only the conventions pertaining to pure gases, solids, and liquids, while those for gas mixtures and solution constituents will be considered in later chapters.

STANDARD STATE FOR GASES

As the standard state of a gas at any given temperature is taken the state in which the fugacity is equal to unity, namely, $f^0 = 1$. On the basis of this definition, the activity of any gas becomes equal to the fugacity, for

$$a = \frac{f}{f^0} = \frac{f}{1} = f \tag{53}$$

and hence for a gas Eq. (51) may be written as

$$F = F^0 + RT \ln f \tag{54}$$

For an ideal gas the fugacity is equal to the pressure. Since, however, any gas can be brought into an ideal state by reducing its pressure to zero, we can complete the definition of the fugacity of any gas by stating that, in general,

$$f = P \text{ as } P \longrightarrow 0 \tag{55a}$$

or that
$$\lim_{P \to 0} \frac{f}{P} = 1 \tag{55b}$$

As long as a gas is ideal the ratio f/P remains equal to unity. However, as soon as a gas deviates from ideal behavior f is no longer equal to P, and the ratio f/P becomes something other than one. The further this ratio is from unity the greater is the nonideality. Consequently, this ratio, called the *activity coefficient* of a gas, and represented by the symbol γ, gives a direct measure of the extent to which any real gas deviates from ideality at any given pressure and temperature.

DETERMINATION OF ACTIVITY COEFFICIENTS OF GASES

Determination of the activity coefficients of gases at any temperature T is based on Eq. (29), namely, $dF = VdP$. Since F^0 is a constant at any given temperature, then differentiation of Eq. (54) yields $dF = RTd \ln f$, and so

$$d \ln f = \frac{V}{RT} dP \tag{56}$$

Let us define now a quantity α as

$$\alpha = V - \frac{RT}{P} \tag{57}$$

and therefore
$$V = \frac{RT}{P} + \alpha \tag{58}$$

Insertion of Eq. (58) into (56) yields

$$d \ln f = d \ln P + \frac{\alpha}{RT} dP$$

or
$$d \ln \gamma = d \ln \frac{f}{P}$$

$$= \frac{\alpha}{RT} dP \tag{59}$$

Equation (59) may be integrated now between the limits $P = 0$, where $\gamma = 1$, and P with its value of γ. We get thus

$$\int_{\gamma = 1}^{\gamma} d \ln \gamma = \frac{1}{RT} \int_{P=0}^{P} \alpha dP$$

and
$$\ln \gamma = \frac{1}{RT} \int_{P=0}^{P} \alpha dP \tag{60}$$

Equation (60) is the expression used for obtaining γ. To do this, when an equation of state for the gas is available, α is substituted as a function of P and the integration performed between $P = 0$ and any desired pressure. If, however, only P-V data are available at a given temperature, then α is plotted against P, and the integration is performed graphically.

The general conclusions deduced from such calculations may be summarized as follows. For all gases at relatively low pressures, pressure may be substituted for fugacity in thermodynamic equations without the introduction of any serious error. The exact pressure up to which this is permissible cannot be specified, for it will depend on the nature of the gas in question, the temperature, and the accuracy required. At higher pressures, however, such a substitution may lead to considerable, and even very large, errors, as may be judged from Table 6-1. An idea of the error

TABLE 6-1. RELATION BETWEEN FUGACITY AND PRESSURE
FOR SOME GASES

	H_2 at 100°C		NH_3 at 200°C		CO_2 at 60°C	
P (atm)	f (atm)	γ	f (atm)	γ	f (atm)	γ
0	0	1.00	0	1.000	0	1.000
25	25.3	1.01	23.9	0.954	23.2	0.928
50	51.5	1.03	45.7	0.913	42.8	0.856
100	105	1.05	84.8	0.848	70.4	0.704
200	222	1.11	144	0.720	91.0	0.455
300	351	1.17	193	0.642	112	0.373
400	492	1.23	—	—	—	—
500	650	1.30	—	—	—	—

that may be involved in the substitution of pressure for fugacity in ΔF calculations can be gathered from the following example.

EXAMPLE: Calculate the free energy change accompanying the compression of 1 mole of CO_2 at 60°C from 25 to 300 atm. Using pressures first, we obtain

$$\Delta F = nRT \ln \frac{P_2}{P_1}$$

$$= 1 \times 1.987 \times 333 \times 2.303 \log_{10} \frac{300}{25}$$

$$= 1640 \text{ cal}$$

Using instead the fugacities given in Table 6-1 and Eq. (52b)

$$\Delta F = nRT \ln \frac{a_2}{a_1} = nRT \ln \frac{f_2}{f_1}$$

$$= 1 \times 1.987 \times 333 \times 2.303 \log_{10} \frac{112}{23.2}$$

$$= 1040 \text{ cal}$$

The approximate calculation gives here, therefore, a result 57.7% too high.

STANDARD STATES FOR SOLIDS AND LIQUIDS

As the standard state of a pure solid or liquid is taken the solid or liquid at 1 atm pressure at each temperature. In this state of the solid or liquid $a = 1$ and $F = F^0$. Since, as we have seen, the free energies of solids and liquids are not very dependent on pressure, for these $a = 1$, to a near approximation, at all temperatures and for wide ranges of pressure.

THE REACTION ISOTHERM

With the free energy of a substance defined in terms of the free energy in the standard state and the activity, we may proceed to deduce the equation for the free energy change involved in any type of transformation, whether physical or chemical. Consider in general the reaction

$$aA + bB + \cdots = cC + dD + \cdots \tag{61}$$

If the activities of A and B at the *start* are a_A and a_B, while the activities of C and D at the *end* of the reaction are a_C and a_D respectively, then the free energies of each of these substances *per mole* at a temperature T are given by the expressions

$$F_A = F_A^0 + RT \ln a_A \tag{62a}$$
$$F_B = F_B^0 + RT \ln a_B \tag{62b}$$
$$F_C = F_C^0 + RT \ln a_C \tag{62c}$$
$$F_D = F_D^0 + RT \ln a_D \tag{62d}$$

where the F^0's with the given subscripts represent the free energies at unit activity of the respective species. By definition the free energy change of the reaction, ΔF, is

$$\Delta F = (cF_C + dF_D + \cdots) - (aF_A + bF_B + \cdots)$$

and hence

$$\begin{aligned}
\Delta F &= (cF_C^0 + cRT \ln a_C + dF_D^0 + dRT \ln a_D + \cdots) \\
&\quad - (aF_A^0 + aRT \ln a_A + bF_B^0 + bRT \ln a_B + \cdots) \\
&= [(cF_C^0 + dF_D^0 + \cdots) - (aF_A^0 + bF_B^0 + \cdots)] \\
&\quad + [(cRT \ln a_C + dRT \ln a_D + \cdots) \\
&\quad\quad - (aRT \ln a_A + bRT \ln a_B + \cdots)] \\
&= [(cF_C^0 + dF_D^0 + \cdots) - (aF_A^0 + bF_B^0 + \cdots)] \\
&\quad\quad\quad\quad\quad\quad\quad + RT \ln \frac{a_C^c a_D^d \cdots}{a_A^a a_B^b \cdots} \tag{63}
\end{aligned}$$

The first term on the right of Eq. (63) gives the free energy change of the reaction in the standard state; i.e., the free energy change involved when the starting materials at *unit* activity react to form products also at *unit* activity. If this free energy change in the standard state be designated by ΔF^0, then Eq. (63) becomes

$$\Delta F = \Delta F^0 + RT \ln \frac{a_C^c a_D^d \cdots}{a_A^a a_B^b \cdots} \tag{64}$$

Equation (64) gives the free energy change of a reaction in terms of the free energy change in the standard state and the starting and final activities at any *constant temperature T*. For this reason this most fundamental thermodynamic equation is referred to as the *reaction isotherm*. When the activities of the starting and final materials are all one, the second term on the right of Eq. (64) is zero, and $\Delta F = \Delta F^0$. On the other hand, when the initial and final activities are other than unity, the second term is not zero, and ΔF is given then by the full Eq. (64).

For any reaction ΔF^0 is constant at any given temperature and completely independent of pressure. The variation of ΔF^0 with temperature, like that of ΔF, is given by Eq. (33) or (37), except that in this case the ΔH's and ΔS's are ΔH^0's and ΔS^0's; i.e., the heat and entropy changes are for the reactions in the standard states. Outside of this one difference, calculations involving ΔF^0's are handled in exactly the same manner as those involving ΔF's.

STANDARD FREE ENERGIES OF FORMATION

Because of the direct relation of ΔF^0 to the conditions prevailing at equilibrium, to be discussed in the next chapter, the evaluation of the standard free energy change of a reaction is a problem of prime concern in physical chemistry. Free energy changes may be obtained from the ΔH^0 and ΔS^0 of a reaction at a particular temperature through Eq. (31), and by other methods to be developed further in the text. At this time will be presented a scheme employed to tabulate free energies of formation of compounds, from which free energy changes for all types of reactions may be calculated.

The standard free energy of formation of a compound is defined as the free energy change accompanying the formation of the compound at unit activity from the elements also at unit activity. The equation

$$H_2(g) + \frac{1}{2} O_2(g) = H_2O(l) \qquad \Delta F^0_{298°K} = -56{,}690 \text{ cal} \tag{65}$$

gives the change in free energy on formation of 1 mole of $H_2O(l)$ at unit activity from $H_2(g)$ and $\frac{1}{2} O_2(g)$, both at unit activity. *Assuming now arbitrarily*, as in heats of formation, *that the free energies of formation of the elements in their standard states at 25°C are zero*, $\Delta F^0_{298°K}$ above becomes the standard free energy of formation of liquid water from the elements, and we may write simply that

$$H_2O(l): \qquad \Delta F^0_{298°K} = -56,690 \text{ cal mole}^{-1} \qquad (66)$$

Table 6–2 lists the standard free energies of formation per mole for a number of compounds at 25°C. These free energies of formation may

TABLE 6–2.   STANDARD FREE ENERGIES OF FORMATION AT 25°C

Substance	ΔF^0 (cal/mole)	Substance	ΔF^0 (cal/mole)
HCl(g)	$-22,770$	Ag$_2$O(s)	$-2,590$
H$_2$O(l)	$-56,690$	HgO(s)	$-13,940$
H$_2$O(g)	$-54,640$	PbSO$_4$(s)	$-193,890$
CO(g)	$-32,810$	CH$_4$(g)	$-12,140$
CO$_2$(g)	$-94,260$	C$_2$H$_6$(g)	$-7,860$
NO(g)	$20,720$	C$_2$H$_4$(g)	$16,280$
NO$_2$(g)	$12,390$	C$_2$H$_2$(g)	$50,000$
N$_2$O$_4$(g)	$23,400$	C$_6$H$_6$(g)	$30,990$
H$_2$S(g)	$-7,890$	C$_6$H$_6$(l)	$29,760$
SO$_2$(g)	$-71,790$	CH$_3$OH(l)	$-39,730$
NH$_3$(g)	$-3,980$	C$_2$H$_5$OH(l)	$-41,770$
NaCl(s)	$-91,790$	HCOOH(l)	$-82,700$
AgCl(s)	$-26,220$	CH$_3$COOH(l)	$-93,800$

be used to calculate ΔF^0 of reactions in the same manner that heats of formation are used to calculate heats of reaction. Thus, for instance, taking the data from the table, we find at 25°C for the standard free energy change of the reaction

$$NO(g) + \frac{1}{2} O_2(g) = NO_2(g)$$

$$\Delta F^0_{25°C} = \Delta F^0_{NO_2} - \left[\Delta F^0_{NO} + \frac{1}{2} \Delta F^0_{O_2} \right]$$
$$= 12,390 - 20,720 - 0$$
$$= -8330 \text{ cal}$$

CRITERIA OF EQUILIBRIUM

Although the establishment of equilibrium in systems has been mentioned a number of times, no thermodynamic criteria for equilibrium have as yet been specified. Actually quite a number of such criteria can be deduced in terms of various thermodynamic functions and the conditions under which the equilibrium is established. Here we shall limit ourselves to a discussion of the three criteria which are of widest interest and utility in handling chemical problems.

A system in equilibrium represents a balance of driving and opposing forces, i.e., a condition of reversibility. Further, no work can be obtained from a system in this state. For such a situation Eq. (17) of the last chapter gives

$$T dS - dE - P dV = 0 \qquad (67)$$

If in such a system the volume and internal energy are kept constant, then $dE = dV = 0$, and hence also $dS = 0$. We may write, therefore, that for equilibrium in a system at constant E and V we must have

$$(\partial S)_{E,V} = 0 \qquad (68a)$$

for an infinitesimal change, or

$$(\Delta S)_{E,V} = 0 \qquad (68b)$$

for a finite change. Equations (68a) and (68b) correspond to a maximum in the entropy of the system for the following reason. Spontaneous processes tend to proceed with an increase in entropy, i.e., with dS positive. In turn, for a nonspontaneous process at constant E and V, dS would be negative. Hence at $dS = 0$, the equilibrium, the entropy must be a maximum.

The criterion for equilibrium at constants T and V is expressed in terms of the Helmholtz free energy. Differentiation of Eq. (1) yields

$$dA = dE - S dT - T dS$$

or

$$T dS - dE = -dA - S dT$$

Substituting the latter expression into Eq. (67) we obtain

$$dA + S dT + P dV = 0 \qquad (69)$$

Consequently at constants T and V

$$(\partial A)_{T,V} = 0 \qquad (70a)$$

for an infinitesimal change, and for a finite change

$$(\Delta A)_{T,V} = 0 \qquad (70b)$$

Equations (70a) and (70b) correspond to a minimum in A for the system, since ΔA is negative for a spontaneous change and positive for one which is nonspontaneous.

Differentiation of Eq. (21) gives

$$dF = dA + P dV + V dP$$

or

$$dA + PdV = dF - VdP$$

Insertion of the latter expression into Eq. (69) yields

$$dF + SdT - VdP = 0 \qquad \text{(71)}$$

and hence at constants T and P we get

$$(\partial F)_{T,P} = 0 \qquad \text{(72a)}$$

for an infinitesimal change, and

$$(\Delta F)_{T,P} = 0 \qquad \text{(72b)}$$

for a finite change. For the same reasons as given for the Helmholtz free energy, the Gibbs free energy of a system is a minimum at equilibrium.

The three criteria of equilibrium developed above are equally important statements of the situation in a system at equilibrium, except that they differ from each other in the constraints imposed, i.e., in the variables held constant. Of the three criteria given, the one in terms of F is of widest applicability in chemistry, since most frequently the independent variables dealt with are temperature and pressure.

PHYSICAL EQUILIBRIA INVOLVING PURE SUBSTANCES

For any pure substance in a single phase, such as liquid or gaseous water, any variation in free energy is given by Eq. (27), namely,

$$dF = -SdT + VdP$$

To have equilibrium in the phase, dF has to be zero at constants T and P. Since $dF = 0$ in the above equation when $dT = dP = 0$, the phase is in equilibrium when the pressure and temperature are constant and uniform throughout the phase.

Consider a more interesting situation, the transition of a pure substance from one phase into another. Examples of such a transformation are transition of a substance from one crystalline form to another, change of a solid to a liquid, sublimation of a solid, and vaporization of a liquid. All such changes can be represented by the equation

$$A_1 = A_2$$

for which ΔF is given by

$$\Delta F = F_2 - F_1 \qquad \text{(73)}$$

where F_2 and F_1 are the molar free energies of the substance in the final and initial states respectively. All such transformations will attain equilibrium when $\Delta F = 0$ at constant temperature and pressure. Imposing this condition on Eq. (73), we see that $F_2 = F_1$; i.e., *all such transformations will be in equilibrium at constant temperature and pressure when the molar free energies of the substance are identical in both phases.*

Suppose that we have two phases in equilibrium and that the pressure of the system is changed by dP. The temperature of the system then will have to change by dT in order to preserve the equilibrium. In such a situation dP and dT can be related as follows: Since $F_2 = F_1$, we have also $dF_2 = dF_1$. But, by Eq. (27) $dF_2 = -S_2dT + V_2dP$ and $dF_1 = -S_1dT + V_1dP$. Equating these expressions we get

$$-S_2dT + V_2dP = -S_1dT + V_1dP$$
$$(V_2 - V_1)dP = (S_2 - S_1)dT$$
$$\frac{dP}{dT} = \frac{(S_2 - S_1)}{(V_2 - V_1)}$$
$$= \frac{\Delta S}{\Delta V} \tag{74}$$

where $\Delta S = S_2 - S_1$ is the change in entropy and $\Delta V = V_2 - V_1$ is the change in volume for the process. Further, Eq. (20) with $\Delta F = 0$ yields $\Delta S = \Delta H/T$, where ΔH is the change in enthalpy for the reversible transformation occurring at temperature T. Substituting this value of ΔS into Eq. (74) we obtain

$$\frac{dP}{dT} = \frac{\Delta H}{T\Delta V} \tag{75}$$

Equation (75), known as the *Clapeyron equation*, relates the change in temperature which must accompany a change in pressure occurring in a system containing two phases of a pure substance in equilibrium. Again, the equation shows that dP/dT is related directly to the enthalpy of transition and inversely to the temperature and the volume change accompanying the transformation.

To integrate Eq. (75), ΔH and ΔV must be known as functions of temperature or pressure. Since such information is usually unavailable, Eq. (75) is used either in differential form, i.e.,

$$\frac{P_2 - P_1}{T_2 - T_1} = \frac{\Delta H}{T\Delta V} \tag{76}$$

where T is taken as the average of T_1 and T_2, or the equation is integrated on the supposition that ΔH and ΔV are constant. We get then

$$\int_{P_1}^{P_2} dP = \frac{\Delta H}{\Delta V} \int_{T_1}^{T_2} \frac{dT}{T}$$
$$P_2 - P_1 = \frac{\Delta H}{\Delta V} \ln \frac{T_2}{T_1} \tag{77}$$

Equation (77) is somewhat more precise than Eq. (76), and so is preferred for calculations involving appreciable temperature changes.

USE OF CLAPEYRON EQUATION

The Clapeyron equation in the form given above is employed primarily for calculations involving condensed phase equilibria, i.e., equilibria involving solids and liquids. Its use can best be illustrated by the following discussion of the fusion of solids. It was shown in Chapter 2 that the fusion of a pure solid occurs at a fixed temperature for a given pressure, and is accompanied by an absorption of heat equal to $\Delta H_f = H_l - H_s$, where the ΔH_f is the molar heat of fusion, and H_l and H_s are, respectively, the molar enthalpies of the liquid and solid. At the same time the process is accompanied by a volume change $\Delta V_f = V_l - V_s$, where V_l and V_s are the molar volumes of the liquid and solid. If we now change the total pressure on the system from P_1 to P_2, then the temperature at which equilibrium melting will occur is changed from T_1 to T_2. To calculate the extent of the change by means of the Clapeyron equation, consider specifically the following example. For acetic acid the melting point at 1 atm pressure is 16.61°C, $\Delta H_f = 2800$ cal/mole, and $\Delta V_f = 9.614$ cc/mole. What will be the melting point at 11 atm pressure? From Eq. (76) we have

$$t_2 - t_1 = \frac{T(P_2 - P_1)\Delta V}{(41.29)\Delta H}$$

where the factor 41.29 converts calories to cc-atm. Upon introducing the data for the acid we get

$$t_2 - t_1 = \frac{289.76(10)(9.614)}{2,800(41.29)} = 0.241°C$$

and hence

$$t_2 = 16.61 + 0.24$$
$$= 16.85°C$$

It will be observed that the effect of pressure on the melting point is not large, and for small pressure variations may be disregarded.

Water is the interesting example of a substance having V_s greater than V_l, and hence the melting point of water is lowered by the application of pressure.

The temperature at which one solid modification of a substance undergoes transformation to another form is, like fusion, a function of the pressure. For any such transformation the variation of the transition temperature T with pressure P is again given by the Clapeyron equation, with ΔH and ΔV now being the enthalpy and volume change accompanying the transformation, i.e., ΔH_t and ΔV_t. For such transitions ΔH_t is always positive when the transformation is from a form stable at lower temperatures to one which is stable at higher ones, and negative for the reverse changes.

THE VAPOR PRESSURE OF LIQUIDS

Attention has already been directed to the fact that a liquid placed in a container will partially evaporate to establish a pressure of vapor above the liquid. The pressure established depends on the nature of the liquid and is, at equilibrium, constant at any given temperature. This constant vapor pressure is called the *saturated vapor pressure* of the liquid at the particular temperature. As long as this vapor pressure is maintained, the liquid exhibits no further tendency to evaporate. At any lower pressure the liquid will evaporate into the gas phase, while at any higher pressure vapor will tend to condense until the equilibrium pressure is re-established.

The vaporization of a liquid is accompanied by absorption of heat. For any liquid at a given temperature the amount of heat required per given weight of liquid is a definite quantity known as the *heat of vaporization* of the liquid. It is the difference in the enthalpies of vapor and liquid respectively, namely, $\Delta H_v = H_v - H_l$, where ΔH_v is the heat of vaporization, H_v the enthalpy of the vapor, and H_l the enthalpy of the liquid. For an evaporation ΔH_v is always positive, while for a condensation ΔH_v is always negative and equal numerically to the heat absorbed in the vaporization. As may be expected from the definition of ΔH, ΔH_v represents the difference in the internal energy of vapor and liquid, $\Delta E_v = E_v - E_l$, and the work involved in the expansion from liquid to vapor; i.e.,

$$\Delta H_v = \Delta E_v + P\Delta V_v \tag{78}$$

where P is the vapor pressure and $\Delta V_v = V_v - V_l$.

The various procedures available for measuring the vapor pressure of a liquid may be classified generally into static and dynamic methods. In the static methods the liquid is permitted to establish its vapor pressure without being disturbed in any way, while in the dynamic methods the liquid either is boiled or has a stream of inert gas passing through it. The line of demarcation between these two classifications is not always sharp, and a particular procedure may actually be a combination of the two.

The isoteniscope method of Menzies and Smith is precise, flexible, and convenient for the measurement of the vapor pressures of a substance over a range of temperatures. A simple laboratory setup is illustrated in Figure 6–2. The isoteniscope bulb B is filled one-half to three-quarters full with the liquid to be studied, and the U-shaped portion of the tube, C, is filled to a depth of 2 or 3 cm with the same liquid. The isoteniscope is then attached to the rest of the apparatus and surrounded by a water bath A, whose temperature is measured by thermometer D. E is a barometric leg for measuring the pressure in the apparatus, while G is a large bottle to smooth out pressure fluctuations in the system. This bottle can be connected alternately to a suction pump or the air. In operation the system is evacuated until the liquid boils vigorously at B to expel all

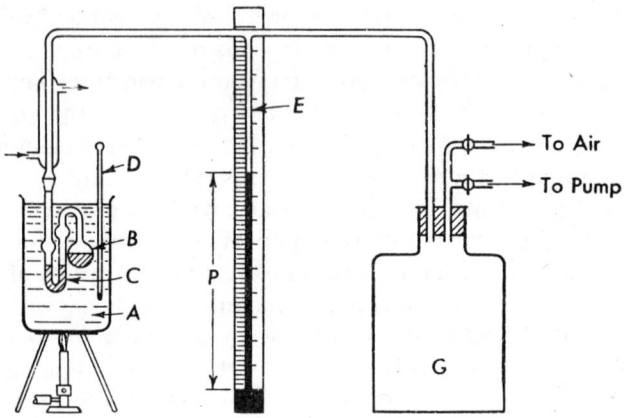

Figure 6–2.   Isoteniscope assembly for determination of vapor pressure.

air from BC. The bath A is then adjusted to the desired temperature, and air is slowly admitted to the system until the liquid levels in the U-tube C are exactly equal. Under these conditions the pressure on either side of the U-tube must be the same. Hence, the vapor pressure in B must be the same as the pressure in the rest of the apparatus and can be obtained from the reading of the barometer and the mercury column at E. The difference between the barometric pressure and that at E is the vapor pressure of the liquid in B at the temperature of the bath. Readings at different temperatures can be obtained by merely changing the temperature of bath A and repeating the operation.

A simple dynamic method is shown in Figure 6–3. The liquid in question, B, is boiled, after deaeration, under a measured external pressure, and the temperature of the condensing vapor is read from thermometer T. The barometric pressure minus the pressure P is the pressure at which the liquid boils, and this is the vapor pressure of the liquid at the temperature T. By changing the pressure P, the liquid may be boiled at different temperatures, and the vapor pressures at these temperatures thus obtained. The function of bottle C is to condense any escaping vapor and thus prevent the distillation of liquid into the mercury manometer.

A more elaborate dynamic method proposed by Walker involves the saturation of some inert gas, such as nitrogen, with vapor by bubbling a measured quantity of the gas through the liquid at constant temperature and subsequently condensing out, or absorbing, and weighing the vapor of the liquid in question. If P_t is the total pressure in the apparatus at saturation, n_g the moles of gas passed through, and $n_v = W_v/M_v$ the number of moles of vapor collected, then the partial pressure of the

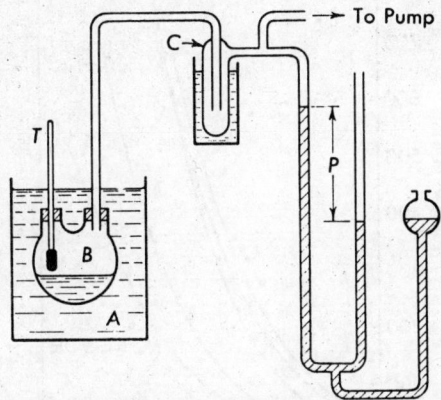

Figure 6–3.   Boiling point method for determination of vapor pressure.

vapor P, which is the same as the vapor pressure of the liquid at saturation, is

$$P = \left(\frac{n_v}{n_g + n_v}\right) P_t \tag{79}$$

This method is as a rule much more tedious than the others mentioned, but with care can be made to yield excellent results. It is especially useful in determinations of partial vapor pressures of mixtures of liquids.

VARIATION OF VAPOR PRESSURE WITH TEMPERATURE

The vapor pressure of a liquid, though constant at a given temperature, increases continuously with increase in temperature up to the critical point of the liquid. Above the critical temperature the liquid no longer exists, and consequently the concept of a saturated vapor pressure is no longer valid. In terms of kinetic theory the increase in vapor pressure with temperature is easily understandable. As the temperature increases, a greater proportion of the molecules acquire sufficient energy to escape from the liquid, and consequently a higher pressure is necessary to establish equilibrium between vapor and liquid. Above the critical temperature the escaping tendency of the molecules is so high that no applied pressure is sufficient to keep any of them in the liquid state, and the whole mass persists as a gas.

The manner in which the vapor pressure varies with temperature is shown graphically in Figure 6–4. The vapor pressure increases slowly at

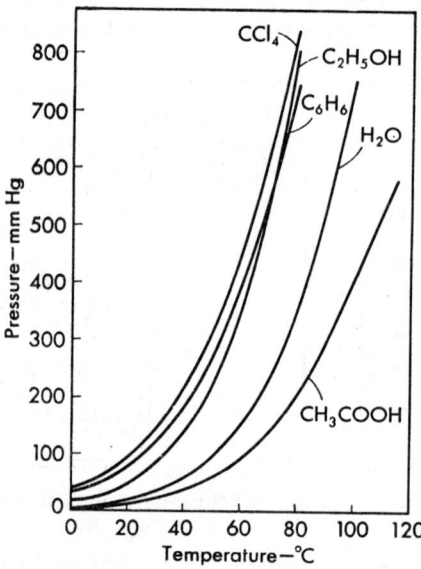

Figure 6–4.   Variation of vapor pressure with temperature.

the lower temperatures, and then quite rapidly, as is shown by the steep rise in the curves. This variation of vapor pressure with temperature can be expressed mathematically by means of the Clausius-Clapeyron equation. For the transition of liquid to vapor P is the vapor pressure at temperature T, $\Delta H = \Delta H_v$ the heat of vaporization of a given weight of liquid, and $V_1 = V_l$ the liquid volume, while $V_2 = V_g$ is the volume of the same weight of vapor. Consequently, for a vaporization the Clapeyron equation can be written as

$$\frac{dP}{dT} = \frac{\Delta H_v}{T(V_g - V_l)} \tag{80}$$

At temperatures not too near the critical, V_l is quite small compared with V_g and may be neglected. Thus, at 100°C, V_g for water is 1671 cc per gram, while V_l is only 1.04 cc per gram. Further, if we assume that the vapor behaves essentially as an ideal gas, then V_g per mole is given by

$V_g = RT/P$, and Eq. (80) becomes

$$\frac{dP}{dT} = \frac{\Delta H_v}{T V_g} = \frac{\Delta H_v P}{RT^2}$$

$$\frac{1}{P}\frac{dP}{dT} = \frac{\Delta H_v}{RT^2}$$

$$\frac{d \ln P}{dT} = \frac{\Delta H_v}{RT^2} \tag{81}$$

Equation (81) is known as the *Clausius-Clapeyron equation*. Before this equation can be integrated, ΔH_v, which is now the heat of vaporization per mole, must be known as a function of temperature. If we assume as an approximation, however, that over the interval in question ΔH_v remains essentially constant, then integration yields

$$\ln P = \frac{\Delta H_v}{R} \int \frac{dT}{T^2} + C'$$

$$= -\frac{\Delta H_v}{R}\left(\frac{1}{T}\right) + C' \tag{82a}$$

or

$$\log_{10} P = -\frac{\Delta H_v}{2.303\,R}\left(\frac{1}{T}\right) + C \tag{82b}$$

where C' and C are constants of integration. Equation (82) predicts that the logarithm of the vapor pressure should be a function of the reciprocal of the absolute temperature. Further, comparison of the equation with the equation of a straight line, namely, $y = mx + b$, suggests that if $\log_{10} P$ for any liquid is plotted against $1/T$, the plot should be a straight line with slope $m = (-\Delta H_v/2.303\,R)$, and y−intercept $b = C$. That this is in accord with the facts may be seen from Figure 6–5, where the same data as were shown in Figure 6–4 are now plotted as $\log_{10} P$ vs. $1/T$. From the slopes of the lines the heats of vaporization of the various liquids follow as

$$\text{slope} = m = \frac{-\Delta H_v}{2.303\,R}$$

and, consequently,

$$\Delta H_v = -2.303\,Rm = -4.576\,m \text{ cal mole}^{-1}$$

The heat of vaporization is obtained in calories per mole when the value of R used is in calories per mole per degree, namely, $R = 1.987$. The heat of vaporization of a liquid thus calculated will be the mean value of the heat of vaporization over the temperature interval considered.

To obtain C in Eq. (82b) it is best to substitute in the equation the calculated value of ΔH_v and a value of $\log_{10} P$ and $1/T$ corresponding to a point on the line, and solve for C. Once ΔH_v and C for a given liquid are known, the vapor pressure of the liquid at any temperature over the range of the equation can easily be calculated by merely substituting the desired value of T.

A word about the units of the various terms in Eq. (82b) is in order. Since ΔH_v and R are both in calories, the first term on the right-hand side of the equation is independent of the units in which P is expressed. C, however, is not, and its magnitude will depend on the units of P. Consequently, in setting up an equation for the vapor pressure, it is

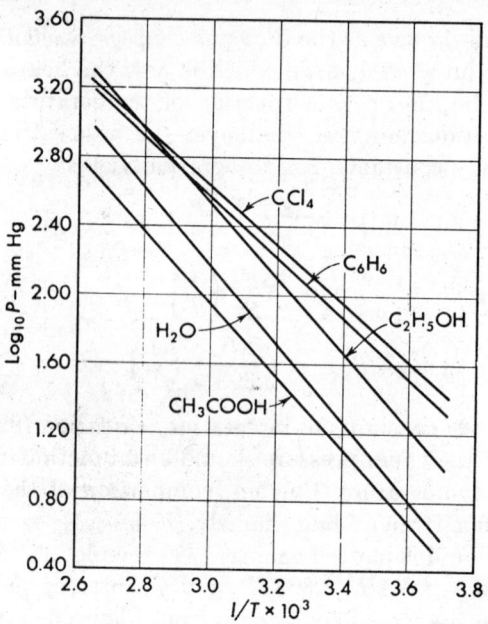

Figure 6–5.   Plot of $\log_{10} P$ vs. $1/T$ for several liquids.

essential to state clearly whether P is expressed in atmospheres, millimeters of mercury, or some other unit. Similarly, in using an equation from some reference source, attention should be paid to the units in which the equation is expressed.

An alternate form of Eq. (82b) may be obtained by integrating Eq. (81) between the limits P_1 and P_2 corresponding to the temperatures T_1 and T_2. Then

$$\int_{P_1}^{P_2} d \ln P = \frac{\Delta H_v}{R} \int_{T_1}^{T_2} \frac{dT}{T^2}$$

$$\ln \frac{P_2}{P_1} = \frac{\Delta H_v}{R} \left[-\frac{1}{T} \right]_{T_1}^{T_2}$$

$$= \frac{\Delta H_v}{R} \left[\frac{T_2 - T_1}{T_1 T_2} \right]$$

or

$$\log_{10} \frac{P_2}{P_1} = \frac{\Delta H_v}{2.303 \, R} \left[\frac{T_2 - T_1}{T_1 T_2} \right] \tag{83}$$

Equation (83) permits the calculation of ΔH_v from the values of the vapor pressure at two temperatures; or, when ΔH_v is known, P at some desired temperature may be calculated from a single available vapor pressure at a given temperature.

EXAMPLE: At 373.6°K and 372.6°K the vapor pressures of $H_2O(l)$ are 1.018 and 0.982 atm respectively. What is the heat of vaporization of water? Employing Eq. (83), we have

$$\log_{10} \frac{P_2}{P_1} = \frac{\Delta H_v}{2.303\,R}\left[\frac{T_2 - T_1}{T_1 T_2}\right]$$

$$\log \frac{1.018}{0.982} = \frac{\Delta H_v}{2.303 \times 1.987}\left[\frac{373.6 - 372.6}{373.6 \times 372.6}\right]$$

$$\Delta H_v = 9790 \text{ cal/mole}$$
$$= 540 \text{ cal/g}$$

The experimentally observed value at 373.21°K is 538.7 cal per gram.

The heats of vaporization of liquids may also be measured directly in a calorimeter by condensing a definite weight of vapor and observing the temperature rise of the calorimeter, or by supplying to a liquid a definite amount of electrical energy and measuring the weight of liquid vaporized thereby. The heat of vaporization of a liquid decreases in general with increase in temperature and becomes zero at the critical temperature. For acetic acid, however, ΔH_v increases at first, goes through a maximum at about 120°C, and then decreases with increase in temperature. In magnitude the heats of vaporization of various liquids differ widely. Especially is the heat of vaporization of water abnormally high, a fact which speaks for the complexity of the liquid.

Equation (82b) will not be strictly valid over wide temperature ranges, particularly because of the assumption of constancy of ΔH_v. When data are available on the variation of ΔH_v with temperature, Eq. (81) may be integrated to give better agreement with experiment and over a much wider temperature range. When such data are not available, vapor pressure–temperature data are usually correlated by empirical equations of the form

$$\log P = A - \frac{B}{T} + C \log T + DT + \cdots \qquad (84)$$

where A, B, C, and D are fitted constants.

THE BOILING POINT OF LIQUIDS

The *normal* boiling point of a liquid is the temperature at which the vapor pressure of the liquid equals 760 mm Hg pressure, or 1 atm. However, a liquid can be made to boil at any temperature between its freezing point and the critical temperature by merely raising or lowering, as the case may be, the external pressure on the liquid. Therefore, it may be stated in general that the boiling point of a liquid is the temperature at which the vapor pressure of a liquid becomes equal to the external pressure acting upon the surface of the liquid. Boiling is characterized by the formation within the liquid of bubbles of vapor which rise and escape into the vapor phase.

The change in boiling point produced by a change in pressure may be calculated with the aid of the Clausius-Clapeyron equation. If ΔH_v for the liquid is known, and if T_1 is the boiling point at pressure P_1, the boiling point T_2 at pressure P_2 follows from Eq. (83). When however, ΔH_v is not known, then its value may be estimated from *Trouton's rule*. This rule states that

$$\frac{\Delta H_v}{T_b} = \text{constant} \tag{85}$$

i.e., *that the ratio of the molar heat of vaporization of a liquid to its normal boiling point on the absolute scale is the same constant for all liquids.*

An alternate way of stating the rule is that the entropy of vaporization

TABLE 6–3.   TROUTON'S CONSTANTS FOR LIQUIDS

Liquid	ΔH_v at Boiling Point (cal mole^{-1})	Normal Boiling Point (°K)	Trouton Constant
Nitrogen	1,338	77.4	17.3
Oxygen	1,636	90.2	18.1
Ammonia	5,570	239.8	23.2
Ethyl ether	6,220	307.8	20.2
Acetone	7,230	329.4	21.9
Methyl acetate	7,270	330.5	22.0
Chloroform	7,040	334.4	21.0
n-Hexane	6,850	341.9	20.0
Carbon tetrachloride	7,140	350.0	20.4
Benzene	7,350	353.3	20.8
Nitrobenzene	9,660	484.1	20.0
Mercury	14,200	629.8	22.5
Zinc	23,700	1,180	20.1
Hydrogen	216	20.5	10.5
Formic acid	5,520	374.0	14.8
Acetic acid	5,810	391.3	14.8
Water	9,710	373.2	26.0
n-Propyl alcohol	9,880	370.4	26.7
Ethyl alcohol	9,410	351.6	26.8
Bismuth	46,100	1,723	26.8
Tin	77,700	2,533	30.7

of all liquids at their normal boiling points is a constant. For $\Delta \bar{H}_v$ in calories per mole and T in °K the value of the constant is usually taken as 21. That this rule is only approximate may be seen from Table 6–3. It holds fairly well for the liquids shown in the first section of the table, but the constants are low for hydrogen and the acids, and high for water, alcohols, and some of the metals.

SUBLIMATION PRESSURE OF SOLIDS

Many solids exhibit discernible and measurable vapor pressures. The equilibrium vapor pressure of a solid, known as the sublimation pressure, is entirely analogous to the saturated vapor pressure of a liquid. For any given solid the vapor pressure is constant at any given temperature, and the process is accompanied by an absorption of heat called the heat of sublimation ΔH_s, and by a volume change $\Delta V_s = V_g - V_s$, where V_s is the molar volume of the solid. If we assume V_s to be negligible compared with V_g and also that the vapor behaves as an ideal gas, then the variation of the sublimation pressure of a solid with temperature is again given by the Clausius-Clapeyron equation. Now T is the sublimation temperature and ΔH_s is the thermal change. For sublimations this equation is handled in exactly the same manner as for vaporizations, and the same kinds of information can be obtained from the equations as for vaporization processes.

At any given temperature the sublimation of a solid is equivalent to the fusion of the solid and vaporization of the resulting liquid. Since ΔH of a process is independent of the path, we must have at any given T that

$$\Delta H_s = \Delta H_f + \Delta H_v \tag{86}$$

Hence, when any two of these quantities are known, the third can be calculated by means of Eq. (86).

REFERENCES

See the references listed at the end of Chapters 3 and 4. Also:

1. A. Weissberger, *Physical Methods of Organic Chemistry*, Interscience Publishers, Inc., New York, 1959, Chaps. 7, 8, and 9.

PROBLEMS

1. The molar volume of $C_6H_6(l)$ is 88.9 cc at 20°C and 1 atm pressure. Assuming the volume to be constant, find ΔF and ΔA for compression of 1 mole of the liquid from 1 to 100 atm.   *Ans.* $\Delta F = 213$ cal mole^{-1}; $\Delta A = 0$.

2. Two moles of an ideal gas are compressed isothermally and reversibly at 100°C from a pressure of 10 to 25 atm. (a) Find the values of ΔF and ΔA for the process. (b) What are the values of ΔE, ΔH, ΔS, q, and w for the process?
Ans. (a) $\Delta F = 1359$ cal.

3. Three moles of an ideal gas are allowed to expand freely at 300°K from a volume of 100 to 1000 liters. (a) What are the values of ΔF and ΔA for the process? (b) What are the values of ΔE, ΔH, ΔS, q, and w?

4. A certain gas obeys the equation

$$PV_m = RT + a_1P + a_2P^2$$

where V_m is the molar volume, and a_1 and a_2 are constants dependent only on the temperature. What will be the expressions for ΔF and ΔA when 1 mole of the gas is compressed from a pressure P_1 to a pressure P_2 at temperature T?

5. Calculate the difference in calories between ΔF and ΔA at 25°C for the reaction

$$H_2 \text{ (g, 1 atm)} + \frac{1}{2} O_2(\text{g, 1 atm}) = H_2O(l)$$

$Ans.$ $\Delta F - \Delta A = -888$ cal.

6. For a certain process $\Delta F = -12,000$ cal and $\Delta H = -17,500$ cal at 400°K. Find for the process at this temperature ΔS, $[\partial(\Delta F)/\partial T]_P$, and $[\partial(\Delta A)/\partial T]_V$.

7. Assuming that the process mentioned in problem 6 is carried out reversibly, how much heat is evolved during its performance?

8. For a certain reaction $\Delta F = 13,580 + 16.1\ T \log_{10} T - 72.59\ T$. Find ΔS and ΔH of the reaction at 25°C. $Ans.$ $\Delta S = 25.74$ eu; $\Delta H = 11,490$ cal.

9. For the reaction $H_2O(l) = H_2O(\text{g, 23.76 mm Hg}) \Delta F_{25°c} = 0$. Assuming the water vapor to behave as an ideal gas, find ΔF at 25°C for the reaction

$$H_2O(l) = H_2O(\text{g, 1 atm})$$

$Ans.$ $\Delta F = 2050$ cal.

10. For the reaction, $H_2(\text{g, 1 atm}) + Cl_2(\text{g, 1 atm}) = 2\ HCl(\text{g, 1 atm}) \Delta F$ at 25°C is $-45,400$ cal. Find ΔF for the process

$$H_2(\text{g, 2 atm}) + Cl_2(\text{g, 1 atm}) = 2\ HCl(\text{g, 0.1 atm})$$

11. For a certain reaction the change in volume as a function of pressure is given by $\Delta V = a_1 + a_2P$, where a_1 and a_2 are constants dependent only on the temperature. What will be the change in free energy produced by a change in pressure from P_1 to P_2 at constant T?

12. At -50°C the molar volumes of N_2(g) at various pressures are as follows:

P (atm)	V (l/mole)
1	18.28
20	0.890
40	0.434
60	0.284
100	0.167
200	0.0879
300	0.0671
400	0.0579
500	0.0526

Determine the activity coefficients of the gas at 100, 200, and 500 atm.

13. Using Table 6–1, find ΔF for the isothermal compression of 1 mole of $H_2(g)$ from 50 to 500 atm at 100°C. Compare the result with the value obtained on assumption that $H_2(g)$ behaves as an ideal gas.

14. A certain gas at temperature T obeys the relation $PV = RT + BP + CP^2$. Deduce the expression for $\ln \gamma$ of the gas as a function of P at this temperature.

15. If in problem 14 $T = 223.2°K$, $B = -3.69 \times 10^{-2}$, and $C = 1.79 \times 10^{-4}$ for volume in liters and pressure in atmospheres, what will be the activity coefficient of the gas at $P = 100$ atm?   *Ans.* $\gamma = 0.858$.

16. From the data given in problem 1 calculate the activity of $C_6H_6(l)$ at 20°C and 100 atm pressure.

17. Derive the expression for the variation of activity with temperature at constant pressure.

18. The standard free energy of formation of $H_2O(l)$ at 25°C is $-56,690$ cal. Find the free energy of formation of H_2O at 25°C from H_2 at a partial pressure of 0.01 atm and O_2 at a partial pressure of 0.25 atm.   *Ans.* $\Delta F = -53,550$ cal.

19. For the formation of $C_2H_5OH(l)$ $\Delta F^0_{25°C} = -41,770$ cal. Find ΔF at 25°C for the reaction

$$2\,C(s) + 3\,H_2(g, f = 50\ \text{atm}) + \frac{1}{2}\,O_2(g, f = 100\ \text{atm}) = C_2H_5OH(l)$$

20. Deduce the criteria of equilibrium for the following conditions:

(a) Constant H and P
(b) Constant S and V
(c) Constant S and P

21. Deduce the criteria of equilibrium for the following conditions:

(a) Constant S and E
(b) Constant S and H

22. From the data of Table 6–2 calculate ΔF^0 for the following reactions at 25°C:

(a) $C_2H_5OH(l) + O_2(g) = CH_3COOH(l) + H_2O(l)$
(b) $2\,CO_2(g) = 2\,CO(g) + O_2(g)$

Which of the above reactions as written is spontaneous in the standard state?
Ans. (a) $\Delta F^0 = -108,720$ cal.

23. Using the absolute entropies from Table 5–2, calculate ΔH^0 at 25°C for the reactions given in the preceding problem.

24. From the following series of reactions find the free energy of formation of $N_2O_4(g)$ at 25°C:

$$\frac{1}{2}\,N_2(g) + \frac{1}{2}\,O_2(g) = NO(g) \qquad \Delta F^0_{298°K} = 20,720\ \text{cal}$$

$$NO(g) + \frac{1}{2}\,O_2(g) = NO_2(g) \qquad \Delta F^0_{298°K} = -8330\ \text{cal}$$

$$2\,NO_2(g) = N_2O_4(g) \qquad \Delta F^0_{298°K} = -1380\ \text{cal}$$

25. From the appropriate absolute entropies and heats of formation calculate the standard free energy change at 25°C for the reaction:

$$CO(g) + H_2O(l) = CO_2(g) + H_2(g)$$

26. For the sublimation $Au(s) = Au(g)$ $\Delta H^0_{298.2°K} = 90,500$ cal/mole and $\Delta F^0_{298.2°K} = 81,000$ cal/mole. Further,

$$Au(g): \quad C_p = 5.00 \text{ cal mole}^{-1} \text{ degree}^{-1}$$
$$Au(s): \quad C_p = 5.61 + 1.44 \times 10^{-3} T \text{ cal mole}^{-1} \text{ degree}^{-1}$$

From these data find an expression for ΔF^0 as a function of T.

$$Ans. \ \Delta F^0 = 90,740 + 1.40 \ T \log_{10} T - 36.23 \ T + 0.72 \times 10^{-3} \ T^2$$

27. For the reaction $MoS_2(s) + 2 H_2(g) = Mo(s) + 2 H_2S(g)$ $\Delta H^0_{25°C} = 46,670$ cal, $\Delta F^0_{25°C} = 38,460$ cal, and ΔC_p is given by

$$\Delta C_p = -12.95 + 3.75 \times 10^{-3} \ T - \frac{0.503 \times 10^5}{T^2}$$

Deduce the expression for ΔF^0 as a function of T.

28. For the reaction $Cu(s) = Cu(g)$

$$\Delta H^0 = 81,730 - 0.47 \ T - 0.731 \times 10^{-3} \ T^2$$
$$\Delta S^0 = 34.94 - 1.08 \log_{10} T - 1.46 \times 10^{-3} \ T$$

Set up the expression for ΔF^0 as a function of T.

29. For the reaction $FeCO_3(s) = FeO(s) + CO_2(g)$

$$\Delta F^0 = 18,660 - 14.42 \ T \log_{10} T - 6.07 \ T + 8.24 \times 10^{-3} \ T^2$$

Find ΔH^0 and ΔS^0 for the reaction at 25°C.

$$Ans. \ \Delta H^0 = 19,790 \text{ cal/mole}; \ \Delta S^0 = 43.1 \text{ eu.}$$

30. For the reaction $C(s, graphite) + S_2(g) = CS_2(g)$

$$\Delta F^0 = -5040 - 7.67 \ T \log_{10} T + 1.51 \times 10^{-3} \ T^2 + \frac{1.106 \times 10^5}{T} + 21.58 \ T$$

Further,

$$C(s, graphite): \quad C_p = 2.673 + 2.617 \times 10^{-3} \ T - \frac{1.169 \times 10^5}{T^2} \text{ cal mole}^{-1} \text{ degree}^{-1}$$

$$CS_2(g): \quad C_p = 13.75 + 0.49 \times 10^{-3} \ T - \frac{3.38 \times 10^5}{T^2} \text{ cal mole}^{-1} \text{ degree}^{-1}$$

From these data find the equation for C_p of $S_2(g)$ as a function of the temperature.

31. For the reaction $A_2(g) + 2B(s) = 2AB(g)$, ΔF^0 is given by

$$\Delta F^0 = A + BT \ln T + CT + DT^2$$

Find the expressions for (a) ΔA^0, (b) ΔH^0, (c) ΔE^0, (d) ΔS^0, (e) ΔC_p^0, and (f) ΔC_v^0 as functions of the temperature.

32. The densities of liquid and solid Hg are respectively 13.70 and 14.19 g/cc at the melting point, $-38.87°C$. The heat of fusion is 566 cal/g atom. Find the change in melting point per atmosphere change in pressure.

Ans. 0.0051°C/atm.

33. *m*-Dinitrobenzene melts at 89.8°C under a pressure of 1 atm, and at 114.8°C under a pressure of 968 atm. If the heat of fusion is 24.7 cal/g, what is the change in volume on fusion?

34. The heat of fusion of ice is 79.7 cal/g at 0°C. The densities of ice and water at the same temperature are, respectively, 0.9168 and 0.9999 g/cc. Calculate the melting point of ice at 325 atm pressure, and compare your answer with the observed value of $-2.5°C$.

35. NH_4NO_3 undergoes a transition from one solid modification to another at 125.5°C at 1 atm, and at 135.0°C at 1000 atm pressure. The form stable at higher temperatures has an average volume of 0.0126 cc/g greater than the other modification over the pressure range studied. From these data calculate the heat of transition.

Ans. 1040 cal/mole.

36. AgI exists in two forms, α and β, which are in equilibrium at 146.5°C at 1 atm pressure. For the change of α to β $\Delta H_t = 1530$ cal/mole, while $\Delta V = -2.2$ cc/mole. Find the pressure at which the transition temperature will be 145.0°C.

37. In measuring the vapor pressure of a liquid by means of the isoteniscope, the height of the Hg in the manometer was found to be 53.32 cm at 40°C, and 39.40 cm at 55°C. The barometric pressure was 741.0 mm. What are the vapor pressures of the liquid at the two temperatures?

38. In measuring the vapor pressure of ethanol by the gas saturation method, the following data were taken:

$$\begin{aligned}
\text{Volume of } N_2 \text{ at 740 mm and } 30°C &= 5.6 \text{ liters}\\
\text{Barometric pressure} &= 740 \text{ mm}\\
\text{Temperature} &= 30°C\\
\text{Loss in weight of alcohol} &= 1.193 \text{ g}
\end{aligned}$$

Find the vapor pressure of ethanol at 30°C.

Ans. 78.2 mm Hg.

39. At the normal boiling point, 61.5°C, the heat of vaporization of $CHCl_3$ is 59.0 cal/g. Assuming that the vapor behaves as an ideal gas and that the volume of the liquid is negligible compared to that of the vapor, what is ΔE per mole for the vaporization process?

40. The vapor pressure of ethanol is 135.3 mm at 40°C and 542.5 mm at 70°C. Calculate the molar heat of vaporization and the vapor pressure of ethanol at 50°C.

Ans. $\Delta H_v = 9880$ cal/mole; $P_{50°C} = 221.0$ mm.

41. The heat of vaporization of ethyl ether is 83.9 cal/g while its vapor pressure at 30°C is 647.3 mm. What will be the vapor pressure at 0°C?

42. The vapor pressure of CH_3Cl between -47 and $-10°C$ can be represented by the equation:

$$\log_{10} P_{mm} = \frac{-1149}{T} + 7.481$$

What is the heat of vaporization of this liquid in calories per gram?

Ans. 104.1 cal/g.

43. A liquid is observed to boil at 120°C under a pressure of 725 mm. Its molar heat of vaporization is 8200 cal/mole. Calculate the normal boiling point of the liquid.

44. The normal boiling point of C_6H_5Br is 156.15°C. Using Trouton's rule, find the vapor pressure at 100°C, and compare with the observed value of 141.1 mm.

45. CCl_4 exhibits the following vapor pressures at the indicated temperatures:

t (°C)	30	50	70	100
P (mm Hg)	142.3	314.4	621.1	1463.0

Set up the equation giving $\log_{10} P$ as a function of the temperature.

46. Beattie and Marple [*J. Am. Chem. Soc.*, **72**, 1450 (1950)] give the following equation for the vapor pressure of 1-butene as a function of the temperature between -75 and 125°C:

$$\log_{10} P \text{ (atm)} = 5.475462 - \frac{1343.516}{T} - 167.515 \times 10^{-5} \, T$$

Find: (a) the expression for ΔH_v as a function of the temperature; (b) ΔH_v at 300°K; (c) the normal boiling point of the liquid.   *Ans.* (c) 267°K.

47. The heat of vaporization of a certain liquid as a function of the temperature is given by the relation

$$\Delta H_v = a + bT + cT^2$$

where a, b, and c are constants. What will be the expression for $\ln P$ as a $f(T)$?

48. To measure the heat of vaporization of a liquid calorimetrically, 13.5200 g of vapor of the liquid, initially at 45.35°C, were passed into a calorimeter and condensed there. During this process the temperature of the calorimeter rose from 25.015 to 26.525°C. If the heat capacity of the calorimeter is 453.25 cal/degree, and if the specific heat of the vapor is 0.180 cal g^{-1} degree^{-1}, what is the heat of vaporization of the liquid in calories per gram?   *Ans.* 47.1 cal/g.

49. The vapor pressure of solid CO_2 is 76.7 mm at $-103°C$, and 1 atm at $-78.5°C$. Calculate the heat of sublimation of CO_2.   *Ans.* 6160 cal/mole.

50. At 630°K the heat of sublimation of $ZrBr_4$ is 25,800 cal/mole, while the vapor pressure is 1 atm. What will be the sublimation pressure at 700°K?

51. At 0°C the heat of sublimation of ice is 675.7 cal/g, while the heat of vaporization of water is 595.9 cal/g. Calculate the rates of change of vapor pressure with temperature for water and ice at 0°C. At this temperature the vapor pressure of water is 4.58 mm Hg.

52. Ice is in equilibrium with *air-free* liquid water at 0.0023°C under a pressure of 1 atm, while under its own vapor pressure the melting point is 0.0075°C higher. Using the data given in problem 34, find the sublimation pressure of the ice.

53. The sublimation pressure of N_2O is given by the relation

$$\log_{10} P_{atm} = \frac{-1294}{T} + 1.405 \log_{10} T - 0.0051\, T + 4.800$$

Find the expression for the heat of sublimation of N_2O as a function of the temperature.

54. The vapor pressure of liquid arsenic is given by the equation

$$\log_{10} P_{mm} = \frac{-2460}{T} + 6.69$$

that of solid arsenic by

$$\log_{10} P_{mm} = \frac{-6947}{T} + 10.8$$

Find the temperature at which the two forms of As have the same vapor pressure. What is the value of this pressure?

55. For a certain liquid the heat of vaporization is $\Delta H_v = 13,500 - 10.0\, T$ cal/mole, while the normal boiling point is 400°K. Assuming the vapor to behave as an ideal gas, find per mole for the vaporization process at 400°K: (a) ΔH, (b) ΔS, (c) ΔF, (d) ΔF^0, (e) q, (f) w, (g) ΔE, (h) ΔA, (i) ΔC_p, and (j) ΔC_v.

7
Chemical Equilibrium

IT IS a familiar and well-established fact that many reactions do not go to completion. They proceed to a certain point and then apparently stop, often leaving considerable amounts of unaffected reactants. Under any given set of conditions of temperature, pressure, and concentration, the point at which any reaction seems to stop is always the same; i.e., there exists at this point among the concentrations of the various reactants and products of any reaction a relationship which is definitely fixed. When a reaction reaches this stage in its course, it is said to be in *equilibrium*.

A state of chemical equilibrium should not be considered the state of a reaction at which all motion ceases. It is much more fruitful to consider the point of equilibrium as the state in which the *rate* at which reactants disappear to form products is exactly equal to the *rate* at which the products interact to reform the reacting substances. Under these conditions no perceptible transformation can be detected in the system, and the net effect is an *apparent* state of complete rest. Such an equilibrium is designated as a *dynamic* one, in contrast to a *static* equilibrium where there is no motion whatsoever. All chemical equilibria and physical equilibria between states are considered to be dynamic in nature.

Chemical equilibria may be classified into two groups, namely, (a) *homogeneous* and (b) *heterogeneous* equilibria. A homogeneous equilibrium is one established in a system in which only one phase occurs, as in a system containing only gases, or a single liquid or solid phase. A heterogeneous equilibrium, on the other hand, is one established in a system in which more than a single phase appears, as equilibrium between solid and gas, liquid and gas, solid and liquid, or solid and solid.

THE THERMODYNAMIC EQUILIBRIUM CONSTANT

For any reaction such as

$$aA + bB + \cdots = cC + dD + \cdots \tag{1}$$

the change in free energy, ΔF, at any temperature T is given by the reaction isotherm, Eq. (64) of the last chapter, namely,

$$\Delta F = \Delta F^0 + RT \ln \left(\frac{a_C^c a_D^d \cdots}{a_A^\alpha a_B^b \cdots} \right) \tag{2}$$

The activities indicated are those of the products at the end and of reactants at the start of the reaction. However, we have seen in the last chapter that the criterion of equilibrium is that $\Delta F = 0$ at constant temperature and pressure. Hence, at equilibrium Eq. (2) becomes

$$0 = \Delta F^0 + RT \ln \left(\frac{a_C^c a_D^d \cdots}{a_A^\alpha a_B^b \cdots} \right)$$

and

$$\Delta F^0 = -RT \ln \left(\frac{a_C^c a_D^d \cdots}{a_A^\alpha a_B^b \cdots} \right) \tag{3}$$

The activities now are those of reactants and products *at equilibrium*. Since at any given temperature ΔF^0, the free energy change in the standard state, is a constant for any given reaction, it follows that the activity ratio in Eq. (3) must also be constant, i.e.,

$$K_a = \frac{a_C^c a_D^d \cdots}{a_A^\alpha a_B^b \cdots} \tag{4}$$

and Eq. (3) may be written as

$$\Delta F^0 = -RT \ln K_a \tag{5}$$

Equation (4) defines K_a, *the thermodynamic equilibrium constant* of a reaction. It also shows that the indicated ratio of the activities of products and reactants at equilibrium must be *constant and independent of all factors except temperature*. Again, Eq. (5) relates directly the thermodynamic equilibrium constant of a reaction at a temperature T to the free energy change in the standard state for the reaction. It permits, therefore, the calculation of K_a from ΔF^0 values, and vice versa, the calculation of ΔF^0 of reactions from their thermodynamic equilibrium constants. This highly important equation makes possible reduction of the free energy change calculations described in the last chapter to equilibrium constants, and allows prediction of the behavior of chemical reactions under various conditions without recourse to direct experiment. Thus, we have seen in the last chapter that for the reaction

$$\frac{1}{2} N_2(g) + \frac{3}{2} H_2(g) = NH_3(g) \qquad \Delta F^0_{298.15°K} = -3,980 \text{ cal}$$

Using Eq. (5), we find for the equilibrium constant of this reaction at 298.15°K,

$$\log_{10} K_a = -\frac{\Delta F^0}{2.303 \, RT}$$
$$= \frac{3980}{2.303 \times 1.987 \times 298.15}$$
$$= 2.917$$
$$K_a = 826.1$$

In writing the expression for the equilibrium constant the *activities of products must always be placed in the numerator* and those of the reactants in the denominator. Inversion of the ratio will not give the constant for the reaction as written, but its reverse; for, as may be seen from Eq. (4), the two constants are reciprocally related to each other, i.e.,

$$K_{\text{direct reaction}} = \frac{1}{K_{\text{reverse reaction}}} \tag{6}$$

K_p AND K_c FOR GASEOUS REACTIONS

Before proceeding it is necessary to relate the thermodynamic equilibrium constant to experimentally measurable quantities. These are the partial pressures of reactants and products at equilibrium in the case of gaseous reactions, and concentrations for reactions in solution. Here we shall discuss reactions involving only gases, or gases and pure condensed phases. Equilibria in nonelectrolyte solutions will be considered in the next chapter and ionic equilibria in Chapter 11.

We saw in the last chapter that the activity of any pure gas, which is identical with the fugacity, is given by $a = P\gamma$, where P is the pressure of the gas and γ is its activity coefficient. The corresponding relation for a gas in a mixture of gases is given by the *Lewis fugacity rule*. This rule states that *the activity of any gas in a mixture is equal to the partial pressure of the gas multiplied by the activity coefficient of the pure gas at the total pressure of the mixture*. In terms of this rule, for any gaseous species i in a mixture

$$a_i = P_i \gamma_i \tag{7a}$$
$$= N_i P_t \gamma_i \tag{7b}$$

where N_i is the mol fraction of i, P_t is the total pressure of the gas mixture, and γ_i is the activity coefficient of pure i at pressure P_t.

When Eq. (7a) is substituted for the activity of each substance in Eq. (4), then we get for K_a of a gaseous reaction at equilibrium

$$K_a = \frac{(P_C \gamma_C)^c (P_D \gamma_D)^d \cdots}{(P_A \gamma_A)^a (P_B \gamma_B)^b \cdots}$$
$$= \left(\frac{P_C^c P_D^d \cdots}{P_A^a P_B^b \cdots} \right) \left(\frac{\gamma_C^c \gamma_D^d \cdots}{\gamma_A^a \gamma_B^b \cdots} \right) \tag{8}$$

The first term on the right in Eq. (8) may be represented by K_p, namely,

$$K_p = \frac{P_C^c P_D^d \cdots}{P_A^a P_B^b \cdots} \tag{9}$$

and the second term by K_γ, where

$$K_\gamma = \frac{\gamma_C^c \gamma_D^d \cdots}{\gamma_A^a \gamma_B^b \cdots} \tag{10}$$

With these definitions Eq. (8) becomes

$$K_a = K_p K_\gamma \tag{11}$$

K_p is the *equilibrium constant of a reaction expressed in pressures.* For evaluating K_p any desired pressure units may be used. However, for substitution in Eq. (11) or Eq. (5) the pressures must be in atmospheres.

In Eq. (11) K_a is a true constant for a reaction at a given temperature. On the other hand, K_γ is a quantity whose magnitude depends on the gases involved and the pressure. For ideal gases, or for real gases at zero pressure, $\gamma = 1$ for each gas, and hence K_γ is also equal to unity. Under such conditions $K_a = K_p$, and the two types of equilibrium constants are identical. However, for nonideal gases at pressures above zero the γ's will deviate from unity, and so will K_γ; in fact, for any given reaction the value of K_γ will be determined by the total pressure of the system and will vary as the latter is changed. Therefore, it must follow from Eq. (11) that, whereas for any given reaction and temperature K_a is a true constant, K_p may not be a constant but a quantity whose value will depend on the total pressure at equilibrium.

The limits within which pressures may be substituted for activities have been discussed in the preceding chapter. In general it may be said that K_p will approximate K_a at low total pressures. Since most of the examples of equilibria involving gases to be cited in the following correspond to the latter situation, we shall proceed here on the supposition that K_a is synonymous with K_p.

Although equilibria involving gases are formulated most frequently in terms of partial pressures to yield K_p, they may also be expressed in concentration terms as

$$K_c = \frac{C_C^c C_D^d \cdots}{C_A^a C_B^b \cdots} \tag{12}$$

where K_c is the concentration equilibrium constant. The values of K_p and K_c thus obtained for a given reaction are generally different numerically. However, a relation between these two at any temperature T can readily be deduced provided the gases involved may be considered to behave ideally. Since for an ideal gas $P = (n/V)RT = CRT$, we obtain on substitution of this relation in Eq. (9),

$$
\begin{aligned}
K_p &= \frac{C_C^c(RT)^c C_D^d(RT)^d \cdots}{C_A^a(RT)^a(C_B^b(RT)^b \cdots} \\
&= \left(\frac{C_C^c C_D^d \cdots}{C_A^a C_B^b \cdots}\right)\frac{(RT)^{c+d+\cdots}}{(RT)^{a+b+\cdots}} \\
&= K_c(RT)^{(c+d+\cdots)-(a+b+\cdots)}
\end{aligned}
\tag{13}
$$

But, $(c + d + \cdots) - (a + b + \cdots)$ represents the change in the total number of moles of gaseous products and reactants during the reaction. Letting this difference be Δn_g, we obtain for the relation between the two constants

$$K_p = K_c(RT)^{\Delta n_g} \tag{14}$$

It is apparent from Eq. (14) that $K_p = K_c$ only when $\Delta n_g = 0$; i.e., when there is no change in volume on reaction. However, when there is a volume change, $K_p \neq K_c$. For an increase in volume on reaction Δn_g is positive, and hence K_p is numerically larger than K_c. On the other hand, when Δn_g is negative, corresponding to a volume decrease, K_p is less than K_c. In using Eq. (14) R must be expressed in the same units as are the pressures and volumes involved in K_p and K_c.

PROPERTIES OF EQUILIBRIUM CONSTANTS

Because of the fundamental importance of equilibrium calculations, it may not be amiss to recapitulate and emphasize the properties of the equilibrium constant of a reaction. In the discussion which follows it will be assumed that K_p or K_c for a reaction is a true constant.

First, the equilibrium constant principle is valid only at *equilibrium*. The relation called for between the concentrations of products and reactants does not apply to all possible concentrations that may be encountered in a reacting system, but only to those which correspond to the point of true equilibrium. Unless true equilibrium concentrations are substituted in Eq. (9) or (12), these equations cannot be expected to yield constants.

Again, the equilibrium constant for any reaction should, at a fixed temperature, be a constant independent of concentration or pressure at all concentrations and pressures. On the other hand, the equilibrium constant of a reaction will not be the same at all temperatures but will vary from temperature to temperature in a manner predictable by thermodynamics.

The magnitude of the equilibrium constant determines the extent to which any particular reaction can proceed under given conditions. A large value of K_p or K_c indicates that the numerator in the equilibrium constant expression is large compared with the denominator, i.e., that the concentrations of products are large compared with those of the reactants and hence that the reaction favors the formation of products. On the other hand, when K_p or K_c is small, the concentrations of reactants are large compared with those of products, and the indications are that the particular reaction does not proceed to any appreciable extent under the given conditions.

Further, the equilibrium constant defines *quantitatively* the effect of concentrations of reactants and products on the extent of reaction. The manner of calculating such effects will be discussed later in the chapter. At present we may deduce qualitatively certain conclusions of general validity. For this purpose consider the reaction

$$H_2(g) + Cl_2(g) = 2\ HCl(g) \tag{15}$$

for which the equilibrium constant is given by

$$\frac{P_{HCl}^2}{P_{H_2}P_{Cl_2}} = K_p \tag{16}$$

Since K_p is a constant at all pressures, the relation among the partial pressures of H_2, Cl_2, and HCl in Eq. (16) should also remain constant under all conditions at a given temperature. If now hydrogen is added to an equilibrium mixture of the three gases, the pressure of this gas is increased, and hence the existing relationship among the partial pressures as given in Eq. (16) is disturbed. To accommodate the added hydrogen without violating the constancy of K_p, the partial pressure of chlorine must decrease, while that of hydrogen chloride must increase. This can be accomplished by further interaction of hydrogen and chlorine to form hydrogen chloride, the process continuing until the relationship called for by K_p is re-established and the gases are again in equilibrium. The same effect may be produced by adding chlorine. Conversely, addition of hydrogen chloride will increase the numerator, and hence the denominator must also be increased to preserve the constancy of K_p. This time the adjustment is brought about by dissociation of HCl into H_2 and Cl_2 until equilibrium is again re-established and K_p returns to its constant value.

From this behavior of an equilibrium mixture on addition of excess reactants or products may be drawn the following two generalizations:

1. Presence of excess of some of the reactants over others tends to drive a reaction further to completion with respect to the reactants not in excess.

2. Initial presence of products tends to decrease the extent of conversion of reactants to products.

EQUILIBRIA IN GASEOUS SYSTEMS

Many direct experimental studies of gaseous equilibria have been made. Several examples of these will be discussed in detail to indicate some of the methods employed, as well as to illustrate the application of equilibrium constants.

THE AMMONIA EQUILIBRIUM

The equilibrium

$$\frac{3}{2} H_2(g) + \frac{1}{2} N_2(g) = NH_3(g) \tag{17}$$

has been extensively investigated by Haber and co-workers, Nernst and Jellinek, and more recently by Larson and Dodge.[1] The last named passed a mixture of nitrogen and hydrogen, in the ratio of 1:3 by volume, through an iron coil immersed in a constant temperature bath. To speed the approach of equilibrium, the coil was lined with finely divided iron to act as a catalyst, i.e., a substance which accelerated the attainment of equilibrium without affecting it. The exit gases were analyzed, then, for hydrogen, nitrogen, and ammonia to determine the composition of the equilibrium mixture. To check their results and to assure themselves that true equilibrium had been attained, Larson and Dodge approached equilibrium also from the ammonia side by passing through the coil mixtures of ammonia, nitrogen, and hydrogen. Some of their results for the direct formation of ammonia from nitrogen and hydrogen at various pressures and temperatures are given in Table 7–1. The first column gives the

[1] Larson and Dodge, *J. Am. Chem. Soc.*, **45**, 2918 (1923).

TABLE 7-1. EQUILIBRIUM CONSTANTS FOR FORMATION OF NH_3

Total Pressure, Atm	$t = 350°C$		$t = 400°C$		$t = 450°C$	
	% NH_3	K_p	% NH_3	K_p	% NH_3	K_p
10	7.35	0.0266	3.85	0.0129	2.04	0.00659
30	17.80	0.0273	10.09	0.0129	5.80	0.00676
50	25.11	0.0278	15.11	0.0130	9.17	0.00690

total equilibrium pressure, the second the mole percentage of ammonia found in the equilibrium mixture at each total pressure and 350°C, while the third gives the value of K_p calculated from

$$K_p = \frac{P_{NH_3}}{P_{H_2}^{3/2} P_{N_2}^{1/2}} \tag{18}$$

The remaining columns give similar information for 400°C and 450°C.

From the percentage of ammonia at equilibrium at a total pressure P, K_p is calculated with the aid of Dalton's law of partial pressures. Taking the data for 30 atm and 400°C, we see that the percentage of ammonia at equilibrium is 10.09, and hence the partial pressure of ammonia is

$$P_{NH_3} = 30 \times 0.1009 = 3.03 \text{ atm}$$

The pressure of hydrogen plus nitrogen is, therefore,

$$P_{H_2} + P_{N_2} = 30.00 - 3.03 = 26.97 \text{ atm}$$

Since, however, the nitrogen and hydrogen are present in the ratio of 1:3, then

$$P_{H_2} = \frac{3}{4} \times 26.97 = 20.22 \text{ atm}$$

$$P_{N_2} = \frac{1}{4} \times 26.97 = 6.75 \text{ atm}$$

and

$$K_p = \frac{P_{NH_3}}{P_{H_2}^{3/2} P_{N_2}^{1/2}}$$

$$= \frac{3.03}{(20.22)^{3/2}(6.75)^{1/2}}$$

$$= 0.0129$$

Table 7-1 clearly substantiates Eq. (18) as the expression for the equilibrium constant of this reaction. Although the variation in total pressure is quite large, 10 to 50 atm, still K_p at any *given temperature* is essentially constant. The total pressure merely determines the relative percentages of hydrogen, nitrogen, and ammonia present at equilibrium, but does not affect the constancy of K_p. On the other hand, variation of temperature produces significant changes both in the percentage of ammonia and in K_p. At all the given pressures increase of temperature operates to decrease the yield of ammonia and hence to decrease K_p.

THE LE CHATELIER - BRAUN PRINCIPLE

To foretell qualitatively the effect of variation in pressure or temperature on a system in equilibrium, use is made of the *Le Chatelier-Braun principle*. This principle states that *whenever stress is placed on any system in a state of equilibrium, the system will always react in a direction which will tend to counteract the applied stress*. Thus, if pressure is applied to a system, the tendency of the stress will be to decrease the volume, and hence that reaction in the system will take place which will favor the smaller volume. In the ammonia equilibrium the combination of nitrogen and hydrogen to form ammonia is attended by a volume *decrease* from two molar volumes to one, and hence according to the Le Chatelier-Braun principle we may expect the formation of ammonia to be favored by an increase in total pressure. That this is actually the case is borne out by Table 7–1.

Again, when a reaction is endothermic, i.e., absorbs heat, addition of heat should favor it, and the reaction should proceed to a greater extent at higher temperatures. Conversely, when a reaction is exothermic we may expect that addition of heat would tend to inhibit the reaction, and hence at higher temperatures the reaction should tend to reverse itself. The latter behavior is shown by the exothermic ammonia synthesis reaction. Table 7–1 indicates, in line with the Le Chatelier-Braun prediction, that the yield of ammonia is higher the lower the temperature.

THE PHOSGENE EQUILIBRIUM

The formation of phosgene,

$$CO(g) + Cl_2(g) = COCl_2(g) \qquad (19)$$

has been fully studied by Max Bodenstein and Heinrich Plaut[2] by a static method, as contrasted to the flow, or dynamic, method employed by Larson and Dodge for the ammonia equilibrium. The apparatus used consisted of a glass reaction bulb set in a cylindrical, electrically heated oven which was kept at a constant temperature. The top of the glass bulb was connected by capillary glass tubing to tanks in which the various gases were stored and to a specially constructed manometer

---

[2] Bodenstein and Plaut, *Z. physik. Chem.*, **110**, 399 (1924).

made of quartz. All pressure measurements were made with this manometer. At the start of an experiment chlorine was first admitted, then carbon monoxide, and the pressure of each was recorded. After equilibrium had been established, the total pressure as recorded on the manometer was also read.

K_p can be calculated directly from the pressure readings. In a typical experiment Bodenstein and Plaut found at 394.8°C that for an initial pressure of chlorine equal to 351.4 mm and carbon monoxide equal to 342.0 mm, the total pressure at equilibrium was 439.5 mm. Since the volume was constant throughout the experiment, the partial pressures are directly proportional to the numbers of moles of each constituent present, and hence we may deal with these directly. If, then, we let x be the drop in the partial pressure of chlorine during the experiment, the drop in partial pressure of carbon monoxide is also x, while the partial pressure of phosgene formed is x. This relation is apparent from the stoichiometry of Eq. (19), where the two gases, chlorine and carbon monoxide, interact mole for mole to form 1 mole of phosgene. We may write, therefore, for the partial pressures of the three gases at equilibrium:

$$P_{Cl_2} = 351.4 - x \text{ mm}$$
$$P_{CO} = 342.0 - x \text{ mm}$$
$$P_{COCl_2} = x \text{ mm}$$

The total pressure at equilibrium must be the sum of these partial pressures and must equal in turn to 439.5 mm as found by experiment. Then

$$P = P_{Cl_2} + P_{CO} + P_{COCl_2} = 439.5 \text{ mm}$$
$$= (351.4 - x) + (342.0 - x) + x = 439.5 \text{ mm}$$
$$= 693.4 - x = 439.5 \text{ mm}$$
$$x = 693.4 - 439.5 = 253.9 \text{ mm}$$

Substituting this value of x into the expressions for the partial pressures, we find

$$P_{Cl_2} = 351.4 - 253.9$$
$$= 97.5 \text{ mm} = 0.128 \text{ atm}$$
$$P_{CO} = 342.0 - 253.9$$
$$= 88.1 \text{ mm} = 0.116 \text{ atm}$$
$$P_{COCl_2} = 253.9 \text{ mm} = 0.334 \text{ atm}$$

Hence at 394.8°C

$$K_p = \frac{P_{COCl_2}}{P_{Cl_2}P_{CO}} = \frac{0.334}{(0.128)(0.116)}$$
$$= 22.5$$

When the equilibrium constant for a forward reaction is known, the equilibrium constant for the reverse reaction is also known through Eq. (6). Thus, since the equilibrium constant for the formation of phosgene from chlorine and carbon monoxide is $K_p = 22.5$ at 394.8°C, the equilibrium constant for the *dissociation* of phosgene into carbon monoxide and chlorine, namely,

$$COCl_2(g) = CO(g) + Cl_2(g) \tag{20}$$

must be

$$K_p' = \frac{P_{CO}P_{Cl_2}}{P_{COCl_2}} = \frac{1}{K_p} \tag{21}$$

$$= \frac{1}{22.5} = 0.0444$$

This new constant may be employed to calculate the extent to which phosgene dissociates into the given products at any specified pressure and 394.8°C. If we start with n moles of phosgene, and let α be the degree of dissociation at equilibrium, i.e., the *fraction of each mole* that dissociates, the number of moles of phosgene undissociated at equilibrium must be $n(1 - \alpha)$, while the number of moles of chlorine and carbon monoxide formed must each be $n\alpha$. Hence the total number of moles of gas present at equilibrium is

$$n_t = n(1 - \alpha) + n\alpha + n\alpha$$
$$= (n + n\alpha)$$
$$= n(1 + \alpha)$$

If the total equilibrium pressure is P, the partial pressures of the three gases, using Dalton's law, must be

$$P_{COCl_2} = \left(\frac{n_{COCl_2}}{n_t}\right) P = \left(\frac{1 - \alpha}{1 + \alpha}\right) P$$

$$P_{Cl_2} = \left(\frac{n_{Cl_2}}{n_t}\right) P = \left(\frac{\alpha}{1 + \alpha}\right) P$$

$$P_{CO} = \left(\frac{n_{CO}}{n_t}\right) P = \left(\frac{\alpha}{1 + \alpha}\right) P$$

and therefore,

$$K_p' = \frac{P_{CO}P_{Cl_2}}{P_{COCl_2}} = \frac{\left(\dfrac{\alpha}{1 + \alpha}\right) P \cdot \left(\dfrac{\alpha}{1 + \alpha}\right) P}{\left(\dfrac{1 - \alpha}{1 + \alpha}\right) P}$$

$$= \frac{\alpha^2 P}{(1 - \alpha)(1 + \alpha)}$$

$$= \frac{\alpha^2 P}{(1 - \alpha^2)} \tag{22}$$

Substituting $K_p' = 0.0444$, and assuming $P = 1$ atm, solution for α yields

$$\frac{\alpha^2}{1 - \alpha^2} = 0.0444$$

$$\alpha = 0.206$$

Consequently, pure phosgene dissociates into chlorine and carbon monoxide to the extent of 20.6 per cent at 1 atm pressure and a temperature of 394.8°C.

This typical calculation illustrates the general procedure followed in setting up expressions for the equilibrium constant in terms of the total pressure and the degree of dissociation.

THE DISSOCIATION OF HYDROGEN SULFIDE

Hydrogen sulfide on being heated dissociates into hydrogen and sulfur according to the equation

$$2\ H_2S(g) = 2\ H_2(g) + S_2(g) \tag{23}$$

To determine the extent of this dissociation, Preuner and Schupp[3] utilized a very novel method. Hydrogen sulfide under a definite pressure was admitted into an elongated porcelain tube where it was allowed to dissociate at constant temperature. In the center of this tube was located a small platinum bulb which acted as a membrane permeable to hydrogen, but not to hydrogen sulfide or sulfur. This bulb was evacuated before each experiment. As the hydrogen sulfide dissociated, the hydrogen formed diffused rapidly into the platinum bulb until it built up within a pressure equal to the equilibrium pressure of hydrogen. This pressure was recorded on a manometer connected directly with the bulb and was used to calculate the degree of dissociation, α, of the hydrogen sulfide. Values of α obtained in this manner at several temperatures for a total equilibrium pressure of 1 atm are shown in Table 7-2.

TABLE 7-2.   DISSOCIATION OF H_2S
AT 1 ATMOSPHERE PRESSURE

$t°C$	α	K_p
750	0.055	0.000091
830	0.087	0.00038
1065	0.247	0.0118
1132	0.307	0.0260

[3] Preuner and Schupp, *Z. physik. Chem.*, **68**, 157 (1909).

To calculate K_p from these values of α we proceed as follows. Of each initial mole of hydrogen sulfide the amount left at equilibrium is $(1 - \alpha)$ and of n moles, $n(1 - \alpha)$. Since for each mole of hydrogen sulfide that dissociates 1 mole of hydrogen and 0.5 mole of sulfur are formed, then from a total of $n\alpha$ moles dissociated, $n\alpha$ moles of hydrogen and $n\alpha/2$ moles of sulfur are obtained. The conditions at equilibrium are, therefore,

$$2\,H_2S \;\; = 2\,H_2 + S_2$$

$$n(1 - \alpha) \qquad n\alpha \qquad \frac{n\alpha}{2}$$

and, instead of the initial n moles of gas, there are present at equilibrium

$$n_t = n(1 - \alpha) + n\alpha + \frac{n\alpha}{2}$$

$$= \frac{n(2 + \alpha)}{2} \text{ moles}$$

For a total equilibrium pressure P the partial pressures are, respectively,

$$P_{H_2S} = \left[\frac{n(1 - \alpha)}{n(2 + \alpha)/2} \right] P = \left[\frac{2(1 - \alpha)}{2 + \alpha} \right] P$$

$$P_{H_2} = \left[\frac{n\alpha}{n(2 + \alpha)/2} \right] P = \left[\frac{2\,\alpha}{2 + \alpha} \right] P$$

$$P_{S_2} = \left[\frac{(n\alpha/2)}{n(2 + \alpha)/2} \right] P = \left[\frac{\alpha}{2 + \alpha} \right] P$$

and, therefore,

$$K_p = \frac{P_{H_2}^2 P_{S_2}}{P_{H_2S}^2} = \frac{[2\,\alpha/(2 + \alpha)]^2 P^2 \cdot [\alpha/(2 + \alpha)]P}{[2(1 - \alpha)/(2 + \alpha)]^2 P^2}$$

$$= \frac{\alpha^3 P}{(2 + \alpha)(1 - \alpha)^2} \tag{24}$$

The values of K_p obtained by substituting $P = 1$ and the α's from Table 7–2 into the above expression are given in column 3 of the table.

Expressions for K_p such as Eq. (22) or (24) apply only to the equilibrium conditions for which they were derived. When the conditions under which equilibrium is established are varied, such as when products as well as reactants are present initially, the expressions are different in form and more complicated. Thus, consider the hydrogen sulfide equilibrium when n_{H_2S} moles of hydrogen sulfide, n_{H_2} moles of hydrogen, and n_{S_2} moles of sulfur are mixed at 1132°C. To obtain the equilibrium relations at a total pressure P, let x be the number of moles of hydrogen sulfide which react to form hydrogen and sulfur. The number of moles of each species at equilibrium is, then,

$$2\,H_2S \;=\; 2\,H_2 \;+\; S_2$$
$$(n_{H_2S} - x) \qquad (n_{H_2} + x) \qquad \left(n_{S_2} + \frac{x}{2}\right)$$

and
$$n_t = (n_{H_2S} - x) + (n_{H_2} + x) + \left(n_{S_2} + \frac{x}{2}\right)$$
$$= \left(n_{H_2S} + n_{H_2} + n_{S_2} + \frac{x}{2}\right)$$

Consequently the partial pressures are

$$P_{H_2S} = \left(\frac{n_{H_2S} - x}{n_{H_2S} + n_{H_2} + n_{S_2} + x/2}\right)P$$

$$P_{H_2} = \left(\frac{n_{H_2} + x}{n_{H_2S} + n_{H_2} + n_{S_2} + x/2}\right)P$$

$$P_{S_2} = \left(\frac{n_{S_2} + x/2}{n_{H_2S} + n_{H_2} + n_{S_2} + x/2}\right)P$$

and the expression for K_p becomes

$$K_p = \frac{P_{H_2}^2 P_{S_2}}{P_{H_2S}^2}$$

$$= \frac{\left(\dfrac{n_{H_2} + x}{n_{H_2S} + n_{H_2} + n_{S_2} + x/2}\right)^2 P^2 \cdot \left(\dfrac{n_{S_2} + x/2}{n_{H_2S} + n_{H_2} + n_{S_2} + x/2}\right)P}{\left(\dfrac{n_{H_2S} - x}{n_{H_2S} + n_{H_2} + n_{S_2} + x/2}\right)^2 P^2}$$

$$= \frac{(n_{H_2} + x)^2(n_{S_2} + x/2)P}{(n_{H_2S} + n_{H_2} + n_{S_2} + x/2)(n_{H_2S} - x)^2} \tag{25}$$

For the *special* case of $n_{H_2S} = n_{H_2} = n_{S_2} = 1$, and $P = 1$ atm, the equilibrium constant expression at 1132°C, where $K_p = 0.0260$, reduces to

$$K_p = \frac{(1 + x)^2(1 + x/2)}{(3 + x/2)(1 - x)^2} = 0.0260 \tag{26}$$

Solution of this equation for x yields $x = -0.526$. The minus sign indicates that hydrogen sulfide does *not* dissociate under the conditions specified, but that hydrogen and sulfur must combine to form hydrogen sulfide before equilibrium can be established in a mixture containing initially 1 mole of each of the participants and at a total equilibrium pressure of 1 atm. At equilibrium there are present, therefore,

$$n_{H_2S} - x = 1 + 0.526 = 1.526 \text{ moles of } H_2S$$
$$n_{H_2} + x = 1 - 0.526 = 0.474 \text{ mole of } H_2$$
$$n_{S_2} + \frac{x}{2} = 1 - 0.263 = 0.737 \text{ mole of } S_2$$

and the partial pressures are, respectively,

$$P_{H_2S} = \frac{1.526}{2.737} \times 1 = 0.558 \text{ atm}$$

$$P_{H_2} = \frac{0.474}{2.737} \times 1 = 0.173 \text{ atm}$$

$$P_{S_2} = \frac{0.737}{2.737} \times 1 = 0.269 \text{ atm}$$

EFFECT OF INERT GASES ON EQUILIBRIUM

All the examples of equilibria considered so far involved only the gases which participate directly in the reaction. Frequently, however, equilibrium mixtures are encountered where gases are present other than those involved in the reaction. The question is: What, if any, will be the effect of gases not entering into the reaction on the extent of reaction? Obviously, the presence of these gases cannot affect in any way the thermodynamic equilibrium constant. But, the presence of these inert gases modifies the γ's, and thereby also K_γ. As a result K_p is changed. However, even if we neglect this effect on K_p, the presence of the inert gases still affects the *partial pressures* of the reactants and products at a given total equilibrium pressure, and hence we may expect a shift in the extent of reaction to permit a redistribution of the partial pressures in accord with the demands of the equilibrium constant.

To illustrate quantitatively the effect of an inert gas on an equilibrium, consider again the reaction

$$COCl_2(g) = CO(g) + Cl_2(g)$$

We have seen that at 394.8°C and at a total equilibrium pressure of 1 atm the degree of dissociation of the phosgene is $\alpha = 0.206$. Suppose now, however, that equilibrium is established in presence of nitrogen gas at a partial pressure of 0.40 atm in a total pressure of 1 atm. Then the sum of the partial pressures of phosgene, chlorine, and carbon monoxide is no longer $P = 1$, but $1 - 0.40 = 0.60$ atm, and from Eq. (22) it follows that

$$K'_p = \frac{P_{CO} \times P_{Cl_2}}{P_{COCl_2}} = \frac{\alpha^2 P}{(1 - \alpha^2)}$$

$$0.0444 = \frac{\alpha^2(0.60)}{(1 - \alpha^2)}$$

$$\alpha = 0.262$$

The addition of nitrogen under the specified conditions leads, therefore, to an increase of 5.6 per cent in the dissociation of phosgene.

Qualitatively, the effect of inert gases on the extent of reaction can be foretold with the aid of the Le Chatelier-Braun principle. Since for any given total pressure the presence of inert gases decreases the partial pressure of reactants and products, the net effect will be the same as if the gases at equilibrium were subjected to a lower total pressure. Or, stated differently, the effect of an inert gas is to *dilute* the concentrations of reactants and products. Consequently, according to the Le Chatelier-Braun principle, presence of inert gases will favor the reaction that results in an *increase* in volume. Thus, when the volume of reactants is greater than that of products, the reaction will be displaced in favor of reactants, while when the opposite is true, the formation of products will be favored, as is substantiated by the above calculation. When there is no change in volume on reaction, however, neither side will be favored, and the equilibrium will be uninfluenced by the introduction of inert gases.

THE EQUILIBRIUM CONSTANT FOR HETEROGENEOUS REACTIONS

Consider the reaction

$$CuO(s) + H_2(g) = Cu(s) + H_2O(g) \qquad (27)$$

The thermodynamic equilibrium constant K_a for this reaction is

$$K_a = \frac{a_{Cu}a_{H_2O}}{a_{CuO}a_{H_2}} \qquad (28)$$

However, it was shown in Chapter 6 that *the activity of a pure solid or liquid may be taken as unity at all temperatures up to fairly high pressures.* Then, so long as the pressures are not too high $a_{Cu} = a_{CuO} = 1$, and

$$K_a = \frac{a_{H_2O}}{a_{H_2}} \qquad (29)$$

Further, when the gases involved may be considered to behave ideally the activity reduces to the pressure, and Eq. (29) becomes

$$K_p = \frac{P_{H_2O}}{P_{H_2}} \qquad (30)$$

Therefore, under the conditions specified the equilibrium constant for a heterogeneous reaction should contain only the activities or pressures of the gaseous constituents and should not include terms for either pure solids or liquids; i.e., *the presence of pure solid or liquid phases can be completely disregarded in writing the expression for the equilibrium constant.* K_p's of heterogeneous reactions are generally referred to as *condensed equilibrium constants.*

Below are given several typical examples of heterogeneous equilibria in order to illustrate some of the methods employed in their study and mathematical treatment.

THE DISSOCIATION OF CUPRIC OXIDE

At elevated temperatures cupric oxide dissociates into cuprous oxide and oxygen according to the reaction

$$4 \, CuO(s) = 2 \, Cu_2O(s) + O_2(g) \tag{31}$$

Since oxygen is the only gaseous constituent involved in this equilibrium, the condensed equilibrium constant K_p for this reaction should be

$$K_p = P_{O_2} \tag{32}$$

i.e., the pressure of oxygen above a mixture of cupric and cuprous oxides should be constant at each given temperature. This conclusion is verified by the experimental results of F. Hastings Smith and H. R. Robert.[4] These investigators placed a charge of pure cupric oxide in a silica tube, evacuated the system, and then heated the tube to the desired temperature in an electric furnace. The equilibrium pressure of oxygen developed as a result of dissociation of the cupric oxide was read on a manometer attached to the silica tube. To make sure that the pressures read were those at true equilibrium, the latter was approached from both lower and higher temperatures, and at each temperature some gas was removed and the equilibrium pressure permitted to reestablish itself. In this manner Hastings Smith and Robert proved that at each temperature the pressure of oxygen is constant in accord with Eq. (32). Some of their results at various temperatures are given in Table 7–3.

TABLE 7–3.   DISSOCIATION PRESSURES
OF CuO AT VARIOUS TEMPERATURES

$t°C$	$K_p = P_{O_2}$ (mm Hg)
900	12.5
940	29.2
980	65.0
1020	137.7
1060	278.0
1080	388.0

[4] F. Hastings Smith and H. R. Robert, *J. Am. Chem. Soc.*, **42**, 2582 (1920).

THE CARBON DISULFIDE EQUILIBRIUM

When gaseous sulfur is passed over carbon at high temperatures, carbon disulfide is formed according to the equation

$$C(s) + S_2(g) = CS_2(g) \tag{33}$$

For this reaction the condensed equilibrium constant is given by

$$K_p = \frac{P_{CS_2}}{P_{S_2}} \tag{34}$$

To study this equilibrium F. Koref[5] employed a dynamic method. Nitrogen gas saturated with sulfur vapor was passed over finely divided carbon kept at the desired temperature in an electric furnace. The exit gases were cooled rapidly to prevent the shift of equilibrium, the sulfur and carbon disulfide condensed out, while the nitrogen was collected in a gasometer. The equilibrium quantities of carbon disulfide and sulfur were then determined by weighing.

TABLE 7-4. K_p AT 1009°C FOR THE
REACTION $C(s) + S_2(g) = CS_2(g)$

V_{CS_2}(cc)	V_{S_2}(cc)	K_p
458	84	5.45
607	109	5.57
738	130	5.68
814	142	5.73
1164	207	5.62
2057	371	5.54

From the data thus obtained the equilibrium constant was calculated as follows: If it is assumed that the sulfur and carbon disulfide vapors behave ideally, the volume of each of these at equilibrium at temperature T and total pressure P is

$$V_{S_2} = \frac{n_{S_2}RT}{P} \qquad V_{CS_2} = \frac{n_{CS_2}RT}{P}$$

where n_{S_2} and n_{CS_2} are the numbers of moles of these substances present in the condensates from the equilibrium mixture. If we let V be the total volume of gases at equilibrium, i.e., the sum of the volumes of sulfur, carbon disulfide, and nitrogen, then the mol fractions of the first

[5] F. Koref, Z. anorg. Chem., **66**, 73 (1910).

two substances are $N_{S_2} = V_{S_2}/V$ and $N_{CS_2} = V_{CS_2}/V$, and hence, according to Dalton's law,

$$P_{S_2} = N_{S_2}P = \left(\frac{V_{S_2}}{V}\right)P \qquad P_{CS_2} = N_{CS_2}P = \left(\frac{V_{CS_2}}{V}\right)P$$

Substituting these expressions for the partial pressures into Eq. (34), we find

$$K_p = \frac{P_{CS_2}}{P_{S_2}} = \frac{(V_{CS_2}P)/V}{(V_{S_2}P)/V}$$
$$= \frac{V_{CS_2}}{V_{S_2}} \tag{35}$$

namely, the equilibrium constant should equal the ratio of the volumes of the two substances at equilibrium. Table 7–4 lists some of Koref's data at 1009°C and the values of K_p calculated from these. The constancy exhibited by K_p is satisfactory.

THE DISSOCIATION OF AMMONIUM CARBAMATE

Ammonium carbamate dissociates even at room temperature as follows:

$$NH_2COONH_4(s) = 2\,NH_3(g) + CO_2(g) \tag{36}$$

The condensed equilibrium constant for this reaction is, therefore,

$$K_p = P^2_{NH_3}P_{CO_2} \tag{37}$$

In studying this equilibrium by a static method T. R. Briggs and V Migidichian[6] introduced solid ammonium carbamate along with definite quantities of ammonia and carbon dioxide into an evacuated glass vessel to which was attached a manometer. The apparatus then was immersed in a water thermostat, the mixture was allowed to reach equilibrium, and the total pressure at equilibrium was read on the manometer.

If we consider the experiments where only ammonia was introduced initially along with the ammonium carbamate, and if we let e_1 be the initial pressure of ammonia and P the total pressure at equilibrium, then the increase in pressure due to dissociation of the solid is $(P - e_1)$. From the stoichiometry of the reaction it follows that of this increase two-thirds must be due to the formation of ammonia, and one-third to formation of carbon dioxide. Consequently,

$$P_{NH_3} = \frac{2}{3}(P - e_1) + e_1 = \left(\frac{2P + e_1}{3}\right)$$

$$P_{CO_2} = \frac{1}{3}(P - e_1)$$

[6] T. R. Briggs and V. Migidichian, *J. Phys. Chem.*, **28**, 1121 (1924).

and Eq. (37) becomes

$$K_p = \left(\frac{2P + e_1}{3}\right)^2 \left(\frac{P - e_1}{3}\right)$$
$$= \frac{(2P + e_1)^2 (P - e_1)}{27} \tag{38}$$

Table 7–5 presents typical data obtained during a series of experiments at

TABLE 7–5.   DISSOCIATION OF AMMONIUM CARBAMATE AT 30°C
(P in mm Hg)

e_1	P	P_{NH_3}	P_{CO_2}	K_p
0	125.0	83.3	41.7	2.89×10^5
13.6	124.9	87.8	37.1	2.86
27.3	125.4	92.7	32.7	2.81
52.5	129.5	103.8	25.7	2.77
141.1	174.2	163.2	11.0	2.93
168.6	194.2	185.7	8.5	2.93

30°C, as well as the equilibrium constants calculated according to Eq. (38). All pressures are given in mm Hg. From the latter equation it is apparent that when $e_1 = 0$, i.e., when no ammonia is present initially, the expression for K_p reduces to

$$K_p = \frac{(2P)^2 P}{27}$$
$$= \frac{4P^3}{27} \tag{39}$$

Other examples of heterogeneous equilibria which can be handled in the manner described are the thermal dissociations of metal carbonates, halides, and sulfides, the oxidation of carbon to carbon monoxide and dioxide, and the reduction of metal oxides by carbon monoxide.

EFFECT OF PRESSURE ON HETEROGENEOUS EQUILIBRIA

As in homogeneous equilibria, the influence of pressure on heterogeneous equilibria can be predicted by means of the Le Chatelier principle. In considering the change in volume accompanying a reaction, the volumes of all condensed phases may be disregarded, since they are negligibly small compared with those of the gases involved.

To illustrate the application of the Le Chatelier principle to a heterogeneous equilibrium, consider again the reaction

$$4\,CuO(s) = 2\,Cu_2O(s) + O_2(g)$$

Since this reaction proceeds with an increase in volume, any increase in the oxygen pressure above its equilibrium value will shift this reaction to the left, i.e., cuprous oxide and oxygen will interact to form cupric oxide until the pressure of oxygen is back to its equilibrium value. On the other hand, when the oxygen pressure is reduced below its equilibrium value, cupric oxide will dissociate, and the process will continue until the requisite oxygen pressure is re-established.

Such equilibrium adjustments are possible only when *all the condensed phases participating in the equilibrium are present*. According to Eq. (32), the pressure of oxygen at equilibrium in this system should be constant at any given temperature. When the equilibrium is approached from the cupric oxide side, and when sufficient cupric oxide is taken to supply the given pressure of the gas in the volume involved, all the phases participating in the equilibrium are present, and the demands of the equilibrium constant can be satisfied. Suppose, however, that the equilibrium is approached from the other side, and that the initial pressure of oxygen is considerably above the equilibrium pressure. Then cuprous oxide will interact with the oxygen to form cupric oxide, and this reaction will proceed until the pressure of the gas has been reduced to the equilibrium pressure. But, this reaction can go this far only if sufficient cuprous oxide is present to react with all the excess oxygen. If such is not the case, all the cuprous oxide will be converted to cupric oxide, and still an excess of oxygen will remain to yield a pressure higher than that demanded by the equilibrium constant. Under such conditions no equilibrium is possible in the system, since no cuprous oxide is present, and hence the oxygen pressure may assume any value dependent on the amount of it present.

These conclusions apply to all heterogeneous equilibria. It must be remembered, therefore, that sufficient amounts of the solid phases involved in the equilibria must be present to permit regulation and adjustment of the conditions in the system necessary for the establishment of true equilibrium. Otherwise true equilibrium is impossible, and the equilibrium constant principle does not apply.

VARIATION OF K_a AND K_p WITH TEMPERATURE

Although K_a for a given reaction is constant at constant T, its magnitude varies appreciably as the temperature is changed. The manner in which K_a for any type of reaction varies with temperature is readily deducible as follows: From Eq. (5) of this chapter we have

$$\ln K_a = -\frac{\Delta F^0}{RT}$$

and, therefore,

$$\frac{d \ln K_a}{dT} = -\frac{1}{R}\left[\frac{d(\Delta F^0/T)}{dT}\right]$$

But

$$\frac{d\,(\Delta F^0/T)}{dT} = -\frac{\Delta H^0}{T^2}$$

and, consequently,

$$\frac{d \ln K_a}{dT} = \frac{\Delta H^0}{RT^2} \tag{40}$$

In the above equations the partial derivatives become totals because neither K_a nor ΔF^0 depends on the pressure. Equation (40) defines the temperature coefficient of $\ln K_a$ in terms of the heat of reaction ΔH^0 and the temperature T. For reactions involving gases and when $K_a = K_p$, Eq. (40) becomes

$$\frac{d \ln K_p}{dT} = \frac{\Delta H^0}{RT^2} \tag{41}$$

For exact integration of Eq. (41) ΔH^0 must be known as a function of T. However, when the temperature interval considered is not large, ΔH^0 may be considered constant over the interval, and

$$\int_{K_{p_1}}^{K_{p_2}} d \ln K_p = \int_{T_1}^{T_2} \frac{\Delta H^0}{RT^2} dT$$

$$\ln K_p \bigg]_{K_{p_1}}^{K_{p_2}} = \frac{\Delta H^0}{R}\left[-\frac{1}{T}\right]_{T_1}^{T_2}$$

$$\ln \frac{K_{p_2}}{K_{p_1}} = \frac{\Delta H^0}{R}\left[\frac{T_2 - T_1}{T_1 T_2}\right] \tag{42}$$

Equation (42) permits the calculation of K_{p_2} at T_2 when K_{p_1} at T_1 and ΔH^0 are available; or, when the equilibrium constants at two different temperatures are known, Eq. (42) may be used to obtain the average heat of the reaction over the temperature range T_1 to T_2.

Instead of integrating Eq. (41) between limits, a solution under the same conditions of constant ΔH^0 may be obtained in the form

$$d \ln K_p = \frac{\Delta H^0}{RT^2} dT$$

$$\ln K_p = -\frac{\Delta H^0}{RT} + C \tag{43}$$

C is an integration constant which can be evaluated for any reaction by substituting a known value of K_p at some given temperature. When several values of K_p at various temperatures are available and ΔH^0 is sought, it is preferable to determine ΔH^0 graphically rather than by use of Eq. (42). According to Eq. (43) a plot of $\log_{10} K_p$ vs. $1/T$ should be a straight line with slope equal to $-\Delta H^0/2.303R$, and hence ΔH^0 follows as

$$\Delta H^0 = -2.303\ R \times \text{slope}$$
$$= -4.576 \times \text{slope} \qquad (44)$$

Figure 7–1 shows such a plot for the homogeneous reaction

$$SO_2(g) + \frac{1}{2} O_2(g) = SO_3(g)$$

between 800 and 1170°K. In agreement with Eq. (43) the plot is a

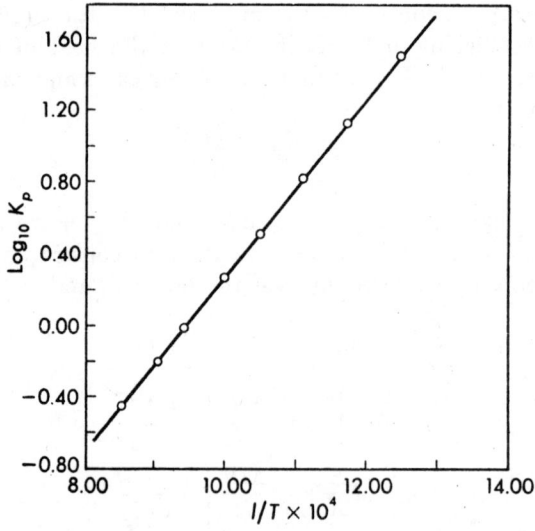

Figure 7–1. Plot of $\log_{10} K_p$ vs. $1/T$ for the reaction $SO_2(g) + \frac{1}{2} O_2(g) = SO_3(g)$.

straight line with slope equal to 4930. Consequently, over this temperature interval

$$\Delta H^0 = -4.576 \times 4930$$
$$= -22,600 \text{ cal}$$

A similar plot of the data given in Table 7–3 for the heterogeneous dissociation of CuO(s) is shown in Figure 7–2. Here again a straight line is obtained for temperatures between 900 and 1080°C, but this time

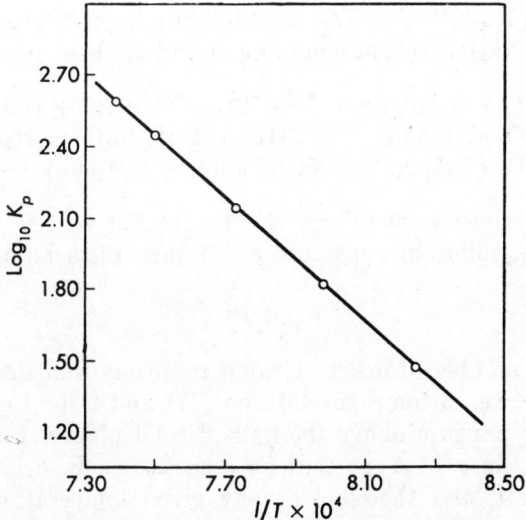

Figure 7–2.   Plot of $\log_{10} K_p$ vs. $1/T$ for the reaction $4\,CuO(s) = 2\,Cu_2O(s) + O_2(g)$.

the slope is negative and equal to $-13{,}200$. Consequently, $\Delta H^0 = 4.576(13{,}200) = 60{,}400$ cal.

VARIATION OF K_c WITH TEMPERATURE

An equation similar to (41) may also be derived for the variation of K_c of a gaseous reaction with temperature, namely,

$$\frac{d \ln K_c}{dT} = \frac{\Delta E^0}{RT^2} \tag{45}$$

where ΔE^0 is the heat of reaction at constant volume. This equation is integrated and handled in exactly the same manner as Eq. (41). As a general rule Eqs. (45) and (41) will yield different results and will be identical only when $K_c = K_p$ and $\Delta H^0 = \Delta E^0$, i.e., when $\Delta n_g = 0$ and there is no volume change on reaction.

EQUILIBRIA IN HYDRATES

An interesting type of heterogeneous equilibrium is that exhibited by hydrates, or in general solvates, of various substances. Under proper conditions these hydrates are found to dissociate into lower hydrates, or the anhydrous substance, and water vapor which establishes a definite vapor pressure above the solid phases. Thus $Na_2HPO_4 \cdot 12\,H_2O$

dissociates into $Na_2HPO_4 \cdot 7\ H_2O$, the latter into $Na_2HPO_4 \cdot 2\ H_2O$, and the dihydrate into Na_2HPO_4. The equilibria attending these dissociations are:

$$Na_2HPO_4 \cdot 12\ H_2O(s) = Na_2HPO_4 \cdot 7\ H_2O(s) + 5\ H_2O(g) \quad (46)$$
$$Na_2HPO_4 \cdot 7\ H_2O(s) = Na_2HPO_4 \cdot 2\ H_2O(s) + 5\ H_2O(g) \quad (47)$$
$$Na_2HPO_4 \cdot 2\ H_2O(s) = Na_2HPO_4(s) + 2\ H_2O(g) \quad (48)$$

These heterogeneous equilibria can be treated by the methods described. The equilibrium constants of all such dissociations are given simply by

$$K_p = P^x \quad (49)$$

where x is the number of moles of vapor resulting from the dissociation of the hydrate, i.e., 5 for Eqs. (46) and (47) and 2 for Eq. (48), while P is the vapor pressure above the pair of solid phases. From Eq. (49) it follows that, since P^x is constant for a particular dissociation, P must also be constant, and therefore at any given temperature the vapor pressure above any *hydrate pair* must be constant as long as both phases are present. This conclusion is in accord with observation. In the presence of any given hydrate and its lower dissociation product the pressure is found to be definite and characteristic of the particular hydrate pair. This is not true, however, when only a single solid phase is present, as under such conditions the vapor pressure can vary within certain limits. It is erroneous, therefore, to speak of the vapor pressure of a hydrate. The vapor pressure is constant only for a hydrate pair or a hydrate and its anhydride.

TABLE 7–6.   AQUEOUS VAPOR PRESSURE OF HYDRATE
PAIRS AT 25°C

Hydrate Pair	P (mm Hg)
$MgSO_4 \cdot 7\ H_2O - MgSO_4 \cdot 6\ H_2O$	11.5
$6\ H_2O - 5\ H_2O$	9.8
$5\ H_2O - 4\ H_2O$	8.8
$4\ H_2O - H_2O$	4.1
$H_2O - MgSO_4$	1.0
$CuSO_4 \cdot 5\ H_2O - CuSO_4 \cdot 3\ H_2O$	7.80
$3\ H_2O - H_2O$	5.60
$H_2O - CuSO_4$	0.8(?)
$Na_2HPO_4 \cdot 12\ H_2O - Na_2HPO_4 \cdot 7\ H_2O$	19.13
$7\ H_2O - 2\ H_2O$	14.51
$2\ H_2O - Na_2HPO_4$	9.80

In Table 7–6 are given the vapor pressures at 25°C for the hydrate pairs of several salts. It will be observed that the vapor pressure is highest for the pair richest in water and decreases as the water content of the solid phase decreases. A clearer appreciation of the relation between the pressures of the various hydrated forms of a substance at a particular temperature can be obtained by plotting the pressure against the number of moles of water in the solid phase, n. Such a plot for disodium phosphate is shown in Figure 7–3. The horizontal portions indicate the values of n over

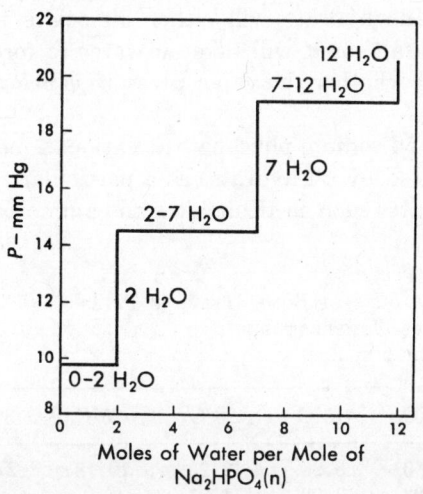

Figure 7–3. Vapor pressures of the hydrates of Na_2HPO_4 at 25°C.

which the particular vapor pressures will remain constant, i.e., between $n = 0$ and 2, $n = 2$ and 7, and $n = 7$ and 12. The vertical portions, on the other hand, give the *vapor pressure ranges* over which the pure solid phases (anhydrous salt, 2 H_2O, 7 H_2O, and 12 H_2O) are stable. Thus between zero and 9.80 mm Hg water vapor pressure anhydrous disodium phosphate does not combine with water vapor. As soon as the latter pressure is reached, however, some $Na_2HPO_4 \cdot 2\,H_2O$ is formed, and the pressure remains constant as long as any unconverted anhydrous salt is present. As soon as all of the disodium phosphate has been converted to the dihydrate, it is found that the aqueous tension can be varied between 9.80 and 14.51 mm Hg without formation of the heptahydrate; i.e., the range of stability of the dihydrate is from 9.80 to 14.51 mm Hg vapor pressure. At 14.51 mm Hg the heptahydrate phase begins to form, and the pressure again becomes constant, this time at 14.51 mm, until all of the dihydrate has been converted to the heptahydrate. The latter in turn remains stable between this pressure and 19.13 mm, at which pressure the 12 H_2O begins to form. Once the solid phase has been converted to the 12 H_2O, the pressure may be increased up to the pressure of the saturated solution of disodium phosphate in water. On reaching this pressure some of the solid will dissolve to form a solution in equilibrium with solid $Na_2HPO_4 \cdot 12\ H_2O$.

On dehydration the phenomena described will be found to take place in reverse order, i.e., the 12 H_2O will break down to the 7 H_2O, the 7 H_2O to 2 H_2O, etc. From this discussion it is apparent that any particular hydrate can be preserved only as long as the aqueous tension lies between the limits of stability exhibited by the hydrate at the given temperature. Outside these limits the hydrate will either dissociate to a lower hydrate or the anhydrous salt, or it will take on water to form a water-richer phase, depending on whether the vapor pressure is below or above the prescribed limits.

This behavior of the hydrates of disodium phosphate at various aqueous tensions is typical of that exhibited by all hydrates at a particular temperature, as well as by other solvates such as alcoholates and ammoniates of various salts.

TABLE 7–7.   VAPOR PRESSURES OF SOME HYDRATE PAIRS
AT VARIOUS TEMPERATURES
(P in mm Hg)

Hydrate Pair	0°C	15°C	20°C	25°C	30°C
$Na_2HPO_4 \cdot 12\ H_2O - 7\ H_2O$*	2.66	8.95	12.93	19.18	27.05
$SrCl_2 \cdot 6\ H_2O - 2\ H_2O$†	1.23	3.99	8.37	—	—
$Na_2SO_4 \cdot 10\ H_2O - Na_2SO_4$†	2.77	9.21	19.20	—	—

* Partington and Winterton, *J. Chem. Soc.*, **132**, 635 (1930); Baxter and Cooper, *J. Am. Chem. Soc.*, **46**, 927 (1924).
† Baxter and Lansing, *J. Am. Chem. Soc.*, **42**, 419 (1920).

The vapor pressures of hydrate and other solvate pairs increase with rise in temperature, as may be seen from Table 7–7, in a manner readily deducible from Eq. (49). Since $K_p = P^x$, then

$$\frac{d \ln P^x}{dT} = \frac{\Delta H^0}{RT^2} \tag{50}$$

and, therefore,

$$\frac{d \ln P}{dT} = \frac{\Delta H^0}{xRT^2} \tag{51}$$

On integration, assuming ΔH^0 constant, this leads to

$$\log_{10} P = \left(\frac{-\Delta H^0}{2.303\ xR}\right)\frac{1}{T} + C \tag{52}$$

Equation (52) is identical with the Clausius-Clapeyron equation **except** that the slope of $\log_{10} P$ vs. $1/T$, instead of being $(-\Delta H^0/2.303\,R)$, **is** given by $(-\Delta H^0/2.303\,xR)$. From the slope of such a plot the heat **of** dissociation of a hydrate or solvate readily follows as

$$\Delta H^0 = -2.303\,xR\,(\text{slope}) \tag{53}$$

REFERENCES

See references listed at end of Chapters 3 and 4. Also:

1. K. G. Denbigh, *The Principles of Chemical Equilibrium*, Cambridge University Press, Cambridge, 1955.
2. K. Jellinek, *Lehrbuch der physikalischen Chemie*, Ferdinand Enke, Stuttgart, 1930, Vol. III.

PROBLEMS

1. State which of the following equilibria are homogeneous and which **are** heterogeneous:

    (a) S (rhombic) = S (monoclinic)
    (b) $Fe_2O_3(s) + CO(g) = CO_2(g) + 2FeO(s)$
    (c) $2\,SO_2(g) + O_2(g) = 2\,SO_3(g)$
    (d) $CaCO_3(s) + H^+(aq) = HCO_3^-(aq) + Ca^{++}(aq)$

2. Formulate the equilibrium constants K_a and K_p for each of the **following** reactions:

    (a) $C_2H_6(g) = C_2H_4(g) + H_2(g)$
    (b) $2\,NO(g) + O_2(g) = 2\,NO_2(g)$
    (c) $NO_2(g) + SO_2(g) = SO_3(g) + NO(g)$
    (d) $3\,O_2(g) = 2\,O_3(g)$

3. A mixture of three gases, whose total pressure is 90 atm at 300°K, **contains** 3 moles of each gas. The activity coefficients of these gases when alone at **90 atm** pressure are, respectively, $\gamma_A = 0.760$, $\gamma_B = 0.920$, and $\gamma_C = 1.16$. What are the fugacities of the gases in the mixture?   *Ans.* $f_A = 22.8$ atm.

4. The standard free energy of formation of HCl(g) at 25°C is $-22,770$ cal/ mole. Calculate the thermodynamic equilibrium constant for the dissociation of HCl into its elements at 25°C.

5. For the synthesis of 1 mole of $NH_3(g)$ from the elements at 600°K, $\log_{10} K_\gamma$ as a function of the total equilibrium pressure P in atmospheres is given by **the** relation [Maron and Turnbull, *Ind. Eng. Chem.*, **33**, 69 (1941)]:

$$\log_{10} K_\gamma = -6.360 \times 10^{-4}\,P - 10.484 \times 10^{-8}\,P^2 + 1.750 \times 10^{-10}\,P^3$$

Determine the ratio of K_p to K_a at (a) 100 atm, (b) 300 atm, **and** (c) **500 atm** total equilibrium pressure.   *Ans.* (a) $K_p/K_a = 1.16$.

6. For the reaction $SO_2(g) + \frac{1}{2} O_2(g) = SO_3(g)$ $K_a = 6.55$ at $900°K$. Find $\Delta F°$ of the reaction at this temperature.

7. Calculate the ratio of K_p to K_c at $300°K$ for each of the reactions given in problem 2.   *Ans.* (a) 24.6.

8. For each of the following equilibria predict qualitatively the effect of increasing the total pressure upon the percentage of products present at equilibrium:

(a) $2 SO_3(g) + heat = 2 SO_2(g) + O_2(g)$
(b) $2 HI(g) = H_2(g) + I_2(g) + heat$
(c) $2 NO_2(g) = N_2O_4(g) + heat$
(d) $CO(g) + H_2O(g) = CO_2(g) + H_2(g) + heat$

9. In the preceding problem, what effect will a decrease in temperature have upon the percentage of product present at equilibrium?

10. At $248°C$ and 1 atm pressure $\alpha = 0.718$ for the dissociation

$$SbCl_5(g) = SbCl_3(g) + Cl_2(g)$$

Calculate K_p for the reaction.   *Ans.* $K_p = 1.07$.

11. At temperature T a compound $AB_2(g)$ dissociates according to the reaction

$$2 AB_2(g) = 2 AB(g) + B_2(g)$$

with a degree of dissociation, α, which is negligibly small compared with unity. Deduce the expression for α in terms of K_p and the total pressure P.

12. At $2155°C$ and 1 atm pressure $H_2O(g)$ is 1.18% decomposed into $H_2(g)$ and $O_2(g)$ in accordance with the equation

$$2 H_2O(g) = 2 H_2(g) + O_2(g)$$

Calculate K_p for the process.   *Ans.* $K_p = 8.38 \times 10^{-7}$.

13. For the reaction

$$2 NO_2(g) = 2 NO(g) + O_2(g)$$

at $184°C$ $K_p = 6.76 \times 10^{-5}$ for pressure in atm. Find the degree of dissociation of the $NO_2(g)$ at a total pressure of 1 atm.

14. $PCl_5(g)$ dissociates according to the reaction

$$PCl_5(g) = PCl_3(g) + Cl_2(g)$$

At $250°C$ $K_p = 1.78$ for P in atm. Find the density of the equilibrium mixture in grams per liter at a total pressure of 1 atm.

15. At $3000°K$ and 1 atm, CO_2 is 40% dissociated into CO and O_2. (a) What will be the degree of dissociation if the pressure is raised to 2 atm? (b) What will be the degree of dissociation when a mixture of 50% CO_2 and 50% O_2 is heated to $3000°K$, the pressure being 1 atm?   *Ans.* (a) 0.335; (b) 0.271.

16. From the average value of K_p in Table 7-1 calculate the mole percentage of NH_3 present at equilibrium at $450°C$ when the total pressure is 100 atm and the molar ratio of H_2 to N_2 is 3 to 1.

17. Repeat the calculation in the preceding problem, assuming that the initial molar ratio of H_2 to N_2 is 1 to 1 and the total equilibrium pressure is 10 atm.

18. At 30°C K_p in atmospheres for the dissociation

$$SO_2Cl_2(g) = SO_2(g) + Cl_2(g)$$

is 2.9×10^{-2}. Calculate the degree of dissociation when the total equilibrium pressure is 1 atm.

19. At 25°C ΔF^0 for the reaction

$$N_2O_4(g) = 2 NO_2(g)$$

is $+1380$ cal. What is the degree of dissociation at 25°C when the total pressure is 10 atm?   *Ans. $\alpha = 0.0494$.*

20. In the preceding problem calculate the degree of dissociation when the total pressure is maintained at 10 atm, but a partial pressure of CO_2 equal to 5 atm is present at equilibrium.

21. NH_4HS dissociates as follows:

$$NH_4HS(s) = NH_3(g) + H_2S(g)$$

At 25°C the dissociation pressure of the pure solid is 500 mm Hg. Calculate (a) K_p and (b) the total pressure at equilibrium when 300 mm of NH_3 are introduced into a flask containing solid NH_4HS.

Ans. (a) 6.25×10^4 mm²; (b) 583.0 mm.

22. For the equilibrium

$$C \text{ (graphite)} + CO_2(g) = 2 CO(g)$$

at 1123°K the mole per cent of CO in the vapor phase at equilibrium is 93.77% at 1 atm pressure. What is (a) K_p, and (b) the mole per cent of CO present at equilibrium when the total pressure is 10 atm?

23. In two experiments solid $NH_2CO_2NH_4$ is introduced at 30°C into a flask containing (a) a partial pressure of 200 mm of NH_3 gas and (b) a partial pressure of 200 mm of CO_2 gas. Using the average value of K_p in Table 7–5, find what will be the total pressure at equilibrium in each case.

24. For the reaction

$$H_2S(g) + I_2(s) = 2 HI(g) + S(s, \text{rhombic})$$

$K_p = 1.33 \times 10^{-5}$ atm at 60°C. What will be the mole fraction of HI in the vapor at 60°C when the total pressure is 1 atm?   *Ans. 0.00365.*

25. At 713°K, K_p for the reaction

$$Sb_2S_3(s) + 3 H_2(g) = 2 Sb(s) + 3 H_2S(g)$$

is 0.429. What is the mole fraction of H_2 in the vapor at 713°K? Will the result depend upon the total pressure? Explain.

26. For the reaction,

$$2 \, CaSO_4(s) = 2 \, CaO(s) + 2 \, SO_2(g) + O_2(g)$$

$K_p = 1.45 \times 10^{-5}$ atm^3 at 1625°K. What is the dissociation pressure of $CaSO_4$ in mm Hg at 1625°K?

27. One mole of H_2 and 1 mole of $Sb_2S_3(s)$ are introduced into a reaction vessel and heated to 713°K. From the data given in problem 25 find the number of moles of Sb formed and the number of moles of H_2 unconverted.

28. The equilibrium constant for the dissociation

$$2 \, H_2S(g) = 2 \, H_2(g) + S_2(g)$$

is $K_p = 0.0118$ at 1065°C, while the heat of dissociation is $\Delta H^0 = 42,400$ cal. Find the equilibrium constant of the reaction at 1200°C.   *Ans.* $K_p = 0.0507$.

29. For the reaction

$$SO_2(g) + \frac{1}{2} \, O_2(g) = SO_3(g)$$

$K_p = 6.55$ at 900°K and $K_p = 1.86$ at 1000°K. Calculate the heat of the reaction over the temperature interval 900 to 1000°K.

30. For the reaction
$$2 \, SO_3(g) = 2 \, SO_2(g) + O_2(g)$$

$\Delta H^0 = 46,980$ cal and $\Delta F^0 = 33,460$ cal at 25°C. Assuming ΔH^0 to be independent of temperature, calculate (a) ΔF^0 and (b) the degree of dissociation, α, of $SO_3(g)$ at 600°K and 0.5 atm total pressure.   *Ans.* (b) $\alpha = 6.3 \times 10^{-3}$.

31. From the heats and free energies of formation given in tables of preceding chapters, find the degree of dissociation of $NO_2(g)$ into $NO(g)$ and $O_2(g)$ at 200°C and 1 atm pressure. Assume ΔH^0 to be independent of temperature.

32. For the reaction

$$CO(g) + H_2O(g) = CO_2(g) + H_2(g)$$

the following are the absolute entropies S^0, and heats of formation ΔH^0, at 25°C:

	S^0(eu)	ΔH^0(cal)
CO	47.30	−26,420
H_2O	45.11	−57,800
CO_2	51.06	−94,050
H_2	31.21	—

Assuming that ΔH^0 is constant with T, calculate K_p for the reaction at 600°K.

33. For the reaction

$$2 \, NaHSO_4(s) = Na_2S_2O_7(s) + H_2O(g)$$

$\Delta H^0 = 19,800$ cal and $\Delta F^0 = 9000$ cal at 25°C. Assuming ΔH^0 to be constant, calculate the dissociation pressure of $NaHSO_4$ at 700°K.   *Ans.* 55.3 atm.

34. Starting with Eq. (14) of this chapter, derive Eq. (45).

35. For the reaction

$$H_2(g) + I_2(g) = 2\ HI(g)$$

$K_c = 50.0$ at $448°C$ and $K_c = 66.9$ at $350°C$. Find ΔE^0 and ΔH^0 of the reaction.

36. One g of $Na_2HPO_4 \cdot 7\ H_2O$ is placed in a 2-liter vessel at $25°C$. What weight of H_2O would have to be added in order to convert practically completely, this hydrate into $Na_2HPO_4 \cdot 12\ H_2O$ and establish a condition of equilibrium in the vessel? Use the data given in Table 7-7. *Ans.* 0.372 g.

37. For the reaction

$$CuSO_4 \cdot 3\ H_2O(s) = CuSO_4 \cdot H_2O(s) + 2\ H_2O(g)$$

the dissociation pressure at $298.2°K$ is 7.37×10^{-3} atm, while $\Delta H^0_{298.2°K} = 27{,}000$ cal. Assuming that ΔH^0 does not vary with T, what is the dissociation pressure in mm Hg at $100°C$?

38. For the reaction

$$(CH_3)_2CHOH(g) = (CH_3)_2CO(g) + H_2(g)$$

K_p at $457.4°K$ is 0.36, $\Delta C_p = 4.0$, and $\Delta H^0_{298.2°K} = 14{,}700$ cal. (a) Derive an expression for $\log_{10} K_p$ as a function of T, and (b) calculate K_p at $500°K$.

$$\textit{Ans. (a) } \log_{10} K_p = -\frac{2950}{T} + 2.01 \log_{10} T + 0.656; \text{ (b) } K_p = 1.42.$$

39. For the reaction $2\ H_2(g) + S_2(g) = 2\ H_2S(g)$

$$\Delta F^0 = -38{,}810 + 15.41\ T \log_{10} T - 2.065 \times 10^{-3}\ T^2 - 25.02\ T$$

Deduce the expressions for $\ln K_p$, ΔH^0, ΔS^0, and ΔC_p^0 of the reaction as a function of the temperature.

40. For the reaction $S_2(g) + 2\ O_2(g) = 2\ SO_2(g)$ $\Delta H^0_{25°C} = -172{,}900$ cal, $\Delta S^0_{25°C} = -33.67$ eu, and ΔC_p^0 as a function of temperature is given by

$$\Delta C_p^0 = -1.49 + 1.424 \times 10^{-3}\ T - \frac{0.336 \times 10^5}{T^2}$$

From these data find the value of K_p of the reaction at $1000°K$.

41. For the reaction $S_2(g) = 2\ S(g)$ $\log_{10} K_p = -16.735$ at $1000°K$, while

$$\Delta H^0 = 102{,}600 + 2.47\ T - 0.444 \times 10^{-3}\ T^2$$

Find the extent to which $S_2(g)$ will be dissociated into atoms at a temperature of $4000°K$ and 1 atm pressure.

42. For the reaction $2\ NaHCO_3(s) = Na_2CO_3(s) + CO_2(g) + H_2O(g)$

$$\Delta H^0 = 29{,}320 + 9.15\ T - 12.75 \times 10^{-3}\ T^2, \text{ and } \Delta F^0_{25°C} = 7080 \text{ cal}$$

What will be the partial pressure of $H_2O(g)$ at $400°K$ above a mixture of the two solids to which is added CO_2 at an initial pressure of 100 mm Hg?

Ans. 1.89 atm.

43. For the reaction $MnCO_3(s) = MnO(s) + CO_2(g)$

$$\Delta F^0 = 27,660 - 14.16\,T\log_{10} T + 10.7 \times 10^{-3}\,T^2 - 10.19\,T$$

Determine the temperature at which the dissociation pressure of $CO_2(g)$ will be 0.5 atm.

44. For the reaction $2\,Mo(s) + CH_4(g) = Mo_2C(s) + 2\,H_2(g)$ $K_p = 3.55$ at $973°K$. What will have to be the initial pressure of methane in order to yield $H_2(g)$ at an equilibrium pressure of 0.75 atm?

45. The integral heat of solution of $MgCl_2(s)$ at $18°C$ is $-35,900$ cal, that of $MgCl_2 \cdot 6\,H_2O(s)$ is -2950 cal, while the heat of vaporization of H_2O is 587 cal/g. Find ΔH at $18°C$ for the reaction

$$MgCl_2 \cdot 6\,H_2O(s) = MgCl_2(s) + 6\,H_2O(g)$$

Ans. 96,350 cal.

46. For the transition

$$HgS\ (red) = HgS\ (black)$$

$\Delta F^0 = 4100 - 6.09\,T$. What is the stable modification of HgS at $100°C$? What is the transition temperature at 1 atm pressure?

8
Solutions

$$W$$

HEN several nonreacting substances are mixed, three possible types of mixtures may be obtained: (a) a coarse mixture, such as that of salt and sugar; (b) a colloidal dispersion, such as results when fine clay is shaken with water; or (c) a true solution, obtained when a substance like sugar dissolves in water. In the coarse mixture, the individual particles are readily discernible and may be separated from each other by mechanical means. Although in a colloidal dispersion the particles are much finer and the heterogeneity is not so readily apparent, the dispersion is, nevertheless, not homogeneous. On the other hand, in the true solution the constituents cannot be separated from each other by mechanical means, and every part of the solution is found to be like every other part; i.e., a true solution constitutes a homogeneous phase.

It is frequently convenient to refer to the substance that dissolves as the *solute* and to the substance in which solution takes place as the *solvent*. For the solubility of solids in liquids, where the liquid is usually present in large excess over the solid, there is no ambiguity in these terms, the solid being the solute, the liquid the solvent. However, when dealing with the solubility of such liquids as acetone and water or dioxane and water, which dissolve in each other in all proportions, it is difficult to differentiate between solute and solvent. Here these terms will be employed only when there is no ambiguity as to meaning.

A solution which contains at a given temperature as much solute as it can hold in presence of the dissolving substance is said to be *saturated*. Any solution which contains less than this amount of solute is *unsaturated*, while if it contains more than this amount it is *supersaturated*. A supersaturated solution can exist only in the absence of dissolving substance and is at best very unstable. Jarring and stirring may, and introduction of solute will, cause the precipitation of excess solute in solution, leading to the formation of a saturated solution. To determine the state of a solution with respect to saturation it is only necessary to introduce some of the dissolving substance. If the substance dissolves, the solution is unsaturated; if no further solubility takes place, the solution is saturated; while if precipitation takes place, the original solution was supersaturated.

FACTORS AFFECTING SOLUBILITY

The extent to which a substance will dissolve in another varies greatly with different substances and depends on the nature of the solute and solvent, the temperature, and the pressure. In general the effect of pressure on solubility is small unless gases are involved. However, the effect of temperature is usually very pronounced. The direction in which the solubility of a substance in a solvent changes with temperature depends on the heat of solution. If a substance dissolves at saturation with evolution of heat, the solubility decreases with rising temperature. On the other hand, if a substance dissolves with absorption of heat, the solubility increases as the temperature is raised.

In general, compounds of similar chemical character are more readily soluble in each other than are those whose chemical character is entirely different. When a similarity of chemical nature exists between two substances, the solution of the two will have an environment not too different from that of the pure substances, and the two can tolerate each other in solution. On the other hand, when the chemical nature of the two substances is considerably different, the substances may not be able to tolerate each other, and hence there may be little tendency to dissolve. Between these two extremes a considerable number of intermediate stages of similarity is possible, and this will account for the wide ranges of solubility of various substances in each other.

These points may be illustrated with the phenomena encountered in the mutual solubility of liquids. When ethyl alcohol and water, which are closely related chemically, are mixed, the two dissolve in each other in all proportions, i.e., there is no saturation limit. Such substances are said to be *completely miscible*. In distinction to these, two liquids such as water and mercury, which are very different chemically, do not dissolve in each other at all and are said to be *completely immiscible*. Between these two limiting types there are liquid pairs, such as ether and water, which dissolve in each other to a limited extent only. Thus, pure ether dissolves a certain amount of water to form a saturated solution of water in ether, while water dissolves a limited amount of ether to form a saturated solution of ether in water. Consequently, with high proportions of one or the other of these liquids, a completely miscible solution can be obtained. When the proportions taken are outside these saturation limits, however, two layers are obtained, one composed of a solution of ether in water, the other of water in ether. Liquid pairs of this sort are said to be *partially miscible.*

TYPES OF SOLUTIONS

Although solutions with many components can be prepared, attention will be confined to binary solutions, i.e., solutions containing two components only. Since the solvent and solute may be either gaseous, liquid, or solid, the number of possible types of binary solutions that may be expected is nine, namely:

1. Solution of a gas in a gas
2. Solution of a liquid in a gas
3. Solution of a solid in a gas
4. Solution of a gas in a solid
5. Solution of a liquid in a solid
6. Solution of a solid in a solid
7. Solution of a gas in a liquid
8. Solution of a solid in a liquid
9. Solution of a liquid in a liquid

Of these types, the last three are common and familiar. The other types require some comment.

All gases are miscible in all proportions, yielding solutions whose physical properties are very nearly additive provided the total pressure is not too high. Under the latter conditions the partial and total pressures are governed by Dalton's law, the partial and total volumes by Amagat's law. Both of these principles have already been discussed.

The vaporization of a liquid and the sublimation of a solid into a gas phase may be considered as solution of these substances in a gas. These processes involve first the conversion of the liquid or solid to vapor, and the subsequent solution of the vapor in the gas. Because the vaporization and sublimation pressures of a substance are fixed at any given temperature, the amounts of liquid and solid that can vaporize into a given volume of gas are limited to the amount necessary to establish the equilibrium pressures.

Gases and liquids may dissolve ih solids to form apparently true homogeneous solutions. Examples are the solubility of hydrogen in palladium and the solubility of liquid benzene in solid iodine. Both solutions formed are solid.

When two solids dissolve in each other, the solutions formed may be completely or partially miscible, depending on the nature of the substances involved and the temperature. Examples of salts forming solid solutions are potassium and ammonium sulfates, copper and ferrous sulfates, and the alums of ammonium and potassium. Many metal pairs likewise form solid solutions, as for instance gold and platinum, gold and palladium, silver and palladium, and copper and nickel. Since the formation of a solid solution would occur extremely slowly while both materials remained solid, it is necessary to resort to crystallization, either from solution in the case of salts or from the molten materials in the case of metals, in order to obtain these solid solutions.

Temperature has no influence on the solubility when both substances involved in the solid solution formation are completely miscible in the solid state. When the two are only partially miscible, however, the extent of solubility depends on the temperature. Some substances may be completely miscible in the solid state at higher temperatures and only partially miscible at lower ones, a transition occurring from one type to the other. There are also other possibilities, but these will be discussed in greater detail in the chapter on the Phase Rule.

THE THERMODYNAMIC PROPERTIES OF A SOLUTION

The total free energy of a solution, F, is given by Eq. (8) of Chapter 3, namely,

$$F = \bar{F}_1 n_1 + \bar{F}_2 n_2 + \cdots \qquad (1)$$

where $\bar{F}_1$, $\bar{F}_2$, etc., are the partial molal free energies of the solution constituents, and n_1, n_2, etc., are the numbers of moles of the various constituents present. Similarly the total entropy and enthalpy are given, respectively, by the relations

$$S = \bar{S}_1 n_1 + \bar{S}_2 n_2 + \cdots \qquad (2)$$
$$H = \bar{H}_1 n_1 + \bar{H}_2 n_2 + \cdots \qquad (3)$$

Since $F = H - TS$, it readily follows from Eqs. (1), (2), and (3) that

$$\bar{F}_1 = \bar{H}_1 - T\bar{S}_1 \qquad (4)$$
$$\bar{F}_2 = \bar{H}_2 - T\bar{S}_2 \qquad (5)$$

and so on for the various solution constituents. Again, Eqs. (28), (29), and (30) of Chapter 6 also apply to the solution constituents, and we get thus for each constituent

$$\left(\frac{\partial \bar{F}_i}{\partial T}\right)_{n,P} = -\bar{S}_i \tag{6}$$

$$\left(\frac{\partial \bar{F}_i}{\partial P}\right)_{n,T} = \bar{V}_i \tag{7}$$

and
$$\left[\frac{\partial(\bar{F}_i/T)}{\partial T}\right]_{n,P} = -\frac{\bar{H}_i}{T^2} \tag{8}$$

In these equations n indicates the constancy of the solution concentration, i.e., n_1, n_2, etc., are all held constant.

Utilizing Eq. (51) of Chapter 6, we can write for the partial molal free energy of any constituent of a solution

$$\bar{F}_i = \bar{F}_i^0 + RT \ln a_i \tag{9}$$

where $\bar{F}_i^0$ is the partial molal free energy of constituent i in some suitably chosen standard state, and a_i is its activity in the solution at any given concentration. The particular state chosen as standard is rather arbitrary. In dealing with solutions of substances which are completely miscible it is customary to take as the standard states for all solution constituents the pure substances. Under these conditions $\bar{F}_i^0$ is identical with F_i^0 of the pure substances, i.e., $\bar{F}_i^0 = F_i^0$, and hence the standard states are the same as those defined in Chapter 6. Again, if we employ mol fractions to express concentration, then a_i can be related to N_i by the equation

$$a_i = N_i \gamma_i' \tag{10}$$

where γ_i' is an activity coefficient converting N_i to a_i. On this basis Eq. (9) for $\bar{F}_i$ becomes

$$\bar{F}_i = F_i^0 + RT \ln N_i \gamma_i' \tag{11}$$

An alternate method frequently employed for defining standard states utilizes the above definition only for the solvent, i.e., for the solvent $\bar{F}_1^0 = F_1^0$, and $a_1/N_1 = \gamma_1' = 1$ as $N_1 \to 1$. In turn, for a nondissociating solute a_2 is taken as

$$a_2 = N_2 \gamma_2' \tag{12}$$

where γ_2' is the activity coefficient of the solute, and the definition is completed by saying that

$$\gamma_2' = \frac{a_2}{N_2} = 1 \quad \text{as } N_2 \to 0 \tag{13}$$

Here the infinitely dilute solution of the solute in the solvent is taken as a point of reference, and in this state we have $a_2 = N_2 = 0$ and $\gamma_2' = 1$. Then, the concentration of solution that turns out to have $a_2 = N_2 \gamma_2' = 1$ becomes the standard state for the solute, and in this state $\bar{F}_2 = \bar{F}_2^0$. The definitions of standard and reference states for dissociating solutes will be discussed in Chapter 11.

Equations similar to (12) and (13) can be used to relate a_2 to concentrations expressed in units other than mol fractions. Thus, if we wish to use moles per liter, C, then we can say that

$$a_2 = C_2 f_2 \tag{14}$$

and

$$f_2 = \frac{a_2}{C_2} = 1 \quad \text{as } C_2 \to 0 \tag{15}$$

Here f_2 is the activity coefficient of the solute for concentration C in moles per liter. Again, if we wish to use molality m and activity coefficient γ_2, then

$$a_2 = m_2 \gamma_2 \tag{16}$$

and

$$\gamma_2 = \frac{a_2}{m_2} = 1 \quad \text{as } m_2 \to 0 \tag{17}$$

On the basis of these various definitions $\gamma_2' = f_2 = \gamma_2 = 1$ in the infinitely dilute solution, but at all other concentrations these various activity coefficients in general will have different values for a given solution concentration.

THE SOLUTION PROCESS

Consider the process

$$n_1 A_1 + n_2 A_2 = \text{Solution} \tag{18}$$

in which n_1 moles of pure substance A_1 are mixed with n_2 moles of pure substance A_2 to form a binary solution. The change ΔG_m in any extensive thermodynamic property of the system at any given temperature and pressure is then

$$\Delta G_m = G - (n_1 G_1^0 + n_2 G_2^0) \tag{19}$$

where G is the property for the solution, and G_1^0 and G_2^0 the values of the property per mole for the two pure constituents. Again, since G is given by

$$G = n_1 \bar{G}_1 + n_2 \bar{G}_2 \tag{20}$$

then insertion of Eq. (20) into Eq. (19) yields

$$\begin{aligned}
\Delta G_m &= (n_1 \bar{G}_1 + n_2 \bar{G}_2) - (n_1 G_1^0 + n_2 G_2^0) \\
&= n_1 (\bar{G}_1 - G_1^0) + n_2 (\bar{G}_2 - G_2^0) \\
&= n_1 \overline{\Delta G_1} + n_2 \overline{\Delta G_2}
\end{aligned} \tag{21}$$

where

$$\overline{\Delta G_1} = \bar{G}_1 - G_1^0 \tag{22}$$

and

$$\overline{\Delta G_2} = \bar{G}_2 - G_2^0 \tag{23}$$

In these equations ΔG_m represents the integral change in some thermo-dynamic property as a result of the mixing process, while $\overline{\Delta G_1}$ and $\overline{\Delta G_2}$ represent the changes in the partial molal quantities for the property involved.

Equation (21) applied to the free energy change attending the mixing process yields for the free energy of mixing

$$\Delta F_m = n_1(\bar{F}_1 - F_1^0) + n_2(\bar{F}_2 - F_2^0) \tag{24}$$

Similarly we obtain for the heat, entropy, and volume change on mixing

$$\Delta H_m = n_1(\bar{H}_1 - H_1^0) + n_2(\bar{H}_2 - H_2^0) \tag{25}$$
$$\Delta S_m = n_1(\bar{S}_1 - S_1^0) + n_2(\bar{S}_2 - S_2^0) \tag{26}$$
$$\Delta V_m = n_1(\bar{V}_1 - V_1^0) + n_2(\bar{V}_2 - V_2^0) \tag{27}$$

Further, at any temperature T, the free energy, entropy, and heat of mixing are related to each other by the equation

$$\Delta F_m = \Delta H_m - T\Delta S_m \tag{28}$$

Equation (24) can be written in a different form by introduction of Eq. (9). On the basis of the pure constituents as standard states, Eq. (9) becomes

$$\bar{F}_i = F_i^0 + RT \ln a_i \tag{29}$$

and hence we get for the constituents of the binary solution

$$\bar{F}_1 - F_1^0 = RT \ln a_1 \tag{30}$$
and $$\bar{F}_2 - F_2^0 = RT \ln a_2 \tag{31}$$

Substituting these expressions into Eq. (24) we get for ΔF_m

$$\Delta F_m = n_1 RT \ln a_1 + n_2 RT \ln a_2 \tag{32}$$

How Eq. (32) is used to find ΔF_m will be explained immediately following the next section.

CONDITION FOR EQUILIBRIUM BETWEEN PHASES

We have seen on page 211 that the condition for equilibrium between phases of a pure substance at constant temperature and pressure is that the molar free energy of the substance be the same in all phases. The question now is: What is the condition for equilibrium between phases when the phases are not pure, i.e., when they are solutions? To answer this question, consider a system composed of several phases, each of which may contain a number of components. The total free energy of each of the phases will be given in general by Eq. (1) of this chapter. Focus attention now on any two of the phases present, and suppose that

dn_1, dn_2, etc., moles of the various constituents are transferred from one of these phases to the other at constants T and P. The decrease in free energy of one of these phases will be then

$$dF = \bar{F}_1 dn_1 + \bar{F}_2 dn_2 + \cdots$$

while the increase in free energy of the second phase will be given by

$$dF' = \bar{F}'_1 dn_1 + \bar{F}'_2 dn_2 + \cdots$$

The net change in the free energy of transfer for the process described, dF, will be thus

$$dF = dF' - dF = (\bar{F}'_1 - \bar{F}_1)dn_1 + (\bar{F}'_2 - \bar{F}_2)dn_2 + \cdots$$

However, if equilibrium is to exist between the two phases, then we must have $(dF)_{T,P} = 0$, or

$$(\bar{F}'_1 - \bar{F}_1)dn_1 + (\bar{F}'_2 - \bar{F}_2)dn_2 + \cdots = 0 \tag{33}$$

Again, since the transfer of each constituent from one phase to another can occur independently, it must follow also that each term in Eq. (33) must be zero; namely

$$(\bar{F}'_1 - \bar{F}_1)dn_1 = (\bar{F}'_2 - \bar{F}_2)dn_2 = (\bar{F}'_i - \bar{F}_i)dn_i = 0 \tag{34}$$

Furthermore, since dn_1, dn_2, etc., are not zero, the other terms must be, and hence $\bar{F}'_1 = \bar{F}_1$, $\bar{F}'_2 = \bar{F}_2$, and so on. Finally, since the same argument can be repeated with any pair of phases chosen, the above result must be valid for all of them, namely, that for equilibrium between phases in a system we must have

$$\bar{F}_1 = \bar{F}'_1 = \bar{F}''_1 \text{ etc.} \tag{35a}$$
$$\bar{F}_2 = \bar{F}'_2 = \bar{F}''_2 \text{ etc.} \tag{35b}$$

or, in general,

$$\bar{F}_i = \bar{F}'_i = \bar{F}''_i \text{ etc.} \tag{35c}$$

Equation (35) states that, for equilibrium at constants T and P in a multicomponent system composed of a number of phases, *the partial molal free energy of each constituent must be the same in all the phases.* If one of the phases happens to be a pure substance, the above deductions still apply, except that now $\bar{F}_i$ for that substance in the pure phase becomes F_i^0.

EQUILIBRIUM BETWEEN A SOLUTION AND ITS VAPOR PHASE

Consider a solution containing components which are volatile. Such a solution, placed in an evacuated space, will establish a vapor phase which will contain the solution constituents. The conditions which must obtain in the system in order to have equilibrium between the solution and the vapor phase can be obtained as follows.

The molar free energy of any constituent i in the gas phase is given by Eq. (54) of Chapter 6, namely,

$$F_{i(g)} = F^0_{i(g)} + RT \ln f_{i(g)} \tag{36}$$

where the subscript g refers to the gas. Again the partial molal free energy of the same constituent in solution is given by Eq. (29) of this chapter

$$\bar{F}_i = F^0_i + RT \ln a_i \tag{37}$$

For equilibrium between the solution and the gas phase we must have $F_{i(g)} = \bar{F}_i$. Consequently, we get from Eqs. (36) and (37)

$$F^0_{i(g)} + RT \ln f_{i(g)} = F^0_i + RT \ln a_i$$

or

$$\ln \frac{f_{i(g)}}{a_i} = \frac{F^0_i - F^0_{i(g)}}{RT} \tag{38}$$

Since at any given temperature the right-hand side of Eq. (38) is a constant, then the quantity within the logarithm must also be constant, and we get thus

$$\frac{f_{i(g)}}{a_i} = K \tag{39}$$

To evaluate K we utilize the fact that, for pure constituent i, $a_i = 1$ and $f_{i(g)} = f^0_{i(g)}$, where $f^0_{i(g)}$ is the fugacity of the vapor above the pure constituent. Inserting these quantities into Eq. (39) we see that $K = f^0_{i(g)}$, and hence the equation becomes

$$a_i = \frac{f_{i(g)}}{f^0_{i(g)}} \tag{40}$$

Equation (40) shows that the activity of any volatile constituent of a solution is given by the ratio of the fugacity of the vapor of the constituent in equilibrium with the solution divided by the fugacity of the vapor in equilibrium with the pure constituent. Again, $f_{i(g)} = P_i \gamma_{i(g)}$, where P_i is the vapor pressure of the given constituent above the solution and $\gamma_{i(g)}$ its activity coefficient. Similarly, $f^0_{i(g)} = P^0_i \gamma^0_{i(g)}$, where P^0_i is the vapor pressure of the pure constituent and $\gamma^0_{i(g)}$ its activity coefficient. Upon inserting these identities into Eq. (40) we get

$$a_i = \frac{P_i \gamma_{i(g)}}{P^0_i \gamma^0_{i(g)}} \tag{41}$$

Finally, if the vapors behave as ideal gases, then $\gamma_{i(g)} = \gamma^0_{i(g)} = 1$, and Eq. (41) becomes simply

$$a_i = \frac{P_i}{P^0_i} \tag{42}$$

Thus, a_i can be obtained from the vapor pressures of the constituent above the solution and when pure.

Substitution of Eq. (41) into Eq. (32) yields for the free energy of mixing of a binary solution

$$\Delta F_m = n_1 RT \ln \frac{P_1 \gamma_{1(g)}}{P_1^0 \gamma_{1(g)}^0} + n_2 RT \ln \frac{P_2 \gamma_{2(g)}}{P_2^0 \gamma_{2(g)}^0} \tag{43}$$

When the gases behave ideally the γ's become unity, and Eq. (43) simplifies to

$$\Delta F_m = n_1 RT \ln \frac{P_1}{P_1^0} + n_2 RT \ln \frac{P_2}{P_2^0} \tag{44}$$

Here all the quantities are directly measurable, and hence ΔF_m can be readily obtained from vapor pressure data.

IDEAL SOLUTIONS

Just as it was found convenient to set up an ideal gas as a criterion of gas behavior, so too is it desirable to define an ideal solution in terms of whose predictable properties can be judged the behavior of real solutions. For this purpose *we shall define an ideal solution as one in which the activity of each constituent is equal to its mol fraction under all conditions of temperature, pressure, and concentration;* i.e., $a_1 = N_1$, $a_2 = N_2$, etc. On this basis Eq. (32) becomes for a binary ideal solution

$$\Delta F_m = n_1 RT \ln N_1 + n_2 RT \ln N_2 \tag{45}$$

Since the right-hand side of Eq. (45) is independent of pressure, we obtain on differentiation of ΔF_m with respect to pressure at constant temperature

$$\left(\frac{\partial \Delta F_m}{\partial P} \right)_T = \Delta V = 0 \tag{46}$$

This result means that on mixing the two pure constituents to form an ideal solution there is no change in volume attending the process, and hence the partial molal volumes of the solution constituents are identical with the molar volumes of the pure constituents. Again, if we divide Eq. (45) by T and differentiate $\Delta F_m/T$ with respect to T at constant P, we get

$$\left[\frac{\partial (\Delta F_m/T)}{\partial T} \right]_P = - \frac{\Delta H_m}{T^2} = 0$$

and hence

$$\Delta H_m = 0 \tag{47}$$

Consequently an ideal solution is formed without any evolution or absorption of heat. Insertion of Eq. (47) into Eq. (28) shows then that for an ideal solution $\Delta F_m = -T\Delta S_m$, and hence ΔS_m follows from Eq. (45) as

$$\Delta S_m = -[n_1 R \ln N_1 + n_2 R \ln N_2] \tag{48}$$

Equation (48) predicts that for an ideal solution ΔS_m is a function of only the concentration and quantity of constituents present. In turn, Eq. (45) shows ΔF_m to be a function of these variables as well as of temperature. Furthermore, neither equation contains any factors specific of the nature of the substances involved. In this respect these equations are analogous to the ideal gas law, which predicts gas behavior to be independent of any characteristics specific of the nature or structure of the gas molecules.

Figures 8–1 and 8–2 show comparisons between ideal solution predictions and actual behavior observed for the ΔF_m, ΔH_m, and $T\Delta S_m$ of two binary solutions of liquids. The calculations given are *per mole of solution*, i.e., $n_1 = N_1$ and $n_2 = N_2$. Further, the vapor pressure data have been corrected for gas nonideality. Inspection of Figure 8–1 shows

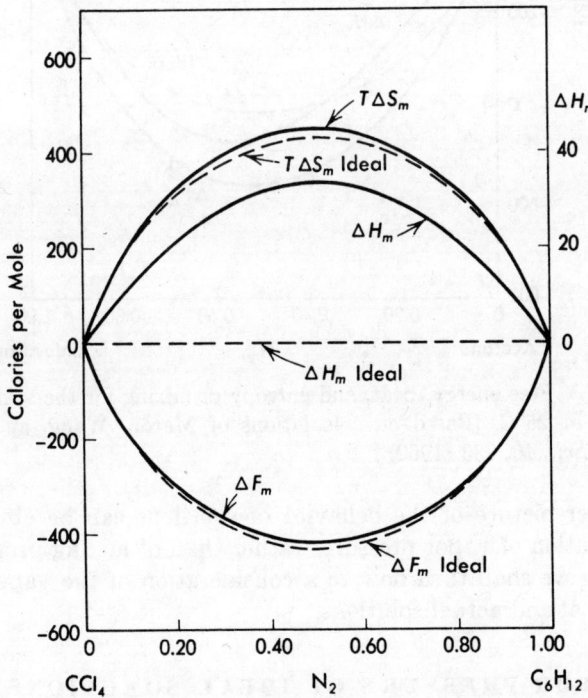

Figure 8–1. Free energy, heat, and entropy of mixing for the system carbon tetrachloride-cyclohexane at 40°C. [Based on calculations of Maron, Wang, and Nakajima, *J. Polymer Sci.*, **46**, 333 (1960).]

that the system carbon tetrachloride-cyclohexane approximates quite closely an ideal solution. However, such is not the case with the system acetone-chloroform, where all the thermodynamic properties of the solution show very large deviations from ideality.

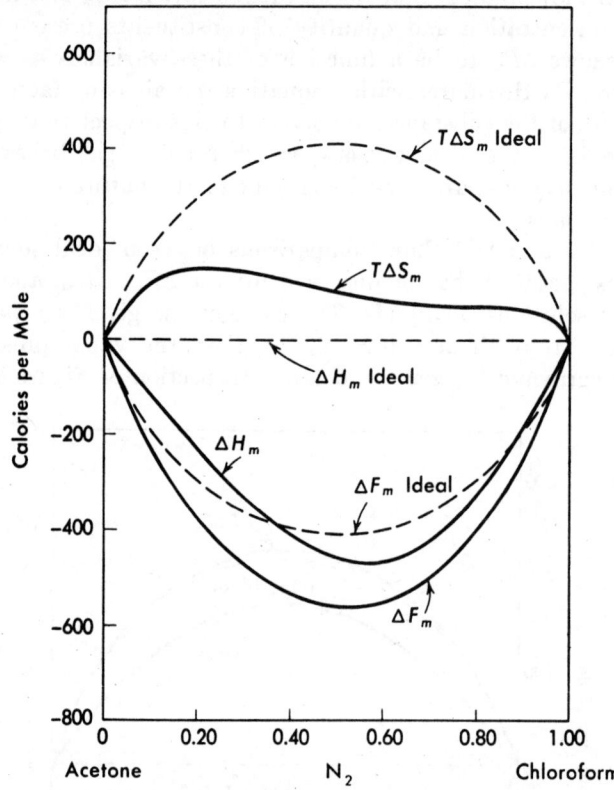

Figure 8–2. Free energy, heat, and entropy of mixing for the system acetone-chloroform at 25°C. [Based on calculations of Maron, Wang, and Nakajima, *J. Polymer Sci.*, **46,** 333 (1960).]

A simpler picture of the behavior of solutions can be obtained from an examination of vapor pressures rather than of mixing properties. For this reason we shall turn now to a consideration of the vapor pressures of both ideal and actual solutions.

THE VAPOR PRESSURE OF IDEAL SOLUTIONS

Suppose that two liquids, A_1 and A_2, are volatile and completely miscible, and suppose further that the two liquids dissolve in each other to form ideal solutions. Since the solutions are ideal, $a_1 = N_1$ and $a_2 = N_2$. Inserting these identities into Eq. (42), we obtain

$$P_1 = N_1 P_1^0 \tag{49}$$

and

$$P_2 = N_2 P_2^0 \tag{50}$$

Equations (49) and (50) are expressions of *Raoult's law*, which states that *the partial vapor pressure of any volatile constituent of a solution is equal to the vapor pressure of the pure constituent multiplied by the mol fraction of that constituent in solution.* From these equations the total vapor pressure, P, above such a solution follows as

$$\begin{aligned} P &= P_1 + P_2 \\ &= P_1^0 N_1 + P_2^0 N_2 \end{aligned} \tag{51}$$

Further, since $N_1 = 1 - N_2$, Eq. (51) can also be written as

$$\begin{aligned} P &= P_1^0(1 - N_2) + P_2^0 N_2 \\ &= (P_2^0 - P_1^0)N_2 + P_1^0 \end{aligned} \tag{52}$$

For any given system and temperature P_1^0 and P_2^0 are constant, and hence a plot of P vs. N_2 should be a straight line with $P = P_1^0$ at $N_2 = 0$ and $P = P_2^0$ at $N_2 = 1$. Such a plot, given by the solid line in Figure 8–3, shows that the total pressures of ideal solutions lie on a straight line joining P_1^0 and P_2^0. In turn, the dotted lines in this figure give plots of

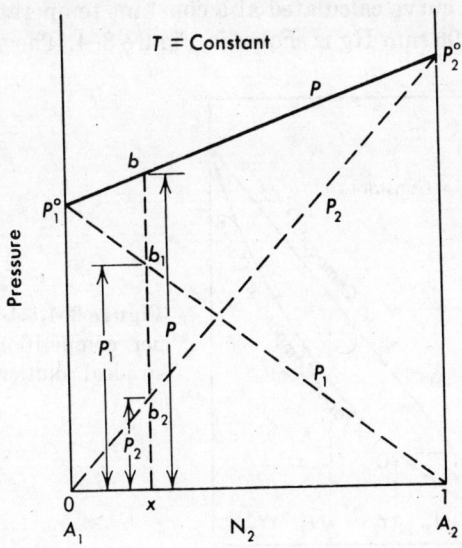

Figure 8–3.   Total and partial vapor pressures of ideal solutions.

Eqs. (49) and (50) for the *partial pressures* of the individual solution components. These vary linearly from $P_1 = P_1^0$ and $P_2 = 0$ at $N_2 = 0$, to $P_1 = 0$ and $P_2 = P_2^0$ at $N_2 = 1$. At all intermediate concentrations the total pressure is the sum of the partial pressure ordinates. Thus, at $N_2 = x$, P_1 is equal to the distance $b_1 x$, $P_2 = b_2 x$, and $P = b_1 x + b_2 x = b x$.

The above relationships demanded by Raoult's law apply to the total and partial vapor pressures as a function of the mol fractions of the constituents *in solution*. To obtain the relation between the composition of a solution and the composition of the vapor above it, let Y_2 be the mol fraction of A_2 *in the vapor* above a solution of composition N_2. Then, according to Dalton's law of partial pressures

$$Y_2 = \frac{P_2}{P} \tag{53}$$

But $P_2 = P_2^0 N_2$, while P is given by Eq. (52). Consequently,

$$Y_2 = \frac{P_2^0 N_2}{(P_2^0 - P_1^0)N_2 + P_1^0} \tag{54}$$

This equation shows that there is a definite composition of vapor corresponding to each composition of solution, and that Y_2 and N_2 will not be the same except in the very special case when $P_1^0 = P_2^0$.

With the aid of Eq. (54) a vapor pressure–vapor composition curve can be constructed for solutions obeying Raoult's law which will show the composition of the vapor corresponding to any particular composition of solution. Such a curve calculated at a constant temperature for $P_1^0 = 147$ mm and $P_2^0 = 396$ mm Hg is shown in Figure 8–4. The straight line gives

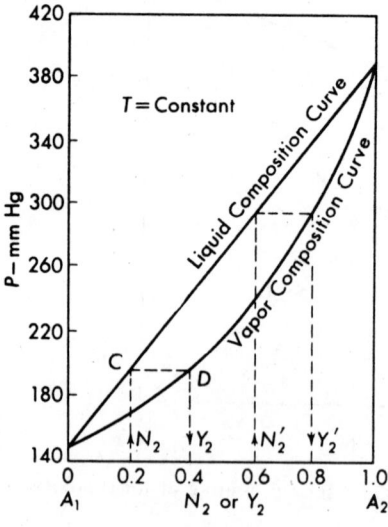

Figure 8–4. Liquid and vapor composition curves for an ideal solution.

the total pressure above the solution as a function of the mol fraction N_2, while the curve lying below it represents the total vapor pressure as a function of the mol fraction of A_2 in the vapor, Y_2. To obtain the composition of vapor corresponding to, say, a solution for which $N_2 = 0.2$, we move vertically to point C to obtain the total vapor pressure of the solution. This same pressure is given also on the vapor composition curve by point D, corresponding to a composition $Y_2 = 0.402$, and hence for the liquids in question when $N_2 = 0.2$, $Y_2 = 0.402$. Similarly, when $N_2' = 0.6$, $Y_2' = 0.803$. It will be observed that the vapor is always richer in A_2 than the solution, i.e., the vapor is richer in the more volatile component.

VAPOR PRESSURE OF ACTUAL LIQUID PAIRS

A few binary miscible liquid systems obey Raoult's law throughout the complete range of concentrations. One of these, the pair ethylene dibromide-propylene dibromide at 85.05°C, is shown in Figure 8–5. Others

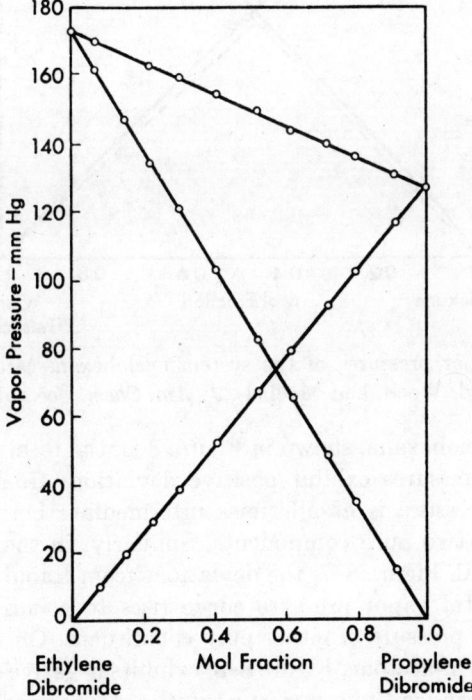

Figure 8–5. Vapor pressures of the system ethylene dibromide-propylene dibromide at 85.05°C.

are the pairs benzene-ethylene dichloride, carbon tetrachloride-stannic chloride, and chlorbenzene-brombenzene. Most systems, however, deviate from Raoult's law to a greater or lesser degree dependent on the nature of the liquids and the temperature.

The character of the deviations from Raoult's law may be judged from Figures 8–6, 8–7, and 8–8, which show the total and partial pressures of several systems in their dependence on the mol fraction. In the pair carbon

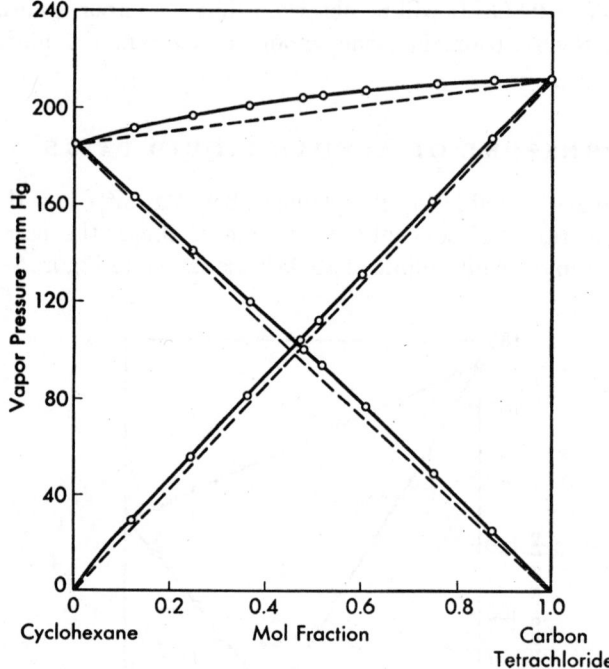

Figure 8–6.   Vapor pressures of the system cyclohexane-carbon tetrachloride at 40°C. [Scatchard, Wood, and Mochels, *J. Am. Chem. Soc.*, **61**, 3208 (1939).]

tetrachloride-cyclohexane, shown in Figure 8–6, the total pressure as well as the partial pressures exhibit positive deviations from Raoult's law, but the total pressure is at all times intermediate between the vapor pressures of the two pure components. Similarly, in the system carbon disulfide-methylal, Figure 8–7, the deviations from Raoult's law are positive, but the total vapor pressure curve rises to a *maximum* which is above the vapor pressure of either pure constituent. On the other hand, the pair chloroform-acetone, Figure 8–8, exhibits negative deviations from Raoult's law which lead to a *minimum* in the total vapor pressure of the system, i.e., the vapor pressures of certain concentrations of the solution are *below* the vapor pressures of either of the pure constituents.

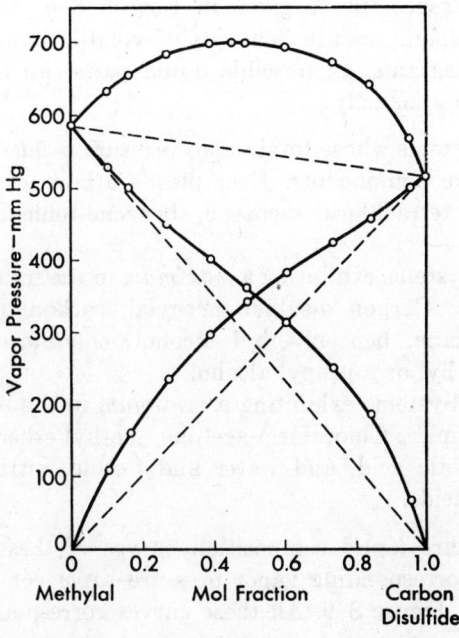

Figure 8–7.   Vapor pressures of the system methylal-carbon disulfide at 35.2°C.

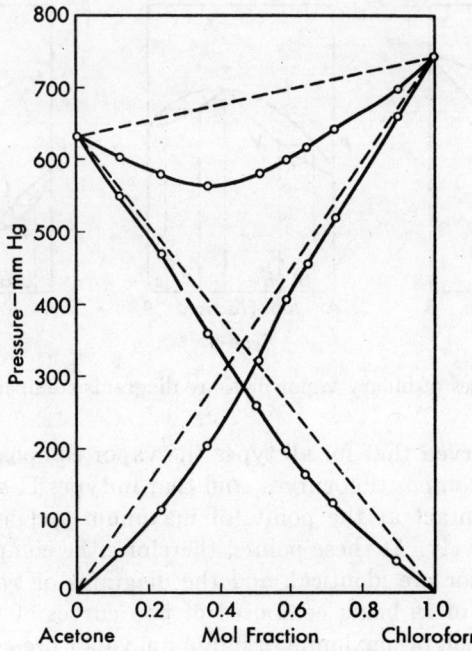

Figure 8–8.   Vapor pressures of the system acetone-chloroform at 55.1°C.

The vapor pressure curves given in Figures 8–6, 8–7, and 8–8, are typical of the vapor pressure behavior of volatile liquid pairs. On the basis of these diagrams, all miscible liquid pairs can be classified into three general types, namely:

T Y P E I :   Systems whose total vapor pressure is *intermediate* between those of the pure components. Examples: Carbon tetrachloride-cyclohexane, carbon tetrachloride-benzene, benzene-toluene, water-methyl alcohol.

T Y P E I I :   Systems exhibiting a *maximum* in the total vapor pressure curve. Examples: Carbon disulfide-methylal, carbon disulfide-acetone, benzene-cyclohexane, benzene-ethyl alcohol, chloroform-ethyl alcohol, and water and ethyl or *n*-propyl alcohol.

T Y P E I I I :   Systems exhibiting a *minimum* in the total vapor pressure curve. Examples: Chloroform-acetone, methyl ether-hydrogen chloride, pyridine-acetic acid, and water and formic, nitric, hydrochloric, or hydrobromic acids.

The vapor pressure–*liquid* composition curves of these various types, along with the corresponding vapor pressure–*vapor* composition curves, are illustrated in Figure 8–9. All these curves correspond to a constant temperature.

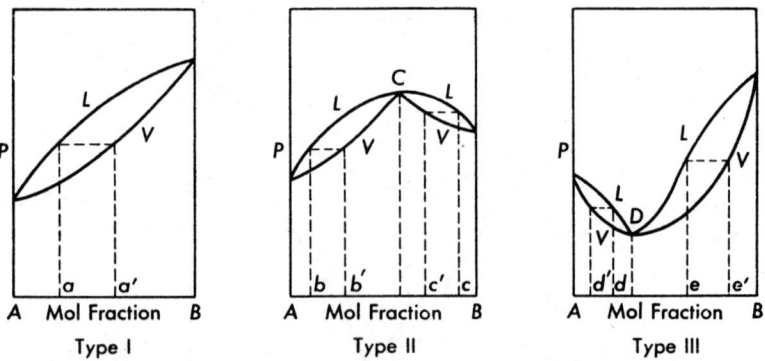

Figure 8–9.   Types of binary vapor pressure diagrams (temperature constant).

It will be observed that for all types the vapor composition curves lie *below* the liquid composition curves, and that in types II and III the two curves are in contact at the points of maximum and minimum vapor pressure respectively. At these points, therefore, the compositions of the solution and vapor are identical, and the diagrams of types II and III may be thought of as being composed of two curves of type I, one for pure A and solution of maximum or minimum vapor pressure C or D, and the other of solution of maximum or minimum vapor pressure C or D and pure B. The position of the vapor composition curves with respect to the liquid curves follows from the fact that out of a mixture of two volatile substances in solution the one of higher vapor pressure will volatilize to a greater extent than the one of lower vapor pressure, yielding a vapor of composition richer in the more volatile constituent than is the solution. Thus, in the liquid pair of type I shown in Figure 8–9, B is more volatile than A, and consequently the vapor above all concentrations of solution must be richer in B than is the solution. Hence, the composition of vapor corresponding to point a on the diagram must lie closer to B than point a, and this is possible only if the vapor composition curve is below the liquid composition curve. As the diagram indicates, the composition of vapor corresponding to a is a', a mixture considerably richer in B than is a. For the same reason all mixtures between A and C in type II must have vapors richer in B than the solution, while all mixtures between C and B must be richer in A than the solution. These conditions are satisfied by the diagram as drawn, for the vapor composition corresponding to a point such as b is b', richer in B than b, while that of composition c is c', richer in A than c. Similarly, the compositions of vapor in type III must be richer in A than the solutions between A and D, while richer in B between D and B. These conditions are again satisfied by the diagram drawn, as may be verified by the two points indicated, d with vapor composition d', and e with vapor composition e'.

BOILING POINT DIAGRAMS OF MISCIBLE BINARY MIXTURES

Because at any given temperature the vapor above any solution is richer in the more volatile substance than the solution, a solution can be made to shift in composition toward the less volatile constituent by removing the vapor above it. Again, if the vapors are condensed, and the new vapor above the condensate removed, the new vapors will be found considerably richer in the more volatile constituent than are the solutions from which they came. By repeating this process it is possible to obtain a concentration of the more volatile constituent in the vapor and a concentration of the less volatile constituent in the solution. Such a process of concentrating the constituents is known as fractional distillation; and, since the process described takes place at constant temperature, it may be designated as an isothermal fractional distillation.

In practice it is much more convenient to conduct a distillation at constant pressure rather than at constant temperature. At a given confining pressure any solution of definite composition will boil at a temperature at which its *total* vapor pressure becomes equal to the confining pressure. If we designate by P the confining pressure, then the condition for boiling may be written as $P = P_A + P_B$. Thus, at atmospheric pressure a solution will boil at the temperature at which its total vapor pressure becomes equal to 760 mm Hg. Since different compositions of a solution have different vapor pressures, it must follow that the various solutions will not reach a total vapor pressure equal to the confining pressure at the same temperature, and therefore solutions of various concentrations will boil at different temperatures. In general, solutions of low vapor pressure will boil at temperatures higher than solutions whose vapor pressure is high, for solutions of high vapor pressure can reach a total pressure equal to the confining pressure at relatively lower temperatures than solutions whose vapor pressure is low.

This latter fact permits the construction of the various types of temperature-composition diagrams which will correspond to the three general types of vapor-pressure-composition diagrams already discussed. These are shown in Figure 8–10. In type I the vapor pressure of A is the lowest pressure in the system and that of B is the highest, while the vapor pressures of all possible compositions of A and B are intermediate between the two. Consequently, at constant pressure the boiling point of A will be the highest in the system and that of B the lowest, while those of all compositions of A and B will be intermediate and will be given by the liquid composition curve in the figure. Since the vapor coming off from any particular composition of solution must be richer in the more volatile constituent B, the vapor composition at any temperature must lie *closer* to B than the corresponding liquid composition, and hence the vapor composition curve must lie now *above* the liquid composition curve, as shown. The same considerations apply to the other two types. In type II the vapor pressure of the system is a maximum for composition C, and hence such a solution will boil at the lowest temperature, leading to a minimum in the boiling point curve. Again, since in type III the solution of composition D has the lowest vapor pressure in the system, it will boil at the highest temperature, and consequently the boiling point curve exhibits a *maximum*. In all cases the vapor composition curves lie above the liquid composition curves for the reasons given. We see, therefore, that any system whose vapor pressures are intermediate between those of the pure constituents will have a distillation diagram with intermediate boiling points, such as type I. On the other hand, any system of the maximum vapor pressure type will give a distillation diagram with minimum boiling point, while any system of the minimum vapor pressure type will have a distillation diagram with a maximum in the boiling point curve.

The concentrations for points C and D in the vapor pressure and temperature diagrams are generally not identical. With change in temperature there is a tendency for the compositions at which the vapor pressure maxima or minima occur to shift toward A or B, depending on the system involved.

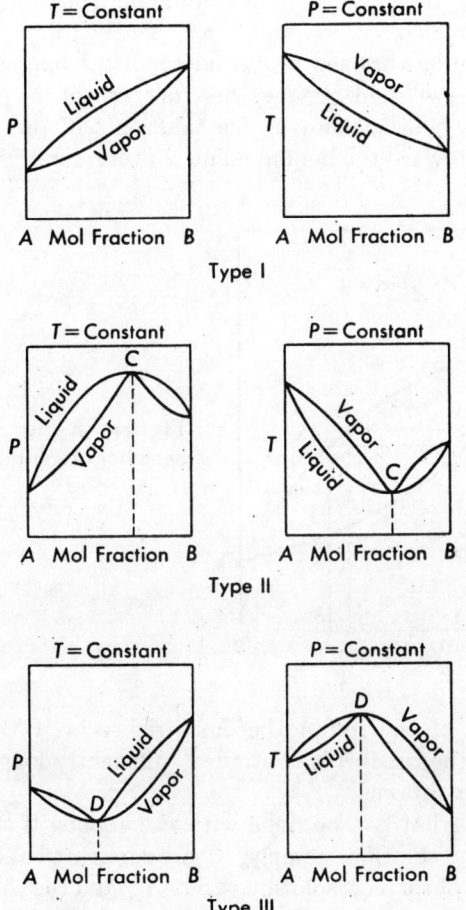

Figure 8-10. Types of distillation diagrams corresponding to various vapor pressure diagrams.

DISTILLATION OF BINARY MISCIBLE SOLUTIONS

Because of the differences in the distillation diagrams of the three types of solutions, the behavior of these on constant pressure distillation will be different. Consider first the behavior of a system of type I (Figure 8-11). If we heat a solution of composition a, no boiling will start until temperature T_a is reached. At this temperature the vapor coming off from a will have the composition a'. Since a' is richer in B than a, the composition of the residue must become richer in A, say b. The new composition of residue, b, cannot boil, however, until temperature T_b is reached, which is higher than T_a. In turn the vapor coming off from b will have the composition b', again richer in B, and consequently the composition of the residue will again be enriched in A, and again the temperature must rise before the residue will boil. We see, therefore, that if the process described is continued the boiling point of the solution will rise from the initial boiling point T_a toward the boiling point, T_A, of pure A. At the same time

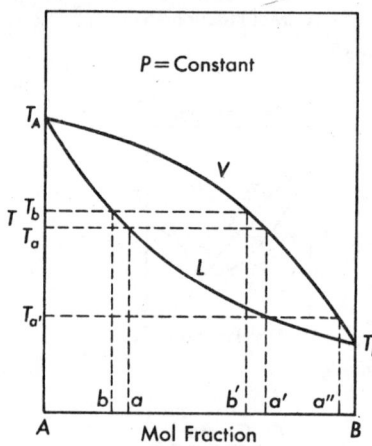

Figure 8–11.   Distillation behavior of solutions of type I.

the composition of the residue becomes richer in A than the original solution, and if the process is continued sufficiently long, a final residue of pure A can be obtained.

Consider now what can be done with the vapors. If the initial vapors obtained from the solution, namely, a', are condensed and again distilled, the boiling point of the new solution will be T'_a, and the composition of the distillate will be given by a''. This distillate is again richer in B than the original. If the process of condensing and redistilling is continued, eventually a vapor can be obtained composed essentially of pure B. Therefore, on distillation of any mixture of type I it is possible to separate eventually the constituents into a residue of the less volatile constituent A and a distillate of the more volatile component B; i.e., *the two constituents forming a solution of type I can be separated by fractional distillation into the pure components.*

Such a separation into the pure components is impossible, however, with solutions of either type II or type III. Consider the distillation behavior of a solution of type II, Figure 8–12, of which the system water (*A*)-ethyl alcohol (*B*) is an outstanding example. If a solution of composition between *A* and *C*, such as *a*, is distilled, the vapor coming off will have the composition *a'* and will be richer in *B* than the residue. Because of this fact the composition of the residue will shift toward *A*, and hence the residue will have to boil at a temperature higher than the original solution *a*. If the distillation is continued, the same argument as was employed for solutions of type I indicates that eventually a residue of pure *A*, boiling at temperature T_A, will be obtained. On the other hand, if the vapors from the original solution, *a'*, are condensed and redistilled repeatedly, a vapor of composition *C* will eventually be obtained. Such

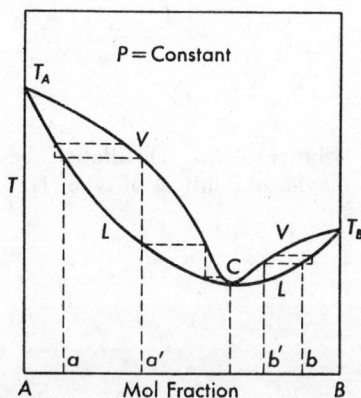

Figure 8–12. Distillation behavior of solutions of type II.

vapor when condensed and distilled will again yield vapor of composition *C*; i.e., the vapors coming off from the solution will have the same composition as the solution, and hence no further separation is possible by distillation. Consequently, *any mixture having a composition between A and C can be separated by fractional distillation only into a residue of pure A and a final distillate of composition C*. No pure *B* can be recovered.

On the other hand, if a solution of composition between *C* and *B* is distilled, for example, *b*, the vapor coming off, *b'*, will be *richer* in *A* than the original solution, and hence on repeated distillation the residue will tend toward pure *B*, while the distillate will tend toward *C*. Such solutions on complete distillation will yield, therefore, pure *B* in the residue and constant boiling mixture *C* in the distillate. No *A* can be recovered by distillation.

The behavior of solutions of type III on distillation will be analogous to that of solutions of type II, with the exception that *the residues tend toward the maximum boiling mixture, while the distillates tend toward the pure constituents* (Figure 8–13). If the starting mixture has a composition between A and D, such as a, the vapor obtained on distillation, a', will be richer in A than the solution. Hence the composition of the residue will shift toward D and will eventually reach it. A redistillation of the vapor, on the other hand, will finally yield a distillate of pure A. A mixture between D and B, such as b, however, will yield on distillation a vapor of composition b', richer in B than the solution. Hence again the residue will shift toward D, while on redistillation the vapors will tend toward pure B. Consequently, complete distillation of a mixture such as b will eventually yield a residue of composition D and a distillate of pure B.

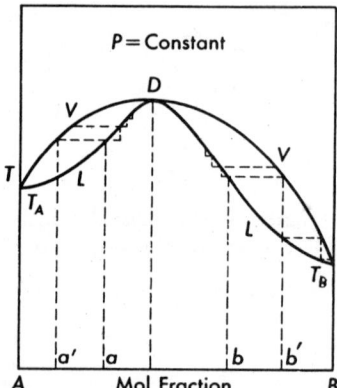

Figure 8–13. Distillation behavior of solutions of type III.

We see, therefore, that any binary system of this type can be separated on complete fractional distillation into a residue of composition D, the constant maximum boiling mixture, and a distillate of either pure A or pure B, depending on whether the starting composition is between A and D or D and B. But a mixture of composition D cannot be separated further by distillation.

AZEOTROPES

The constant boiling mixtures described above are called *azeotropes*. The composition of azeotropes is remarkably constant at any given confining pressure. However, when the total pressure is changed, both the boiling point and the composition of the azeotrope change, as may be seen from Table 8–1. Consequently, azeotropes are not definite compounds, whose composition must remain constant over a range of temperatures and pressures, but rather are mixtures resulting from the interplay of intermolecular forces in solution.

TABLE 8–1. EFFECT OF PRESSURE
ON COMPOSITION OF AZEOTROPE
IN SYSTEM H₂O-HCl

Pressure (mm Hg)	Weight % of HCl in Constant Boiling Mixture
730	20.314
740	20.290
750	20.266
760	20.242
770	20.218

Table 8–2 lists the boiling points and compositions of some azeotropes, all for a total pressure of 760 mm Hg.

TABLE 8–2. BOILING POINTS AND COMPOSITIONS OF AZEOTROPIC MIXTURES
($P = 760$ mm Hg)

Type	A	B	Boiling Point (°C)	Weight % of B in Azeotrope
Minimum boiling point	Water	Ethyl alcohol	78.15	95.57
	Water	n-Propyl alcohol	88.1	71.8
	Ethyl alcohol	Benzene	67.8	67.6
	Acetic acid	Benzene	80.05	98.0
	Carbon disulfide	Ethyl acetate	46.1	3.0
	Pyridine	Water	92.6	43.0
Maximum boiling point	Water	Nitric acid	121.0	68.5
	Water	Hydrochloric acid	108.6	20.24
	Water	Hydrobromic acid	126.0	47.5
	Water	Hydriodic acid	127.0	57.0
	Water	Hydrofluoric acid	114.4	35.6
	Water	Formic acid	107.1	77.5
	Chloroform	Acetone	64.7	20.0
	Pyridine	Formic acid	149.0	18.0

THE FRACTIONATING COLUMN

The type of distillation described heretofore, in which the vapor removed is in equilibrium with the total mass of boiling liquid, is designated as *equilibrium distillation*. The process of separating mixtures by distillation would be extremely complicated and tedious if it had to be performed by repeated distillations and condensations in a discontinuous manner. Instead, the separation is performed in a continuous operation, known as *fractional distillation*, utilizing a distilling apparatus called a *fractionating column*, Figure 8–14. The fractionating column consists essentially of three parts: a heated still A; the column proper D, composed of a series of plates whose detailed construction is shown in the figure; and a condenser F. The preheated mixture to be distilled is admitted through E onto one of these plates, and overflows through 2 to the plate below. On this lower

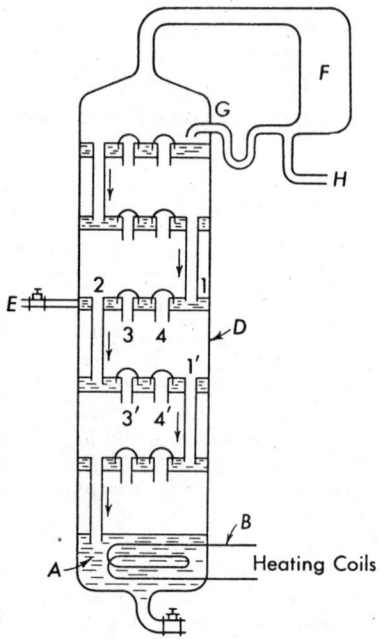

Figure 8–14. Schematic diagram of fractionating column.

plate the liquid comes in contact with vapor moving upward from the still through the "bubble caps" 3' and 4'. These caps are so designed that the vapor must bubble through the layer of liquid on each plate before it can escape. In doing so, part of the less volatile constituent is condensed out of the vapor, and part of the more volatile constituent is vaporized out of the liquid. The vapor moving on to the next higher plates through 3 and 4 is richer, then, in the more volatile constituent than the vapor which approached the plate from below, while the liquid overflowing to the next lower plate through 1' is richer in the less volatile constituent than the liquid which reached the plate from above. The net result of the interaction between vapor and liquid at the plate is, therefore, a redistribution in favor of the more volatile constituent in the vapor and the less volatile constituent in the liquid; i.e., each plate acts essentially as a miniature still.

Since this process repeats itself at each plate, it is possible with a sufficient number of plates to separate the mixture into two end fractions, a residue of the less volatile component running into still A, where it can be drawn off, and a vapor passing from the top of the column containing essentially the more volatile constituent. This vapor is fed into a condenser F, where it is liquefied. Part of this liquid is drawn off through H, while part, the reflux, is returned to the column through G in order to maintain the stock of essentially pure distillate on the upper plates.

RATIO OF DISTILLATE TO RESIDUE

For various calculations it is frequently necessary to know the ratio of weight of distillate to weight of residue at each stage of an equilibrium distillation. For a binary mixture this information is readily available from the distillation diagram.

Consider a binary system whose distillation diagram is given by Figure 8–15. The abscissa is expressed now as weight per cent rather than mol

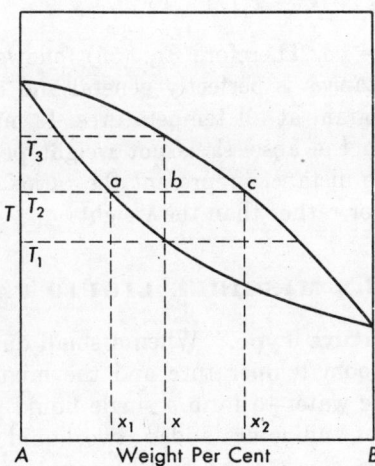

Figure 8–15. Ratio of distillate to residue on distillation.

fraction. If we start with a mixture whose weight per cent is x, the system will be entirely liquid until temperature T_1 is reached. At T_1 the solution will begin to boil, and since the liquid becomes less concentrated in B, the boiling point will gradually rise as distillation proceeds. At a temperature such as T_2, if no vapor is removed from the system, the latter will still have the same over-all composition as the starting mixture x, but it will be composed now of liquid of composition x_1 and vapor of composition x_2. Under such conditions the weight of liquid present, W_1, is proportional to the linear distance bc, while the weight of vapor present, W_2, is proportional to the distance ab; i.e.,

$$\frac{W_1}{W_2} = \frac{bc}{ab} \tag{55}$$

The argument leading to Eq. (55) is as follows. If we let W be the total weight of mixture of composition x, then

$$W = W_1 + W_2$$

Again, a material balance on constituent B yields

$$Wx = W_1 x_1 + W_2 x_2$$

and hence we get on substitution for W,

$$(W_1 + W_2)x = W_1 x_1 + W_2 x_2$$
$$W_1(x - x_1) = W_2(x_2 - x)$$
$$\frac{W_1}{W_2} = \frac{(x_2 - x)}{(x - x_1)}$$

But $(x_2 - x) = bc$ and $(x - x_1) = ab$. Therefore Eq. (55) follows. This relation between weights and distances is perfectly general and can be applied to any portion of the diagram at all temperatures. It must be remembered, however, that, when the abscissa is not weight per cent but mol fraction, the ratios of the distances represent the ratios of the number of *moles* of liquid and vapor rather than the weights.

SOLUBILITY OF PARTIALLY MISCIBLE LIQUID PAIRS

Maximum Solution Temperature Type. When a small quantity of aniline is added to water at room temperature and the mixture is shaken, the aniline dissolves in the water to form a single liquid phase. However, when larger quantities of aniline are added, two liquid layers are formed. One of these, the lower, consists of a small amount of water dissolved in the aniline, while the upper consists of a small amount of aniline dissolved in water. Further addition of aniline to the system causes the water-rich layer to diminish in size until finally it disappears, leaving only a single liquid phase composed of water in the aniline.

If this experiment is performed at constant temperature, it is found that the compositions of the two layers, although different from each other, remain constant as long as two phases are present. The addition of small amounts of either aniline or water merely changes the relative volumes of the two layers, not their composition. As the temperature is raised, this behavior is found to persist except that the mutual solubility of the two liquids increases. When the temperature finally reaches 168°C, the compositions of the two layers become identical, and thereafter the two liquids are completely miscible. In other words, at 168°C and above aniline and water dissolve in each other in all proportions and yield only a single liquid layer on mixing.

This variation of the mutual solubility of water and aniline with temperature is illustrated in Figure 8–16. At a temperature such as 100°C,

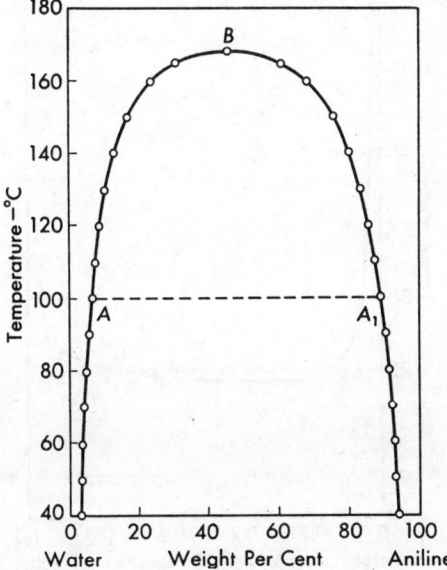

Figure 8–16. Mutual solubility of water and aniline at various temperatures.

point A represents the composition of the water-rich layer and point A_1 the composition of the aniline-rich layer in equilibrium with A. Between A and A_1 all mixtures yield two layers of compositions A and A_1. Outside these compositions the two liquids are mutually soluble at 100°; i.e., all compositions between pure water and A yield a solution of aniline in water, while all compositions between A_1 and pure aniline yield a solution of water in aniline. Since the same argument holds at other temperatures, it must follow that the dome-shaped area represents the range of existence of two liquid phases, the area outside the dome that of a single liquid layer. The temperature corresponding to point B, i.e., the temperature at which solubility first becomes complete, is called either the *critical solution temperature* or the *consolute temperature*.

Minimum Solution Temperature Type.   Figure 8–17 shows the effect of temperature on the mutual solubility of triethylamine and water. The two liquids are completely miscible at or below 18.5°C, but only partially miscible above this temperature. Thus at 30°C, for instance, a solution of 5.6 per cent triethylamine in water is in equilibrium with one containing 4 per cent water in triethylamine. The temperature at which the two liquids become completely miscible is called in this case the

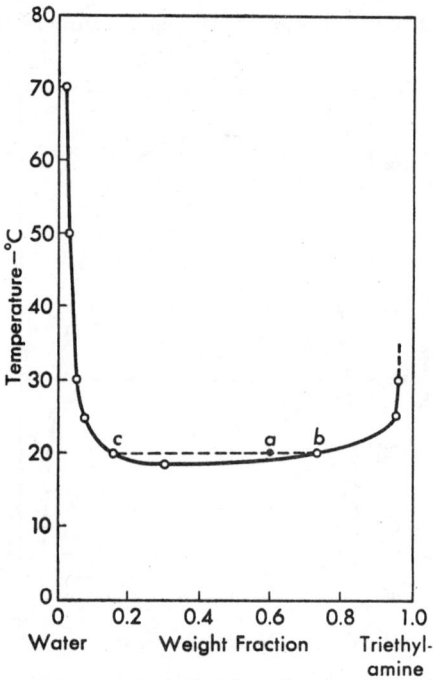

Figure 8–17.   Solubility of triethylamine in water at various temperatures.

minimum critical solution temperature, since the curve confining the area of partial miscibility exhibits a minimum.

Maximum and Minimum Solution Temperature Type.   The system nicotine-water exhibits two critical solution temperatures, an upper and a lower, as may be seen from Figure 8–18. Within the enclosed area the liquids are only partially miscible, while outside the enclosed area they are completely miscible. The upper, or maximum solution temperature, point C, is 208°C, while the lower, or minimum solution temperature, point C', is 60.8°C. The compositions corresponding to C and C' are the same, 34 per cent nicotine. At 94 to 95°C, point A, nicotine is least soluble in water, while water is least soluble in nicotine at 129 to 130°C, point B.

It has been found that on applying external pressure to this system the upper and lower critical solution temperatures approach each other, until a pressure is finally reached at which the two liquids become completely miscible.

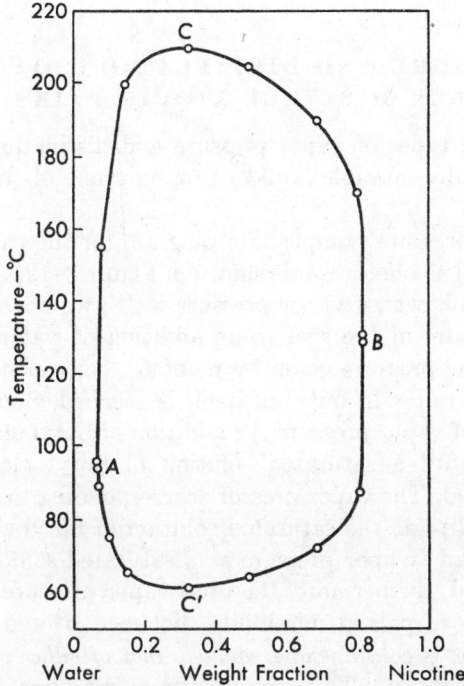

Figure 8–18.   Solubility of nicotine in water at various temperatures.

Type Without Critical Solution Temperature.   A final variation of these types is exhibited by the system ethyl ether-water, which has neither an upper nor a lower critical solution temperature. The two liquids are therefore only partially soluble in each other at all the temperatures over which the solution exists.

It is always possible to deduce from solubility diagrams such as those given in Figures 8–16, 8–17, and 8–18 the proportions by weight of the two layers present at equilibrium at the various temperatures. Consider specifically a system composed of 60 g of triethylamine·and 40 g of water present in equilibrium at 20°C. Since the percentage of amine by weight is 60 per cent, the over-all composition is represented in Figure 8–17 by point a. But this system is composed of solutions of compositions b and c, and hence, by the method used before,

$$\frac{\text{Wt amine layer}}{\text{Wt water layer}} = \frac{\text{distance } ca}{\text{distance } ab}$$

$$= \frac{60 - 15.5}{73 - 60}$$

$$= 3.42$$

VAPOR PRESSURE AND DISTILLATION DIAGRAMS OF PARTIALLY MISCIBLE LIQUID PAIRS

Although three types of vapor pressure and distillation diagrams are possible for partially miscible liquid pairs, only one of these will be discussed here.

A total vapor pressure–composition diagram for the partially miscible liquid pair n-butyl alcohol-water is shown in Figure 8–19(a). Starting with pure butyl alcohol, whose vapor pressure is D', we observe an increase in the total pressure of the system on addition of water. This increase continues until the pressure given by point C', corresponding to a saturated solution of water in butyl alcohol, is reached. Similarly, starting with pure water of vapor pressure A', addition of butyl alcohol raises the vapor pressure until a saturated solution of butyl alcohol in water, point B', is reached. The vapor pressure corresponding to C' is *exactly the same* as that of B'; i.e., the saturated solution of butyl alcohol in water has exactly the same vapor pressure as a saturated solution of water in butyl alcohol. And, furthermore, the total vapor pressure above the system in the region of partial miscibility, between B' and C', where two layers are present, is *constant and equal to that of either of the two layers and not the sum of the two*. This state of affairs is a direct consequence of the thermodynamic conditions of equilibrium in such a system. For equilibrium to exist in a system involving the distribution of a particular component between two phases, such as butyl alcohol between the upper and lower layers, the vapor pressure of the particular constituent above each of the layers must be the same, and the vapor pressure above both must be that of any one. Consequently, the vapor pressure of butyl alcohol above B' must be equal to that above C', and the total must be the same as *either* that of B' or C'. Since the same considerations apply to the water in the two phases, the total pressure above both layers is the same as that of B' or C' and is constant as long as both phases are present.

The curves $A'H'$ and $D'H'$ give the compositions of vapor in equilibrium with $A'B'$ and $D'C'$ respectively. For all compositions of liquid between B' and C' the composition of vapor is constant and equal to H'.

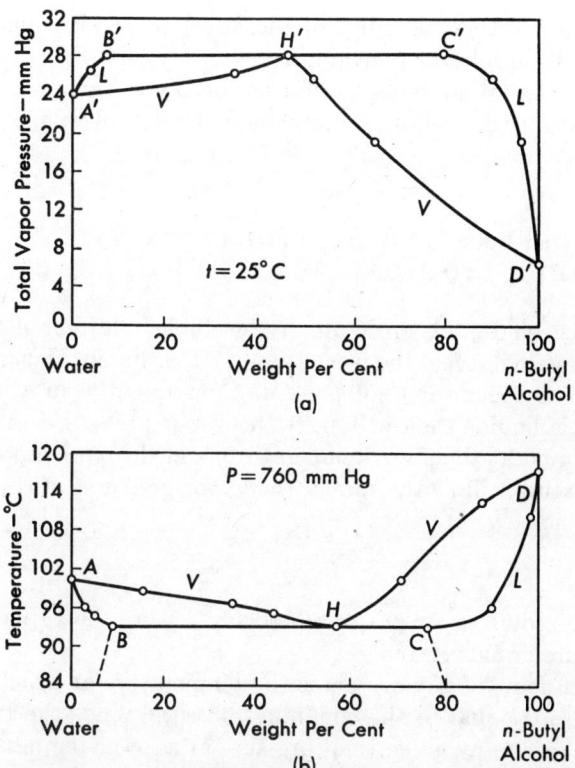

Figure 8–19. Vapor and distillation diagrams for the system *n*-butyl alcohol-water.

The distillation diagram which corresponds to a system exhibiting the vapor pressure behavior shown by butyl alcohol-water is shown in Figure 8–19(b). The curve *CD* gives the boiling points of all solutions of water in butyl alcohol, while *HD* gives the compositions of vapor corresponding to the various compositions of the liquid phase. Similarly, *AB* gives the boiling points of all solutions of butyl alcohol in water, while *AH* gives the corresponding vapor compositions. The boiling points of all over-all compositions that yield two layers, namely, between *B* and *C*, are given by *BC*, while the composition of vapor that corresponds to *BC* is given by point *H*; i.e., as long as the two saturated layers are present, the boiling point of the system is *constant*, and the composition of the vapor coming off is also *constant and independent of the over-all composition*.

The dotted lines emanating from *B* and *C* in Figure 8–19(b) indicate the variation of the mutual solubility of the two liquids with temperature, and are part of the solubility diagram.

Other examples of systems exhibiting behavior similar to that of butyl alcohol-water are aniline-water, *iso*-butyl alcohol-water, and ethyl acetate-water.

VAPOR PRESSURE AND DISTILLATION OF IMMISCIBLE LIQUIDS

Since immiscible liquids are mutually insoluble, addition of one liquid to the other does not affect the properties of either liquid. Hence each will behave as if the other were not present. Consequently, in a mixture of two immiscible liquids each will exert the vapor pressure corresponding to the pure liquid at the given temperature, and the *total* vapor pressure above the mixture will be the sum of the vapor pressures of the two pure constituents, namely,

$$P = P_A^0 + P_B^0 \qquad (56)$$

where P is the total vapor pressure, and P_A^0, P_B^0 are the vapor pressures of the two pure liquids A and B.

The boiling point of any system is the temperature at which the total vapor pressure is equal to the confining pressure. Since the two liquids together can reach any given total pressure at a lower temperature than either liquid alone, it must follow that any mixture of two immiscible liquids must boil at a temperature *lower* than the boiling point of *either* of the two liquids. Furthermore, since at any given temperature there is no change in total vapor pressure with change in over-all composition, the boiling point of all possible mixtures of the two must remain *constant* as long as both liquids are present. As soon as one of the liquids is boiled away, however, the boiling temperature will rise abruptly from that of the mixture to either T_A or T_B, depending on whether one or the other is boiled away first.

At any boiling temperature for the mixture, T, the partial vapor pressures of the two constituents are P_A^0 and P_B^0 corresponding to the given temperature. If we let N_A' and N_B' be the mol fractions of the two constituents in the vapor, then $P_A^0 = N_A'P$, $P_B^0 = N_B'P$, and hence,

$$\frac{P_A^0}{P_B^0} = \frac{N_A'P}{N_B'P} = \frac{N_A'}{N_B'} \qquad (57)$$

But $N_A' = n_A/(n_A + n_B)$, and $N_B' = n_B/(n_A + n_B)$, where n_A and n_B are the number of moles of A and B in any given volume of vapor.

Consequently,

$$\frac{P_A^0}{P_B^0} = \frac{n_A}{n_B} \tag{58}$$

and, since the ratio of the partial pressures at T is constant, n_A/n_B must also be constant; i.e., the composition of the vapor is at all times constant as long as both liquids are present. Further, since $n_A = W_A/M_A$, and $n_B = W_B/M_B$, where W_A, W_B are the weights in any given volume and M_A, M_B are the molecular weights of A and B respectively, Eq. (58) becomes

$$\frac{P_A^0}{P_B^0} = \frac{n_A}{n_B} = \frac{W_A}{W_B} \cdot \frac{M_B}{M_A}$$

and
$$\frac{W_A}{W_B} = \frac{M_A P_A^0}{M_B P_B^0} \tag{59}$$

Equation (59) relates directly the weights of the two constituents distilled from a mixture of two immiscible liquids to the molecular weights and the vapor pressures of the two pure constituents. It will be observed that the weight of any constituent distilled over depends on both its vapor pressure and molecular weight, and hence the effect of a low vapor pressure is counteracted by a high molecular weight as far as weight of a particular substance distilled over is concerned.

Distillation of immiscible liquids is utilized industrially and in the laboratory for the purification of organic liquids which either boil at high temperatures or tend to decompose when heated to their normal boiling point. The other liquid frequently is water, and the whole process is generally referred to as *steam distillation*. The immiscible mixture of the liquid and water is heated either directly or by injection of steam, and the vapors coming off are condensed and separated. In this manner it is possible to distill many liquids of high boiling point at temperatures below 100°C, the boiling point of water.

Distillation of immiscible liquids can also be utilized in determining the approximate molecular weight of one of the liquids involved. When the vapor pressures and weight ratios of distillates of two liquids are determined, and the molecular weight of one of the liquids is known, the molecular weight of the other can readily be calculated from Eq. (59). The manner in which the necessary data are obtained can best be explained through an example. When the two immiscible liquids chlorbenzene and water are boiled at a pressure of 734.4 mm Hg, the boiling point is 90°C, while the ratio of weight of chlorbenzene to water collected in the distillate is 2.47. Since at 90°C the vapor pressure of water is 526.0 mm Hg, the vapor pressure of chlorbenzene must be 734.4 − 526.0 = 208.4 mm Hg. Therefore, letting A be chlorbenzene and B water, and applying Eq. (59),

$$M_A = \left(\frac{W_A}{W_B}\right)\left(\frac{P_B^0}{P_A^0}\right) M_B$$

$$= 2.47 \times \frac{526.0}{208.4} \times 18.02$$

$$= 112.3$$

The molecular weight of chlorbenzene calculated from atomic weights is 112.6

SOLUBILITY OF GASES IN LIQUIDS

Gases dissolve in liquids to form true solutions. The degree of solubility depends on the nature of the gas, the nature of the solvent, the pressure, and the temperature. Gases like nitrogen, hydrogen, oxygen, and helium dissolve in water only to a slight extent, while gases like hydrogen chloride and ammonia are very soluble. The large solubility in the latter cases is accounted for by the chemical reaction of these gases with the solvent to form hydrochloric acid and ammonium hydroxide respectively. Again, nitrogen, oxygen, and carbon dioxide are much more soluble in ethyl alcohol than they are in water at the same pressure and temperature, while hydrogen sulfide and ammonia are more soluble in water than in ethyl alcohol. Frequently, chemical similarity between solute and solvent leads to a higher solubility, as is evidenced by the fact that hydrocarbon vapors dissolve more readily in hydrocarbon and other organic solvents than they do in water. Still, chemical similarity is not an infallible criterion of solubility. Thus acetylene, which is quite different in chemical characteristics from water, dissolves to a greater extent in water at 0°C than does oxygen.

The effect of pressure on the solubility of a given gas in a particular liquid at constant temperature can be obtained readily by viewing the process in reverse; i.e., by considering the gas as a solute which vaporizes to establish a vapor pressure above the solution. For the latter situation Eq. (39) applies, namely, that

$$\frac{f_{2(g)}}{a_2} = K \tag{60}$$

where $f_{2(g)}$ is the fugacity of the gas above the solution and a_2 is the activity of the gas in solution. If both the gas phase and the solution behave ideally, then $f_{2(g)} = P_2$, $a_2 = N_2$, and Eq. (60) becomes

$$\frac{P_2}{N_2} = K$$

or
$$N_2 = K'P_2 \tag{61}$$

Equation (61) is known as *Henry's law*, which states that *at constant temperature the solubility of a gas in a liquid is directly proportional to the pressure of the gas above the liquid.* The proportionality factor K' is called Henry's law constant. Its magnitude depends on the nature of the gas and solvent, temperature, and the units in which P_2 is expressed.

When several gases are being dissolved simultaneously in a solvent, Eq. (61) is valid for each gas independently, provided N_2 is the concentration and P_2 the *partial pressure* of each gas. We may say, therefore, that *the solubility of each gas from a mixture of gases is directly proportional to the partial pressure of the gas in the mixture.* The proportionality constant K' will, of course, be different for each gas.

The validity of Henry's law is illustrated by the data given in Table 8-3 for the solubility of oxygen in water at 25°C. If the law is correct, K' should be a constant independent of pressure. This is actually the case, as may be seen from column 3 of the table.

TABLE 8–3.   SOLUBILITY OF OXYGEN IN WATER
AT 25°C

P_2 (Atm)	N_2	$K' = \dfrac{N_2}{P_2}$
0.230	0.537×10^{-5}	2.33×10^{-5}
0.395	0.904	2.29
0.545	1.24	2.28
0.803	1.84	2.29
1.000	2.30	2.30

The strict applicability of Henry's law is limited to the lower pressures. At high pressures the law becomes less exact, and the proportionality constants exhibit considerable variation. Generally, the higher the temperature and the lower the pressure, the more closely is the law obeyed. Furthermore, the law as given above is not applicable where the dissolved gas reacts with the solvent, or where the dissolved gas ionizes. When the ionization in solution is complete, the law breaks down altogether. The deviations in case of chemical reaction and partial dissociation can be readily understood and corrected for when it is realized that Henry's law is valid only when it is applied to the concentration in solution of the *same molecular species* as exists in the gas phase and not to the total concentration in solution. Thus, when ammonia dissolves in water, part of the dissolved gas reacts to form ammonium hydroxide, which in turn dissociates partially to $NH_4^+ + OH^-$. The reactions involved may be written as

$$NH_3 \text{ (gas)} = NH_3 \text{ (dissolved)}$$
$$NH_3 \text{ (dissolved)} + H_2O = NH_4OH$$
$$NH_4OH = NH_4^+ + OH^-$$

In the light of the above limitation, Henry's law, to be applicable to the solubility of ammonia in water, must be expressed not as $N_{total}/P_{NH_3} = K'$, but as $N_{NH_3}/P_{NH_3} = K'$, where N_{NH_3} is the mol fraction of ammonia in solution present as NH_3.

The solubility of most gases in liquids decreases with increase in temperature, and consequently the Henry's law constants have smaller values at the higher temperatures, as may be seen from Table 8–4. Because of the decrease in solubility at higher temperatures, liquids containing many types of dissolved gases may be purged of these by boiling. But this is not always the case. Some gases are more soluble at higher temperatures than at lower, and hence these are not readily removable by heating. In fact, dilute solutions of hydrogen chloride in water become more concentrated on boiling, until eventually a solution containing about 20 per cent hydrogen chloride is attained.

TABLE 8–4.   HENRY'S LAW CONSTANTS FOR SOLUBILITY
OF GASES IN WATER

Gas	$K' \times 10^5$ (Atm^{-1})				
	0°C	20°C	40°C	60°C	80°C
H_2	1.72	1.46	1.31	1.31	1.33
N_2	1.86	1.32	1.00	0.874	—
O_2	3.98	2.58	1.84	1.57	1.44
C_2H_4	20.5	10.1	6.18	—	—

THE NERNST DISTRIBUTION LAW

Iodine is soluble in both water and carbon tetrachloride. When a solution of iodine in water is shaken with carbon tetrachloride, which is immiscible with water, it is found that the iodine distributes itself between the water and carbon tetrachloride layers in such a way that at equilibrium the ratio of the concentrations of iodine in the two layers is a constant at any given temperature. Such distribution of a solute between two immiscible or only slightly miscible solvents can be accomplished with any solute for which a pair of immiscible solvents can be found.

The above behavior is a direct consequence of the thermodynamic requirements for equilibrium. To show this, consider a pair of immiscible solvents in contact, A and B, both containing the same substance in solution. The partial molal free energy of the solute in liquid A, $\bar{F}_A$, can be represented by

$$\bar{F}_A = \bar{F}_A^0 + RT \ln a_A \tag{62}$$

where $\bar{F}_A^0$ is the standard free energy and a_A the activity of the solute in solvent A. Similarly, the partial molal free energy of the solute in the second liquid, $\bar{F}_B$, can be written as

$$\bar{F}_B = \bar{F}_B^0 + RT \ln a_B \tag{63}$$

where all the quantities have the same significance as in Eq. (62) except that they refer now to liquid B. Since for equilibrium between the layers we must have $\bar{F}_A = \bar{F}_B$ at constant temperature and pressure, it follows that

$$\bar{F}_B^0 + RT \ln a_B = \bar{F}_A^0 + RT \ln a_A$$

and
$$\ln \frac{a_B}{a_A} = \frac{\bar{F}_A^0 - \bar{F}_B^0}{RT} \tag{64}$$

However, at any given temperature $\bar{F}_A^0$ and $\bar{F}_B^0$ are constants for a given substance in the particular solvents. Hence

$$\ln \frac{a_B}{a_A} = \text{constant}$$

and, therefore,

$$\frac{a_B}{a_A} = K \tag{65}$$

Equation (65) is a mathematical statement of the *Nernst distribution law*, which states that a substance will distribute itself between two solvents until at equilibrium the *ratio* of the activities of the substance in the two layers is constant at any given temperature. When the solutions are dilute, or when the solute behaves ideally, the activity is essentially equal to the concentration C, and Eq. (65) reduces to

$$\frac{C_B}{C_A} = K \tag{66}$$

The constant K is called either the *distribution* or *partition coefficient* of the solute between the two solvents.

The applicability of the simplified distribution law, Eq. (66), may be judged from the data given in Table 8–5. The essential constancy of the distribution coefficients for low concentrations shows that in dilute solutions Eq. (66) is valid. However, the last two values for iodine and the last for boric acid indicate that in more concentrated solutions activities must be employed instead of concentrations to obtain a true constant for the partition coefficient. Furthermore, K depends on the nature of the solute and the liquids involved. Other factors affecting the magnitude of this constant are the temperature and the manner in which the constant is written, i.e., C_A/C_B or C_B/C_A.

TABLE 8–5.   DISTRIBUTION COEFFICIENTS AT 25°C
(C in moles/liter)

I_2 Between H_2O and CCl_4			H_3BO_3 Between H_2O and Amyl Alcohol		
C_{H_2O}	C_{CCl_4}	$K = \dfrac{C_{H_2O}}{C_{CCl_4}}$	C_{H_2O}	C_A	$K = \dfrac{C_{H_2O}}{C_A}$
0.000322	0.02745	0.0117	0.02602	0.00805	3.24
0.000503	0.0429	0.0117	0.05104	0.01545	3.31
0.000763	0.0654	0.0117	0.1808	0.0540	3.35
0.00115	0.1010	0.0114	0.3012	0.0857	3.52
0.00134	0.1196	0.0112			

Other examples of distribution which may be cited are the distributions of iodine between water and carbon disulfide, chloroform, and ethylene glycol, bromine between water and carbon disulfide or bromoform, hydrogen peroxide between water and various organic solvents, and phenol between water and amyl alcohol.

Walter Nernst[1] first called attention to the fact that the above statement of the distribution law is valid only when the solute undergoes no change such as dissociation or association. If a solute does dissociate into ions or simpler molecules or if it associates into more complex molecules, then the distribution law does not apply to the total concentrations in the two phases, but only to the concentrations of the particular species *common* to both. Thus, if a substance A dissolves in one solvent without any change in molecular form, and in another with partial association into, say, A_2, the partition coefficient for the distribution will not be given by the ratio of the total concentrations in the two phases, but rather by the total concentration in the first solvent divided by the concentration of unassociated molecules in the second solvent; i.e., by the ratio of the concentrations of the molecules having identical molecular weights in the two solvents. This limitation to the applicability of the distribution law was previously pointed out in conjunction with the use of Henry's law for gases that form new molecular or ionic species in solution.

To illustrate how the distribution law may be utilized in handling more complicated cases, consider the distribution of benzoic acid between water and chloroform, for which data are given in Table 8–6. Column 1 gives the

[1] W. Nernst, *Z. physik. Chem.*, **8**, 110 (1891).

TABLE 8-6. DISTRIBUTION OF BENZOIC ACID BETWEEN WATER AND
CHLOROFORM AT 40°C
(C in moles/liter)

C_W	C_C	$\dfrac{C_W}{C_C}$	C_{W_1}	C_{C_1}	$K = \dfrac{C_{W_1}}{C_{C_1}}$
0.00211	0.00721	0.292	0.00178	0.00404	0.441
0.00268	0.01084	0.247	0.00231	0.00523	0.442
0.00353	0.01686	0.210	0.00310	0.00701	0.442
0.00725	0.05700	0.127	0.00662	0.01497	0.442
0.01272	0.16733	0.076	0.01188	0.02687	0.442

total concentration of benzoic acid in the water layer, column 2 the same data for the chloroform layer at equilibrium, and column 3 shows the ratio of the two concentrations. Obviously there is no semblance of constancy in the ratio of total concentrations of benzoic acid in the two solvents. However, the observed results can be explained on the basis that the acid is partially *dissociated* in water into benzoate and hydrogen ions and *associated* in chloroform into double molecules $(C_6H_5COOH)_2$. Since the distribution law can be applied only to the common species present in the two phases, i.e., the single molecules of the acid, correction must be applied for the dissociation of the acid in the aqueous phase and for its association in the chloroform phase. When this is done, we obtain the concentrations of the single molecules given in columns 4 and 5 of the table, and these lead to the very constant value of K shown in the last column.

Other systems to which such analysis has been applied with success are the distribution of benzoic acid between water and benzene, salicylic acid between water and benzene or chloroform, and acetic acid between water and various organic solvents.

Distribution coefficients, like other equilibrium constants, vary with temperature. Thus K for the distribution of benzoic acid between water and chloroform is 0.564 at 10°C and 0.442 at 40°C. The variation of these constants with temperature is given by Eq. (40), Chapter 7, where ΔH is now the heat of transfer per mole of the solute from one solution to the other.

The distribution law has been applied to the study of problems of both theoretical and practical interest, such as extraction, analysis, and determination of equilibrium constants. Extraction is a subject of great importance both in the laboratory and in industry. In the laboratory occasion frequently arises for the removal of a dissolved substance from, say, a water solution, with solvents such as ether, chloroform, carbon tetrachloride, or benzene. Again, in industry extraction is used to remove various undesirable constituents of a product, such as harmful ingredients in petroleum oils, by treating the product with an immiscible solvent in which the impurity is also soluble. In all such processes it is important to know how much solvent and how many treatments are necessary in order to accomplish a particular degree of separation.

When a substance distributes itself between two solvents without the complications of association, dissociation, or reaction with the solvent, it is possible to calculate the weight of substance which can be removed in a series of extractions. Suppose we have a solution containing W g of a substance in V_1 cc of solution, and suppose that this solution is shaken repeatedly with V_2 cc samples of pure immiscible second solvent until distribution equilibrium is attained. Then at the end of n extractions the weight W_n of solute remaining unextracted will be

$$W_n = W \left(\frac{KV_1}{KV_1 + V_2} \right)^n \tag{67}$$

and therefore the weight extracted will be

$$W - W_n = W - W \left(\frac{KV_1}{KV_1 + V_2} \right)^n$$
$$= W \left[1 - \left(\frac{KV_1}{KV_1 + V_2} \right)^n \right] \tag{68}$$

Here $K = C_1/C_2$. When K is known, Eq. (67) may be employed to estimate the number of extractions necessary to reduce W to some given value W_n. Another important deduction which can be made from Eq. (67) is that if a given volume V of a solvent is available for extraction, greater extracting efficiency can be obtained if this volume is utilized in a number of separate extractions than if it were all used once. In other words, greater extracting efficiency is obtained by keeping V_2 small and n large than the other way around, and hence it is better to extract with small volumes of solvent several times than once with a large volume. The same conclusions apply to washing of precipitates, in which case the process may be considered as the distribution of the impurity between the wash liquid and the precipitate.

Another application of distribution coefficients is in analysis. Suppose a substance is present in a solvent A, in which analysis for the substance is difficult, and suppose, further, that analysis in another solvent B is readily possible. Then a distribution of the substance between the two solvents can be carried out and the substance analyzed for in solvent B. From the results of analysis in B, the volumes of the two solvents used, and the distribution coefficient, K, for the substance between the two solvents, the weight of it present originally in A can be obtained.

SOLUTIONS OF SOLIDS IN LIQUIDS

The extent to which solids dissolve in liquids varies greatly with the nature of the solid and liquid, the temperature, and to a much lesser degree the pressure on the system. In all cases the limit of solubility is the *saturated* solution. For any particular solute and solvent the concentration of the saturated solution at any given temperature and pressure is constant and does not depend on the manner in which the solution is prepared.

The concentrations of various solutes in a solvent necessary for saturation range over wide limits. Thus, at 20°C 100 g of water dissolve 192 g of ammonium nitrate, 6.5 g of mercuric chloride, and only 8.4×10^{-6} g of silver bromide. In ethyl alcohol, on the other hand, the order of solubility of mercuric chloride and ammonium nitrate is reversed, 100 g of the solvent dissolving 47.6 g of mercuric chloride and only 3.8 g of ammonium nitrate. As a rule most inorganic substances are more soluble in water than in organic solvents, while the reverse holds true for organic substances. There are, however, many exceptions.

The influence of temperature on the solubility of a solute in a particular solvent is, in general, quite pronounced, as may be seen from Table 8-7. Because most substances absorb heat on solution they tend to become

TABLE 8-7. SOLUBILITY OF SOLIDS IN WATER AT VARIOUS TEMPERATURES
(Grams of anhydrous salt/100 g H_2O)

Solid	0°C	20°C	40°C	60°C	100°C
NH_4Cl	29.4	37.2	45.8	55.2	77.3
$CaSO_4 \cdot 2\ H_2O$	0.176	—	0.210	0.205	0.162
$CuSO_4 \cdot 5\ H_2O$	14.3	20.7	28.5	40.0	75.4
KCl	27.6	34.0	40.0	45.5	56.7
KNO_3	13.3	31.6	63.9	110.0	246.
$AgNO_3$	122.	222.	376.	525.	952.
NaCl	35.7	36.0	36.6	37.3	39.8
Na_2SO_4	—	—	48.8	45.3	42.5

more soluble at higher temperatures. On the other hand, when the solution process is exothermic, a decrease of solubility with temperature may be expected, as is the case with sodium sulfate, However, for some substances the solubility behavior is not so regular. Thus the solubility of $CaSO_4 \cdot 2\ H_2O$ increases up to 40°C, passes through a maximum at this temperature, and then decreases at higher temperatures.

When the solubility of any substance is plotted against temperature, the curve obtained is continuous as long as there is no change in the nature of the saturating solid phase. As soon as the solid phase changes, however, a break in the solubility curve appears, and a new solubility curve, originating at the point of break, is obtained, which gives now the solubility of the solid phase formed as a function of the temperature. For substances exhibiting such changes in solid phase the temperatures at which the breaks in the solubility curve occur are definite and characteristic of the substances involved; they represent the temperatures at which the original solid phase and the new phase are in equilibrium with the same solution. In other words, such a break occurs at the temperature at which the solution is saturated with respect to both solid phases.

The particular change in the nature of the solid phase involved may be a transformation of one crystalline form to another, a change from a hydrate to the anhydrous salt, or a transformation of one hydrate to another. These changes will be discussed more fully in Chapter 10. At present only one example will be given, namely, the transformation of β-rhombic ammonium nitrate to the γ-rhombic solid modification, which leads to a discontinuity in the water solubility curve at 32°C. Each of these forms of ammonium nitrate has its own solubility curve, but each form is stable over a different temperature interval. Below 32°C the β-form is stable and is the saturating phase, while above 32°C the γ-modification is the stable form. At 32°C, the transition temperature, both forms are stable. Hence the solubility curves of the two intersect, and the solution is saturated with respect to both. Any break in a solubility curve may be viewed, therefore, as the intersection of two distinct solubility curves, each corresponding to and characteristic of its particular saturating phase.

The effect of pressure on the solubility of solids in liquids is generally quite small. A change of 500 atm in pressure increases the solubility of sodium chloride in water by only 2.3 per cent and decreases the solubility of ammonium chloride by only 5.1 per cent. It has also been observed that the solubility of a solid in a liquid is increased when the particle size of the saturating phase becomes very small. Thus, when the particle size of calcium sulfate is decreased from 2 to 0.3 micron the solubility in water at 25°C goes up from 2.085 to 2.476 g per liter. This fact explains why it is necessary in analytical procedures to digest a precipitate in order to increase the particle size and thereby decrease the solubility. Any solution saturated with respect to fine particles will be supersaturated with respect to any coarser ones, and hence the tendency will be for solute to precipitate onto the coarser particles. The result will be that the crystal size of the larger particles will increase, while the finer particles will disappear by solution.

CHEMICAL EQUILIBRIA IN SOLUTION

We have seen in the last chapter that for any reaction such as

$$aA + bB + \cdots = cC + dD + \cdots \qquad (69)$$

the thermodynamic equilibrium constant, K_a, is given by

$$K_a = \frac{a_C^c a_D^d \cdots}{a_A^a a_B^b \cdots} \qquad (70)$$

Equation (70) is also applicable to chemical equilibria occurring in solution, except that now the activities refer to solution constituents. If we express the concentration of the species involved in moles per liter, C, then $a = Cf$ by Eq. (14), and Eq. (70) thus becomes

$$K_a = \frac{(C_C f_C)^c (C_D f_D)^d \cdots}{(C_A f_A)^a (C_B f_B)^b \cdots}$$

$$= \left(\frac{C_C^c C_D^d \cdots}{C_A^a C_B^b \cdots} \right) \left(\frac{f_C^c f_D^d \cdots}{f_A^a f_B^b \cdots} \right) \qquad (71)$$

If we write now

$$K_c = \frac{C_C^c C_D^d \cdots}{C_A^a C_B^b \cdots} \qquad (72)$$

and

$$K_f = \frac{f_C^c f_D^d \cdots}{f_A^a f_B^b \cdots} \qquad (73)$$

then Eq. (71) becomes

$$K_a = K_c K_f \qquad (74)$$

Equation (72) defines K_c, the *concentration equilibrium constant* of a reaction.

In Eq. (74) K_a is again a true constant for a given reaction and temperature, while K_f is a factor dependent on the nature of the dissolved substances and their concentration. For ideal solutions, or real solutions at low concentrations, $f = 1$, $K_f = 1$, and therefore $K_a = K_c$. However, for nonideal solutions at higher concentrations K_f may not equal unity, and hence K_c may not be identical with K_a. Then, since K_f is a function of the concentration, K_c will vary with the concentration of the solution at equilibrium. Generally such deviations are not large with solutions of nonelectrolytes. If they are, then K_a's may be obtained from K_c's either by introduction of the activity coefficients, or by plotting the observed K_c's vs. the concentration of one of the reacting species and extrapolating to zero concentration. In the latter limit K_c becomes K_a.

As an example of an equilibrium in a liquid solution may be taken the dissociation of the amyl ester of dichloracetic acid into the acid and amylene, namely,

$$CHCl_2COOC_5H_{11} = CHCl_2COOH + C_5H_{10} \qquad (75)$$

for which the equilibrium constant K_c is given by

$$K_c = \frac{(C_{acid})(C_{amylene})}{C_{ester}} \qquad (76)$$

In investigating this equilibrium Nernst and Hohmann[2] mixed various proportions of amylene and dichloracetic acid, sealed the mixtures in glass tubes, and kept the tubes at 100°C until equilibrium was established. The tubes were chilled then to "freeze" the equilibrium, opened, and the contents analyzed for the amount of ester present.

TABLE 8–8. DECOMPOSITION OF THE AMYL ESTER
OF DICHLORACETIC ACID AT 100°C
($\alpha = 1$ mole)

b (moles)	V (liters)	x	K_c
1.05	0.215	0.455	3.31
2.61	0.401	0.615	3.12
4.45	0.640	0.628	3.54
5.91	0.794	0.658	3.44
7.30	0.959	0.650	3.73
8.16	1.062	0.669	3.49
11.33	1.439	0.688	3.35
13.80	1.734	0.700	3.24
15.36	1.829	0.703	3.39

If we let α equal the initial number of moles of acid, b the initial number of moles of amylene, x the number of moles of ester at equilibrium, and V the total volume of mixture in liters, then the concentrations of the various substances at equilibrium in moles per liter are

$$C_{ester} = \frac{x}{V}, \quad C_{acid} = \frac{\alpha - x}{V}, \quad C_{amylene} = \frac{b - x}{V}$$

and, therefore,

$$K_c = \frac{\left(\dfrac{\alpha - x}{V}\right)\left(\dfrac{b - x}{V}\right)}{\left(\dfrac{x}{V}\right)}$$

$$= \frac{(\alpha - x)(b - x)}{xV} \qquad (77)$$

[2] Nernst and Hohmann, *Z. physik. Chem.*, **11**, 352 (1893).

In all of Nernst and Hohmann's experiments α was kept at 1 mole, while b was varied. The values of b, V, and x for a series of runs are shown in the first three columns of Table 8–8, while the fourth column shows the values of K_c calculated from these data by Eq. (77). The constancy in K_c is fairly satisfactory.

REFERENCES

See references listed at end of Chapter 3. Also:

1. Hala, Pick, Fried, and Vilim, *Vapor-Liquid Equilibrium*, Pergamon Press, New York, 1958.
2. Hildebrand and Scott, *Solubility of Non-Electrolytes*, Reinhold Publishing Corporation, New York, 1950.
3. Robinson and Gilliland, *Elements of Fractional Distillation*, McGraw-Hill Book Company, Inc., New York, 1950.
4. A. Weissberger, *Physical Methods of Organic Chemistry*, Interscience Publishers, Inc., New York, 1959, Vol. I, Chaps. 9 and 11.
5. A. Weissberger, *Distillation*, Interscience Publishers, Inc., New York, 1951.

PROBLEMS

1. At 20°C diethyl ether exhibits a vapor pressure of 442.2 mm Hg. At the same temperature a solution of a nonvolatile solute in diethyl ether gives a vapor pressure of 413.5 mm Hg. Assuming the vapors to behave ideally, find (a) the activity of the solvent in the given solution, and (b) its partial molal free energy of mixing.   *Ans.* (a) 0.935; (b) -39.17 cal mole^{-1}.

2. At 300°K liquid A has a vapor pressure of 280.0 mm Hg and liquid B a vapor pressure of 170.0 mm Hg. When a solution is prepared containing 2 moles of each of the liquids, the vapor pressure above the solution is found to be 380.0 mm Hg, and the vapor contains 60.0 mole per cent of A. Assuming the vapors to be ideal, find:

(a) The activities of A and B in the solution
(b) The activity coefficients of A and B in solution
(c) The free energy of mixing for the solution
(d) The free energy of mixing to be expected for a corresponding ideal solution

3. Show that, on the basis of the infinitely dilute solution as a point of reference for solute, the relation between the molar activity coefficient f_2 and the molal activity coefficient γ_2 is given by the relation

$$f_2 = \left(\frac{m_2 \rho_0}{C_2} \right) \gamma_2$$

where m_2 is the molality of a given solution, C_2 the molarity, and ρ_0 the density of the pure solvent.

4. On the same basis as in problem 3, show that

$$\gamma_2' = (1 + 0.001 \, m_2 M_1) \gamma_2$$

where γ_2' is the activity coefficient of the solution for concentration in mol fractions, and M_1 is the molecular weight of the solvent.

5. Adcock and McGlashan [*Proc. Roy. Soc.* (London), **A226**, 266 (1954)] found that for the system carbon tetrachloride-cyclohexane ΔH_m between 10 and 55°C can be represented by the relation

$$\Delta H_m = (281 - 0.468\ T)N_1N_2\ \text{cal mole}^{-1}$$

where T is the absolute temperature, N_1 the mol fraction of carbon tetrachloride, and N_2 the mol fraction of cyclohexane. Deduce from this equation the partial molal heats of mixing of the two solution constituents.

$\qquad\qquad\qquad\qquad\qquad Ans.\ \overline{\Delta H}_1 = (281 - 0.468\ T)N_2^2.$

6. From the equations in problem 5, deduce the total and partial heat capacities for the mixing process, i.e., ΔC_{p_m}, $\overline{\Delta C}_{p_1}$, and $\overline{\Delta C}_{p_2}$.

7. A mixture of $C_6H_5CH_3$ and C_6H_6 contains 30% by weight of $C_6H_5CH_3$. At 30°C the vapor pressure of pure $C_6H_5CH_3$ is 36.7 mm Hg while that of pure C_6H_6 is 118.2 mm. Assuming that the two liquids form ideal solutions, calculate the total pressure and the partial pressure of each constituent above the solution at 30°C. $\qquad Ans.\ P_{C_6H_6} = 86.7$ mm; $P_{total} = 96.5$ mm.

8. At 60°C the vapor pressure of ethyl alcohol is 352.7 mm Hg and that of methyl alcohol 625 mm Hg. A mixture of the two, which may be assumed to be ideal, contains 50% by weight of each constituent. What will be the composition of the vapor above the solution at 60°C?

9. At 140°C the vapor pressure of C_6H_5Cl is 939.4 mm and that of C_6H_5Br is 495.8 mm. Assuming that these two liquids form an ideal solution, what will be the composition of a mixture of the two which boils at 140°C under 1 atm pressure? What will be the composition of the vapor at this temperature?

10. Solutions of two volatile liquids, A and B, obey Raoult's law. At a certain temperature it is found that when the total pressure above a given solution is 400 mm Hg, the mol fraction of A in the vapor is 0.45 and in the liquid, 0.65. What are the vapor pressures of the two pure liquids at the given temperature?

11. Kretschmer and Wiebe [*J. Am. Chem. Soc.*, **71**, 3176 (1949)] give the following liquid-vapor equilibrium data for ethanol-methylcyclohexane solutions at 55°C:

Mol fraction of ethanol in		Total pressure
Liquid	Vapor	(mm Hg)
0.0000	0.0000	168.1
0.0528	0.4837	319.8
0.1251	0.5375	352.8
0.2205	0.5645	368.0
0.3621	0.5846	376.3
0.5071	0.5988	379.8
0.6832	0.6244	380.1
0.7792	0.6528	375.8
0.9347	0.7879	337.5
1.0000	1.0000	279.9

Prepare a plot giving the total and partial pressures of the constituents as a function of the mol fraction of ethanol in the liquid phase, and determine the mol fraction at which the vapor pressure is a maximum. Assume that the vapor behaves as an ideal gas.

12. Using the data given in the preceding problem, plot the mol fractions of the liquid phase against those of the vapor with which they are in equilibrium. From the plot determine the mol fraction of liquid phase corresponding to the point of maximum vapor pressure.

13. What weight of HCl-H_2O azeotrope prepared at 740 mm Hg pressure will have to be added to water in order to prepare 2 liters of 0.50 molar HCl solution?

14. At 30°C a mixture of C_6H_5OH and H_2O is made up containing 60% by weight of H_2O. The mixture splits into two layers, the C_6H_5OH layer containing 70% by weight of C_6H_5OH and the H_2O layer containing 92% by weight of H_2O. Calculate the relative weights of the two layers.

$Ans.$ $W_{H_2O}/W_{Ph} = 0.934.$

15. A mixture of aniline and H_2O is made up containing 30% by weight of aniline. From Figure 8–16 determine graphically the relative weights of the two layers which form and the proportion of the total amount of aniline present in each layer at 40°C.

16. Using data from a suitable handbook, plot on the same graph the vapor pressures of H_2O, C_6H_6, and the total pressure of a mixture of the two against temperature between 40 and 80°C. From this plot, find the boiling point of the immiscible system C_6H_6-H_2O under 1 atm pressure.

17. A totally immiscible liquid system composed of H_2O and an organic liquid boils at 90°C when the barometer reads 734 mm Hg. The distillate contains 73% by weight of the organic liquid. What is the molecular weight and vapor pressure at 90°C of the organic liquid? $Ans.$ 122.9 g/mole; 208.2 mm Hg.

18. Naphthalene may be steam distilled at 99.3°C under atmospheric pressure. What weight of steam will be required to carry 2 lb of naphthalene into the distillate at atmospheric pressure?

19. If the specific heat of steam is 0.5 cal/g and the heat of vaporization of the liquid in problem 17 is 78 cal/g, calculate the total amount of heat and the minimum amount of steam, delivered at 99°C, required to steam-distill 500 g of the liquid at 90°C.

20. By use of Table 8–4 calculate the weight of ethylene which will dissolve in 1000 g of H_2O under an ethylene pressure of 2 atm at 20°C. $Ans.$ 0.314 g.

21. Dry air contains 21 mole % of O_2 and 79 mole % of N_2. What will be the composition of the air in water when the latter is equilibrated at 20°C with air at a pressure of 1 atm?

22. A mixture of H_2 and N_2 is equilibrated with 100 g of water at 40°C. At equilibrium it is found that the total pressure of the gas phase is 790.0 mm Hg and that the gas, after drying, consists of 40.0% H_2 by volume. Assuming that the vapor pressure of water above the solution is the same as that of pure water, namely, 55.3 mm Hg at 40°C, calculate the weights of dissolved H_2 and N_2.

23. At 20°C SO_2 was permitted to distribute itself between 200 cc of $CHCl_3$ and 75 cc of H_2O. When equilibrium was established, the $CHCl_3$ layer contained 0.14 mole of SO_2 and the H_2O layer 0.05 mole. What is the distribution coefficient of SO_2 between H_2O and $CHCl_3$ at 20°C?   *Ans.* $C_{H_2O}/C_{CHCl_3} = 0.953$.

24. Using the average value of the distribution coefficients given in Table 8-5, calculate the number of moles of H_3BO_3 which may be extracted from 50 cc of a 0.2 molar aqueous solution (a) by a single extraction with 150 cc of amyl alcohol, and (b) by three extractions with 50-cc portions of amyl alcohol.

25. (a) Show that the following data for the distribution of benzoic acid between H_2O and C_6H_6 at 20°C obey quite well the relation $C_{H_2O}^2/C_{C_6H_6} = K$.

C_{H_2O}	$C_{C_6H_6}$
0.0150	0.242
0.0195	0.412
0.0289	0.970

(b) Show that this relation can be obtained on the assumptions (1) that the dissociation of benzoic acid in water is slight, and (2) that the acid is practically completely associated into double molecules in C_6H_6.

26. Using the results of the preceding problem, calculate the number of moles of benzoic acid which may be extracted from 100 cc of a 0.2 molar aqueous solution by 10 cc of C_6H_6 at 20°C.

27. At 25°C the distribution coefficient of H_2S between H_2O and C_6H_6, defined as $[H_2S]_{H_2O}/[H_2S]_{C_6H_6}$, is 0.167. What is the minimum volume of C_6H_6 necessary at 25°C to extract in a single step 90% of the H_2S from 1 liter of a 0.1 molar aqueous solution of H_2S?   *Ans.* 1.50 liters.

28. Use the data of the preceding problem to find what total volume of C_6H_6 would be necessary to remove 90% of the H_2S from the given aqueous solution in three separate extractions using equal volumes of C_6H_6 in each.

29. At 25°C the distribution coefficient of C_2H_5OH between CCl_4 and H_2O, $K = [C_2H_5OH]_{CCl_4}/[C_2H_5OH]_{H_2O}$, is 0.0244. How will 1 g of C_2H_5OH distribute itself between 20 cc of H_2O and 50 cc of CCl_4?

30. Expressed in grams of anhydrous salt per 100 g H_2O, the solubilities of $MnSO_4 \cdot 5 H_2O$ and $MnSO_4 \cdot 4 H_2O$ in water as a function of the temperature are as follows:

	Solubility	
$t°C$	$MnSO_4 \cdot 5 H_2O$	$MnSO_4 \cdot 4 H_2O$
10	59.5	—
20	62.9	64.5
30	67.8	66.4
40	—	68.8
50	—	72.6

From a plot of solubilities vs. temperature ascertain the transition temperature of $MnSO_4 \cdot 5 H_2O$ to $MnSO_4 \cdot 4 H_2O$. Which of these phases is stable at 25°C?

31. For the reaction

$$C_2H_5OH(l) + CH_3COOH(l) = CH_3COOC_2H_5(l) + H_2O(l)$$

let α be the number of moles of alcohol present initially per mole of acid and x the number of moles of acid esterified after equilibrium is established. Then from the following data

α	x
0.5	0.420
1.0	0.665
1.5	0.779

(a) calculate K_c in each case, and (b) from the average value of K_c find x when α is 0.1 mole.

9
Colligative
Properties
of Solutions

$\mathbf{I}$N THIS chapter we shall consider four properties of solutions containing *nonvolatile* solutes, namely: (a) the vapor pressure lowering of the solvent, (b) the freezing point lowering, (c) the boiling point elevation, and (d) the osmotic pressure of the solution. These properties of a solution are referred to as the *colligative properties*. A colligative property is any property which depends only on the number of particles in solution and not in any way on the nature of these. As we shall see, this is the essential attribute of the four phenomena mentioned above, at least in dilute solutions.

On theoretical grounds it is convenient to subdivide solutions into (a) solutions of nonelectrolytes and (b) solutions of electrolytes. In nonelectrolytic solutions the solute dissolved in the solvent persists in molecular, uncharged form and exhibits no tendency to dissociate into electrically charged ions. For such solutions certain general laws have been developed which will be discussed below. In electrolytic solutions, on the other hand, the solute dissociates to a greater or lesser degree into ions, increasing thereby the total number of particles in solution. The behavior of the solution with respect to certain properties, therefore, is changed, and the simple laws deduced for nonelectrolytic solutions require modification. For this reason these two types of solutions will be discussed in this chapter in separate sections.

312

SOLUTIONS OF NONELECTROLYTES

The colligative properties of nonelectrolytic solutions deserve serious consideration because they supply valuable methods for estimating the molecular weight of the dissolved substance and for evaluating a number of highly important thermodynamic quantities. Here attention will be confined primarily to the presentation of the basic principles involved and their use for arriving at the molecular weights of the solutes.

LOWERING OF VAPOR PRESSURE OF SOLVENT

A dissolved solute lowers the vapor pressure of a liquid solvent in which it is dissolved. The vapor pressure lowering suffered by the solvent can be readily understood in terms of Raoult's law as developed in the preceding chapter. Let N_1 be the mol fraction of the solvent, N_2 the mol fraction of the solute, P^0 the vapor pressure of the pure solvent, and P the vapor pressure of the solvent above a given solution. Then according to Raoult's law P is given by

$$P = P^0 N_1 \tag{1}$$

Since N_1 in any solution is always less than unity, P must always be less than P^0. Consequently, *solution of a solute in a solvent leads to a lowering of the vapor pressure of the latter below that of the pure solvent.* Furthermore, when the solute is nonvolatile it does not contribute to the total vapor pressure, and hence Eq. (1) gives as well the total vapor pressure above the solution, which in this case is due to solvent only and is always less than P^0.

The extent of the vapor pressure lowering, ΔP, is

$$\begin{aligned} \Delta P &= P^0 - P = P^0 - P^0 N_1 \\ &= P^0(1 - N_1) \\ &= P^0 N_2 \end{aligned} \tag{2}$$

According to Eq. (2), the vapor pressure lowering of the solvent depends both on the vapor pressure of the solvent and the mol fraction of solute in solution. In other words, it depends on the nature of the solvent and on the concentration of solute, but not on the nature of the latter. However, if we consider the *relative vapor pressure lowering*, i.e., the ratio $\Delta P/P^0$, then from Eq. (2)

$$\frac{\Delta P}{P^0} = \frac{P^0 - P}{P^0} = N_2 \tag{3}$$

and the relative vapor pressure lowering of the solvent depends only on the mol fraction of solute and is *completely independent* of either the nature of solute or solvent. Equation (3), the form Raoult's law takes for solutions of nonvolatile solutes, shows that the relative vapor pressure lowering of a solvent is a colligative property, because it depends only upon the concentration of the solute and upon nothing else.

The validity of Raoult's law applied to vapor pressure lowering may be judged from the data given in Table 9–1 for water solutions of mannite at 20°C. Considering the difficulties involved in measuring small vapor pressure differences, the concordance between theory and experiment is satisfactory.

TABLE 9–1.   VAPOR PRESSURE LOWERING FOR
AQUEOUS MANNITE SOLUTIONS AT 20°C
$(P° = 17.51$ mm Hg)

Moles of Mannite per 1000 g H_2O	ΔP Observed (mm Hg)	ΔP Calculated (mm Hg)
0.0984	0.0307	0.0311
0.1977	0.0614	0.0622
0.2962	0.0922	0.0931
0.4938	0.1536	0.1547
0.6934	0.2162	0.2164
0.8922	0.2792	0.2775
0.9908	0.3096	0.3076

Equation (2) or (3) may be employed to calculate the vapor pressure lowering and the vapor pressure of solutions of nonvolatile solutes; or, knowing the vapor pressure lowering, they may be utilized to calculate the molecular weight of the dissolved substance. Consider the problem of calculating the vapor pressure of solvent above a solution containing 53.94 g of mannite (molecular weight = 182.11) per 1000 g of water at 20°C. At this temperature the vapor pressure of water is 17.51 mm Hg. According to Eq. (2),

$$\Delta P = P^0 - P = P^0 N_2$$
$$= P^0 \left[\frac{W_2/M_2}{W_1/M_1 + W_2/M_2} \right] \tag{4}$$

where W_2 and M_2 are the weight and molecular weight respectively of the solute, while W_1 and M_1 are the same quantities for the solvent. Substituting the given data into Eq. (4), we obtain

$$\Delta P = 17.51 \left[\frac{53.94/182.11}{1000/18.016 + 53.94/182.11} \right]$$
$$= 17.51 \times 0.0053$$
$$= 0.0929 \text{ mm Hg}$$

The vapor pressure of the solution is, therefore,

$$P = P^0 - \Delta P = 17.51 - 0.09 = 17.42 \text{ mm Hg}$$

When the vapor pressure lowering is known, the calculation may be reversed and Eq. (4) utilized to calculate M_2. Further, for very dilute solutions W_2/M_2 is small compared with W_1/M_1, and Eq. (4) can be reduced to

$$\frac{P^0 - P}{P^0} = \frac{W_2 M_1}{W_1 M_2} \tag{5}$$

BOILING POINT ELEVATION OF SOLUTIONS

Solutions containing nonvolatile solutes boil at temperatures *higher* than the boiling point of the pure solvent. The difference between the boiling points of the solution and pure solvent at any given constant pressure is referred to as the *boiling point elevation* of the solution. The boiling point elevation of a solution depends on the nature of the solvent and the concentration of solute, but is independent, at least in dilute solutions, of the nature of the solute as long as the latter is not ionized.

This elevation of the boiling point is readily understood in terms of the vapor pressure lowering and is a direct consequence of it. Consider the vapor pressure–temperature diagram shown in Figure 9–1. In this diagram curve AB represents the vapor pressure of the pure solvent as a function of temperature. Since the vapor pressure of the solution is at all

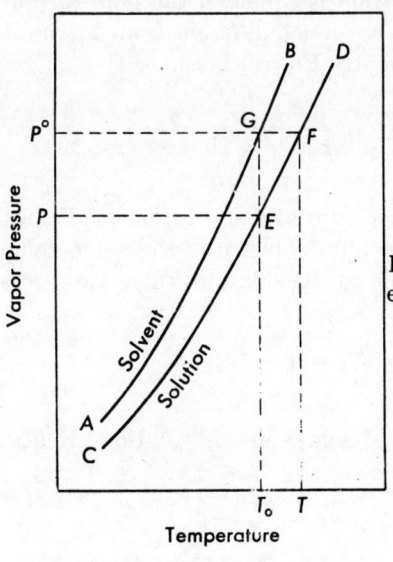

Figure 9–1. Boiling point elevation due to solutes.

temperatures lower than that of the solvent, the vapor pressure–temperature curve of the solution must lie below that of the pure solvent and hence must be represented by some curve such as CD in the figure. In order to reach the boiling point at some given external pressure P^0, the solvent and the solution must be heated to temperatures at which their respective vapor pressures become equal to the given confining pressure. As the diagram indicates, the solvent can attain the pressure P^0 at the temperature T_0, but the solution must be raised to temperature T, higher than T_0, before the same pressure is reached. Consequently, at the same external pressure the solution must boil at a temperature higher than the pure solvent; and, the elevation of the boiling point of the solution, ΔT_b, is given by $\Delta T_b = T - T_0$. These considerations are perfectly general and apply to any solution of a nonvolatile solute.

By applying the Clausius-Clapeyron equation and Raoult's law to the conditions depicted in Figure 9–1, it is possible to deduce a relation between the boiling point elevation of the solution and its concentration. Since points E and F lie on the vapor pressure curve of the solution, they are both given by the Clausius-Clapeyron equation

$$\ln \frac{P^0}{P} = \frac{\Delta H_v}{R}\left(\frac{T - T_0}{TT_0}\right) \tag{6}$$

where P is the vapor pressure of the solution at temperature T_0, while P^0 is the vapor pressure at temperature T. ΔH_v is heat of vaporization per mole of the solvent from the solution. When the solution is dilute, this is essentially the heat of vaporization per mole of the pure solvent. Again, when the solution is dilute T is not much different from T_0, and hence we may write $TT_0 = T_0^2$. Consequently Eq. (6) becomes

$$\ln \frac{P^0}{P} = -\ln \frac{P}{P^0} = \frac{\Delta H_v}{R} \cdot \frac{\Delta T_b}{T_0^2} \tag{7}$$

In Eq. (7) P is the vapor pressure of the solution at T_0, while P^0 is also the vapor pressure of the pure solvent at the same temperature. When Raoult's law is applicable to the solution, these two pressures are related through

$$\frac{P}{P^0} = N_1 = 1 - N_2 \tag{8}$$

where N_2 is the mol fraction of solute in the solution. Hence Eq. (7) becomes

$$\ln (1 - N_2) = -\frac{\Delta H_v}{R} \cdot \frac{\Delta T_b}{T_0^2} \tag{9}$$

Further, expansion of $\ln (1 - N_2)$ in series yields the expression

$$\ln (1 - N_2) = -N_2 - \frac{N_2^2}{2} - \frac{N_2^3}{3} - \cdots$$

and since the solution was already specified to be dilute, N_2 must be small, and all terms in the expansion beyond the first can be considered negligible. Writing, then, $-N_2$ for $\ln (1 - N_2)$ in Eq. (9), we have

$$-N_2 = -\frac{\Delta H_v}{R} \cdot \frac{\Delta T_b}{T_0^2}$$

and, therefore,

$$\Delta T_b = \frac{R T_0^2}{\Delta H_v} \cdot N_2 \tag{10}$$

Equation (10) gives the boiling point elevation of a solution in terms of the boiling point and heat of vaporization of the solvent, and the mol fraction of the solute in solution. Since for any given solvent T_0 and ΔH_v are constant, the boiling point elevation for dilute solutions is seen to be directly proportional to the mol fraction of the solute only and is in no way dependent on the nature of the solute. The boiling point elevation of a solution is thus a colligative property.

The common practice in boiling point elevation work is to express the concentration not in mol fractions but in *moles of solute per 1000 g of solvent*, i.e., the molality m. If we now let n_1 be the number of moles of solvent in 1000 g, then,

$$N_2 = \frac{m}{n_1 + m} \approx \frac{m}{n_1}$$

since for dilute solutions m is small compared with n_1 and may be disregarded. Therefore,

$$\Delta T_b = \left[\frac{R T_0^2}{\Delta H_v n_1} \right] m \tag{11}$$

For any given solvent all the quantities in the brackets of Eq. (11) are constant, and hence the whole term is constant. Writing

$$K_b = \frac{R T_0^2}{\Delta H_v n_1} \tag{12}$$

Eq. (11) finally reduces to

$$\Delta T_b = K_b m \tag{13}$$

According to Eq. (13) the boiling point elevation of any dilute solution is directly proportional to the *molality of the solution*. The proportionality constant K_b is called either the *molal boiling point elevation constant*, or *the ebullioscopic constant*, and signifies the rise in boiling point for a 1 molal solution of a solute in a solvent provided the laws of dilute solutions are applicable to such a concentration. Actually it is the boiling point elevation per mole calculated by proportion from the boiling point elevation of much more dilute solutions.

The validity of Eq. (13) may be tested in several ways. First, the equation demands that for a given solvent the boiling point elevation be proportional to the molality of the solute irrespective of its nature. Second, the equation demands that for any given solvent the proportionality constant K_b be independent of the nature or concentration of the solute. In both these respects the equation is in good agreement with experimental results on dilute solutions. A third, and more crucial, test can be made by comparing the observed values of K_b with those predicted by Eq. (12). If Eq. (12) is valid, it should be possible to calculate the molal boiling point elevation constant from a knowledge of the normal boiling point and heat of vaporization of the solvent. For instance, for water $T_0 = 373.2°K$, while the heat of vaporization at the boiling point is 539 cal per gram. Applying Eq. (12), the molal boiling point elevation constant for water as a solvent follows as

$$K_b = \frac{RT_0^2}{\Delta H_v n_1}$$
$$= \frac{1.987 \times (373.2)^2}{(18.02 \times 539)(1000/18.02)}$$
$$= 0.513°$$

This value compares very well with the experimentally observed $K_b = 0.52°$.

Table 9–2 lists the normal boiling points and observed ebullioscopic constants for a number of solvents. The molal elevation constants for these solvents calculated by Eq. (12) are also included for comparison. The agreement between calculated and observed values is highly satisfactory when uncertainties involved in boiling point determinations are con-

TABLE 9–2.   MOLAL BOILING POINT ELEVATION CONSTANTS

Solvent	Boiling Point (°C)	K_b (obs.)	K_b (calc.)
Acetone	56.5	1.72	1.73
Carbon tetrachloride	76.8	5.0	5.02
Benzene	80.1	2.57	2.61
Chloroform	61.2	3.88	3.85
Ethyl alcohol	78.4	1.20	1.19
Ethyl ether	34.6	2.11	2.16
Methyl alcohol	64.7	0.80	0.83
Water	100.0	0.52	0.51

sidered. In view of this concordance between theory and experiment, Eq. (12) may be employed also to evaluate heats of vaporization of the solvent from experimental values of K_b.

CALCULATION OF MOLECULAR WEIGHTS FROM BOILING POINT ELEVATION

If we let ΔT_b be the boiling point elevation for a solution containing W_2 g of solute of molecular weight M_2 dissolved in W_1 g of solvent, then the weight of solute per 1000 g of solvent is

$$\frac{W_2 \times 1000}{W_1}$$

and hence m, the molality of the solution, is

$$m = \frac{W_2 \times 1000}{W_1 M_2} \tag{14}$$

Equation (13) in terms of Eq. (14) is, therefore,

$$\Delta T_b = K_b \left(\frac{1000\, W_2}{W_1 M_2} \right) \tag{15}$$

When the ebullioscopic constant of a solvent is known, a determination of the boiling point elevation of a solution containing the unknown solute in definite concentration is sufficient to yield the molecular weight of the solute. When the ebullioscopic constant is unknown, however, an independent determination of ΔT_b must be made with a solute of known molecular weight. The calculations involved can best be seen from the following example. A solution containing 0.5126 g of naphthalene (molecular weight = 128.17) in 50.00 g of carbon tetrachloride yields a boiling point elevation of 0.402°C, while a solution of 0.6216 g of an unknown solute in the same weight of solvent gives a boiling point elevation of 0.647°C. Find the molecular weight of the unknown solute.

For finding K_b of carbon tetrachloride, the given data are

$$W_1 = 50.00 \text{ g} \qquad \Delta T_b = 0.402°C$$
$$W_2 = 0.5126 \text{ g} \qquad M_2 = 128.17$$

Substituting these into Eq. (15) and solving for K_b, we obtain

$$K_b = \frac{\Delta T_b\, W_1 M_2}{1000\, W_2}$$
$$= \frac{0.402 \times 50.00 \times 128.17}{1000 \times 0.5126}$$
$$= 5.03°C/\text{mole}/1000 \text{ g of solvent}$$

Using now the found value of K_b along with the information for the unknown,

$$W_1 = 50.00 \text{ g} \qquad W_2 = 0.6216 \text{ g} \qquad \Delta T_b = 0.647°C$$

Eq. (15) yields for M_2 of the unknown,

$$M_2 = \frac{1000 \ W_2 K_b}{W_1 \cdot \Delta T_b}$$

$$= \frac{1000 \times 0.6216 \times 5.03}{50.00 \times 0.647}$$

$$= 96.7 \text{ g mole}^{-1}$$

FREEZING POINT LOWERING OF SOLUTIONS

When a dilute solution is cooled, a temperature is eventually reached at which *solid solvent* begins to separate from solution. The temperature at which this separation begins is called *the freezing point of the solution*. More generally the freezing point of a solution may be defined as the temperature at which a particular solution is in equilibrium with solid solvent.

Solutions freeze at *lower* temperatures than the pure solvent. The freezing point lowering of a solution is again a direct consequence of the vapor pressure lowering of the solvent by dissolved solute. To appreciate this, consider the vapor pressure–temperature diagram shown in Figure 9–2. In this diagram AB is the sublimation curve of the solid solvent, while CD is the vapor pressure curve of pure liquid solvent. At the freezing point of the pure solvent the solid and liquid phases are in equilibrium, and consequently they must both have at this temperature identical vapor pressures. The only point on the diagram at which the two forms of the pure solvent have the same vapor pressure is B, the intersection of AB and CD, and therefore T_0, the temperature corresponding to B, must be the freezing point of the pure solvent. When a solute is dissolved in the solvent, however, the vapor pressure of the latter is lowered, and equilibrium can no longer exist at T_0. To ascertain the new point of equilibrium between the solution of the solute and the solid solvent, the temperature must be found at which the vapor pressure of the solution becomes equal to that of the solid, i.e., the temperature at which the vapor pressure curve of the solution intersects the sublimation curve, and this will be the freezing point of the solution. Since the vapor pressure curve of the solution, EF, always lies below that of the pure solvent, the intersection of EF and AB can occur only at a point such as E for which the temperature is lower than T_0. Hence any solution of the solute in the solvent must have a freezing point, T, lower than that of the solvent, T_0.

The freezing point depression of a solution is defined as $\Delta T_f = T_0 - T$ and represents the number of degrees by which the freezing point of a solution is lower than that of the pure solvent. The magnitude of ΔT_f depends both on the nature of the solvent and the concentration of the solution. For dilute solutions of various solutes in a given solvent ΔT_f varies linearly with concentration irrespective of the nature of the solute. The proportionality constant of this concentration variation is, however, a function of the solvent and varies considerably for different solvents.

To relate mathematically the freezing point depression of a solution to the factors mentioned, consider again Figure 9–2. Let P_s be the vapor

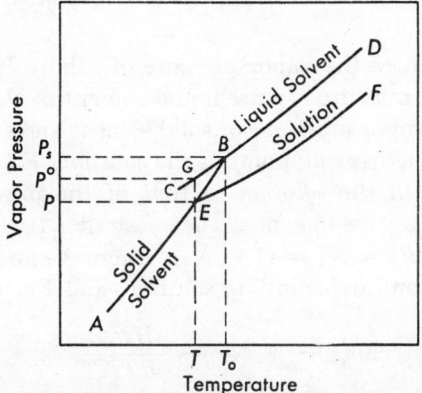

Figure 9–2. Depression of freezing point by solutes.

pressure of solid and pure liquid solvent at T_0 and P be the vapor pressure of both solid solvent and solution at temperature T. Again, let P^0 be the vapor pressure of pure supercooled liquid solvent at T, point G. Then, since points G and B lie on the same vapor pressure curve, they must both be related by the Clausius-Clapeyron equation

$$\ln \frac{P_s}{P^0} = \frac{\Delta H_v(T_0 - T)}{RT_0 T} \tag{16}$$

where ΔH_v is the heat of vaporization of the pure solvent. Similarly, since points E and B lie on the same sublimation curve, they must be given by the equation

$$\ln \frac{P_s}{P} = \frac{\Delta H_s(T_0 - T)}{RT_0 T} \tag{17}$$

where ΔH_s is the heat of sublimation of the solid solvent. Subtracting now

Eq. (17) from Eq. (16), we obtain

$$\ln P_s - \ln P^0 - \ln P_s + \ln P = \frac{\Delta H_v(T_0 - T)}{RT_0T} - \frac{\Delta H_s(T_0 - T)}{RT_0T}$$

$$\ln P - \ln P^0 = -\frac{(\Delta H_s - \Delta H_v)(T_0 - T)}{RT_0T}$$

$$\ln \frac{P}{P^0} = -\frac{(\Delta H_s - \Delta H_v)(T_0 - T)}{RT_0T} \tag{18}$$

But $(\Delta H_s - \Delta H_v) = \Delta H_f$, the heat of fusion of the solvent. Therefore,

$$\ln \frac{P}{P^0} = -\frac{\Delta H_f(T_0 - T)}{RT_0T} = -\frac{\Delta H_f \Delta T_f}{RT_0T} \tag{19}$$

Equation (19) relates the vapor pressure of solid solvent at temperature T to the vapor pressure of pure liquid solvent at the same temperature. But since the vapor pressures of solid solvent and solution are equal at temperature T, the freezing point of the solution, Eq. (19) also relates the vapor pressure of the *solution* to that of the pure solvent at the temperature T. If we assume now that Raoult's law is applicable to the solution, the $P/P^0 = N_1 = (1 - N_2)$, where N_1 and N_2 are the mol fractions of the solvent and solute in solution, and Eq. (19) becomes

$$\ln (1 - N_2) = -\frac{\Delta H_f \Delta T_f}{RT_0T} \tag{20}$$

When N_2 is small, i.e., when the solution is dilute, $\ln (1 - N_2)$ is equal essentially to $-N_2$ and T_0T to T_0^2. Hence

$$-N_2 = -\frac{\Delta H_f \Delta T_f}{RT_0^2}$$

and
$$\Delta T_f = \left(\frac{RT_0^2}{\Delta H_f}\right) N_2 \tag{21}$$

Finally, designating by m the *molality* of the solution and by n_1 the number of moles of solvent in 1000 g, $N_2 = m/n_1$ (approximately), and

$$\Delta T_f = \left(\frac{RT_0^2}{\Delta H_f n_1}\right) m$$
$$= K_f m \tag{22}$$

where
$$K_f = \frac{RT_0^2}{\Delta H_f n_1} \tag{23}$$

Equation (22) is the fundamental relation of cryoscopy and is directly analogous to Eq. (13) for boiling point elevation. K_f, called the *molal freezing point lowering* or *cryoscopic constant* of a solvent, is defined in terms of quantities characteristic of the solvent only and in no way depends on either the concentration or nature of the solute. It is the freezing point depression counterpart of the boiling point elevation constant K_b. Since for any given solvent K_f is a constant, the freezing point depression of a solution is determined by the concentration of solute only, and hence the freezing point depression, like the vapor pressure lowering and boiling point elevation, is a colligative property.

An important test of Eq. (22) is the constancy of K_f with concentration for a given solvent. Table 9–3 shows some experimental data for

TABLE 9–3. FREEZING POINT DEPRESSIONS
FOR SOLUTIONS OF UREA IN WATER

m	ΔT_f	$K_f = \Delta T_f/m$
0.000538	0.001002	1.862
0.004235	0.007846	1.851
0.007645	0.01413	1.849
0.012918	0.02393	1.850
0.01887	0.03496	1.853
0.03084	0.05696	1.848
0.04248	0.07850	1.848
		1.852

the freezing point lowering of solutions of urea in water and the value of $K_f = \Delta T_f/m$ calculated therefrom. As the last column indicates, K_f is constant throughout the given concentration range and equal to 1.85. Essentially identical values for the cryoscopic constant of water have been obtained with numerous other solutes. Furthermore, the observed value of this constant is in good agreement with the value predicted for water as a solvent by Eq. (23). For water, $T_0 = 273.2$, $\Delta H_f = 79.71 \times 18.02$, and $n_1 = 1000/18.02$. Therefore,

$$K_f = \frac{1.987(273.2)^2}{(79.71 \times 18.02)(1000/18.02)}$$
$$= 1.857°C/mole/1000 \text{ g } H_2O$$

Similar concordance between theory and experiment has been obtained with many other solvents. Table 9–4 lists a number of solvents, their freezing points, and their cryoscopic constants. It will be observed that, of those listed, the cryoscopic constant of water is the lowest. Hence, for a given concentration of solute, more pronounced freezing point depressions may be obtained in the various other solvents, and this fact is of importance in the practical use of these solvents for molecular weight determination.

TABLE 9–4.   CRYOSCOPIC CONSTANTS
FOR VARIOUS SOLVENTS

Solvent	Freezing Point (°C)	K_f
Acetic acid	16.7	3.9
Benzene	5.5	5.12
Bromoform	7.8	14.4
Camphor	178.4	37.7
Cyclohexane	6.5	20.0
1,4-Dioxane	10.5	4.9
Naphthalene	80.2	6.9
Phenol	42	7.27
Tribromophenol	96	20.4
Triphenylphosphate	49.9	11.76
Water	0.00	1.86

Because of the assumptions made in its derivation, Eq. (22) can be expected to be valid only in dilute solutions. In more concentrated solutions K_f deviates considerably from constancy, and the deviation is the more pronounced the more concentrated the solution. Thus the cryoscopic constant of benzene as evaluated from freezing point determinations of carbon tetrachloride in the solvent is 5.09°C per mole in a 0.1184 molal solution and only 4.82°C per mole in a 1.166 molal solution.

In view of the ease with which fairly precise freezing point data can be obtained, such data are particularly suitable for determining molecular weights of solutes in solution. The calculations involved are exactly analogous to those made in conjunction with boiling point elevation, as are also the equations and data required. If the expression for m from Eq. (14) is substituted into Eq. (22), we get

$$\Delta T_f = K_f \left(\frac{1000 \, W_2}{W_1 M_2} \right) \tag{24}$$

from which the molecular weight follows as

$$M_2 = K_f \left(\frac{1000 \, W_2}{\Delta T_f \, W_1} \right) \tag{25}$$

Therefore, to calculate the molecular weight, K_f for the solvent must be known, and ΔT_f, W_1 and W_2 must be measured. If K_f is not known, it can be either calculated by means of Eq. (23) or determined by first making a freezing point determination with a solution of a solute of known molecular weight in the same solvent.

SEPARATION OF SOLID SOLUTIONS ON FREEZING

These freezing point depression considerations and the equations deduced are valid only when the solid separating from solution is the pure solvent. Occasionally cases are encountered, such as solutions of iodine or thiophene in benzene, where the solid crystallizing out contains solute dissolved in it in the form of solid solutions. For such cases Eq. (22), and all others based on it, are no longer applicable. It can be shown by theoretical argument that when the separating solid phase is a solid solution, Eq. (22) must be replaced by

$$\Delta T_f = K_f (1 - k)m \qquad (26)$$

where the symbols have the same significance as before, and k is the ratio of mol fraction of solute in the solid to mol fraction of solute in solution. When the solid phase is pure solvent, k is zero, and Eq. (26) reduces to Eq. (22). When the solid phase is not pure, however, two conditions may be distinguished. If the solute is more soluble in the liquid solvent than in the solid, k is a positive fraction, and hence $(1 - k)$ is less than unity. The effect will be, therefore, to give a freezing point depression less than would be anticipated for separation of pure solid solvent. If, on the other hand, the solute is more soluble in the solid phase than in the liquid, $k > 1$ and $(1 - k)$ is negative. Under these conditions ΔT_f is also *negative;* i.e., a freezing point *elevation* rather than a lowering is observed for the solution. Such behavior is rarely encountered under ordinary circumstances, but it is not at all unusual with metals and salts that form solid solutions.

OSMOSIS AND OSMOTIC PRESSURE

When a solution of a solute is separated from pure solvent by a semipermeable membrane, i.e., a membrane that permits the passage of solvent but not of solute, it is observed that solvent tends to pass through the membrane into the solution, and thereby to dilute it. The phenomenon, called *osmosis*, was first reported by the Abbé Nollet in 1748. For low molecular weight solutes in water the best semipermeable membrane known is a film of copper ferrocyanide, $Cu_2Fe(CN)_6$, prepared by contacting a solution of a cupric salt with one of potassium ferrocyanide. For high molecular weight solutes in organic solvents the membranes used most frequently are thin films of either cellulose or cellulose nitrate.

Before proceeding it is necessary to define a quantity called the osmotic pressure, Π. For this purpose consider the diagram shown in Figure 9-3. A is a chamber open at one end and fitted at the other with a movable piston B. The chamber is divided by means of a semipermeable membrane

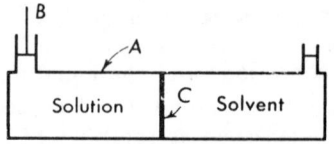

Figure 9–3.   Osmotic pressure of solutions.

C into two sections, of which the right one is filled with pure solvent, the other with some solution of a solute in the solvent. Because of osmosis, the solvent will tend to pass through the membrane into the solution and displace the piston upward. The motion of the piston and osmosis of the solvent can be prevented, however, by the application of pressure to the piston in order to keep it in its original position. The mechanical pressure which must be applied on a solution to prevent osmosis of the solvent into the solution through a semipermeable membrane is called the *osmotic pressure* of the solution. This pressure for a given solution depends on a number of factors, as we shall see later, but it does not depend on the nature of the membrane so long as the membrane is truly semipermeable. Hence the osmotic pressure of a solution must be considered as a measure of some real difference, expressible in pressure units, in the natures of the pure solvent and the solution, rather than as a phenomenon for which the membrane is responsible. The membrane is merely the artifice by which this difference is made manifest.

The most extensive series of measurements on the osmotic pressures of aqueous solutions were made by the Earl of Berkeley and Hartley (1906–1909) in England, and by Frazer, Morse, and their collaborators (1901–1923) in America. A schematic diagram of the apparatus used by Berkeley and Hartley is shown in Figure 9–4. A is a porous tube on

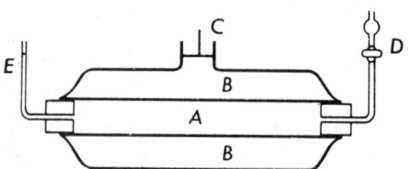

Figure 9–4. Osmotic pressure apparatus of Berkeley and Hartley.

the outside of which was deposited by a special technique a layer of copper ferrocyanide. This tube was mounted by means of water-tight joints within an outer metal jacket B, which carried an attachment C through which pressure could be applied. A was filled through D with pure water up to a definite mark on the capillary E, B with the solution under test. The apparatus was then immersed in a thermostat. As a result of osmosis the liquid level in E tended to drop. By applying pressure through C the liquid level at E could be restored to its initial position, and the pressure necessary to accomplish this restoration was taken as the osmotic pressure of the solution. With this method equilibrium is established rapidly, and the concentration of the solution is not changed by dilution with solvent.

Some accurate data on the osmotic pressures of aqueous sucrose solutions at various temperatures are shown in Table 9–5. They are based on the measurements of Berkeley and Hartley and of Morse, Frazer, and their co-workers and are typical of the results obtained with nonelectrolytes in water solutions.

TABLE 9–5.   OSMOTIC PRESSURES OF AQUEOUS SUCROSE SOLUTIONS

m (moles/1000 g H_2O)	Osmotic Pressure (atm)				
	0°	20°	40°	60°	80°
0.100	2.46	2.59	2.66	2.72	
0.200	4.72	5.06	5.16	5.44	
0.300	7.09	7.61	7.84	8.14	
0.400	9.44	10.14	10.60	10.87	
0.500	11.90	12.75	13.36	13.67	
0.600	14.38	15.39	16.15	16.54	
0.700	16.89	18.13	18.93	19.40	
0.800	19.48	20.91	21.80	22.33	23.06
0.900	22.12	23.72	24.74	25.27	25.92
1.000	24.83	26.64	27.70	28.37	28.00

RELATION OF OSMOTIC PRESSURE TO VAPOR PRESSURE

It is possible to derive from purely thermodynamic considerations a relation between osmotic pressure and the vapor pressure lowering of a solution. At any constant temperature and external pressure of 1 atm, transfer of solvent into solution occurs because the molar free energy of the pure solvent, F_1^0, is greater than the partial molal free energy of the solvent in solution, $\bar{F}_1$. To bring about equilibrium between the two and thus stop osmosis, it is necessary to increase the value of $\bar{F}_1$ to $\bar{F}_1'$ by raising the external pressure on the solution from 1 atm to a pressure P. If this increase in free energy is ΔF, then the condition for osmotic equilibrium must be

$$F_1^0 = \bar{F}_1 + \Delta F$$

and hence

$$\bar{F}_1 - F_1^0 = -\Delta F \tag{27}$$

But, by Eqs. (30) and (42) of the last chapter, $\bar{F}_1 - F_1^0 = RT \ln a_1 = RT \ln P_1/P_1^0$ when the vapor phase of the solvent behaves ideally, and hence

Eq. (27) becomes

$$RT \ln P_1/P_1^0 = -\Delta F$$

or

$$\Delta F = RT \ln \frac{P_1^0}{P_1} \tag{28}$$

To find ΔF we utilize Eq. (7) of Chapter 8, which leads to the expression

$$d\bar{F}_1 = \bar{V}_1 dP$$

where $\bar{V}_1$ is the partial molal volume of the solvent in solution. If we integrate this equation now between the limits $\bar{F}_1 = \bar{F}_1$ at $P = 1$ atm and $\bar{F}_1 = \bar{F}_1'$ at pressure P, we obtain

$$\int_{\bar{F}_1}^{\bar{F}_1'} d\bar{F}_1 = \int_1^P \bar{V}_1 dP$$

$$\bar{F}_1' - \bar{F}_1 = \int_1^P \bar{V}_1 dP \tag{29}$$

But $\bar{F}_1' - \bar{F}_1 = \Delta F$. Further, if we take $\bar{V}_1$ to be independent of pressure, then Eq. (29) becomes

$$\Delta F = \bar{V}_1(P - 1)$$
$$= \bar{V}_1 \Pi \tag{30}$$

where $\Pi = (P - 1)$ is the pressure which must be applied to stop osmosis, i.e., the osmotic pressure. Substitution of Eq. (30) into Eq. (28) thus yields

$$\Pi = \frac{RT}{\bar{V}_1} \ln \frac{P_1^0}{P_1} \tag{31}$$

which is the relation sought between osmotic pressure and vapor pressure lowering. When the solutions involved are not too concentrated, $\bar{V}_1$ may be taken to be identical with V_1^0, the molar volume of the solvent, and

TABLE 9–6. COMPARISON OF OBSERVED AND CALCULATED
OSMOTIC PRESSURES OF SUCROSE SOLUTIONS AT 30°C

m	$\Pi_{obs.}$ (atm)	$\Pi_{calc.}$	
		Eq. (32)	Eq. (33)
0.100	2.47	2.44	2.40
1.000	27.22	27.0	20.4
2.000	58.37	58.5	35.1
3.000	95.16	96.2	45.5
4.000	138.96	138.5	55.7
5.000	187.3	183.0	64.5
6.000	232.3	231.0	—

then Eq. (31) becomes

$$\Pi = \frac{RT}{V_1^0} \ln \frac{P_1^0}{P_1} \tag{32}$$

Equation (32) allows the calculation of the osmotic pressure of a solution from its vapor pressure and that of the pure solvent at any constant temperature T. Further, it reproduces the observed osmotic pressures up to fairly high concentrations, as may be seen from the second and third columns of Table 9–6.

VAN'T HOFF EQUATION FOR OSMOTIC PRESSURE

Equation (32) can be reduced to a simpler form for the special case of dilute solutions obeying Raoult's law. For such solutions $P_1/P_1^0 = N_1 = 1 - N_2$, and hence

$$\Pi V_1^0 = -RT \ln (1 - N_2)$$

If $\ln (1 - N_2)$ be expanded now in series as before, then for dilute solutions all terms beyond the first can be neglected, and $\ln (1 - N_2)$ becomes $-N_2 = - n_2/n_1$, where n_2 is the number of moles of solute in n_1 moles of solvent. Hence

$$\Pi V_1^0 = \frac{RTn_2}{n_1}$$

and

$$\Pi(V_1^0 n_1) = n_2 RT$$

But $V_1^0 n_1$ is the total volume of solvent containing n_2 moles of solute, which for dilute solutions is essentially the volume V of the solution. Consequently,

$$\Pi V = n_2 RT \tag{33}$$

or, alternately,

$$\Pi = CRT \tag{34}$$

where C is the molarity of the solution.

Equation (33) is known as *van't Hoff's law* for ideal solutions. This equation is identical in form with the ideal gas law, except that Π replaces the gas pressure P. Its validity can be judged from the data given in Table 9–7 and the last column of Table 9–6. As expected, the equation applies only to dilute solutions, i.e., below about 0.2 molar.

The van't Hoff equation can be used to find molecular weights of dissolved solutes from measured osmotic pressures in the same manner as the ideal gas law is used to find the molecular weights of gases. However, this method is rarely employed, except in high polymer work, because of the extreme difficulty in obtaining accurate osmotic pressure

TABLE 9–7. OSMOTIC PRESSURE OF
AQUEOUS SUCROSE SOLUTIONS AT 14°C

C	Π (Atm)	
	Obs.	Eq. (34)
0.0588	1.34	1.39
0.0809	2.00	1.91
0.1189	2.75	2.80
0.1794	4.04	4.23

data for dilute solutions. The same information can be obtained more readily by, say, a freezing point lowering determination with an accuracy that can hardly be attained in osmotic pressure measurements.

SOLUTIONS OF ELECTROLYTES

Solutions of nonelectrolytes in water or other solvents do not conduct electricity. Such solutions exhibit the colligative behavior described in the preceding sections and they obey in dilute solutions the various relations deduced there. These are also the solutions which yield for the dissolved substances normal molecular weights, i.e., the molecular weights expected from their chemical formulas, or, occasionally, some simple multiple of it. On the other hand, there are substances, particularly salts and inorganic acids and bases, which when dissolved in water or other appropriate solvents yield solutions which conduct electricity to a greater or lesser extent. Such solutions are said to be electrolytes. Solutions of electrolytes also exhibit the colligative properties of vapor pressure lowering, boiling point elevation, freezing point lowering, and osmotic pressure, but they do not obey the simple relations deduced for nonelectrolytes. The colligative effects observed are always *greater* than those to be expected from the concentration. Stated differently, solutions of electrolytes behave as if the solute had in solution a molecular weight *lower* than the simplest formula weight of the substance. The remainder of this chapter will be devoted to a discussion of the colligative properties of electrolytic solutions and to an exposition of some of the theories which have been advanced to explain their behavior. Other aspects of electrolyte behavior will be elaborated on in later chapters.

COLLIGATIVE PROPERTIES OF ELECTROLYTES

As was indicated above, the freezing point lowering, boiling point elevation, vapor pressure lowering, and osmotic pressure of solutions of electrolytes all are higher than the corresponding effects for solutions of nonelectrolytes of the same total concentration. The freezing point data shown in Table 9–8 may be taken as typical of the nature of the deviations. This table lists the ratios of observed freezing point lowering, ΔT_f, to molality, m, at various concentrations for a number of electrolytes in water solution. According to the arguments developed above, this ratio should approach for dilute aqueous solutions the value of K_f for water, namely, 1.86° per mole per 1000 g of solvent. Inspection of the table reveals, however, that the limiting values approached by the various

TABLE 9–8. $\Delta T_f/m$ FOR AQUEOUS SOLUTIONS OF ELECTROLYTES

m	HCl	HNO₃	NH₄Cl	CuSO₄	H₂SO₄	CoCl₂	K₂SO₄	K₃Fe(CN)₆
0.0005	—	—	—	—	—	—	—	7.3
0.0010	3.690	—	—	—	—	—	5.280	7.10
0.0020	3.669	—	—	—	—	5.35	—	6.87
0.0025	—	—	—	3.003	5.052	—	5.258	—
0.0050	3.635	3.67	3.617	2.871	4.814	5.208	5.150	6.53
0.0100	3.601	3.64	3.582	2.703	4.584	5.107	5.010	6.26
0.0500	3.532	3.55	3.489	2.266	4.112	4.918	4.559	5.60
0.1000	3.523	3.51	3.442	2.08	3.940	4.882	4.319	5.30
0.2000	3.54	3.47	3.392	1.91	3.790	4.946	4.044	5.0
0.4000	—	3.46	—	—	3.68	5.170	3.79	—
1.0000	3.94	3.58	3.33	1.72	4.04	6.31	—	—
2.0000	4.43	3.79	3.34	—	5.07	8.51	—	—
4.0000	5.65	4.16	3.35	—	7.05	—	—	—

electrolytes are considerably higher than 1.86°. Furthermore, the limit approached is not the same throughout, but varies from approximately $2 \times 1.86 = 3.72°$ for substances like hydrochloric acid and ammonium chloride to $3 \times 1.86 = 5.58°$ for cobalt chloride and $4 \times 1.86 = 7.44°$ for potassium ferricyanide.

To represent the colligative properties of electrolytes by means of the relations for nonelectrolytes, van't Hoff suggested the use of a factor i, which is defined as *the ratio of the colligative effect produced by a concentration m of electrolyte divided by the effect observed for the same concentration of nonelectrolyte.* Applying this definition of i to freezing point depression of electrolyte solutions, it follows that

$$i = \frac{\Delta T_f}{(\Delta T_f)_0} \tag{35}$$

where ΔT_f is the freezing point lowering for the electrolyte and $(\Delta T_f)_0$ is the freezing point depression for a nonelectrolyte of the same concentration. Since according to Eq. (22) $(\Delta T_f)_0 = K_f m$, then

$$\Delta T_f = i K_f m \tag{36}$$

The values of i must be calculated from experimental data for each electrolyte at various concentrations. It has been found, however, that once i is known for a particular concentration of an electrolyte for one of the colligative properties, the same value of i, within a small temperature correction where necessary, is essentially valid for the other properties at the same concentration. Consequently, we may write

$$i = \frac{\Delta T_f}{(\Delta T_f)_0} = \frac{\Delta T_b}{(\Delta T_b)_0} = \frac{\Delta P}{(\Delta P)_0} = \frac{\Pi}{(\Pi)_0} \tag{37}$$

where the quantities without subscripts refer to the electrolyte and those with subscripts to the nonelectrolyte of the same concentration. On substitution of Eqs. (2), (13), and (33) into Eq. (37), the expressions for the vapor pressure lowering, boiling point elevation, and osmotic pressure for solutions of electrolytes become:

$$\Delta P = i(\Delta P)_0 = i P^0 N_2 \tag{38}$$
$$\Delta T_b = i(\Delta T_b)_0 = i K_b m \tag{39}$$
$$\Pi = i(\Pi)_0 = \frac{i n_2 R T}{V} \tag{40}$$

These equations may be expected to be applicable only to dilute solutions.

Table 9–9 lists the values of i calculated from Table 9–8 by means of Eq. (36). A perusal of this table reveals that in dilute solutions i increases as the molality is lowered and approaches a limit of two for electrolytes such as hydrochloric acid, nitric acid, ammonium chloride, and copper

TABLE 9–9.   VAN'T HOFF FACTORS, i, FOR VARIOUS ELECTROLYTES

m	HCl	HNO$_3$	NH$_4$Cl	CuSO$_4$	H$_2$SO$_4$	CoCl$_2$	K$_2$SO$_4$	K$_3$Fe(CN)$_6$
0.0005	—	—	—	—	—	—	—	3.92
0.0010	1.98	—	—	—	—	—	2.84	3.82
0.0020	1.97	—	—	—	—	2.88	—	3.70
0.0025	—	—	—	1.61	2.72	—	2.83	—
0.0050	1.95	1.97	1.95	1.54	2.59	2.80	2.77	3.51
0.0100	1.94	1.96	1.92	1.45	2.46	2.75	2.70	3.31
0.0500	1.90	1.91	1.88	1.22	2.21	2.64	2.45	3.01
0.1000	1.89	1.89	1.85	1.12	2.12	2.62	2.32	2.85
0.2000	1.90	1.87	1.82	1.03	2.04	2.66	2.17	2.69
0.4000	—	1.86	—	—	1.98	2.78	2.04	—
1.0000	2.12	1.92	1.79	0.93	2.17	3.40	—	—
2.0000	2.38	2.04	1.80	—	2.73	4.58	—	—
4.0000	3.04	2.24	1.80	—	3.79	—	—	—

sulfate, a limit of three for electrolytes of the type sulfuric acid, cobalt chloride, and potassium sulfate, and a limit of four for potassium ferricyanide. In more concentrated solutions, on the other hand, i passes through a minimum and then increases, the increase frequently rising above the limits established in dilute solutions.

THE ARRHENIUS THEORY OF ELECTROLYTIC DISSOCIATION

The colligative properties of electrolytes and the fact that solutions of electrolytes conduct electricity led Svante Arrhenius to propose in 1887 his celebrated *theory of electrolytic dissociation*. The essential points of this theory are already familiar to the student. Arrhenius postulated that electrolytes in solution are dissociated into electrically charged particles, called ions, in such a manner that the total charge on the positive ions is equal to the total charge on the negative ions. The net result is, therefore, that the solution as a whole is neutral in spite of the presence of electrically charged particles in it. The presence of these ions accounts for the electrical conductivity of the solutions.

Arrhenius pointed out further that an electrolyte in solution need not necessarily be completely dissociated into ions; instead it may be only partially dissociated to yield ions in equilibrium with unionized molecules of the substance. It may then be anticipated from the laws of chemical equilibrium that the extent of dissociation will vary with concentration, becoming greater as the concentration of dissolved substance becomes lower. In view of this, complete dissociation may be expected to take place only in infinitely dilute solutions. At finite concentrations, however, the electrolyte will be only partially ionized to a degree dependent on the nature of the substance and the concentration.

This idea of partial electrolytic dissociation was employed by Arrhenius to explain the colligative behavior of solutions of electrolytes. The colligative properties of a dilute solution depend on the number of particles, irrespective of kind, present in a given quantity of solvent. When a substance in solution dissociates into ions, the number of particles in solution is increased. If it is assumed now that an ion acts with respect to the colligative properties in the same manner that an unionized molecule does, the increase in the number of particles in solution should cause an increase in the colligative effects. Thus, since the molal freezing point depression for a nonelectrolyte in water is 1.86°C, it may be anticipated that for an electrolyte which yields on complete dissociation two ions for every molecule the molal freezing point depression should be twice as great, namely, 3.72°C. Similarly, for an electrolyte yielding three ions the molal freezing point depression should be 5.58°C, while for one with four ions, 7.44°C. These molal lowerings for the types of electrolytes mentioned are the ones to be anticipated on complete dissociation, and hence these are the values to be observed only in extremely dilute solutions. That such is actually the case is evident from an inspection of Table 9–8.

In terms of this theory any observed values of molal freezing point depressions higher than 1.86°C and lower than the limits which would be reached on complete dissociation are to be accounted for by partial dissociation of the electrolyte. If this be the case, it is possible to calculate the degree of ionization of an electrolyte from the observed colligative data or from the values of i calculated from these. Consider an electrolyte A_xB_y which dissociates into x ions of A, each of charge z_+, and y ions of B, each of charge z_-, according to the equation

$$A_xB_y = xA^{z+} + yB^{z-} \tag{41}$$

If the original molality of the electrolyte is m, and if we let α be the degree of dissociation, then the number of moles of A_xB_y that dissociate is $m\alpha$, and the number of moles that remain unionized is $m - m\alpha = m(1 - \alpha)$. But for each mole of A_xB_y that dissociates, x moles of positive ions and y moles of negative ions are obtained. Consequently, for $m\alpha$ moles dissociating, $x(m\alpha)$ moles of A^{z+} and $y(m\alpha)$ moles of B^{z-} are obtained. The total number of moles, m_t, of substances of all types present in solution is then

$$m_t = m(1 - \alpha) + x(m\alpha) + y(m\alpha)$$
$$= m[1 + \alpha(x + y - 1)] \tag{42}$$

Designating by ν the *total number of ions yielded by a molecule of the electrolyte*, $\nu = x + y$, and Eq. (42) becomes

$$m_t = m[1 + \alpha(\nu - 1)] \tag{43}$$

Now, for a total molality m_t the freezing point depression must be given by $\Delta T_f = K_f m_t$, and hence, in terms of Eq. (43),

$$\Delta T_f = K_f m[1 + \alpha(\nu - 1)] \tag{44}$$

Solving for α we obtain

$$\alpha = \frac{(\Delta T_f/K_f m - 1)}{(\nu - 1)}$$
$$= \frac{\Delta T_f - K_f m}{(\nu - 1)(K_f m)} \tag{45}$$

A more general relation for α follows by comparing Eq. (44) with Eq. (36). Since $\Delta T_f = iK_f m$, i must be given by

$$i = 1 + \alpha(\nu - 1) \tag{46}$$

and, therefore,

$$\alpha = \frac{i-1}{\nu-1} \tag{47}$$

Equation (47) is applicable to any of the colligative properties and can just as readily be derived from boiling point elevation, vapor pressure lowering, or osmotic pressure considerations. It gives the degree of dissociation of an electrolyte from a knowledge of i and the type of electrolyte in question.

CLASSIFICATION OF ELECTROLYTES

Calculations of the degree of dissociation for various electrolytes in aqueous solutions show that practically all salts are highly dissociated into ions. The same is true for strong acids and bases. As may be expected from their high degree of dissociation, aqueous solutions of these substances are good conductors of electricity. On the other hand, there are many substances whose aqueous solutions exhibit relatively poor conductivity and whose colligative behavior indicates that they are only slightly dissociated even at fairly low concentrations. Among these are included a large number of organic acids such as acetic, propionic, and benzoic; inorganic acids such as carbonic, hydrosulfuric, hydrocyanic, orthoarsenic, boric, and hypochlorous acids; and bases such as ammonium, zinc, and lead hydroxides. Solutions of substances that show good conductance and which indicate a high degree of dissociation in solution are designated as *strong electrolytes*. In turn, solutions of substances that exhibit only poor conductance and a low degree of dissociation are called *weak electrolytes*.

However, not all electrolytes can be classified clearly as being either strong or weak. There are some electrolytes, such as aqueous solutions of o-chlorbenzoic, o-nitrobenzoic, 3,5-dinitrobenzoic, and cyanoacetic acids, whose behavior indicates that they are intermediate in properties between the strong and weak electrolytes. Still, the number of substances of intermediate strength is not large, and hence such electrolytes will not be considered further.

It is frequently convenient to subdivide strong electrolytes further according to the charge of the ions produced. An electrolyte that yields two singly charged ions is called an electrolyte of 1-1 type. Again, an electrolyte that yields univalent positive ions and bivalent negative ions, such as potassium sulfate or sulfuric acid, is called an electrolyte of 1-2 type, while one that yields bivalent positive ions and univalent negative ions, such as barium chloride or magnesium nitrate, is called an electrolyte of 2-1 type. Similarly, copper sulfate is an electrolyte of 2-2 type, while potassium ferricyanide is an electrolyte of 1–3 type. In this method of classification of strong electrolytes the charge of the positive ions is given first followed by the charge of the negative ions.

CRITICISM OF THE ARRHENIUS THEORY

Applications of the Arrhenius theory to the colligative behavior, electrical conductance, and ionic equilibria of weak electrolytes have shown that the theory is essentially satisfactory for these. However, in attempting to apply this theory to strong electrolytes so many anomalies and inconsistencies have been encountered that serious questioning arose early as to the validity of some of Arrhenius's postulates in respect to the nature of strong electrolytes.

The degree of dissociation of an electrolyte may be determined not only from the colligative properties but also from conductance measurements by means that will be described in Chapter 11. The values of α obtained by these two methods agree quite well for weak electrolytes, but for strong electrolytes the agreement is not what is to be expected from the accuracy of the measurements. Again, application of the laws of equilibrium to the partial dissociation postulated for electrolyte solutions by Arrhenius shows that these are obeyed quite well by weak electrolytes but not at all by strong electrolytes.

Another factor which militates against the simple dissociation theory when applied to strong electrolytes is the fact that Arrhenius considered solutions containing ions to behave essentially as ideal solutions containing the same number of neutral molecules. Yet such an assumption is hardly plausible. Whereas the forces between neutral molecules in fairly dilute solutions may be relatively small, the electrostatic attractions between electrically charged particles in solutions of electrolytes may exercise a significant effect on the motion and distribution of ions. Consequently such solutions can hardly be ideal in behavior, and ions can hardly be expected to act as if they were neutral and ideal molecules. Still, the effect produced by interionic attractions may be quite small in solutions of weak electrolytes where the number of ions is not large. Such solutions may behave, therefore, in line with Arrhenius's expectations. But, in strong electrolytes, where the number of ions is large, the effect of interionic attractions should be appreciable and should be the more pronounced the more concentrated the solution and the higher the valence of the ions.

These considerations, and others, point to the conclusion that the Arrhenius theory, although essentially valid for weak electrolytes, does not represent the true situation in solutions of strong electrolytes. The consensus is that solutions of strong electrolytes are *completely ionized* even at moderate concentrations, and that the α's calculated for strong electrolytes merely give an indication of the interionic forces of attraction operating in such solutions.

THE DEBYE-HÜCKEL THEORY OF INTERIONIC ATTRACTION

In 1923 P. Debye and E. Hückel[1] published a theory of interionic attraction in dilute solutions of electrolytes which occupies a dominant position in all considerations involving electrolytes and their kinetic and thermodynamic behavior.[2] At this time attention will be confined to the presentation of the salient qualitative details of the theory, without any attempt at mathematical formulation. Like Arrhenius, Debye and Hückel postulate that strong electrolytes exist in solution as ions of the types mentioned, but, unlike Arrhenius, they believe that strong electrolytes, at least in dilute solutions, are completely ionized and that the effects observed are due to the unequal distribution of ions resulting from interionic attraction. Debye and Hückel showed that, because of electrostatic attractions between charged ions, each positive ion in solution must be surrounded on an average with more negative ions than ions of like charge; and, conversely, each negative ion must be surrounded on an average with more positive than negative ions. In other words, each ion in solution is surrounded by an *ionic atmosphere whose net charge is opposite to that of the central ion.* They showed, further, that the properties of the electrolyte are determined by the interaction of the central ion with its atmosphere. Since the nature of the atmosphere is determined by the valences of the ions in solution, their concentration, the temperature, and the dielectric constant of the medium, it must follow that these are also the factors controlling the thermodynamic properties of the electrolyte. At any given temperature and in any given solvent, the temperature and dielectric constant are fixed, and hence the properties of the electrolyte should depend only on the charges of the ions and their concentration and not at all on the specific nature of each electrolyte. These conclusions are strictly valid only for very dilute solutions. This limitation arises from the fact that Debye and Hückel were forced to make certain mathematical simplifications which in essence reduce the applicability of their equations to such solutions.

In the Debye-Hückel theory the effect of the concentration of the ions

[1] Debye and Hückel, *Physikalische Zeitschrift*, **24**, 185 (1923).
[2] Although this theory will be discussed briefly at this point and will be elaborated on at several other stages in the book, no complete and rigid exposition is possible in an elementary text. Any student interested in further and more complete details is advised to read some of the references mentioned at the end of the chapter.

enters through a quantity called the *ionic strength* of the solution. This quantity, is a measure of the electrical environment in solution and plays in this theory a role analogous to concentration in the Arrhenius theory. It is defined as

$$\mu = \frac{1}{2}(C_1 z_1^2 + C_2 z_2^2 + C_3 z_3^2 + \cdots)$$

$$= \frac{1}{2}\Sigma C_i z_i^2 \tag{48}$$

where μ is the ionic strength of the solution, C_1, C_2, C_3, $\cdots$ the concentrations of the various ions in gram ionic weights per liter, while z_1, z_2, z_3, $\cdots$ are the valences of the respective ions. The ionic strength of a solution is equal to the molarity only in the case of 1–1 electrolytes. In all other instances the two are not equal. This may be illustrated with the calculation of the ionic strength of C molar solutions of potassium chloride, barium chloride, and lanthanum sulfate. In potassium chloride, $C_+ = C_- = C$, $z_+ = z_- = 1$, and, therefore,

$$\mu_{KCl} = \frac{1}{2}[C(1)^2 + C(1)^2]$$

$$= \frac{2\,C}{2} = C$$

For barium chloride, $C_+ = C$, $C_- = 2\,C$, $z_+ = 2$, $z_- = 1$. Hence

$$\mu_{BaCl_2} = \frac{1}{2}[C(2)^2 + 2\,C(1)^2]$$

$$= \frac{6\,C}{2} = 3\,C$$

Finally, for lanthanum sulfate, $C_+ = 2\,C$, $C_- = 3\,C$, $z_+ = 3$, $z_- = 2$, and, therefore,

$$\mu = \frac{1}{2}[2\,C(3)^2 + 3\,C(2)^2]$$

$$= \frac{30\,C}{2} = 15\,C$$

These examples show that the ionic strength of a solution is determined not only by the stoichiometric concentration of the electrolyte but also by the valences of its ions. It should also be emphasized that in calculating the ionic strength of a solution the summation called for in Eq. (48) must include any and all ionic species present, no matter what their source. Thus the ionic strength of a solution containing $C = 0.10$ potassium chloride in presence of $C = 0.01$ barium chloride is

$$\mu = \frac{1}{2} [0.1(1)^2 + 0.1(1)^2 + 0.01(2)^2 + 2(0.01)(1)^2]$$
$$= 0.13$$

Various indications of the ability of the Debye-Hückel theory to explain the behavior of dilute solutions of strong electrolytes will be given at appropriate stages in this book. At present it is sufficient to point out that this theory can account for the values of i observed in dilute solutions of strong electrolytes. The theory predicts that the values of i for such solutions at 0°C should be given by the relation

$$i = \nu(1 - 0.375 \, z_+z_- \, \sqrt{\mu}) \tag{49}$$

where ν is the number of ions yielded by a given molecule of electrolyte. The extent to which this prediction is verified may be seen from Table 9–10, where observed values of i are compared with those calculated by means of Eq. (49). The agreement between theory and experiment is good in the very dilute solutions, but past $m = 0.01$ the deviations become rapidly significant. These deviations are least pronounced with the 1–1 electrolyte HCl and most marked with the 2–2 electrolyte $CuSO_4$, i.e., the deviations are the greater the larger the product z_+z_-.

TABLE 9–10. COMPARISON OF OBSERVED AND CALCULATED VAN'T HOFF FACTORS AT 0°C

	HCl		K_2SO_4		$CuSO_4$	
m	$i_{obs.}$	$i_{calc.}$	$i_{obs.}$	$i_{calc.}$	$i_{obs.}$	$i_{calc.}$
0.0010	1.98	1.98	2.84	2.88	—	—
0.0025	—	—	2.83	2.81	1.61	1.70
0.0050	1.95	1.95	2.77	2.73	1.54	1.58
0.0100	1.94	1.92	2.70	2.65	1.45	1.40
0.0500	1.90	1.83	2.45	2.13	1.22	0.66
0.1000	1.89	1.63	2.32	1.77	1.12	0.10

In conclusion, the present status of the problem of electrolytes may be summarized as follows. For weak electrolytes the Arrhenius theory of partial dissociation is, within minor corrections, adequate. On the other hand, for strong electrolytes this theory is not satisfactory, for in these interionic attraction is the dominant factor. On the quantitative and theoretical side the Debye-Huckel theory supplies the explanation for the thermodynamic properties of very dilute and moderately dilute solutions, but this theory has not been extended as yet to concentrated solutions. It is quite probable that for concentrated solutions the treatment presented by the Debye-Hückel theory is considerably oversimplified. In order to understand the properties of concentrated solutions it may be necessary to consider not only interionic attraction, but also such phenomena as interaction between ions resulting in association, interaction of the solvent with ions, and the change in the nature of the solvent as a result of the presence of charged particles. Although the importance of these factors is appreciated, and although some of them have been investigated both theoretically and experimentally, no complete analysis of the complex problems involved here has as yet been made.

REFERENCES

See references listed at end of Chapter 3. Also:
1. H. Falkenhagen, *Electrolytes*, Oxford University Press, New York, 1934.
2. Harned and Owen, *The Physical Chemistry of Electrolytic Solutions*, Reinhold Publishing Corporation, New York, 1950.
3. G. Scatchard, *Chem. Rev.*, **13**, 7 (1933).
4. Shedlovsky, Brown, and MacInnes, *Trans. Am. Electrochem. Soc.*, **66**, 237 (1934).
5. A. Weissberger, *Physical Methods of Organic Chemistry*, Interscience Publishers, Inc., New York, 1959, Vol. I, Chaps. 7, 8, 9, 11, and 15.

PROBLEMS

1. A solution contains 5 g of urea ($M_2 = 60.05$) per 100 g of water. What will be the vapor pressure of this solution at 25°C? The vapor pressure of pure H_2O at this temperature is 23.756 mm.   *Ans.* 23.40 mm.

2. At 25°C 10.50 liters of pure N_2, measured at 760 mm Hg, are passed through an aqueous solution of a nonvolatile solute, whereby the solution loses 0.2455 g in weight. If the total pressure above the solution is also 760 mm, what is the vapor pressure of the solution and the mol fraction of solute?

3. At 50°C the vapor pressures of pure water and ethyl alcohol are, respectively, 92.5 and 219.9 mm Hg. If 6 g of a nonvolatile solute of molecular weight 120 are dissolved in 150 g of each of these solvents, what will be the relative vapor pressure lowerings in the two solvents?

4. A solution composed of 10 g of a nonvolatile organic solute in 100 g of diethyl ether has a vapor pressure of 426.0 mm at 20°C. If the vapor pressure of the pure ether is 442.2 mm at the same temperature, what is the molecular weight of the solute?

5. Derive Eq. (12) of this chapter by starting with the free energy condition for equilibrium between vapor and solution at the normal boiling point. Assume both the vapor and solution to be ideal.

6. If 30 g of diphenyl are dissolved in 250 g of benzene, what will be the boiling point of the resulting solution under atmospheric pressure?   *Ans.* 82.1°C.

7. A solution consisting of 5.00 g of an organic solute per 25.00 g of CCl_4 boils at 81.5°C under atmospheric pressure. What is the molecular weight of the solute?

8. From the data given in Table 9–2 calculate the molar heat of vaporization of ethyl alcohol.   *Ans.* 9420 cal/mole.

9. A certain weight of a nonvolatile solute dissolved in chloroform gives a boiling point elevation of 3.00°C at 760 mm pressure. Assuming the solution to be ideal, calculate the mol fractions of solute in the solution by Eqs. (9), (10), and (13). Which of these results would be nearest the correct value?

10. In deriving Eq. (9) it has been assumed that ΔH_v is a constant. Suppose, however, that ΔH_v in calories/mole is not constant, but varies with temperature according to the equation

$$\Delta H_v = A + BT + CT^2 + DT^3$$

where A, B, C, and D are constants. Starting with the unintegrated Clausius-Clapeyron equation, and without making any assumptions other than that Raoult's law is valid, derive the relation between $\ln(1 - N_2)$, T, and T_0.

11. For water the constants in the heat of vaporization equation given in problem 10 are: $A = 13,425$, $B = -9.81$, $C = 7.5 \times 10^{-5}$, and $D = 4.46 \times 10^{-7}$. Using these constants and the equation derived in problem 10, calculate the mol fraction of solute in solution when the boiling point elevation is 0.50°C. Compare the result with those given by Eqs. (9), (10), and (13).

12. Derive Eq. (22) of this chapter by starting with the free energy condition for equilibrium between pure solid solvent and solution. Assume the solution to be ideal.

13. What weight of glycerol would have to be added to 1000 g of water in order to lower its freezing point 10°C?   *Ans.* 495 g.

14. An aqueous solution contains 5% by weight of urea and 10% by weight of glucose. What will be its freezing point?

15. Compare the weights of methanol and glycerol which would be required to lower the freezing point of 1000 g of water 1°C.

16. From the data given in Table 9–4, calculate the heat of fusion per mole of phenol.

17. A sample of CH_3COOH is found to freeze at 16.4°C. Assuming that no solid solution is formed, what is the concentration of impurities in the sample?

18. A mixture which contains 0.550 g of camphor and 0.045 g of an organic solute freezes at 157.0°C. The solute contains 93.46% of C and 6.54% by weight of H. What is the molecular formula of the compound?   *Ans.* $C_{12}H_{10}$.

19. When dissolved in 100 g of a solvent whose molecular weight is 94.10 and whose freezing point is 45.0°C, 0.5550 g of a solute of molecular weight 110.1

gave a freezing point depression of 0.382°C. Again, when 0.4372 g of solute of unknown molecular weight was dissolved in 96.50 g of the same solvent, the freezing point lowering was found to be 0.467°C. From these data find (a) the molecular weight of the unknown solute, (b) the cryoscopic constant of the solvent, and (c) the heat of fusion of the solvent per mole.

20. In the derivation of Eq. (20) it has been assumed that ΔH_v and ΔH_s are constant. Suppose, however, that both are dependent upon the temperature, and given by the relations

$$\Delta H_v = A + BT + CT^2 + DT^3$$
$$\Delta H_s = a + bT + cT^2 + dT^3$$

where A, B, C, D, a, b, c, and d are constants. Starting with the unintegrated Clausius-Clapeyron equation, derive the relation between $\ln (1 - N_2)$, T, and T_0. Make no assumptions other than that Raoult's law is valid.

21. For water the constants A, B, C, and D are given in problem 11. Again, for this substance $a = 11,260$, $b = 7.66$, $c = -15.6 \times 10^{-3}$, and $d = 4.46 \times 10^{-7}$. Using these constants and the equation derived in problem 20, calculate the mol fraction of solute in solution when the freezing point lowering is 1.000°C. Compare the result with those given by Eqs. (20), (21), and (22).

22. Derive Eq. (26) on the assumption that both the liquid and solid solutions behave ideally.

23. An aqueous solution contains 20 g of glucose per liter. Assuming the solution to be ideal, calculate its osmotic pressure at 25°C.   *Ans.* 2.72 atm.

24. The osmotic pressure of an aqueous solution containing 45.0 g of sucrose per liter of solution is 2.97 atm at 0°C. Find the value of the gas constant and compare the result with the accepted value.

25. A solution of 1.00 g of antipyrine ($C_{11}H_{12}N_2O$) in 100 cc of aqueous solution gave an osmotic pressure of 1.18 atm at 0°C. Calculate the molecular weight of the compound and compare the result with that expected from the given formula.

26. An aqueous solution freezes at -1.50°C. Calculate (a) the normal boiling point, (b) the vapor pressure at 25°C, and (c) the osmotic pressure at 25°C of the given solution.

27. The average osmotic pressure of human blood is 7.7 atm at 40°C. (a) What should be the total concentration of various solutes in the blood? (b) Assuming this concentration to be essentially the same as the molality, find the freezing point of blood.

28. The vapor pressure of an aqueous solution at 25°C is 23.45 mm. Using Eq. (30), calculate its osmotic pressure given that the vapor pressure of pure H_2O is 23.756 mm at 25°C.

29. A 0.2 molal aqueous solution of KCl freezes at -0.680°C. Calculate i and the osmotic pressure at 0°C. Assume volume to be that of pure H_2O.

Ans. $i = 1.83$; $\Pi = 8.2$ atm.

30. A 0.4 molal aqueous solution of K_2SO_4 freezes at $-1.52°C$. Assuming that i is constant with temperature, calculate the vapor pressure at $25°C$ and the normal boiling point of the solution.

31. A solution of HCl, 0.72% by weight, freezes at $-0.706°C$. Calculate the apparent molality and the apparent molecular weight of the HCl.

32. A 0.1 molal solution of a weak electrolyte ionizing into two ions freezes at $-0.208°C$. Calculate the degree of dissociation. *Ans.* 0.118.

33. A 0.01 molal solution of $K_3Fe(CN)_6$ freezes at $-0.062°C$. What is the apparent percentage of dissociation? *Ans.* 78%.

34. A 2.00 molal HCl solution freezes at $-8.86°C$. Calculate the apparent percentage of dissociation and explain your answer.

35. At $25°C$ a 0.1 molal solution of CH_3COOH is 1.35% dissociated. Calculate the freezing point and osmotic pressure of the solution. Compare your results with those expected under conditions of no dissociation.

36. Compare the ionic strengths of 0.1 N solutions of HCl, $SrCl_2$, $AlCl_3$, $ZnSO_4$, and $Fe_2(SO_4)_3$.

37. A solution is 0.5 molar in $MgSO_4$, 0.1 molar in $AlCl_3$, and 0.2 molar in $(NH_4)_2SO_4$. What is the total ionic strength? *Ans.* 3.2.

38. Using the Debye-Hückel equation, calculate the values of i at $0°C$ for 0.0005 molar aqueous solutions of HCl, $BaCl_2$, H_2SO_4, $CuSO_4$, and $La(NO_3)_3$.

10
The Phase
Rule

U<small>P TO</small> this point various types of heterogeneous equilibria were considered from a number of different points of view. Thus, heterogeneous equilibria such as vaporization, sublimation, fusion, transition of one solid phase to another, solubility of solids, liquids, and gases in each other, vapor pressure of solutions, chemical reaction between solids or liquids and gases, and distribution of solutes between phases all have been approached by methods suitable for each particular type of equilibrium. These involved the Clapeyron and Clausius-Clapeyron equations, Raoult's law, Henry's law, equilibrium constants, and the distribution law. However, it is possible to treat all heterogeneous equilibria from a unified standpoint by means of a principle called the *phase rule*. With this principle the number of variables to which each and every type of heterogeneous equilibrium is subject may be defined under various experimental conditions. By this definition the phase rule does not invalidate or supersede some of the methods of attack described for the quantitative study of such equilibria. The phase rule is merely able to fix the number of variables involved, but the quantitative relations among the variables must be established through supplementary expressions such as some of those mentioned above. The significance of this statement will become clearer as soon as the nature of the phase rule and the manner in which it is used are developed.

DEFINITIONS

Before proceeding to a statement of the phase rule, it will be necessary to define and explain in some detail certain terms which are employed frequently in this connection. These are true, metastable, and unstable equilibrium, number of components, and degrees of freedom of a system. The meanings of system and phase were given in Chapter 3.

A state of *true equilibrium* is said to exist in a system when the same state can be realized by approach from either direction. Thermodynamically speaking, true equilibrium is attained when the free energy content of the system is at a minimum for the given values of the variables. An instance of such an equilibrium is ice and liquid water at 1 atm pressure and 0°C. At the given pressure, the temperature at which the two phases are in equilibrium is the same whether it is attained by partial melting of the ice or a partial freezing of the water. On the other hand, water at −5°C can be obtained by careful cooling of the liquid, but not by fusion of ice. Water at −5°C is said to be in a state of *metastable equilibrium*. Such a state can be realized only by careful approach from one direction and may be preserved provided the system is not subjected to sudden shock, stirring, or "seeding" by solid phase. As soon as a crystal of ice is introduced, solidification sets in rapidly, and the temperature rises to 0°C.

A state of *unstable equilibrium* is said to exist when the approach to equilibrium in a system is so slow that the system *appears* to undergo no change with time. An instance of such a situation is sodium chloride dissolving into a solution which is very nearly saturated with the salt. Insufficient time of observation might make it appear that equilibrium had been reached, whereas actually the process is still proceeding very slowly toward true saturation. It must be realized that, although a metastable equilibrium represents a state of at least partial stability, unstable equilibrium does not involve any equilibrium at all, but only a process of very slow change.

The *number of components* of a system is the *smallest number* of independently variable constituents, in terms of whose formulas equations may be written expressing the composition of each of the possible phases that may occur. The quantity desired here is the smallest number, and it is immaterial which particular constituents are chosen to express the compositions of the various phases. This point will become clearer from the following examples. In the system "water" the phases that occur are ice, liquid water, and water vapor. The composition of each of these phases can be expressed in terms of the single constituent water, and hence this is a one-component system. The variable could equally well be hydrogen or oxygen, for the specification of one of these automatically fixes the other through the formula H_2O. Similar considerations show that the minimum number of constituents necessary to describe the composition of all phases in the system sodium sulfate-water is two, and hence this is a two-component system. In this system the various phases that may occur are Na_2SO_4, $Na_2SO_4 \cdot 7\ H_2O$, $Na_2SO_4 \cdot 10\ H_2O$, solutions of Na_2SO_4 in water, ice, and water vapor. The composition of each of these phases in terms of the two components sodium sulfate and water may be stated as follows:

$$Na_2SO_4: \quad Na_2SO_4 + 0\ H_2O$$
$$Na_2SO_4 \cdot 7\ H_2O: \quad Na_2SO_4 + 7\ H_2O$$
$$Na_2SO_4 \cdot 10\ H_2O: \quad Na_2SO_4 + 10\ H_2O$$
$$Na_2SO_4(aq.): \quad Na_2SO_4 + x\ H_2O$$
$$H_2O(s),\ H_2O(l),\ H_2O(g): \quad 0\ Na_2SO_4 + H_2O$$

It will be noted that the composition of certain phases may be stated in terms of only one of these constituents, whereas certain others necessitate a knowledge of the amounts of both present in order to specify unambiguously the composition of the phase. Since two components are the smallest number by which the compositions of *all* the phases can be defined, sodium sulfate-water must be a two-component system.

The essential fact to remember in deciding upon the number of components of a system is that the particular constituents chosen as independent variables do not matter, but their *smallest* number does. If the number chosen is not the smallest, certain of these will not be independent of the others. Again, in writing the composition of a phase in terms of the components selected, plus, minus, and zero coefficients in front of a component are permissible. Thus, in a system in which solid magnesium carbonate dissociates according to

$$MgCO_3(s) = MgO(s) + CO_2(g)$$

the compositions of the various phases may be represented in terms of magnesium carbonate and magnesium oxide as follows:

$$MgCO_3: \quad MgCO_3 + 0\ MgO$$
$$MgO: \quad 0\ MgCO_3 + MgO$$
$$CO_2: \quad MgCO_3 - MgO$$

Finally, by the *degrees of freedom* or the *variance* of a system is meant the *smallest number* of independent variables (such as pressure, temperature, concentration) that must be specified in order to define completely the remaining variables of the system. The significance of the degrees of freedom of a system may be gathered from the following examples. In order to specify unambiguously the density of liquid water, it is necessary to state also the temperature and pressure to which this density corresponds; i.e., the density of water is 0.99973 g per milliliter at 10°C and 1 atm pressure. A statement of the density at 10°C without mention of pressure does not define clearly the state of the water, for at 10°C the water may exist at any and all possible pressures above its own vapor pressure. Similarly, mention of the pressure without the temperature leaves ambiguity. Hence, for complete description of the state of the water, two variables must be given, and this phase, when present alone in a system, possesses two degrees of freedom, or the system is said to be *bivariant*. When liquid and solid water exist in equilibrium, however, the temperature and the densities of the phases are determined only by the pressure, and a statement of some arbitrary value of the latter is sufficient to define all the other variables. Thus, if we know that ice and water are at equilibrium at 1 atm pressure, the temperature can be only 0°C and the densities are also established. The same applies to the choice of temperature as the independent variable. At each arbitrarily chosen temperature (within the range of existence of the two phases) equilibrium is possible only at a given pressure, and once again the system is defined in terms of one variable. Under these conditions the system possesses only one degree of freedom, or it is *monovariant*.

THE GIBBS PHASE RULE

That there is a definite relation in a system between the number of degrees of freedom, the number of components, and the number of phases present was first established by J. Willard Gibbs in 1876. This relation, known as the *phase rule*, is a principle of the widest generality, and its validity is in no way dependent on any concepts of atomic or molecular constitution. Credit is due to Ostwald, Roozeboom, van't Hoff, and others for showing how this generalization can be utilized in the study of problems in heterogeneous equilibrium.

To arrive at a formulation of the phase rule, consider in general a system of C components in which P phases are present. The problem now is to determine the total number of variables upon which such a system depends. First, the state of the system will depend upon the pressure and the temperature. Again, in order to define the composition of each phase, it is necessary to specify the concentration of $(C - 1)$ constituents of the phase, the concentration of the remaining component being determined by difference. Since there are P phases, the total number of concentration variables will be $P(C - 1)$, and these along with the temperature and pressure constitute a total of $[P(C - 1) + 2]$ variables.

The student will recall from his study of algebra that when an equation in n independent variables occurs, n independent equations are necessary in order to solve for the value of each of these. Similarly, in order to define the $[P(C-1)+2]$ variables of a system, this number of equations relating these variables would have to be available. The next question is then: How many equations involving these variables can possibly be written from the conditions obtaining in the system? To answer this query recourse must be had to thermodynamics. Thermodynamics tells us that equilibrium between the various phases in a system is possible only provided the partial molal free energy of each constituent of a phase is equal to the partial molal free energy of the *same* constituent in every other phase. Since the partial molal free energy of the constituent of a phase is a function of the pressure, temperature, and $(C-1)$ concentration variables, it readily follows that the thermodynamic condition for equilibrium makes it possible to write *one equation* among the variables *for each constituent distributed between any two phases*. When P phases are present, $(P-1)$ equations are available for each constituent, and for C constituents a total of $C(P-1)$ equations.

If this number of equations is equal to the number of variables, the system is completely defined. However, generally this will not be the case, and the number of variables will exceed the number of equations by F, where

$$F = \text{Number of variables} - \text{Number of equations}$$
$$= [P(C-1)+2] - [C(P-1)]$$
$$= C - P + 2 \tag{1}$$

Equation (1) is the celebrated phase rule of Gibbs. F is the number of degrees of freedom of a system and gives the number of variables whose values must be specified arbitrarily before the state of the system can be completely and unambiguously characterized. According to the phase rule the number of degrees of freedom of a system is determined by the difference between the number of components and the number of phases present, i.e., by $(C-P)$.

In this derivation it was assumed that each component is present in every phase. If a component is missing from a particular phase, however, the number of concentration variables is decreased by one. But at the same time the number of possible equations is also decreased by one. Hence the value of $(C-P)$, and therefore F, remains the same whether each constituent is present in every phase or not. This means that the phase rule is not restricted by the assumption made, and is generally valid under all conditions of distribution provided that equilibrium exists in the system.

The principal value of Eq. (1) is in its use as a check in the construction of various types of plots for the representation of the equilibrium conditions existing in heterogeneous systems. Before proceeding to a discussion of some specific systems and the application of the phase rule to these, it is convenient to classify all systems according to the number of components present. Thus we may have one-, two-, three-, etc., component systems. The advisability of this approach will become apparent from what follows.

ONE - COMPONENT SYSTEMS

The complexity of one-component systems depends on the number of solid phases that can exist in the system. The simplest case is one in which only a single solid phase occurs, as in water at the lower pressures and carbon dioxide. When more than one solid phase appears in a system the number of possible equilibria is considerably enhanced, and hence the phase diagram, or the plot showing the various equilibria, becomes more involved. The possibilities in such systems and their phase relations can best be brought out by the consideration of several specific examples.

The System Water.   Above about $-20°C$ and below 2000 atm pressure there is only one solid phase in this system, namely, ordinary ice. This solid phase, liquid water, and water vapor constitute the *three* possible single phases in the system. These phases may be involved in *three* possible two-phase equilibria, namely, liquid-vapor, solid-vapor, and solid-liquid, and *one* three-phase equilibrium, solid-liquid-vapor. Applying the phase rule to the system when only a single phase is present, we see that $F = 2$, and therefore each single phase at equilibrium possesses two degrees of freedom. If temperature and pressure are chosen as the independent variables, the phase rule predicts that both of these must be stated in order to define the condition of the phase. Since two independent variables are necessary to locate any point in an area, it must follow that each phase on a $P–T$ diagram occupies an area; and, as three single phases are possible in this system, we may anticipate three such areas on the plot, one for each phase.

For two phases in equilibrium the phase rule predicts that $F = 1$. Since a single variable determines a line, we may expect for each two-phase equilibrium a line on the $P–T$ plot. As three such equilibria may occur in the system, the diagram will be characterized by the existence of three lines separating the various areas from each other. Finally, for the three-phase equilibrium $F = 0$, i.e., no variables need be specified. This must mean that when all three phases coexist the temperature and pressure are fixed, and the position of this equilibrium on the diagram is characterized by the intersection of the three lines at a common point.

Although the phase rule makes it possible to predict the general appearance of the diagram, the exact positions of all lines and points can be determined only by experiment. An inspection of the possible equilibria in this system shows that the data necessary for the construction of a

P–T plot are: (a) the vapor pressure curve of water (liquid-vapor equilibrium), (b) the sublimation curve of ice (solid-vapor equilibrium), (c) the melting point curve of ice as a function of pressure (solid-liquid equilibrium), and (d) the position of the solid-liquid-vapor equilibrium point. These experimental data for the system water are shown in Figure 10–1. In this phase diagram line AO gives the sublimation curve of ice, line OB the vapor pressure curve of liquid water, and OC the line along which equilibria between ice and liquid water occur at various pressures. O is the *triple point* at which ice, water, and water vapor are in equilibrium. This equilibrium is possible only at 0.010°C and 4.58 mm pressure. As predicted by the phase rule, there are three areas on the diagram, disposed as shown, one such area corresponding to each of the single phases.

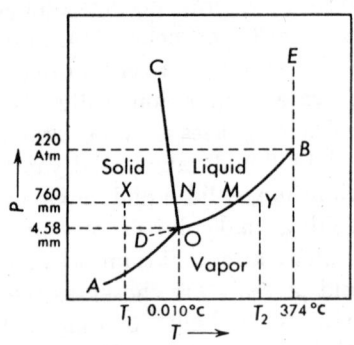

Figure 10–1.   The system water at moderate pressures (schematic).

The vapor pressure curve of water, line OB, extends from the triple point O up to the critical point B, corresponding to 374°C and 220 atm pressure. However, under certain conditions it is possible to supercool water below point O to yield the metastable liquid-vapor equilibria shown by the dotted line OD. The fact that OD lies above AO shows that at temperatures below that of the triple point liquid water has a vapor pressure higher than the sublimation pressure of ice, and hence the supercooled liquid is unstable at these temperatures with respect to the ice. The sublimation curve of ice, line AO, may extend from absolute zero up to O. No superheating of ice beyond O has ever been realized. Line CO runs from O up to a point corresponding to 2000 atm pressure and about -20°C; at this point ordinary ice, type I, in equilibrium with water undergoes a transformation into another solid modification, type III, in equilibrium with the liquid. The slope of this line indicates that the melting point of ice is *lowered* by increase in pressure in accord with the Le Chatelier principle and the fact that ice has a larger specific volume than liquid water. The slopes of the lines AO, OB, and OC are determined at each point by the Clapeyron equation or one of its modifications as applicable in each instance. From the slopes and this equation it is possible to evaluate the heats of vaporization from OB, the heats of sublimation from AO, and the heats of fusion from OC.

Since no liquid may exist above the critical temperature, dotted line BE has been inserted in the diagram to separate the liquid from vapor areas above the critical temperature. Consequently the vapor area lies below and to the right of $AOBE$, the liquid area above OB and between the lines OC and BE, while the solid area extends to the left of OC and above AO.

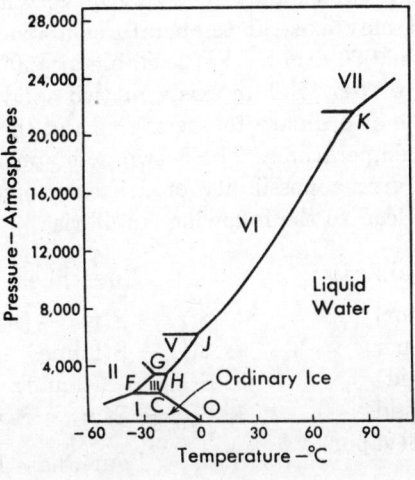

Figure 10–2. The system water at high pressures.

The manner in which a diagram such as Figure 10–1 may be used to follow the changes that occur in the system with a change in the variables may be seen from the following example. Suppose it is desired to know the behavior of the system on heating ice at a pressure of 760 mm and temperature T_1, corresponding to point X in the diagram, to a point corresponding to point Y at the same pressure but temperature T_2. Starting with ice at X and heating it slowly at constant pressure, the system follows line XN with increase in the temperature of the ice. However, once N is reached the ice begins to melt, the temperature remains constant until the fusion is complete, and only then does the temperature begin to rise again along NM. Between N and M the only change is an increase in the temperature of the liquid. But at M vaporization sets in, and the temperature again is constant until all the liquid is converted to vapor. On complete transformation of liquid to vapor, any further addition of heat results merely in an increase of the temperature of the vapor along

MY until the latter point is reached. In the same way it is possible with the aid of Figure 10–1 to predict and outline any changes that may take place in this system with a variation of temperature, pressure, or both.

At high pressures a number of other solid modifications besides the ordinary form have been observed.[1] The phase diagram of the system under these conditions is shown in Figure 10–2. Figures 10–1 and 10–2 are parts of the same diagram, the latter figure being merely the high-pressure portion of the former. The manner in which the two diagrams tie in can be judged from the line OC, which is the same in both plots. It is of interest to observe in Figure 10–2 that at very high pressures ice of types VI and VII may exist at temperatures above 0°C. In fact, at a pressure of about 40,000 atm ice VII is stable at 190°C.

The System Sulfur. Sulfur exists in two solid modifications, the *rhombic* form stable at ordinary temperatures and the *monoclinic* variety stable at higher temperatures. These two solid phases along with the liquid and vapor give a possibility of existence of four single phases, which in turn can lead to the following equilibria:

Two-Phase Equilibria	Three-Phase Equilibria
1. S(r) − S(vapor)	1. S(r) − S(m) − S(liquid)
2. S(m) − S(vapor)	2. S(r) − S(liquid) − S(vapor)
3. S(r) − S(liquid)	3. S(m) − S(liquid) − S(vapor)
4. S(m) − S(liquid)	4. S(r) − S(m) − S(vapor)
5. S(liquid) − S(vapor)	
6. S(r) − S(m)	Four-Phase Equilibria
	1. S(r) − S(m) − S(liquid) − S(vapor)

Applying the phase rule to these possible equilibria, we may anticipate four separate divariant single-phase areas, six monovariant two-phase equilibrium lines, and four invariant three-phase equilibrium points. Since the maximum number of phases that may be present in equilibrium is given by the phase rule with $F = 0$, it follows that $P = 3$ for a one-component system, and hence the four-phase equilibrium cannot exist in this or any other one-component system.

Figure 10–3 shows the schematic phase diagram for this system. The four single-phase areas are disposed as indicated. Lines OP and PK are the sublimation curves of rhombic and monoclinic sulfur, respectively, while KU is the vapor pressure curve of liquid sulfur. At point P rhombic sulfur undergoes a transition to monoclinic sulfur, and hence this is an invariant point corresponding to the equilibrium S(r) − S(m) − S(vapor). At point

---

[1] Tammann, *Zeit. physik. Chem.*, **72**, 609 (1910); Bridgman, *Proc. Am. Acad. Sci.*, **47**, 441 (1912); *J. Chem. Phys.*, **5**, 964 (1937).

_K_ monoclinic sulfur melts, and thus this point corresponds to the three-phase equilibrium S(m) − S(liquid) − S(vapor). Line _PS_ shows the variation of the transition point with pressure, while line _KS_ shows in a like manner the variation of the melting point of monoclinic sulfur with the same variable. These two lines intersect at _S_ to yield the three-phase equilibrium S(r) − S(m) − S(liquid). Finally, the line _SW_ gives the melting point of rhombic sulfur. These are all the *stable* equilibria that occur. The monoclinic sulfur area is enclosed by the lines _PS_, _PK_, and _KS_, and therefore no monoclinic sulfur can exist in stable condition outside this area. Furthermore, no vapor can exist stably at pressures above those

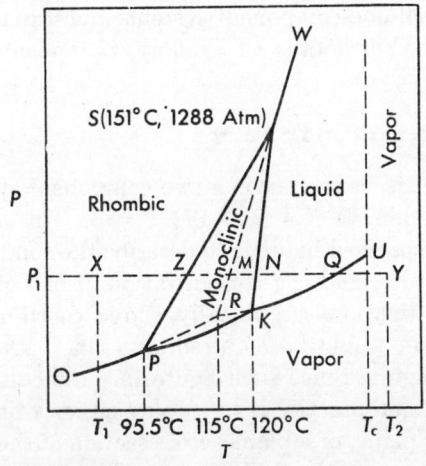

Figure 10–3. The system sulfur (schematic).

given by the lines _OP_, _PK_, and _KU_ below the temperature of the critical point _U_. Above this temperature, however, no liquid is possible. Hence the liquid area is terminated along the vertical dotted line through _U_, and vapor may exist thereafter at high pressures.

The remaining equilibria in this system are all *metastable*. By rapid heating it is possible to superheat rhombic sulfur along the extension of _OP_, the line _PR_, which is the equilibrium line for superheated S(r) with S(vapor). Similarly, it is possible to supercool liquid sulfur along the extension of _UK_ to _R_. At point _R_ the metastable S(r) − S(vapor) line intersects the metastable S(liquid) − S(vapor) line to yield the metastable invariant point S(r) − S(liquid) − S(vapor). The line _RS_ shows the variation of this metastable point with pressure and is, therefore, the melting point line for rhombic sulfur as a function of pressure. This line is an extension of the stable _WS_ line into the metastable range. It should be clearly understood that, when these various metastable equilibria occur, monoclinic sulfur does not appear; instead rhombic sulfur is transformed directly to liquid along RS or to vapor along PR without passing through the monoclinic stage.

Other One-Component Systems. The system *carbon dioxide* is very similar to the system water as shown in Figure 10–1, except that the triple point occurs at $-56.4°C$ and a pressure of about 5 atm. Furthermore, the solid-liquid line slopes to the right instead of to the left as in the water diagram. Since the 1 atm line in this system cuts only the sublimation curve of the solid, solid carbon dioxide must change directly to vapor at this pressure without passing through the liquid state. Liquefaction of the solid can be attained only under pressures of about 5 atm or higher.

Other examples of one-component systems investigated are phosphorus and benzophenone. For details see Findlay, Campbell and Smith.[2]

TWO - COMPONENT SYSTEMS

When a single phase is present in a two-component system, the number of degrees of freedom is $F = 2 - 1 + 2 = 3$. This means that three variables must be specified in order to describe the condition of the phase: pressure, temperature, and the concentration of one of the components. To present these relations graphically, three coordinate axes at right angles to each other would be required, and the diagram thus resulting would be a solid figure. Since such figures are difficult to construct and use, the more common practice is to employ either a projection of such a solid diagram on a plane, or a planar cross section of the figure for a given constant value of one of the variables. In this manner it is possible to present the various relations in two-component systems in a two-dimensional plot of any two of the three variables mentioned.

Moreover, the discussion of two-component systems can be simplified further by considering the various possible types of equilibria separately. Thus the usual practice is to study liquid-gas, solid-gas, liquid-liquid, and solid-liquid equilibria individually and, when necessary, to combine the diagrams. Since the first three types of equilibria have already been considered in various places in the text, attention here will be devoted exclusively to an application of the phase rule to the very important category of solid-liquid equilibria.

Solid-liquid equilibria are of great importance because of their connection with all crystallization problems. Such equilibria are characterized generally by the absence of a gas phase and by the fact that they are little affected by small changes in pressure. Systems where the gas phase

---

[2] Findlay, Campbell, and Smith, *The Phase Rule and Its Applications*, Dover Publications, New York, 1951.

is absent are called *condensed systems*, and it is with condensed systems that we shall be concerned.

Measurements on solid-liquid equilibria in condensed systems are usually carried out at atmospheric pressure. Because of the relative insensitivity of such systems to small variations in pressure, the latter may be considered constant, and for such systems the phase rule takes the form

$$F = C - P + 1 \tag{2}$$

For two-component systems Eq. (2) becomes

$$
\begin{aligned}
F &= 2 + 1 - P \\
&= 3 - P
\end{aligned} \tag{3}
$$

where the only remaining variables are temperature and the concentration of one of the constituents. Solid-liquid equilibria are represented, therefore, on temperature-composition diagrams. For limited ranges of concentration any scheme of expressing concentration will do, but where the range may extend from 100 per cent of one constituent to 100 per cent of the other, it is preferable to employ as an abscissa either weight or mol per cent, as in distillation diagrams.

DETERMINATION OF SOLID - LIQUID EQUILIBRIA

Of the many experimental procedures employed for the determination of equilibrium conditions between solid and liquid phases, the two of widest utility and applicability are the *thermal analysis* and *saturation or solubility methods*. These methods, supplemented when necessary by an investigation of the nature of the solid phases occurring in a system, can cover between them the study of any system which may be encountered.

The *thermal analysis method* involves a study of the cooling rates, i.e., temperature–time curves, of various compositions of a system during solidification. From such curves it is possible to ascertain the temperature of initial and final solidification of a mixture and to detect the temperatures at which various transformations and transitions occur. Although thermal analysis is applicable under all temperature conditions it is particularly suitable for equilibrium investigations at temperatures considerably above and below that of the room.

In order to illustrate the experimental steps involved in this procedure, the interpretation of the curves, and the plotting of the final diagram, consider specifically the problem of determining the condensed phase diagram for the binary system bismuth-cadmium. The first step involves the preparation of a number of mixtures of the two metals ranging in over-all composition from 100 per cent bismuth to 100 per cent cadmium.

These mixtures may be spaced at 10 per cent intervals and should all be preferably of equal weight. Each of these mixtures of solids is placed in an inert crucible of, say, fireclay or graphite and is then melted in an electric furnace. To prevent oxidation of the metals it is advisable to maintain an inert or reducing atmosphere over them by passing hydrogen, nitrogen, or carbon dioxide through the furnace. A molten flux, such as borax, or a layer of powdered graphite may be used to cover the crucible charge as an added precaution. After melting and thorough agitation a thermocouple is inserted in the melt, and the furnace and contents are allowed to cool slowly. Temperature and time readings are taken until the charge in the crucible is completely solidified. Finally, plots of the temperatures thus obtained against time are prepared. If a check on the composition is desired, the solidified alloys are removed and carefully analyzed.

Figure 10–4 shows a set of cooling curves thus obtained for various compositions of bismuth-cadmium mixtures. The explanation of these curves is as follows. When a body is cooled slowly and uniformly, a smooth cooling curve is obtained and the temperature of the body approaches that of the room as a limit. However, when some transformation that liberates heat occurs during cooling, the slope of the cooling curve is reduced suddenly. The nature of the reduction depends on the degrees of freedom of the system. A single phase with $F = 2$ exhibits a continuous cooling curve. When a new phase appears, the variance of the system is reduced to one, and the heat generated by the formation of the new phase results in a discontinuity in the curve due to change of slope for the cooling of one phase to a lesser slope corresponding to the cooling of two phases. Again, when still a third phase appears, $F = 0$, and the temperature of the system must remain constant until one of the phases disappears. The result is a flat portion on the cooling curve. Finally, when solidification is complete, the system regains a degree of freedom, and the cooling curves once again exhibit continuous variation of temperature with time. In light of these facts a "break" or arrest in a cooling curve indicates the appearance of a second phase, usually the separation of a solid from the melt, while a horizontal portion indicates the coexistence of three phases. The third phase may result from the separation of two solids from the melt, the interaction of the melt with a solid to form another solid, or the separation of a solid from two liquid phases. The nature of the particular change occurring can be ascertained from an inspection of the final phase diagram and an analysis of the solids in the system.

With these considerations in mind we may conclude from the cooling curves in Figure 10–4 that the arrests indicated by t_i signify the appearance of a second phase in the system, while the horizontal portions result from the coexistence of three phases. In this system the only solid phases are pure bismuth and pure cadmium, and hence the horizontal portions are the result of simultaneous occurrence of these solids and melt. However, in curves (a) and (h) the horizontal portions are due to two phases, since these are one-component systems.

To construct the equilibrium diagram from the cooling curves, the initial and final solidification temperatures, t_i and t_f, are taken off the cooling curves for the various overall concentrations and are plotted on a temperature-composition diagram. Smooth curves are drawn then through all the t_i and t_f temperatures to yield the diagram shown in Figure 10–5. Curve AB indicates the temperatures at which bismuth

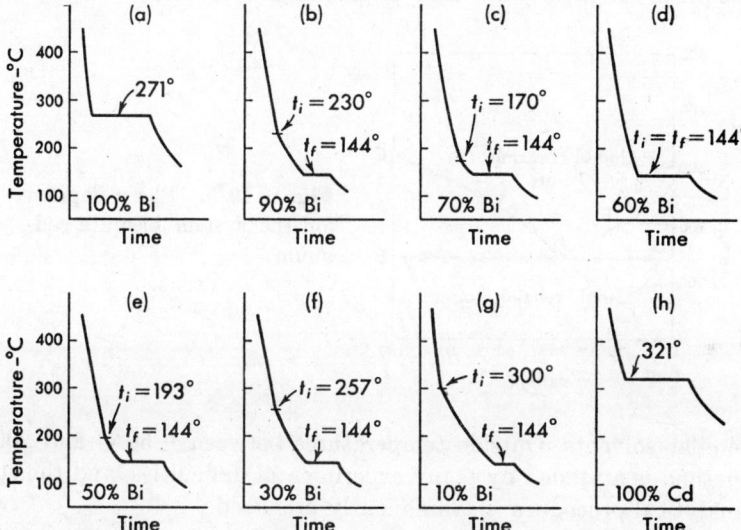

Figure 10–4.   Cooling curves for the system bismuth-cadmium.

begins to separate from various concentrations of melt, while BC gives the same information for initial separation of cadmium. Line DE indicates the temperature at which all mixtures become completely solid. Further details of this type of diagram will be discussed later.

In Figure 10–5 line AB may be looked upon not only as the initial freezing point curve for bismuth, but also as the *solubility* curve of bismuth in molten cadmium. Points on this curve represent then the solubilities of bismuth in the molten cadmium at various temperatures. Similarly, curve BC gives the solubilities at various temperatures of cadmium in molten bismuth, while at B the solution is saturated with respect to both solids.

In the *saturation method* the solubilities of one substance in another are determined at various constant temperatures, and the solubilities are then plotted as a function of the temperature. To obtain the composition of a solution of cadmium saturated with bismuth at say 200°C, point F in Figure 10–5, excess powdered bismuth may be added to molten cadmium, the mixture brought to 200°C, and the mass agitated until equilibrium is attained. The excess solid bismuth is now filtered off, and the saturated solution is analyzed for both constituents. By repeating this operation at various temperatures between 144 and 271°C, curve AB may be traced out. By a similar procedure, but employing molten bismuth and excess solid cadmium, curve BC may be obtained between 144 and 321°C. Although this method is rarely applied to the study of metallic systems, it is the principal means employed in systems containing water

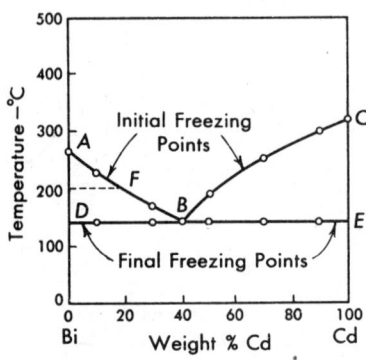

Figure 10–5. Phase diagram for the system bismuth-cadmium.

and similar solvents. Outside temperatures between −50°C and 200°C this method is attended by many experimental difficulties, and the thermal analytical procedure is consequently preferred.

DETERMINATION OF NATURE OF SOLID PHASES

For a complete interpretation of the phase diagram it is essential to know the nature and composition of the solid phases which appear during crystallization and in the final solid. These may be:

1. *Pure components*, such as bismuth or cadmium in the system discussed above.
2. *Compounds* formed by reaction between the pure constituents. Examples are $MgZn_2$ in the system Mg-Zn, $Na_2SO_4 \cdot 10\ H_2O$ in the system Na_2SO_4-H_2O, and $FeCl_3 \cdot 6\ H_2O$ in the system $FeCl_3$-H_2O. Such compounds have definite composition, are stable over definite temperature and solution concentration intervals, possess definite melting or transition temperature, and have a definite crystal structure.

3. *Solid solutions*, homogeneous solids whose composition may vary within certain concentration intervals and is determined by the composition of the solution from which the solid crystallizes. Such variability in composition differentiates the solid solution from a compound.
4. *Mixtures* of solids which may be pure components, compounds, or solid solutions.

Frequently it is possible to deduce the nature of the solid phases from the shape of the phase diagram. However, in some instances it is necessary to subject the solids to a more careful scrutiny. To do this the completely solidified mass may be inspected under a microscope to determine the number of solids present and, if possible, their identity. Another possibility is the use of X rays. Direct separation of solids from a solution and their subsequent analysis may also be resorted to, but this procedure is attended by some uncertainty because of difficulty in obtaining a solid free from contamination by saturated solution. This difficulty may be circumvented sometimes by the use of a "telltale." The "telltale" is a substance that is soluble in the solution but not in the solid phase. On adding a definite small quantity of this substance to the solution and determining the amount of it present in the wet solid, it is possible to ascertain the quantity of solution adhering to the solid and thereby to arrive at the composition of the pure solid phase.

CLASSIFICATION OF TWO - COMPONENT SOLID - LIQUID EQUILIBRIA

Every condensed phase diagram may be considered to be composed of a combination of a number of simple types of diagrams. In some systems only a single type occurs; in others a number of these simple types may occur combined to yield a more complicated complete diagram. In either case the phase relations in a system can readily be understood when the significance of the elementary types of diagrams is manifest.

All condensed two-component systems may be classified first according to the miscibility of the liquid phases, and these in turn according to the nature of the solid phases which crystallize from the solution. On this basis the important elementary types are:

Class A. The two components are completely miscible in the liquid state.

      Type I.   The pure components only crystallize from the solution.

      Type II. The two constituents form a solid compound stable up to its melting point.

Type III. The two components form a solid compound which decomposes before attaining its melting point.

Tpye IV. The two constituents are completely miscible in the solid state and yield thereby a complete series of solid solutions.

Type V. The two constituents are partially miscible in the solid state and form stable solid solutions.

Type VI. The two constituents form solid solutions which are stable only up to a transition temperature.

Class B. The two components are partially miscible in the liquid state.

Type I. Pure components only crystallize from solution.

Class C. The two components are immiscible in the liquid state.

Type I. Pure components only crystallize from solution.

CLASS A: TYPE I. SIMPLE EUTECTIC DIAGRAM

Two-component condensed systems belonging to this class have a diagram of the general form shown in Figure 10–6. They are characterized by the fact that two constituents A and B are completely miscible in the liquid state, and such solutions yield only pure A or pure B as solid phases. In this figure points D and E are the melting points of pure A and pure B respectively. Line DG gives the concentrations of solutions saturated with A at temperatures between D and F, or the freezing points of the solutions which yield A as a solid phase. Similarly, line EG gives the concentrations of solutions saturated with solid B at temperatures between E and F. At G the solution is saturated with both A and B, i.e., at G three phases are in equilibrium. It follows, therefore, that the lines DG and EG represent monovariant two-phase equilibria, while G is an invariant point. At this point the temperature F and the composition C of the solution must remain constant as long as three phases coexist. The temperature can be lowered below F only when one of the phases has disappeared, and on cooling this must be the saturated solution. In other words, at F solution G must solidify completely. Temperature F must consequently be the *lowest* temperature at which a liquid phase may exist in the system A–B; below this temperature the system is completely solid. Temperature F is called the *eutectic temperature*, composition C the *eutectic composition*, and point G the *eutectic point* of the system.

Above the lines DG and GE is the area in which unsaturated solution or melt exists. In this area only one phase is present and the system is divariant. In order to define any point in this area both the temperature and composition must be specified. The significance of the remaining portions of the diagram can be made clear by considering the behavior on cooling of several mixtures of A and B. Take first a mixture of over-all composition a. If such a mixture is heated to point a''' an unsaturated solution is obtained. On cooling this solution nothing beyond a drop of temperature of the liquid phase occurs until point a'', corresponding to temperature x'', is reached. At this point the solution becomes saturated with A; or, in other words, a'' is the freezing point of the solution at temperature x''. As cooling continues A keeps on separating out, and the composition of the saturated solution changes along the line $a''G$. Thus at a temperature such as x' solid A is in equilibrium with saturated solution of composition y', and so on. It is seen, therefore, that for any over-all composition falling in area DFG solid A is in equilibrium with various compositions of solution given by the curve DG at each temperature. However, at temperature F another solid phase, B, appears, and the system becomes

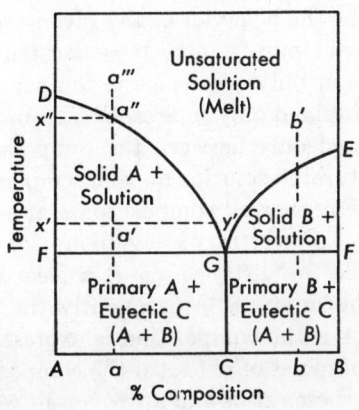

Figure 10-6. Simple eutectic diagram.

invariant. On extracting heat from this system A and B must crystallize from the saturation solution in the fixed ratio C, and this crystallization will proceed until the solution has been completely solidified. Once this process is over and nothing but a *mixture* of solid A and solid B remains, the system becomes monovariant, and the cooling may be continued below F into area $FACG$, the area of coexistence of the two solids A and B.

Inspection of the solid in area $FACG$ under a microscope would reveal that it is composed of relatively large crystals of A which have had a chance to grow from temperature x'' to the temperature below F, and an intimate *mixture* of finer crystals of A and B which crystallized in definite proportion C at temperature F. The larger crystals of A are called *primary* crystals, since they appeared first. Area $FACG$ may be marked, therefore, as containing primary crystals of A and an intimate eutectic mixture of fine crystals of A and B.

Similar considerations applied to over-all compositions lying between

C and B, such as b, for instance, show that in area EFG solid B is in equilibrium with saturated solutions along the line EG. At temperature F solid A also appears, the system becomes invariant, and it remains so until the solution at G solidifies. Once the solidification is complete, the mixture passes into area $FBCG$ where now primary B and eutectic mixture of over-all composition C are present. Finally, cooling a mixture of over-all composition C will yield no solid until point G is reached. At this point solids A and B appear simultaneously, and the system solidifies at constant temperature to yield only the eutectic mixture. In this respect composition C behaves like a pure substance on freezing. However, the result is not a single solid, but a mixture of two.

Once a phase diagram such as Figure 10–6 is available for a binary system, it is possible to specify the conditions under which particular solid phases may be obtained and to describe the behavior of any given over-all mixture on cooling. Thus it is seen that pure A may be separated only from mixtures falling in area DFG, and only between the temperatures D and F. Similarly, pure B may be obtained only in area EFG from over-all compositions between C and B, and only between the temperatures E and F. The proportion of solid to saturated solution at each temperature can be estimated from the diagram. For over-all composition a at a temperature such as x', the distance $x'a'$ is a measure of the amount of saturated solution of composition y', while the distance $a'y'$ is a measure of the amount of solid A present in the mixture. Consequently the ratio $x'a'/a'y'$ is the weight ratio of y' to A if the composition is expressed in weight percentage, or of the number of moles of y' to A if the composition is expressed in mol percentage. From these ratios and any over-all weight it is possible to calculate the yield of solid phase to be anticipated at any given temperature.

Examples of systems exhibiting simple eutectic diagrams of the type shown in Figure 10–6 are aluminum-tin, bismuth-cadmium, potassium chloride-silver chloride, sodium sulfate-sodium chloride, and benzene-methyl chloride. The behavior of all these systems on cooling will be similar to that described for the system bismuth-cadmium.

CLASS A: TYPE II. FORMATION OF COMPOUND WITH CONGRUENT MELTING POINT

When the two pure components react to form a compound stable up to its melting point, the phase diagram takes on the typical form shown in Figure 10–7 for the system cuprous chloride-ferric chloride. If the compound, in this case $CuCl \cdot FeCl_3$, is considered as a separate component, the whole diagram may be thought of as being composed of two diagrams of the simple eutectic type, one for $CuCl$-$CuCl \cdot FeCl_3$, and the other for

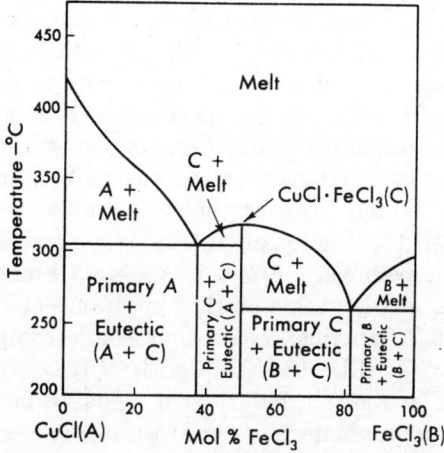

Figure 10–7.   Formation of compound with congruent melting point.

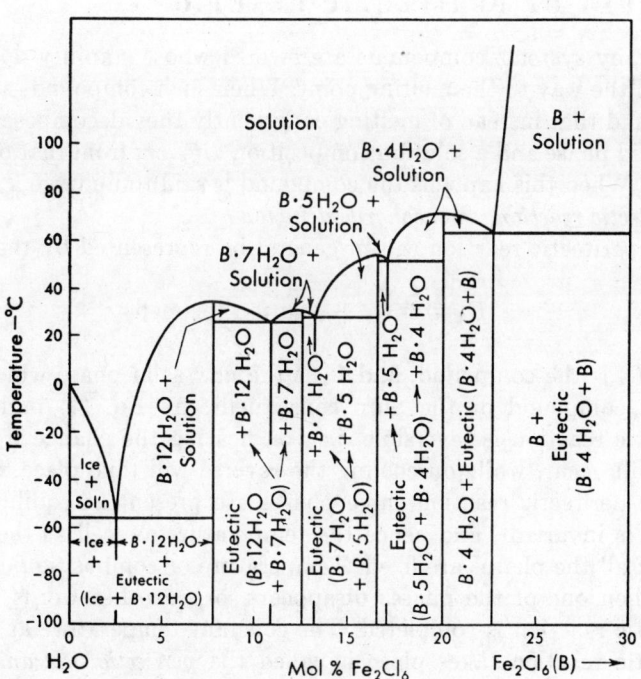

Figure 10–8.   The system Fe_2Cl_6-H_2O.

$CuCl \cdot FeCl_3\text{-}FeCl_3$. The discussion of type I diagrams can now be applied to each portion with the results shown in the figure.

A compound such as $CuCl \cdot FeCl_3$ which melts at a constant temperature to yield a liquid of the same composition as the solid compound is said to have a *congruent* melting point. Compounds with congruent melting points also appear in such binary systems as gold-tellurium ($AuTe_2$), aluminum-selenium (Al_2Se_3), calcium chloride-potassium chloride ($CaCl_2 \cdot KCl$), urea-phenol ($1:1$), and many others. When several compounds with congruent melting points are formed in a system a maximum is obtained for each, and the diagram takes on the appearance of Figure 10–8 for the system $Fe_2Cl_6\text{-}H_2O$.[3] In this system four stable compounds have been observed, namely, $Fe_2Cl_6 \cdot 12\ H_2O$, $Fe_2Cl_6 \cdot 7\ H_2O$, $Fe_2Cl_6 \cdot 5\ H_2O$, and $Fe_2Cl_6 \cdot 4\ H_2O$. The occurrence of these compounds increases the number of areas and eutectic points but otherwise introduces nothing new. By employing the artifice of considering each compound as a constituent, the significance of the various areas, lines, and points can readily be deduced to be as indicated.

CLASS A: TYPE III. COMPOUND FORMATION AS RESULT OF PERITECTIC REACTION

In many systems compounds are formed whose stability does not extend all the way to the melting point. When such compounds are heated, it is found that instead of melting congruently they decompose to yield a new solid phase and a solution composition *different* from that of the solid phases. When this happens the compound is said to undergo a *transition*, or *peritectic reaction*, or *incongruent fusion*.

Any peritectic reaction can in general be represented by the equation

$$C_2 \rightleftharpoons C_1 + \text{solution (or melt)} \qquad (4)$$

where C_2 is the compound, and C_1 is the new solid phase which may be itself a compound or the pure constituent. As Eq. (4) indicates, the peritectic reaction is reversible, i.e., on heating the change from left to right will occur, while on cooling the reverse will take place. Since during the peritectic reaction three phases are present at equilibrium, the system is invariant, and hence the temperature as well as the compositions of all the phases are fixed. Temperature or composition can change only when one of the phases disappears, or, in other words, when the peritectic reaction is completed. The constant temperature at which the peritectic reaction takes place is called the *peritectic* or *transition tem-*

[3] B. Roozeboom, *Z. physik. Chem.*, **10,** 477 (1892).

perature, and for this temperature there will be obtained a horizontal portion on the cooling curves analogous to the eutectic portion.

The relations in a system in which a peritectic reaction occurs can be illustrated with the diagram for the binary condensed system calcium fluoride-calcium chloride shown in Figure 10–9. Line AB gives the concentrations of melt in equilibrium with solid calcium fluoride. When point B, corresponding to 737°C, is reached, a peritectic reaction between the

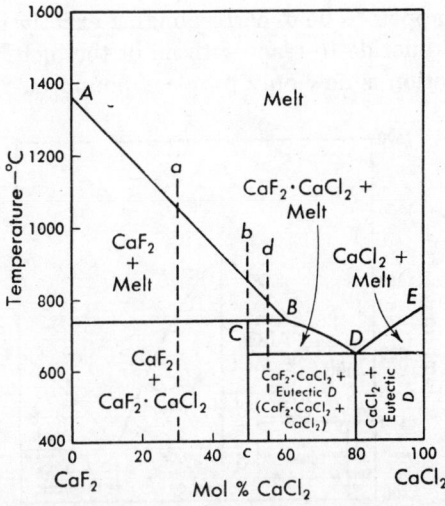

Figure 10–9.   Phase diagram for the system CaF_2-$CaCl_2$.

melt of composition B and solid calcium fluoride sets in to form the double compound $CaF_2 \cdot CaCl_2$ according to the equation

$$CaF_2(s) + \text{melt } (B) = CaF_2 \cdot CaCl_2(s) \tag{5}$$

This reaction proceeds isothermally to form compound C until either all the calcium fluoride or all of the melt has been consumed. Whether one or the other of these phases disappears depends on the over-all composition of the mixture. If the over-all composition lies between calcium fluoride and C, as a in the figure, there is more calcium fluoride present than is necessary to react with all of the melt B, and hence on completion of the reaction all the melt will be converted to $CaF_2 \cdot CaCl_2$, while the excess calcium fluoride will remain as such. Consequently, on passing into the area between the calcium fluoride axis and C below 737°C, only the two solid phases, CaF_2 and $CaF_2 \cdot CaCl_2$, will be present. If, however, the over-all composition of the mixture lies between C and B, such as d, more melt is present than is required to react with all of the solid calcium fluoride, and hence the products of the peritectic reaction will be C and unreacted melt. On passing into the area immediately below CB we shall have, therefore, solid C and melt in equilibrium with the concentrations of the saturated solutions given by the line BD. It should be observed that the lines AB and BD are not continuous but show a break at B. This means that the solid phases calcium fluoride and $CaF_2 \cdot CaCl_2$ each have separate and distinct solubility curves which intersect at B, the concentration of solution saturated with both phases. Finally, should the overall composition happen to be b, corresponding exactly to C, there is just sufficient calcium fluoride to react with all of the melt, and the result of the peritectic reaction is now only pure compound C.

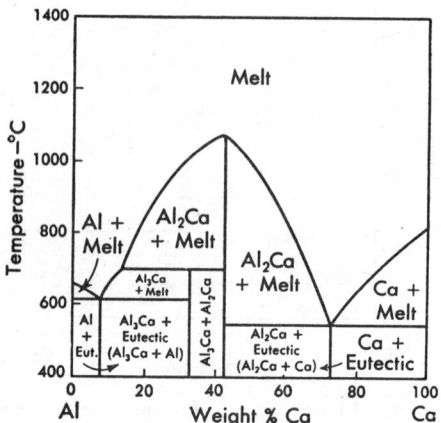

Figure 10–10.   Phase diagram for the system aluminum-calcium.

The remainder of the diagram introduces nothing new. $CBDE$ is merely a simple eutectic-type diagram involving the constituents C and calcium chloride. From the interpretation of such a diagram the area designations shown in the figure readily follow. It will be observed that the eutectic mixture is composed of calcium chloride, $CaF_2 \cdot CaCl_2$, and melt, with no pure calcium fluoride appearing.

Peritectic reactions have been observed in many binary systems. As examples may be quoted the following, along with the compounds formed: gold-antimony ($AuSb_2$), potassium chloride-cupric chloride ($2\ KCl \cdot CuCl_2$), picric acid-benzene ($1:1$), sodium chloride-water ($NaCl \cdot 2\ H_2O$), and sodium sulfate-water ($Na_2SO_4 \cdot 10\ H_2O$). In some systems several compounds occur, some of which have congruent melting points while others do not. Thus in the system aluminum-calcium the compound Al_2Ca has a congruent melting point, while the compound Al_3Ca is formed by peritectic reaction. The phase diagram for this system, shown in Figure 10–10, may be considered typical of the relations encountered under such conditions. Finally, in many systems several compounds are formed, none of which has a congruent melting point. Instances of this type are potassium sulfate-cadmium sulfate ($K_2SO_4 \cdot 2\ CdSO_4$, $K_2SO_4 \cdot 3\ CdSO_4$) and magnesium sulfate-water ($MgSO_4 \cdot H_2O$, $MgSO_4 \cdot 6\ H_2O$, $MgSO_4 \cdot 7\ H_2O$, $MgSO_4 \cdot 12\ H_2O$). The water-end portion of the diagram for the latter system is shown in Figure 10–11. Study of this diagram will reveal that its interpretation involves merely an extension

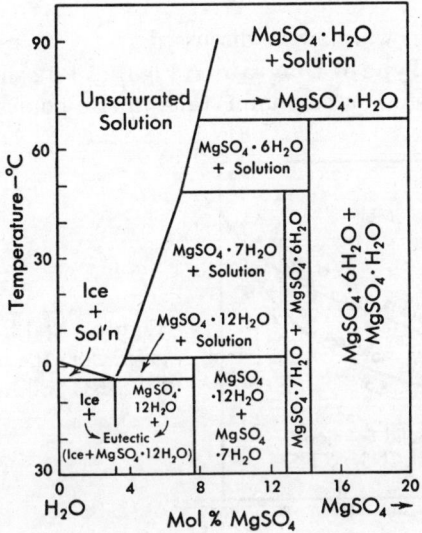

Figure 10–11.   Phase diagram for the system $MgSO_4$-H_2O.

of the principles employed for interpreting the simple diagram shown in Figure 10–9.

CLASS A: TYPE IV. COMPLETE MISCIBILITY IN SOLID STATE

Just as two liquids may dissolve in each other to form a liquid solution, so one solid may dissolve in another to form a *solid solution*. Solid, like liquid, solutions are homogeneous and may vary in composition within wide limits. In the latter respect they differ from solid compounds whose composition is always fixed and definite. X-ray examination of the lattices of solid solutions reveals that one constituent enters the lattice of the other and is uniformly distributed through it. This uniformity of distribution differentiates a solid solution from a mixture of solids, for in the latter instance each constituent preserves its own characteristic crystal structure.

The condensed phase diagrams of binary systems where both the liquid and solid phases are completely miscible fall into three groups, for which:

1. The melting points of all solutions are intermediate between those of the pure constituents.
2. The melting point curve exhibits a minimum.
3. The melting point curve exhibits a maximum.

These various cases will now be discussed.

Intermediate Type of Diagram.   Figure 10–12 shows the phase diagram for the system ammonium thiocyanate-potassium thiocyanate,

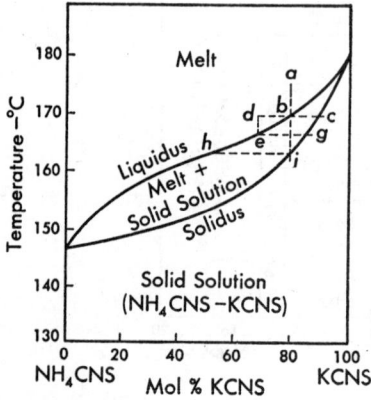

Figure 10–12. Phase diagram for the system NH₄-CNS-KCNS.

where the two constituents are completely miscible in both the solid and liquid states. In this diagram the upper or *liquidus* curve gives the temperature of initial solidification, i.e., the compositions of melt saturated with solid solutions at various temperatures, while the lower or *solidus* curve gives the temperatures at which final solidification takes place. To follow the changes involved in the solidification of any given composition of melt, consider a point such as *a* in the liquid phase. On cooling this solution, no solid phase will separate until the temperature corresponding to point *b* is reached. At this point a small amount of *solid solution* of composition *c* will separate and change thereby the composition of the melt toward, say, *d*. On further cooling *d* will start freezing at *e*, and the composition of the solid phase will adjust itself to *g* by solution of some ammonium thiocyanate from the melt. At all times the composition of the liquid and solid solutions at equilibrium is given by the intersections of the horizontal isothermal lines with the liquidus and solidus curves. As cooling is continued the compositions of the melt will move along the liquidus curve toward ammonium thiocyanate, while the compositions of the solid solution will move in the same direction along the solidus curve. When point h is reached, the solid solution in equilibrium with the small amount of melt remaining has the over-all composition of the mixture, and hence at this point the mixture solidifies completely to yield a solid solution of composition i equal to the original composition of the melt. Further cooling will result merely in lowering the temperature of i.

The same considerations apply to the cooling of any other over-all composition in the diagram. Although the cooling process descrived above has been considered to be stepwise, actually the changes in the compositions of solid and melt involve a continuous adjustment of the concentrations to the equilibrium values called for by the liquidus and solidus curves at each temperature. It should be pointed out, however, that this mode of solidification is obtained only if the cooling rate is sufficiently slow and there is sufficiently good contact between the phases to allow equilibrium to be reached. With rapid cooling and little agitation solid solutions of varying composition may separate, and the final congealing temperature may be lower than hi.

Since in this system no more than two phases are present at any time in equilibrium, there are no invariant points, and hence there are no horizontal portions on the cooling curves. The only discontinuities are an arrest when the liquidus curve is reached and solid starts separating, and another arrest when the over-all composition line crosses the solidus curve and the melt disappears; i.e., for a composition such as a arrests are observed at temperatures corresponding to points b and i.

Continuous solid solution formation of the type discussed is fairly common. Systems besides the above that exhibit such behavior are lead chloride-lead bromide, silver chloride-sodium chloride, copper-nickel, cobalt-nickel, silver-gold, and naphthalene–β-naphthol.

Minimum Type of Diagram. A variation of Figure 10–12 is the diagram shown in Figure 10–13 for the system p-iodochlorbenzene (A)–p-dichlorbenzene (B). Here, as before, A and B form a complete series of solid solutions, but the melting point curve now exhibits a minimum. Again the upper curve is the liquidus, while the lower curve is the solidus. Except for the fact that the melt composition corresponding to the minimum freezes at constant temperature to yield a solid solution, the relations in this system are similar to those described for ammonium thiocyanate-potassium thiocyanate.

Some other systems falling within this category are sodium carbonate-potassium carbonate, potassium chloride-potassium bromide, mercuric bromide-mercuric iodide, potassium nitrate-sodium nitrate, silver-antimony, and copper-gold.

Maximum Type of Diagram. Though rarely encountered, a third possible type of diagram is one in which the melting point curve of the system exhibits a maximum. Figure 10–14 shows such a diagram for the system d-carvoxime–l-carvoxime ($C_{10}H_{14}NOH$). Again the liquidus is the

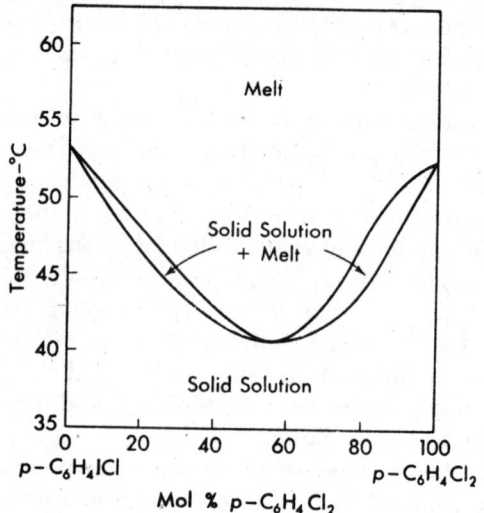

Figure 10–13. Phase diagram for the system p-C_6H_4ICl–p-$C_6H_4Cl_2$.

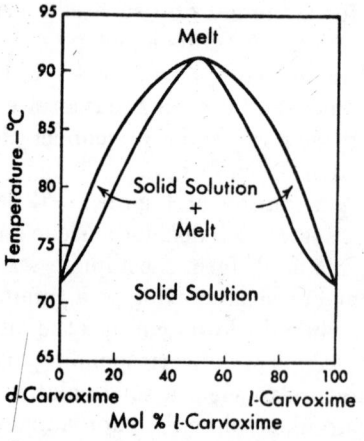

Figure 10–14. Phase diagram for the system d-carvoxime–l-carvoxime.

upper curve, the solidus the lower. As in the preceding type, there is just one composition of melt which solidifies at constant temperature to yield a solid solution, the one corresponding to the maximum. All other over-all compositions behave as described.

CLASS A: TYPE V. PARTIAL MISCIBILITY IN SOLID STATE WITH EUTECTIC

When two solids are soluble in each other to a limited degree only, A will dissolve a given amount of B to yield a saturated solution of B in A, while B will dissolve some A to yield a saturated solution of A in B. As long as these limiting concentrations are not exceeded, the solid phase is homogeneous and constitutes a single solid solution. If the range of miscibility is exceeded, however, two solid phases result, each composed of a saturated solution of one constituent in the other. It follows from the phase rule, therefore, that for equilibria between a single solid solution and melt the system will be monovariant, while for equilibria between the two solid solutions and melt the system will be invariant. Any processes involving the coexistence of the two solid solutions and melt will have to take place then isothermally.

Systems in which partial miscibility in the solid state occurs exhibit two types of diagrams. In the first of these, type V, the cooling of melt within certain composition limits results in the appearance of a eutectic involving melt and the two solid solutions. The other possibility, type VI, is observed when the two solid solutions are not stable within certain concentration ranges, and one of these is transformed to the other through a peritectic reaction.

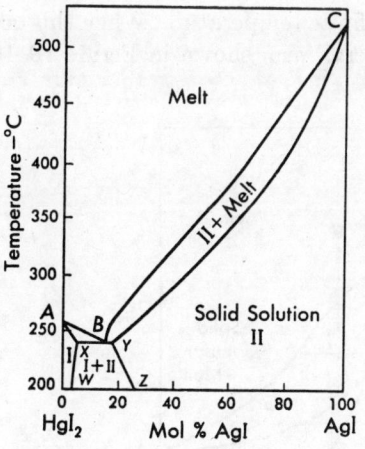

Figure 10–15. Phase diagram for the system mercuric iodide-silver iodide.

Figure 10–15 shows a typical phase diagram for a system in which the partial miscibility in the solid state leads to a eutectic B. Line AB is the liquidus curve for solid solutions of silver iodide in mercuric iodide (I), while line AX is the solidus curve for the same solutions. Similarly, line BC is the liquidus curve for solid solutions of mercuric iodide in silver iodide (II), while line YC is the solidus curve for these solutions. The area enclosed by the lines AX, XB, and BA contains then solid solutions I and melt and the area enclosed by the lines BY, YC, and CB solid solutions II and melt. Below $AXYC$ no melt is present, only solid solutions. To the right of YZ and below CY the solid phase present is solid solution II, while to the left of WX and below AX the solid phase is solid solution I. The area below XY and between the lines WX and YZ is the range of partial miscibility of the solid solutions. In this area I and II coexist, with the lines WX and YZ showing the compositions of I and II respectively, at each temperature. From the directions of these lines it is apparent that the partial miscibility of the solid iodides decreases as the temperature is lowered. The eutectic at B involves two solid solutions with the fixed compositions given by X and Y.

Other systems exhibiting diagrams of this type are silver chloride-cuprous chloride, potassium nitrate-thallium nitrate, azobenzene-azoxybenzene, naphthalene-monochloracetic acid, and the metal pairs lead-antimony, silver-copper, lead-tin, and cadmium-zinc.

CLASS A: TYPE VI. PARTIAL MISCIBILITY IN SOLID STATE WITH PERITECTIC

Instead of exhibiting a eutectic, two solid solutions may undergo a peritectic reaction in which a solid solution of one type is transformed to a solid solution of another type at a definite temperature. When this occurs, the phase diagram takes on the general form shown in Figure 10–16. In

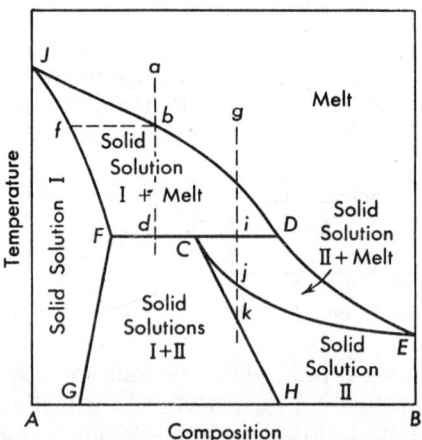

Figure 10–16. Phase diagram for partially miscible solid solutions with peritectic.

this diagram JD and JF are the liquidus and solidus curves, respectively, of solid solutions of B in A, while DE and CE are the corresponding curves for solid solutions of A in B. As may be seen from the diagram, any overall compositions lying between pure A and point F yield only solid solutions I; those lying between pure B and point D yield only solid solutions II. However, the solid phases resulting from mixtures between F and D depend on whether the over-all composition lies between F and C, or C and D. Cooling any mixture between F and C, such as a, will yield first solid solution I of composition f when point b is reached. Further cooling will result in further separation of I of compositions along fF, while the melt composition adjustment will proceed along bD. But when the temperature corresponding to line FD is reached, point d, solid solution I of composition F becomes unstable and begins to react with melt of composition D to form solid solution II of composition C. Since this *peritectic* reaction between the two solid solutions and melt involves the coexistence of three phases, the process must occur isothermally until the melt is all consumed and only the two solid phases remain. Once the melt has disappeared the temperature begins to fall again, and we pass into the area of partial miscibility of the two solid solutions. The lines FG and CH give the compositions of I and II respectively at various temperatures below the peritectic.

For any composition falling between C and D, such as g, the behavior on cooling will be similar to that of a up to point i, the peritectic temperature. At i solid solution F and melt D again interact to form solid solution II of composition C; but, since now there is more melt present than is necessary to react with all of F to form C, F must disappear, and the end of the peritectic reaction must result in the presence of solid solution II and melt. This mixture of melt and II will eventually solidify at j to leave only solid solution II.

In the diagram under discussion the solid solution of composition j will be stable only between the temperatures corresponding to j and k. When the latter temperature is reached, j passes into the range of partial miscibility and breaks up into two solid solutions.

The appearance of peritectics involving solid solutions is found in such binary systems as silver chloride-lithium chloride, silver nitrate-sodium nitrate, cobalt-iron, indium-thalium, and p-iodo-chlorbenzene-p-diiodobenzene.

CLASS B: PARTIAL MISCIBILITY IN LIQUID STATE

Although the discussion has centered so far on systems exhibiting a single liquid phase, there are systems in which the melt is only partially miscible over certain temperature and concentration ranges. When this separation of a liquid into two layers occurs, the number of phases is increased, and the phase relations are thereby modified. The case to be considered here is that in which the melt is partially miscible but the solid phases are the pure constituents, Figure 10–17. This figure is essentially a simple eutectic-type diagram with an area of partial miscibility of the melt super-imposed upon it. Outside the dome-shaped area and above the solid lines a single liquid phase is present. Within the dome-shaped area and above line DE two liquid phases coexist whose compositions at each temperature are given by horizontal tie-lines such as

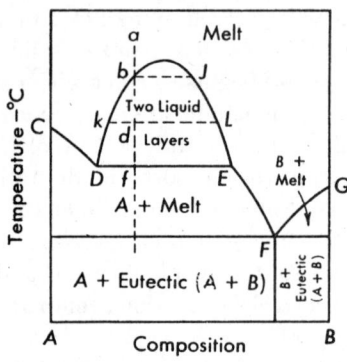

Figure 10–17. Phase diagram for system with partially miscible melt.

bJ or kL. Since the relations below line DE have been discussed, we need elaborate only the behavior at and above this line.

For this purpose consider specifically a melt of composition a lying between D and E. Cooling of this melt will yield no new phases until at b a small amount of a second liquid layer J appears. Further cooling of these two layers to a point such as d merely changes the relative amounts of the two layers and brings about an adjustment in the concentrations along bk and JL. However, when the layer compositions reach D and E, corresponding to the over-all composition point f, layer D becomes saturated with A, and the latter begins to deposit according to the scheme

$$\text{Liquid } D \longrightarrow A \text{ (solid)} + \text{Liquid } E \qquad (6)$$

During this crystallization of A and transformation of liquid layer D into E, three phases are present in equilibrium, the system is invariant, and hence the temperature remains constant until all of D has disappeared. Once D is gone the system regains a degree of freedom, and the temperature falls to yield A in equilibrium with a single melt of composition given by EF. Final solidification takes place at F to yield a eutectic mixture of A and B.

For compositions between C and D the behavior will be somewhat different. Cooling of any such mixture will result first in a separation of A when line CD is crossed. Further cooling will shift the melt composition toward D, and when this point is reached a separation of the melt into D and E will occur. From this stage on, separation of A will proceed under isothermal conditions until D is gone, and thereafter the mixture of A and melt along EF will cool in the manner described. Since to the right of E only a single liquid phase is present the cooling behavior will be the same as that of any simple eutectic system.

In many systems the line CD is either extremely short or nonexistent. In the latter instance points D and C coincide; i.e., at the melting point of A the A-rich layer is practically pure A. Again, the eutectic point F may be displaced so close to the B axis as to obliterate practically the line FG and the $(B + \text{melt})$ area. Examples of all these various modifications of Figure 10–17 may be found in the condensed binary pairs lithium-sodium, bismuth-zinc, bismuth-cobalt, chromium-copper, copper-lead, benzoic acid-water, and phenol-water.

CLASS C: IMMISCIBILITY IN SOLID AND LIQUID STATES

When two constituents are completely immiscible in both the solid and liquid states, each of these substances will melt and freeze independently of the other. An example of such behavior is shown in Figure 10–18

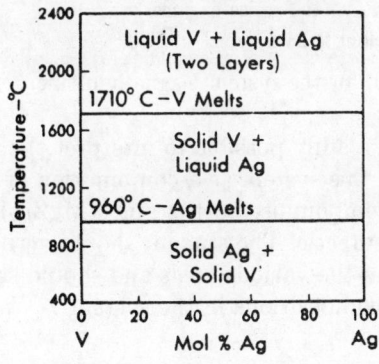

Figure 10–18. Phase diagram for the system vanadium-silver.

for the system vanadium-silver. Below 960°C the two elements exist as two solid phases. At 960°C the silver melts sharply to yield the liquid, which then coexists with the solid vanadium up to the melting point of the latter. At 1710°C the vanadium melts, and thereafter the system contains only the two pure liquids in two layers

Similar behavior is observed also with the metal pairs bismuth-chromium, chromium-iron, aluminum-sodium, aluminum-lead, gallium-mercury, potassium-magnesium, and others. In practically all instances the melts become partially miscible at temperatures above the fusion point of the higher melting constituent.

COMPOSITE DIAGRAMS

As a rule binary solid-liquid diagrams are not of the simple types described but have the elements of several types combined into a single diagram. Such a more complicated phase relation for the system magnesium-zinc is shown in Figure 10–19. By applying the general principles

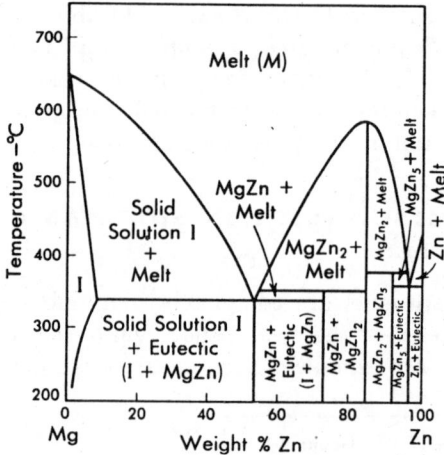

Figure 10–19.   Phase diagram for the system magnesium-zinc.

developed for the simple types it is readily possible to interpret the more complicated equilibria occurring in this system as a combination of solid solution formation, formation of compounds, both stable ($MgZn_2$) and unstable ($MgZn$ and $MgZn_5$), and eutectic. The student should verify for himself the validity of the legends in the various areas and should sketch cooling curves for various over-all compositions in the system.

THREE - COMPONENT SYSTEMS

In three-component systems a single phase possesses four degrees of freedom, namely, temperature, pressure, and the compositions of two out of the three components. This number of variables poses great difficulty in the graphical presentation of the phase relations. For this reason data in ternary systems are generally presented at some fixed pressure, such as atmospheric, and at various constant temperatures. Under these conditions it is possible to show the concentration relations among the three components at any given temperature on a planar diagram. By combining such planar diagrams for various temperatures it is then possible, if desired, to construct a solid model having concentrations as a base and temperature as a vertical axis.

For a three-component system the phase rule takes the form $F = 5 - P$. At a fixed pressure and temperature the number of degrees of freedom is reduced by two, so that $F = 3 - P$, and the maximum number of phases which can occur simultaneously is thus three. This is the same number as is possible in two-component systems under a constant pressure only. Therefore, under the specified conditions an area will again indicate divariance, a line monovariance, and a point invariance.

METHOD OF GRAPHICAL PRESENTATION

Several schemes for plotting two-dimensional equilibrium diagrams for ternary systems have been proposed. Of these the equilateral triangle method suggested by Stokes and Roozeboom is most generally employed and will be used here. In this method concentrations of the three components at any given temperature and pressure are plotted on an equilateral triangle such as that shown in Figure 10–20. Each apex of the

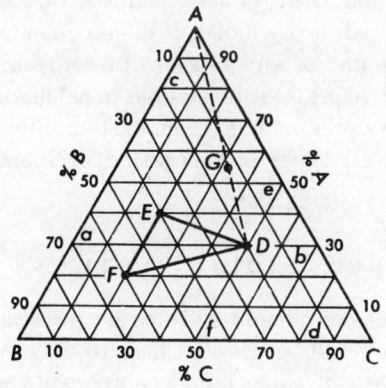

Figure 10–20. Graphical representation of three-component systems.

triangle is taken as 100 per cent of the component with which it is designated. To obtain percentages other than 100 for A, the sides AB and AC are divided into 10 (or sometimes 100) equal parts, and lines parallel to the side BC are drawn. Each of these lines represents then a definite percentage of A ranging from zero on line BC to 100 at A. Similarly, lines dividing the sides BA and BC and parallel to AC give various percentages of B, while the lines along the sides CA and CB and parallel to AB represent various percentages of C. To plot any point on the diagram such as D, having 30 per cent A, 20 per cent B, and 50 per cent C, we locate first the 30 per cent A line, namely ab, and next the 20 per cent B line, or cd. The intersection of these two lines yields the desired point. This point should lie also on the 50 per cent C line, or ef, and this fact may be utilized as a check on the accuracy of location of the point.

From the nature of the diagram it is readily apparent that the sides of the triangle represent various proportions of constituents in two-component systems. Thus lines AB, BC, and AC give the concentration relations in the binary systems A–B, B–C, and A–C, respectively, and any point on these lines refers only to these binary systems. On the other hand, any mixture composed of A, B, and C must lie within the diagram. In fact, the argument may be extended to show that similar relations apply also to any line or smaller triangle within the diagram. Thus all mixtures that can be prepared from D and E will lie on DE, those prepared from E and F will lie on EF, while those composed of F and D will lie on the line FD; and all possible compositions that can be prepared from D, E, and F will fall within the small triangle DEF. From the same considerations it also follows that if a point such as G lies on a straight line connecting D and A, then D must lie on an extension of the straight line through G linking A and G.

Another relation to remember is this: Any mixture such as G, composed of A and D, will contain A and D in the length ratios DG:AG by weight, if weight percentage is plotted, or by moles, if mole percentage is plotted. By knowing the total amount of any mixture present, and by determining these lengths from the diagram, it is possible to calculate the weights (or moles) of various phases present in a given system. Such calculations find wide application in all types of separation problems involving three components.

PARTIALLY MISCIBLE THREE-LIQUID SYSTEMS

Although many categories of three-component systems are possible and have been observed, prime attention will be devoted here to only two of these, namely, (a) systems composed of three liquid components which exhibit partial miscibility and (b) systems composed of two solid components and a liquid.

Systems composed of three liquids which show partial miscibility may be classified as follows:

Type I.   Formation of one pair of partially miscible liquids
Type II.   Formation of two pairs of partially miscible liquids
Type III. Formation of three pairs of partially miscible liquids

These three types will be discussed now in turn.

TYPE I. ONE PARTIALLY MISCIBLE PAIR

Consider a pair of liquids, B and C, that are partially soluble in each other at a given temperature and pressure. If we mix relative amounts of the two so as to exceed the mutual solubility limits, two layers will be obtained, one composed of a solution of B in C, the other of C in B. Suppose we add now to the two-layer mixture a third liquid A, which is completely miscible with both B and C. Experience shows that A will distribute itself between the two layers and will promote thereby a greater miscibility of B and C. The increase in miscibility brought about by A depends on the amount of it added and on the amounts of B and C present. If sufficient A is introduced, the two layers can be changed into a single solution composed of the three liquids.

The changes in miscibility produced by progressive additions of A to mixtures of B and C can be followed on the diagram shown in Figure 10–21. Points a and b designate the compositions of the two liquid layers

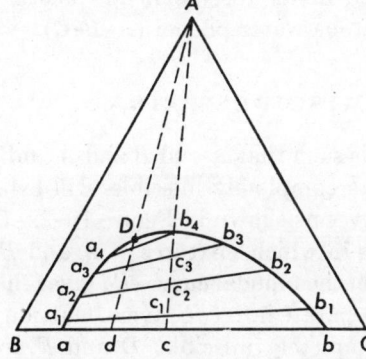

Figure 10–21.   Three liquids with one partially miscible pair.

resulting from mixing B and C in some arbitrary over-all proportion such as c, while line Ac shows the manner in which the over-all composition of c is changed by addition of A. When enough A is added to change c to c_1, the compositions of the two layers are shifted from a and b to a_1 and b_1. The line a_1b_1 through c_1 connects the compositions of the two layers in equilibrium, and is called a *tie-line*. In a similar manner the compositions are changed to a_2, a_3 and b_2, b_3 when the over-all compositions reach c_2 and c_3. Finally, at point b_4 sufficient A has been added to form only a single layer of this composition, and thereafter only a single solution is obtained.

The tie-line for b_4 shows that the composition of the B-rich layer at the point of complete miscibility is not identical with b_4 but is equal to a_4. This fact indicates that complete miscibility is brought about not by coalescence of the two layers into one, but rather by the disappearance of the B-rich layer. Complete miscibility by the merging of the two layers into one can occur only at one point on the diagram, D. At this point the compositions of the two layers become identical, and the two solutions coalesce into a single liquid phase of constant composition. Point D is called either the *isothermal critical point* of the system, or the *plait point*, and can be obtained only by adding A to a single mixture of B and C, namely, d.

From the preceding discussion it follows that any mixture of A, B, and C of over-all composition falling within the dome-shaped area will yield two liquid layers of compositions given by the appropriate tie-line through the composition of the mixture. On the other hand, any mixture of over-all composition outside this area will yield only a single homogeneous solution of the three liquids. Curve aDb is frequently referred to as a *binodal* curve. In general the plait point D will fall off the maximum of the binodal curve. Furthermore, since as a rule component A will not distribute itself equally between the two layers a and b, the tie-lines will not be parallel either to BC or to each other.

Examples of systems of the type under discussion are acetic acid–chloroform–water $(A–B–C)$, and acetone–water–phenol $(A–B–C)$.

TYPE II. TWO PARTIALLY MISCIBLE PAIRS

A system composed of three liquids such that A and B and A and C are partially miscible, while B and C are completely miscible, will exhibit a phase diagram with two binodal curves as shown in Figure 10–22. Curve aDb gives the range of compositions in which mixtures of A and B containing C are partially miscible. Again the binodal curve cFd gives the area within which C and A containing B separate into two layers. Outside these areas the three components are completely miscible. D and F are the respective plait points of the two heterogeneous regions, while the indicated tie-lines show the compositions of the various layers in equilib-

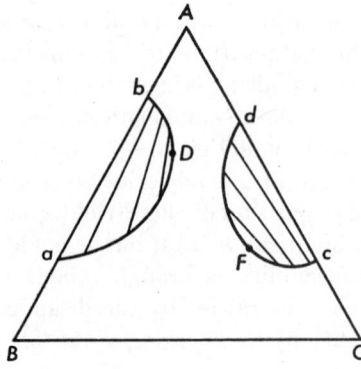

Figure 10–22. Partially miscible liquids with two binodal curves.

rium. This type of diagram is shown by the liquids succinic nitrile-water-ethyl alcohol (*A–B–C*) between 18.5 and 31°C.

Although some ternary liquid systems exhibit this type of diagram at elevated temperatures, at lower temperatures, when the miscibility decreases, the two binodal curves may intersect to form the "band" type of diagram shown in Figure 10–23. Here the partial miscibility area *abdc* extends across the width of the diagram, with *bd* giving the composition of one layer and *ac* that of the other. The indicated tie-lines join solutions in equilibrium. Examples of "band" formation are found in the systems water-phenol-aniline (*A–B–C*) and water-ethyl acetate–*n*-butyl alcohol (*A–B–C*).

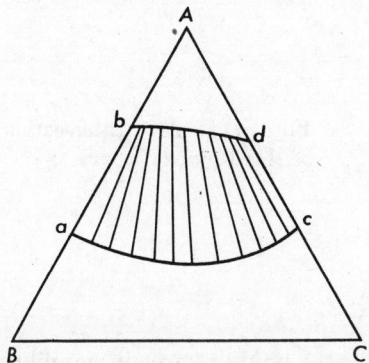

Figure 10–23. Partially miscible liquids with intersecting binodal curves.

TYPE III. THREE PARTIALLY MISCIBLE PAIRS

When all three liquids are partially soluble in each other, three binodal curves (Figure 10–24) result, provided the temperature is sufficiently high to prevent intersections. Here again the dome-shaped areas indicate two-phase liquid regions, while outside these only a single phase is present.

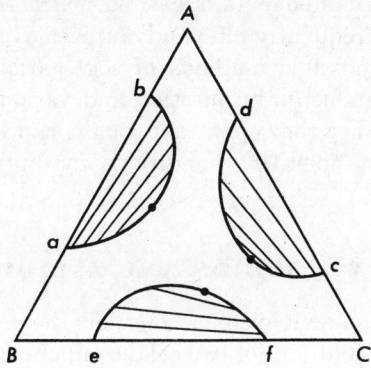

Figure 10–24. Partially miscible liquids with three binodal curves.

However, when two of the binodal curves intersect, as may occur at lower temperatures, the diagram takes on the appearance of Figure 10–25. Here in the areas designated as 1 only a single phase exists, while in the areas marked 2 two liquid phases coexist with the equilibrium concentrations given by the connecting tie-lines. But the area marked 3 contains now *three* liquid phases. Since for three phases in equilibrium the system must be invariant at constant temperature and pressure, the compositions of the three layers must be fixed and independent of the over-all composition—as long as it falls within this area. These constant compositions for the three liquid layers in equilibrium are given by the points D, E, and F.

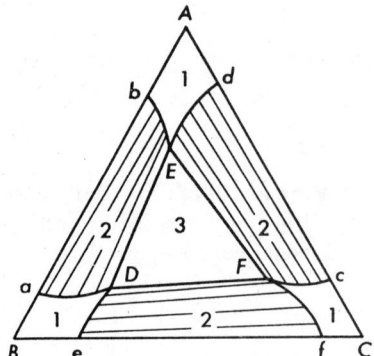

Figure 10–25. Intersection of three binodal curves.

An example of a system yielding three liquid phases in equilibrium is succinic nitrile-water-ether. At higher temperatures this system goes over to the type shown in Figure 10–24.

APPLICATION OF TERNARY LIQUID DIAGRAMS

Ternary liquid diagrams are of considerable value in various types of separation and solvent extraction problems. In certain binary mixtures where separation of the individual components is desired, extraction of one component by a third solvent frequently offers advantages over distillation, crystallization, or other possible methods of segregation. By studying the diagrams for the components in question and various solvents, it is possible to deduce whether the separation sought can be accomplished and to define the best operating conditions for optimum results.

SYSTEMS COMPOSED OF TWO SOLIDS AND A LIQUID

This important class of three-component systems considers the crystallization of various solid phases from solutions of two solid components in a liquid solvent. Although the solvent may be any suitable liquid in which the solid components may be soluble, attention will be focused primarily on solutions in water, which is by far our most important crystallization medium. However, the relations developed will apply equally to any other systems of the general type under discussion.

All the systems to be described below will exhibit only a single liquid phase. For one liquid phase occurring in a system, the behavior of the solid phases may be classified as follows:

Type I.   Crystallization of pure components only
Type II.   Formation of binary compounds
Type III. Formation of ternary compounds
Type IV. Formation of complete series of solid solutions
Type V.   Partial miscibility of solid phases

TYPE I. CRYSTALLIZATION OF PURE COMPONENTS ONLY

When in a system composed of water and two solid components B and C only the two pure solid components appear on crystallization, the *isothermal* phase diagram has the general form shown in Figure 10-26. In this

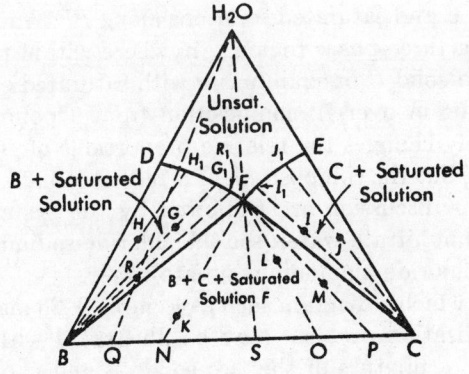

Figure 10-26.   Crystallization of two solid components from solution.

diagram points D and E are solubilities in water of pure B and C respectively at the given temperature. As C is added to the solution saturated with B, the concentration of the latter changes and follows the line DF. Similarly, when B is dissolved in the water solution saturated with C, the composition of the solution changes along line EF. The line DF is, therefore, the saturation *solubility curve* of B in water containing C, while EF is the corresponding solubility curve of C in water containing B. At F, the point of intersection of the two solubility curves, the solution becomes saturated with both B and C. Since the system contains now three phases in equilibrium, B, C, and saturated solution of composition F, there are no degrees of freedom left, and hence the composition at this point must be constant. For this reason F is called the *isothermal invariant point*.

The area above the lines DF and FE contains only unsaturated solution. However, area DFB is a two-phase region in which solid B is in equilibrium with saturated solutions of compositions lying along DF. The particular concentration of saturated solution resulting from a given over-all composition, such as G, can readily be determined from the diagram. Any mixture G of B, C, and water within the indicated area will yield at equilibrium pure solid B and a saturated solution somewhere along DF. From the relations of the equilateral triangle it follows, therefore, that the concentration of solution sought must lie on a straight line from B through G, namely, G_1. By the same token any mixture such as H will have H_1 as a saturated solution. Lines such as G_1B and H_1B which connect the concentrations of the saturated solutions with the solid phases in equilibrium with them are called tie-lines. The points of convergence of various sets of these tie-lines on a diagram determine the nature of the solid phases with which various solutions are saturated.

Corresponding to area DFB is the area EFC in which the saturating phase is C. Any mixtures falling within this area, such as I or J, yield at equilibrium solid C and saturated solutions along EF. On the other hand, the area BFC is a three-phase region. Anywhere within this area will be found solid B and solid C in equilibrium with saturated solution of composition F. A shift in over-all composition from a point K to L or M in this area merely changes the relative proportions of B and C present from N to O or P, but the composition of F remains unaffected. Diagrams of the type under discussion are exhibited by the systems ammonium chloride-ammonium nitrate-water, sodium chloride-sodium nitrate-water, and ammonium chloride-ammonium sulfate-water.

The manner in which a diagram such as Figure 10–26 may be utilized in solving a crystallization problem may be illustrated with the following example. Suppose a mixture of the two solids B and C of over-all composition Q is available and it is desired to separate from this mixture pure B. In order to recover this solid it is necessary to bring the over-all composition into area DFB. This may be accomplished by the addition of water which will make the composition change along line Q–H_2O. When water is first added we enter area BFC in which both B and C are in equilibrium with solution F. However, as soon as line BF is crossed to a point such as R, all the C dissolves, and a solution R_1 is obtained saturated with solid B which can be filtered off, washed, and dried to obtain pure B. The amount of water necessary to dissolve all of C, and the yield of B, can be calculated with the aid of the diagram. From the distances QR and $R-H_2O$ it follows that the ratio of the weight of Q to the weight of water to be added to reach R is

$$\frac{\text{Weight of } Q}{\text{Weight of water}} = \frac{R-H_2O}{RQ}$$

Knowing the original weight of Q, the weight of water to be added and the total weight of the mixture at R may be determined. Furthermore, since R is composed of B and R_1, we find in the same manner that

$$\frac{\text{Weight of } B}{\text{Weight of } R_1} = \frac{RR_1}{BR}$$

By using this ratio and the total weight of the mixture at R, the weight of B which can be recovered may be predicted.

The maximum amount of B is recovered when R is very close to the line BF. Farther along the line $R-H_2O$ the proportion of saturated solution to B is greater, and hence the yield of B is less. Another significant fact is that B can be recovered by the addition of pure water only when the mixture of B and C does not exceed the composition S. Once this composition is exceeded toward C, the saturation area of C is entered, and consequently only this substance may be recovered from the mixture.

THE SCHREINEMAKERS "WET RESIDUE" METHOD

Experimental methods employed for the determination of phase diagrams of ternary systems containing solid and liquid phases are in general similar to those described for binary systems. Thermal analysis may be used, but it usually yields cooling curves that are more difficult to interpret than those encountered in binary mixtures. For this reason the saturation method is used almost exclusively in the study of equilibria at and near room temperatures. The procedure for this method is briefly as follows: Mixtures of various proportions of the solid components with water are prepared and agitated in a thermostat until equilibrium is established. The liquid phase is separated then from the wet crystals, and both are weighed and analyzed carefully. The compositions thus obtained for saturated solution and the wet residue are plotted finally on a triangular diagram.

Figure 10–27 shows a series of points arrived at in this manner. S_1, S_2, etc., are the compositions of saturated solutions, while R_1, R_2, etc., are the compositions of the corresponding wet residues. To ascertain the nature of the solid phases in equilibrium with the various solutions and present in the residues, a graphical scheme is employed known as the *Schreine-*

makers method of wet residues. This method is based on the fact that any wet residue composed of a given solid phase and a saturated solution must lie on a straight line joining the composition of the solid phase and that of the saturated solution. Consequently a tie-line drawn between any corresponding pair of R and S points must pass, on extension past R, through the composition of the solid phase. Moreover, as several solutions may have the same solid phase, all the tie-lines for such solutions

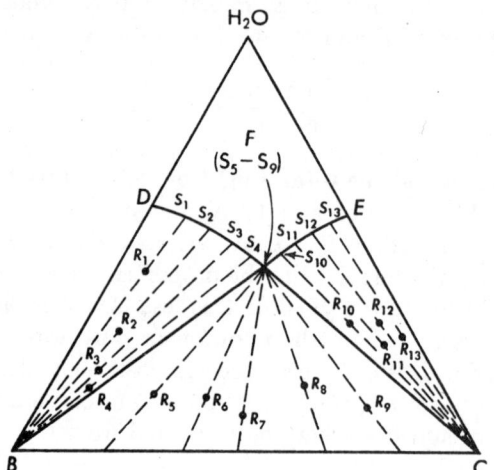

Figure 10–27. The Schreinemakers method of wet residues.

must intersect at a common point, which is the composition of the common solid phase. By this reasoning we deduce that the solid phase for all solutions between D and F is B, for those between F and E is C, while F is saturated with various proportions of B and C

TYPE II. FORMATION OF BINARY COMPOUNDS

In many systems composed of two salts and water compounds may be formed at a given temperature either between the salts and water, namely, hydrates, or between the two solid salts. Although a number of possible combinations can be envisaged, only several typical cases will be discussed to indicate the nature of the diagrams obtained under these conditions.

Hydrate Formation. Figure 10–28 shows the phase diagram for a system in which one of the components, B, forms a hydrate. Since the hydrate is composed of B and water, its composition must lie along the line B–H_2O and is given by D. E is the solubility of the hydrate in pure water at the given temperature, while line EF gives the solubility of the hydrate in solutions containing C. Within area DEF the hydrate D coexists with saturated solutions given by line *EF*. The area *FGC* contains again pure *C* in equilibrium with saturated solutions along *FG*. However, at *F*, the isothermal invariant point, the solution is saturated with both *D* and *C*. As all possible mixtures of the latter two lie along line *DC*, the tie-lines within the area *DFC* must terminate along this line, as shown. Since all of the liquid phase disappears when line *DC* is reached, below this line only a mixture of solids *D*, *B*, and *C* can exist. The system sodium sulfate-

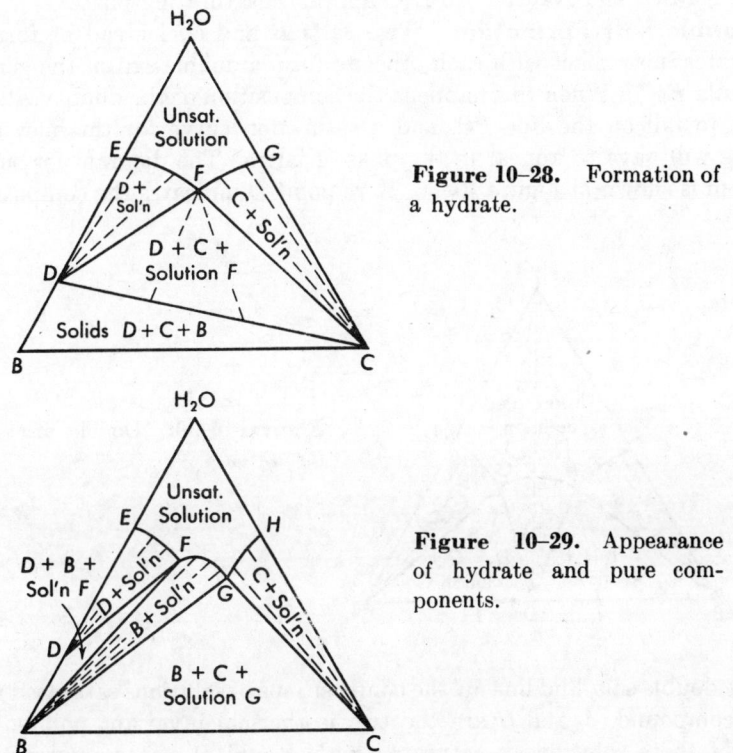

Figure 10-28.   Formation of a hydrate.

Figure   10-29.   Appearance of hydrate and pure components.

sodium chloride-water at 15°C is an example of this type of behavior, the hydrate formed being $Na_2SO_4 \cdot 10\ H_2O$.

A system in which both solid components form a hydrate will have a diagram similar to Figure 10-28 except that the tie-lines, instead of intersecting at *C*, will intercept the *C*–H_2O line at some point between *G* and *C*. The line *DC* will similarly be elevated above *C*. An example of such a system is magnesium chloride-calcium chloride-water at 0°C, where the hydrates $MgCl_2 \cdot 6\ H_2O$ and $CaCl_2 \cdot 6\ H_2O$ occur.

Under some temperature conditions in certain systems not only the hydrate of a component but also the anhydrous salt appears. The phase diagram then has the appearance of Figure 10-29. Three saturation curves are now obtained, one for each of the solid phases, and two invariant points, F and G. The solution at the invariant point F is in equilibrium with D and B, and any mixture within area DFB will yield these three phases. On the other hand, any mixture falling within area BGC will have B and C in equilibrium with the solution at the isothermal invariant point G. A diagram similar to Figure 10–29 is again exhibited by the system sodium sulfate-sodium chloride-water, but at 25°C. At this temperature both Na_2SO_4 and $Na_2SO_4 \cdot 10\ H_2O$ appear as saturating phases.

Double Salt Formation.   Two salts B and C, instead of forming hydrates, may react with each other to form a double salt of the general formula B_nC_m. When this happens the composition of the double salt will have to fall on the line BC, and a saturation curve for this new solid phase will have to appear in the phase triangle. The diagram for such a system is shown in Figure 10–30. Here point D indicates the composition

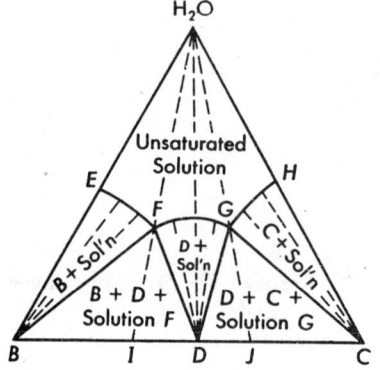

Figure 10–30.   Double salt formation.

of the double salt, and line FG the compositions of solutions saturated with this compound. F and G are the two isothermal invariant points. The first of these solutions is saturated with B and D, the second with D and C.

The stability of the double salt in presence of water depends on where along BC the composition of the salt falls. If point D falls between the points I and J, it is possible to prepare stable solutions of the double salt by adding water to it, as is indicated by the line D–H_2O. Such a salt is said to be *congruently saturating*. On the other hand, if the composition of the salt falls either to the left of I or to the right of J, it is impossible to prepare a saturated solution of the salt in water by adding water to D. A line from D lying between B and I to the water apex will miss curve FG to yield a mixture either within areas BDF or BEF Similarly, for D between J and C a line to the water point will miss FG to form either a mixture of D, C, and saturated solution G, or a mixture of C and saturated solution. In either eventuality the double salt undergoes partial or complete decomposition. Double salts behaving in this manner are said to be *incongruently saturating*.

As an example of a congruently saturating double salt may be mentioned $NH_4NO_3 \cdot AgNO_3$ in the system ammonium nitrate-silver nitrate-water at 30°C. However, in the system potassium nitrate-silver nitrate-water at the same temperature the double salt formed, $KNO_3 \cdot AgNO_3$, decomposes on addition of water and hence is incongruently saturating.

TYPE III. FORMATION OF TERNARY COMPOUNDS

In certain systems not only can binary compounds be formed but ternary compounds involving all three components as well. Figure 10–31

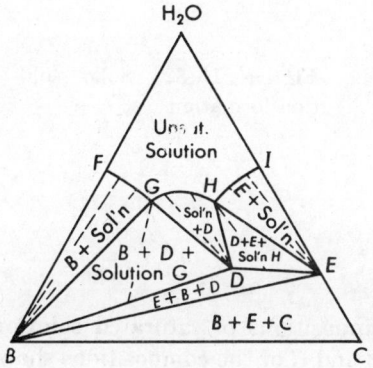

Figure 10–31. Formation of ternary compound.

shows the phase diagram for a system in which appear a binary compound, hydrate E, and ternary compound D, composed of B, E, and water. Within area GHD compound D is in equilibrium with saturated solutions along GH, and may be recovered from these. G and H are the isothermal invariant points. G is saturated with mixtures of B and D, H with the solids D and E. All mixtures are completely solid below the lines BD and DE. Within the triangle BDE the solid phases are B, D, and E; but once line BE is crossed into area BEC, the solid phases become B, E, and C.

The ternary compound D shown in Figure 10–31 is evidently of the incongruently saturating type and decomposes on addition of water. Examples of such a compound are $CaCl_2 \cdot MgCl_2 \cdot 12 H_2O$ (tachydrite) in the system calcium chloride-magnesium chloride-water at 25°C, and $MgSO_4 \cdot Na_2SO_4 \cdot H_2O$ in the system magnesium sulfate-sodium sulfate-water at the same temperature. On the other hand, ternary compounds such as the alums, whose general formula is $B_2SO_4 \cdot C_2(SO_4)_3 \cdot 24\ H_2O$, form congruently saturating salts which are stable in presence of water.

TYPE IV. FORMATION OF SOLID SOLUTIONS

When two solid components B and C are completely soluble in each other in the solid phase, a series of solid solutions ranging in composition from pure B to pure C can be recovered from a solution of these in water. Since under these conditions only two phases appear in the system, the solid solution and the liquid saturated solution, no invariant point is observed. Figure 10–32 shows the phase diagram for such a system. In

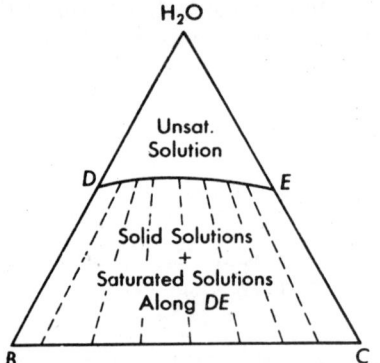

Figure 10–32. Solid solution formation.

this diagram line DE gives the compositions of saturated solutions in equilibrium with solid solutions of B and C of the compositions shown by the various tie-lines. In the area above DE only unsaturated solutions can be obtained. Below this line two phases occur, the saturated solutions along DE and the solid solutions in equilibrium with them.

TYPE V. PARTIAL MISCIBILITY OF SOLID PHASES

Finally, Figure 10–33 shows the phase diagram for a system in which the solid phases B and C are only partially soluble in each other. Under these conditions two sets of solid solutions result, one of C in B, lying between points B and D, and another of B in C, between points E and C. Line FG gives the compositions of saturated solutions in equilibrium with the first series of these solid solutions and line GH those in equilibrium with the second. Between points D and E mixtures of B and C yield two solid phases of which one has the composition D, the other E. G is an isothermal invariant point such that any over-all composition falling

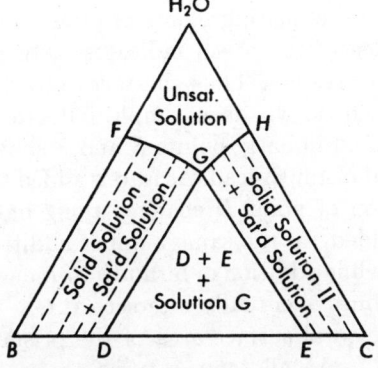

Figure 10–33. Partial miscibility of solid phases.

within area DGE yields this solution and the two solid solutions, D and E, in equilibrium with it.

THE THREE-DIMENSIONAL PHASE DIAGRAM

The various triangular diagrams described heretofore are actually isothermal sections of a solid figure consisting of an equilateral triangular base, along which concentrations are plotted, and a vertical axis giving temperature. Such a solid diagram can be constructed from a study of a given ternary system at various temperatures and subsequent assembly of the isothermal sections into a three-demensional model. The result of this kind of study for the system bismuth-tin-lead is shown schematically in Figure 10–34. The relations in this system are the simplest that can be encountered, for here only the pure components appear as solid phases.

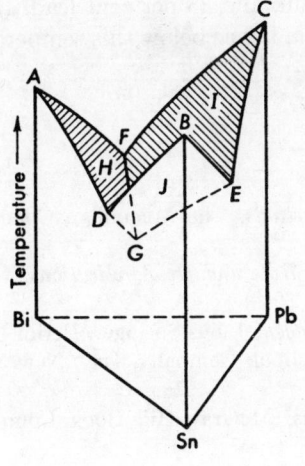

Figure 10–34. The system Bi-Sn-Pb at various temperatures.

In Figure 10–34 points A, B, C give the melting points of pure bismuth, tin, and lead respectively. Each face of the prism indicates in turn the phase behavior of a two-component system. Thus the face ADB-Sn-Bi shows that the binary pair bismuth-tin yields a simple eutectic-type diagram with the eutectic point at D. Similarly points E and F show the eutectics for the systems tin-lead and bismuth-lead. As lead is added to the binary pair bismuth-tin, the position of point D changes along line DG according to the amount of lead added. In the same manner addition of tin to F changes its locus along FG, while addition of bismuth to E changes this point along line EG. All these lines eventually intersect at G.

Lines FG, DG, and GE divide the total surface area of the prism into three distinct surfaces H, I, and J. At all temperatures above these surfaces only melt exists. As soon as a surface is reached, however, a solid phase begins to separate. On H the solid phase is bismuth, on I lead, on J tin. These surfaces represent, therefore, two-phase equilibria involving a pure component and melt. Since FG is the line of intersection of surface H, involving an equilibrium between solid bismuth and melt, and surface I, involving solid lead and melt, this line must represent three phases in equilibrium, namely, bismuth, lead, and melt. By the same reasoning along line GE, tin, lead, and melt, and along line DG bismuth, tin, and melt, are in equilibrium. When these lines eventually meet at G, four phases coexist, the three solids and melt. As for four phases in equilibrium a three-component system at constant pressure possesses zero degrees of freedom, point G must be the absolute invariant point of the system. This invariant point can occur only at a fixed temperature and must correspond to a constant composition of the three components. For the system bismuth-tin-lead the coordinates of G are 97°C and approximately 51 per cent bismuth, 16 per cent tin, and 33 per cent lead. Below 97°C no melt can appear in this system and hence below this temperature only the three solid phases coexist.

REFERENCES

1. S. T. Bowden, *Phase Rule and Phase Reactions*, The Macmillan Company, New York, 1939.
2. Findlay, Campbell, and Smith, *The Phase Rule and Its Applications*, Dover Publications, New York, 1951.
3. M. Hansen, *Der Aufbau der Zweistofflegierungen*, Julius Springer, Berlin, 1936.
4. *International Critical Tables*, McGraw-Hill Book Company, Inc., New York, 1923, Vol. II.
5. J. S. Marsh, *Principles of Phase Diagrams*, McGraw-Hill Book Company, Inc., New York, 1935.

6. Purdon and Slater, *Aqueous Solutions and Phase Diagrams*, Edward Arnold & Co., London, 1946.

7. J. E., Ricci, *The Phase Rule and Heterogeneous Equilibrium*, D. Van Nostrand Company, Inc., Princeton, 1951.

PROBLEMS

1. State how many componen.s are present in each of the following systems:

  (a) $H_2(g) + N_2(g)$
  (b) $NH_3(g)$
  (c) A solution of $Ca(NO_3)_2$ in water
  (d) An aqueous solution of $NaCl + Na_2SO_4$
  (e) An aqueous solution of $KCl + Na_2SO_4$ *Ans.* (a) Two; (b) one.

2. How many degrees of freedom will each of the systems enumerated in problem 1 possess?

3. How many degrees of freedom will each of the following systems possess:

  (a) $NaCl(s)$ in equilibrium with its saturated solution at 25°C and 1 atm pressure.

  (b) $I_2(s)$ in equilibrium with its vapor

  (c) $I_2(s)$ in equilibrium with its vapor at 50°C

  (d) $HCl(g)$ and $NH_3(g)$ in equilibrium with $NH_4Cl(s)$ when the equilibrium is approached by starting with the two gases only

  (e) $HCl(g)$ and $NH_3(g)$ in equilibrium with $NH_4Cl(s)$ when the equilibrium is approached by starting only with the solid

4. (a) Referring to Fig. 10–3, describe all the changes which will occur in the system when sulfur is heated *slowly* at pressure P_1 from point X at temperature T_1 to point Y at temperature T_2. Which of the equilibria encountered will be stable ones?

  (b) Describe the changes which will occur in this system in going from point X to point Y when the sulfur is heated *rapidly*. Which of the equilibria encountered will be stable ones?

5. From the following information sketch and label the phase diagram for CH_3COOH. (a) The melting point of the solid is 16.6°C under its own vapor pressure of 9.1 mm; (b) solid CH_3COOH exists in two modifications, I and II, both of which are more dense than the liquid, and I is the stable modification at low pressure; (c) phases I, II, and liquid are in equilibrium at 55.2°C under a pressure of 2000 atm; (d) the transition temperature from I to II decreases as the pressure is decreased.

6. Suppose that a one-component system exhibits a gas phase, a liquid phase, and three solid modifications. How many one-, two-, three-, and four-phase equilibria are possible in the system?

7. The following data are given by Andrews [*J. Phys. Chem.*, **29,** 1041 (1925)] for the system *o*-dinitrobenzene–*p*-dinitrobenzene:

Mol % of Para Compound	Initial Melting Point, °C
100	173.5
90	167.7
80	161.2
70	154.5
60	146.1
50	136.6
40	125.2
30	111.7
20	104.0
10	110.6
0	116.9

Construct a temperature-composition diagram for the system, and determine therefrom the eutectic temperature and composition.

8. Using the plot constructed in the preceding problem, find graphically the maximum percentage of *p*-dinitrobenzene which can be recovered pure by crystallization from mixtures of the two compounds containing originally 95%, 75%, and 45% of the para compound.   *Ans.* 98.4% for first.

9. From the phase diagram of the system Cd-Bi given in the text, estimate the solubility of Cd in Bi at 200°C.   *Ans.* 110.5 g/100 g Bi.

10. The system *n*-heptane and 2,2,4-trimethyl pentane exhibits a simple eutectic point at $-114.4°C$ corresponding to 24 mole per cent of *n*-heptane [Smittenberg, Hoog, and Henkes, *J. Am. Chem. Soc.*, **60,** 17 (1938)]. Determine analytically the maximum mole per cents of *n*-heptane which can be recovered by crystallization from mixtures of the two compounds containing 80, 90, and 95 mole per cent of *n*-heptane.

11. In the system NaCl-H_2O a simple eutectic is observed at $-21.1°C$ for a solution containing 23.3% by weight NaCl where NaCl $\cdot$ 2 H_2O and ice crystallize out of the mixture. At a composition of 27% by weight of NaCl and at $-9°C$ a peritectic point exists where the dihydrate decomposes to form anhydrous NaCl. The temperature coefficient of solubility of anhydrous NaCl is very small and positive. Make a rough sketch for the system showing clearly the phases in equilibrium in the various areas and along the various curves of the diagram.

12. Using Figure 10–7, sketch and interpret the cooling curves which would result when melts corresponding to 20, 40, and 60 mole per cent of $FeCl_3$ are cooled.

13. By referring to Figure 10–9, state the conditions of temperature and composition which must be met in order to crystallize in the pure state the compound $CaF_2 \cdot CaCl_2$ from the CaF_2-$CaCl_2$ system.

14. By referring to Figure 10–11, explain how you would proceed to obtain the optimum amount of pure $MgSO_4 \cdot 6 H_2O$ crystals from a dilute aqueous solution of $MgSO_4$.

15. The system SO_3-H_2O exhibits congruent melting points at compositions by weight of 68.96%, 81.63%, and 89.89% SO_3. What are the formulas of the corresponding compounds? *Ans.* $H_2SO_4 \cdot H_2O$ for first.

16. Complete and interpret the phase diagram, Figure 10–35, for the system Al-Ni, explaining what phases are in equilibrium under the various conditions represented by the areas and curves.

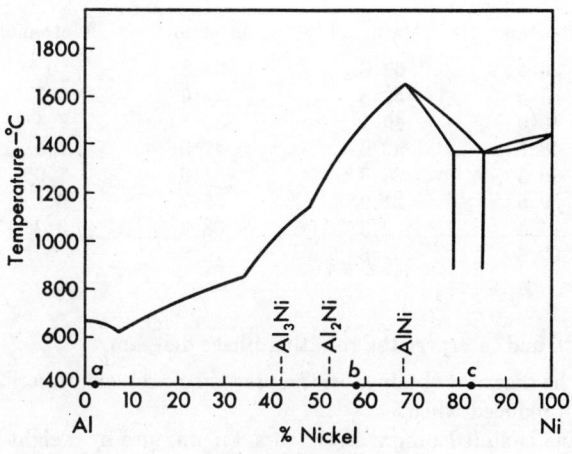

Figure 10–35.

17. In the preceding problem sketch cooling curves showing the complete solidification of melts having compositions a, b, and c.

18. Will the addition of a small amount of impurity to a pure compound or element always lower its melting point? Explain your answer.

19. Iodine dissolves in both water and CCl_4, but the latter two are practically immiscible. Apply the phase rule to the system H_2O-CCl_4-I_2, and explain what variables must be specified in order to determine the state of the system.

20. The following data are for the system water-alcohol-benzene at 25°C. The first two columns give the percentages by weight of alcohol and benzene in one layer, while the third column gives the percentage by weight of water in the layer conjugate to these.

Layer I		Layer II
% C_6H_6	% C_2H_5OH	% H_2O
1.3	38.7	
9.2	50.8	
20.0	52.3	3.2
30.0	49.5	5.0
40.0	44.8	6.5
60.0	33.9	13.5
80.0	17.7	34.0
95.0	4.8	65.5

Construct the phase diagram for the system, and draw in the tie-lines.

21. An aqueous solution contains 46% by weight of ethyl alcohol. Using the diagram of the preceding problem, find how much alcohol would be extracted from 25 g of this solution by 100 g of C_6H_6.   *Ans.* 5.05 g.

22. A. W. Francis [*J. Am. Chem. Soc.*, **76**, 393 (1954)] obtained the following data for the ternary system nitrobenzene-methanol-isooctane at 20°C:

Weight %		Weight %	
Isooctane	Methanol	Isooctane	Methanol
30.4	69.6	23.8	0
36.5	53.8	27.6	3.7
40.0	46.1	35.1	7.4
48.7	32.0	47.0	7.5
50.3	28.8	51.6	7.0*
59.5	21.9*	55.2	6.4
84.4	9.8	68.3	3.2
92.5	7.5	76.2	0

* Plait point.

Plot these data and interpret the resulting phase diagram.

23. Using the diagram obtained in the preceding problem, describe the effects which will be produced when:

(a) Isooctane is added progressively to a 1:1 mixture by weight of methanol and nitrobenzene

(b) Nitrobenzene is added to a 1:1 mixture by weight of methanol and isooctane

(c) Methanol is added to a 1:1 mixture by weight of nitrobenzene and isooctane

(d) A 1:1 mixture of nitrobenzene and isooctane is added to a 1:1 mixture of methanol and isooctane

24. The following data are given by Prutton, Brosheer, and Maron [*J. Am. Chem. Soc.*, **57**, 1656 (1935)] for the system NH_4Cl-NH_4NO_3-H_2O at 25°C:

Saturated Solution		Wet Residue	
% NH_4NO_3	% NH_4Cl	% NH_4NO_3	% NH_4Cl
67.73	0	—	—
66.27	2.00	88.20	0.79
64.73	3.82	88.00	1.34
62.24	5.58	90.25	1.65
61.68	6.97	87.65	2.28
53.49	11.08	23.31	62.22
36.99	15.80	13.63	66.29
19.05	21.81	7.09	72.75
0	28.33	—	—

Construct and interpret the phase diagram for the system, and determine the ternary composition. Is there evidence of hydrate or double salt formation in this system?

25. Using the plot constructed in the preceding problem, determine the maximum theoretical recovery of NH_4Cl from a dry salt mixture of NH_4Cl-NH_4NO_3 containing 80% by weight of NH_4Cl.   *Ans.* 96.1%.

26. For the system KNO_3-$NaNO_3$-H_2O a ternary point exists at 5°C at which the two anhydrous salts are in equilibrium with a saturated solution containing 9.04% by weight of KNO_3 and 41.01% $NaNO_3$. Determine analytically the maximum weight of KNO_3 which could be recovered pure from a salt mixture containing 70 g of KNO_3 and 30 g of $NaNO_3$ by crystallization from an aqueous solution at 5°C.   *Ans.* 90.5%.

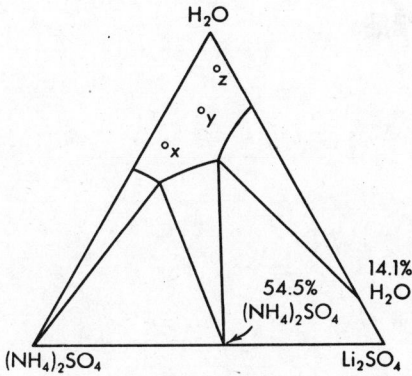

Figure 10–36.

27. Complete and interpret the phase diagram, Figure 10–36, for the system $(NH_4)_2SO_4$-Li_2SO_4-H_2O at 25°C.

28. In the diagram of the preceding problem, what will be the composition of the first crystals formed by the evaporation of H_2O from solutions of compositions x, y, and z at 25°C?

29. The following hydrates may under proper conditions be crystallized from the system $Ba(OH)_2$-$NaOH$-H_2O at 30°C: $Ba(OH)_2 \cdot 8 H_2O$, $Ba(OH)_2 \cdot 3 H_2O$, $Ba(OH)_2 \cdot H_2O$, and $NaOH \cdot H_2O$. Sketch and interpret the phase diagram for this system.

11

Conductance and Ionic Equilibria

ELECTROCHEMISTRY is the branch of physical chemistry that concerns itself with the interrelation of chemical phenomena and electricity. It deals with the study of the electrical properties of solutions of electrolytes and with the elucidation of the relation between chemical action and electricity in such systems. The phenomena encountered in electrochemistry are of such theoretical and practical importance that this and the next chapter are to be devoted to an exposition of various aspects of the subject.

OHM'S LAW AND ELECTRICAL UNITS

The strength of an electric current flowing through a conductor, i.e., the quantity of electricity flowing per second, is determined by the difference in potential applied across the conductor and by the resistance offered by the conductor to the current. According to Ohm's law the relation among these three quantities is given by

$$I = \frac{\varepsilon}{R} \tag{1}$$

where I is the strength of a current flowing through a resistance R under an applied potential ε. From this equation it is evident that the current strength is directly proportional to the difference in potential and inversely proportional to the resistance. By an appropriate choice of units the constant of proportionality is made unity.

Electrical quantities may be expressed in cgs *electromagnetic units (emu)*, based on the law of attraction or repulsion of magnets, in cgs *electrostatic units (esu)*, based on the Coulomb law of force between electric charges, or in *absolute units* derived from the emu units. As of January 1, 1948, the United States Bureau of Standards has adopted the last-named system as the official one for this country.[1] Before this date the official system of units was the one known as the *international*.

In the absolute system the units of current, potential, and resistance are respectively the absolute *ampere, volt,* and *ohm*. The corresponding units in the international system were the international *ampere, volt,* and *ohm.* The international ampere was defined as the invariable current of such strength that on passage through a water solution of silver nitrate it will deposit 0.00111800 g of silver in 1 sec. In turn, the international ohm was defined as the resistance at 0°C of a column of mercury of uniform cross section, 106.300 cm long, and containing 14.4521 g of mercury. From these two units and Ohm's law the international volt followed as the potential difference required to send a current of 1 amp through a resistance of 1 ohm.

Depending on the system, the unit of quantity of electricity is either the absolute or international *coulomb.* The coulomb is the quantity of electricity transported by a current of 1 amp in 1 sec. Since the quantity of electricity carried by a current must equal rate of transport times the time, the charge Q carried by a current I in t sec must be

$$Q = It \tag{2}$$

coulombs. Another unit of quantity of electricity which we shall employ frequently is the faraday, $\mathfrak{F}$. A faraday is equal to 96,490 absolute coulombs.

The electrical work w performed when a current of strength I passes for t sec through a resistance across which the potential drop is $\mathcal{E}$ is given by Joule's law, namely,

$$w = \mathcal{E}It = \mathcal{E}Q \tag{3}$$

where w is expressed in *joules*. The *joule* is the electrical unit of energy and is defined as the amount of work performed by a current of 1 amp flowing for 1 sec under a potential drop of 1 volt. The work in joules is readily convertible to other energy units through the relations

$$1 \text{ joule(abs.)} = 1 \times 10^7 \text{ ergs} = 0.2390 \text{ cal} \tag{4}$$

[1] National Bureau of Standards Circular C 459 (1947).

Finally, the rate at which work is being done by an electric current is expressed in *watts*. A *watt* is work performed at the rate of 1 joule per second, and is obviously a unit of electrical power. From Eq. (3) the power in watts p delivered by a current follows as

$$p = \varepsilon I = \frac{\varepsilon Q}{t} \tag{5}$$

A larger unit of power is the kilowatt, which is equal to 1000 watts.

In Table 11-1 is given a comparison of the magnitudes of the electrical units in the various systems. Since the difference between the values of the units for a given quantity in the international and absolute systems is generally negligible for our purposes, no distinction will be made between these except where necessary.

TABLE 11-1. COMPARISON OF ELECTRICAL UNITS IN VARIOUS SYSTEMS
(Each quantity in the first column is equal to all others on line with it)

Absolute	International	Electromagnetic (cgs emu)	Electrostatic (cgs esu)
1 volt	0.999670 volt	1×10^8	$\frac{1}{300}$
1 ampere	1.000165 ampere	1×10^{-1}	2.9978×10^9
1 ohm	0.999505 ohm	1×10^9	$\frac{1}{9} \times 10^{-11}$
1 coulomb	1.000165 coulomb	1×10^{-1}	2.9978×10^9
1 watt	0.999835 watt	1×10^7	0.9993×10^7
1 joule	0.999835 joule	1×10^7	0.9993×10^7

ELECTROLYTIC CONDUCTION

Flow of electricity through a conductor involves a transfer of electrons from a point of higher negative potential to one of lower. However, the mechanism by which this transfer is accomplished is not the same for all conductors. In *electronic* conductors, of which solid and molten metals and certain solid salts (cupric sulfide, cadmium sulfide) are examples, conduction takes place by direct migration of electrons through the conductor under the influence of an applied potential. Here the atoms or ions composing the conductor are not involved in the process, and, except for a vibration about their mean positions of equilibrium, they remain stationary. On the other hand, in *electrolytic* conductors, which include solutions of strong and weak electrolytes, fused salts, and also some solid salts such as sodium chloride and silver nitrate, electron transfer takes place by a *migration of ions*, both positive and negative, toward the electrodes. This migration involves not only a transfer of electricity from one electrode to the other, but also a transport of matter from one part of the conductor to another. Further, current flow in electrolytic conductors is always accompanied by chemical changes at the electrodes which are quite characteristic and specific for the substances composing the conductor and the electrodes. Finally, while the resistance of electronic conductors increases with temperature, that of electrolytic conductors always decreases as the temperature is raised.

The mechanism by which an electric current passes through a solution can best be understood from a specific example. For this purpose consider a cell, Figure 11-1, composed of two inert electrodes, in this case platinum,

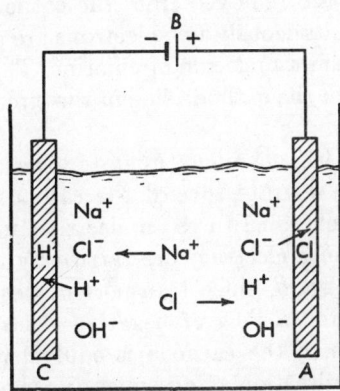

Figure 11-1. Electrolytic conduction.

connected to a source of current B and dipping into an aqueous solution of sodium chloride. The electrode C, connected to the negative side of B, is called the *cathode*. This is the electrode by which electrons from B, say a battery, enter the solution. In turn, electrode A, connected to the positive side of the battery, is termed the *anode*. It is the electrode through which the electrons leave the solution to return to B. In solution we have sodium and chloride ions, and also some hydrogen and hydroxyl ions due to the very slight ionization of the water. Now, when the circuit is closed and a current passes through the solution, it is found that chlorine gas escapes at the anode and hydrogen gas at the cathode, while sodium hydroxide forms in the solution immediately adjacent to the cathode. The explanation of these changes is as follows: Electrons enter the solution at the cathode C by combining with hydrogen ions in solution to form monatomic hydrogen. Two atoms of hydrogen thus deposited on the electrode combine then to form a molecule of hydrogen which escapes from the electrode as a gas. The reactions involved can be represented by

$$2 \, H^+ + 2 \ominus \text{(electrons)} = 2 \, H$$
$$2 \, H = H_2(g)$$

Again, electrons leave the solution at the anode by the discharge of chloride ions, with each chloride ion giving up one electron to the electrode and becoming a chlorine atom. The electrons thus liberated flow through the external circuit from the electrode to the source of potential, while the chlorine atoms combine with each other to form gaseous chlorine which escapes. The reactions here are

$$2\ Cl^- = 2\ Cl + 2\ \ominus$$
$$2\ Cl = Cl_2(g)$$

We see, therefore, that two electrons are removed from the cathode to form a molecule of hydrogen, and simultaneously two electrons are given up to the anode by chloride ions to form a molecule of chlorine. The net result is a transfer of two electrons from the cathode side of the circuit to the anode side.

When the circuit is closed negative ions, or *anions*, migrate toward the anode, while positive ions, or *cations*, migrate toward the cathode. As these particles are charged, their motion constitutes an electric current. The anions move to the anode, and hence electrons are carried by these ions from the cathode to the anode. Again, since transport of positive electricity to the *left* may be considered a flow of negative electricity to the *right*, the migration of cations to the cathode is equivalent to a flow of electrons in the opposite direction. Consequently, the net result of the migration is a flow of electrons through the solution in the direction of the current, with each ion carrying part of the current and thus contributing its share to the transport of electricity through the solution.

The formation of sodium hydroxide in the cathode portion of the cell is understood when we remember that from this part of the solution hydrogen ions have been removed by discharge on the electrode, leaving in solution an excess of hydroxyl ions. Since these have no existence independent of positive ions, sodium ions migrate into the cathode compartment in quantity just sufficient to give an electrically neutral solution. When this solution of sodium and hydroxyl ions is evaporated, solid sodium hydroxide is obtained.

The process of current passage through an electrolytic conductor with all the accompanying chemical and migratory changes is called *electrolysis*. From the above discussion the mechanism of electrolysis may be summarized by saying that (a) electrons enter and leave the solution through chemical changes at the electrodes, and (b) electrons pass through the solution by migration of ions. Just as many electrons pass through the solution and leave it as entered it, no more, no less. The evidence for this statement follows from Faraday's laws of electrolysis.

FARADAY'S LAWS OF ELECTROLYSIS

The chemical reaction which occurs during electrolysis at the anode need not be necessarily a deposition of ions, but may be any *oxidation* reaction, as solution of a metal or oxidation of ferrous to ferric ions. Similarly, the reaction at the cathode may be any *reduction* possible under the circumstances, as solution of iodine to form iodide ions, or reduction of stannic to stannous tin. Still, no matter what the nature of the reaction may be, Michael Faraday found that the *mass of a substance involved in reaction at the electrodes is directly proportional to the quantity of electricity passed through the solution*. This statement is known as *Faraday's first law of electrolysis*. The law has been shown to hold very rigidly provided the passage of electricity takes place entirely by electrolytic conduction. It applies to molten electrolytes as well as to solutions of electrolytes and is independent of temperature, pressure, or the nature of the solvent, as long as the latter can promote ionization of the solute.

From Faraday's first law of electrolysis the quantity of electricity necessary to deposit one equivalent weight of silver may readily be calculated. One absolute coulomb deposits 0.00111797 g of silver, and since the mass deposited is directly proportional to the quantity of electricity, the number of coulombs required for deposition of 1 g atomic weight of silver, 107.870 g, must be

$$\frac{107.870}{0.00111797} = 96,487 \text{ absolute coulombs}$$

The question that immediately arises is: What mass of other substances will this quantity of electricity deposit or form? If two cells, one composed of silver electrodes in silver nitrate, the other of copper electrodes in copper sulfate, are connected in series, any current passed through one cell must also pass through the other. With such a setup it can be shown that the weight of copper deposited per coulomb of electricity is 0.0003293 g. Therefore the quantity of electricity necessary to deposit 1 g atomic weight of copper, 63.54 g, is

$$\frac{63.54}{0.0003293} = 193,000 = 2(96,500) \text{ coulombs}$$

However, since copper is divalent, an equivalent is 63.54/2 g. Consequently, to deposit *one equivalent* of copper only one-half the above amount of electricity is required, or 96,500 coulombs.

From a series of such experiments Faraday arrived at his *second law of electrolysis*, namely, that *the masses of different substances produced during electrolysis are directly proportional to their equivalent weights*. Another way of stating this law is that *the same quantity of electricity will produce chemically equivalent quantities of all substances* resulting from the process. Moreover, since 96,487 coulombs will yield one equivalent of silver, a direct consequence of Faraday's second law is that *during electrolysis 96,487 coulombs of electricity will yield one equivalent weight of any substance.* The name faraday (symbol $\mathcal{F}$) has been given to this quantity of electricity. For ordinary calculations 1 faraday = 96,500 coulombs will be sufficiently exact.

It is essential to realize that 1 faraday will produce one equivalent of *each* of the primary products of electrolysis. A primary product is one formed directly by the current rather than by subsequent chemical action. To the latter Faraday's laws do not apply. For instance, 1 faraday passed through a sodium chloride cell will produce one equivalent each of chlorine, hydrogen, and sodium hydroxide. These are the primary products of the electrolysis. However, should the chlorine happen to diffuse into the cathode compartment and react with the sodium hydroxide to form sodium hypochlorite and sodium chlorate, the amounts of these formed will not depend on the quantity of electricity passed through the cell, but rather on the operating conditions. In fact, with proper care the formation of these *secondary* products can be avoided, but not that of the primary.

By applying Faraday's laws, the weight of primary products formed in any electrolytic process may be calculated very simply from a knowledge of the strength of the current and its time of passage. Thus, suppose a solution of silver nitrate is electrolyzed between silver electrodes with a current of 0.20 amp flowing for 30 min. The quantity of electricity delivered to the cell is $0.2 \times 30 \times 60 = 360$ coulombs, and hence the weight of silver deposited is 360/96,500 of an equivalent, or

$$\frac{107.87 \times 360}{96,500} = 0.4024 \text{ g}$$

Like the first law of electrolysis, the second law holds very rigidly for electrolytes both in solution and when fused. Its validity is again independent of temperature, pressure, and the nature of the solvent. The reason for the exactness of these laws, and an insight into their significance, may be gathered from the following simple calculation. The charge on the electron is equal to 1.602×10^{-19} absolute coulomb of electricity. Consequently the number of electrons in 1 faraday is

$$\frac{96,487}{1.602 \times 10^{-19}} = 6.023 \times 10^{23} \text{ electrons}$$

which is exactly Avogadro's number. Hence 1 faraday of electricity is associated with 6.023×10^{23} particles of unit charge, or, in general, with one equivalent of a chemical substance. One equivalent of positive ions lacks this number of electrons, while one equivalent of negative ions has this number of electrons in excess. When 1 faraday of electricity is passed through a solution, Avogadro's number of electrons is removed from the cathode by reduction of one equivalent of substance, and exactly the same number of electrons is donated to the anode as a result of some oxidation. Through the solution the equivalent of Avogadro's number of electrons is carried by the migration of positive and negative ions to the appropriate electrodes. In this way the whole process of electrolysis reduces itself to the transport of a given number of electrons through the electrolytic conductor. Since a given number of electrons does not depend on pressure, temperature, or other such factors, neither should the process of electrolysis, and this is the case.

As Faraday's laws are obeyed so rigidly, they may be utilized to determine the quantity of electricity passing through a circuit by observing the chemical changes produced by the same current in a suitable electrolytic cell. A cell used for this purpose is called a *coulometer*. The coulometer is placed in the circuit in series with any other apparatus and is allowed to remain there as long as the current is flowing. It is then removed, and the chemical changes produced by the current are determined by some appropriate means.

The silver coulometer is commonly employed for precise work. This coulometer consists of a platinum dish serving as both cathode and cell vessel, and pure silver as anode. The electrolyte is an aqueous solution of purified silver nitrate. The dish is weighed before electrolysis, and the cell is assembled. After electrolysis the electrolyte is carefully decanted, the deposit of silver on the dish is thoroughly washed with distilled water, and the dish plus silver are dried and weighed. From the increase in weight is calculated then the quantity of electricity passed through the coulometer. With care such coulometers can give results accurate to 0.05 per cent or better.

The iodine coulometer[2] can also yield high precision. In this coulometer the iodine liberated from a solution of potassium iodide by passage of current is estimated by titration with sodium thiosulfate or arsenious acid. For less accurate work copper coulometers, consisting of copper electrodes in a solution of copper sulfate, are quite suitable. Here the copper deposited is estimated by weighing.

TRANSFERENCE AND TRANSFERENCE NUMBERS

Although current is transported through a solution by migration of positive and negative ions, the fraction of the total current carried by each

---

[2] For details on this and other coulometers see the book by MacInnes listed at end of the chapter.

is not necessarily the same. Thus in dilute solutions of magnesium sulfate the magnesium ion carries about 0.38 of the total current, while the sulfate ion carries the balance, or 0.62. Similarly, in dilute nitric acid solutions the nitrate ion carries only 0.16 of the total current, the hydrogen ion 0.84. The sulfate and hydrogen ions transport the greater fraction of the total current because in their respective solutions they move faster than the other ions present. If both ions in a solution moved with the same speed, each would transfer past any fixed plane in the solution the same quantity of electricity in any given time. However, when the speeds of two ions are not the same, in any given period of time the faster ion will carry past any plane a greater fraction of the current, and hence will perform a greater percentage of the total work involved in current transfer.

The quantitative relation between the fraction of the current carried by an ion and its speed can be established as follows. Consider two parallel plates d cm apart, Figure 11–2, across which is applied a potential of ε

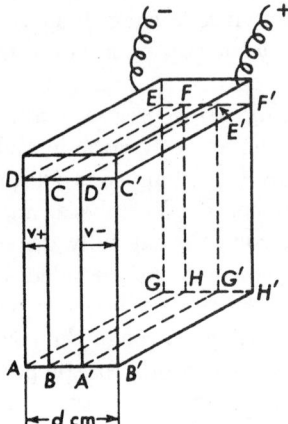

Figure 11–2. Relation of current to ionic velocities.

volts and between which is contained some volume of an electrolyte. Let the average migration velocity of the cation in this solution be v_+ cm per second, the charge of the ion be z_+, and the number of these ions n_+. Similarly, let the velocity of the anion be v_-, its charge z_-, and the number of these n_-. Then the quantity of electricity transported by the cation in 1 sec, i.e., the current due to the cation, is going to be all the electricity possessed by all the cations that lie within a distance v_+ of the negative plate, or, in other words, all those which lie within the volume $ABCDEFGH$. The number of these ions is obviously the fraction v_+/d of the total or n_+v_+/d. Again since the charge of each ion is z_+, and the quantity of electricity associated with unit charge is e, the electronic charge, the current carried by the positive ions, must be

$$I_+ = \frac{n_+ v_+ z_+ e}{d} \tag{6}$$

In a like manner, the current carried by the anions to the positive plate must be that corresponding to all the electricity possessed by the anions in the volume $A'B'C'D'E'F'G'H'$, namely,

$$I_- = \frac{n_- v_- z_- e}{d} \tag{7}$$

Consequently the total current carried by both ions is

$$I = I_+ + I_- = \frac{n_+ v_+ z_+ e + n_- v_- z_- e}{d} \tag{8}$$

But the condition for electroneutrality of the solution demands that the total charge of the cations must be equal to that of the anions, namely,

$$n_+ z_+ = n_- z_- \tag{9}$$

Therefore,

$$I = \frac{n_+ z_+ e v_+ + n_+ z_+ e v_-}{d}$$

$$= \frac{n_+ z_+ e (v_+ + v_-)}{d} \tag{10}$$

From Eqs. (6) and (10) the fraction of the total current carried by the cations, t_+, follows as

$$t_+ = \frac{I_+}{I} = \frac{n_+ v_+ z_+ e}{n_+ z_+ e (v_+ + v_-)}$$

$$= \left(\frac{v_+}{v_+ + v_-} \right) \tag{11}$$

while the fraction of the total current carried by the anions, t_-, follows from Eqs. (7) and (10) as

$$t_- = \frac{I_-}{I} = \frac{n_- v_- z_- e}{n_+ z_+ e (v_+ + v_-)} = \frac{n_+ z_+ e v_-}{n_+ z_+ e (v_+ + v_-)}$$

$$= \left(\frac{v_-}{v_+ + v_-} \right) \tag{12}$$

where t_+ and t_- are the *transport* or *transference numbers* of the cation and anion respectively. These numbers give the fraction of the total current carried by a given ion in a solution. On dividing Eq. (11) by Eq. (12) we see that

$$\frac{t_+}{t_-} = \frac{v_+}{v_-} \tag{13}$$

and hence the transport numbers of the ions, and therefore the fractions of the total current they carry, are directly proportional to their absolute velocities. When these are equal, $t_+ = t_-$ and both ions contribute equally to the transport of current. However, when v_+ does not equal v_-, t_+ will not equal t_-, and the two ions will carry different proportions of the total current. Still, no matter what the ratio between t_+ and t_- may be, since the two ions carry between them all of the current, the sum of the two transference numbers must be unity, i.e.,

$$t_+ + t_- = 1 \tag{14}$$

HITTORF'S RULE

As a result of passage of current through a solution concentration changes directly related to the velocities of the ions occur in the vicinity of the electrodes. To understand the nature of these changes and their dependence on the ionic speeds, consider the cell illustrated under I in Figure 11–3. Let this cell be divided by the imaginary planes AA' and

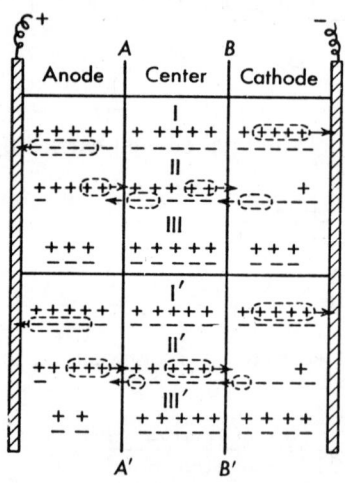

Figure 11–3. Concentration changes due to transference.

BB' into three compartments, anode, center, and cathode, and let each compartment contain five equivalents of positive ions and the same number of equivalents of negative ions. Assume, further, that the speeds of the two ions are equal. Now, if 4 faradays of electricity are passed through the cell, four equivalents of positive ions in the cathode compartment will accept electrons from the electrode and deposit. Again four equivalents of negative ions in the anode compartment will give up their electrons to the anode and will also deposit. The result of these changes is summarized under II. But while these changes at the electrodes are taking place, the ions migrate through the solution. Since in this case both ions move with the same speed, each must carry one-half the current; i.e., past the planes AA' and BB' the cations must transport 2 faradays of electricity from left to right, the anions 2 faradays of electricity from right to left. Consequently, two equivalents of positive ions must move from the anode compartment into the center, and the same number of equivalents from the center into the cathode compartment. Simultaneously, two equivalents of negative ions must move past these planes from the cathode compartment to the center, and also two of these from the center into the anode compartment. When these migrations are added to the changes at the electrodes, the final result is III. From III it is evident that the concentration in the central compartment has not been affected by the passage of current. On the other hand, the concentrations in the cathode and anode compartments have both decreased, but the decrease, two equivalents, is the *same* for both.

However, the situation is different when the speeds of the ions are not the same. Brackets I', II', III' in Figure 11–3 illustrate what happens when the speed of the cation is three times that of the anion, the quantity of electricity being again 4 faradays. As before, four equivalents of each ion are deposited at the electrodes. But, since the cation carries here three times as much current as the anion, three equivalents of positive ions must migrate from anode to center, and from center to cathode compartments. At the same time only one equivalent of anions leaves the cathode section, and only one equivalent leaves the center for the anode compartment. From III', where the final state of the cell is given, it is seen that again there is no change in concentration in the central compartment. Moreover, as before there are concentration changes at the two electrodes, but this time the two are not the same. In fact, the anode compartment has suffered a concentration change equal to three times that for the cathode. By repeating the above argument for various ionic speed ratios, it can readily be shown that the changes in concentration at the electrodes as a result of electrolysis will be equal only when the ionic speeds are the same. When these are different, so are the concentration changes at the electrodes.

An inspection of Figure 11–3 reveals further that for the case of *equal* speeds the loss in concentration of cations due to migration from the anode compartment is equal to the loss in concentration of anions due to migration from the cathode compartment. On the other hand, when $v_+:v_- = 3:1$, the loss in concentration of cations from the anode compartment due to migration is *three times* that of anions from the cathode section. This parallelism between concentration loss due to migration and the velocity of the ion responsible for it leads to *Hittorf's rule*, namely,

$$\frac{\text{Loss in cation equivalents at anode due to migration}}{\text{Loss in anion equivalents at cathode due to migration}} = \frac{v_+}{v_-} = \frac{t_+}{t_-} \quad (15)$$

Since the total current passed through the cell, expressed in equivalents, is proportional to $t_+ + t_- = 1$, direct consequences of this rule are also the statements:

$$\frac{\text{Loss in cation equivalents at anode due to migration}}{\text{Equivalents of current passed}} = \frac{t_+}{1} = t_+ \quad (16)$$

$$\frac{\text{Loss in anion equivalents at cathode due to migration}}{\text{Equivalents of current passed}} = \frac{t_-}{1} = t_- \quad (17)$$

DETERMINATION OF TRANSFERENCE NUMBERS

Experimentally, transference numbers may be determined by three different methods: (a) the *Hittorf method*, based on observation of the changes in concentration about the electrodes due to migration, (b) the *moving boundary method*, and (c) from *electromotive force measurements*. The first two of these will be described here. A typical setup for determination of transference numbers by the Hittorf method is shown in Figure 11–4. The apparatus consists of a transport cell A, in series with a silver

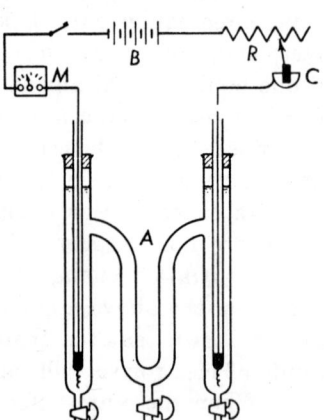

Figure 11–4. Transference apparatus for Hittorf method.

coulometer C, both connected to a battery B through the variable resistance R. The milliammeter M in the circuit permits adjustment of the current to any desired value and makes possible a rough estimation of the quantity of electricity passed through the cell. The cell is filled with the electrolyte to be investigated, the current is turned on, and the solution is electrolyzed sufficiently long to give an appreciable change in concentration about the electrodes. The current is then stopped, and the solution is drained from one or both of the electrode compartments, weighed, and analyzed. The quantity of electricity passed through the cell is obtained from the increase in weight of the cathode in the coulometer. If the original concentration is known, the solution in the central compartment may be ignored or analyzed to see whether any diffusion has taken place. If it is not known, the central compartment must be drained, weighed, and analyzed to yield the original composition of the solution. The following example illustrates the steps in a typical calculation of transport numbers from observed data. It is based on the assumption usually made that only the ions migrate, but not the water.

EXAMPLE: To determine the transport numbers of the ions in an exactly 0.2000 molal solution of copper sulfate, the cell was filled with the solution and electrolyzed between copper electrodes for some time. The cathode solution from the cell was found to weigh 36.4340 g and to contain 0.4415 g of copper. Further, the cathode in the coulometer showed an increase in weight of 0.0405 g due to deposited silver. From these data it is required to calculate the transference numbers of the copper and sulfate ions.

The 36.4340 g of final cathode solution contained 0.4415 g of copper, which is equivalent to $0.4415 \times CuSO_4/Cu = 1.1090$ g of copper sulfate. The weight of water in this solution was, then, $36.4340 - 1.1090 = 35.3250$ g. Now, since before electrolysis the solution was 0.2000 molal, each gram of water had associated with it $0.2000 \times CuSO_4/1000$ g of the salt. Hence 35.3250 g of water had associated with them initially

$$35.3250 \left(0.2000 \times \frac{CuSO_4}{1000} \right) = \frac{35.3250 \times 0.2000 \times 159.60}{1000}$$

$$= 1.1276 \text{ g } CuSO_4$$

Therefore, the loss in weight of the copper sulfate in cathode compartment is equal to $1.1276 - 1.1090 = 0.0186$ g, or $2 \times 0.0186/159.60 = 0.000233$ equivalent.

The total current passed through the cell is given by 0.0405 gram of silver, or $0.0405/107.87 = 0.000375$ equivalent. Consequently, the loss in copper sulfate due to deposition of copper on cathode should have been this number of equivalents. But, the actual loss was only 0.000233 equivalent, and hence the difference, $0.000375 - 0.000233 = 0.000142$ equivalent, must have migrated into this compartment. As the ion that migrates toward the cathode is Cu^{++}, this number of equivalents must have been transported by this ion, and therefore its transport number is

$$t_{Cu^{++}} = \frac{0.000142}{0.000375} = 0.379$$

while that of SO_4^{--} is

$$t_{SO_4^{--}} = 1 - 0.379 = 0.621$$

An independent check on these values could have been obtained by analyzing the anode solution and calculating the transport numbers therefrom. In this connection it should be remembered that copper dissolved at the anode, and hence an actual increase in copper concentration would have been observed there. However, owing to migration of copper ions out of the compartment, the increase will not be as large as may be anticipated from the quantity of electricity passed through the cell.

In the moving boundary method the motion of ions under the influence of an applied potential is observed directly rather than through the concentration changes at the electrodes. To understand the principle of this method, consider specifically the determination of the transference number of the hydrogen ion in hydrochloric acid. The apparatus required, Figure 11–5, is the same as in the Hittorf method except for the cell.

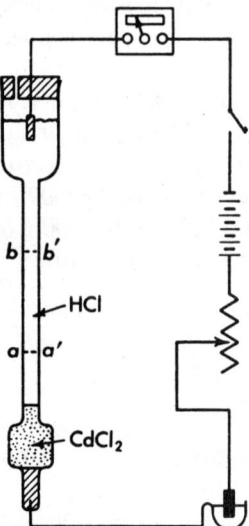

Figure 11–5. Transference numbers by moving boundary.

This time the cell consists of a tube mounted vertically and filled with cadmium chloride and the acid in the manner shown, so as to leave a sharp boundary between the two. The cathode, inserted at the top, is a platinum wire, while the anode at the bottom is a stick of cadmium metal. When current is turned on cadmium dissolves at the anode, hydrogen is evolved at the cathode, and hydrogen ions migrate upward through the cell. As these ions move toward the cathode their place is taken by cadmium ions, and hence the boundary between the two solutions also moves upward. By observing the volume swept out by the moving boundary for a given quantity of electricity passed through the cell, it is possible to calculate the transport number of the hydrogen ion as follows:

Suppose that the volume swept out by the boundary in moving from aa' to bb' during the passage of Q faradays of electricity, as measured in the coulometer, is V cc. Then, if the concentration of the acid is C equivalents per liter, the number of equivalents of acid carried toward the cathode by the current is $V \times C/1000$ equivalents. Since this number of equivalents is carried toward the cathode by hydrogen ions, and as the total current carried is Q equivalents, the transport number of the hydrogen ion, t_+, is given by

$$t_+ = \frac{V \times C}{1000\, Q} \tag{18}$$

Although as described here the method appears to be very simple, actually certain conditions must be met in choosing the solutions to be employed in connection with the one to be studied, and various corrections must be applied. For details the student is referred to the excellent discussion given by MacInnes,[3] who with Longsworth has made very valuable contributions to the technique and precision of this method.

In Table 11–2 are given the transference numbers of the cations for a

TABLE 11–2. TRANSFERENCE NUMBERS OF CATIONS

	Temp. (°C)	Concentration—Equivalents per Liter							
		0.005	0.01	0.02	0.05	0.10	0.20	0.50	1.00
HCl	0	0.847	0.846	0.844	0.839	0.834	—	—	—
	18	0.832	0.833	0.833	0.834	0.835	0.837	0.840	0.844
	25	—	0.825	0.827	0.829	0.831	0.834	—	—
HNO₃	20	0.839	0.840	0.841	0.844	—	—	—	—
H₂SO₄	20	—	—	0.822	0.822	0.822	0.820	0.816	0.812
NH₄Cl	25	—	0.491	0.491	0.491	0.491	0.491	—	—
AgNO₃	25	—	0.465	0.465	0.466	0.468	—	—	—
LiCl	25	—	0.329	0.326	0.321	0.317	0.311	0.300	0.287
KCl	25	0.490	0.490	0.490	0.490	0.490	0.489	0.489	0.488
	30	0.498	0.498	0.498	0.498	0.497	0.496	—	—
NaCl	25	—	0.392	0.390	0.388	0.385	0.382	—	—
NaOH	25	—	0.203	—	0.189	0.183	0.177	0.169	0.163
NaC₂H₃O₂	25	—	0.554	0.555	0.557	0.559	0.561	—	—
CaCl₂	25	—	0.426	0.422	0.414	0.406	0.395	—	—
CdSO₄	18	—	0.389	0.384	0.374	0.364	0.350	0.323	0.294
CuSO₄	18	—	—	0.375	0.375	0.373	0.361	0.327	—

[3] See reference at end of chapter.

number of electrolytes in aqueous solution. As far as can be observed, these numbers are not affected by current strength. They vary somewhat with concentration, but the variation is not large. As a rule the transference numbers that are large in dilute solution increase with rise in concentration, while those which are small decrease. However, there are exceptions to this rule. With increase in temperature the transport numbers of the cation and anion tend to equalize and approach a value of 0.5. Hence transference numbers greater than 0.5 decrease as the temperature is raised, while those less than 0.5 increase. This variation is illustrated in the data for hydrochloric acid and potassium chloride given in the table.

ELECTROLYTIC CONDUCTANCE

The resistance of an electrolytic conductor to current passage can be determined by the application of Ohm's law to such conductors. However, instead of resistance, it is customary here to speak of the *conductance*, which is merely the *reciprocal* of the electrical resistance.

As is well known, the resistance of any conductor is proportional directly to its length and inversely to its cross-sectional area, namely,

$$R = \rho \, \frac{l}{A} \tag{19}$$

where R is the resistance in ohms, l the length in centimeters, A the area in square centimeters, and ρ the *specific resistivity*. The value of ρ depends on and is characteristic of the nature of the conductor. From Eq. (19) the expression for the corresponding conductance L follows as

$$L = \frac{1}{R} = \frac{1}{\rho}\left(\frac{A}{l}\right)$$
$$= L_s\left(\frac{A}{l}\right) \tag{20}$$

where $L_s = 1/\rho$ is the *specific conductance* of the conductor. This quantity may be considered to be the conductance of 1 cm cube (not cc) of material, and is expressed in reciprocal ohms or *mhos* per centimeter.

Although the specific conductance is a property of the conducting medium, in dealing with solutions of electrolytes a quantity of greater significance is the *equivalent conductance*, Λ. The equivalent conductance of an electrolyte is defined as the conductance of a volume of solution containing one equivalent weight of dissolved substance when placed between two parallel electrodes 1 cm apart, and large enough to contain between them all of the solution. Λ is never determined directly, but is calculated from the specific conductance. If C is the concentration of a solution in gram equivalents per liter, then the concentration per cubic centimeter is $C/1000$, and the volume containing one equivalent of the solute is, therefore, $1000/C$ cc. Since L_s is the conductance of a centimeter cube of the solution, the conductance of $1000/C$ cc, and hence Λ, will be

$$\Lambda = \frac{1000\,L_s}{C} \tag{21}$$

Equation (21) is the defining expression for the equivalent conductance. It must be remembered that C in this equation is in *equivalents of solute per liter of solution*.

DETERMINATION OF CONDUCTANCE

The problem of obtaining Λ reduces itself to a determination of the specific conductance of the electrolyte, and this, in turn, to a measurement of the resistance of the solution and use of Eq. (20). For measuring resistances of electrolytic solutions the Wheatstone bridge method is employed, a schematic diagram of which is shown in Figure 11–6. R_x, the

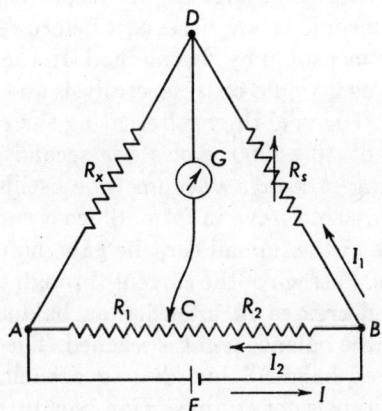

Figure 11–6. Principle of direct-current Wheatstone bridge.

unknown resistance whose value is to be determined, is placed in one arm of the bridge, a variable known resistance R_s in the other. AB is a uniform slide wire across which moves a contact point C. To balance the bridge the contact is moved along this resistance until no current from the battery E flows through the galvanometer G. When this condition is reached, R_s, the resistance from A to C, R_1, and that from C to B, R_2, are read. R_x is then calculated from these according to the following considerations.

When the current I from the battery reaches point B it divides into two parallel paths and into the two currents, I_1 and I_2. These currents lead to potential drops across the resistances they traverse. The purpose of balancing the bridge is to find a point along AB, namely C, such that the potential drop from B to C is equal to that from B to D. When this point is found, D and C are at the same potential and no current flows through the branch containing the galvanometer. The condition for bridge balance is, then,

$$R_s I_1 = R_2 I_2$$

But, when these IR drops are equal, those from D to A and from C to A must also be equal. Hence

$$R_x I_1 = R_1 I_2$$

Dividing now the second of these equalities by the first, we obtain

$$\frac{R_x}{R_s} = \frac{R_1}{R_2}$$

and, therefore,

$$R_x = R_s \left(\frac{R_1}{R_2}\right) \qquad (22)$$

Although the principle of the Wheatstone bridge as just given remains the same, several modifications in technique are necessary before resistances of electrolytic solutions can be measured by this method. In the first place, direct current cannot be used, as it would cause electrolysis and concentration changes at the electrodes. To avoid these, alternating current is employed, usually at a frequency of 500–2000 cycles per second, and supplied by either a vibrating tuning fork or a vacuum tube oscillator. This current should be close to a pure sine wave in form. Since a current of this frequency is within the range of the human ear, the galvanometer can be replaced by a set of earphones. Passage of the current through these produces a buzzing sound which decreases in intensity as balance is approached and is a minimum when the balance point is reached. Theoretically the sound should be zero at balance, but due to capacitance introduced by the cell, such an ideal state is not attained. However, by placing a variable condenser across the standard resistance, it is possible to sharpen the balancing by adjusting the condenser to the capacitance of the cell, thus neutralizing the two to a degree. An alternate detecting device employed at present is the cathode-ray oscilloscope. By its use a balance of both resistance and capacitance can be obtained.

The cells employed for conductance work are of various types and shapes, depending on the purpose and on the accuracy required. They are constructed of glass, with electrodes of either platinum or gold. To overcome imperfections in the current and other effects at the electrodes, the latter are coated electrolytically from a solution of chloroplatinic acid with a thin layer of finely divided platinum, called *platinum black* because of its color. The distance apart the electrodes are placed in a cell is determined by the conductance of the solution to be measured. For solutions of high conductance the electrodes are widely spaced, while for poorly conducting solutions the electrodes are mounted near each other. In Figure 11–7 are shown several cells commonly encountered in the laboratory.

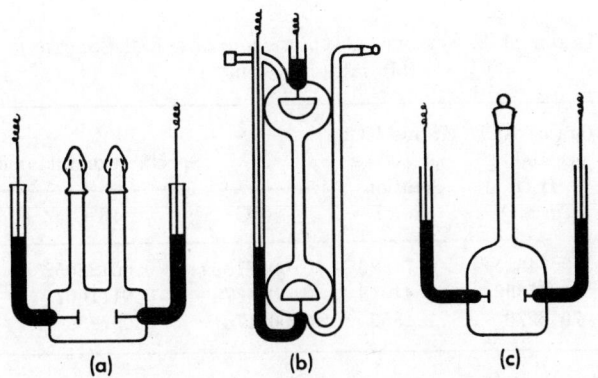

(a)   (b)   (c)

Figure 11–7.   Types of conductivity cells.

According to Eq. (20), the specific conductance of any electrolytic conductor is given by

$$L_s = \left(\frac{l}{A}\right) L = \left(\frac{l}{A}\right) \frac{1}{R} \tag{23}$$

and hence before L_s can be calculated from the measured resistance the ratio (l/A) for the particular cell used is required. For any given cell this ratio is a fixed quantity called the *cell constant K*. To obtain the value of the cell constant it is not necessary to determine l and A. Instead, a solution of known L_s is placed in the cell, the resistance is measured, and K is calculated. Once K is available, the specific conductance of any other solution whose resistance is observed in the same cell follows from

$$L_s = \frac{K}{R} \tag{24}$$

To determine cell constants either 1, 0.1, or 0.01 *demal* solutions of potassium chloride are used. A demal solution is a solution containing 1 g mole of salt per *cubic decimeter* of solution at 0°C, or 76.6276 g of potassium chloride in 1000 g of water, both weighed in air. The 0.1 and 0.01 demal solutions contain, respectively, 7.47896 and 0.74625 g of potassium chloride per 1000 g of water. The conductances of these solutions have been measured with great accuracy in cells with electrodes of definite area and definite spacing, and their specific conductances are well known. The most concentrated solution is used only for cells with electrodes far apart, the other two for electrodes close together or intermediate in spacing. Table 11–3 gives the specific conductances of these solutions at several temperatures as determined by Jones and Bradshaw.[4] The calculation of cell constants, and specific and equivalent conductances from experimental data may be illustrated with the following example.

TABLE 11–3. SPECIFIC CONDUCTANCES OF KCl SOLUTIONS
(In int. ohm^{-1} cm^{-1})

Demal Conc.	Grams KCl per 1000 g H$_2$O (in air)	Grams KCl per 1000 g Solution (in air)	Specific Conductance		
			0°C	18°C	25°C
0.01	0.74625	0.74526	0.00077364	0.00122052	0.00140877
0.10	7.47896	7.41913	0.0071379	0.0111667	0.0128560
1.00	76.6276	71.1352	0.065176	0.097838	0.111342

EXAMPLE: In a particular cell a 0.01 demal solution of potassium chloride gave a resistance of 150.00 ohms at 25°C, while a 0.01 N solution of hydrochloric acid gave a resistance of 51.40 ohms at the same temperature. At 25°C the specific conductance of 0.01 demal potassium chloride is 0.0014088, and hence the cell constant for the given cell is

$$K = L_s R = 0.0014088 \times 150.00$$
$$= 0.21132$$

From this K and the resistance of the hydrochloric acid solution the specific conductance of the latter follows as

$$L_s = \frac{K}{R} = \frac{0.21132}{51.40}$$
$$= 0.004111 \text{ mho cm}^{-1}$$

and the equivalent conductance as

$$\Lambda = \frac{1000 \, L_s}{C}$$
$$= \frac{0.004111 \times 1000}{0.01}$$
$$= 411.1 \text{ mho cm}^2 \text{ equiv}^{-1}$$

[4] Jones and Bradshaw, *J. Am. Chem. Soc.*, **55**, 1780 (1933).

VARIATION OF CONDUCTANCE WITH CONCENTRATION

Both the specific and equivalent conductances of a solution vary with concentration. For strong electrolytes at concentrations up to several equivalents per liter, the specific conductance increases sharply with increase in concentration. In contrast, the specific conductances of weak electrolytes start at lower values in dilute solutions and increase much more gradually. In both instances the increase in conductance with concentration is due to the increase in the number of ions per unit volume of solution. In strong electrolytes the number of ions per cubic centimeter increases in proportion to the concentration. In weak electrolytes, however, the increase cannot be quite so large because of the changing partial ionization of the solute, and consequently the conductance does not go up so rapidly as in strong electrolytes.

TABLE 11–4.   EQUIVALENT CONDUCTANCES OF ELECTROLYTES IN AQUEOUS SOLUTION AT 25°C

C Equiv./ liter	KCl	HCl	$AgNO_3$	$\frac{1}{2} H_2SO_4$	$\frac{1}{2} BaCl_2$	$HC_2H_3O_2$
0.0000	149.86	426.16	133.36	429.6	139.98	390.7
0.0001	—	—	—	—	—	134.7
0.0005	147.81	422.74	131.36	413.1	135.96	67.7
0.001	146.95	421.36	130.51	399.5	134.34	49.2
0.005	143.55	415.80	127.20	364.9	128.02	22.9
0.01	141.27	412.00	124.76	336.4	123.94	16.3
0.02	138.34	407.24	121.41	308.0	119.09	11.6
0.05	133.37	399.09	115.24	272.6	111.48	7.4
0.10	128.96	391.32	109.14	250.8	105.19	—
0.20	123.9	379.6	101.8	234.3	98.6	—
0.50	117.2	359.2	—	222.5	88.8	—
1.00	111.9	332.8	—	—	80.5	—

Unlike the specific conductance, the equivalent conductance Λ of both strong and weak electrolytes increases *with dilution*. The reason for this is that the decrease in specific conductance is more than compensated by the increase in the value of $1/C$ on dilution, and hence Λ goes up. The manner in which Λ varies with concentration can be judged from Table 11–4, and from the plot of Λ vs. $\sqrt{C}$ given in Figure 11–8. It may be seen from these that the Λ's for strong and weak electrolytes behave differently on decrease of concentration. On dilution of a strong electrolyte Λ rapidly approaches already in 0.001 or 0.0001 N solutions a value close to the limiting value of the conductance at zero concentration, Λ_0. On the other hand, although the equivalent conductance of weak electrolytes increases rapidly on dilution, at the concentrations mentioned it is still very far from its limit. For instance, Λ at 25°C for 0.001 N sodium chloride is 123.7 as against a Λ_0 of 126.5. At the same concentration and temperature Λ for acetic acid is only 49.2, while Λ_0 is 390.7. Because of this fundamental difference in the behavior of the equivalent conductances of strong and weak electrolytes on dilution, quite diverse procedures must be used for obtaining their limiting equivalent conductances.

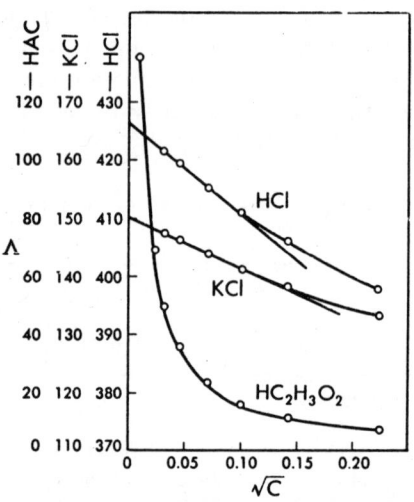

Figure 11-8.   Plot of Λ vs. $\sqrt{C}$ for strong and weak electrolytes.

EQUIVALENT CONDUCTANCES AT INFINITE DILUTION

Kohlrausch was the first to point out that when Λ for *strong electrolytes* is plotted against $\sqrt{C}$ the curve approaches linearity in dilute solutions, i.e., in dilute solutions

$$\Lambda = \Lambda_0 - b\sqrt{C} \tag{25}$$

where b is a constant. The validity of this finding may be seen from the plots for hydrochloric acid and potassium chloride in Figure 11-8. Consequently, to obtain Λ_0 of such electrolytes the curve may be extrapolated to $\sqrt{C} = 0$ and the intercept read, or the slope of the linear portion of the curve may be obtained from the plot, and Λ_0 solved for from Eq. (25). Some Λ_0 values are given in the first horizontal column of Table 11-4.

The above method for evaluation of Λ_0 cannot be used for weak electrolytes. As may be seen from Figure 11–8, the plot of Λ vs. $\sqrt{C}$ for the weak electrolyte acetic acid does not approach linearity in solutions as dilute as 0.0001 N. Instead, Λ exhibits a very rapid increase with decrease in concentration. Again, it is not practicable to carry the measurements to concentrations much lower than 0.0001 N, for in such dilute solutions the conductance of the water becomes an appreciable part of the total. Although the conductance of the water may be subtracted from the specific conductance of the solution to yield that of the electrolyte, such corrections are not always satisfactory and introduce uncertainty into the final result.

Equivalent conductances at infinite dilution for weak electrolytes are obtained by application of *Kohlrausch's law of independent migration of ions*. This law states that at infinite dilution, where dissociation for all electrolytes is complete and where all interionic effects disappear, each ion migrates independently of its co-ion, and contributes to the total equivalent conductance of an electrolyte a definite share which depends only on its own nature and not at all on that of the ion with which it is associated. If this be the case, then Λ_0 of any electrolyte should be the sum of the equivalent conductances of the ions composing it, provided, of course, that the solvent and temperature are the same. Evidence for the validity of this statement is given in Table 11–5. According to this law the difference between the Λ_0's of electrolytes containing a common ion should

TABLE 11–5. KOHLRAUSCH'S LAW OF INDEPENDENT MIGRATION OF IONS

Electrolyte	Λ_0 (25°C)	Difference	Electrolyte	Λ_0 (25°C)	Difference
KCl	149.9		HCl	426.2	
LiCl	115.0	34.9	HNO$_3$	421.3	4.9
KNO$_3$	145.0		KCl	149.9	
LiNO$_3$	110.1	34.9	KNO$_3$	145.0	4.9
KOH	271.5		LiCl	115.0	
LiOH	236.7	34.8	LiNO$_3$	110.1	4.9

be a constant equal to the difference in the equivalent conductances of the ions not in common. In line with this requirement we see that, irrespective of the nature of the co-ion, the difference between the conductances of K^+ and Li^+ is constant, and the same is true for the difference in conductance between the Cl^- and NO_3^- ions. The law has been tested on other ions as well, and with the same results.

From the law of independent migration of ions it follows that Λ_0 for any electrolyte may be written as

$$\Lambda_0 = l_+^0 + l_-^0 \tag{26}$$

where l_+^0 and l_-^0 are the *equivalent ionic conductances at infinite dilution* of the cation and anion respectively. Moreover, since the fraction of the total current carried by any ion is given by its transport number, this number must represent as well the fraction of the total conductance due to the ion. Consequently l_+^0 and l_-^0 are also related to Λ_0 by the relations

$$l_+^0 = t_+^0 \Lambda_0 \tag{27}$$

$$l_-^0 = t_-^0 \Lambda_0 \tag{28}$$

t_+^0 and t_-^0 being the transference numbers at infinite dilution as obtained by extrapolation. These equations permit ready calculation of the limiting ionic conductances from transference numbers and Λ_0 values of strong electrolytes. For example, Λ_0 for hydrochloric acid at 25°C is 426.16, while t_+^0 of the hydrogen ion is 0.821. Therefore,

$$l_{H^+}^0 = 0.821 \times 426.16 = 349.9$$
$$l_{Cl^-}^0 = 0.179 \times 426.16 = 76.3$$

The equivalent ionic conductances of other ions have similarly been evaluated. These are summarized in Table 11–6. The ionic conductances of ions of weak acids or bases were deduced from the Λ_0's of their salts, which are strong electrolytes.

Through Eq. (26) Λ_0 values of *strong and weak electrolytes* follow from Table 11–6 on addition of the appropriate ionic conductances of the cation and anion. Thus we find for Λ_0 of acetic acid, a weak electrolyte,

$$
\begin{aligned}
\Lambda_{0(HAc)} &= l_{H^+}^0 + l_{Ac^-}^0 \\
&= 349.8 + 40.9 \\
&= 390.7 \text{ mhos cm}^2 \text{ equiv}^{-1}
\end{aligned}
$$

In arriving at the equivalent ionic conductances given in Table 11–6, and from these at the Λ_0's of electrolytes, transference numbers were used. However, it is possible to obtain Λ_0's of electrolytes by direct addition and subtraction of appropriate Λ_0 values without using transport numbers. Thus, if Λ_0 for hydrochloric acid is added to that for sodium acetate, and that for sodium chloride subtracted, the result is Λ_0 for acetic acid,

$$
\begin{aligned}
\Lambda_{0(NaAc)} + \Lambda_{0(HCl)} - \Lambda_{0(NaCl)} &= l_{Na^+}^0 + l_{Ac^-}^0 + l_{H^+}^0 + l_{Cl^-}^0 - l_{Na^+}^0 - l_{Cl^-}^0 \\
&= l_{H^+}^0 + l_{Ac^-}^0 \\
&= \Lambda_{0(HAc)}
\end{aligned}
$$

Similarly, Λ_0 of nitric acid follows from the Λ_0's of potassium nitrate, potassium chloride, and hydrochloric acid. This method of calculating Λ_0

is particularly valuable with weak electrolytes for whose salt solutions transport numbers are not available, and neither are the equivalent ionic conductances at infinite dilution for both ions. From the example cited it is evident that all that is necessary to obtain Λ_0 for a weak acid for which no data are extant is Λ_0 of its sodium salt. Since the salt is a strong electrolyte, its Λ_0 can be evaluated from the measured equivalent conductances without any special difficulty. By combining this Λ_0 with

TABLE 11-6. EQUIVALENT IONIC CONDUCTANCES
AT INFINITE DILUTION
(25°C)

Cations	l_+^0	Anions	l_-^0
K^+	73.52	Cl^-	76.34
Na^+	50.11	Br^-	78.4
Li^+	38.69	I^-	76.8
NH_4^+	73.4	NO_3^-	71.44
H^+	349.82	HCO_3^-	44.48
Ag^+	61.92	OH^-	198
Tl^+	74.7	$Acetate^-$	40.9
$\frac{1}{2} Ca^{++}$	59.50	$Chloracetate^-$	39.7
$\frac{1}{2} Ba^{++}$	63.64	ClO_4^-	68.0
$\frac{1}{2} Sr^{++}$	59.46	$\frac{1}{2} SO_4^{--}$	79.8
$\frac{1}{2} Mg^{++}$	53.06	$\frac{1}{3} Fe(CN)_6^{---}$	101.0
$\frac{1}{3} La^{+++}$	69.6	$\frac{1}{4} Fe(CN)_6^{----}$	110.5

that of hydrochloric acid and sodium chloride, as shown above, the Λ_0 for the acid follows immediately. Similarly, Λ_0 for a weak base like ammonium hydroxide can be calculated from Λ_0 of ammonium chloride, sodium hydroxide, and sodium chloride.

EFFECT OF OTHER FACTORS ON CONDUCTANCE

The conductance of all electrolytes increases with temperature. The variation of Λ_0 with temperature can be represented by the equation

$$\Lambda_{0_{(t)}} = \Lambda_{0_{(25°C)}}[1 + \beta(t - 25)] \tag{29}$$

where $\Lambda_{0_{(t)}}$ is the limiting equivalent conductance at $t°C$, $\Lambda_{0_{(25°C)}}$ that at 25°C, and β a constant. For salts β is usually 0.022 to 0.025, for acids 0.016 to 0.019. Similar behavior is exhibited by the equivalent conductances of strong electrolytes in finite concentrations. However, with weak electrolytes the variation of Λ with temperature is not so regular, for in these not only do the velocities of the ions and the interionic forces change, but also the degree of dissociation.

The conductance behavior observed in nonaqueous solvents depends pretty much on the dielectric constant of the medium. The dielectric constant of water is high, 78.6 at 25°C, whereas that of most other solvents is considerably lower. Thus the dielectric constants at 25°C for methyl alcohol, ethyl alcohol, and dioxane are, respectively, 31.5, 24.3, and 2.2. As the dielectric constant of a solvent is lowered, the conductance of an electrolyte in the medium also decreases. Beyond this conductance drop, and beyond the fact that some electrolytes that are strong in water may be weak in other solvents, the conductance behavior of these substances is not very different in nonaqueous solvents from that in water, provided the dielectric constant is above about 25. For example, the halides and nitrates of alkali metals, the thiocyanates of the alkali and alkaline earth metals, and the tetraalkyl ammonium salts are strong electrolytes in ethyl and methyl alcohols and behave pretty much as do strong electrolytes in water. Again, picric acid, various substituted acetic acids, and phenols are weak electrolytes in methyl alcohol, as are also hydrochloric, hydrobromic, and hydriodic acids in ethyl alcohol.

In solvents of dielectric constant less than 25, the dependence of the equivalent conductance on the concentration becomes complex. Plots of $\log \Lambda$ vs. $\log C$, instead of being linear or slightly curved as they are in solvents of higher dielectric constant, contain minima which appear at lower concentrations the lower the dielectric constant. To explain these minima and the curves in general it has been suggested[5] that in these solvents ions exhibit a tendency to associate into complexes such as A^+B^-, $A^+B^-A^+$, and $B^-A^+B^-$, which decrease the number of ions available to carry current, and hence the conductance. These theories seem to account fairly well for the observed phenomena.

INTERIONIC ATTRACTION THEORY OF CONDUCTANCE

The decrease in the equivalent conductance with increase in concentration for weak electrolytes can be explained as due essentially to a decrease in the degree of ionization. However, such an explanation cannot apply to strong electrolytes, for these, at least in the more dilute solutions, are completely dissociated. Consequently, to account for the variation of Λ with concentration in strong electrolytes some other explanation must be sought, and this is found in the Debye-Hückel-Onsager theory of conductance.

According to the Debye-Hückel theory of interionic attraction, as developed briefly in Chapter 9, each ion in solution is surrounded by

[5] Kraus and Fuoss, *J. Am. Chem. Soc.*, **55**, 21 (1933); **55**, 476, 1019, 2387 (1933); **57**, 1 (1935).

an atmosphere of other ions whose net charge is on the average opposite to that of the central ion. When the ions have no external force applied upon them, this atmosphere is spherically and symmetrically distributed about the ion. However, when an external force is imposed, as when a potential is applied across two electrodes immersed in the solution during conductance, the ions are set in motion, and as a consequence certain effects and changes in the ionic atmosphere arise which result in a decrease in the speeds of the ions. Debye and Huckel[6] first pointed out that these effects are twofold, namely, (a) the *relaxation of the ionic atmosphere* due to an applied potential, and (b) the *electrophoretic effect*. The first of these arises from the fact that any central ion and its atmosphere are oppositely charged, i.e., when the central ion is positively charged the atmosphere is negative, and vice versa. Because of this difference in sign of the atmosphere and central ion, a potential applied across the combination will tend to move the central ion in one direction, the atmosphere in the other. Thus a central positive ion will tend to move toward the cathode while its ionic atmosphere will tend toward the anode. The symmetry of the atmosphere about an ion is destroyed by these opposing tendencies, and the atmosphere becomes distorted. In this state the force exerted by the atmosphere on the ion is no longer uniform in all directions, but is greater *behind* the ion than in front of it. Consequently, the ion experiences a retarding force opposite to the direction of its motion, and the ion is slowed down by these interionic attractions.

The electrophoretic effect arises from the fact that an ion, in moving through the solution, does not travel through a stationary medium, but through one that moves in a direction opposite to that of the ion. Ions are generally solvated, and when these move, they carry with them solvent. Any positive ion migrating toward the cathode has, then, to thread its way through medium moving with the negative ions toward the anode. Similarly, negative ions have to migrate through molecules of solvent carried by positive ions in the opposite direction. These countercurrents make it more difficult for the ion to move through the solution, and thus slow it down in the same way as swimming against the current in a river would slow down a swimmer.

Debye and Hückel showed that both of these retarding effects on an ion produce a decrease in equivalent conductance dependent on the concentration. Their mathematical treatment was subsequently extended by Lars Onsager[7] to include not only the relaxation and electrophoretic effects, but also the natural Brownian movement of the ions. Onsager obtained the following equation for the dependence of the equivalent

[6] P. Debye and E. Huckel, *Physik. Zeit.*, **24**, 185, 305 (1923).
[7] L. Onsager, *Physik. Zeit.*, **27**, 388 (1926); **28**, 277 (1927).

conductance of a binary strong electrolyte on the concentration

$$\Lambda = \Lambda_0 - \left[\frac{0.9834 \times 10^6}{(DT)^{3/2}} w\Lambda_0 + \frac{28.94(z_+ + z_-)}{\eta(DT)^{1/2}} \right] \sqrt{(z_+ + z_-)C} \quad (30)$$

where

$$w = z_+ z_- \left(\frac{2q}{1 + \sqrt{q}} \right)$$

$$q = \frac{z_+ z_- \Lambda_0}{(z_+ + z_-)(z_+ l_-^0 + z_- l_+^0)}$$

In this equation Λ, Λ_0, C, l_+^0, and l_-^0 have the same significance as before, z_+ and z_- are the charges of the two ions, T is the absolute temperature, and D and η are the dielectric constant and viscosity of the solvent, respectively. Because of simplifications in derivation this equation is applicable only to very dilute solutions, and it must be considered as essentially a limiting equation for conductance. *For the special case of 1-1 electrolytes in water at 25°C*, for which $z_+ = z_- = 1$, $D = 78.55$, and $\eta = 0.008949$ poise, Eq. (30) reduces to

$$\Lambda = \Lambda_0 - [\theta \Lambda_0 + \sigma] \sqrt{C} \quad (31)$$

where θ and σ are constants with values $\theta = 0.2273$, and $\sigma = 59.78$. Since the quantity in brackets is a constant, Eq. (31) is identical in form with Eq. (25) proposed by Kohlrausch on empirical grounds.

Equation (31) has the correct form for the dependence of Λ on $\sqrt{C}$. The question remaining is whether the experimental slope is in accord with that predicted by the Onsager equation, namely, slope = $[\theta \Lambda_0 + \sigma]$. Exhaustive tests by Shedlovsky, MacInnes, and others indicate that in very dilute solutions the Onsager equation is in agreement with observation not only for 1-1 strong electrolytes, but also for electrolytes of higher valence types, such as calcium chloride and lanthanum chloride. The concordance obtained may be judged from the fact that for potassium chloride Λ_0 found by graphical extrapolation is 149.86, while that calculated by Eq. (31) from the experimental data is 149.98.

The Onsager equation is applicable not only to aqueous solutions, but also to strong electrolytes in other solvents. However, the concordance usually is not as good as in water. In some instances the slopes predicted for the Λ vs. $\sqrt{C}$ plots deviate from the experimental quite appreciably and may be off as much as 100 per cent or over, as 126 per cent for silver nitrate in ethyl alcohol at 25°C.

ABSOLUTE VELOCITIES OF IONS

The absolute velocity with which any ion moves through a solution depends on the nature of the ion, the concentration of the solution, the temperature, and the applied potential drop per centimeter of conducting path. Such velocities can be measured directly by application of the moving boundary method. However, the velocities may also be calculated from conductance measurements, and to this end we turn now our attention.

Consider a pair of parallel electrodes d cm apart, such as those illustrated in Figure 11–2, across which is applied a potential difference ε, and between which is contained a volume of solution containing 1 mole of an electrolyte A_xB_y. For the sake of generality, consider this electrolyte to be ionized to an extent α according to the equation

$$A_xB_y = xA^{z_+} + yB^{z_-}$$
$$1(1 - \alpha) \qquad x\alpha \qquad y\alpha \qquad\qquad (32)$$

where z_+ and z_- are the valences of the ions, while x and y are the numbers of these obtained from one molecule of A_xB_y. The quantities of the various species are, then, as indicated in Eq. (32), and the number of positive ions present between the two electrodes is $n_+ = x\alpha N$, where $N = $ Avogadro's number. Now, according to Eq. (10) of this chapter, the current I flowing between the plates is given by

$$I = \frac{n_+z_+e(v_+ + v_-)}{d}$$

Substituting for n_+ its equivalent $x\alpha N$, we have

$$I = \frac{x\alpha Nz_+e(v_+ + v_-)}{d}$$
$$= \frac{\alpha\mathfrak{F}(xz_+)(v_+ + v_-)}{d} \qquad\qquad (33)$$

where $\mathfrak{F}$, the value of the faraday, was substituted for Ne.

One mole of A_xB_y corresponds to xz_+ equivalents of substance, i.e., the number of positive ions per molecule multiplied by the valence. Therefore, if the concentration of the solution is C equivalents per 1000 cc of solution, the volume per equivalent is $1000/C$, and per mole $(xz_+1000)/C$ cc. The latter volume is also equal to Ad, where A is the area of one of the parallel plates. Consequently,

$$Ad = \frac{1000xz_+}{C}$$

and

$$\frac{AC}{1000} = \frac{xz_+}{d} \qquad\qquad (34)$$

Now, the conductance of the solution is given by Eq. (20), namely, $L = L_sA/d$. Again, by Eq. (21) $L_s = C\Lambda/1000$. Hence we may write for L

$$L = \frac{L_s A}{d} = \left(\frac{AC}{1000}\right)\frac{\Lambda}{d}$$

$$= \frac{x z_+ \Lambda}{d^2} \tag{35}$$

in view of Eq. (34). Further, by Ohm's law $\mathcal{E}/R = \mathcal{E}L = I$, and so on substitution of Eq. (35) for L and Eq. (33) for I we get

$$\frac{\mathcal{E}(x z_+)\Lambda}{d^2} = \frac{\alpha \mathcal{F}(x z_+)(v_+ + v_-)}{d}$$

or

$$\left(\frac{\mathcal{E}}{d}\right)\Lambda = \alpha \mathcal{F}(v_+ + v_-) \tag{36}$$

If we set now $\mathcal{E}/d = \mathcal{E}'$, $\mu_+ = v_+/\mathcal{E}'$, and $\mu_- = v_-/\mathcal{E}'$, then Eq. (36) becomes

$$\Lambda = \alpha \mathcal{F}(\mu_+ + \mu_-) \tag{37}$$

The quantity $\mathcal{E}'$ is the *voltage gradient*, i.e., the potential drop per centimeter of path between the electrodes. Again, the quantities μ_+ and μ_-, called the *ionic mobilities*, represent the velocities of the ions in centimeters per second when the potential gradient is 1 volt per centimeter.

At infinite dilution $\alpha = 1$, $\Lambda = \Lambda_0$, $\mu_+ = \mu_+^0$, and $\mu_- = \mu_-^0$. Equation (37) for infinite dilution becomes thus

$$\Lambda_0 = \mathcal{F}\mu_+^0 + \mathcal{F}\mu_-^0 \tag{38}$$

Again, for infinite dilution Kohlrausch's law of independent migration of ions yields

$$\Lambda_0 = l_+^0 + l_-^0 \tag{26}$$

and hence we get on comparison of the two expressions

$$\mu_+^0 = \frac{l_+^0}{\mathcal{F}} \tag{39}$$

and

$$\mu_-^0 = \frac{l_-^0}{\mathcal{F}} \tag{40}$$

i.e., *at infinite dilution the velocity of any ion in centimeters per second, under a potential drop of 1 volt per centimeter, is given by the limiting equivalent conductance of the ion divided by the value of the faraday.*

The limiting mobilities of a number of ions calculated with the aid of Eqs. (39) and (40) from the equivalent ionic conductances are given in Table 11–7. It will be observed that the velocities are unusually low. Further, with the exception of the hydrogen and hydroxyl ions, the speeds are not very different for the various ions. The hydrogen and hydroxyl ions are unique in their high mobility, and this fact is the basis of several applications of conductance measurements.

TABLE 11-7.   ABSOLUTE MOBILITIES OF IONS AT 25°C
(cm² volt⁻¹ sec⁻¹)

Cation	μ_+^0	Anion	μ_-^0
K⁺	0.000762	Cl⁻	0.000791
Na⁺	0.000520	Br⁻	0.000812
Li⁺	0.000388	I⁻	0.000796
NH₄⁺	0.000760	NO₃⁻	0.000740
H⁺	0.003620	HCO₃⁻	0.000461
Ag⁺	0.000642	OH⁻	0.002050
Tl⁺	0.000774	C₂H₃O₂⁻	0.000424
Ca⁺⁺	0.000616	C₃H₅O₂⁻	0.000411
Ba⁺⁺	0.000659	ClO₄⁻	0.000705
Sr⁺⁺	0.000616	SO₄⁻⁻	0.000827
Mg⁺⁺	0.000550	Fe(CN)₆⁻⁻⁻	0.001040
La⁺⁺⁺	0.000721	Fe(CN)₆⁻⁻⁻⁻	0.001140

If Eq. (37) is rewritten

$$\Lambda = \alpha\mathfrak{F}\mu_+ + \alpha\mathfrak{F}\mu_- \tag{37a}$$

the new form suggests that Λ at *finite* concentration can also be represented by

$$\Lambda = l_+ + l_- \tag{41}$$

where l_+ and l_- are now the equivalent ionic conductances for the concentration to which Λ corresponds. Analogously to Eqs. (27) and (28) these equivalent conductances of the ions can be evaluated from Λ and the transport numbers by the relations

$$l_+ = t_+\Lambda \tag{42}$$
$$l_- = t_-\Lambda \tag{43}$$

By comparison of Eqs. (37a) and (41), the mobilities of the ions at any given concentration follow as

$$\mu_+ = \frac{l_+}{\alpha\mathfrak{F}} \tag{44}$$

$$\mu_- = \frac{l_-}{\alpha\mathfrak{F}} \tag{45}$$

for weak electrolytes, and

$$\mu_+ = \frac{l_+}{\mathfrak{F}} \tag{46}$$

$$\mu_- = \frac{l_-}{\mathfrak{F}} \tag{47}$$

for strong electrolytes, since these are completely ionized.

DEGREE OF IONIZATION AND CONDUCTANCE

Arrhenius ascribed the decrease in Λ with increase in concentration entirely to variation of the degree of dissociation of the electrolyte. If this is correct, it can readily be shown that the degree of dissociation α must be given by the ratio Λ/Λ_0. From Eqs. (37) and (38), Λ/Λ_0 for any electrolyte follows as

$$\frac{\Lambda}{\Lambda_0} = \frac{\alpha(\mu_+ + \mu_-)}{(\mu_+^0 + \mu_-^0)} \tag{48}$$

Assuming, further, that the mobilities of the ions at any finite concentration are the same as at infinite dilution, then $(\mu_+ + \mu_-) = (\mu_+^0 + \mu_-^0)$, and, therefore,

$$\alpha = \frac{\Lambda}{\Lambda_0} \tag{49}$$

Accordingly the degree of dissociation of any electrolyte is unity at infinite dilution. On the other hand, as the concentration is increased Λ decreases, and so does the value of α.

Equation (49) cannot apply to strong electrolytes, for in these the decrease in Λ with concentration is due to interionic attraction and not partial dissociation. However, within a small correction, Λ/Λ_0 does give the degree of dissociation of weak electrolytes. In weak electrolytes the variation of Λ with concentration is due to two factors, namely, (a) the partial dissociation of the electrolyte and (b) the interionic attractions between ions present. Because of the relatively low degree of dissociation of these electrolytes, the concentrations of ions present in solution are quite low, and so are the interionic effects. Consequently most of the decrease in equivalent conductance is due to decrease in α, and within a fairly close approximation Eq. (49) holds true.

In a weak electrolyte such as acetic acid at a concentration C and degree of dissociation α, the concentration of dissociated acid is $C\alpha$. Because of this concentration of dissociated acid interionic forces are present that tend to decrease the value of Λ from Λ_0 to Λ_e, where Λ_e is the equivalent conductance of the completely dissociated acid at a concentration $C\alpha$. Consequently, in order to obtain the degree of dissociation of the electrolyte corrected for the interionic effects it would be more nearly correct to calculate α not as Λ/Λ_0, but as

$$\alpha = \frac{\Lambda}{\Lambda_e} \tag{50}$$

i.e., as the ratio of the actual equivalent conductance to the equivalent conductance of the same electrolyte when completely dissociated at the same ionic concentration as is present in the solution. The evaluation of Λ_e for use in Eq. (50) may be illustrated with the following example. At $C = 0.1000$, Λ for acetic acid at 25°C is 5.201, while $\Lambda_0 = 390.71$. From these data we find provisionally that

$$\alpha = \frac{\Lambda}{\Lambda_0} = \frac{5.201}{390.71} = 0.0133 = 1.33\%$$

and hence $C\alpha = 0.00133$ equivalent per liter. Now, on the supposition that acetic acid is completely dissociated, and on the assumption that Kohlrausch's law is applicable, it can be calculated from the Λ's of sodium acetate, sodium chloride, and hydrochloric acid, all at $C = 0.00133$, that for acetic acid Λ_e would be 385.40 rather than 390.71. Dividing now the observed Λ by the calculated Λ_e, we find for the true degree of dissociation

$$\alpha = \frac{\Lambda}{\Lambda_e} = \frac{5.201}{385.40} = 0.0135 = 1.35\%$$

Consequently, the true degree of dissociation is 1.35 per cent as against 1.33 per cent estimated from the ratio Λ/Λ_0.

Although such small corrections are found necessary in precise calculations, for our purposes the simpler approximate Eq. (49) will be sufficient and will be employed hereafter.

APPLICATION OF CONDUCTANCE MEASUREMENTS

Conductance measurements find extensive application in chemistry and chemical industry for obtaining important information concerning the behavior of electrolytes, for analysis, and for concentration control. A discussion of some of these applications follows.

Solubility of Difficultly Soluble Salts.   Conductance offers a very simple and convenient means for determining the solubility of difficultly soluble salts such as barium sulfate or silver chloride. The procedure involved is this. A saturated solution of the salt in water of known specific conductance, $L_{s_{(H_2O)}}$, is prepared. Next, the specific conductance of the saturated solution, L_s, is measured. This specific conductance is due to both the salt and the water, and hence the specific conductance of the salt alone, $L_{s(salt)}$, is

$$L_{s(salt)} = L_s - L_{s_{(H_2O)}} \tag{51}$$

From $L_{s(salt)}$ the equivalent conductance follows as

$$\Lambda = \frac{1000\, L_{s(salt)}}{C}$$

where C is the concentration of the salt in equivalents per liter, and hence the solubility. Since the solution is a st very dilute, and since salts are strong electrolytes, Λ must be equal essentially to Λ_0. Making this substitution, C is obtained as

$$C = \frac{1000 \, L_{s(\text{salt})}}{\Lambda_0} \tag{52}$$

By looking up the value of Λ_0 in a table of limiting equivalent conductances, C can readily be calculated through Eq. (52) from the measured specific conductance of the saturated solution.

Determination of Degree of Ionization. The determination of the degree of ionization of weak electrolytes is a problem of importance in physical and analytical chemistry, because from such information can be calculated the ionization constants of the electrolytes. More extensive reference to this subject will be made later in the chapter. Here attention will be directed only to the estimation of the degree of ionization of water.

TABLE 11–8.   SPECIFIC CON-
DUCTANCE OF VERY
PURE WATER

$t°C$	L_s (mho cm^{-1})
0	0.14×10^{-7}
18	0.40
25	0.58
34	0.89
50	1.76

Kohlrausch and Heydweiller[8] found that, no matter how long and how carefully purified, water still exhibits a definite though small conductance at each temperature, as may be seen from the data given in Table 11–8. From these results it must be concluded that water is a weak electrolyte and that it ionizes according to the equation

$$H_2O = H^+ + OH^- \tag{53}$$

To find the degree of ionization α at, say, 25°C, Λ is required. Since $\Lambda = L_s \times V_e$, where V_e is the volume in cubic centimeters containing one equivalent of water, V_e must be the molecular weight of water divided by the density of the water at 25°C. Therefore,

$$\Lambda = (0.58 \times 10^{-7}) \frac{18.016}{0.9971}$$
$$= 1.05 \times 10^{-6}$$

[8] Kohlrausch and Heydweiller, *Z. physik. Chem.*, **14**, 317 (1894).

Again, Λ_0 for water is the sum of the equivalent ionic conductances of H^+ and OH^- ions, or $\Lambda_0 = 349.8 + 198 = 547.8$. Consequently,

$$\alpha = \frac{\Lambda}{\Lambda_0} = \frac{1.05 \times 10^{-6}}{547.8}$$
$$= 1.9 \times 10^{-9}$$

i.e., water is ionized to the extent of 1.9×10^{-7} per cent at 25°C. This degree of ionization, though extremely small, is sufficient to account for many phenomena encountered in aqueous solutions.

Conductometric Titrations. Conductance measurements may be employed to determine the end points of various titrations. Consider first the titration of a strong acid like hydrochloric with a strong base like sodium hydroxide. Before base is added, the acid solution has a high content of the highly mobile hydrogen ions which give the solution a high conductance. As alkali is added the hydrogen ions are removed to form water, and their place is taken by the much slower cations of the base. Consequently the conductance of the solution decreases and keeps on falling with addition of base until the equivalent point is reached. Further addition of alkali introduces now an excess of the fast hydroxyl ions, and these cause the conductance to rise again. When this variation of the conductance of the solution is plotted against the volume of alkali added, the result is curve ABC in Figure 11-9. The descending branch of this curve gives the conductances of mixtures of acid and salt, the ascending branch conductances of mixtures of the salt and excess base. At the minimum, point B, there is no excess present of either acid or base, and hence it is the end point.

Titration curves of the type just described are obtained only on neutralization of strong acids by strong bases. When the acid is weak, say acetic acid, and the base is strong, the titration curve has the general form of $A'B'C'$ in Figure 11-9. Since the acid is weak, its conductance is

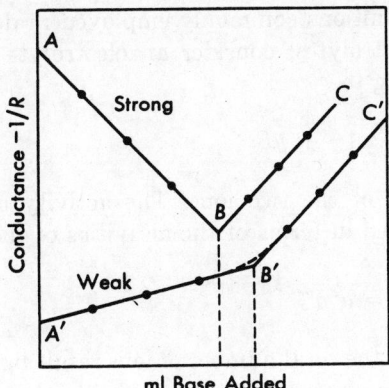

Figure 11-9. Conductometric titration of acids with a strong base.

correspondingly low. As base is added, the poorly conducting acid is converted to highly ionized salt, and consequently the conductance goes up along $A'B'$. Once the acid is neutralized, addition of excess base causes another sharp increase in conductance, and the curve rises along $B'C'$. The equivalence point is again the intersection of the two straight lines. Actually the intersection of $A'B'$ and $B'C'$ is not as sharp as shown but has the form indicated by the dotted lines. This rounding of the intersection is due to hydrolysis of the salt formed during the neutralization reaction. However, this rounding introduces no particular difficulty, for the straight-line portions of the curve may be extended, as in the figure, to give the correct end point.

The conductometric titration curves described apply only to the conditions specified above. When the relative strengths of acid and base are changed, so may be the titration curves. Suffice it to say that many titrations can be run conductometrically in water or mixed solvents which would be difficult or impossible with indicators, and that the method can be applied to mixtures of weak and strong acids, weak and strong bases, precipitation, oxidation-reduction, and other types of reactions.

ACTIVITIES AND ACTIVITY COEFFICIENTS OF STRONG ELECTROLYTES

Equilibrium constants for gaseous equilibria and for nonionic reactions in solution can be expressed in terms of concentrations to a fairly good approximation at relatively low pressures and concentrations. However, in dealing with ionic equilibria, with certain aspects of kinetics in solution, and with electromotive force studies, substitution of concentrations for activities is frequently not possible. For this reason it is essential to consider how ionic concentrations may be converted to activities, and how such activities can be evaluated.

In order to introduce some definitions commonly employed in dealing with the activities of strong electrolytes, consider an electrolyte A_xB_y which ionizes in solution according to

$$A_xB_y = xA^{z_+} + yB^{z_-} \qquad (54)$$

where z_+ and z_- are the charges of the two ions. The activity of the electrolyte as a whole, a_2, is defined in terms of the activities of the two ions a_+ and a_- as

$$a_2 = a_+^x a_-^y \qquad (55)$$

If we designate now by $\nu = x + y$ the total number of ions resulting from 1 molecule of the electrolyte, *the geometric mean activity of the electrolyte,* or more simply the *mean activity*, written $a_{\pm}$, is defined as

$$a_{\pm} = \sqrt[\nu]{a_2} = \sqrt[\nu]{a_+^x a_-^y} \tag{56}$$

To relate the activities of the ions to their concentrations we write

$$a_+ = C_+ f_+ \tag{57a}$$
$$a_- = C_- f_- \tag{57b}$$

where C_+ and C_- are the gram ionic weights per liter of the two ions in solution, while f_+ and f_- are the *activity coefficients* of the two ions. These activity coefficients are appropriate factors which when multiplied by the concentrations of the respective ions yield their activities. Introducing Eqs. (57a) and (57b) into Eq. (55), we obtain for a_2

$$a_2 = (C_+ f_+)^x (C_- f_-)^y$$
$$= (C_+^x C_-^y)(f_+^x f_-^y) \tag{58}$$

and for the mean activity from Eq. (56),

$$a_{\pm} = \sqrt[\nu]{a_2} = \sqrt[\nu]{(C_+^x C_-^y)(f_+^x f_-^y)}$$
$$= (C_+^x C_-^y)^{1/\nu}(f_+^x f_-^y)^{1/\nu} \tag{59}$$

The factor $(f_+^x f_-^y)^{1/\nu}$ is called the *mean activity coefficient* of the electrolyte, f; i.e.,

$$f = (f_+^x f_-^y)^{1/\nu} \tag{60}$$

Similarly, the factor $(C_+^x C_-^y)^{1/\nu}$ is defined as the *mean molarity* of the electrolyte, $C_{\pm}$; or

$$C_{\pm} = (C_+^x C_-^y)^{1/\nu} \tag{61}$$

In terms of the mean molarity and mean activity coefficient, Eqs. (58) and (59) may be written simply as

$$a_{\pm} = a_2^{1/\nu} = C_{\pm} f \tag{62}$$
$$a_2 = a_{\pm}^\nu = (C_{\pm} f)^\nu \tag{63}$$

Finally, since for any electrolyte of molarity C we have $C_+ = xC$ and $C_- = yC$, Eqs. (62) and (63) become also

$$a_{\pm} = a_2^{1/\nu} = [(xC)^x (yC)^y]^{1/\nu} f$$
$$= (x^x y^y)^{1/\nu} C f \tag{64}$$
$$a_2 = a_{\pm}^\nu = (x^x y^y) C^\nu f^\nu \tag{65}$$

Equations (64) and (65) are the expressions needed for converting activities to molarities, or vice versa. Although these expressions may appear complicated, they are actually not so when applied to specific cases. Thus for a 1–1 electrolyte, such as sodium chloride, of molarity C

we have $x = 1$, $y = 1$, $\nu = 2$, and, therefore,

$$a_{\pm} = (1 \times 1)^{1/2} Cf = Cf$$
$$a_2 = a_{\pm}^2 = C^2 f^2$$

Again, for an electrolyte of the 2–1 type, such as barium chloride, we obtain $x = 1$, $y = 2$, $\nu = 3$, and hence

$$a_{\pm} = (1 \times 2^2)^{1/3} Cf$$
$$= \sqrt[3]{4}\, Cf$$
$$a_2 = a_{\pm}^3 = 4\, C^3 f^3$$

In Table 11–9 are summarized the relations connecting C, f, $a_{\pm}$, and a_2 for a number of different types of electrolytes. It will be observed that the expressions resulting from Eqs. (64) and (65) depend on the electrolyte type and are identical for the 1–1, 2–2, 3–3 types; for the 1–2 and 2–1 types; for the 1–3 and 3–1 types; and for the 2–3 and 3–2 types.

TABLE 11–9. RELATION OF $a_{\pm}$ AND a_2 TO C AND f FOR VARIOUS ELECTROLYTES

Electrolyte Type	Example	x	y	ν	$C_{\pm}$	$a_{\pm} = C_{\pm}f$	$a_2 = a_{\pm}^{\nu}$
1–1	NaCl	1	1	2	C	Cf	$C^2 f^2$
2–2	CuSO$_4$	1	1	2	C	Cf	$C^2 f^2$
3–3	AlPO$_4$	1	1	2	C	Cf	$C^2 f^2$
1–2	Na$_2$SO$_4$	2	1	3	$\sqrt[3]{4}\ C$	$\sqrt[3]{4}\ Cf$	$4\,C^3 f^3$
2–1	BaCl$_2$	1	2	3	$\sqrt[3]{4}\ C$	$\sqrt[3]{4}\ Cf$	$4\,C^3 f^3$
1–3	Na$_3$PO$_4$	3	1	4	$\sqrt[4]{27}\ C$	$\sqrt[4]{27}\ Cf$	$27\,C^4 f^4$
3–1	La(NO$_3$)$_3$	1	3	4	$\sqrt[4]{27}\ C$	$\sqrt[4]{27}\ Cf$	$27\,C^4 f^4$
2–3	Ca$_3$(PO$_4$)$_2$	3	2	5	$\sqrt[5]{108}\ C$	$\sqrt[5]{108}\ Cf$	$108\,C^5 f^5$
3–2	La$_2$(SO$_4$)$_3$	2	3	5	$\sqrt[5]{108}\ C$	$\sqrt[5]{108}\ Cf$	$108\,C^5 f^5$

The definitions for the ionic and mean activity coefficients have been expressed in terms of concentration in moles per liter of solution. In electrochemical work quite frequently concentrations are expressed on a molality basis, m. When this is the case, the activities of the ions are defined analogously to Eq. (57) by

$$a_+ = m_+ \gamma_+ \tag{66a}$$
$$a_- = m_- \gamma_- \tag{66b}$$

By repeating the above argument with these new definitions, it can readily be shown that now

$$a_{\pm} = (x^x y^y)^{1/\nu} m\gamma \tag{67}$$
$$a_2 = a_{\pm}^{\nu} = (x^x y^y) m^{\nu} \gamma^{\nu} \tag{68}$$

where γ is the *mean activity coefficient of the electrolyte* for concentration in molalities, and is given by

$$\gamma = (\gamma_+^z \gamma_-^y)^{1/\nu} \tag{69}$$

With Eqs. (67) and (68) the expressions for the mean molality, $m_\pm$, i.e.,

$$m_\pm = (m_+^z m_-^y)^{1/\nu} \tag{70}$$

and a_2 in Table 11-9 have exactly the same form as for molarities, except that C and f are replaced by m and γ. Since C and m are not identical, neither will f as a rule be equal to γ. In fact, the two activity coefficients are related by the expression

$$f = \gamma \left(\frac{\rho_0 m}{C} \right) \tag{71}$$

where ρ_0 is the density of the pure solvent, and m/C is given by

$$\frac{m}{C} = \frac{1 + 0.001 \, mM_2}{\rho} \tag{72}$$

Here M_2 is the molecular weight of the electrolyte, while ρ is the density of the solution. From Eqs. (71) and (72) it can be shown that for dilute aqueous solutions f will be essentially equal to γ; however, in more concentrated solutions the two will have different values.

DETERMINATION OF ACTIVITY COEFFICIENTS

Equations (64) and (65) or (67) and (68) indicate that for conversion of molarities or molalities to activities the mean activity coefficients for various concentrations of an electrolyte must be known. Such mean activity coefficients can be determined from vapor pressure, freezing point lowering, boiling point elevation, osmotic pressure, distribution, solubility, and electromotive force measurements by well-known thermodynamic methods. Although in appropriate places in this book the solubility and electromotive force methods for evaluating activity coefficients will be explained, a discussion of the other methods is beyond our scope. For these, books on chemical thermodynamics should be consulted.

All evaluations of activity coefficients are made on the supposition that in the infinitely dilute solution $\gamma = 1$ or $f = 1$, depending on which concentration scale is employed; i.e., that

$$\frac{m_\pm}{a_\pm} = \gamma = 1 \qquad \text{as } m \to 0 \tag{73}$$

or that

$$\frac{a_\pm}{C_\pm} = f = 1 \qquad \text{as } C \to 0 \tag{74}$$

TABLE 11–10. MEAN ACTIVITY COEFFICIENTS, γ, OF ELECTROLYTES AT 25°C

Molality m	HCl	KCl	NaOH	CaCl$_2$	H$_2$SO$_4$	ZnSO$_4$
0.000	1.000	1.000	1.000	1.000	1.000	1.000
0.001	0.966	0.965	—	0.888	0.830	0.734
0.005	0.929	0.927	—	0.789	0.639	0.477
0.01	0.905	0.902	0.899	0.732	0.544	0.387
0.02	0.876	0.869	0.860	0.669	0.453	0.298
0.05	0.830	0.817	0.818	0.584	0.340	0.202
0.10	0.796	0.769	0.766	0.531	0.265	0.148
0.20	0.767	0.719	0.719	0.482	0.209	0.104
0.50	0.757	0.651	0.693	0.457	0.154	0.063
1.00	0.809	0.606	0.679	0.509	0.131	0.044
1.50	0.896	0.585	0.683	0.628	0.124	0.037
2.00	1.009	0.576	0.698	0.807	0.124	0.035
3.00	1.316	0.571	0.774	—	0.141	0.041
4.00	1.762	0.579	0.888	—	0.171	—

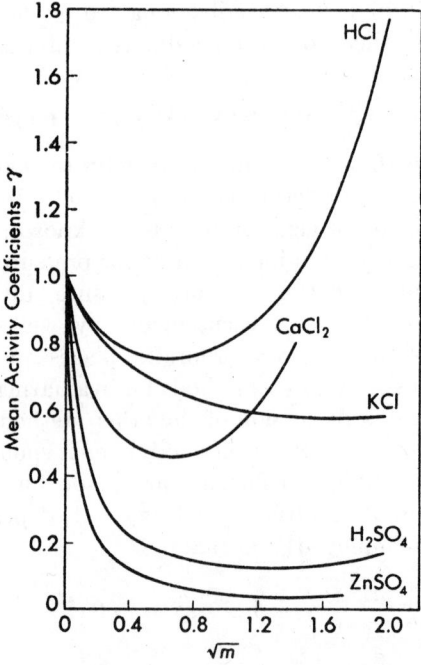

Figure 11–10. Mean activity coefficients of various electrolytes at 25°C.

These definitions are equivalent to the statement that the activity of an ion is equal to its concentration in the infinitely dilute solution. On this basis the activity coefficients of all electrolytes are unity at zero concentration. As the concentration is increased the activity coefficients decrease at first below unity, pass through a minimum, and then increase again to values which may rise considerably above one. In Table 11–10 are listed the γ's for a number of electrolytes at 25°C, while in Figure 11–10 is shown a plot for some of these as function of $\sqrt{m}$. Although at the higher concentrations the behavior of the γ's is highly individualized, at the lower concentrations the curves converge in a manner dependent on the type of electrolyte. This fact is of great theoretical significance in explaining the thermodynamic behavior of dilute solutions of strong electrolytes.

THE DEBYE-HÜCKEL THEORY OF ACTIVITY COEFFICIENTS

A theory which attempts to account for the activity coefficients of electrolytes in terms of electrostatic attractions operating between ions in a solution is that of Debye and Hückel. These authors start by assuming that electrically charged particles in solution, like other charges, are subject to the Coulomb law of force, namely, that the forces of attraction or repulsion between charges vary directly as the product of the charges, q_1 and q_2, and inversely as the square of the distance r between them

$$\text{Force} = \frac{1}{D}\left(\frac{q_1 q_2}{r^2}\right) \tag{75}$$

The proportionality constant D, called the *dielectric constant*, is determined by the medium in which the charges are immersed, in our case the solvent. As a result of these forces the distribution of ions throughout a solution is not random but is such that any central positive ion is surrounded on an average by an atmosphere of other ions whose net charge is negative, while any central negative ion is surrounded by an atmosphere of net positive charge. Owing to the presence of an atmosphere about an ion there is a potential ε_i established at its surface whose magnitude Debye and Hückel showed to be

$$\varepsilon_i = \frac{-z_i e \kappa}{D(1 + \kappa a_i)} \tag{76}$$

In this equation z_i is the valence of the central ion, e the electronic charge, a_i the ionic diameter, and κ is

$$\kappa = \sqrt{\frac{4\pi e^2 \Sigma n_i z_i^2}{DkT}} \tag{77}$$

where k is the gas constant per *molecule*, i.e., R/N, T the absolute temperature, and n_i the number of any given kind of ions per *cubic centimeter* of solution. The summation Σ in Eq. (77) must be carried out over all the ions present in a given solution. κ has the dimensions of a reciprocal length and may be looked upon as the reciprocal of the average thickness of the ionic atmosphere about an ion.

The presence of a potential $\mathcal{E}_i$ at the surface of an ion due to the ionic atmosphere can be shown to give the solution an electrical free energy F_e

$$F_e = \frac{-z_i^2 e^2 \kappa}{2\, D(1 + \kappa a_i)} \tag{78}$$

in excess of what the solution would have if the ionic atmosphere resulting from electrostatic attraction were not present. This excess free energy F_e is thermodynamically related to the activity coefficient of the ion f_i by

$$F_e = kT \ln f_i \tag{79}$$

and hence from Eqs. (78) and (79) we obtain

$$\ln f_i = \frac{-z_i^2 e^2 \kappa}{2\, kTD(1 + \kappa a_i)} \tag{80}$$

On introducing into Eq. (80) the value of κ from Eq. (77), and the concentration of the ions C_i in gram ionic weights per liter from the identity

$$C_i = \frac{n_i}{N} \times 1000$$

Eq. (80) becomes

$$\log_{10} f_i = \frac{-A z_i^2 \sqrt{\mu}}{1 + B a_i \sqrt{\mu}} \tag{81}$$

For any given temperature and solvent, A and B in this equation are constants defined by

$$A = \frac{e^3}{2.303(DkT)^{3/2}} \sqrt{\frac{2\,\pi N}{1000}} \tag{82}$$

$$B = \sqrt{\frac{8\,\pi N e^2}{1000\, DkT}} \tag{83}$$

while μ is the ionic strength of the solution (see page 338), i.e.,

$$\mu = \frac{1}{2} \Sigma C_i z_i^2 \tag{84}$$

Equation (81) gives the *activity coefficient of an ion* as a function of the ionic strength of the solution. Since experimentally we evaluate not the ionic activity coefficient but the mean activity coefficient of the electrolyte f, Eq. (81) must be transformed with the aid of Eq. (60) to the logarithm of the mean activity coefficient. The result is

$$\log_{10} f = \frac{-A z_+ z_- \sqrt{\mu}}{1 + B a_i' \sqrt{\mu}} \tag{85}$$

where a_i' is now an average ionic diameter. The constants A and B for this equation are given for *water as a solvent* at various temperatures in Table 11–11. The mean ionic diameter a_i' must be assumed to fit the data. However, *for very dilute solutions* $B a_i' \sqrt{\mu}$ is small compared with unity

TABLE 11–11. DEBYE-HÜCKEL CONSTANTS
A AND B FOR WATER AS SOLVENT

t (°C)	A	B
0	0.4883	0.3241×10^8
15	0.5002	0.3267
25	0.5091	0.3286
40	0.5241	0.3318
55	0.5410	0.3353
70	0.5599	0.3392

and may be neglected. We obtain, then,

$$\log_{10} f = -A z_+ z_- \sqrt{\mu} \tag{86}$$

which is the *limiting equation of Debye and Hückel for the activity coefficients of strong electrolytes*.

Because of certain simplifications necessitated by mathematical complexities of the derivation, the Debye-Hückel theory, as epitomized in Eq. (85), can be expected to be applicable only to *dilute solutions*. Further, Eq. (86) must be looked upon as a limiting law for the behavior of activity coefficients in *very dilute solutions*. According to Eq. (86) the activity coefficients of all strong electrolytes at high dilution should be determined, in a given solvent and at a given temperature, only by the ionic strength of the solution and the valence type ($z_+ z_-$) of the electrolyte, and not at all by the nature of the electrolyte. Further, a plot of $-\log f$ vs. $\sqrt{\mu}$ for all electrolytes should yield straight lines through the origin, but the slope of the lines should depend on $z_+ z_-$. For the 1–1 electrolytes this slope should be A, for 1–2 or 2–1 electrolytes $2A$, for 1–3 or 3–1 electrolytes $3A$, etc. That these requirements are met by experimented data in very dilute solutions may be seen from Figure 11–11. In this figure are plotted $-\log f$ values of various complex cobalt ammines in aqueous solution as a function of the square root of the ionic strength. The points are the experimentally observed $-\log f$ values as obtained by Bronsted and LaMer[9] from solubility measurements, while the solid lines give the plots predicted by theory. Considering the difficulties involved in the determination of solubilities in such dilute solutions, the agreement between theory and experiment must be considered satisfactory. Similar tests have been carried out for other substances with the same result. We may conclude, therefore, that for strong electrolytes in extremely dilute solutions the limiting law of Debye and Huckel does represent the behavior of activity coefficients to $\sqrt{\mu} = 0.1$ or less for electrolytes

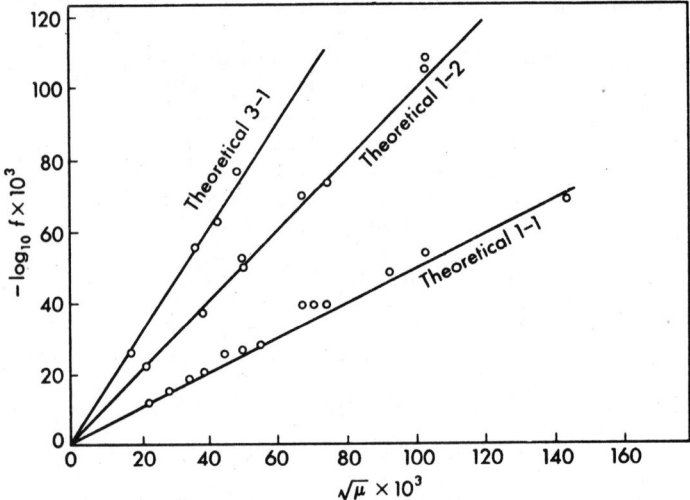

Figure 11-11.   Test of Debye-Huckel limiting law at 15°C.

of $z_+ z_-$ up to three. With valence types higher than $z_+ z_- = 3$ the agreement usually does not extend to $\sqrt{\mu} = 0.1$.

For reproducing the course of activity coefficients with $\sqrt{\mu}$ at ionic strengths higher than $\sqrt{\mu} = 0.1$, the more complete Eq. (85) must be used. By an appropriate choice of a_i', which must be of the order of atomic diameters, i.e., several angstroms, the activity coefficients of 1–1 electrolytes can be reproduced to $C = 0.1$, and of 2–1 electrolytes to $C = 0.05$ or so. Huckel and others had pointed out that Eq. (85) can be extended even further by the introduction of another term linear in μ, but for our purposes this modification is not necessary and will not be discussed.

Although the Debye-Hückel theory was originally proposed for strong electrolytes, nothing in the theory precludes its application to the *ionic*

[9] Brönsted and LaMer, *J. Am. Chem. Soc.*, **46**, 555 (1924).

portion of weak electrolytes. This extension of the theory to weak electrolytes will be illustrated later in the chapter.

IONIZATION CONSTANTS OF WEAK ACIDS

The equilibrium constant principle cannot be applied to the ionization of strong electrolytes because, as we have seen, they are already completely ionized. Consequently, we need concern ourselves only with the ionization equilibria of weak electrolytes, namely, the ionization of weak acids and bases.

The ionization equilibrium of any weak monobasic acid, such as acetic acid, can be represented by the equation

$$HA = H^+ + A^-$$

for which the thermodynamic ionization constant K_a is given by

$$K_a = \frac{a_{H^+} a_{A^-}}{a_{HA}} \tag{87}$$

Introducing the activity coefficients and concentrations for the activities, Eq. (87) becomes

$$K_a = \frac{(C_{H^+} f_{H^+})(C_{A^-} f_{A^-})}{(C_{HA} f_{HA})}$$

$$= \left(\frac{C_{H^+} C_{A^-}}{C_{HA}}\right)\left(\frac{f_{H^+} f_{A^-}}{f_{HA}}\right) \tag{88}$$

Now, for a total acid concentration C with degree of ionization α the concentration of HA will be $C_{HA} = C(1 - \alpha)$, while that of the ions will be $C_{H^+} = C_{A^-} = C\alpha$. On substitution of these into Eq. (88), we have

$$K_a = \frac{(C\alpha)(C\alpha)}{C(1 - \alpha)}\left(\frac{f_{H^+} f_{A^-}}{f_{HA}}\right)$$

$$= \left(\frac{C\alpha^2}{1 - \alpha}\right)\left(\frac{f_{H^+} f_{A^-}}{f_{HA}}\right)$$

$$= K_a' K_f \tag{89}$$

where

$$K_a' = \frac{C\alpha^2}{1 - \alpha} \tag{90}$$

is the ionization constant of the acid in terms of concentrations, while

$$K_f = \frac{f_{H^+} f_{A^-}}{f_{HA}} \tag{91}$$

is the ratio of the activity coefficients. For dilute solutions K_f will be close to unity, and for such K_a will be essentially equal to K_a'. However, in more concentrated solutions a correction to K'_a for the deviation of the activity coefficients from unity will be in order.

Use of the above equations to calculate K'_a and K_a may be illustrated with data on the ionization of acetic acid obtained from conductance measurements by MacInnes and Shedlovsky.[10] The first column in Table 11–12 lists the concentrations of the acid, while the second column gives the degree of ionization of the acid as estimated from Eq. (50), i.e., $\alpha = \Lambda/\Lambda_e$. The concentration ionization constants K'_a, calculated from

TABLE 11–12.   The Ionization Constant of Acetic Acid at 25°C

C	α	K'_a	K_a
0.00002801	0.5393	1.77×10^{-5}	1.75×10^{-5}
0.0001114	0.3277	1.78	1.75
0.0002184	0.2477	1.78	1.75
0.001028	0.1238	1.80	1.75
0.002414	0.08290	1.81	1.75
0.005912	0.05401	1.82	1.75
0.009842	0.04222	1.83	1.75
0.02000	0.02988	1.84	1.74
0.05000	0.01905	1.85	1.72
0.10000	0.01350	1.85	1.70

C and α by means of Eq. (90), are given in the third column. It will be observed that although the K'_a's show a fairly satisfactory constancy, there is a consistent increase in K'_a with concentration. This drift is due to the neglect of K_f in Eq. (89). To take this factor into account we proceed as follows. On taking logarithms of this equation, we have

$$\log K_a = \log K'_a + \log K_f$$
$$= \log K'_a + \log \frac{f_{H^+} f_{A^-}}{f_{HA}} \tag{92}$$

Since f_{HA} is the activity coefficient of an unionized molecule, it cannot be far from unity. Again, since $f_{H^+} f_{A^-} = f^2$, then

$$\log f_{H^+} f_{A^-} = 2 \log f$$

and on introduction of the limiting law of Debye and Hückel, Eq. (86), we find for $2 \log f$ at 25°C with $z_+ = z_- = 1$

$$2 \log f = -2 A z_+ z_- \sqrt{\mu}$$
$$= -1.018 \sqrt{\mu} \tag{93}$$

[10] MacInnes and Shedlovsky, *J. Am. Chem. Soc.*, **54**, 1429 (1932).

The ionic strength of the solution is in this instance one-half the total *ionic concentration*, namely, $\mu = C\alpha$. Inserting this value of μ into Eq. (93), and substituting Eq. (93) into Eq. (92), we obtain finally, for log K_a,

$$\log K_a = \log K'_a - 1.018 \sqrt{C\alpha} \qquad (94)$$

The values of K_a calculated with Eq. (94) are shown in the last column of Table 11–12. It will be observed that within the range of validity of the limiting law of Debye and Hückel the values of K_a are now constant and do not exhibit the drift shown by K'_a. Consequently 1.75×10^{-5} is the thermodynamic ionization constant of acetic acid at 25°C.

Although strictly speaking the ionization constants should always be expressed in activities, for many purposes no appreciable error is introduced by using K_a for K'_a or vice versa. Hence, we shall hereafter follow the practice of expressing the ionization constants in concentrations rather than activities for calculation purposes.

Acids possessing more than one ionizable hydrogen do not ionize in a single step, but in successive stages. Thus, sulfuric acid ionizes in two steps, namely,

$$H_2SO_4 = H^+ + HSO_4^-$$
$$HSO_4^- = H^+ + SO_4^{--}$$

while phosphoric acid dissociates in three steps

$$H_3PO_4 = H^+ + H_2PO_4^-$$
$$H_2PO_4^- = H^+ + HPO_4^{--}$$
$$HPO_4^{--} = H^+ + PO_4^{---}$$

In any dibasic acid the primary ionization is more complete than the second, while in any tribasic acid the primary dissociation is greater than the second, and the second greater than the third. Each of the dissociation stages constitutes a true equilibrium, and for each of these there is an ionization constant. In sulfuric acid the first ionization stage is complete, i.e., H_2SO_4 is a strong acid. However, the ionization of HSO_4^- is incomplete, and hence for this ionization stage we have the constant

$$K_a = \frac{a_{H^+} a_{SO_4^{--}}}{a_{HSO_4^-}}$$

In phosphoric acid all three stages involve partial ionization, and we have thus for this acid three ionization constants. At 25°C the magnitudes of the three constants are $K_{a_1} = 7.5 \times 10^{-3}$, $K_{a_2} = 6.2 \times 10^{-3}$, and $K_{a_3} = 4.8 \times 10^{-13}$, indicating that at any given acid concentration the extent of dissociation of each stage decreases markedly from the primary to the tertiary. In fact, it can be calculated from the above ionization constants that in a 0.0233 molar solution of phosphoric acid, where the concentration of hydrogen ions is approximately 0.01, the concentrations of the various molecular and ionic species present are

$$C_{H_3PO_4} = 0.0133 \qquad C_{HPO_4^{--}} = 6 \times 10^{-8}$$
$$C_{H_2PO_4^-} = 0.01 \qquad C_{PO_4^{---}} = 2.9 \times 10^{-18}$$

These figures show that even dilute solutions of pure phosphoric acid consist primarily of undissociated H_3PO_4, H^+, and $H_2PO_4^-$ and that in such solutions practically no PO_4^{---} ions are present.

IONIZATION CONSTANTS OF WEAK BASES

Any weak base BOH, such as ammonium hydroxide, ionizes according to

$$BOH = B^+ + OH^- \tag{95}$$

For this process the thermodynamic ionization constant of the base K_b is

$$K_b = \frac{a_{B^+} a_{OH^-}}{a_{BOH}} \tag{96}$$

while the concentration ionization constant K_b' is

$$K_b' = \frac{C_{B^+} C_{OH^-}}{C_{BOH}} \tag{97}$$

As in weak acids, the two constants K_b and K_b' are identical only in dilute solutions. Otherwise,

$$K_b = K_b' K_f \tag{98}$$

where K_f is again the activity coefficient ratio for the various species. We shall employ K_b' expressed in concentration units as a sufficiently close approximation to K_b.

DETERMINATION OF IONIZATION CONSTANTS

Ionization constants of weak acids and bases can be determined from conductance data, and measurements of the hydrogen ion concentration of a given solution of the acid or base in presence of its salt. The conductance method has already been discussed in connection with the ionization constant of acetic acid. We turn our attention, then, to the estimation of ionization constants from hydrogen ion concentration measurements.

For this purpose consider specifically a solution of acetic acid of concentration C containing also sodium acetate at concentration C'. If in this solution the concentration of hydrogen ions is C_{H^+}, then the concentration of unionized acid is $C_{HA} = C - C_{H^+}$, while the concentration of acetate ions is that due to ionization plus that contributed by the completely ionized salt. Since on ionization of the acid equivalent quantities of hydrogen and acetate ions result, the acetate ion concentration resulting from ionization is C_{H^+}; and this concentration plus C' from the

TABLE 11-13. IONIZATION CONSTANTS OF WEAK ACIDS AT 25°C

Acid	Formula	K_{a_1}	K_{a_2}	K_{a_3}
Arsenic	H_3ASO_4	5.0×10^{-3}	8.3×10^{-8}	6×10^{-10}
Boric	H_3BO_3	5.80×10^{-10}		
Carbonic	H_2CO_3	4.52×10^{-7}	4.69×10^{-11}	
Hydrocyanic	HCN	7.2×10^{-10}		
Iodic	HIO_3	1.67×10^{-1}		
Phosphoric	H_3PO_4	7.52×10^{-3}	6.23×10^{-8}	4.8×10^{-13}
Phosphorous	H_3PO_3	1.6×10^{-2}	7×10^{-7}	
Sulfuric	H_2SO_4	strong	1.01×10^{-2}	
Sulfurous	H_2SO_3	1.72×10^{-2}	6.24×10^{-8}	
Formic	HCOOH	1.77×10^{-4}		
Acetic	CH_3COOH	1.75×10^{-5}		
Propionic	C_2H_5COOH	1.34×10^{-5}		
Chloracetic	$CH_2ClCOOH$	1.38×10^{-3}		
Dichloracetic	$CHCl_2COOH$	5×10^{-2}		
Benzoic	C_6H_5COOH	6.29×10^{-5}		
Oxalic	$(COOH)_2$	5.02×10^{-2}	5.18×10^{-5}	
Phenol	C_6H_5OH	1.20×10^{-10}		

salt gives $C_{A^-} = (C_{H^+} + C')$. In terms of these quantities, the ionization constant of the acid becomes

$$K_a' = \frac{C_{H^+}C_{A^-}}{C_{HA}}$$

$$= \frac{C_{H^+}(C_{H^+} + C')}{C - C_{H^+}} \tag{99}$$

Therefore, as C and C' are known, measurement of C_{H^+} of the solution by some suitable means is all that is necessary to yield K_a'.

Similar considerations apply to the determination of the ionization constants of bases. For the base ammonium hydroxide at concentration C in the presence of ammonium chloride at concentration C', the ionization constant K_b' takes the form

$$K_b' = \frac{C_{OH^-}(C_{OH^-} + C')}{C - C_{OH}} \tag{100}$$

As will be shown presently, the concentrations of hydrogen and hydroxyl ions in water solution are related in a definite manner, so that as soon as

C_{H^+} is measured, C_{OH^-} is also known. Hence by measuring the *hydrogen ion* concentration in the solution of the base and its salt, C_{OH^-} is also determined, and K_b' follows from Eq. (100).

In Table 11–13 are listed the ionization constants of a number of weak acids at 25°C, while in Table 11–14 are given the ionization constants of some weak bases at the same temperature.

TABLE 11–14.   IONIZATION CONSTANTS OF WEAK BASES
AT 25°C

Base	Formula	K_b
Ammonia	NH_4OH	1.81×10^{-5}
Silver hydroxide	$AgOH$	1.1×10^{-4}
Methyl amine	$(CH_3)NH_2$	4.38×10^{-4}
Dimethyl amine	$(CH_3)_2NH$	5.12×10^{-4}
Trimethyl amine	$(CH_3)_3N$	5.21×10^{-5}
Ethyl amine	$(C_2H_5)NH_2$	5.6×10^{-4}
Aniline	$(C_6H_5)NH_2$	3.83×10^{-10}
Hydrazine	$NH_2 \cdot NH_2$	3×10^{-6}
Pyridine	C_6H_5N	1.4×10^{-9}
Urea	$CO(NH_2)_2$	1.5×10^{-14}

CALCULATIONS INVOLVING IONIZATION CONSTANTS

When the ionization constant of an acid or base is known, it is readily possible to calculate the degree of ionization and the concentrations of the species present in solution under various given conditions. Two such calculations will be presented as examples of the procedure.

Calculation of the Degree of Ionization of Pure Acids or Bases.
Suppose that it is desired to calculate the degree of ionization and the concentrations of the various species in a solution of a monobasic acid at concentration C. If α is the degree of ionization, then according to Eq. (90)

$$K_a' = \frac{C\alpha^2}{1 - \alpha} \tag{90}$$

from which α follows as

$$\alpha = \frac{-K_a' + \sqrt{(K_a')^2 + 4\,K_a'C}}{2\,C} \tag{101}$$

When K_a' is small so will be α, and the denominator in Eq. (90) will be essentially unity. Under these conditions the equation for α reduces to

$$\alpha = \sqrt{\frac{K_a'}{C}} \tag{102}$$

Once α is known, C_{HA}, C_{H^+}, and C_{A^-} are readily obtained. In an exactly analogous manner may be calculated α and the concentrations in the ionization of bases.

Degree of Ionization in Presence of a Common Ion. When the solution of a weak acid or base contains as well a substance possessing an ion in common with the weak electrolyte, the degree of ionization of the latter is invariably repressed. The generality of this statement may be illustrated with the following example. Suppose that we have a C molar solution of propionic acid, and suppose, further, that this solution contains also sodium propionate at concentration C'. If α is again the degree of ionization, then $C_{HA} = C(1 - \alpha)$, and $C_{H^+} = C\alpha$. But the concentration of propionate ions is now that due to ionization, $C\alpha$, plus that contributed by the salt, C', or, $C_{A^-} = (C\alpha + C')$. Inserting these concentrations into the expression for K_a', we have

$$
\begin{aligned}
K_a' &= \frac{C_{H^+}C_{A^-}}{C_{HA}} \\
&= \frac{(C\alpha)(C\alpha + C')}{C(1 - \alpha)} \\
&= \frac{\alpha(C\alpha + C')}{(1 - \alpha)}
\end{aligned}
\tag{103}
$$

and hence

$$
\alpha = \frac{-(C' + K_a') + \sqrt{(C' + K_a')^2 + 4\,K_a'C}}{2\,C}
\tag{104}
$$

When we can take $(1 - \alpha) = 1$, Eq. (104) becomes

$$
\alpha = \frac{-C' + \sqrt{(C')^2 + 4\,K_a'C}}{2\,C}
\tag{105}
$$

Further, with low values of K_a' no significant error is introduced when not only α is disregarded in the denominator of Eq. (103) but also $C\alpha$ compared with C' in the numerator. Equation (103) yields then

$$
K_a' = C'\alpha
$$

and

$$
\alpha = \frac{K_a'}{C'}
\tag{106}
$$

Comparison of Eqs. (102) and (106) shows α to be smaller in the latter instance.

THE ION PRODUCT OF WATER

Conductance measurements and other evidence definitely indicate that water ionizes according to the equation

$$
H_2O(l) = H^+ + OH^-
\tag{107}
$$

For this ionization, the equilibrium constant is

$$K = \frac{a_{H^+}a_{OH^-}}{a_{H_2O}} \tag{108}$$

However, since the ionization is at best very slight, the activity of the water in any aqueous solution will be constant and may be included in K. Equation (108) becomes then

$$K_w = Ka_{H_2O} = a_{H^+}a_{OH^-} \tag{109}$$

K_w is called the *ion product of water*. It indicates that in any aqueous solution both hydrogen and hydroxyl ions must be present and that at all times the product of the activities of the two ions must be a constant. In terms of concentrations this constant may be written as

$$K'_w = C_{H^+}C_{OH^-} \tag{110}$$

where K'_w differs from K_w by the product of the activity coefficients of the ions. In solutions of low ionic strength K'_w is essentially equal to K_w.

The value of K'_w may be calculated from Kohlrausch and Heydweiller's data on the conductance of pure water given in Table 11–8. These data yield for the degree of ionization of water at 25°C $\alpha = 1.9 \times 10^{-9}$. Now, as 1 liter of water at 25°C weighs 997.07 g, and as the molecular weight of water is 18.016, the molar concentration C of the water is $C = 997.07/18.016 = 55.34$ molar. From C and α the concentrations of the ions follow as

$$C_{H^+} = C_{OH^-} = C\alpha$$
$$= 55.34 \times 1.9 \times 10^{-9}$$
$$= 1.05 \times 10^{-7}$$

and therefore K'_w at 25°C is

$$K'_w = C_{H^+}C_{OH^-}$$
$$= (1.05 \times 10^{-7})^2$$
$$= 1.10 \times 10^{-14}$$

In Table 11–15 are given the present best values of K_w at a number of temperatures. One point of interest in this table is that the ion product of water is 1×10^{-14} only at 25°C. At other temperatures the constant has different values, and consequently the concentration of hydrogen and hydroxyl ions in pure water at these temperatures will *not* be the same as at 25°C, namely, 1×10^{-7} g ionic weights per liter.

As the ion product principle must be valid in any solution in which

TABLE 11-15. ION PRODUCT OF WATER,
K_w, AT VARIOUS TEMPERATURES

Temperature °C	K_w
0	0.114×10^{-14}
10	0.292
25	1.008
40	2.919
60	9.614

water is present, this constant may be employed to calculate the concentrations of hydrogen and hydroxyl ions present in such solutions. For instance, in 0.001 molar aqueous sodium hydroxide at 25°C $C_{OH^-} = 0.001$, and therefore the concentration of hydrogen ions must be

$$C_{H^+} = \frac{K'_w}{C_{OH^-}}$$
$$= \frac{1.01 \times 10^{-14}}{1 \times 10^{-3}}$$
$$= 1.01 \times 10^{-11}$$

Similarly may be calculated the concentration of hydroxyl ions in any acid solution.

pH AND pOH

In passing from acid to alkaline solutions the concentration of hydrogen ions can vary within very wide limits. To permit a convenient means of expressing the concentration of hydrogen ions without involving negative exponents, Sørensen suggested the use of the *pH* (puissance d'hydrogen) *scale*. On this scale the pH of any solution is defined as

$$\text{pH} = -\log_{10} a_{H^+} \tag{111}$$

i.e., the hydrogen ion activity of a solution is equal to 10^{-pH}. Thus, for a solution of pH = 4, $a_{H^+} = 10^{-4}$, while for a solution of pH = 12, $a_{H^+} = 10^{-12}$. Similarly, the activity of hydroxyl ions can be expressed on a *pOH scale* by the definition

$$\text{pOH} = -\log_{10} a_{OH^-} \tag{112}$$

Since in any aqueous solution $a_{H^+}a_{OH^-} = K_w$, we have

$$-\log_{10} a_{H^+} - \log_{10} a_{OH^-} = -\log_{10} K_w$$

and, therefore,

$$pH + pOH = - \log_{10} K_w \tag{113}$$

From this equation it follows that the sum of pH and pOH for any aqueous solution must always be a constant equal to $- \log_{10} K_w$, and hence when the pH of a solution goes up the pOH must decrease, and vice versa. For neutrality the concentrations of hydrogen and hydroxyl ions must be equal, and so must pH and pOH. Then

$$pH = pOH = \frac{- \log_{10} K_w}{2} \tag{114}$$

Any solution having a pH lower than $-\frac{1}{2} \log_{10} K_w$ will thus be acid, while any solution having a pH higher than this value will be alkaline.

In the specific case of aqueous solutions *at 25°C*, $K_w = 1 \times 10^{-14}$, and $- \log_{10} K_w = 14$. For this temperature, then, pH + pOH = 14, and the neutral solution has a pH of seven. Any solution of pH lower than seven will be acid, while any solution of pH above seven will be alkaline.

BUFFER SOLUTIONS

Solutions composed of an acid and one of its salts, or of a base and one of its salts, possess the ability to resist to a greater or lesser degree changes in pH when some acid or base is added to them. Such solutions exhibiting the property of opposing a change in their pH are called *buffer solutions*, or simply *buffers*. Particularly effective in this respect are mixtures in which the acid or base involved is weak.

The buffering action of a solution of a weak acid HA in presence of one of its salts is explainable as follows. In any mixture of acid and salt the ionization equilibrium of the acid determines the hydrogen ion concentration of the solution. When hydrogen ions are added to this solution in the form of some acid, the equilibrium is disturbed by the presence of excess hydrogen ions. To remove this excess and to re-establish the ionization equilibrium, hydrogen ions combine with anions of the salt to form molecules of unionized acid HA, and thereby the pH of the solution reverts to a value not far different from what it was originally. Again, when base is added to the mixture of acid and salt, the excess of hydroxyl ions disturbs the ionization equilibrium of the water. To re-establish this equilibrium hydrogen and hydroxyl ions combine to form water. As this reaction removes hydrogen ions, the ionization equilibrium of the acid is also disturbed, and to re-establish the latter some of the acid HA ionizes to yield the requisite hydrogen ions. The result of these changes is that the excess hydroxyl ions are essentially neutralized, and the solution reverts to a pH close to its original value. These considerations apply only when the amount of acid or base added to a buffer solution is not so large as to change to any great extent the ratio of acid to salt.

In a like manner can be explained the buffering action of solutions of a weak base BOH and one of its salts. The particular pH at which a buffer solution is effective is determined by the ratio of acid or base to salt present and by the magnitude of the ionization constant of the acid or base. Thus, for a weak acid the hydrogen ion activity of the buffer is determined from the ionization constant by

$$a_{H^+} = K_a \frac{a_{HA}}{a_{A^-}} \tag{115}$$

while for a weak base the hydroxyl ion activity is controlled by

$$a_{OH^-} = K_b \frac{a_{BOH}}{a_{B^+}} \tag{116}$$

When the activity ratios of acid to salt or of base to salt are unity, these equations reduce to $a_{H^+} = K_a$ and $a_{OH^-} = K_b$. However, when these ratios are not unity, a_{H^+} and a_{OH^-} must be estimated from Eq. (115) or (116).

Buffers are used whenever solutions of known pH are required, or when-ever it is necessary to keep the pH of a solution constant. MacIlvaine has given instructions for preparing buffer solutions of definite pH rang-ing from pH = 3.4 to pH = 8.0 in 0.2 unit steps. These are based on the use of disodium acid phosphate and citric acid solutions in various propor-tions. Another system of buffer mixtures ranging from pH = 1 to pH = 10 in 0.2 unit steps has been proposed by Clark and Lubs. Details for preparing these buffers may be found in Lange's Handbook.[11]

GENERALIZED CONCEPTS OF ACIDS AND BASES

The term *acid* is ordinarily taken to mean any substance that yields hydrogen ions in solution, and *base* any substance that yields hydroxyl ions. Although these narrow definitions of acid and base may be satis-factory for some purposes in aqueous solutions, they are altogether insuf-ficient to cover all observed phenomena either in water or in nonaqueous solvents.

A hydrogen ion has heretofore been considered to be a hydrogen atom with the electron removed, i.e., a *proton*. However, Brönsted[12] has shown that the free energy change of the reaction

$$H^+ + H_2O(l) = H_3O^+$$

[11] N. A. Lange, *Handbook of Chemistry*, Handbook Publishers, Inc., Sandusky, Ohio, 1956.

[12] Brönsted, *Chem. Rev.*, **5,** 231 (1928).

is extremely large and negative, and hence the equilibrium constant of this reaction must also be very large. H^+ ions as such must be, then, practically nonexistent in aqueous media, and what we think of as the hydrogen ion is actually the hydrated proton, H_3O^+. The latter ion has variously been designated as the *hydronium, hydroxonium,* or *oxonium* ion. Likewise, certain studies of the glass electrode by Dole[13] and of the kinetics of reactions subject to acid catalysis have led to the conclusion that the proton is solvated.

These and other considerations have led Brönsted[14] and Lowry[15] to a redefinition of the concept of acid and base. They define an *acid* as *any substance that can donate a proton to any other substance*. Again, they define a *base* as *any substance that can accept a proton from an acid*. In other words, an acid is any substance, whether charged or uncharged, that can act as a *proton donor*, while any substance, whether charged or uncharged, that can act as a *proton acceptor* is a base. Inherent in these new definitions of acid and base is the significant fact that when an acid gives off a proton there must be a base to receive it; and, vice versa, no base can act as such unless there is an acid present to donate protons to it.

The differences between this *generalized concept of acids and bases* and the older, more restricted concept can best be brought out with several examples. According to both points of view acetic acid is an acid. However, whereas the older concept represents the ionization of this acid by the process

$$CH_3COOH = H^+ + CH_3COO^- \tag{117a}$$

the generalized concept represents the ionization as

$$CH_3COOH + H_2O(l) = H_3O^+ + CH_3COO^- \tag{117b}$$

In the latter equation the acetic acid donates a proton to a water molecule, which acts as a base, to form a hydronium ion and an acetate ion. Furthermore, since H_3O^+ can donate a proton to the acetate ion to form acetic acid and water, H_3O^+ itself must be an acid, while CH_3COO^- must be a base. Consequently, any interaction of an acid and a base must always result in the formation of another acid and another base; i.e.,

$$Acid_1 + Base_1 = Acid_2 + Base_2 \tag{118}$$

$Base_2$, which results from $Acid_1$, is said to be the base conjugate to $Acid_1$. Similarly, $Acid_2$, which results from $Base_1$, is said to be conjugate to $Base$,

[13] Dole, *J. Am. Chem. Soc.*, **54**, 2120, 3095 (1932).
[14] Brönsted, *Chem. Rev.*, **5**, 231 (1928); *Rec. trav. chim.*, **42**, 718 (1923).
[15] Lowry, *J. Chem. Soc.*, **123**, 848 (1923).

By extending this argument, we obtain the following formulations for the ionization in water of a number of generalized acids:

Acid$_1$		Base$_1$		Acid$_2$		Base$_2$
HCl	+	H_2O	=	H_3O^+	+	Cl^-
HCOOH	+	H_2O	=	H_3O^+	+	$HCOO^-$
HSO_4^-	+	H_2O	=	H_3O^+	+	SO_4^{--}
NH_4^+	+	H_2O	=	H_3O^+	+	NH_3
$C_6H_5NH_3^+$	+	H_2O	=	H_3O^+	+	$C_6H_5NH_2$
H_3O^+	+	H_2O	=	H_3O^+	+	H_2O
H_2O	+	H_2O	=	H_3O^+	+	OH^-
H_2SO_3	+	H_2O	=	H_3O^+	+	HSO_3^-
HSO_3^-	+	H_2O	=	H_3O^+	+	SO_3^{--}

From these examples it may be seen that besides substances ordinarily considered to be acids, H_3O^+, H_2O, NH_4^+, $C_6H_5NH_3^+$, and other proton donors are also acids. Again, bases are not only substances which possess hydroxyl ions, but also anions of acids, water, ammonia, aniline, HSO_3^-, and other proton acceptors. It will be observed that water may act as both acid or base, depending on the conditions and the reaction, i.e., water is *amphoteric*. Further, an ion like HSO_3^- is also amphoteric, for it may act as an acid or base, depending on circumstances of the reaction.

The strength of a given acid in the new theory is measured by its ability to donate protons to the solvent and is expressible by the ionization constant of the acid. On the other hand, the strength of a given base K_B is defined as

$$K_B = \frac{1}{K_a} \tag{119}$$

where K_a is the ionization constant of the acid conjugate to the base. For *aqueous* solutions it can also be shown that

$$K_B = \frac{K_b}{K_w} \tag{120}$$

where K_b is the ionization constant of a base as defined by Eq. (96).

The generalized concept of acids and bases does not contradict the older views of these substances, but rather extends them. And, although the ideas involved may appear strange to one not accustomed to thinking in these terms, we shall see in a subsequent chapter that these ideas are very fruitful and permit the correlation and explanation of certain kinetic phenomena in aqueous solutions which would otherwise appear to be very puzzling.

A theory of acids and bases even broader in scope than the preceding has been proposed by G. N. Lewis. According to Lewis, *a base is any substance which donates a pair of electrons* to the formation of a coordinate bond. In turn, *an acid is any substance which accepts a pair of electrons* to form such a bond. This theory embraces not only the acids and bases in the Brönsted-Lowry theory, but many other substances which are usually not classified in this category. For details the student must be referred to the literature.[16]

HYDROLYSIS

The reaction between an acid and a base in aqueous solution always results in the formation of water and a salt. If this reaction were complete, exact neutralization would always give an exactly neutral solution; i.e., one containing no excess of hydrogen or hydroxyl ions. This is very nearly the case when a strong acid like hydrochloric is neutralized by a strong base like sodium hydroxide. However, when a weak acid like acetic is neutralized by sodium hydroxide, it is found that the final solution is not neutral but basic. Again, when a weak base like ammonium hydroxide is neutralized by hydrochloric acid, the final solution is acid. The basicity of the final solution in one case and acidity in the other are due to the tendency of the salt formed by neutralization to react with water and thereby reverse the neutralization. This tendency of salts when dissolved in water to react with the solvent and thereby reverse the neutralization process is called *hydrolysis*.

In considering the hydrolytic behavior of various salts, four cases may be distinguished, namely, (a) salts of strong acids and strong bases, (b) salts of weak acids and strong bases, (c) salts of strong acids and weak bases, and (d) salts of weak acids and weak bases. Each of these categories will be discussed now in turn.

Salts of Strong Acids and Strong Bases.   Sodium chloride may be taken as an example of the hydrolytic behavior of a salt of a strong acid and a strong base. This salt exists in aqueous solution as sodium and chloride ions. If these two ions were to react with the water, the products would be hydrochloric acid and sodium hydroxide. However, since the latter two are also strong electrolytes, the products would again be sodium and chloride ions, and the hydrogen and hydroxyl ions would recombine to form water. In other words, the products of hydrolysis would be identical with the reactants, and there would be no change in the nature of the species in solution. We may say, therefore, that a salt of a strong acid and a strong base does not hydrolyze, and the solution of such a salt is essentially neutral.

Salts of Weak Acids and Strong Bases.   When a salt of a weak acid and a strong base, such as sodium acetate, is dissolved in water, the

---

[16] See Luder and Zuffanti, *The Electronic Theory of Acids and Bases*, John Wiley & Sons, Inc., New York, 1946.

cation of the base, i.e., the sodium ion, will not undergo hydrolysis for the reason given above. However, the anion of the weak acid, i.e., the acetate ion, will react with water to form sufficient unionized acetic acid for the ionization constant of the acid to hold. The result is that the acetate ion undergoes the hydrolytic reaction

$$CH_3COO^- + H_2O = CH_3COOH + OH^-$$

which leads also to the formation of hydroxyl ions, and the solution becomes alkaline.

In general, the hydrolysis of any salt BA of a weak acid HA and a strong base BOH is due to the hydrolysis of the anion of the acid. This hydrolysis of the anion can be represented by the equation

$$A^- + H_2O = HA + OH^- \tag{121}$$

for which the equilibrium constant K_h, again including the activity of the water in the constant, is given by

$$K_h = \frac{a_{HA}a_{OH^-}}{a_{A^-}} \tag{122}$$

The constant K_h, called the *hydrolytic constant* of the ion A^-, determines the extent to which the ion A^- will react with water to form HA and OH^-. The magnitude of this constant, in turn, depends on the ionization constant of the acid HA, K_a, and the ion product of water K_w, as may be shown as follows. If the numerator and denominator of the right-hand side of Eq. (122) are multiplied by a_{H^+}, we obtain

$$K_h = \left(\frac{a_{HA}}{a_{A^-}a_{H^+}}\right)(a_{H^+}a_{OH^-})$$

But $\left(\dfrac{a_{HA}}{a_{A^-}a_{H^+}}\right) = \dfrac{1}{K_a}$, and $K_w = a_{H^+}a_{OH^-}$. Therefore,

$$K_h = \frac{K_w}{K_a} \tag{123}$$

and the hydrolytic constant may be calculated from the ion product of water and the ionization constant of the weak acid.

Salts of Strong Acids and Weak Bases. In salts of this class, of which ammonium chloride is an example, the anion of the strong acid will suffer no hydrolysis. But the cation B^+ of the weak base BOH will undergo hydrolysis according to the reaction

$$B^+ + H_2O = BOH + H^+ \tag{124}$$

This time the hydrolysis yields molecules of the unionized base and hydrogen ions, and hence the solution of the salt in water is acid. The hydrolytic constant is now given by

$$K_h = \frac{a_{BOH}a_{H^+}}{a_{B^+}} \tag{125}$$

If the numerator and denominator of this expression are multiplied by a_{OH^-}, K_h becomes

$$K_h = \left(\frac{a_{BOH}}{a_{B^+}a_{OH^-}} \right) (a_{H^+}a_{OH^-})$$

But the first quantity in parentheses is $1/K_b$, where K_b is the ionization constant of the weak base BOH. Again, $K_w = a_{H^+}a_{OH^-}$. Therefore,

$$K_h = \frac{K_w}{K_b} \tag{126}$$

and the hydrolytic constant of a salt of a strong acid and a weak base can be calculated from the ion product of water and the ionization constant of the weak base BOH.

Salts of Weak Acids and Weak Bases.   When the salt BA is the product of the interaction of a weak acid HA and a weak base BOH, such as ammonium acetate, both the cation and anion of the salt undergo hydrolysis. The reaction is

$$B^+ + A^- + H_2O = BOH + HA \tag{127}$$

and whether the solution of the salt in water is acid or basic is determined by the relative strengths of the acid and base. In this instance the hydrolytic constant is defined by

$$K_h = \frac{a_{BOH}a_{HA}}{a_{B^+}a_{A^-}} \tag{128}$$

If now the numerator and denominator of Eq. (128) are multiplied by $a_{H^+}a_{OH^-}$, K_h becomes

$$K_h = \left(\frac{a_{BOH}}{a_{B^+}a_{OH^-}} \right) \left(\frac{a_{HA}}{a_{H^+}a_{A^-}} \right) (a_{H^+}a_{OH^-})$$

But $\left(\dfrac{a_{BOH}}{a_{B^+}a_{OH^-}} \right) = \dfrac{1}{K_b}$, $\left(\dfrac{a_{HA}}{a_{H^+}a_{A^-}} \right) = \dfrac{1}{K_a}$, $(a_{H^+}a_{OH^-}) = K_w$. Therefore,

$$K_h = \frac{K_w}{K_b K_a} \tag{129}$$

This time the ionization constants of both the weak acid and the weak base are involved in the expression for the hydrolytic constant of the salt.

CALCULATIONS INVOLVING HYDROLYTIC CONSTANTS

Hydrolytic constants of salts may be calculated from known K_a, K_b, and K_w values by means of Eq. (123), (126), or (129), or they may be determined experimentally. In the latter instance the usual practice is to dissolve a given amount of the salt in water and measure the hydrogen ion concentration of the solution. Once this quantity and the original concentration of the salt are known, the hydrolytic constant may be calculated. In all the calculations which follow it will be assumed that concentrations may be substituted for activities. To illustrate the calculation of K_h from measurements of the hydrogen ion concentration of a salt in water, consider the following example.

EXAMPLE: An 0.02 molar solution of sodium acetate in water at 25°C is found to have a hydrogen ion concentration of 3.0×10^{-9} gram ionic weights per liter. What is the hydrolytic constant of the salt? The hydrolytic reaction of this salt is given by

$$CH_3COO^- + H_2O = CH_3COOH + OH^-$$

Since $C_{H^+} = 3.0 \times 10^{-9}$, and since $K_w = 1.01 \times 10^{-14} = C_{H^+}C_{OH^-}$, the concentration of hydroxyl ions must be

$$C_{OH^-} = \frac{K_w}{C_{H^+}} = \frac{1.01 \times 10^{-14}}{3.0 \times 10^{-9}}$$
$$= 3.37 \times 10^{-6}$$

This must also be the concentration of the acetic acid, as the latter is formed in quantity equivalent to the hydroxyl ions during the hydrolysis. Finally, the concentration of acetate ions at hydrolytic equilibrium must be the original concentration of the salt minus the amount reacted, or

$$C_{Ac^-} = 0.02 - 3.37 \times 10^{-6}$$
$$= 0.02$$

to a very near approximation. Therefore,

$$K_h = \frac{C_{HA}C_{OH^-}}{C_{Ac^-}}$$
$$= \frac{(3.37 \times 10^{-6})(3.37 \times 10^{-6})}{0.02}$$
$$= 5.68 \times 10^{-10}$$

Calculated from $K_w = 1.01 \times 10^{-14}$ and $K_a = 1.75 \times 10^{-5}$ for acetic acid at 25°C, $K_h = 5.77 \times 10^{-10}$.

Once K_h is known, it may be employed to estimate the degree of hydrolysis of the salt under other conditions. Thus, suppose it is desired to know the degree of hydrolysis of sodium acetate in 0.01 molar solution at 25°C. If we let α be the degree of hydrolysis of the acetate ion, then $C_{Ac^-} = 0.01(1 - \alpha)$, and $C_{HAc} = C_{OH^-} = 0.01\ \alpha$. Consequently,

$$5\ 77 \times 10^{-10} = \frac{(0.01\ \alpha)(0.01\ \alpha)}{0.01(1 - \alpha)}$$

As K_h is small, so will be α, and we may write $1 - \alpha = 1$. Then

$$\frac{(0.01)^2 \alpha^2}{0.01} = 5.77 \times 10^{-10}$$

$$\alpha = 2.40 \times 10^{-4}$$

i.e., the acetate ion is hydrolyzed to the extent of 0.024 per cent in 0.01 molar solution at 25°C.

THE SOLUBILITY PRODUCT

A particularly important type of heterogeneous ionic equilibrium is involved in the solubility of difficultly soluble salts in water. When a difficulty soluble salt, such as barium sulfate, is agitated with water until the solution is *saturated*, the equilibrium established between the solid phase and the completely ionized salt in solution is given by

$$BaSO_4(s) = Ba^{++} + SO_4^{--} \tag{130}$$

For this process the equilibrium constant is

$$K_s = a_{Ba^{++}} a_{SO_4^{--}} \tag{131}$$

i.e., *in any solution saturated with barium sulfate the product of the activities of barium and sulfate ions is a constant equal to K_s.* The constant K_s is called the *solubility product constant*, or simply the *solubility product*, of the salt.

Every difficultly soluble salt has a solubility product constant of its own. For salt yielding only two ions, such as barium sulfate or silver chloride, the solubility product is merely the product of the activities of the two ions. However, for salts yielding more than two ions the solubility product expression is a little more complicated. The solubility equilibrium of any salt $A_x B_y$ yielding x positive ions of A and y negative ions of B is given by

$$A_x B_y(s) = xA + yB \tag{132}$$

and hence the most general expression for the solubility product is

$$K_s = a_A^x a_B^y \tag{133}$$

Thus, for a salt such as silver carbonate, for which $x = 2$ and $y = 1$, $K_s = a_{Ag^+}^2 a_{CO_3^{--}}$, while for calcium fluoride $x = 1$, $y = 2$, and $K_s = a_{Ca^{++}} a_F^2_{-}$. The solubility products are readily evaluated from the saturation solubilities of salts in pure water. These solubilities may be obtained by direct analysis of the saturated solutions, by conductance measurements as described in this chapter, or by electromotive force measurements. The calculations involved may be illustrated with the following example. At 25°C the solubility of silver carbonate in water is 1.16×10^{-4} mole per liter. Hence the concentration of carbonate ions in the saturated solution is $C_{CO_3^{--}} = 1.16 \times 10^{-4}$, while that of the silver ions is twice that of the carbonate, or $C_{Ag^+} = 2.32 \times 10^{-4}$ gram ionic weight per liter. For silver carbonate K_s is given by

$$K_s = a_{Ag^+}^2 a_{CO_3^{--}}$$
$$= (C_{Ag^+}^2 C_{CO_3^{--}})(f_{Ag^+}^2 f_{CO_3^{--}})$$

If we assume now that the activity coefficients are unity in the very dilute solution involved, then K_s reduces to K_s', the stoichiometric solubility product. Therefore,

$$K_s = K_s' = C_{Ag^+}^2 C_{CO_3^{--}}$$
$$= (2.32 \times 10^{-4})^2 (1.16 \times 10^{-4})$$
$$= 6.2 \times 10^{-12}$$

TABLE 11–16. SOLUBILITY PRODUCTS FOR VARIOUS SUBSTANCES AT 25°C

Substance	K_s	Substance	K_s
Al(OH)$_3$	3.7×10^{-15}	HgI$_2$	3.2×10^{-29}
BaSO$_4$	1.08×10^{-10}	AgBr	7.7×10^{-13}
CaCO$_3$	8.7×10^{-9}	AgCl	1.56×10^{-10}
CuS (18°)	8.5×10^{-45}	AgI	1.5×10^{-16}
Fe(OH)$_3$ (18°)	1.1×10^{-36}	Ag$_2$CO$_3$	6.15×10^{-12}
Fe(OH)$_2$ (18°)	1.64×10^{-14}	Ag$_2$CrO$_4$	9×10^{-12}
PbI$_2$	1.39×10^{-8}	Ag$_2$S	1.6×10^{-49}
Mg(OH)$_2$ (18°)	1.2×10^{-11}	SrCO$_3$	1.6×10^{-9}
HgBr$_2$	8×10^{-20}	TlCl	2.02×10^{-4}

In Table 11–16 are listed the solubility products at 25°C for a number of salts as well as for several difficultly soluble hydroxides to which the solubility product principle is applicable. From these constants the solubility of these substances in water can readily be obtained by reversing the calculations for the solubility product given above.

SALT EFFECTS AND SOLUBILITY

From the formulation of the solubility product in terms of concentrations it is to be expected that addition of salts possessing an ion in common with the dissolving salt should decrease its solubility. Such is actually the case, as may be seen from Figure 11–12 for the effects of added $TlNO_3$ and KCl on the solubility of TlCl at 25°C. However, these effects are less than calculated from the relation

$$K'_s = C_{Tl^+}C_{Cl}$$

for the common ion effect (bottom curve in the figure). Furthermore, the upper three curves in the figure show that electrolytes without a common ion actually increase the solubility.

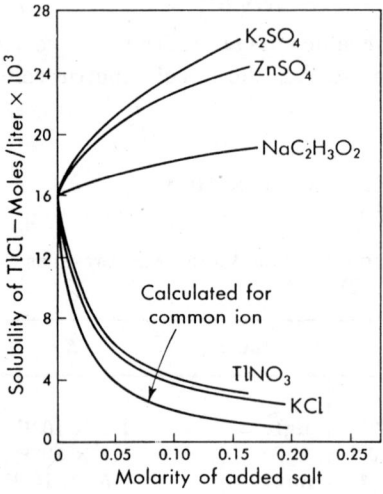

Figure 11–12. Effect of salts on solubility of TlCl at 25°C.

Although inexplicable from the concentration approach, this marked increase in solubility with electrolyte concentration is just what is to be expected from the more exact activity formulation of the solubility product. The thermodynamic solubility product of thallous chloride is given by

$$\begin{aligned} K_s &= a_{Tl^+}a_{Cl^-} \\ &= (C_{Tl^+}C_{Cl^-})(f_{Tl^+}f_{Cl^-}) \\ &= K'_s f^2 \end{aligned} \tag{134}$$

where f is the mean activity coefficient of the salt and K'_s the concentration solubility product. From Eq. (134) it follows that, since K_s must be a true constant, K'_s cannot be constant and equal to K_s unless $f = 1$. However, f is not unity except at zero ionic strength and varies with ionic strength as the latter is changed. Consequently, K'_s must also be a function of the ionic strength and must vary with it according to the dependence of f^2 on μ in the expression

$$K_s' = \frac{K_s}{f^2} \tag{135}$$

Now, at not too excessively high ionic strengths the activity coefficients are less than one, and consequently at these ionic strengths K_s' must be greater than K_s. Since the solubility is equal to $\sqrt{K_s'}$, the solubility in presence of added salt must also be greater. We see, therefore, that whereas the thermodynamic solubility product K_s remains constant, theory predicts that the concentration solubility product K_s' should increase with increase in ionic strength, and so should the solubility, as is the case. This argument accounts also for the fact that the observed solubilities in presence of common ions are greater than predicted when no allowance is made for activity coefficients. This variation of solubility with concentration of added electrolytes may be employed to evaluate the activity coefficients of the *dissolving salt* at various ionic strengths. The procedure may again be illustrated with thallous chloride. If we represent the solubility of this salt in pure water by S_0, in presence of added electrolyte with no common ion by S, and let f_0 and f be the mean activity coefficients of the salt in the two solutions, then, in view of the constancy of K_s, we have the relation

$$K_s = S_0^2 f_0^2 = S^2 f^2 \tag{136}$$

and, therefore,

$$Sf = S_0 f_0 \tag{137}$$

Taking logarithms, we have

$$\log S = \log S_0 f_0 - \log f \tag{138}$$

According to this expression f can be evaluated from the solubilities as soon as the product $S_0 f_0$ is available. To obtain this product $\log S$ is plotted against the square root of the total ionic strength of the solution, i.e., of dissolved salt and added electrolyte, and the plot extrapolated to $\sqrt{\mu} = 0$. As at zero ionic strength $f = 1$, the intercept on the $\log S$ axis gives immediately $\log S_0 f_0$. Once this quantity is known, subtraction of $\log S_0 f_0$ from $\log S$ yields $-\log f$, and hence f, at various ionic strengths. Or, on evaluating $S_0 f_0$, f follows from Eq. (137) by dividing $S_0 f_0$ by S.

This procedure for evaluating f is illustrated with the data given in Table 11-17 for the solubility of thallous chloride at 25°C in water and in presence of various concentrations of potassium nitrate. Column 1 gives the concentration C of added salt, column 2 the observed solubility S of thallous chloride in moles per liter, and the third column $\mu = (C + S)$.

TABLE 11-17. SOLUBILITY OF TlCl IN PRESENCE OF KNO_3 AT 25°C

Conc. KNO_3 Added C	Solubility TlCl S	$\mu = (C + S)$	Mean Activity Coefficient of TlCl f	$K_s = S^2 f^2$
0	0.01607	0.01607	0.885	2.02×10^{-4}
0.02	0.01716	0.03716	0.829	2.02
0.05	0.01826	0.06826	0.779	2.02
0.16	0.01961	0.11961	0.725	2.02
0.30	0.02312	0.32313	0.615	2.02
1.00	0.03072	1.03072	0.463	2.02

From a plot of log S against $\sqrt{\mu}$ it is found that $S_0 f_0 = 0.01422$. Dividing now $S_0 f_0$ by S at the various ionic strengths, the activity coefficients in column 4 result. Finally, the last column gives the thermodynamic solubility product of thallous chloride as calculated from $K_s = S^2 f^2$. As it should be, K_s is constant throughout.

IONIC EQUILIBRIA AND TEMPERATURE

Equilibrium constants of ionic reactions, like all other equilibrium constants, vary with temperature. As before, the variation with temperature is given by

$$\frac{d \ln K}{dT} = \frac{\Delta H^0}{RT^2} \tag{139}$$

and this expression may be employed to evaluate the heats of various ionic reactions or to estimate K at one temperature from that at another when ΔH^0 is known.

REFERENCES

1. Creighton and Koehler, *Principles and Applications of Electrochemistry*, John Wiley & Sons, Inc., New York, 1943, Vol. I.
2. P. Delahay, *Advances in Electrochemistry*, John Wiley & Sons, Inc., New York, 1961–1963, Vol. I and III.
3. S. Glasstone, *Introduction to Electrochemistry*, D. Van Nostrand Company, Inc., Princeton, N.J., 1942.
4. Harned and Owen, *The Physical Chemistry of Electrolytic Solutions*, Reinhold Publishing Corporation, New York, 1958.
5. D. A. MacInnes, *The Principles of Electrochemistry*, Reinhold Publishing Corporation, New York, 1939.
6. Robinson and Stokes, *Electrolyte Solutions*, Academic Press, New York, 1959.

7. A. Weissberger, *Physical Methods of Organic Chemistry*, Interscience Publishers, Inc., New York, 1960. Vol. IV, Chapters 45 and 46.

PROBLEMS

Note: In problems where insufficient data are given for ascertaining activities of ions of strong electrolytes, assume ionic activity equal to ionic concentration.

1. A direct current of 0.5 int. amp flows through a circuit for 10 min under an applied potential of 30 int. volts. Find the quantity of electricity transported by the current in (a) int. coulombs, (b) abs. coulombs, (c) emu, and (d) esu.

Ans. (a) 300.0 int. coulombs.

2. What is the rate of dissipation of energy by the current mentioned in the preceding problem in (a) int. watts, (b) abs. watts, (c) ergs sec^{-1}, (d) cal sec^{-1}?

3. When a potential of 110 volts d-c is applied to the terminals of an electric lamp, a current of 2 amp is found to flow. (a) What is the resistance of the lamp, and (b) how many calories of heat are dissipated per hour?

Ans. (a) 55 ohms; (b) 189,300 cal.

4. A constant direct current flows through an iodine coulometer for a period of 2 hours. At the end of this time it is found that the coulometer contains 0.0020 equivalents of liberated I_2. What was the current passing through the coulometer?

Ans. 0.027 amp.

5. The platinum crucible used in a silver coulometer gains 0.500 g in a certain electrolysis. What would be the gain in weight of a copper cathode in a cell filled with potassium cuprocyanide $[KCu(CN)_2]$ placed in the same circuit?

6. What volume of O_2 would be liberated from an aqueous solution of NaOH by a current of 2 amp flowing for $1\frac{1}{2}$ hr? The temperature is 27°C and the total pressure is 1 atm. *Ans.* 688.8 cc.

7. (a) How long would it take a current of 1 amp to reduce completely 80 cc of 0.1 molar $Fe_2(SO_4)_3$ to $FeSO_4$? (b) How many cc of 0.1 molar $K_2Cr_2O_7$ could be reduced to chromic sulfate, $Cr_2(SO_4)_3$, by the same quantity of electricity?

8. What quantity of electricity would be required to reduce 10 g of nitrobenzene, $C_6H_5NO_2$, completely to aniline, $C_6H_5NH_2$? If the potential drop across the cell is 2 volts, how much energy, in calories, is consumed in the process?

Ans. 47,070 coulombs; 22,510 cal.

9. A $AgNO_3$ solution containing 0.00739 g of $AgNO_3$ per gram of H_2O is electrolyzed between silver electrodes. During the experiment 0.078 g of Ag plate out at the cathode. At the end of the experiment the anode portion contains 23.14 g of H_2O and 0.236 g of $AgNO_3$. What are the transport numbers of Ag^+ and NO_3^- ions? *Ans.* $t_+ = 0.47$.

10. A 4 molal solution of $FeCl_3$ is electrolyzed between platinum electrodes. After the electrolysis the cathode portion, weighing 30 g, is 3.15 molal in $FeCl_3$ and 1.00 molal in $FeCl_2$. What are the transport numbers of Fe^{+++} and Cl^- ions?

11. The transference numbers of the ions in 1.000 N KCl were determined by the moving boundary method using 0.80 N BaCl$_2$ as the following solution. With a current of 0.0142 amp, the time required for the boundary to sweep through a volume of 0.1205 cc was 1675 sec. What are the transport numbers of K$^+$ and Cl$^-$ ions?

Ans. $t_+ = 0.49$.

12. The cathode, center, and anode chambers of an electrolytic cell contain each 10 milliequivalents of HCl in aqueous solution. What will be the number of milliequivalents of HCl in each compartment after the passage of 5 milliequivalents of electricity through the cell? Assume that t_+ is 0.8 and that H$_2$ is given off at the cathode and Cl$_2$ at the anode.

13. A conductivity cell filled with 0.10 demal KCl solution gives at 25°C a resistance of 910 ohms. What will be the resistance when this cell is filled with a solution whose specific conductance at 25°C is 0.00532 mho?

14. A conductivity cell filled with 0.01 demal KCl solution gives at 0°C a resistance of 11,210 ohms. The distance between the electrodes in the cell is 6 cm. Find (a) the cell constant and (b) the average cross-sectional area of the electrodes.

Ans. (a) 8.673.

15. At 25°C the equivalent conductances of dilute NaI solutions are as follows:

Molarity	Λ
0.0005	125.36
0.0010	124.25
0.0050	121.25

Find Λ_0 of NaI at 25°C.

16. At 25°C a cell filled with 0.01 demal KCl solution gave a resistance of 484.0 ohms. The following data for NaCl solutions were then taken in the same cell at 25°C:

Normality	Resistance (ohms)
0.0005	10,910
0.0010	· 5,494
0.0020	2,772
0.0050	1,129

(a) Calculate Λ for NaCl at each concentration, and (b) evaluate Λ_0 by plotting Λ against $\sqrt{C}$ and extrapolating to infinite dilution.

17. From the following equivalent conductances at infinite dilution and 18°C find Λ_0 for NH$_4$OH:

$$Ba(OH)_2: \quad \Lambda_0 = 228.8$$
$$BaCl_2: \quad \Lambda_0 = 120.3$$
$$NH_4Cl: \quad \Lambda_0 = 129.8$$

18. In measuring the mobility of H$^+$ ions by the moving boundary procedure, it is observed that the boundary moves a distance of 4.0 cm in 12.52 min. The voltage drop across the cell before the formation of the boundary is 16.0 volts. The distance between electrodes is 9.6 cm. Calculate from these data the mobility and equivalent ionic conductance of hydrogen ions.

Ans. $\mu_+ = 0.0032$ cm^2 volt^{-1} sec^{-1}.

19. At 25°C the equivalent conductance of a 0.020 molar AgNO$_3$ solution is 128.7, while the transport number of Ag$^+$ is 0.477. Calculate the equivalent ionic conductances and the ionic mobilities of Ag$^+$ and NO$_3^-$ in a 0.020 molar solution of AgNO$_3$.

20. At 18°C the mobility at infinite dilution of the ammonium ion is 0.00066 while that of the chlorate ion is 0.00057 cm^2 volt^{-1} sec^{-1}. Calculate Λ_0 of ammonium chlorate and the transport numbers of the two ions.

21. For AgNO$_3$ in aqueous solution $\Lambda_0 = 133.36$ at 25°C. Using the Onsager equation find Λ for a 0.0010 molar solution at the same temperature, and compare the result with the experimentally observed value of $\Lambda = 130.5$.

22. At 25°C the resistance of a cell filled with 0.01 demal KCl solution is 525 ohms. The resistance of the same cell filled with 0.1 N NH$_4$OH is 2030 ohms. What is the degree of dissociation of NH$_4$OH in this solution?

Ans. $\alpha = 0.0134$.

23. What will be the resistance of the cell used in the preceding problem when it is filled with H$_2$O having a specific conductance of 2.00×10^{-6} mho cm^{-1}?

24. The specific conductance at 25°C of a saturated aqueous solution of SrSO$_4$ is 1.482×10^{-4}, while that of the H$_2$O used is 1.5×10^{-6} mho cm^{-1}. Determine at 25°C the solubility in grams per liter of SrSO$_4$ in water. *Ans.* 0.0967 g/liter.

25. In the titration of 25.0 cc (diluted to 300 cc) of a NaC$_2$H$_3$O$_2$ solution with 0.0972 N HCl solution the following data were found:

Volume of HCl Used	Conductance $\times 10^4$
10.0	3.32
15.0	3.38
20.0	3.46
45.0	4.64
50.0	5.85
55.0	7.10

What is the concentration of the NaC$_2$H$_3$O$_2$ solution in moles per liter?

26. Sketch the conductance against volume of reagent curves for the following titrations: (a) C$_6$H$_5$OH with NaOH; (b) CuSO$_4$ with NaOH; (c) K$_2$CrO$_4$ with AgNO$_3$.

27. For each of the following solutions evaluate the mean molality, the mean ionic activity, and the activity of the salt:

	Molality	Mean Activity Coefficient
K$_3$Fe(CN)$_6$	0.010	0.571
CdCl$_2$	0.100	0.219
H$_2$SO$_4$	0.050	0.397

28. The density of a 1.190 molar solution of $CaCl_2$ in water is 1.100 g/cc at 20°C. What is the molality of the solution and the ratio of f to γ?

Ans. 1.228 molal; $f/\gamma = 1.029$.

29. Using the Debye-Hückel limiting law, calculate the mean ionic activity coefficient at 25°C of a 0.001 molar solution of $K_3Fe(CN)_6$, and compare with the observed value of 0.808.

30. A solution is 0.002 molar in $CoCl_2$ and 0.002 molar in $ZnSO_4$. Calculate the activity coefficient of Zn^{++} ions in the solution using the Debye-Hückel limiting law.

31. Ascertain whether Eq. (85) of this chapter predicts maxima or minima in the variation of $\log_{10} f$ with $\sqrt{\bar{\mu}}$.

32. A solution of a strong electrolyte A_xB_y has a molality m, activity of solute a_2, and mean activity coefficient γ_2 at temperature T. (a) Write down the expression for the partial molal free energy of the solute in this solution. (b) Using this expression, derive the equation for the variation of a_2 with temperature at constant m. (c) From the expression obtained in (b), deduce the equation for the variation of γ_2 with temperature at constant m.

33. The equivalent conductance of a 0.0140 N solution of chloracetic acid is 109.0 at 25°C. If Λ_0 is 389.5, what is the ionization constant of chloracetic acid?

Ans. 1.52×10^{-3}.

34. Assuming that the conductance measurements give the true degree of dissociation in problem 33, use the Debye-Hückel limiting law to calculate the thermodynamic dissociation constant of chloracetic acid.

35. For propionic acid $K_a = 1.34 \times 10^{-5}$ at 25°C. Find for a 0.01 molar solution of the acid (a) the degree of dissociation, (b) the hydrogen ion concentration, and (c) the pH of the solution. *Ans.* (a) $\alpha = 0.0364$.

36. An aqueous solution at 25°C is 0.01 molar in propionic acid and 0.02 molar in sodium propionate. Find for this solution (a) the degree of ionization of the acid, (b) the hydrogen ion concentration, and (c) the pH.

37. A solution at 25°C contains 0.01 molar propionic acid and 0.03 molar hydrochloric acid. Find for this solution (a) the degree of ionization of the weak acid, (b) the hydrogen ion concentration, and (c) the pH.

38. A solution composed of 0.05 molar benzoic acid and 0.10 molar sodium benzoate gives a pH of 4.50 at 25°C. Find the ionization constant of the acid.

39. Calculate the pH and pOH of each of the following solutions at 25°C: (a) 0.001 molar H_2SO_4; (b) 0.001 molar $NaHSO_4$; and (c) 0.01 molar NH_4OH.

Ans. (a) pH = 2.73.

40. Calculate the H^+, H_3PO_4, $H_2PO_4^-$, HPO_4^{--}, and PO_4^{---} concentrations in a 0.10 molar H_3PO_4 solution at 25°C.   *Ans.* $C_{H^+} = 0.0239$; $C_{H_3PO_4} = 0.0761$.

41. If the final volume is to be 1 liter, how many moles of HCl will have to be added to 500 cc of 0.1 molar Na_2CO_3 in order to adjust the pH to 10.0?

42. Give the formulas of the acids conjugate to (a) methyl alcohol, (b) aniline, (c) dimethyl ether, and of the bases conjugate to (a) methyl alcohol, and (b) phenol.

43. Calculate the hydrolytic constants for each of the following salts at 25°C: (a) urea hydrochloride, (b) amnonium carbonate, (c) disodium acid phosphate, and (d) sodium bicarbonate.   *Ans.* (a) 0.672; (b) 11.9.

44. Calculate the degree of hydrolysis and the OH^- ion concentration at 25°C in (a) a 0.50 molar KCN solution, and (b) a 0.01 molar Na_2CO_3 solution.

45. Assuming the degrees of hydrolysis to be the same for both ions, calculate the extent to which a 0.10 molar solution of aniline acetate, $C_6H_5NH_3C_2H_3O_2$, will hydrolyze at 25°C. What will be the pH of the solution?

46. (a) Calculate at 25°C the hydrogen ion concentration in a 0.05 molar CH_3COOH solution on the basis that $K_a = K_a'$.

(b) Using now the limiting law of Debye and Hückel to obtain K_a', find the more precise value of the hydrogen ion concentration in the solution.

47. (a) Starting with Eq. (123) of this chapter, deduce the relation among the standard free energy changes of the reactions represented by the three equilibrium constants involved.

(b) What is the relation among the standard heats of these reactions?

(c) What is the relation among the standard entropies of these reactions?

48. Formulate equilibrium constants for each of the following equilibria:

(a) $AgBr(s) + 2\ NH_4OH = Ag(NH_3)_2^+ + Br^- + 2\ H_2O(l)$

(b) $IO_3^- + 5\ I^- + 6\ H^+ = 3\ H_2O(l) + 3\ I_2(s)$

(c) $H^+ + HCO_3^- = H_2O(l) + CO_2(g)$

49. For the reaction $Sn(s) + Pb^{++} = Pb(s) + Sn^{++}$, the following data have been obtained at 25°C:

Equilibrium Concentration
(moles/liter)

$C_{Pb^{++}}$	$C_{Sn^{++}}$
0.0233	0.0704
0.0235	0.0692
0.0275	0.0821

Calculate the equilibrium constant of the reaction.

50. For the reaction $Ag(CN)_2^- = Ag^+ + 2CN^-$ the equilibrium constant at 25°C is 4.0×10^{-19}. Calculate the silver ion concentration in a solution which was originally 0.10 molar in KCN and 0.03 molar in $AgNO_3$.

Ans. 7.5×10^{-18} equiv./liter.

–51. The solubility of CaF_2 in water at 18°C is 2.04×10^{-4} mole/liter. Calculate (a) the solubility product and (b) the solubility in 0.01 molar NaF solution.

Ans. (a) 3.4×10^{-11}; (b) 3.4×10^{-7} mole/liter.

52. Calculate the solubility of PbI_2 in (a) pure H_2O, (b) 0.04 molar KI, and (c) 0.04 molar $Pb(NO_3)_2$ at 25°C.

53. At 25°C the solubility product of $AgBrO_3$ is 5.77×10^{-5}. Using the Debye-Hückel limiting law, calculate its solubility in (a) pure H_2O and (b) 0.01 molar $KBrO_3$.   *Ans.* (a) 0.0084; (b) 0.0051 mole/liter.

54. At 25°C the solubility product of FeC_2O_4 is 2.1×10^{-7}. Using the Debye-Hückel limiting law, calculate the solubility of FeC_2O_4 in (a) a solution 0.002 molar in $MgSO_4$ and 0.005 molar in KNO_3, (b) pure H_2O.

55. Using data given in problem 50, predict whether or not AgCl would be precipitated from a solution which is 0.02 molar in NaCl and 0.05 molar in $KAg(CN)_2$.

56. The solubility product of CaF_2 at 18°C is 3.4×10^{-11} while that of $CaCO_3$ is 9.5×10^{-9}. What will be the nature of the first precipitate when a solution of $CaCl_2$ is added to a solution which is 0.05 molar in NaF and 0.02 molar in Na_2CO_3? In a 0.02 molar solution of Na_2CO_3 what is the minimum concentration of NaF at which CaF_2 and $CaCO_3$ will precipitate simultaneously?

57. The solubility product of PbI_2 is 7.47×10^{-9} at 15°C and 1.39×10^{-8} at 25°C. Calculate (a) the molar heat of solution of PbI_2 and (b) the solubility in moles per liter at 75°C.   *Ans.* (a) 10,600 cal/mole; (b) 0.00357 mole/liter.

12
Electrochemical Cells

ELECTROCHEMICAL CELLS may be used to perform two functions namely, (a) to convert chemical energy into electrical and (b) to convert electrical energy into chemical. In the common dry cell and the lead storage battery we have converters of chemical into electrical energy, while in the charging of the storage battery and in the electrolytic purification of copper electrical energy is used to bring about chemical action. We shall divide our discussion of cells along these functional lines.

Before proceeding, a distinction should be made between the terms *cell* and *battery*. A cell is a single arrangement of two electrodes and an electrolyte capable of yielding electricity due to chemical action within the cell, or of producing chemical action due to passage of electricity through the cell. A battery, on the other hand, is a combination of two or more cells arranged in series or parallel. Thus, the ordinary 6-volt lead storage battery is a combination of three 2-volt cells connected in series.

REVERSIBLE AND IRREVERSIBLE CELLS

In dealing with the energy relations of cells thermodynamic principles find very extensive application. However, the use of these principles is subject to one very important restriction, namely, that the processes to which they are applied be *reversible*. It will be recalled that the conditions for thermodynamic reversibility are (a) that the driving and opposing forces be only infinitesimally different from each other, and (b) that it should be possible to reverse any change taking place by applying a force infinitesimally greater than the one acting. When these requirements are satisfied by a cell, the cell is *reversible*, and its potential difference measured under appropriate conditions may be substituted into the relevant thermodynamic relations. When these conditions are not satisfied, the cell is said to be *irreversible*, and the thermodynamic equations do not apply.

The difference between reversible and irreversible cells may be illustrated with the following two examples. Consider first a cell composed of a zinc and a silver-silver chloride electrode, both dipping into a solution of zinc chloride. When the two electrodes are connected externally through a conductor, electrons flow through the outer circuit from the zinc to the silver-silver chloride. During this passage of current zinc dissolves at one electrode to form zinc ions, while at the other electrode the reaction

$$AgCl(s) = Ag(s) + Cl^- \tag{1}$$

takes place. The net reaction for the cell is, therefore,

$$\frac{1}{2} Zn(s) + AgCl(s) = Ag(s) + \frac{1}{2} Zn^{++} + Cl^- \tag{2}$$

and this process continues as long as the external opposing potential is infinitesimally smaller than that of the cell. However, as soon as the opposing potential becomes slightly larger than that of the cell, the direction of current flow is reversed, and so is the cell reaction. Now zinc ions go to form zinc at one electrode, silver chloride is formed from silver and chloride ions at the other, and the over-all cell reaction becomes

$$Ag(s) + \frac{1}{2} Zn^{++} + Cl^- = \frac{1}{2} Zn(s) + AgCl(s) \tag{3}$$

From this description it is obvious that the cell in question meets the second condition of reversibility. Again, the first condition can be satisfied by drawing from or passing through the cell a very minute current. Hence, this cell is reversible, and it may be treated by thermodynamic methods without any ambiguity.

Consider now instead a cell composed of zinc and silver electrodes immersed in a solution of sulfuric acid. When the two electrodes are short-circuited, zinc dissolves with evolution of hydrogen to form zinc sulfate according to the scheme

$$Zn(s) + H_2SO_4 = ZnSO_4 + H_2(g) \tag{4}$$

However, when the cell is connected with an external source of potential slightly greater than its own, silver dissolves at one electrode, hydrogen is evolved at the other, and the cell reaction becomes

$$2 \, Ag(s) + H_2SO_4 = Ag_2SO_4 + H_2(g) \tag{5}$$

From Eqs. (4) and (5) it is readily evident that, even though this cell may be made to satisfy the first condition of reversibility, the second does not hold, and consequently the cell cannot be reversible. The potential of such a cell does not have the definite thermodynamic significance which can be ascribed to the potentials of reversible cells.

There are a number of other types of irreversibility to which reference will be made later. For the present, suffice it to point out that in theoretical study of the potentials of cells it is the reversible ones which are of importance, and it is to these that we shall confine our prime attention.

ELECTROMOTIVE FORCE AND ITS MEASUREMENT

When a cell is connected in series with a galvanometer and the circuit is closed, the galvanometer is deflected, indicating that a current is flowing through the circuit. The passage of current from one electrode to the other is evidence for the existence of a potential difference between them, for without the presence of a potential difference no electricity can flow from one point to another. *This difference of potential which causes a current to flow from the electrode of higher potential to the one of lower is called the electromotive force*, abbreviated emf, *of the cell* and is expressed in volts.

The most common method of determining the potential difference between any two points in an electric circuit is to connect a voltmeter across the two points. The potential difference or voltage is read then directly from the instrument. A serious objection to the use of the voltmeter for accurate measurement of cell potentials is that it draws some current from the cell, causing thereby a change in the emf due to formation of reaction products at the electrodes and changes in the concentration of the electrolyte around the electrodes. Again, with appreciable current flow part of the emf will have to be utilized to overcome the internal resistance of the cell, and hence the potential measured on the voltmeter will not be the total cell emf. For these reasons precise emf's of cells are never determined with voltmeters. Instead *potentiometers* are used which require extremely small currents at balance.

Potentiometers for emf measurements operate on the Poggendorff compensation principle. In this method the unknown emf is opposed by another known emf until the two are equal. The setup and the conditions at balance may be understood from the diagram shown in Figure 12-1.

In this diagram A is a cell of known emf ε_A, whose potential is impressed across a uniform resistance ab. Connected with A in such a way that the two emf's oppose each other is the source X of unknown potential ε_X. To find ε_X the sliding contact C is moved along ab until a position S is found at which the galvanometer G gives no deflection. From ε_A, and the distances ab and aS, the unknown emf ε_X is found as follows: Since ε_A is impressed across the full length ab, for any given current passing through the resistance ε_A must be proportional to ab. Again, as ε_X is impressed

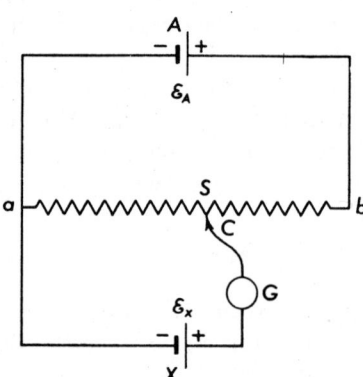

Figure 12–1. Poggendorff compensation method for measuring emf.

only across the distance aS, it must be proportional to this length. Consequently, on dividing ε_X by ε_A we obtain

$$\frac{\varepsilon_X}{\varepsilon_A} = \frac{aS}{ab}$$

and, therefore,

$$\varepsilon_X = \left(\frac{aS}{ab}\right)\varepsilon_A \tag{6}$$

The only requirements that need be met here are that ε_A be larger than ε_X, that the wire ab be uniform, and that the galvanometer be sufficiently sensitive to allow a balance of the potentials without appreciable current flow. Once these requirements are satisfied, the unknown emf can be obtained through Eq. (6) under conditions where the cell suffers no disturbance due to passage of current, and hence under conditions approximating very closely true reversibility.

To permit *direct reading* of voltage and to conserve the standard source of potential A, the setup shown in Figure 12–1 is modified in practice to that given in Figure 12–2. In this diagram W is the *working cell* whose emf can be impressed across ab, which is calibrated in volts, through the variable resistance R. Against this cell may be applied either the unknown emf X or the standard cell S.C. through the double-pole double-throw switch D. Before the emf of X can be measured, the slide wire must be standardized to read emf as follows. First pointer C is set at a point S' along ab corresponding to the value of the emf of the standard cell, say 1.0183 volts. Next, switch D is thrown to the S.C. side, the tapping key is closed gently, and F is moved along resistance R until the galvanometer G shows no deflection. When this balance is established, the current flowing through ab is of such a magnitude as to make the potential drop between a

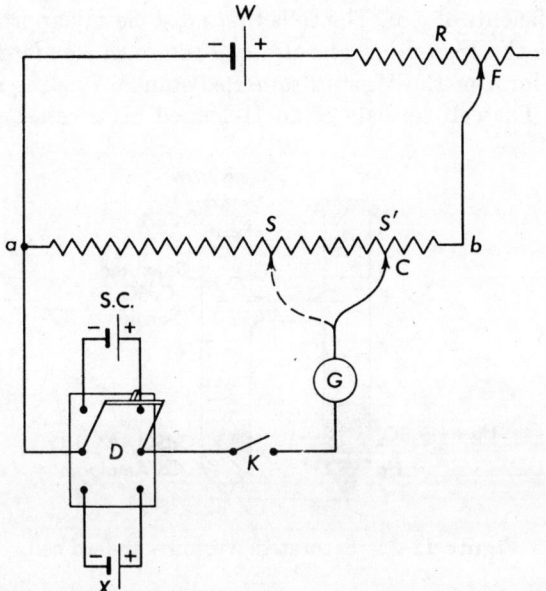

Figure 12–2. Principle of direct reading potentiometer.

and S' exactly 1.0183 volts and the voltage drop anywhere along ab identical with the voltage markings on the slide wire. In other words, the potentiometer has been standardized to read voltage directly.

Now R is left undisturbed, the switch D is shifted to the unknown emf side, the key is closed, and this time C is moved along ab until a point S is found at which the galvanometer G again shows no deflection. The reading of the slide wire at S gives then the voltage of X directly.

In laboratory potentiometers the slide wire ab is usually not a single unit, but consists first of a series of coils of nominal 0.1 volt values, and an extended slide wire covering *in toto* a 0.1 volt range. By this means simple potentiometers can be used to measure potentials up to 1.6 volts to 1×10^{-4} volt. In more precise potentiometers the precision can be extended to 1×10^{-5} volt, and for small values of emf down to 1×10^{-7}

volt. With the latter instruments galvanometers of very high sensitivity are required.

STANDARD CELLS

It is essential to have available as standards cells whose potentials are reproducible, constant with time, and well known. Also, such cells should be reversible, should not be subject to permanent damage due to passage of current through them, and should preferably have low temperature coefficients of emf. The cells that most closely approximate these properties are the *Weston unsaturated and saturated standard cells.*

The usual form of the Weston *saturated* standard cell is illustrated in Figure 12–3. The cell consists of an H-shaped glass vessel containing in

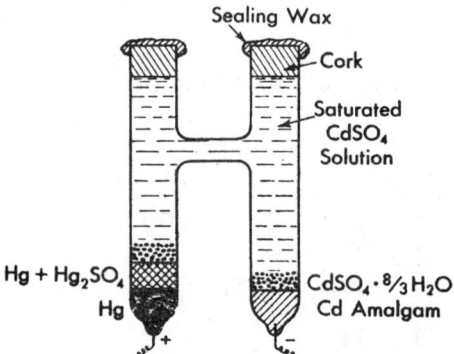

Figure 12–3.   Saturated Weston standard cell.

each arm one of the electrodes and filled throughout with the electrolyte. Contact with the active material is made through short platinum wires sealed into the bottoms of the arms. The positive electrode consists of mercury covered with a paste of mercurous sulfate and mercury. The negative electrode, in turn, consists of a cadmium amalgam containing 12 to 14 per cent cadmium by weight. Over both electrodes are sprinkled some crystals of solid $CdSO_4 \cdot \frac{8}{3} H_2O$, the entire cell is filled with a saturated solution of cadmium sulfate, and the cell is closed with either corks and sealing wax, or the arms are drawn off. The purpose of the solid crystals of $CdSO_4 \cdot \frac{8}{3} H_2O$ is to keep the electrolyte saturated with this phase at all temperatures.

The operation of the Weston saturated standard cell depends on the reversible reaction

$$Cd(s) + Hg_2SO_4(s) + \frac{8}{3} H_2O(l) = CdSO_4 \cdot \frac{8}{3} H_2O(s) + 2 Hg(l) \quad (7)$$

The reaction as written occurs when the cell is acting as a source of current, while the reverse reaction takes place when current is passed through the cell.

The *unsaturated* Weston cell is similar to the saturated, except that the crystals of solid $CdSO_4 \cdot \frac{8}{3} H_2O$ are omitted, and the electrolyte is a solution of cadmium sulfate *saturated only at 4°C*. At all temperatures above 4°C the electrolyte in this cell is unsaturated with respect to $CdSO_4 \cdot \frac{8}{3} H_2O$.

The Weston standard cells when carefully prepared and not subjected to abuse will preserve their potentials excellently for many years. The Weston *saturated cell*, or as it is frequently called the Weston normal cell, is generally used in laboratories as a standard of emf. Its potential at any temperature $t°C$, in *international volts*,[1] is given by the equation

$$\mathcal{E}_t = 1.01830 - 4.06 \times 10^{-5}(t - 20) - 9.5 \times 10^{-7}(t - 20)^2 \qquad (8)$$

The potential of the cell is thus 1.01830 volts at 20°C and 1.01807 volts at 25°C. The unsaturated cell is employed only as a secondary standard, possesses a lower temperature coefficient than the saturated, and has a potential close to 1.0186 volts at 20°C.

CELL REACTION AND EMF

In studying cells it is necessary to determine not only the emf of the cell but also the reaction responsible for it. The nature of the reaction proceeding in a cell can be deduced from the manner in which the electrodes must be connected to the standard cell in order to obtain a balance of the potentiometer. Since such a balance is possible only when the cells are connected so as to oppose each other, it must follow that the *electrode connected to the negative side of the standard cell is the negative electrode*, while the one connected to the positive side is the positive electrode. And since a negative electrode contains electrons in excess of the number present on the positive electrode, *electrons will have to flow from the negative to the positive electrode in the external circuit*.

Once the direction of current flow is known, the processes responsible for emission of electrons at the negative electrode and the taking on of the electrons at the positive electrode can be ascertained from the nature of the electrode materials on the basis of our ordinary concepts of oxidation and reduction. The student is aware that any reaction that gives off electrons must be an oxidation, while any reaction that involves a

---

[1] Since practically all emf data in the chemical literature are still expressed in international volts, no attempt will be made at present to convert the values given in this book to absolute volts.

taking on of electrons with consequent decrease in positive valence must be a reduction. Hence, *an oxidation must occur at the negative electrode* where electrons are given off, and *a reduction must take place at the positive electrode* where electrons enter. By *adding* the oxidation reaction at the negative electrode to the reduction at the positive electrode we obtain the cell reaction.

This manner of arriving at the cell reaction from the direction of electron flow may be illustrated with an example. Consider a cell composed of a zinc electrode dipping in a solution of zinc ions at unit activity and a cadmium electrode dipping into a solution of cadmium ions at the same activity; in other words, the cell

$$\text{Zn} \mid \text{Zn}^{++}(a = 1) \mid\mid \text{Cd}^{++}(a = 1) \mid \text{Cd} \tag{9}$$

At 25°C it is found with the aid of a potentiometer that this cell has an emf of $\varepsilon_{25°C} = 0.3590$ volt and that the zinc electrode is negative. Hence the electron flow through the external circuit must be from the zinc to the cadmium electrodes, and consequently an oxidation occurs at the zinc electrode, a reduction at the cadmium electrode. The only oxidation process possible at the zinc electrode is, obviously,

$$\text{Zn}(\text{s}) = \text{Zn}^{++}(a = 1) + 2\ominus$$

Again, the reduction at the cadmium electrode is

$$\text{Cd}^{++}(a = 1) + 2\ominus = \text{Cd}(\text{s})$$

The over-all cell reaction must be, then,

$$\text{Zn}(\text{s}) + \text{Cd}^{++}(a = 1) = \text{Zn}^{++}(a = 1) + \text{Cd}(\text{s}) \tag{10}$$

and this reaction yields an emf of $\varepsilon = 0.3590$ volt at 25°C.

It is important to realize that an emf without the reaction responsible for it is as meaningless as the age of an undesignated individual. For absolute clarity each emf must be accompanied by the reaction to which it refers, as well as by a complete statement of the nature of the phases, their concentrations, and the temperature.

CONVENTION REGARDING SIGN OF EMF

The net electrical work performed by a reaction yielding an emf ε and supplying a quantity of electricity Q is $Q\varepsilon$. For each equivalent reacting Q is equal to the faraday $\mathfrak{F}$, and hence for n equivalents reacting

$Q = n\mathfrak{F}$. Therefore, the electrical work obtained from any reaction supplying $n\mathfrak{F}$ coulombs of electricity at a potential ε is

$$\text{Net electrical work} = n\mathfrak{F}\varepsilon$$

But any work performed by a cell can be accomplished only at the expense of a *decrease in free energy* occurring within the cell. Further, when the electrical work is a *maximum*, as when the cell operates *reversibly*, the decrease in free energy, $-\Delta F$, must equal the electrical work done. We obtain, therefore,

$$\Delta F = -n\mathfrak{F}\varepsilon \tag{11}$$

from which we see that the reversible emf of any cell is determined by the free energy change of the reaction going on in the cell. Equation (11) is the "bridge" between thermodynamics and electrochemistry. Through it the evaluation of various thermodynamic properties of reactions becomes possible from emf measurements.

It was shown in Chapter 6 that for any spontaneous reaction at constant temperature and pressure ΔF is negative, for any nonspontaneous one ΔF is positive, while for equilibrium $\Delta F = 0$. In view of this, it may be deduced from Eq. (11) that for any spontaneous reaction ε will have to be positive, for any nonspontaneous reaction ε will be negative, while for any reaction in equilibrium ε will have to be zero. These relations between the spontaneity of a reaction and the signs of ΔF and ε are summarized in Table 12–1. They indicate that in order to have conformity between emf and ΔF it is necessary to prefix each and every emf with the appropriate sign depending on the spontaneity of the reaction as written.

TABLE   12–1.   RELATION   BETWEEN
SIGNS OF ΔF AND ε

Reaction	ΔF	ε
Spontaneous	−	+
Nonspontaneous	+	−
Equilibrium	0	0

The emf's of all cells under discussion here result from the spontaneous reactions occurring within the cells, and hence for all of them ε is positive. However, it is not always possible to tell *a priori* which of two processes possible, namely the forward and reverse reactions, is the spontaneous one. Consequently it is necessary to have some means of ascertaining which reaction results in the measured positive emf. For this purpose we lay down the following rule: *If a cell is written on paper such that the negative electrode is on the left and the positive electrode on the right*, in other words, *so that electrons flow from left to right through the external circuit, the reaction deduced on the basis of oxidation-reduction described above is the spontaneous one, and the emf of the cell is positive.* By observing this rule the spontaneous reaction in any cell may be arrived at, and the correct sign ascribed to the emf corresponding to it.

We may illustrate the application of this rule with the cell given in Eq. (9). The cell is written with the negative electrode on the left, and hence the reaction deduced for it, Eq. (10), is the spontaneous one. For this reaction, then, $\varepsilon_{25°C} = +0.3590$ volt. It follows, therefore, that the reverse of Eq. (10), namely,

$$Cd(s) + Zn^{++}(a = 1) = Zn(s) + Cd^{++}(a = 1) \qquad (12)$$

is not spontaneous, and for this reaction the emf at 25°C will be $\varepsilon_{25°C} = -0.3590$. This means that in order to carry out the reaction in Eq. (12) at least 0.3590 volt will have to be applied to the cell; and, instead of obtaining electrical work from the cell, work will have to be done to get the reaction to go.

SINGLE ELECTRODE POTENTIALS

Any electrochemical reaction is the sum of the two electrode reactions, of which one is an oxidation, the other a reduction. Similarly, every cell emf may be thought of as being composed of two individual *single electrode potentials*, such that their *algebraic sum* is equal to the total emf of the cell.

Only differences in potential between two electrodes can be measured experimentally. To determine directly single electrode potentials, it would be necessary to couple an electrode with another whose potential is zero. However, as no electrode of true zero potential is known, we must fall back upon an arbitrary standard electrode to which all other electrode potentials can be referred. For this reason *we define as the reference of emf the standard hydrogen electrode whose potential at all temperatures is taken as zero.* A standard hydrogen electrode consists of a piece of platinized platinum foil surrounded by hydrogen gas at 1 atm pressure and immersed in a solution containing hydrogen ions at unit activity. Details on the theory and operation of this electrode will be given later.

Because of experimental difficulties involved in the preparation and use of the standard hydrogen electrode, secondary reference electrodes have been compared with the former and are widely used. Among these are the three *calomel electrodes*, whose emf depends on the process spontaneous with respect to the hydrogen electrode reaction

$$Hg_2Cl_2(s) + 2 \ominus = 2 Hg(l) + 2 Cl^-(C = x) \qquad (13)$$

Here x is 0.1 N and 1 N potassium chloride for the 0.1 N and 1 N cal-omel electrodes, and a saturated potassium chloride solution for the saturated calomel electrode. A form of these electrodes is shown in Figure 12–4. The electrode consists of a glass vessel A to which are attached the sidearm B for making electrical contact and the arm C for insertion in any desired solution. Into the bottom of A is sealed a platinum wire over which are placed in turn a layer of specially purified mercury, a paste of mercury and calomel, and then the appropriate potassium chloride solution saturated with calomel so as to fill the cell and the arm C. In the 0.1 N

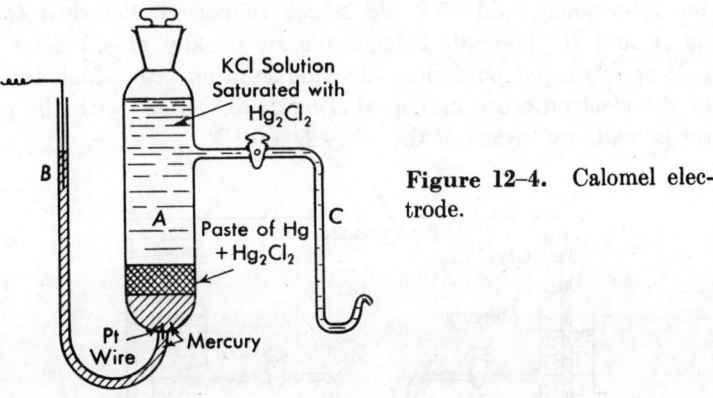

Figure 12–4.   Calomel electrode.

calomel this solution is 0.1 N potassium chloride, in the 1 N calomel it is 1 N potassium chloride, and in the saturated calomel it is an aqueous solution saturated with both potassium chloride and mercurous chloride. In the latter electrode some crystals of potassium chloride are also placed over the mercury-mercurous chloride paste in order to keep the solution saturated at all temperatures.

The potentials of the various calomel electrodes depend at each temperature on the concentration of potassium chloride with which the electrode vessels are filled. In Table 12–2 are given the potentials of these electrodes as a function of temperature, the emf at 25°C, and the corresponding electrode reactions. These emf's are highest for the 0.1 N calomel and decrease as the concentration is increased. Further, the 0.1 N calomel has the smallest temperature coefficient of emf, while the saturated calomel electrode has the highest.

The manner in which these auxiliary reference electrodes are combined with other electrodes to form cells is shown in Figure 12–5. Here A is the

TABLE 12-2.   POTENTIALS OF CALOMEL ELECTRODES

Electrode	Symbol	Emf	$\mathcal{E}_{25°C}$	Reaction
0.1 N calomel	Hg | Hg$_2$Cl$_2$(s), KCl(0.1 N)	$\mathcal{E} = 0.3338 -$ $7 \times 10^{-5}(t - 25)$	0.3338	Hg$_2$Cl$_2$(s) + 2 $\ominus$ = 2 Hg(l) + 2 Cl$^-$ (0.1 N)
1 N calomel	Hg | Hg$_2$Cl$_2$(s), KCl(1 N)	$\mathcal{E} = 0.2800 -$ $2.4 \times 10^{-4}(t - 25)$	0.2800	Hg$_2$Cl$_2$(s) + 2 $\ominus$ = 2 Hg(l) + 2 Cl$^-$(1 N)
Saturated calomel	Hg | Hg$_2$Cl$_2$(s), KCl (sat'd.)	$\mathcal{E} = 0.2415 -$ $7.6 \times 10^{-4}(t - 25)$	0.2415	Hg$_2$Cl$_2$(s) + 2 $\ominus$ = 2 Hg(l) + 2 Cl$^-$(sat'd.)

reference calomel electrode, B the other single electrode whose potential is to be determined, and C a *salt bridge* to permit electrical contact between A and B. The salt bridge consists usually of a beaker filled with 1 N or saturated potassium chloride solution into which the side-arms of the electrodes are immersed. Electrical contact with the potentiometer is made by means of the wires D and D'.

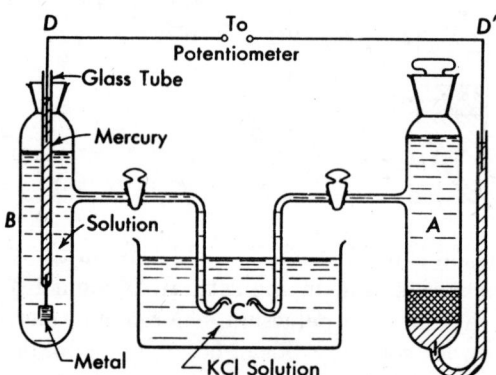

Figure 12-5.   Determination of single electrode potentials.

CALCULATION OF SINGLE ELECTRODE POTENTIALS

From emf's of cells involving various electrodes in combination with reference electrodes, single electrode potentials are readily calculated. To illustrate the procedure consider first the cell

$$\overline{\text{Cd} \mid \text{Cd}^{++}(a = 1) \parallel \text{KCl}(1 \ N), \text{Hg}_2\text{Cl}_2(\text{s}) \mid \text{Hg}} \qquad (14)$$

consisting of a cadmium-cadmium ion electrode and a 1 N calomel. For this cell it is found that at 25°C $\varepsilon = 0.6830$ volt and that the cadmium electrode is negative. Consequently, the cadmium electrode undergoes the oxidation reaction

$$Cd(s) = Cd^{++}(a = 1) + 2 \ominus$$

while the calomel electrode undergoes the reduction

$$Hg_2Cl_2(s) + 2 \ominus = 2 Hg(l) + 2 Cl^-(1 N)$$

For the latter process Table 12-2 shows that the emf at 25°C is $\varepsilon_C = +0.2800$ volt. Hence

$$\varepsilon = \varepsilon_{Cd} + \varepsilon_C$$
$$0.6830 = \varepsilon_{Cd} + 0.2800$$

and

$$\varepsilon_{Cd} = 0.4030 \text{ volt}$$

Therefore the reaction and the corresponding emf at 25°C for the Cd | Cd^{++}($a = 1$) electrode are

$$Cd(s) = Cd^{++}(a = 1) + 2 \ominus \qquad \varepsilon_{25°C} = 0.4030 \text{ volt}$$

As a second example let us take the cell

$$Hg \mid Hg_2Cl_2(s), KCl(1 N) \parallel Cu^{++}(a = 1) \mid Cu \qquad (15)$$

consisting of a 1 N calomel and a copper electrode dipping in a solution of cupric ions at unit activity. At 25°C the emf of this cell is $\varepsilon = 0.0570$ volt, and *the calomel electrode is negative*. Since the latter electrode is negative, the reaction occurring in it must be an oxidation, or

$$2 Hg(l) + 2 Cl^-(1 N) = Hg_2Cl_2(s) + 2 \ominus$$

and the emf on oxidation must be $\varepsilon_C = -0.2800$ volt. The corresponding reduction at the copper electrode involves the reaction

$$Cu^{++}(a = 1) + 2 \ominus = Cu(s)$$

with a single electrode potential ε_{Cu}. We obtain, then,

$$\varepsilon = \varepsilon_C + \varepsilon_{Cu}$$
$$0.0570 = -0.2800 + \varepsilon_{Cu}$$
$$\varepsilon_{Cu} = 0.3370 \text{ volt}$$

and so for the Cu | Cu^{++}($a = 1$) electrode

$$Cu^{++}(a = 1) + 2 \ominus = Cu(s) \qquad \varepsilon_{25°C} = 0.3370 \text{ volt}$$

CALCULATION OF CELL EMF'S FROM SINGLE ELECTRODE POTENTIALS

When the single electrode potentials and reactions are available, the calculations may be reversed to predict cell emf's and cell reactions. Suppose it is desired to know the reaction and the emf at 25°C for the cell

$$\text{Cd} \mid \text{Cd}^{++}(a = 1) \parallel \text{Cu}^{++}(a = 1) \mid \text{Cu} \tag{16}$$

Assume that the cell as written is correct, i.e., that the cadmium electrode is negative. Then, for the oxidation at the cadmium electrode we have

$$\text{Cd(s)} = \text{Cd}^{++}(a = 1) + 2 \ominus \qquad \mathcal{E}_{25°C} = 0.4030 \text{ volt}$$

Again, for the reduction at the copper electrode we get

$$\text{Cu}^{++}(a = 1) + 2 \ominus = \text{Cu(s)} \qquad \mathcal{E}_{25°C} = 0.3370 \text{ volt}$$

Adding now the single electrode reactions and the single electrode potentials, we find for the cell reaction

$$\text{Cd(s)} + \text{Cu}^{++}(a = 1) = \text{Cd}^{++}(a = 1) + \text{Cu(s)} \tag{17a}$$

and for the emf

$$\begin{aligned}
\mathcal{E}_{cell} &= \mathcal{E}_{Cd} + \mathcal{E}_{Cu} \\
&= 0.4030 + 0.3370 \\
&= +0.7400 \text{ volt at } 25°C
\end{aligned} \tag{17b}$$

Since the calculated emf is positive, the assumption made with respect to the polarity of the electrodes is correct, and the cell reaction as given is the spontaneous one.

On the other hand, had we assumed the copper electrode in Eq. (16) to be negative, the latter would have to undergo the oxidation

$$\text{Cu(s)} = \text{Cu}^{++}(a = 1) + 2 \ominus$$

with emf equal to $\mathcal{E}_{Cu} = -0.3370$ volt; the cadmium electrode would have to suffer the reduction

$$\text{Cd}^{++}(a = 1) + 2 \ominus = \text{Cd(s)}$$

with emf $\mathcal{E}_{Cd} = -0.4030$ volt; and we would have found for the cell reaction and emf

$$\text{Cu(s)} + \text{Cd}^{++}(a = 1) = \text{Cu}^{++}(a = 1) + \text{Cd(s)}$$
$$\mathcal{E}_{cell} = -0.7400 \text{ volt at } 25°C \tag{18}$$

Equation (18) reveals that since $\mathcal{E}_{cell}$ is negative the reaction as written is not spontaneous, and consequently the wrong assumption was made with respect to the polarity of the electrodes. All that need be done to rectify the situation is to *reverse the reaction* written in Eq. (18) and *to change the sign of* ε_{cell}. The result is then the spontaneous reaction and emf of the cell as given in Eqs. (17a) and (17b).

SUMMARY OF RULES

Before proceeding we shall summarize the various rules introduced above with respect to cell reactions and emf's. These are:

1. Any cell reaction is the *sum* of the single electrode reactions *as they occur in the cell.*
   (a) At the *negative electrode* the reaction is an *oxidation.*
   (b) At the *positive electrode* the reaction is a *reduction.*
2. The total cell emf is the algebraic sum of the single electrode potentials provided each emf is affixed with the sign corresponding to the reaction as it actually takes place at the electrode.
3. If any cell is written down with the *negative electrode on the left*, so that electrons flow through the external circuit from left to right, the cell reaction deduced by rule (1) will be the *spontaneous* process, and the emf deduced by rule (2) for the cell will be *positive.*
4. If the wrong assumption be made with respect to the polarity of the electrodes, rule (3) will yield a *negative* emf corresponding to the *nonspontaneous* reaction. To obtain the spontaneous reaction and its emf, all that need be done is *to reverse the reaction and change the sign of the emf without changing its magnitude.*

By adhering rigidly to these rules the student can avoid confusion in regard to signs or reactions.

THERMODYNAMICS AND EMF: ΔH AND ΔS FROM EMF DATA

According to Eq. (36) of Chapter 6 the change in free energy, ΔF, is related to the change in enthalpy, ΔH, by the Gibbs-Helmholtz equation

$$\Delta F - \Delta H = T\left[\frac{\partial(\Delta F)}{\partial T}\right]_P \tag{19}$$

Again, the fundamental relation between emf and ΔF is

$$\Delta F = -n\mathfrak{F}\varepsilon \tag{11}$$

which on differentiation with respect to temperature gives

$$\left[\frac{\partial(\Delta F)}{\partial T}\right]_P = -n\mathfrak{F}\left(\frac{\partial\varepsilon}{\partial T}\right)_P \tag{20}$$

If we substitute now Eqs. (20) and (11) into Eq. (19) we find for ΔH

$$-n\mathfrak{F}\mathcal{E} - \Delta H = -n\mathfrak{F}T\left(\frac{\partial \mathcal{E}}{\partial T}\right)_P$$

$$\Delta H = -n\mathfrak{F}\mathcal{E} + n\mathfrak{F}T\left(\frac{\partial \mathcal{E}}{\partial T}\right)_P$$

$$= n\mathfrak{F}\left[T\left(\frac{\partial \mathcal{E}}{\partial T}\right)_P - \mathcal{E}\right] \tag{21}$$

Equation (21) permits the calculation of the heat of a reaction from the measured emf and temperature coefficient of emf of the reaction, or $(\partial \mathcal{E}/\partial T)_P$ from ΔH and $\mathcal{E}$. In using this equation, ΔH follows in *joules* when $\mathcal{E}$ and $\mathfrak{F}$ are in volts and coulombs, respectively, while ΔH must be substituted in joules to obtain $\mathcal{E}$ in volts, or $(\partial \mathcal{E}/\partial T)_P$ in volts per °K.

Substitution of Eq. (20) into Eq. (19) yields

$$\Delta F - \Delta H = -n\mathfrak{F}T\left(\frac{\partial \mathcal{E}}{\partial T}\right)_P \tag{22}$$

Therefore, for any electrochemical reaction ΔF can equal ΔH only when $(\partial \mathcal{E}/\partial T)_P = 0$. Further, since

$$\Delta F - \Delta H = -T\Delta S \tag{23}$$

comparison of Eqs. (22) and (23) shows that

$$-T\Delta S = -n\mathfrak{F}T\left(\frac{\partial \mathcal{E}}{\partial T}\right)_P$$

and

$$\Delta S = n\mathfrak{F}\left(\frac{\partial \mathcal{E}}{\partial T}\right)_P \tag{24}$$

With the aid of Eq. (24) the entropy change of a reaction can readily be calculated from the temperature coefficient of emf.

THERMODYNAMICS OF ELECTRODE POTENTIALS

Single electrode and cell potentials are determined not only by the nature of the constituents composing the electrodes, but also by the temperature and the activities of the solutions employed. The dependence of the emf's on the latter variables is deducible from thermodynamics. For any reaction such as

$$aA + bB + \cdots = cC + dD + \cdots \tag{25}$$

the change in free energy, ΔF, as a function of the *initial* activities of the reactants and the *final* activities of the products, is given by Eq. (64) of Chapter 6 as

$$\Delta F = \Delta F^0 + RT \ln \frac{a_C^c a_D^d \cdots}{a_A^a a_B^b \cdots} \qquad (26)$$

Here the a's in the numerator are the activities of the products, those in the denominator the activities of the reactants, and ΔF^0, the standard free energy change, is the free energy change attending the reaction when the activities of products and reactants are all unity. If we substitute into Eq. (26) $\Delta F = -n\mathfrak{F}\mathcal{E}$, and define $\mathcal{E}^0$ by

$$\Delta F^0 = -n\mathfrak{F}\mathcal{E}^0 \qquad (27)$$

where $\mathcal{E}^0$ is the value of $\mathcal{E}$ corresponding to ΔF^0, Eq. (26) becomes

$$-n\mathfrak{F}\mathcal{E} = -n\mathfrak{F}\mathcal{E}^0 + RT \ln \frac{a_C^c a_D^d \cdots}{a_A^a a_B^b \cdots}$$

and hence $\mathcal{E}$ follows as

$$\mathcal{E} = \mathcal{E}^0 - \frac{RT}{n\mathfrak{F}} \ln \frac{a_C^c a_D^d \cdots}{a_A^a a_B^b \cdots}$$

$$= \mathcal{E}^0 - \frac{2.3026\, RT}{n\mathfrak{F}} \log_{10} \frac{a_C^c a_D^d \cdots}{a_A^a a_B^b \cdots} \qquad (28)$$

Equation (28) gives the potential of any electrode or cell as a function of the initial and final activities. It shows that the emf is determined by the activities of the reacting species, the temperature, and by the quantity $\mathcal{E}^0$. Since ΔF^0 is a constant for any reaction at constant temperature, $\mathcal{E}^0$ must also be a constant at constant temperature and characteristic of the electrode or cell. It is in fact the emf of the electrode or cell when the activities are all unity. $\mathcal{E}^0$ is called the *standard potential* of the electrode or cell in question.

The *standard electrode potentials* at 25°C for a number of electrodes, evaluated by methods to which reference will be made later, are given in Table 12-3. The corresponding electrode reactions are also included. In every instance the sign of the emf refers to the *oxidation* reaction. Before using these $\mathcal{E}^0$ values it must first be ascertained whether the electrode reaction *as it actually occurs in the particular cell* is the oxidation or the reduction. For oxidation the $\mathcal{E}^0$'s tabulated are used directly. However, for reduction the sign of each and every $\mathcal{E}^0$ given must be reversed. Thus, for a process such as $K(s) = K^+ + \ominus$ $\mathcal{E}^0_{25°C} = +2.9241$, but for $K^+ + \ominus = K(s)$ $\mathcal{E}^0_{25°C} = -2.9241$ volts. Similarly, for $Cu(s) = Cu^{++} + 2\ominus$ $\mathcal{E}^0_{25°C} = -0.337$ volt, while for $Cu^{++} + 2\ominus = Cu(s)$ $\mathcal{E}^0_{25°C} = +0.337$ volt.

TABLE 12–3. STANDARD ELECTRODE POTENTIALS AT 25°C FOR OXIDATION REACTIONS

Electrode	Electrode Reaction	ε^0 (volts)
Li | Li$^+$	Li(s) = Li$^+$ + $\ominus$	+3.045
K | K$^+$	K(s) = K$^+$ + $\ominus$	+2.9241
Ca | Ca^{++}	Ca(s) = Ca^{++} + 2 $\ominus$	+2.87
Na | Na$^+$	Na(s) = Na$^+$ + $\ominus$	+2.7146
Zn | Zn^{++}	Zn(s) = Zn^{++} + 2 $\ominus$	+0.7618
Fe | Fe^{++}	Fe(s) = Fe^{++} + 2 $\ominus$	+0.441
Cd | Cd^{++}	Cd(s) = Cd^{++} + 2 $\ominus$	+0.403
Pb | PbSO$_4$(s), SO$_4^{--}$	Pb(s) + SO$_4^{--}$ = PbSO$_4$(s) + 2 $\ominus$	+0.3546
Tl | Tl$^+$	Tl(s) = Tl$^+$ + $\ominus$	+0.3363
Ni | Ni^{++}	Ni(s) = Ni^{++} + 2 $\ominus$	+0.236
Ag | AgI(s), I$^-$	Ag(s) + I$^-$ = AgI(s) + $\ominus$	+0.1522
Sn | Sn^{++}	Sn(s) = Sn^{++} + 2 $\ominus$	+0.140
Pb | Pb^{++}	Pb(s) = Pb^{++} + 2 $\ominus$	+0.1265
H$_2$ | H$^+$	H$_2$(g, 1 atm) = 2 H$^+$ + 2 $\ominus$	±0.0000
Ag | AgBr(s), Br$^-$	Ag(s) + Br$^-$ = AgBr(s) + $\ominus$	−0.0711
Hg | Hg$_2$Br$_2$(s), Br$^-$	2 Hg(l) + 2 Br$^-$ = Hg$_2$Br$_2$(s) + 2 $\ominus$	−0.1385
Pt | Sn^{++}, Sn^{++++}	Sn^{++} = Sn^{++++} + 2 $\ominus$	−0.14
Ag | AgCl(s), Cl$^-$	Ag(s) + Cl$^-$ = AgCl(s) + $\ominus$	−0.2225
Hg | Hg$_2$Cl$_2$(s), Cl$^-$	2 Hg(l) + 2 Cl$^-$ = Hg$_2$Cl$_2$ + 2 $\ominus$	−0.2680
Cu | Cu^{++}	Cu(s) = Cu^{++} + 2 $\ominus$	−0.337
I$_2$ | I$^-$	2 I$^-$ = I$_2$(s) + 2 $\ominus$	−0.5355
Hg | Hg$_2$SO$_4$(s), SO$_4^{--}$	2 Hg(l) + SO$_4^{--}$ = Hg$_2$SO$_4$(s) + 2 $\ominus$	−0.6141
Pt | Fe^{++}, Fe^{+++}	Fe^{++} = Fe^{+++} + $\ominus$	−0.771
Ag | Ag$^+$	Ag(s) = Ag$^+$ + $\ominus$	−0.7991
Br$_2$ | Br$^-$	2 Br$^-$ = Br$_2$(l) + 2 $\ominus$	−1.0652
Pt | Tl$^+$, Tl^{+++}	Tl$^+$ = Tl^{+++} + 2 $\ominus$	−1.250
Cl$_2$ | Cl$^-$	2 Cl$^-$ = Cl$_2$(g, 1 atm) + 2 $\ominus$	−1.3595
Pt | Ce^{+++}, Ce^{++++}	Ce^{+++} = Ce^{++++} + $\ominus$	−1.61
Pt | Co^{++}, Co^{+++}	Co^{++} = Co^{+++} + $\ominus$	−1.82

In Eq. (28) the factor $(2.3026\,RT)/\mathfrak{F}$ is a constant for any given temperature. To evaluate it R must be taken in joules, namely,

$$\frac{2.3026\,RT}{\mathfrak{F}} = \frac{2.3026 \times 8.3143\,T}{96,487}$$
$$= 1.9841 \times 10^{-4}\,T$$

Values of this quantity for a number of temperatures are given in Table 12–4.

Once the ε^0's for various electrodes are available, they may be utilized to calculate the single electrode potentials. Specifically, suppose the emf of the zinc electrode

$$\text{Zn} \mid \text{Zn}^{++}(a = 0.1)$$

is sought at 25°C in a solution of zinc ions where the activity is 0.1. For *oxidation* the reaction at this electrode is

$$\text{Zn(s)} = \text{Zn}^{++}(a = 0.1) + 2 \ominus$$

and hence on applying Eq. (28) we obtain for the single electrode potential, ε_{Zn},

$$\varepsilon_{\text{Zn}} = \varepsilon_{\text{Zn}}^0 - \frac{2.3026\,RT}{n\mathfrak{F}} \log_{10} \frac{a_{\text{Zn}^{++}}}{a_{\text{Zn}}}$$

For this reaction $n = 2$, $a_{\text{Zn}^{++}} = 0.1$, and $a_{\text{Zn}} = 1$, since zinc is a solid.

TABLE 12–4. VALUES OF
$(2.3026\,RT)/\mathfrak{F}$ AT VARIOUS
TEMPERATURES

$t°C$	$(2.3026\,RT)/\mathfrak{F}$
0	0.054195
10	0.056180
15	0.057172
20	0.058164
25	0.059156
30	0.060148

Again, at 25°C, $(2.3026\,RT)/\mathfrak{F} = 0.05916$, while for oxidation $\varepsilon_{\text{Zn}}^0 = +0.7618$ volt. Substituting these quantities into the expression for ε_{Zn}, we find that

$$\varepsilon_{\text{Zn}} = +0.7618 - \frac{0.05916}{2} \log_{10} \frac{0.1}{1}$$

$$= 0.7618 + 0.0296$$

$$= 0.7914 \text{ volt}$$

However, when the reaction at the electrode is a *reduction*, namely,

$$\text{Zn}^{++}(a = 0.1) + 2 \ominus = \text{Zn(s)}$$

the equation yields for ε_{Zn}

$$\varepsilon_{\text{Zn}} = \varepsilon_{\text{Zn}}^0 - \frac{0.05916}{2} \log_{10} \frac{a_{\text{Zn}}}{a_{\text{Zn}^{++}}}$$

$$= \varepsilon_{\text{Zn}}^0 - \frac{0.05916}{2} \log_{10} \frac{1}{0.1}$$

Further, for reduction $\varepsilon_{Zn}^0 = -0.7618$, and, therefore,

$$\begin{aligned}\varepsilon_{Zn} &= -0.7618 - 0.0296 \\ &= -0.7914 \text{ volt}\end{aligned}$$

i.e., the potential is the same as before but opposite in sign.

EQUATION (28) AND CELL EMF'S

Consider now the cell

$$Zn \mid Zn^{++}(a_{Zn^{++}}) \parallel Cl^-(a_{Cl^-}), \; Hg_2Cl_2(s) \mid Hg \qquad (29)$$

In this cell the zinc electrode is negative, and undergoes the oxidation reaction

$$Zn(s) = Zn^{++}(a_{Zn^{++}}) + 2 \ominus \qquad (30a)$$

for which the electrode emf is given by

$$\varepsilon_{Zn} = \varepsilon_{Zn}^0 - \frac{RT}{2\,\mathfrak{F}} \ln a_{Zn^{++}} \qquad (30b)$$

Again, for the reduction at the calomel electrode the reaction is

$$Hg_2Cl_2(s) + 2 \ominus = 2 \; Hg(l) + 2 \; Cl^-(a_{Cl^-}) \qquad (31a)$$

and the single electrode potential is

$$\varepsilon_C = \varepsilon_C^0 - \frac{RT}{2\,\mathfrak{F}} \ln a_{Cl^-}^2 \qquad (31b)$$

On adding Eqs. (30a) and (31a) we find the cell reaction to be

$$Zn(s) + Hg_2Cl_2(s) = 2 \; Hg(l) + Zn^{++}(a_{Zn^{++}}) + 2 \; Cl^-(a_{Cl^-}) \qquad (32a)$$

and similarly on adding Eqs. (30b) and (31b) the cell emf follows as

$$\begin{aligned}\varepsilon &= \varepsilon_{Zn} + \varepsilon_C \\ &= \left(\varepsilon_{Zn}^0 - \frac{RT}{2\,\mathfrak{F}} \ln a_{Zn^{++}}\right) + \left(\varepsilon_C^0 - \frac{RT}{2\,\mathfrak{F}} \ln a_{Cl^-}^2\right) \\ &= (\varepsilon_{Zn}^0 + \varepsilon_C^0) - \frac{RT}{2\,\mathfrak{F}} \ln (a_{Zn^{++}}a_{Cl^-}^2)\end{aligned} \qquad (32b)$$

Instead of approaching the cell emf through the single electrode potentials, we may apply Eq. (28) directly to the cell reaction given in Eq. (32a). If this be done, we find that the cell emf is

$$\varepsilon = \varepsilon_{cell}^0 - \frac{RT}{2\,\mathfrak{F}} \ln (a_{Zn^{++}}a_{Cl^-}^2) \qquad (33)$$

where ε_{cell}^0 is the *standard potential of the cell*. Comparison of Eqs. (32b) and (33), which must be identical, shows that $\varepsilon_{cell}^0 = \varepsilon_{Zn}^0 + \varepsilon_C^0$, i.e., that the ε^0 *value of the cell is the sum of the* ε^0 *values of the electrodes composing it.* This relation between the standard electrode and cell potentials, namely,

$$\varepsilon_{\text{cell}}^0 = \varepsilon_1^0 + \varepsilon_2^0 \tag{34}$$

where ε_1^0 and ε_2^0 are the ε^0's for the two electrodes, applies to any cell. Consequently, as soon as the standard emf's of the electrodes are known, that of the cell composed of these is also available, and Eq. (28) may be applied directly to the over-all reaction of the cell.

In employing Eq. (34) each standard electrode potential must be prefixed by the sign corresponding to the reaction as it occurs in the cell. Thus the reaction at the zinc electrode in Eq. (29) is an oxidation, and therefore $\varepsilon_{\text{Zn}}^0 = +0.7618$ volt at 25°C. Again, the reaction at the calomel electrode is a reduction, and hence $\varepsilon_C^0 = +0.2680$ volt. Consequently, $\varepsilon_{\text{cell}}^0$ is

$$\varepsilon_{\text{cell}}^0 = 0.7618 + 0.2680$$
$$= 1.0298 \text{ volts}$$

STANDARD POTENTIALS AND EQUILIBRIUM CONSTANTS

Standard electrode and cell emf's may also be utilized for obtaining equilibrium constants. According to Eq. (27), ΔF^0 for any electrochemical process is given by

$$\Delta F^0 = -n\mathfrak{F}\varepsilon^0$$

But ΔF^0 is also related to the equilibrium constant K_a of the process by the equation

$$\Delta F^0 = -RT \ln K_a$$

If these two expressions for ΔF^0 are equated, we obtain

$$-n\mathfrak{F}\varepsilon^0 = -RT \ln K_a$$

and hence

$$\varepsilon^0 = \frac{RT}{n\mathfrak{F}} \ln K_a \tag{35}$$

With Eq. (35) K_a can be evaluated from ε^0, or, when K_a is available, ε^0 may be calculated. Thus, for the reduction of stannous ions by thallium, $\varepsilon^0 = 0.196$ volt at 25°C. For this process the equilibrium constant is, therefore,

$$0.196 = \frac{0.05916}{1} \log_{10} K_a$$

$$\log_{10} K_a = \frac{0.196}{0.05916} = 3.31$$

$$K_a = 2.0 \times 10^3$$

Since K_a for this reaction is

$$K_a = \frac{a_{Tl^+}}{a_{Sn^{++}}^{1/2}}$$

where the activities are those *at equilibrium*, this means that if thallium were added to a solution of stannous ions, the reaction would proceed and equilibrium would not be established until the activity of thallous ions became equal to

$$a_{Tl^+} = K_a \, a_{Sn^{++}}^{1/2}$$
$$= 2.0 \times 10^3 \, a_{Sn^{++}}^{1/2}$$

Once this condition is attained the system will be in equilibrium, and no further chemical action will take place.

CLASSIFICATION OF ELECTRODES

In electrochemical work the cells encountered involve various electrodes depending on the purpose at hand. The more important of these may be grouped into the following five types, namely:

1. Metal-metal ion electrodes
2. Amalgam electrodes
3. Gas electrodes
4. Metal-insoluble salt electrodes
5. Oxidation-reduction electrodes

We proceed now to a discussion of each of these.

METAL-METAL ION ELECTRODES

Electrodes of this type involve a metal in equilibrium with a solution of its ions. They have already been described in some detail. Examples are the zinc, copper, cadmium, and sodium electrodes. All these electrodes operate on the general reaction

$$M = M^{+n} + n \ominus \tag{36}$$

for which the electrode potential equation is given by

$$\varepsilon_M = \varepsilon_M^0 - \frac{RT}{n\mathfrak{F}} \ln a_{M^{+n}} \tag{37}$$

Each of these electrodes is said to be *reversible* to its own ions; i.e., the potential of each of these electrodes is sensitive to and is determined by the activities of its own ions in the solution in which the metal is immersed. Thus, the zinc electrode is reversible to zinc ions, the iron electrode to ferrous ions, the tin electrode to stannous ions, etc.

AMALGAM ELECTRODES

It is quite common practice to substitute for the pure metals in metal-metal ion electrodes solutions of the metal in mercury, namely, *amalgams.* Amalgams of metals more active than mercury behave essentially as do the pure metals, the only difference being that the activity of the metal is lowered somewhat by dilution by the mercury. These amalgam electrodes are preferred frequently because equilibrium can as a rule be established much more rapidly with them than with the pure metals, and because they are more readily reversible. Again, with metals such as sodium, potassium, or calcium the activity in aqueous solutions is too great for direct use. However, by converting these metals to amalgams the activity can be moderated sufficiently to permit measurements in presence of water. Still another factor in favor of the amalgams is that small quantities of impurities, which may cause erratic behavior in pure metals, are often diluted enough by amalgamation to yield satisfactory and reproducible results.

As an example of electrodes in this class may be taken the lead amalgam electrode, consisting of the lead amalgam, written $Pb(Hg)$, dipping into a solution of plumbous ions, namely, $Pb(Hg) \mid Pb^{++}(a_{Pb^{++}})$. For this electrode the reaction is

$$Pb(Hg) = Pb^{++}(a_{Pb^{++}}) + 2 \ominus \tag{38}$$

and hence the electrode potential ε_a is given by

$$\varepsilon_a = \varepsilon_{Pb}^0 - \frac{RT}{2\,\mathfrak{F}} \ln \frac{a_{Pb^{++}}}{a_{Pb}} \tag{39}$$

where ε_{Pb}^0 is the standard electrode potential of the pure lead electrode, $a_{Pb^{++}}$ is the activity of plumbous ions in solution, and a_{Pb} is the activity of *metallic lead in the amalgam.* In general a_{Pb} is not unity. The usual procedure is to write Eq. (39) as

$$\varepsilon_a = \left(\varepsilon_{Pb}^0 + \frac{RT}{2\,\mathfrak{F}} \ln a_{Pb} \right) - \frac{RT}{2\,\mathfrak{F}} \ln a_{Pb^{++}}$$

$$= \varepsilon_a^0 - \frac{RT}{2\,\mathfrak{F}} \ln a_{Pb^{++}} \tag{40}$$

where ε_a^0 is the standard potential of the given amalgam, and to evaluate ε_a^0 first. Then, in order to transform ε_a^0 to ε_{Pb}^0, the emf of the amalgam is measured against pure lead when both are immersed in a solution of plumbous ions of the *same* concentration. Since for pure lead the single electrode potential equation is

$$\varepsilon_{Pb} = \varepsilon_{Pb}^0 - \frac{RT}{2\,\mathfrak{F}} \ln a_{Pb^{++}}$$

while for the amalgam it is Eq. (40), the difference in potential ε between the lead and amalgam electrodes follows as

$$\varepsilon = \varepsilon_{Pb} - \varepsilon_a$$
$$= \left(\varepsilon_{Pb}^0 - \frac{RT}{2\,\mathfrak{F}} \ln a_{Pb^{++}} \right) - \left(\varepsilon_a^0 - \frac{RT}{2\,\mathfrak{F}} \ln a_{Pb^{++}} \right)$$
$$= \varepsilon_{Pb}^0 - \varepsilon_a^0 \tag{41}$$

Consequently, with ε and ε_a^0 known ε_{Pb}^0 may be evaluated, and the emf data obtained with the amalgam may be reduced to the pure metal. Thus Carmody[2] found that for the particular lead amalgam he employed in his studies the standard electrode potential of $Pb(Hg) \mid Pb^{++}$ was 0.1207 volt at 25°C, while the emf of a cell composed of this amalgam and pure lead in a solution of plumbous ions was 0.0058 volt. The standard potential of the $Pb \mid Pb^{++}$ electrode at 25°C is then

$$\varepsilon_{Pb}^0 = \varepsilon + \varepsilon_a^0$$
$$= 0.0058 + 0.1207$$
$$= 0.1265 \text{ volt}$$

GAS ELECTRODES

Gas electrodes consist of a gas bubbling about an inert metal wire or foil immersed in a solution containing ions to which the gas is reversible. The function of the metal wire or foil, which usually is platinized platinum, is to facilitate establishment of equilibrium between the gas and its ions and to serve as the electric contact for the electrode.

Among the gas electrodes are the hydrogen electrode, which is reversible to hydrogen ions, the chlorine electrode, reversible to chloride ions, and the oxygen electrode, whose emf depends on the activity of hydroxyl ions. However, though the first two electrodes can be made reversible, no suitable electrode material has so far been found which can catalyze satisfactorily the establishment of equilibrium between oxygen and hydroxyl ions. Whatever precise information is available on the latter electrode has been obtained not by direct emf measurements, but by calculation from suitable standard free energy data obtained from various other sources.

The fundamental reaction for the hydrogen electrode is

$$\frac{1}{2} H_2(g, P_{H_2}) = H^+(a_{H^+}) + \ominus \tag{42}$$

Since the activity of hydrogen gas at relatively low pressures is equal to the pressure P_{H_2} of the gas in atmospheres, the single electrode potential of the hydrogen electrode must be determined both by the pressure of

[2] Carmody, *J. Am. Chem. Soc.*, **51**, 2905 (1929).

the hydrogen about the electrode, and the activity of hydrogen ions in solution; namely,

$$\varepsilon_{H_2} = \varepsilon_{H_2}^0 - \frac{RT}{\mathcal{F}} \ln \frac{a_{H^+}}{P_{H_2}^{1/2}} \tag{43}$$

But $\varepsilon_{H_2}^0$, i.e., the emf of the hydrogen electrode at 1 atm hydrogen pressure and at unit activity of hydrogen ions, is the reference of all single electrode emf measurements and is taken by definition to be zero at all temperatures. Consequently, Eq. (43) becomes

$$\begin{aligned}
\varepsilon_{H_2} &= - \frac{RT}{\mathcal{F}} \ln \frac{a_{H^+}}{P_{H_2}^{1/2}} \\
&= - \frac{RT}{\mathcal{F}} \ln a_{H^+} + \frac{RT}{\mathcal{F}} \ln P_{H_2}^{1/2}
\end{aligned} \tag{44}$$

Further, when the pressure of hydrogen is 1 atm, $\ln P_{H_2}^{1/2} = 0$. Then

$$\varepsilon_{H_2} = - \frac{RT}{\mathcal{F}} \ln a_{H^+} \tag{45}$$

and the *hydrogen electrode becomes strictly dependent only on the activity of hydrogen ions in solution*, i.e., the pH. This use of the hydrogen electrode for pH measurements will be discussed later.

The chlorine electrode shows similar behavior. For this electrode the reduction reaction is

$$\frac{1}{2} Cl_2(g, P_{Cl_2}) + \ominus = Cl^-(a_{Cl^-}) \tag{46}$$

and hence the emf is given by

$$\varepsilon_{Cl_2} = \varepsilon_{Cl_2}^0 - \frac{RT}{\mathcal{F}} \ln \frac{a_{Cl^-}}{P_{Cl_2}^{1/2}} \tag{47}$$

However, $\varepsilon_{Cl_2}^0$ is not zero, but is in fact equal to 1.3595 volts at 25°C.

METAL-INSOLUBLE SALT ELECTRODES

Electrodes of this type are extremely important in electrochemistry and they are encountered very frequently. In this category fall the various calomel electrodes, the silver-silver chloride electrode, the lead-lead sulfate electrode, the silver-silver bromide electrode, and others.

The common characteristic of electrodes in this class is that they all consist of a metal in contact with one of its difficultly soluble salts and a solution containing the ion present in the salt *other than the metal*. For instance, a silver-silver chloride electrode is composed of a silver wire coated with silver chloride and immersed in a solution of chloride ions. Again, the lead-lead sulfate electrode may be either pure lead or a lead amalgam covered with crystals of lead sulfate and surrounded by solutions of sulfates. In all instances *these electrodes are reversible to the ions other than those of the metal present in the insoluble salt.* Thus the calomel and silver-silver chloride electrodes are reversible to chloride ions, the lead-lead sulfate electrode to sulfate ions, the silver-silver bromide electrode to bromide ions, etc. Why this is so may be seen from the following discussion of the silver-silver chloride electrode, a typical member of this class.

Consider a silver electrode dipping into a solution containing chloride ions and saturated with silver chloride. If we imagine this electrode to act as an ordinary metal-metal ion electrode, silver ions will pass from the electrode into the solution according to the reaction

$$Ag(s) = Ag^+ + \ominus \qquad \textbf{(a)}$$

However, since the solution is saturated with silver chloride, the introduction of any silver ions will upset the requirements of the solubility product principle for this salt. Consequently, in order to re-establish the solubility equilibrium, silver ions will have to combine with chloride ions to precipitate solid silver chloride according to the relation

$$Ag^+ + Cl^- = AgCl(s) \qquad \textbf{(b)}$$

Moreover, as the over-all electrode reaction is the sum of all the processes occurring at the electrode, the over-all electrode reaction will have to be the sum of (a) and (b), or

$$Ag(s) + Cl^- = AgCl(s) + \ominus \qquad \textbf{(48)}$$

For this reaction the electrode potential equation is

$$\mathcal{E}_{Ag-AgCl} = \mathcal{E}^0_{Ag-AgCl} - \frac{RT}{\mathcal{F}} \ln \frac{1}{a_{Cl^-}} \qquad \textbf{(49)}$$

and hence this electrode is reversible to *chloride ions.*

In an analogous manner may be deduced the reactions for other electrodes of this type. Thus we find for the calomel electrode

$$2\,Hg(l) + 2\,Cl^- = Hg_2Cl_2(s) + 2 \ominus \qquad \textbf{(50)}$$

for the lead-lead sulfate electrode

$$Pb(s) + SO_4^{--} = PbSO_4(s) + 2 \ominus \qquad \textbf{(51)}$$

and for the silver-silver bromide electrode

$$Ag(s) + Br^- = AgBr(s) + \ominus \qquad \textbf{(52)}$$

OXIDATION-REDUCTION ELECTRODES

Although every electrode reaction involves an oxidation or a reduction, the term *oxidation-reduction electrodes* is used to designate electrodes in which the emf results from the presence of *ions* of a substance in two different stages of oxidation. When a platinum wire is inserted into a solution containing both ferrous and ferric ions it is found that the wire acquires a potential. The same is true of solutions of cerous-ceric ions, stannous-stannic ions, manganous-permanganate ions, etc. These electrode emf's arise from the tendency of ions in one state of oxidation to pass over into a second more stable state. The function of the platinum wire is merely to "pick up" the potential corresponding to this tendency toward a free energy decrease, and to serve as the electrical contact of the electrode.

The general reaction for all electrodes of the oxidation-reduction type may be written as

$$A^{n_1}(a_1) + n\ominus = A^{n_2}(a_2) \tag{53}$$

where n_1 is the valence in the higher stage of oxidation, n_2 that in the lower, while $n = n_1 - n_2$ is the *valence change* attending the electrode process. From Eq. (53) the general emf equation follows as

$$\varepsilon = \varepsilon^0 - \frac{RT}{n\mathfrak{F}} \ln \frac{a_2}{a_1} \tag{54}$$

i.e., the emf of oxidation-reduction electrodes depends on the *ratio of the activities* of the two ions. Thus, for the ferric-ferrous electrode, designated symbolically as $Pt \mid Fe^{+++}, Fe^{++}$, the electrode reaction is

$$Fe^{+++}(a_{Fe^{+++}}) + \ominus = Fe^{++}(a_{Fe^{++}}) \tag{55}$$

and the electrode emf is given by

$$\varepsilon = \varepsilon^0 - \frac{RT}{\mathfrak{F}} \ln \frac{a_{Fe^{++}}}{a_{Fe^{+++}}} \tag{56}$$

In Table 12–3 are given ε^0 values for several electrodes of this type at 25°C.

ELECTROCHEMICAL CELLS

When various electrodes are combined, two general classes of cells may result, namely: (a) *chemical cells,* in which the emf is due to a *chemical reaction* occurring within the cell, and (b) *concentration·cells,* in which the emf is due to the free energy decrease attending the *transfer of matter* from one part of the cell to another. Further, each of these types of cells may or may not involve a liquid junction; or, as it is usually said, the cell may or may not have *transference*. On this basis we arrive at the following classification of electrochemical cells:

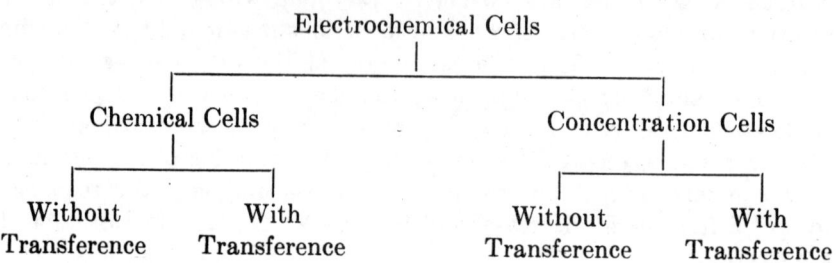

Electrochemical Cells

Chemical Cells   Concentration Cells

Without   With   Without   With
Transference   Transference   Transference   Transference

CHEMICAL CELLS WITHOUT TRANSFERENCE

To construct a chemical cell with no liquid junction or transference, two electrodes and an electrolyte must be selected such that *one of the electrodes is reversible to the cation, the other to the anion of the electrolyte.* For instance, if we desire a cell without transference employing hydrochloric acid as the electrolyte, we must use an electrode reversible to hydrogen ions, namely, the hydrogen electrode, and an electrode reversible to chloride ions. The latter may be either a silver-silver chloride, a mercury-mercurous chloride, or chlorine electrode, all of which are reversible to this ion. Again, for an electrolyte such as zinc bromide we shall have to employ zinc as one of the electrodes, and either bromine, mercury-mercurous bromide, or silver-silver bromide as the electrode for the bromide ions. Where a choice is possible, the electrode that can be handled most conveniently is selected. Although the particular electrode chosen determines the over-all cell reaction and the value of ε^0, it is characteristic of all such possibilities that they give the same dependence of the emf of the cell on the activity of the electrolyte.

As a typical example of chemical cells without transference we shall take the cell

$$H_2(g, P_{H_2}) \mid HCl(a_{HCl}), AgCl(s) \mid Ag \qquad (57)$$

consisting of the hydrogen and silver-silver chloride electrodes in hydrochloric acid as the electrolyte. Since these electrodes are reversible to the ions of the electrolyte, they may be immersed directly in the acid to yield a cell with no liquid junction. The entire emf of the cell is composed then only of the emf's existing at the electrode-solution interfaces.

In cell (57) the hydrogen electrode is negative, and so we obtain for the oxidation at this electrode

$$\frac{1}{2} H_2(g, P_{H_2}) = H^+(a_{H^+}) + \ominus \tag{58a}$$

and
$$\mathcal{E}_{H_2} = -\frac{RT}{\mathcal{F}} \ln \frac{a_{H^+}}{P_{H_2}^{1/2}} \tag{58b}$$

Again, for the reduction at the silver-silver chloride electrode we have

$$AgCl(s) + \ominus' = Ag(s) + Cl^-(a_{Cl^-}) \tag{59a}$$

and
$$\mathcal{E}_{Ag-AgCl} = \mathcal{E}^0_{Ag-AgCl} - \frac{RT}{\mathcal{F}} \ln a_{Cl^-} \tag{59b}$$

Adding now Eqs. (58a) and (59a), the cell reaction follows as

$$\frac{1}{2} H_2(g, P_{H_2}) + AgCl(s) = Ag(s) + H^+(a_{H^+}) + Cl^-(a_{Cl^-}) \tag{60a}$$

while the cell emf on combination of Eqs. (58b) and (59b) is

$$\mathcal{E}_{cell} = \mathcal{E}^0_{Ag-AgCl} - \frac{RT}{\mathcal{F}} \ln \frac{(a_{H^+}a_{Cl^-})}{P_{H_2}^{1/2}}$$

$$= \mathcal{E}^0_{Ag-AgCl} - \frac{RT}{\mathcal{F}} \ln \frac{a_{HCl}}{P_{H_2}^{1/2}} \tag{60b}$$

In the last equation a_{HCl}, the activity of the electrolyte as a whole, has been substituted for its equivalent $a_{H^+}a_{Cl^-}$.

Equation (60a) indicates that the emf of this cell results from a chemical reaction, namely, the reduction of silver chloride by hydrogen gas to form solid silver and hydrochloric acid ($H^+ + Cl^-$) in *solution*. Since this cell possesses as well no liquid junction, it constitutes a chemical cell without transference. Further, Eq. (60b) shows that the emf of this cell depends on the activity of the acid in solution and on the pressure of hydrogen gas. When the latter is 1 atm, as it usually is under experimental conditions, Eq. (60b) reduces to

$$\mathcal{E}_{cell} = \mathcal{E}^0_{Ag-AgCl} - \frac{RT}{\mathcal{F}} \ln a_{HCl} \tag{61}$$

and the emf becomes dependent only on the activity of the hydrochloric acid in solution.

By following the above procedure the student can readily verify that

$$H_2(g, P_{H_2}) \mid H_2SO_4(a_{H_2SO_4}), Hg_2SO_4(s) \mid Hg \tag{62a}$$

$$Cd \mid CdSO_4(a_{CdSO_4}), Hg_2SO_4(s) \mid Hg \tag{63a}$$

are chemical cells without transference for which the reactions are

$$H_2(g, P_{H_2}) + Hg_2SO_4(s) = 2 Hg(l) + H_2SO_4(a_{H_2SO_4}) \tag{62b}$$

$$Cd(s) + Hg_2SO_4(s) = 2 Hg(l) + CdSO_4(a_{CdSO_4}) \tag{63b}$$

Finally, the emf for (62b) at 1 atm hydrogen pressure is

$$\mathcal{E} = \mathcal{E}^0_{\text{Hg}-\text{Hg}_2\text{SO}_4} - \frac{RT}{2\,\mathcal{F}} \ln a_{\text{H}_2\text{SO}_4} \qquad (62c)$$

while for (63b) $\mathcal{E}$ is

$$\mathcal{E} = \mathcal{E}^0_{\text{cell}} - \frac{RT}{2\,\mathcal{F}} \ln a_{\text{CdSO}_4} \qquad (63c)$$

where $\mathcal{E}^0_{\text{cell}}$ is the sum of the standard electrode potentials of the cadmium and mercury-mercurous sulfate electrodes.

Chemical cells without transference find extensive application for the evaluation of standard electrode potentials of cells and electrodes and for the determination from emf data of the activity coefficients of various electrolytes. To illustrate the method, let us take specifically the cell represented in Eq. (57), for which the cell emf at 1 atm hydrogen pressure is given by Eq. (61). The problem in this instance is: How can $\mathcal{E}^0_{\text{Ag}-\text{AgCl}}$ and the activity coefficients of hydrochloric acid solutions of various molalities be evaluated from a series of emf measurements made on cells such as Eq. (57) with different concentrations of the acid?

First, the activity of hydrochloric acid is related at any molality m to the mean activity coefficient γ by the expression $a_{\text{HCl}} = m^2\gamma^2$. Substituting this value of a_{HCl} into Eq. (61) and rearranging terms, we obtain

$$\mathcal{E}_{\text{cell}} = \mathcal{E}^0_{\text{Ag}-\text{AgCl}} - \frac{RT}{\mathcal{F}} \ln (m^2\gamma^2)$$

$$= \mathcal{E}^0_{\text{Ag}-\text{AgCl}} - \frac{2\,RT}{\mathcal{F}} \ln m - \frac{2\,RT}{\mathcal{F}} \ln \gamma$$

$$\left(\mathcal{E}_{\text{cell}} + \frac{2\,RT}{\mathcal{F}} \ln m\right) = \mathcal{E}^0_{\text{Ag}-\text{AgCl}} - \frac{2\,RT}{\mathcal{F}} \ln \gamma \qquad (64)$$

All the quantities on the left-hand side of Eq. (64) are experimentally available, and hence γ could be calculated if $\mathcal{E}^0_{\text{Ag}-\text{AgCl}}$ were known. To determine the latter quantity $[\mathcal{E}_{\text{cell}} + (2\,RT/\mathcal{F}) \ln m]$ is plotted against $\sqrt{\mu}$, i.e., the square root of the ionic strength of the solution which in this case is identical with $\sqrt{m}$, and the plot is extrapolated to $\sqrt{\mu} = 0$. To make this extrapolation with certainty, emf data for dilute solutions are necessary. As at $m = 0$, $\gamma = 1$, the last term on the right in Eq. (64) becomes thus zero, and the value of the extrapolated ordinate yields immediately $\mathcal{E}^0_{\text{Ag}-\text{AgCl}}$. Once this constant is known, the difference between $[\mathcal{E}_{\text{cell}} + (2\,RT/\mathcal{F}) \ln m]$ and $\mathcal{E}^0_{\text{Ag}-\text{AgCl}}$ at each value of m gives immediately $(-2\,RT/\mathcal{F}) \ln \gamma$, and hence γ.

In Table 12–5 column 1 gives the molalities of a number of dilute hydrochloric acid solutions and column 2 the emf's at 25°C obtained for these with cells of the type under discussion. By calculating from these data the quantity $[\mathcal{E}_{\text{cell}} + (2\,RT/\mathcal{F}) \ln m] = (\mathcal{E}_{\text{cell}} + 0.11831 \log_{10} m)$, and plotting it against $\sqrt{m}$, it is found that on extrapolation to $\sqrt{m} = 0$ the ordinate is 0.2225 volt. Consequently, $\mathcal{E}^0_{Ag-AgCl}$ at 25°C is 0.2225 volt. Utilizing this value of $\mathcal{E}^0_{Ag-AgCl}$ and the various values of $\mathcal{E}_{cell}$ and m in Eq. (64), the activity coefficients of hydrochloric acid are found to be those listed in column 3 of the table.

TABLE 12–5.   ACTIVITY COEFFICIENTS OF HCl FROM
EMF'S OF THE CELL $H_2 \mid HCl, AgCl(s) \mid Ag$ AT 25°C*

m_{HCl}	$\mathcal{E}_{cell}$ (volts)	Activity Coefficient γ
0.003215	0.52053	0.942
0.004488	0.50384	0.933
0.005619	0.49257	0.926
0.007311	0.47948	0.917
0.009138	0.46860	0.909
0.011195	0.45861	0.903
0.013407	0.44974	0.895
0.01710	0.43783	0.884
0.02563	0.41824	0.866
0.05391	0.38222	0.829
0.1238	0.34199	0.788

*From D. A. MacInnes, *Principles of Electrochemistry*, Reinhold Publishing Corporation, New York, 1939 p. 187.

CHEMICAL CELLS WITH TRANSFERENCE

In chemical cells with transference the emf again results from a chemical reaction occurring within the cell, but this time assembly of the electrodes leads to a liquid junction between solutions of different electrolytes. Cells of this type are very common; in fact, of those described heretofore, such cells as

$$Zn \mid Zn^{++} \parallel Cd^{++} \mid Cd$$
$$Hg \mid Hg_2Cl_2(s), KCl(1\ N) \parallel Cu^{++} \mid Cu$$
$$Tl \mid Tl^+ \parallel Sn^{++} \mid Sn$$

belong to this category. In treating these we have assumed that the total measured emf is the sum of the two electrode potentials. Although this may be the case to a fairly close approximation, it is not exactly true, and consequently chemical cells where transference is present require further consideration.

Contact between two solutions of different concentrations, or different ions, or both, leads to a junction potential $\mathcal{E}_j$. This potential arises from diffusion of ions across the boundary between the two solutions. Because of a concentration gradient existing across the boundary, ions tend to diffuse from the side of higher to that of lower concentration. If the two ions of an electrolyte migrated with equal velocities, such diffusion would cause no complications. However, as in general this is not the case, the faster ion moves across the boundary ahead of the slower, and a separation of charges results. This separation of charges causes the establishment of a junction potential which is measured experimentally along with the two electrode potentials and appears in the total emf of the cell. In other words, whenever a junction potential ε_j is present, the emf of a cell is no longer $\varepsilon = \varepsilon_1 + \varepsilon_2$, where ε_1 and ε_2 are the two electrode potentials, but instead

$$\varepsilon = \varepsilon_1 + \varepsilon_2 + \varepsilon_j \tag{65}$$

In most instances junction potentials cannot be measured separately. Consequently various attempts have been made to calculate junction potentials,[3] and to arrive thereby through Eq. (65) at the sum of the electrode potentials corresponding to the cell reaction. However, the progress made in this direction is not sufficient to permit the calculation of $\varepsilon_1 + \varepsilon_2$ from ε without some ambiguity. For this reason chemical cells with transference are not considered suitable for *exact* evaluation of the thermodynamic properties of such cells. Nevertheless, under suitable experimental conditions these cells can be utilized to yield valuable information, and they are employed a great deal for measurements of pH, electrometric titrations, etc.

The simplest means of forming a liquid junction between two solutions is to bring the two together in a tube without causing undue mixing. Although when carefully prepared such *static junctions* may remain sharp and give reproducible potentials, a more satisfactory arrangement is the *flowing junction*. In the latter the two solutions are brought together as two streams which merge in a common tube and flow out. By this artifice the boundary between the two solutions can be kept continually renewed and sharp.

However, the type of junction employed most frequently is the *salt bridge*. In this scheme a fairly concentrated solution of a salt, usually $1\ N$ or saturated potassium chloride, is interposed between the two solutions in the manner indicated in Figure 12–5. This salt bridge is supposed to minimize or reduce the junction potential. Exactly how this reduction is accomplished is not quite clear, but it is supposed to be associated with the fact that the two ions in potassium chloride possess about equal velocities, and these operate to yield junction potentials between the two

---

[3] See D. A. MacInnes, *Principles of Electrochemistry*, Reinhold Publishing Corporation, New York, 1939, Chapter XIII.

solutions and the bridge which are opposite in sign, and hence cancel each other to a degree. Whether such a reduction in junction potentials with the aid of salt bridges actually occurs is problematical. It is still more questionable whether such bridges can reduce the junction potentials to a point where they are negligible. Nevertheless, effective or not, the fact remains that salt bridges have come into extensive use and are encountered in various electrochemical assemblies.

In calculations involving these cells it is assumed that a salt bridge eliminates completely the junction potential and that the measured cell emf is merely the sum of the two electrode potentials. When $\mathcal{E}_j$ is disregarded in this manner, it is customary to write two vertical lines between the two solutions, namely,

$$Zn \mid Zn^{++} \parallel Cd^{++} \mid Cd$$

On the other hand, when the presence of the liquid junction is recognized and taken into account, only one line is interposed, as

$$Zn \mid Zn^{++} \mid Cd^{++} \mid Cd$$

It is characteristic of chemical cells with transference that they involve in their emf equations the *activities of ions* rather than mean activities of the electrolyte as a whole. This would suggest that such cells may be suitable for the evaluation of activities and activity coefficients of individual ions. However, this is not quite the case because of the junction potential uncertainty. In fact, it is generally agreed that there is no exact thermodynamic method known at present by which anything more than the geometric mean activity coefficient of an electrolyte can be determined. Therefore, when the emf's of chemical cells with transference are used to estimate activity coefficients of ions, this is done with the understanding that the quantities evaluated possess only quasi-thermodynamic significance. Again, in calculating the emf's of such cells from molalities and activity coefficients, the usual practice is to assume that the activity coefficient of any ion is essentially equal to the geometric mean activity coefficient of the electrolyte, and to use the latter. With these remarks in mind we are ready to consider a specific calculation.

EXAMPLE: Suppose that it is desired to estimate at 25°C the emf of the cell

$$Zn \mid ZnCl_2(m = 0.5) \parallel CdSO_4(m = 0.1) \mid Cd$$

where the m's are the molalities of the two solutions. For this cell

$$\mathcal{E} = \mathcal{E}^0 - \frac{RT}{2\,\mathfrak{F}} \ln \frac{a_{Zn^{++}}}{a_{Cd^{++}}}$$

Since $a_{Zn^{++}} = m_{Zn^{++}}\gamma_{Zn^{++}}$, and $a_{Cd^{++}} = m_{Cd^{++}}\gamma_{Cd^{++}}$, where γ's are the activity coefficients of the ions in their respective solutions, the above equation may be written as

$$\varepsilon = \varepsilon^0 - \frac{RT}{2\,\mathfrak{F}} \ln \frac{m_{Zn^{++}}\gamma_{Zn^{++}}}{m_{Cd^{++}}\gamma_{Cd^{++}}}$$

Now, at 25°C $\varepsilon^0 = 0.359$ volt for this cell. Again, for $0.5m$ $ZnCl_2$ $\gamma = 0.376$, while for $0.1m$ $CdSO_4$ $\gamma = 0.137$. Assuming that these γ's are also the ionic activity coefficients, we find for ε

$$\varepsilon = 0.359 - \frac{0.05916}{2} \log_{10} \frac{(0.5)(0.376)}{(0.1)(0.137)}$$
$$= 0.359 - 0.034$$
$$= 0.325 \text{ volt at } 25°C$$

CONCENTRATION CELLS

Unlike chemical cells, whose emf arises from a chemical reaction, *concentration cells depend for their emf on a transfer of material from one electrode to the other due to a concentration difference between the two.* This difference in concentration may arise from the fact that two like electrodes dipping in the same solution may be at different concentrations, as two hydrogen electrodes at unequal gas pressures immersed in the same solution of hydrogen ions, namely,

$$H_2(P_{H_2} = P_1) \mid H^+ \mid H_2(P_{H_2} = P_2)$$

or two amalgam electrodes of different concentration dipping in a solution of the metal ions, as

$$Cd(Hg)(C_{Cd} = C_1) \mid Cd^{++} \mid Cd(Hg)(C_{Cd} = C_2)$$

Again, the difference in concentration may not be in the electrodes, but in the solutions with which they are in contact, as in the cells

$$H_2(g, 1 \text{ atm}) \mid H^+(a_1) \mid H^+(a_2) \mid H_2(g, 1 \text{ atm})$$
$$Ag \mid Ag^+(a_1) \mid Ag^+(a_2) \mid Ag$$

In the first two cells mentioned there is no liquid junction present, and hence they are *concentration cells without transference.* On the other hand, the latter two cells involve a liquid junction between two solutions of the same kind but unlike concentration, and therefore these constitute *concentration cells with transference.*

ELECTRODE CONCENTRATION CELLS
WITHOUT TRANSFERENCE

To see why an emf arises where two like electrodes are at different concentrations, but the electrolyte is the same for both, consider first the cell

$$H_2(P_{H_2} = P_1) \mid H^+(a_{H^+}) \mid H_2(P_{H_2} = P_2) \tag{66}$$

For the electrode on the left, the oxidation reaction yields

$$\frac{1}{2} H_2(P_1) = H^+(a_{H^+}) + \ominus \tag{67a}$$

and, therefore, since $\varepsilon_{H_2}^0 = 0$,

$$\varepsilon_1 = -\frac{RT}{\mathfrak{F}} \ln \frac{a_{H^+}}{P_1^{1/2}} \tag{67b}$$

Similarly, for the reduction at the right-hand electrode the process is

$$H^+(a_{H^+}) + \ominus = \frac{1}{2} H_2(P_2) \tag{68a}$$

and

$$\varepsilon_2 = -\frac{RT}{\mathfrak{F}} \ln \frac{P_2^{1/2}}{a_{H^+}} \tag{68b}$$

On adding Eqs. (67a) and (68a), the cell reaction follows as

$$\frac{1}{2} H_2(P_1) = \frac{1}{2} H_2(P_2) \tag{69a}$$

while the cell emf, on adding Eqs. (67b) and (68b), is

$$\varepsilon = -\frac{RT}{\mathfrak{F}} \ln \frac{a_{H^+}}{P_1^{1/2}} - \frac{RT}{\mathfrak{F}} \ln \frac{P_2^{1/2}}{a_{H^+}}$$

$$= -\frac{RT}{2\mathfrak{F}} \ln \frac{P_2}{P_1} \tag{69b}$$

Equation (69a) shows that the cell reaction involves merely the transfer of 0.5 mole of hydrogen gas from a pressure P_1 atm at one electrode to a pressure P_2 at the other; i.e., that the cell reaction for the spontaneous process is an expansion of hydrogen gas from a pressure P_1 to a pressure P_2. Again, Eq. (69b) shows that the emf resulting from this expansion depends only on the two pressures and is *independent of the activity of the hydrogen ions* in which the electrodes are immersed.

Consider next another cell of this type, namely,

$$Zn(Hg)(a_{Zn} = a_1) \mid Zn^{++}(a_{Zn^{++}}) \mid Zn(Hg)(a_{Zn} = a_2) \tag{70}$$

consisting of two zinc amalgams with activities of zinc equal to a_1 and a_2 immersed in a solution of zinc ions of activity $a_{Zn^{++}}$. Analogous to Eqs.

(38) and (39), the electrode reaction on the left is

$$Zn(Hg)(a_1) = Zn^{++}(a_{Zn^{++}}) + 2 \ominus \tag{71a}$$

with ε_1 given by

$$\varepsilon_1 = \varepsilon_{Zn}^0 - \frac{RT}{2\,\mathfrak{F}} \ln \frac{a_{Zn^{++}}}{a_1} \tag{71b}$$

Again, for the reduction on the right, we have

$$Zn^{++}(a_{Zn^{++}}) + 2 \ominus = Zn(Hg)(a_2) \tag{72a}$$

and

$$\varepsilon_2 = \varepsilon_{Zn}^{0\prime} - \frac{RT}{2\,\mathfrak{F}} \ln \frac{a_2}{a_{Zn^{++}}} \tag{72b}$$

where $\varepsilon_{Zn}^{0\prime} = -\varepsilon_{Zn}^0$. We obtain, therefore, for the cell reaction

$$Zn(Hg)(a_1) = Zn(Hg)(a_2) \tag{73a}$$

and for the cell emf

$$\begin{aligned} \varepsilon &= \varepsilon_{Zn}^0 - \frac{RT}{2\,\mathfrak{F}} \ln \frac{a_{Zn^{++}}}{a_1} + (-\varepsilon_{Zn}^0) - \frac{RT}{2\,\mathfrak{F}} \ln \frac{a_2}{a_{Zn^{++}}} \\ &= -\frac{RT}{2\,\mathfrak{F}} \ln \frac{a_2}{a_1}. \end{aligned} \tag{73b}$$

From Eq. (73a) it is evident that the emf of this cell is due to a transfer of zinc from the amalgam where its activity is a_1 to the amalgam where its activity is a_2. Further, Eq. (73b) shows that this emf depends only on the ratio of the zinc activities in the two amalgams, and not at all on the activity of the zinc ions in the solution. In the final cell emf equation ε^0 does not appear. This is true of all concentration cells. We may conclude, therefore, *that for concentration cells ε^0 is zero*, and the emf equation takes on the simplified form

$$\varepsilon_{cell} = -\frac{RT}{n\mathfrak{F}} \ln \frac{a_2}{a_1} \tag{74}$$

ELECTROLYTE CONCENTRATIONS CELLS WITHOUT TRANSFERENCE

Consider a chemical cell without transference, such as

$$H_2(g,\ 1\ atm)\ |\ HCl(a_1),\ AgCl(s)\ |\ Ag \tag{75a}$$

For this cell the reaction is

$$\frac{1}{2} H_2(g,\ 1\ atm) + AgCl(s) = Ag(s) + HCl(a_1) \tag{75b}$$

and the cell emf, Eq. (61),

$$\varepsilon_1 = \varepsilon^0 - \frac{RT}{\mathfrak{F}} \ln a_1 \tag{75c}$$

Again, for the same cell but with a different activity of hydrochloric acid, namely,

$$H_2(g, 1\ atm)\ |\ HCl(a_2),\ AgCl(s)\ |\ Ag \tag{76a}$$

the cell reaction will be

$$\frac{1}{2}\ H_2(g,\ 1\ atm)\ +\ AgCl(s)\ =\ Ag(s)\ +\ HCl(a_2) \tag{76b}$$

and the cell emf

$$\mathcal{E}_2\ =\ \mathcal{E}^0\ -\ \frac{RT}{\mathfrak{F}}\ \ln a_2 \tag{76c}$$

If these two cells are connected so as to *oppose* each other, i.e.,

$$H_2(g,\ 1\ atm)\ |\ HCl(a_1),\ AgCl(s)\ |\ Ag\text{-}Ag\ |\ AgCl(s),$$
$$HCl(a_2)\ |\ H_2(g,\ 1\ atm) \tag{77a}$$

the over-all reaction of the combination will be the *difference* between Eqs. (75b) and (76b), namely,

$$\frac{1}{2}\ H_2(g,\ 1\ atm)\ +\ AgCl(s)\ -\ \frac{1}{2}\ H_2(g,\ 1\ atm)\ -\ AgCl(s)\ =$$
$$Ag(s)\ +\ HCl(a_1)\ -\ Ag(s)\ -\ HCl(a_2)$$

or
$$HCl(a_2)\ =\ HCl(a_1) \tag{77b}$$

Similarly the emf of (77a) will be the *difference* between Eqs. (75c) and (76c). Hence

$$\begin{aligned}
\mathcal{E}\ &=\ \mathcal{E}_1\ -\ \mathcal{E}_2\\
&=\ \left(\mathcal{E}^0\ -\ \frac{RT}{\mathfrak{F}}\ \ln a_1\right)\ -\ \left(\mathcal{E}^0\ -\ \frac{RT}{\mathfrak{F}}\ \ln a_2\right)\\
&=\ -\frac{RT}{\mathfrak{F}}\ \ln \frac{a_1}{a_2}
\end{aligned} \tag{77c}$$

According to Eq. (77b) the over-all reaction resulting from the combination given in equation (77a) is a *transfer* for each faraday drawn from the cell of 1 mole of hydrochloric acid from the solution where the activity is a_2 to the solution where the activity is a_1. Consequently, whereas each of the individual cells constituting Eq. (77a) is a chemical cell, the combination of the two is a *concentration cell without transference in which the emf arises from different concentrations of the electrolyte.* For this emf to be positive, Eq. (77c) shows that a_2 must be greater than a_1; i.e., the transfer process is spontaneous for passage of electrolyte from the more concentrated to the more dilute solution.

Concentration cells of this type may be assembled from any chemical cells without transference. For instance, a sodium chloride concentration cell results from the combination

$$Na(Hg) \mid NaCl(a_1), AgCl(s) \mid Ag\text{-}Ag \mid AgCl(s), NaCl(a_2) \mid Na(Hg)$$

and a zinc sulfate concentration cell from

$$Zn(Hg) \mid ZnSO_4(a_1), PbSO_4(s) \mid Pb(Hg)\text{-}Pb(Hg) \mid PbSO_4(s),$$
$$ZnSO_4(a_2) \mid Zn(Hg)$$

Such cells are suitable for the determination of the activity coefficients of the electrolytes involved. For this purpose the usual practice is to keep one of the electrolyte concentrations constant, and to measure the emf of the cells with varying concentrations of the second solution. If for a_1 and a_2 in Eq. (77c) we make the substitutions $a_1 = m_1^2\gamma_1^2$ and $a_2 = m_2^2\gamma_2^2$, the expression for $\mathcal{E}$ becomes

$$\begin{aligned}
\mathcal{E} &= -\frac{RT}{\mathcal{F}} \ln \frac{m_1^2\gamma_1^2}{m_2^2\gamma_2^2} \\
&= \frac{2\,RT}{\mathcal{F}} \ln \frac{m_2\gamma_2}{m_1\gamma_1} \\
&= \frac{2\,RT}{\mathcal{F}} \ln \frac{m_2}{m_1} + \frac{2\,RT}{\mathcal{F}} \ln \frac{\gamma_2}{\gamma_1}
\end{aligned} \tag{78}$$

As the molalities are known, Eq. (78) permits evaluation of the ratio of the activity coefficients of the electrolyte at any molality m_2 to that at the reference m_1. Hence when γ_1 corresponding to m_1 is available, the various activity coefficients follow from these ratios.

CONCENTRATION CELLS WITH TRANSFERENCE

A typical concentration cell with transference is

$$H_2(g, 1\ atm) \mid HCl(a_1) \mid HCl(a_2) \mid H_2(g, 1\ atm) \tag{79}$$

consisting of two identical hydrogen electrodes immersed in two hydrochloric acid solutions of different concentrations. For this cell the total emf is composed of the two single electrode potentials and the potential at the junction, $\mathcal{E}_j$. Further, the over-all cell reaction is the sum of the two electrode processes and any material transfer taking place across the junction. In a case such as this, where the electrolytes constituting the junction are identical except for their concentrations, these various processes can be analyzed, and an equation for the cell can be arrived at which takes into account the junction potential.

To understand how this is possible, assume that the electrode on the left in Eq. (79) is negative, which will be true when $a_2 > a_1$. For a faraday of electricity drawn from this cell, the reaction at the negative electrode will be

$$\frac{1}{2} H_2(g, 1 \text{ atm}) = H^+(a_1) + \ominus$$

that at the positive

$$H^+(a_2) + \ominus = \frac{1}{2} H_2(g, 1 \text{ atm})$$

and hence the sum of the two electrode reactions is

$$H^+(a_2) = H^+(a_1) \tag{80a}$$

However, when electrons flow externally from left to right, they must complete the circuit by passing *through the cell from right to left;* i.e., electrons must pass across the liquid junction from right to left. This current in the cell is composed, of course, not of free electrons, but of negative ions, namely Cl^-, moving from *right to left,* and positive ions, or H^+, moving across the junction from *left to right.* If t_- is the transference number of the chloride ions, then for every faraday passing through the cell t_- equivalents of chloride ions will be transported from the solution of activity a_2 to the solution where the activity is a_1; i.e.,

$$t_-Cl^-(a_2) = t_-Cl(a_1) \tag{80b}$$

Again, $t_+ = 1 - t_-$ equivalents of hydrogen ions will be transferred from the solution of activity a_1 to the solution of activity a_2, namely,

$$(1 - t_-)H^+(a_1) = (1 - t_-)H^+(a_2) \tag{80c}$$

Therefore, in order to obtain the *net transfer* of material we must add Eqs. (80b) and (80c) to (80a). We thus get for the over-all process of the cell

$$H^+(a_2) + t_-Cl^-(a_2) + (1 - t_-)H^+(a_1)$$
$$= H^+(a_1) + t_-Cl^-(a_1) + (1 - t_-)H^+(a_2)$$

$$t_-H^+(a_2) + t_-Cl^-(a_2) = t_-H^+(a_1) + t_-Cl^-(a_1)$$

$$t_-HCl(a_2) = t_-HCl(a_1) \tag{81}$$

Equation (81) shows that in the concentration cell with transference t_- equivalents of hydrochloric acid are transferred from the solution of activity a_2 to the solution of activity a_1 for every faraday of electricity. This may be contrasted with the same process in a concentration cell without transference, Eq. (77b), where for 1 faraday the transfer of a full equivalent of the electrolyte is accomplished.

Application of the emf equation to the process in (81) gives

$$\varepsilon = -\frac{RT}{\mathcal{F}} \ln \frac{a_1^{t-}}{a_2^{t-}}$$
$$= \frac{t_-RT}{\mathcal{F}} \ln \frac{a_2}{a_1} \tag{82a}$$

On insertion of the molalities and activity coefficients, ε becomes

$$\varepsilon = \frac{t_-RT}{\mathcal{F}} \ln \frac{m_2^2 \gamma_2^2}{m_1^2 \gamma_1^2}$$
$$= \frac{2\,t_-RT}{\mathcal{F}} \ln \frac{m_2 \gamma_2}{m_1 \gamma_1} \tag{82b}$$

and hence the emf of such a cell can be calculated from a knowledge of the molalities, activity coefficients, and the *transport number of the ion other than the one to which the electrodes are reversible.*

Equations analogous to Eqs. (81), (82a), and (82b) are applicable to any concentration cell with transference in which the electrodes are reversible to the *cation*. However, when the electrodes are reversible to the *anion*, as in the cell

$$\text{Ag} \mid \text{AgCl(s), HCl}(a_1) \mid \text{HCl}(a_2), \text{AgCl(s)} \mid \text{Ag} \tag{83}$$

repetition of the above reasoning shows that the cell reaction is

$$t_+\text{HCl}(a_1) = t_+\text{HCl}(a_2) \tag{84}$$

and the emf

$$\varepsilon = -\frac{RT}{\mathcal{F}} \ln \frac{a_2^{t+}}{a_1^{t+}}$$
$$= \frac{t_+RT}{\mathcal{F}} \ln \frac{m_1^2 \gamma_1^2}{m_2^2 \gamma_2^2}$$
$$= \frac{2\,t_+RT}{\mathcal{F}} \ln \frac{m_1 \gamma_1}{m_2 \gamma_2} \tag{85}$$

Now a_1 must be greater than a_2 for the reaction to be spontaneous.

Since the above concentration cells with transference involve transport numbers in their emf equations, such cells may be utilized also for obtaining these quantities from emf data. Equations (82b) and (85) indicate that we need for this purpose the emf's of the respective cells and the activity coefficients of the electrolytes involved. However, in precise determinations of transference numbers from emf measurements corrections have to be made for the variation of transport numbers with concentration. Means of making such corrections have been developed, and with these it is possible to obtain results in excellent accord with those given by other methods.

THE JUNCTION POTENTIAL

The emf given by Eq. (82b) for cell (79) is the sum of the two electrode potentials and ε_j. It is of interest to ascertain how much of this total emf is due to the electrodes and how much to the liquid junction. According to Eq. (80a), the sum of the two electrode reactions is

$$H^+(a_2) = H^+(a_1)$$

and hence the sum of the two electrode potentials, $\varepsilon_1 + \varepsilon_2$, is

$$\varepsilon_1 + \varepsilon_2 = -\frac{RT}{\mathcal{F}} \ln \frac{(a_{H^+})_1}{(a_{H^+})_2}$$
$$= \frac{RT}{\mathcal{F}} \ln \frac{(m_{H^+}\gamma_{H^+})_2}{(m_{H^+}\gamma_{H^+})_1} \tag{86}$$

If Eq. (86) be subtracted from Eq. (82b), the result is ε_j, namely,

$$\varepsilon_j = \varepsilon - (\varepsilon_1 + \varepsilon_2)$$
$$= \frac{2\,t_-RT}{\mathcal{F}} \ln \frac{m_2\gamma_2}{m_1\gamma_1} - \frac{RT}{\mathcal{F}} \ln \frac{(m_{H^+}\gamma_{H^+})_2}{(m_{H^+}\gamma_{H^+})_1}$$

But $(m_{H^+})_2 = m_2$ and $(m_{H^+})_1 = m_1$. Further, if we take the activity coefficient of an ion equal to the mean activity coefficient of the electrolyte, $(\gamma_{H^+})_2 = \gamma_2$ and $(\gamma_{H^+})_1 = \gamma_1$. ε_j follows thus as

$$\varepsilon_j = \frac{2\,t_-RT}{\mathcal{F}} \ln \frac{m_2\gamma_2}{m_1\gamma_1} - \frac{RT}{\mathcal{F}} \ln \frac{m_2\gamma_2}{m_1\gamma_1}$$
$$= (2\,t_- - 1)\frac{RT}{\mathcal{F}} \ln \frac{m_2\gamma_2}{m_1\gamma_1} \tag{87a}$$

Again, since $(t_+ + t_-) = 1$, $(2\,t_- - 1) = (t_- - t_+)$, and Eq. (87a) may also be written as

$$\varepsilon_j = (t_- - t_+)\frac{RT}{\mathcal{F}} \ln \frac{m_2\gamma_2}{m_1\gamma_1} \tag{87b}$$

We see, therefore, that besides the activities of the two solutions constituting the junction, ε_j is determined also by the difference between the two transport numbers of the electrolyte. When these are equal, $(t_- - t_+) = 0$, and so is the junction potential. This fact is the basis for the use of potassium chloride as a bridge in cell measurements involving liquid junction, for in this electrolyte the two transference numbers are very nearly identical. In turn, when $t_- > t_+$, ε_j is positive and adds to $\varepsilon_1 + \varepsilon_2$, while when $t_- < t_+$, ε_j is negative and operates to yield for ε of the cell a potential less than $\varepsilon_1 + \varepsilon_2$.

Similar considerations, applied to concentration cells with transference containing electrodes reversible to the *anion*, give for ε_j

$$\varepsilon_j = (2\,t_+ - 1)\frac{RT}{\mathcal{F}} \ln \frac{m_1\gamma_1}{m_2\gamma_2} \tag{88a}$$

or
$$\varepsilon_j = (t_+ - t_-)\frac{RT}{\mathcal{F}} \ln \frac{m_1\gamma_1}{m_2\gamma_2} \tag{88b}$$

Here m_1 is the more concentrated solution. From Eq. (88b) it is clear that again $\varepsilon_j = 0$ for $t_+ = t_-$, but this time ε_j is positive for $t_+ > t_-$, and negative for $t_+ < t_-$.

SOLUBILITY PRODUCTS AND EMF

The saturation solubility of any difficultly soluble salt such as silver bromide is given by the equation

$$AgBr(s) = Ag^+ + Br^- \tag{89}$$

where the product of the activities of the two ions is equal to the solubility product. Since this solubility product K_s is an equilibrium constant, it must be related to the ε^0 value of a cell in which the above reaction occurs. Hence, by finding two single electrodes whose reactions lead to the over-all process given in Eq. (89), and by appropriately combining their ε^0 values, it should be possible to calculate from the latter the solubility product.

Inspection of Eq. (89) suggests that the requisite combination in this case is the silver and the silver-silver bromide electrodes. For the silver electrode, the reaction is

$$Ag(s) = Ag^+ + \ominus$$

for the silver-silver bromide electrode i' is

$$AgBr(s) + \ominus = Ag(s) + Br^-$$

and therefore the sum of the two, namely,

$$AgBr(s) = Ag^+ + Br^- \tag{90}$$

is identical with Eq. (89). From Table 12–3 ε^0 for the silver electrode at 25°C is seen to be $\varepsilon^0_{Ag} = -0.7991$ volt, ε^0 of the silver-silver bromide electrode for reduction $\varepsilon^0_{Ag-AgBr} = +0.0711$ volt, and hence ε^0 for (89) follows as

$$\varepsilon^0 = \varepsilon^0_{Ag} + \varepsilon^0_{Ag-AgBr}$$
$$= -0.7991 + 0.0711$$
$$= -0.7280 \text{ volt}$$

Substituting this value of ε^0 into Eq. (35), we obtain for the solubility product of silver bromide

$$\varepsilon^0 = \frac{RT}{n\mathfrak{F}} \ln K_s$$

$$-0.7280 = 0.05916 \log_{10} K_s$$

$$K_s = 4.9 \times 10^{-13}$$

This value of K_s may be compared with $K_s = 7.7 \times 10^{-13}$ given in **Table 11–16** of Chapter 11. In a like manner may be calculated the solubility products of other difficultly soluble salts.

A less exact method for estimating the solubility product of a difficultly soluble salt involves the measurement of the single electrode potential of an electrode reversible to one of the ions of the salt and immersed in a solution saturated with this salt. To explain this method let us consider specifically the determination of the solubility product of silver chloride. For this purpose a solution of a chloride is taken, say 0.1 m potassium chloride, and the solution saturated with silver chloride by addition of a few drops of silver nitrate. A silver wire is inserted then into this solution, and the silver electrode thus formed is combined with a reference calomel electrode to form the cell

$$\text{Ag} \mid \text{KCl}(0.1\ m,\ \text{sat'd. with AgCl}) \mid \text{Calomel electrode}$$

When the reference electrode is the 0.1 N calomel, the emf of this cell at 25°C is 0.0494 volt. Since ε_C of the 0.1 N calomel for reduction is 0.3338 volt at 25°C, we have that

$$\varepsilon = \varepsilon_{Ag} + \varepsilon_C$$
$$0.0494 = \varepsilon_{Ag} + 0.3338$$
$$\varepsilon_{Ag} = -0.2844 \text{ volt}$$

But ε_{Ag} is also given by

$$\varepsilon_{Ag} = \varepsilon_{Ag}^0 - \frac{RT}{\mathfrak{F}} \ln a_{Ag^+}$$

and hence the activity of silver ions in this solution follows as

$$-0.2844 = -0.7991 - 0.05916 \log_{10} a_{Ag^+}$$
$$\log_{10} a_{Ag^+} = -\frac{0.5147}{0.05916} = -8.700$$
$$a_{Ag^+} = 2.00 \times 10^{-9}$$

Now, for the activity of the chloride ions we may take the **product of** the molality, 0.1, and the mean activity coefficient of potassium **chloride** at this concentration, 0.769. We obtain thus for K_s of silver chloride **at**

25°C

$$K_s = a_{Ag^+} a_{Cl^-}$$
$$= (2.00 \times 10^{-9})(0.1 \times 0.769)$$
$$= 1.54 \times 10^{-10}$$

The thermodynamic value of K_s for this salt at 25°C is 1.76×10^{-10}.

POTENTIOMETRIC DETERMINATION OF pH

One of the most extensive applications of emf measurements is in determining the pH of various solutions. In all potentiometric pH work the procedure followed is very much the same. First a cell is assembled in which one of the electrodes is reversible to hydrogen ions and dips into the solution whose pH is to be determined, while the other electrode is usually one of the calomels. Junction between the two is made either through a salt bridge or by immersing the reference electrode directly into the solution. Next the emf of the combination, ε, is measured with a suitable potentiometer. From ε is subtracted the emf of the calomel electrode to yield the single potential of the electrode reversible to hydrogen ions, and from the latter, in turn, the pH is calculated by means of the emf equation applicable to the particular electrode used.

Of several electrodes more or less suitable for potentiometric pH determinations, we shall discuss two, namely, (a) the hydrogen electrode, and (b) the glass electrode. The hydrogen electrode is the standard of all pH measurements and is the electrode against which all the others are checked. According to Eq. (44), the emf of this electrode is given by

$$\varepsilon_{H_2} = -\frac{RT}{\mathcal{F}} \ln a_{H^+} + \frac{RT}{\mathcal{F}} \ln P_{H_2}^{1/2} \tag{44}$$

Since by definition pH $= -\log_{10} a_{H^+}$, Eq. (44) in terms of pH is

$$\varepsilon_{H_2} = -\frac{2.3026\,RT}{\mathcal{F}} \log_{10} a_{H^+} + \frac{2.3026\,RT}{\mathcal{F}} \log_{10} P_{H_2}^{1/2}$$
$$= \frac{2.3026\,RT}{\mathcal{F}} \text{pH} + \frac{2.3026\,RT}{\mathcal{F}} \log_{10} P_{H_2}^{1/2} \tag{91}$$

When P_{H_2} is exactly 1 atm, Eq. (91) reduces to

$$\varepsilon_{H_2} = \left(\frac{2.3026\,RT}{\mathcal{F}}\right) \text{pH} \tag{92}$$

and hence under these conditions *the emf of the hydrogen electrode is linearly dependent only on the pH of the solution.*

It was mentioned already that the hydrogen electrode consists of a piece of platinized platinum foil immersed in the solution under test around which is bubbled hydrogen gas. For satisfactory operation the platinized coat must be freshly deposited from a solution of chloroplatinic acid, and the hydrogen gas must be carefully purified to remove impurities, particularly oxygen. With all details carefully controlled this electrode can give very accurate results over the entire pH range, provided the solution does not contain oxidizing agents that may react with the hydrogen or metals that may be thrown out of solution by the gas. However, because of the ease with which this electrode is poisoned even by oxygen of the air, and because of the rather elaborate setup required, its use is pretty much confined to the laboratory.

To illustrate the calculation of pH from data obtained with a hydrogen electrode, let us take the following example. Suppose it is found that the emf of the cell

$$H_2 \mid \text{solution (pH} = x) \mid 1\ N \text{ calomel}$$

is $\varepsilon = 0.5164$ volt at 25°C when the corrected barometric pressure is 754.1 mm Hg. Subtracting from this emf the potential of the calomel electrode, we obtain for the emf of the hydrogen electrode

$$\begin{aligned}
\varepsilon_{H_2} &= \varepsilon - \varepsilon_C \\
&= 0.5164 - 0.2800 \\
&= 0.2364 \text{ volt}
\end{aligned}$$

This emf is due to both the pH of the solution and the pressure of the hydrogen gas. The total barometric pressure at which the gas escapes from solution is composed of the pressure of the hydrogen and the pressure of water vapor with which this gas becomes saturated in bubbling through the solution; i.e.,

$$P_{H_2} + P_{H_2O} = P_{\text{barometer}}$$

For P_{H_2O} we may take without any error the vapor pressure of water at 25°C, 23.8 mm. P_{H_2} is then

$$\begin{aligned}
P_{H_2} &= P_{\text{barometer}} - P_{H_2O} \\
&= 754.1 - 23.8 \\
&= 730.3 \text{ mm Hg} = 0.961 \text{ atm}
\end{aligned}$$

Substituting now $\varepsilon_{H_2} = 0.2364$ and $P_{H_2} = 0.961$ atm into Eq. (91), the pH of the solution follows as

$$0.2364 = 0.05916\ \text{pH} + \frac{0.05916}{2} \log_{10} 0.961$$

$$\begin{aligned}
\text{pH} &= \frac{0.2364 + 0.0005}{0.05916} \\
&= 4.00
\end{aligned}$$

It will be observed that the pressure correction to the emf is small and amounts in this case to only 0.01 of a pH unit.

Fritz Haber and Z. Klemensiewicz[4] first showed that, when two solutions of different pH are separated by a glass membrane, a potential is established across the membrane whose magnitude depends on the difference in pH of the two solutions. If the pH of one of these solutions is held constant while that of the second is varied, the emf of the *glass electrode* follows the equation

$$\varepsilon_G = \varepsilon_G^0 - \frac{RT}{\mathfrak{F}} \ln a_{H^+}$$

$$= \varepsilon_G^0 + \left(\frac{2.3026\ RT}{\mathfrak{F}}\right) pH \qquad (93)$$

where ε_G^0 is a constant determined by the magnitude of the fixed pH and the particular arrangement employed. Equation (93) indicates that the glass electrode should be suitable for pH measurements.

A glass electrode assembly frequently utilized for pH work is illustrated in Figure 12–6. Here A is the glass electrode, B is the solution under test,

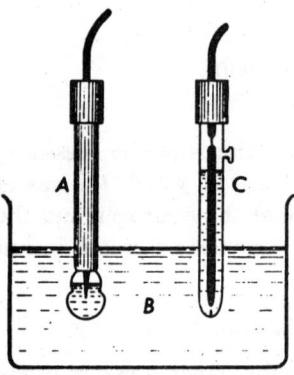

Figure 12–6.   Glass electrode assembly for pH.

and C is a calomel electrode used to complete the cell arrangement. The glass electrode, immersed directly in the solution whose pH is to be determined, consists of a bulb constructed of a special glass, on the inside of which is placed, say, 0.1 N hydrochloric acid and a silver-silver chloride electrode. This entire combination, namely,

$$Ag \mid AgCl(s),\ 0.1\ N\ HCl \mid Glass \mid$$

constitutes the glass electrode. When combined with the calomel the resulting cell is

$$Ag \mid AgCl(s),\ 0.1\ N\ HCl \mid Glass \mid Solution\ (pH = x) \mid Calomel \qquad (94)$$

[4] F. Haber and Z. Klemensiewicz, *Zeit. physik. Chem.*, **67**, 385 (1909).

from whose measured emf the single electrode potential of the glass electrode can be deduced, and therefrom the pH.

Because of the high internal resistance of the glass electrode, which may amount to as much as 100 million ohms, ordinary potentiometers cannot be used to measure the emf of cell (94). Either quadrant electrometers or vacuum tube voltmeters which require practically no current for their operation must be employed. For this purpose vacuum tube circuits have been developed which not only are sensitive to 0.01 pH unit or better, but also are portable and very rugged.

Because of an "asymmetry potential," not all glass electrodes in a particular assembly have the same value of $\mathcal{E}_G^0$. For this reason it is best to determine $\mathcal{E}_G^0$ for each electrode before use. To do this a solution of definite pH is placed in B, the emf of the cell is measured, and $\mathcal{E}_G^0$ is evaluated. Then the first solution is discarded, the solution under test substituted, the measurement repeated, and the pH calculated with the $\mathcal{E}_G^0$ found above. In this manner any "asymmetry potential" is incorporated into $\mathcal{E}_G^0$ of the electrode, and errors are avoided.

A solution frequently used for calibration of glass electrode assemblies is 0.05 molar potassium acid phthalate. This solution has a pH of 4.00 between 10°C and about 30°C, and a pH of 4.02 at 38°C.

The glass electrode is the nearest approach to a universal pH electrode known at present. It is not poisoned easily, nor is it affected by oxidizing or reducing agents or by organic compounds. Further, it can be used on quantities of solution as small as a fraction of a cubic centimeter. Its only limitations arise in strongly alkaline solutions where the glass is attacked, and in solutions of pH = 9 and above where presence of various cations, particularly sodium, leads to appreciable errors. However, new glasses have been developed with which good results can be obtained up to pH = 13 or 14.

POTENTIOMETRIC TITRATIONS

In Figure 12-7(a) is shown a characteristic plot for the variation of pH with volume of added base during the titration of an acid such as hydrochloric with a base like sodium hydroxide. It will be observed that the pH of the solution rises gradually at first, then more rapidly, until at the equivalence point there is a very sharp increase in pH for a very small quantity of added base. Past the equivalence point the curve again tapers off, indicating that the pH increases only slightly on the addition of excess base.

These changes in pH during a titration may be followed potentiometrically by immersing in the solution being titrated an electrode reversible to hydrogen ions, and coupling it with a suitable reference electrode.

Since the potential of the latter remains constant, the emf of such a cell will vary only with the pH of the solution. Further, since the emf of any electrode reversible to hydrogen ions is proportional to the pH, the cell emf will exhibit a course parallel to the curve shown in Figure 12–7(a). Consequently, by measuring this emf at each stage of the titration and plotting it against the volume of base, we may deduce from the plot the equivalence point. This is the principle of all potentiometric acid-base titrations.

A more sensitive and satisfactory means of deducing the end point is to plot not ε of the cell against the volume, but the slope of the curve, i.e., $\Delta\varepsilon/\Delta cc$ vs. cc, Figure 12–7(b). As the slope is greatest at the equivalence point, the latter plot exhibits a maximum at a volume corresponding

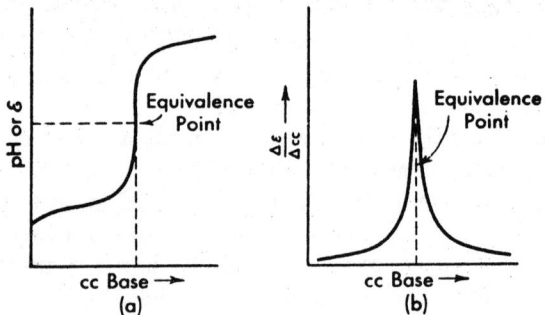

Figure 12–7.   Potentiometric titration of an acid with a base.

to the end point of the titration. In practice a known volume of the acid is placed in a beaker, the electrodes are inserted, and the base is added with stirring in rather large increments at the start. After each addition the emf is read. When the latter begins to show a tendency to increase appreciably, the increments are made smaller and smaller until near the end point only 0.1 cc or less are added at a time. By such a procedure it is possible to obtain a sharp maximum and hence within narrow limits the volume of base equivalent to the acid taken.

Potentiometric acid-base titrations possess a number of distinct advantages over the ordinary methods involving indicators. In the first place, indicators cannot be employed when solutions have a strong color of their own, whereas potentiometric titrations are not subject to this limitation. Second, indicators for any acid-base titration must be chosen so that the pH at which the indicator changes color corresponds more or less to the pH of the solution at the equivalence point. Hence some information is required *a priori* concerning the relative strengths of the acid and base involved. This is not true of potentiometric titrations. These

always yield the equivalence point whether this point comes exactly at the neutral point, or on the acid or basic side.

Still a third advantage of the potentiometric method appears in the titration of polybasic acids or of mixtures of a strong and a weak acid with a base. With indicators it is as a rule impossible or difficult to titrate a polybasic acid in steps corresponding to different stages of neutralization. This is also the case with mixtures of strong and weak acids unless one indicator can be found whose color change corresponds to the neutralization of the strong acid, and another to that of the weak. However, when the ionization constants of the different stages of a polybasic acid or of the different acids in a mixture differ by a factor of at least 1000, the potentiometric titration yields directly a number of distinct steps for the various neutralizations. This may be seen from Figure 12–8, where the plot for the potentiometric titration of phosphoric acid

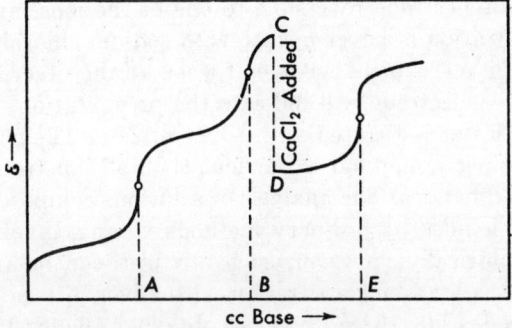

Figure 12–8. Stepwise potentiometric titration of H_3PO_4 with NaOH.

with sodium hydroxide is shown. Point A indicates the neutralization of the first ionizable hydrogen, B of the second. The third cannot be thus obtained, because beyond B the solution is too alkaline for the sodium hydroxide to produce any significant difference in pH. To get around this difficulty calcium chloride is added to the solution at a point such as C. The calcium chloride reacts with the disodium phosphate present to precipitate calcium phosphate and to produce a quantity of hydrochloric acid exactly equal to the acid content of the disodium phosphate. Because of the appearance of this strong acid the pH of the solution falls from C to a point D, thus making possible further titration from D until the third break E, corresponding to complete neutralization, results.

By such procedures it is possible to analyze mixtures of, say, phosphoric acid, sodium dihydrogen phosphate, and disodium phosphate. Likewise other combinations may be titrated in stages. But such stepwise neutralization cannot be obtained with sulfuric acid. In this acid partial neutralization of the first hydrogen leads to appreciable ionization of the second, and the two titrate as if sulfuric acid were a monobasic acid.

Oxidation-reduction titrations can be carried out potentiometrically in a manner identical to acid-base neutralizations by substituting for the electrode reversible to hydrogen ions an inert metal, such as a platinum wire. This metal acts as an oxidation-reduction electrode whose emf is determined by the activity ratio of the substance being oxidized or reduced. The variation of the emf of the cell with volume of reagent added follows essentially a curve similar to Figure 12–7(a). As the sharp change in emf occurs at the equivalence point, these plots may again be utilized to detect the end points. In this way may be titrated various reducing agents with oxidants, or vice versa, without dependence on color change or oxidation-reduction indicators. Further, in certain instances several substances may be estimated in the same solution by a single titration.

The same type of behavior is observed also in many precipitation reactions when an electrode reversible to one of the ions involved is used. Thus, in the titration of silver nitrate with sodium chloride a silver electrode will exhibit a sharp change in emf when all the silver is precipitated. Likewise a silver electrode will indicate the precipitation of all the chloride when the latter is titrated with silver nitrate. The power and speed of the potentiometric method is particularly well illustrated in the estimation of mixed halides. The analysis of solutions composed of chlorides, bromides, and iodides by ordinary methods is long, involved, and tedious. Potentiometrically, however, such mixtures can be analyzed under proper conditions by a single titration with silver nitrate in presence of a silver electrode. The three "breaks" obtained give first the precipitation of the least soluble silver iodide, next the intermediately soluble silver bromide, and finally the precipitation of the most soluble silver chloride. From the volumes of silver nitrate required to produce these "breaks," the amount of each halide present in the mixture can readily be calculated.

ELECTROLYSIS AND POLARIZATION

When an electrochemical cell is employed to convert chemical energy into electrical, the cell yields electrical energy as a result of a free energy decrease accompanying a spontaneous reaction within the cell. Under reversible conditions the electrical work obtained is exactly equal to the free energy decrease. However, if the cell is operated irreversibly, the electrical energy available for doing work is less than the free energy decrease, and the difference becomes dissipated as heat.

Of necessity it must follow that reactions accompanied by a free energy increase cannot be utilized for doing work. Instead, to make such reactions possible energy must be supplied to raise the free energy content of the products above that of the reactants. Electrochemically this addition of energy is accomplished by passing electricity under a suitable applied potential through a cell in which the particular reaction can take place. By means of such *electrolysis* chemical changes are produced at the electrodes, and we obtain chemical action at the expense of the electrical energy.

From thermodynamic considerations it may be anticipated that the *minimum* electrical energy required to carry out a nonspontaneous reaction would be equal to the free energy increase accompanying the change; and this ΔF would in turn be equal, but opposite in sign, to the free energy decrease attending the reverse spontaneous process. This would be true when the electrolysis is carried out reversibly, i.e., when the electrodes are completely reversible and when only a very minute current is passed through the cell. When these conditions of reversibility are not satisfied, the energy required will be the theoretical minimum plus the energy necessary to overcome the irreversibility. This means that in presence of any irreversibility the potential to be applied for electrolysis will have to be greater than the reversible emf of the cell. A cell thus requiring voltage in excess of the theoretical is said to be *polarized*. The excess voltage is called *polarization voltage*, while the phenomenon in general is referred to as *polarization*.

APPLIED POTENTIAL AND ELECTROLYSIS

To understand what happens when an external voltage is applied to a cell, consider the cell

$$\text{Tl} \mid \text{Tl}^+(a = 0.5) \parallel \text{Sn}^{++}(a = 0.01) \mid \text{Sn}$$

for which the spontaneous reaction is

$$\text{Tl(s)} + \frac{1}{2}\text{Sn}^{++}(a = 0.01) = \frac{1}{2}\text{Sn(s)} + \text{Tl}^+(a = 0.5) \tag{95}$$

and the potential 0.155 volt at 25°C. To reverse this reaction an outside potential must be applied to the cell of a magnitude sufficient to overcome the emf of the cell itself. For this purpose an external source of emf must be connected to the cell such that electrons enter the thallium electrode and leave through the tin electrode; i.e., the negative pole of the external source of emf must be connected to the negative electrode of the cell, the positive pole to the positive electrode. Electrons will then flow externally from the thallium to the tin as long as the applied potential is less than 0.155 volt. When the latter becomes equal to 0.155 volt, the two emf's are balanced, and no current flows in either direction. Finally, when the applied potential exceeds that of the cell, electrons from the external battery will enter the thallium electrode and exit through the tin electrode. Since now the direction of current flow is opposite to that of the spontaneous action of the cell, the reactions at the electrodes are also reversed, and we obtain thus the process opposite to that given in Eq. (95).

From this description it is evident that not all of the applied emf is effective in passing current through the cell. Part of it, in this case 0.155 volt at 25°C, has to be utilized in overcoming the spontaneous emf of the cell itself. In general, if we let ε be the applied potential and ε_b be the potential exerted by the cell itself, the net voltage operative in current passage is $\varepsilon - \varepsilon_b$. With this net voltage the current I resulting in a cell whose internal resistance is R follows from Ohm's law as

$$I = \frac{\varepsilon - \varepsilon_b}{R} \tag{96}$$

ε_b is referred to as the *back emf* of the cell in question. Under reversible conditions ε_b is nothing more than the reversible emf of the cell. However, under irreversible conditions, i.e., when polarization is present, ε_b is greater than the reversible emf, and a higher potential is required to yield a given current.

Ordinarily the cells encountered in electrolysis are not of the type discussed above. Rather they involve a *single electrolyte* into which dip a pair of electrodes. For instance, for the electrolysis of water the cell may consist of a pair of platinum electrodes immersed in a solution of an acid or base. Such a cell possesses no initial back emf, since both the electrodes and the electrolyte are the same. It may appear, therefore, that in such a cell all of the applied emf will go toward passage of electricity, with none of it necessary for overcoming back emf. However, this is not the case, for it is found experimentally that a potential of about 1.7 volts must be applied to the cell before there sets in a continuous evolution of hydrogen and oxygen. When a potential of less than 1.7 volts is applied, there is observed a momentary surge of current which rapidly falls to practically zero and stays there. The reason for this behavior is that, owing to the initial surge of current, small amounts of hydrogen are generated at the negative electrode and of oxygen at the positive electrode. These gases adhere to the electrode surfaces and convert these into hydrogen and oxygen electrodes with a back emf which operates to oppose the applied potential. At each potential below 1.7 volts the amounts of the two gases formed at the electrodes are just sufficient to make the back emf equal to the applied potential, and hence the first surge of current is rapidly reduced to practically zero. But, once the applied potential exceeds 1.7 volts, enough hydrogen and oxygen are formed to permit these gases to escape from the electrodes against the confining atmospheric pressure, and *continuous electrolysis* becomes possible. As the minimum voltage at which continuous evolution of hydrogen and oxygen sets in is about 1.7 volts, any voltage above this value is available for current passage. Therefore, by increasing the voltage above 1.7 volts higher and higher currents can be realized.

The significant fact in the above description is that *products of electrolysis convert even inert electrodes into active electrodes which can exercise a back emf.* This is true not only of gases but of other substances as well. Thus, as soon as some zinc is deposited on a piece of platinum, the latter is transformed into an active zinc electrode. Consequently, even where there is no initial back emf present, one is generated on attempted electrolysis by formation of a cell involving the products of the electrolysis.

DECOMPOSITION POTENTIALS

Since even inert electrodes are converted on electrolysis into active electrodes with back emf, it becomes necessary to ascertain the minimum voltage required to produce continuous electrolysis of an electrolyte. This minimum voltage, called the *decomposition potential* of an electrolyte, can be determined by means of the setup illustrated in Figure 12–9. In this diagram A is the electrolytic cell containing the electrolyte to be studied, the two electrodes E and E', and the stirrer S. Connected to these electrodes through a variable resistance C is an external source of emf B and a milliammeter M for measuring the current passing through the

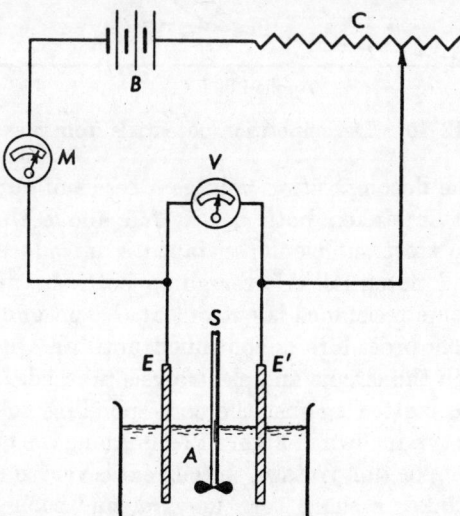

Figure 12–9. Determination of decomposition potentials.

circuit. By means of the variable resistance C it is possible to regulate the voltage applied to the cell, and this voltage in turn can be measured by means of the voltmeter V inserted across the two electrodes. By using a high resistance voltmeter, all of the current indicated by the milliammeter may be considered to pass through the cell. Otherwise the reading of the milliammeter has to be corrected for the current passing through the voltmeter. The latter is given by the reading of the voltmeter divided by the internal resistance of the instrument.

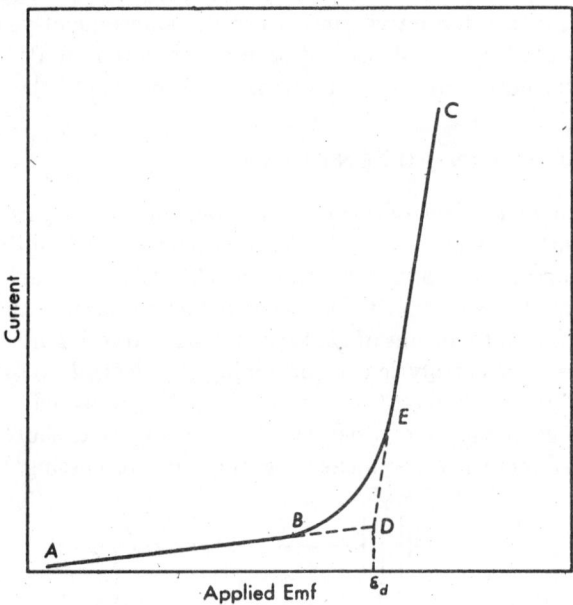

Figure 12-10.   Decomposition potentials from I vs. ε plots.

To measure the decomposition voltage a series of current vs. applied voltage readings are taken both below and above the decomposition potential. At the start sufficient resistance is introduced at C to yield only a very small potential. After reading both the milliammeter and the voltmeter, some resistance is cut out at C, and another set of readings is taken. This procedure is continued until an appreciable current is passing through the circuit and electrolysis proceeds freely. If the current readings are plotted against the corresponding voltages, the curve resulting is of the type shown in Figure 12–10. Along the branch AB, which is below the decomposition voltage, the current is very small. Past B, however, the curve takes a sharp turn upward, and soon continuous electrolysis may be observed in the cell. Since the curve AB lies below the decomposition voltage while EC is above it, the voltage at which decomposition sets in must lie at the intersection of the two curves. To find this point the usual practice is to extend lines AB and CE until they intersect, as at D, and the voltage at this intersection, ε_d, is taken as the decomposition voltage.

The appearance of current below the decomposition potential arises from the fact that the products of electrolysis deposited on the electrodes tend to diffuse slowly back into the solution. To compensate for this loss by diffusion a small current must pass through the cell in order to regenerate the material lost. As the diffusion is continuous, the current must also be supplied continuously, and this accounts for the small but steady current observed below the decomposition potential.

In column 2 of Table 12-6 are given the decomposition potentials obtained with platinum electrodes in 1 N solutions of various electrolytes at room temperature. In each instance the products of electrolysis are given, as well as the reversible decomposition potentials calculated for the various electrolytes. The manner of arriving at the latter and the significance of the observed results will now be explained.

TABLE 12-6. DECOMPOSITION POTENTIALS IN 1 N SOLUTIONS (PT ELECTRODES)

Electrolyte	Dec. Potential ε_d (volts)	Products of Electrolysis	Rev. Decomp. Potential ε_r	$\varepsilon_d - \varepsilon_r$
HNO_3	-1.69	$H_2 + O_2$	-1.23	-0.46
$CH_2ClCOOH$	-1.72	$H_2 + O_2$	-1.23	-0.49
H_2SO_4	-1.67	$H_2 + O_2$	-1.23	-0.44
H_3PO_4	-1.70	$H_2 + O_2$	-1.23	-0.47
$NaOH$	-1.69	$H_2 + O_2$	-1.23	-0.46
KOH	-1.67	$H_2 + O_2$	-1.23	-0.44
NH_4OH	-1.74	$H_2 + O_2$	-1.23	-0.51
HCl	-1.31	$H_2 + Cl_2$	-1.37	$+0.06$
HBr	-0.94	$H_2 + Br_2$	-1.08	$+0.14$
HI	-0.52	$H_2 + I_2$	-0.55	$+0.03$
$CoCl_2$	-1.78	$Co + Cl_2$	-1.69	-0.09
$NiCl_2$	-1.85	$Ni + Cl_2$	-1.64	-0.21
$ZnBr_2$	-1.80	$Zn + Br_2$	-1.87	$+0.07$
$Cd(NO_3)_2$	-1.98	$Cd + O_2$	-1.25	-0.73
$CoSO_4$	-1.92	$Co + O_2$	-1.14	-0.78
$CuSO_4$	-1.49	$Cu + O_2$	-0.51	-0.98
$Pb(NO_3)_2$	-1.52	$Pb + O_2$	-0.96	-0.56
$NiSO_4$	-2.09	$Ni + O_2$	-1.10	-0.99
$AgNO_3$	-0.70	$Ag + O_2$	-0.04	-0.66
$ZnSO_4$	-2.55	$Zn + O_2$	-1.60	-0.95

CALCULATION OF REVERSIBLE DECOMPOSITION POTENTIALS

The reversible decomposition potential of an electrolyte is the sum of the reversible emf's established at the electrodes by the products of electrolysis. Consequently, as soon as the products of electrolysis, the concentration of the electrolyte, and the direction of the electrode reactions, i.e., oxidation or reduction, are known, the reversible decomposition potential can be calculated by the methods already described.

The direction of any electrode reaction in an electrolytic cell is readily ascertained from the polarity of the electrodes. In any electrolytic cell electrons enter through the electrode connected to the *negative* side of the external source of emf. At this electrode, therefore, called the *cathode*, the reaction must always be a *reduction*. On the other hand, electrons leave the cell through the electrode connected to the positive side of the external battery, and hence at this electrode, called the *anode*, the reaction must always be an *oxidation*.

In order to calculate the reversible decomposition potentials in electrolytes where oxygen is evolved, the ε^0 value of the oxygen-hydroxyl ion electrode is required. Since the oxygen electrode is highly irreversible, this quantity cannot be obtained directly from emf measurements, but must be calculated from free energy data from other sources. For this purpose we have the free energy change of the reaction

$$H_2(g) + \frac{1}{2} O_2(g) + H_2O(l) = 2\,H^+ + 2\,OH^- \quad \Delta F^0_{25°C} = -18,600 \text{ cal} \quad (96)$$

As 2 faradays are required to accomplish the reaction, the ε^0 corresponding to the ΔF^0 of this process is

$$\Delta F^0 = -n\mathfrak{F}\varepsilon^0$$
$$-18,600 \times 4.184 = -2 \times 96,500\,\varepsilon^0$$
$$\varepsilon^0 = 0.403 \text{ volt}$$

However, for the reaction $H_2(g) = 2\,H^+$, $\varepsilon^0 = 0$. Subtracting this reaction from Eq. (96), we arrive then at the standard electrode potential for the solution of oxygen to form hydroxyl ions, namely,

$$\frac{1}{2} O_2(g) + H_2O(l) + 2\ominus = 2\,OH^- \qquad \varepsilon^0_{25°C} = 0.403 \text{ volt} \quad (97)$$

For the converse process, i.e., the deposition of oxygen at 1 atm pressure from a solution containing hydroxyl ions at unit activity, the standard potential of the oxygen electrode follows, therefore, as

$$2 \, OH^- = \frac{1}{2} O_2(g) + H_2O(l) + 2 \ominus \qquad \mathcal{E}^0_{25°C} = -0.403 \text{ volt} \qquad (98)$$

With the $\mathcal{E}^0$ of the oxygen electrode thus available, the reversible decomposition potentials for electrolytes involving evolution of oxygen on electrolysis may be calculated. As an initial example let us take the electrolysis of 1 N silver nitrate, which yields silver at the cathode and oxygen at the anode. For the reduction at the silver electrode thus formed at the cathode, we have the reaction $Ag^+(N = 1) + \ominus = Ag(s)$. Hence the corresponding emf is

$$\begin{aligned}
\mathcal{E}_{Ag} &= \mathcal{E}^0_{Ag} - \frac{0.059}{1} \log_{10} \frac{1}{C_{Ag^+} f_{Ag^+}} \\
&= 0.799 + 0.059 \log_{10} (1 \times 0.396) \\
&= 0.78 \text{ volt}
\end{aligned}$$

Again, for the oxidation at the oxygen electrode formed at the anode we have the process

$$2 \, OH^- = \frac{1}{2} O_2(g) + H_2O(l) + 2 \ominus$$

where, since the solution is essentially neutral, the concentration of hydroxyl ions may be taken as 10^{-7} gram ionic weights per liter. The potential of the oxygen electrode is, then,

$$\begin{aligned}
\mathcal{E}_{O_2} &= \mathcal{E}^0_{O_2} - \frac{0.059}{2} \log_{10} \frac{1}{a^2_{OH^-}} \\
&= -0.403 + 0.059 \log_{10} (10^{-7}) \\
&= -0.82 \text{ volt}
\end{aligned}$$

Adding the two, the reversible decomposition voltage of the silver nitrate follows as

$$\begin{aligned}
\mathcal{E} &= \mathcal{E}_{Ag} + \mathcal{E}_{O_2} \\
&= 0.78 - 0.82 \\
&= -0.04 \text{ volt}
\end{aligned}$$

This example shows that the reversible decomposition potentials depend on the concentration of the electrolyte. However, this is not true when water is electrolyzed to yield hydrogen and oxygen. In the latter instance the over-all cell reaction is given by the converse of Eq. (96), for which the emf follows as

$$\begin{aligned}
\mathcal{E} &= \mathcal{E}^0 - \frac{0.059}{2} \log_{10} \frac{1}{a^2_{H^+} a^2_{OH^-}} \\
&= -0.403 + \frac{0.059}{1} \log_{10} (a_{H^+} a_{OH^-})
\end{aligned}$$

But, *in any aqueous solution* the product $a_{H^+}a_{OH^-}$ is a constant equal to 1×10^{-14} at 25°C. Hence

$$\varepsilon = -0.403 + 0.059 \log_{10} (1 \times 10^{-14})$$
$$= -0.403 - 0.826$$
$$= -1.23 \text{ volts}$$

and this should be, therefore, the reversible decomposition potential of any cell in which the essential process is the electrolysis of water. Further, this potential should be independent of the concentration of the electrolyte or its nature. This deduction accounts for the fact that the reversible decomposition potentials in Table 12–6 for all the acids and bases yielding oxygen and hydrogen are the same.

SIGNIFICANCE OF OBSERVED DECOMPOSITION POTENTIALS

With this explanation of the calculation of reversible decomposition potentials, we are in a position to consider the significance of the experimental results given in Table 12–6. First, a comparison of the potentials given in columns 2 and 4 of the table for electrolytes that do not yield oxygen on electrolysis of 1 N solutions reveals that for these the observed decomposition potentials are not far different from those calculated theoretically. From this it may be concluded that the ions in these substances, i.e., hydrogen, the metals, and the halogens, are deposited on platinum electrodes under conditions which approximate more or less to reversibility. On the other hand, in electrolytes in which oxygen is evolved, as in the acids and bases at the top of the table and in some of the salts, the potentials required for decomposition are considerably greater than those calculated on the assumption of reversibility. Moreover, since the metals and hydrogen are deposited essentially reversibly on platinum, this discrepancy indicates that the observed irreversibility must occur primarily in the evolution of oxygen at the platinum surfaces. In other words, oxygen electrodes, even at platinum surfaces, are appreciably polarized. In the case of the acids and bases the extent of this polarization, which for the moment may be ascribed totally to the oxygen, amounts to about 0.45 volt. However, in the deposition of oxygen from salt solutions the polarization is greater and may be as much as a volt.

Another interesting point in Table 12–6 is that acids and bases yielding oxygen and hydrogen have essentially the same decomposition potentials, namely, about −1.7 volts. Further, whereas the decomposition potential of 1 N hydrochloric is −1.31 volts, the potential increases as the acid is diluted, eventually reaching a value of −1.7 volts. From this it may be concluded that, although in the more concentrated solutions the products of electrolysis are hydrogen and chlorine, in dilute solutions they are hydrogen and oxygen. Between the two extremes both chlorine and oxygen are evolved at the anode in varying proportions, giving this electrode a potential intermediate between that of the oxygen and chlorine electrodes.

TYPES OF POLARIZATION

Measurements of decomposition potentials are not suitable for extended polarization studies. First, such data yield information on the sum of the polarization voltages at the two electrodes rather than on that at any one. Again, the results apply only to the conditions obtaining when continuous electrolysis sets in, and they tell us nothing about what happens at higher current densities. Consequently, in order to ascertain the extent of polarization at each electrode separately and at various current densities, more suitable methods are required.

However, before turning to these it is necessary to point out that any observed polarization can be of two types, namely, (a) *concentration polarization* and (b) *overvoltage*. The first arises from the fact that, with the changes in the concentration of the electrolyte about the electrodes during electrolysis, cells may arise which can develop an emf opposite to that applied. On the other hand, overvoltage is a polarization potential whose source lies in some process at the electrode which takes place irreversibly and is thus a phenomenon intimately associated with the nature of the electrode and the processes occurring at its surface. Under ordinary conditions the total polarization observed will be the sum of both effects. But since the overvoltage is of particular electrochemical interest, it becomes important to minimize or eliminate concentration polarization in order to arrive at the overvoltage. Unless the concentration changes at electrodes become very large, the polarizations developed by them are at best rather small and can hardly account for the large polarization potentials observed in many cells. Further, some forces operating in solution, such as electrolytic conduction by the current and thermal diffusion, tend to equalize the concentration differences. As increase in temperature favors electrolytic conduction and diffusion, this equalization can be promoted by raising the temperature. However, a much more effective method of overcoming concentration polarization is efficient stirring of the electrolyte. In electrolytic cells where the electrolyte is thoroughly agitated the concentration polarization can be practically eliminated, and hence any polarization voltage observed under such conditions must be essentially overvoltage due to irreversibility in the electrode processes.

MEASUREMENT OF OVERVOLTAGE

Direct measurements of overvoltage attending the deposition of electrolytic products at various electrodes are made by observing the potentials of the electrodes when current is passing, and under conditions where there is no concentration polarization, i.e., in stirred solutions. For this purpose the setup generally employed is shown in Figure 12–11. The

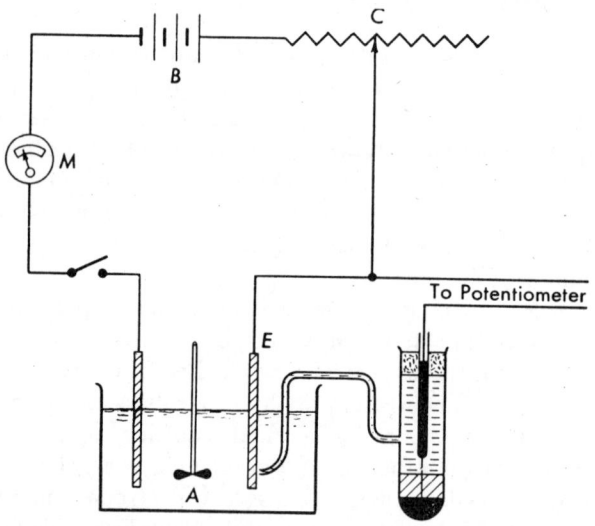

Figure 12–11. Determination of overvoltage.

electrolytic cell A, containing electrodes of known cross-sectional area, the electrolyte, and a stirrer, is connected in series with an external battery B through a variable resistance C, an ammeter M, and a switch. Any desired current strength can thus be sent through the electrolyzing circuit by varying the resistance at C. To obtain now the potential of, say, electrode E, the latter is made one-half of a new cell, the other half of which is a suitable reference electrode, such as a calomel, and the two are connected to a potentiometer. In order to avoid inclusion of any appreciable IR drop through the solution in measuring the potential of E against the reference electrode, the arm of the latter is usually drawn out to a fine tip and placed flush against the electrode.

To make a measurement the stirrer is started, the switch closed, and the current through the electrolyzing circuit, as indicated by M, is adjusted to any desired value by means of C. After the current has become steady, the emf of E against the reference electrode is read on the potentiometer, as well as the current given by M. The latter is then readjusted to some other value, and the whole operation is repeated. In this manner a series of emf readings can be obtained for various values of the current passing through the cell. On dividing the current strength at each point by the area of electrode E, the current density follows, while the emf of E at each stage is obtained by subtracting from the measured emf the potential of the reference electrode. Finally, in order to arrive at the overvoltage the reversible potential of E must be subtracted from the observed electrode potentials. These reversible emf's may be calculated, or they can be measured by taking the potential of E against a reference electrode when no current is flowing through the cell. In the latter instance an electrode material must be used at which the particular electrode reaction is reversible.

OVERVOLTAGE IN CATHODIC DEPOSITION OF METALS

Overvoltage studies on the cathodic deposition of metals indicate that the deposition of metals takes place with relatively small overvoltage. In line with theoretical requirements, metals do not begin to plate on the cathode until the potential of the given metal in the particular solution is exceeded. Thereafter the potential of the metal rises slowly with current density, indicating some polarization, but the overvoltage hardly ever exceeds several tenths of a volt at the higher current densities. Metals such as zinc, copper, and cadmium exhibit particularly low overvoltage, while metals of the iron group, namely, iron, cobalt, and nickel, show higher overvoltages. Just what is the cause of metal overvoltage is not yet clear. Various theories have been proposed, but at present too little information is available to make a choice among them.

The overvoltage of a particular metal is not the same for all salt solutions of the metal, but varies slightly with the nature of the co-ion. It is affected also by the presence of acids, colloids, and other addends. Finally, the overvoltage of metals is increased when these are deposited from solutions in which the particular metal can exist as a complex, as silver in cyanide solutions.

HYDROGEN OVERVOLTAGE

Of particular importance is the overvoltage exhibited by hydrogen in its deposition on various cathodic surfaces. *On platinized platinum and at zero current density the overvoltage of hydrogen is zero*, since these are the conditions under which the hydrogen electrode is reversible. However, as the current density is increased the overvoltage also goes up, but at no time does it become particularly large. In fact, at a current density of

1.5 amp per square centimeter it is only 0.05 volt at 25°C. On the other hand, at other cathodic surfaces the overvoltage is much higher, as may be seen from Figure 12–12. These data, obtained in 2 N sulfuric acid at 25°C, indicate that the overvoltage of hydrogen is low only on platinized platinum. On smooth platinum surfaces the overvoltage is low at very small current densities, but increases gradually with the latter to approach a limiting value at high current densities. This is true also of gold, graphite, tellurium, and palladium cathodes. However, on silver, mercury, bismuth, cadmium, tin, iron, zinc, nickel, and lead surfaces the hydrogen overvoltage is already quite large at zero current density. These electrodes, therefore, cannot function reversibly under any conditions, and it is not

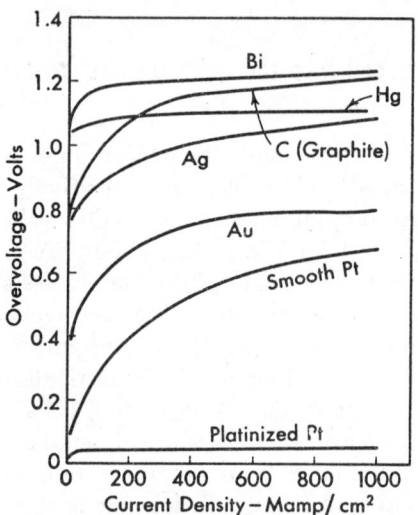

Figure 12–12.   Hydrogen overvoltage on various metals (2 N H_2SO_4 at 25°C).

possible to deposit hydrogen on them at anywhere near the reversible potentials. Another characteristic of these metals with high hydrogen overvoltage is that they exhibit at low current densities a very rapid increase in overvoltage to values which remain essentially constant at the higher densities.

The variation of hydrogen overvoltage with current density at constant temperature can frequently be represented by the equation

$$\varepsilon_0 = a + b \log_{10} I \tag{99}$$

where ε_0 is the overvoltage, I the current density, and a and b are constants. According to this equation a plot of ε_0 against $\log_{10} I$ should be linear. This expectation has been realized in a number of instances, and over a fairly wide range of current densities. Furthermore, many metals give approximately a slope, b, which corresponds very closely to $(2 \times 2.303 \, RT)/\mathfrak{F}$. Study of the effect of temperature on overvoltage has shown that the slope of the ε_0-$\log_{10} I$ curves actually increases linearly with the absolute temperature in accord with the relation

$$\varepsilon_0 = a + \frac{(2 \times 2.303 \, RT)}{\mathfrak{F}} \log_{10} I \qquad (100)$$

Increase of temperature has been found to decrease hydrogen overvoltage, although, as we have seen, the slope of the overvoltage–current density curve is steeper at higher temperatures. Again, measurements of hydrogen overvoltage at various pressures, both above and below atmospheric, indicate that at higher pressures the overvoltage is changed only slightly, but at the lower pressures it rises sharply on copper, nickel, and mercury cathodes. Overvoltage is also influenced by the presence of various impurities in the cathode materials.

Available data show that the composition and pH of the solution exert little effect on the hydrogen overvoltage. On the other hand, the nature of the cathode surface is very important. On smooth, shiny, and polished surfaces the overvoltage is invariably greater than on rough, pitted, or etched surfaces. This difference is particularly well illustrated with the hydrogen overvoltage on smooth and platinized platinum. Analogous differences, though not so large, have been observed also on other metals.

OVERVOLTAGE IN ANODIC PROCESSES

The limited data on anodic deposition indicate that evolution of the halides at electrodes such as platinum and graphite takes place with relatively small overvoltages. Thus, at 1000 milliamperes per square centimeter and 25°C the overvoltage of chlorine is about 0.07 volt on platinized platinum, 0.24 volt on smooth platinum, and 0.5 volt on graphite. At very small current densities the overvoltages are practically negligible, and, therefore, the deposition of the halogens takes place essentially reversibly.

However, this is not true of oxygen evolution. Although precise oxygen overvoltages are difficult to obtain because of attack of the electrodes by the oxygen, the data available indicate that they are invariably high even at low current densities. This may be seen from Table 12–7, where the oxygen overvoltages on various metals at several current densities are given. Another important fact revealed by this table is that oxygen overvoltages attain higher values on metals such as platinum and gold than on metals such as copper and nickel, which is just the opposite to the

TABLE 12-7.   OXYGEN OVERVOLTAGES AT 25°C

Current Density (ma/cm²)	Overvoltage (volts)						
	Plat'd. Pt	Smooth Pt	Au	Graphite	Cu	Ag	Ni
1	0.40	0.72	0.67	0.53	0.42	0.58	0.35
10	0.52	0.85	0.96	0.90	0.58	0.73	0.52
50	0.61	1.16	1.06	—	0.64	0.91	0.67
100	0.64	1.28	1.24	1.09	0.66	0.98	0.73
500	0.71	1.43	1.53	1.19	0.74	1.08	0.82
1000	0.77	1.49	1.63	1.24	0.79	1.13	0.85
1500	0.79	1.38	1.68	1.28	0.84	1.14	0.87

hydrogen overvoltages on these substances. Otherwise oxygen overvoltage behaves essentially analogously to that of hydrogen. In acid solutions the overvoltage appears to be independent of pH, increase of temperature leads to a decrease in overvoltage, and the dependence of overvoltage on current density can again be represented by Eq. (100) with exactly the same expression for the slope of the curve. In alkaline solutions, however, the overvoltage appears to vary with the hydroxyl ion concentration.

THEORIES OF OVERVOLTAGE

When the various stages involved in any over-all electrode reaction proceed rapidly, equilibrium can readily be established between the ions in solution and the electrode, and the latter can thus behave reversibly and exhibit its equilibrium potential. On the other hand, if any of the intermediate processes for one reason or another are slow, or if any complicating reactions take place between the electrode and some of the substances appearing in the electrode reaction, equilibrium cannot be established, and hence the electrode cannot behave reversibly. Utilizing this idea, many theories have been proposed to account for overvoltage, especially that of hydrogen. However, no theory has as yet been proposed which can account satisfactorily for all the observed phenomena. A review of these theories has been given by Bockris.[5]

METAL DEPOSITION AND HYDROGEN OVERVOLTAGE

Deposition of most metals from aqueous solutions takes place with very little overvoltage and hence at essentially reversible potentials.

[5] J. Bockris, *Chem. Rev.*, **43**, 525 (1948).

Were the metal ions the only cations present, these potentials would alone determine the deposition. However, in any aqueous solution hydrogen ions are also present, and consequently two cathodic reactions are possible, the deposition of the metal and the evolution of hydrogen. The question arises, therefore: Under what conditions can a metal be deposited from aqueous solution before evolution of hydrogen sets in?

The answer to this question is simply this: If the potential required for metal deposition is less than that for the hydrogen, the product of electrolysis at the cathode will be the metal. If the reverse is true, cathodic liberation of hydrogen will occur. However, whereas the cathodic deposition of metals involves essentially only the reversible metal electrode potentials, the liberation of hydrogen involves not only the reversible potential of the hydrogen electrode in the particular solution, but also the overvoltage of hydrogen at the material composing the cathode. In fact, it is largely because of the latter factor that many metals can be deposited from aqueous solutions under conditions which in absence of overvoltage would invariably yield hydrogen.

To illustrate the above points, consider first the electrolysis of an aqueous solution of silver nitrate in which the activity of the salt is unity. In this solution the deposition potential of the silver will be the standard electrode potential of the metal, or $\varepsilon_{Ag}^0 = +0.80$ volt at 25°C. Again, in this same solution the activity of hydrogen ions is 10^{-7}, and hence the potential for the reversible evolution of hydrogen would be

$$\varepsilon_{H_2} = -\frac{0.059}{1} \log_{10} \frac{1}{10^{-7}}$$
$$= -0.41 \text{ volt}$$

Since the deposition of silver requires a lesser expenditure of energy than the evolution of hydrogen, as the former process is spontaneous whereas the latter is not, silver will plate out in preference to hydrogen. In this case, therefore, silver will be the electrolytic product at the cathode whether hydrogen overvoltage is present or not. The same will be true also of copper deposition. Further, although the difference between the silver or copper potentials and that of hydrogen can be reduced by making the solutions acid, still the potential of the latter cannot be reduced below those of the metals mentioned, and these will plate out whether the solution is acid or neutral.

On the other hand, the deposition of most metals above hydrogen in the electromotive series would be impossible were it not for overvoltage. For a solution of cadmium ions at unit activity the potential of the cadmium electrode will be 0.40 volt, and hence the potential required for deposition will be -0.40 volt. This potential is so close to that of the hydrogen electrode in neutral solution, -0.41 volt, that both cadmium and hydrogen may be expected to deposit. Again, if the solution were acid, the potential of the hydrogen electrode would be less negative, and in such solutions it would appear to be impossible to plate cadmium at all. Still, this is not the case, for evolution of hydrogen does not take place reversibly on a cadmium cathode; in fact, at a current density as low as 1 milliampere per square centimeter the overvoltage of hydrogen on cadmium is already about a volt. Consequently, in order to plate hydrogen from neutral solutions of a cadmium salt a potential of about -1.4 volts will be necessary, while from strongly acid solutions, about -1.0 volt. As both of these potentials are considerably higher than that required for cadmium plating, -0.40 volt, the latter process takes place rather than hydrogen evolution.

In this manner may be explained the electrolytic separation of such metals as zinc, iron, nickel, tin, and lead from their aqueous solutions. Were it not for the presence of hydrogen overvoltage, the deposition of some of these metals would be complicated by appearance of hydrogen, while in other cases, as with zinc, hydrogen would be the exclusive product. Particularly illustrative in this connection is the fact that even sodium may be deposited on a mercury cathode in preference to hydrogen. With an overvoltage of about a volt on mercury, a potential near -1.4 volts is required to liberate hydrogen on this metal. On the other hand, deposition of sodium requires a potential of over 2 volts. However, when sodium is deposited on mercury, the former dissolves in the latter to form an amalgam in which the activity of the sodium is moderated. When the amalgam is dilute, the sodium potential is only about 1.2 volts, and hence the deposition of sodium on the amalgam can take place at -1.2 volts. As this potential is less than the -1.4 volts for hydrogen evolution on mercury, sodium can be recovered from aqueous solutions as an amalgam in preference to hydrogen. This fact is utilized commercially in the Castner cell for the electrolysis of sodium chloride, where the sodium is first deposited cathodically as an amalgam and is allowed then to react with water to yield hydrogen and sodium hydroxide.

ELECTROLYTIC SEPARATION OF METALS

Not only may metals be separated electrolytically from hydrogen ions in solution when sufficient differences in potential exist, but also various metals may be separated from each other. To accomplish clean-cut separation of one metal from another it is desirable that a difference of about 0.2 volt exist between the potential of one metal at near complete removal and the potential of the next metal at the inception of its deposition. When such conditions obtain, various metals can be removed from solution and separated from each other quantitatively with sufficient accuracy for analytical purposes. In this manner silver may be separated from copper, copper from cadmium, and cadmium from zinc. In a solution containing all of the metals mentioned the procedure is to adjust the electrolytic potential first at a point higher than that required to deposit silver, but not high enough to yield copper. At such a potential appreciable current will flow as long as any silver is present in solution, and deposition takes place. However, as soon as the silver is removed the potential is not sufficiently high to permit the next metal in the series to plate, and hence the current falls to a low value, indicating the end of the silver separation. At this stage the cathode is removed, a new one is inserted, the potential is adjusted to a value intermediate between the copper and cadmium, and electrolysis is again initiated. By repeating this sequence of potential adjustment and electrolysis, the copper can be removed next, then the cadmium, and finally the zinc.

Occasionally, in order to avoid gassing at the cathode due to hydrogen evolution, the acidity of the solution may require adjustment before separation of metals higher in the series. Again, frequently a greater difference in separation potentials can be obtained by converting some of the ions to complexes, or by changing the temperature. Various such artifices and other experimental details may be found described in books on quantitative electroanalysis.

SIMULTANEOUS DEPOSITION OF METALS

In order to deposit two metals simultaneously at a cathode it is necessary that the two have the same deposition potentials. This means that the concentrations of the two ions in solution must be so adjusted that the potentials of the two metals, which are ordinarily different, are brought together. Such adjustment is usually not possible with solutions of simple salts of the metals, but may be accomplished by converting the metal ions to complexes. The outstanding example of such behavior is the direct formation of brasses by simultaneous deposition of copper and zinc. In aqueous solutions of their simple salts these metals have deposition potentials that differ by over a volt. However, when the cations of these metals are converted to their cyanide complexes, the difference in potential between them is considerably reduced until at a given concentration of cyanide they become essentially the same. From such solutions, therefore, both metals can be plated simultaneously as a brass. Again, by varying the proportions of zinc, copper, and cyanide in the electrolyte, and by controlling the current density, deposits can be obtained varying in metal content from pure copper to pure zinc; i.e., brasses of any desired composition can be obtained. Further, various addition agents and the temperature alter the composition of the deposit.

Increase in temperature invariably leads to an increase in copper content, so that at sufficiently high temperatures the product is exclusively this metal.

COMMERCIAL CELLS

We shall conclude this chapter with a description of electrochemical cells commonly used as sources of electrical energy. These cells are of two types. In the *primary cells* electrical energy can be obtained at the expense of chemical reactivity only as long as the active materials are still present. Once these have been consumed, the cell cannot be profitably or readily rejuvenated and must be discarded. On the other hand, in *secondary* or *storage cells* the cell once used can be recharged by passage of current through it, and it may hence be used over and over again. A cell of the first type is the Leclanché or dry cell, while the acid lead cell is of the second type. These cells will now be described in turn.

The Leclanché or Dry Cell. The Leclanché primary cell consists of a negative zinc electrode, a positive carbon electrode surrounded by manganese dioxide, and a paste of ammonium chloride and zinc chloride as an electrolyte. The reaction at the negative electrode is solution of zinc to form zinc ions, while that at the positive electrode appears to be

$$2\,MnO_2(s) + H_2O(l) + 2\,\ominus = Mn_2O_3(s) + 2\,OH^-$$

From these the over-all process for the cell follows as

$$Zn(s) + 2\,MnO_2(s) + H_2O(l) = Zn^{++} + 2\,OH^- + Mn_2O_3(s) \quad \textbf{(101)}$$

The hydroxyl ions generated by the action of the cell liberate ammonia from the ammonium chloride, which in turn combines with the zinc ions to precipitate the difficulty soluble salt $Zn(NH_3)_2Cl_2$; i.e.,

$$2\,NH_4Cl + 2\,OH^- = 2\,NH_3 + 2\,Cl^- + 2\,H_2O \quad \textbf{(102a)}$$
$$Zn^{++} + 2\,NH_3 + 2\,Cl^- = Zn(NH_3)_2Cl_2 \quad \textbf{(102b)}$$

However, the latter are secondary processes not involved directly in the electrode reactions, and so they do not contribute to the potential of the cell.

The Leclanché cell develops a potential of about 1.5 volts on open circuit. With usage this voltage tends to fall until eventually the cell has to be discarded. At best these cells have a low capacity, and they are not intended for heavy duty.

The Lead Storage Cell. In this cell the negative electrode is lead, the positive electrode lead impregnated with lead dioxide, and the electrolyte is a solution of approximately 20 per cent sulfuric acid with a specific gravity of about 1.15 at room temperature. On drawing current from this cell lead dissolves at the negative electrode to form lead ions, which combine then with sulfate ions in solution to precipitate lead sulfate. The over-all electrode process is, therefore,

$$Pb(s) + SO_4^{--} = PbSO_4(s) + 2 \ominus$$

which shows that actually this electrode is a lead-lead sulfate electrode reversible to sulfate ions. The action of the positive electrode, in turn, can be accounted for by the process

$$PbO_2(s) + 4 H^+ + SO_4^{--} + 2 \ominus = PbSO_4(s) + 2 H_2O(l)$$

from which we see that this is essentially an oxidation-reduction electrode dependent on both hydrogen and sulfate ions for its potential. From these equations the over-all reaction in the lead storage cell on discharge follows as

$$Pb(s) + PbO_2(s) + 2 H_2SO_4 = 2 PbSO_4(s) + 2 H_2O(l) \qquad \textbf{(103)}$$

According to Eq. (103) the potential of the lead storage cell should depend at any given temperature only on the activity of sulfuric acid in solution. This is actually the case. Thus, at 25°C and at a concentration of 7.4 per cent sulfuric acid the potential is 1.90 volts, at 21.4 per cent 2.00 volts, and at 39.2 per cent 2.14 volts. Further, since sulfuric acid is consumed in the operation of the cell, the specific gravity of the electrolyte should fall on discharge, as it does. Finally, the heats of reaction (103) for different concentrations of sulfuric acid obtained electrochemically are in excellent accord with thermally obtained results, confirming thus, if not the mechanism of the electrode processes, at least the validity of the over-all cell reaction.

To recharge a lead storage cell, reaction (103) taking place on discharge must be reversed. For this purpose a potential higher than that of the cell must be applied externally. During this electrolysis lead is deposited on the cathode, lead dioxide is formed at the anode, and sulfuric acid is regenerated in the cell. Owing to this increase in acid concentration the potential of the cell rises, and so does the specific gravity. As long as lead ions are present in solution, no gassing will occur on charging. However, if the electrolysis is permitted to proceed until all the lead is removed, evolution of hydrogen at the lead surface and oxygen at the lead dioxide surface will take place, and there is danger of injury to the electrode plates. For this reason lead cells should be charged only until the specific gravity is up to the desired point and no further.

The voltage of a lead cell does not depend on the size of the cell or of the electrodes. Small or large, a lead cell always has the same potential for a given acid concentration. The only advantage of large electrodes is that they give a cell a higher capacity to deliver electrical energy. Finally, the lead cell is quite efficient in its operation, yielding on discharge 90 to 95 per cent of the electrical energy put into it on charge.

REFERENCES

See list at end of Chapter 11. Also:

1. R. G. Bates, *Electrometric pH Determinations*, John Wiley & Sons, Inc., New York, 1954.
2. Kortum and Bockris, *Textbook of Electrochemistry*, Elsevier Publishing Company, New York, 1951.
3. W. M. Latimer, *The Oxidation States of the Elements and Their Potentials in Aqueous Solutions*, Prentice-Hall, Inc., Englewood Cliffs, N.J., 1952.
4. A. Weissberger, *Physical Methods of Organic Chemistry*, Interscience Publishers, Inc., New York, 1960, Vol. IV, Chapters 44, 48, and 49.

PROBLEMS

Note: In the following problems the temperature is to be taken as 25°C unless otherwise specified.

1. A silver-silver ion electrode is measured against a normal calomel electrode at 25°C. The calomel electrode is negative, and the observed emf is 0.2360 volt. Write the cell reaction and calculate the potential of the silver-silver ion electrode.

$$Ans. \ \varepsilon = 0.5160 \ volt.$$

2. A lead-lead ion electrode is measured against a 0.1 N calomel electrode at 25°C. The lead electrode is found to be negative, and the observed emf is 0.4603 volt. Write the cell reaction and calculate the potential of the lead-lead ion electrode.

3. Suppose that the lead electrode in problem 2 is coupled with the silver electrode in problem 1. (a) What will be the cell reaction? (b) Which electrode will be negative? (c) What will be the emf of the cell?

4. Write the individual electrode reactions and the total cell reaction for each of the following cells. In each cell the negative electrode is at the left.

(a) $Pb \mid PbSO_4(s), SO_4^{--} \parallel Cu^{++} \mid Cu$
(b) $Cd \mid Cd^{++} \parallel H^+ \mid H_2(g)$
(c) $Zn \mid Zn^{++} \parallel Fe^{+++}, Fe^{++} \mid Pt$

5. Assuming that the substances present in each of the cells of problem 4 are in their standard states, calculate the emf of each cell and the free energy change accompanying the cell reaction.

6. In each of the following cells write the cell reaction, and designate which reactions are spontaneous as written.

Sign of emf

(a) $Ag \mid AgCl(s), Cl^- \parallel I^-, AgI(s) \mid Ag$ --

(b) $K \mid K^+ \parallel Cl^-, Hg_2Cl_2(s) \mid Hg$ +

(c) $Pt \mid Tl^+, Tl^{+++} \parallel Cu^{++} \mid Cu$ −

(d) $Ni \mid Ni^{++} \parallel Pb^{++} \mid Pb$ +

7. For the reaction

$$Zn(s) + 2\,AgCl(s) = ZnCl_2(0.555\ m) + 2\,Ag(s)$$

$\mathcal{E}_{0°C} = 1.015$ volt while $(\partial\mathcal{E}/\partial T)_P = -4.02 \times 10^{-4}$ volt per degree. Find ΔF, ΔH, and ΔS of the reaction.   *Ans.* (c) $\Delta S = -18.55$ eu.

8. From Eq. (8) of this chapter for the emf of the saturated Weston cell as a function of temperature, calculate ΔF, ΔH, and ΔS for the cell reaction at 25°C.

9. At 25°C the free energy of formation of $H_2O(l)$ is $-56,700$ cal/mole, while that of its ionization to hydrogen and hydroxyl ions is 19,050 cal/mole. What will be the reversible emf at 25°C of the cell

$$H_2(g,\ 1\ atm) \mid H^+ \parallel OH^- \mid O_2(g,\ 1\ atm)$$

Ans. (a) 0.403 volt.

10. The heat of formation at 25°C of $H_2O(l)$ is $-68,320$ cal/mole, while its heat of ionization is 13,600 cal/mole. Calculate the temperature coefficient of emf at 25°C for the cell given in problem 9.

11. Calculate the oxidation potentials of each of the following single electrodes:

(a) $Ag \mid AgCl(s), Cl^-(a = 0.0001)$

(b) $Pt \mid I_2(s), I^-(a = 1.5)$

(c) $Sn \mid Sn^{++}(a = 0.01)$

(d) $H_2(g,\ 1\ atm) \mid H^+(a = 15)$

12. Calculate the potentials, and specify the electrode polarity for the spontaneous reaction, for the cells obtained by combining (a)–(d), (b)–(d), and (a)–(c) in problem 11.   *Ans.* 0.3897 volt for first cell.

13. Calculate the potential of the cell

$$Fe \mid Fe^{++}(a = 0.6) \parallel Cd^{++}(a \quad 0.001) \mid Cd$$

What must be the polarity of the Cd electrode in order for the cell to serve as a source of energy?

14. Devise a cell in which the reaction

$$\frac{1}{2} Br_2(l) + Fe^{++} = Br^- + Fe^{+++}$$

will take place. What will be the standard emf of the cell and the standard free energy change in calories accompanying the reaction? Is the reaction as written spontaneous?

15. For the cell

$$Ag \mid AgBr(s), Br^-(a = 0.10) \parallel Cl^-(a = 0.01), AgCl(s) \mid Ag$$

write the cell reaction, and calculate the potential at 25°C. Is the reaction as written spontaneous?

16. Calculate the potential at 25°C for the cell

$$H_2(P = 1 \text{ atm}) \mid HBr(a_\pm = 0.2), Hg_2Br_2(s) \mid Hg$$

Ans. 0.2212 volt.

17. Write the cell reaction and calculate the potential of the cell

$$H_2(g, 0.4 \text{ atm}) \mid HCl(a_\pm = 3.0), AgCl(s) \mid Ag$$

18. Write the cell reaction and calculate the potential of the following cell:

$$Zn \mid Zn^{++}(a = 0.01) \parallel Fe^{++}(a = 0.001), Fe^{+++}(a = 0.1) \mid Pt$$

19. The potential of the cell

$$Cd \mid CdI_2(a_2), AgI(s) \mid Ag$$

is 0.2860 volt at 25°C. Calculate the mean ionic activity of the ions in the solution, and the activity of the electrolyte.   *Ans.* $a_\pm = 0.403$; $a_{CdI_2} = 0.0656$.

20. The potential of the cell

$$Zn(s) \mid ZnCl_2(m = 0.01021), AgCl(s) \mid Ag$$

was found to be 1.1566 volts. What is the mean ionic activity coefficient of $ZnCl_2$ in this solution?   *Ans.* 0.705.

21. Write the cell reaction and calculate the potential of the cell

$$H_2(g, 1 \text{ atm}) \mid H_2SO_4(m = 0.05, \gamma = 0.340), Hg_2SO_4(s) \mid Hg$$

22. Indicate the cell in which the equilibrium constant for the following reaction can be measured:

$$\frac{1}{2} H_2(g, 1 \text{ atm}) + AgI(s) = H^+ + I^- + Ag(s)$$

What is the equilibrium constant for this reaction at 25°C?

Ans. $K = 2.67 \times 10^{-3}$.

23. From the results of the preceding problem, calculate the activity of HI in equilibrium with AgI(s), Ag(s), and $H_2(g, 1 \text{ atm})$ at 25°C.

24. Finely divided Ni is added to a solution in which the molality of Sn^{++} is 0.1. What will be the activities of Ni^{++} and Sn^{++} when equilibrium is established? Assume that the molality is equal to activity for this calculation.

Ans. $a_{Ni^{++}} = 0.10$; $a_{Sn^{++}} = 5.7 \times 10^{-5}$.

25. An excess of solid AgCl is added to a 0.1 normal solution of Br^- ions. Assuming activity equal to ionic concentration, calculate the concentration of Cl^- and Br^- ions at equilibrium.

26. Explain what cell may be employed to measure the mean ionic activity coefficients of CdCl$_2$. Derive an expression for the emf of such a cell.

27. It is desired to determine the $\mathcal{E}^0$ value of the Cu | Cu^{++} electrode and the mean ionic coefficients of CuSO$_4$. Explain what cells may be assembled, what measurements made, and derive an expression for the emf of the cells chosen. Assume that the $\mathcal{E}^0$ value of the second electrode used is known.

28. For the cells

$$\text{Na(s)} \left| \begin{array}{c} \text{NaI in} \\ \text{C}_2\text{H}_5\text{NH}_2 \end{array} \right| \text{Na(Hg)(0.206\%)}$$

$$\text{Na(Hg)(0.206\%)} \mid \text{NaCl}(m = 1.022, \gamma = 0.650), \text{Hg}_2\text{Cl}_2\text{(s)} \mid \text{Hg}$$

the potentials are 0.8453 and 2.1582 volts respectively. Write the reaction for each cell, and find from the data the $\mathcal{E}^0$ value for the sodium electrode.

29. Keston [*J. Am. Chem. Soc.*, **57**, 1671 (1935)] found that for cells of the type H$_2$(g, 1 atm) | HBr(m), AgBr(s) | Ag(s) the emf's at various molalities of HBr are:

m	0.0003198	0.0004042	0.0008444	0.001355	0.001850	0.002396	0.003719
$\mathcal{E}$	0.48469	0.47381	0.43636	0.41243	0.39667	0.38383	0.36173

Obtain $\mathcal{E}^0$ for the cell by a graphical method, and then calculate the activity coefficients at each molality.

30. The potential of the cell

$$\text{Hg} \mid \text{Hg}_2\text{Cl}_2\text{(s)}, \text{KCl (sat'd.)} \mid \text{Cl}_2(P = 0.283 \text{ atm})$$

was observed to be 1.0758 volts. Using the $\mathcal{E}^0$ value for the calomel electrode given in the text, calculate $\mathcal{E}^0$ for the chlorine electrode.

31. For the cell
$$\text{Zn(Hg)} \mid \text{ZnSO}_4, \text{PbSO}_4\text{(s)} \mid \text{Pb(Hg)}$$

having an $\mathcal{E}^0$ value of 0.4109 volt, the following data were obtained by Cowperthwaite and LaMer [*J. Am. Chem. Soc.*, **53**, 4333 (1931)]:

Molality of ZnSO$_4$	Emf
0.0005	0.61144
0.002	0.58319
0.01	0.55353
0.05	0.52867

Calculate the mean ionic activity coefficient of ZnSO$_4$ in each of these solutions.

32. Write the cell reaction and calculate the potential of the cell

$$\text{Cl}_2(P = 0.9 \text{ atm}) \mid \text{NaCl (sol'n.)} \mid \text{Cl}_2(P = 0.1 \text{ atm})$$

Will the cell reaction be spontaneous as written?   *Ans.* $\mathcal{E} = -0.0282$ volt.

33. Given the cells:
(a) H$_2$($P = 1$ atm) | HBr($a_{\pm} = 0.001$), AgBr(s) | Ag
(b) same as (a) with $a_{\pm} = 2.5$

(c) same as (a) with $a_{\pm} = 0.01$

show how cell (a) may be combined with either (b) or (c) to give a positive emf, and calculate the emf in each case.

34. What will be the cell reaction and the potential of the cell

$$Zn(s) \mid ZnCl_2(m = 0.02, \gamma = 0.642), AgCl(s) \mid Ag\text{-}Ag \mid AgCl(s),$$
$$ZnCl_2(m = 1.50, \gamma = 0.290) \mid Zn(s)$$

Ans. 0.1359 volt.

35. Find the cell reaction and calculate the potential of the cell

$$Cd(s) \mid CdSO_4(m = 0.01, \gamma = 0.383), PbSO_4(s) \mid Pb\text{-}Pb \mid PbSO_4(s),$$
$$CdSO_4(m = 1.00, \gamma = 0.042) \mid Cd(s)$$

36. Find the cell reaction and calculate the potential of the following cell with transference:

$$Pb(s) \mid PbSO_4(s), CuSO_4(m = 0.2, \gamma = 0.110) \mid CuSO_4(m = 0.02,$$
$$\gamma = 0.320), PbSO_4(s) \mid Pb$$

The transport number of Cu^{++} is 0.370. Is the cell reaction spontaneous as written?　　　　　　　　　　　　　　　　　　　　*Ans.* 0.0117 volt; yes.

37. For the cell with transference

$$H_2(P = 1 \text{ atm}) \mid H_2SO_4(m = 0.005, \gamma = 0.643) \mid H_2SO_4(m = 2.00,$$
$$\gamma = 0.125) \mid H_2(P = 1 \text{ atm})$$

$\mathcal{E} = 0.0300$ volt at 25°C. Find the cell reaction and the transport number of the hydrogen ions.

38. The emf of the following cell

$$Ag(s) \mid AgCl \text{ (sat'd.)}, KCl(m = 0.05, \gamma = 0.817) \mid KNO_3 \mid AgNO_3(m = 0.1,$$
$$\gamma = 0.723) \mid Ag(s)$$

is 0.4312 volt. Calculate from these data the solubility product of AgCl.

Ans. 1.52×10^{-10}

39. Show how the proper $\mathcal{E}^0$ values from Table 12–3 may be used to calculate the thermodynamic solubility product of $PbSO_4$.

40. Using $\mathcal{E}^0$ values listed in Table 12–3, calculate $\mathcal{E}^0$ and ΔF^0 for the electrode process

$$Fe = Fe^{+++} + 3 \ominus$$

Ans. $\mathcal{E}^0 = 0.037$ volt; $\Delta F^0 = -2560$ cal.

41. The emf of the cell

$$H_2(g) \mid Buffer \parallel Normal \text{ calomel electrode}$$

is 0.6885 volt at 40°C when the barometric pressure is 725 mm. What is the pH of the solution?　　　　　　　　　　　　　　　　　　　　*Ans.* 6.60.

42. The buffer in the cell of the preceding problem is replaced with a standard buffer of pH = 4.025, while all other conditions remain the same. What should then be the cell potential?

43. The cell

$$Ag(s) \mid AgCl(s), \ HCl(0.1 \ N) \mid Glass \mid Buffer \parallel Sat'd. \ calomel \ electrode$$

gave an emf of 0.1120 volt when the pH of the buffer used was 4.00. When a buffer of unknown pH was used, the potential was 0.3865 volt. What is the pH of the unknown buffer?

44. What would be the potential of the cell used in the preceding problem if a buffer of pH = 2.50 were used?   *Ans.* 0.0232 volt.

45. One hundred cc of an HCl solution, on potentiometric titration with 0.1000 N NaOH, showed the following pH values on addition of various volumes of base:

cc NaOH	pH	cc NaOH	pH
9.00	3.00	10.00	7.00
9.50	3.27	10.10	9.70
9.75	3.66	10.25	10.40
9.90	4.15	10.50	10.76

Determine (a) the equivalence point of the titration, (b) the pH of the solution at this point, and (c) the original normality of the HCl solution.

46. One hundred cc of 0.0100 N CH$_3$COOH are titrated with 0.1000 N NaOH. (a) What will be the volume of base required to reach the equivalence point? (b) What will be the pH of the solution at this point?

47. One hundred cc of 0.0100 N KCl are titrated with 0.1000 N AgNO$_3$. Calculate the potential of a silver electrode in the solution at the equivalence point on the supposition that ionic activity is equal to ionic concentration. The solubility product of AgCl is 1.56×10^{-10}.

48. Which of the following cells, as *written*, are electrolytic and which are primary cells:
  (a) $Cd(s) \mid CdSO_4(a = 1), \ PbSO_4(s) \mid Pb(s)$
  (b) $Au(s) \mid H_2SO_4(dilute \ solution) \mid Pt(s)$
  (c) $Pt(s) \mid Na_2SO_4 \ solution \mid Pt(s)$
  (d) $Ag(s) \mid AgNO_3(m = 0.5) \parallel AgNO_3(m = 0.005) \mid Ag(s)$
  (e) $Pb(s) \mid Pb^{++}(a = 1) \parallel Ag^+(a = 1) \mid Ag(s)$

49. Write the electrode and cell reactions which will occur in the above cells when a current is passed through them.

50. A 2.00-volt storage battery is used in an electrolysis in which a back emf of 1.45 volts is developed. If the resistance of the entire circuit is 10 ohms, (a) what is the magnitude of the current, and (b) what quantity of heat is generated by the current flow per faraday of electricity?

                    Ans. (a) 0.055 amp; (b) 12,690 cal.

51. Under reversible conditions what applied potential will be necessary to initiate steady electrolysis in each of the following cells?

(a) Pb(s) | PbSO$_4$(s), H$_2$SO$_4$, PbSO$_4$(s) | Pb(s)

(b) Pt(s) | Fe^{++}($a = 1$), Fe^{+++}($a = 1$) ‖ Br$^-$($a = 1$), AgBr(s) | Ag(s)

(c) Ni(s) | NiSO$_4$($a = 1$), PbSO$_4$(s) | Pb(Hg)

(d) Ag(s) | AgCl(s), ZnCl$_2$($a = 1$) | Zn(s)

The $\mathcal{E}^0$ value for the Pb(Hg) | PbSO$_4$(s), SO$_4^{--}$($a = 1$) electrode is 0.3505 volt.

52. Repeat the preceding problem for each of the following cells:

(a) Pt(s) | HI($a = 1$) | Pt(s)

(b) Pt(s) | CdCl$_2$($a = 1$) | Pt(s)

(c) Pt(s) | NiSO$_4$($a = 1$) | Pt(s)

53. Calculate the reversible decomposition potentials of the following cells:

(a) Pt(s) | HBr($m = 0.05$, $\gamma = 0.860$) | Pt(s)

(b) Ag(s) | AgNO$_3$($m = 0.5$, $\gamma = 0.526$) ‖ AgNO$_3$($m = 0.01$,

$$\gamma = 0.902) \,|\, \text{Ag(s)}$$
$$\textit{Ans. (a) } -1.2269 \text{ volts}$$

54. In the cell

$$\text{Pt(s)} \left| \begin{array}{c} \text{CdCl}_2(m\gamma = 1) \\ \text{NiSO}_4(m\gamma = 1) \end{array} \right| \text{Pt(s)}$$

what will be the cell reaction and the reversible decomposition potential when electrolysis is initiated?

55. The overvoltage of H$_2$ on Pb was determined in a 0.1 molal H$_2$SO$_4$ solution ($\gamma = 0.265$) by measuring during electrolysis the potential of the lead cathode against a normal calomel celectrode. The potential thus observed was 1.0685 volts. What is the overvoltage of H$_2$ on Pb?   *Ans.* 0.713 volt.

56. The overvoltages of H$_2$ and O$_2$ on Ag at a current density of 0.1 amp/cm^2 are, respectively, 0.87 and 0.98 volt. What will be the decomposition potential of a dilute NaOH solution between Ag electrodes at this current density?

57. (a) Using the data of Table 12–7, plot overvoltage against the log of the current density for Ag, and evaluate the constants a and b in Eq. (99). (b) At what current density will the overvoltage be 0.80 volt?

58. The overvoltage of H$_2$ on polished Pt is 0.24 volt when the current density is 1 ma/cm^2. What will be the deposition potential of H$_2$ on this electrode from a solution of pH $= 3$?   *Ans.* 0.42 volt.

59. The overvoltage of H$_2$ on Zn is 0.72 volt. If it is desired to deposit Zn^{++} down to a concentration of 10^{-4} mole/liter without deposition of H$_2$ occurring, what must be the lowest permissible pH of the solution?

60. The overvoltage of H$_2$ on Fe is 0.40 volt at a current density of 1 ma/cm^2. In a solution containing Fe^{++} ions at unit activity, what is the lowest pH at which deposition of Fe is possible under these conditions?

61. A solution contains Fe^{++} and Zn^{++} ions at unit activity. If the overvoltage of H$_2$ on Fe is 0.40 volt, what must be the maximum pH of the solution in order to plate Fe but obtain "gassing" at the cathode prior to Zn deposition? What will be the activity of Fe^{++} ions when "gassing" begins?

$$Ans. \text{ pH} = 6.11; a_{Fe++} = 1.41 \times 10^{-11}.$$

62. A solution contains $Ag^+(a = 0.05)$, $Fe^{++}(a = 0.01)$, $Cd^{++}(a = 0.001)$, $Ni^{++}(a = 0.1)$, and $H^+(a = 0.001)$. The overvoltage of H_2 on Ag is 0.20 volt, on Ni 0.24 volt, on Fe 0.18 volt, on Cd 0.30 volt. Predict what will happen at a cathode immersed in this solution as the potential applied is gradually increased from zero.

63. A solution contains Fe^{++} ions at unit activity. (a) Under reversible conditions what must be the activity of Cd^{++} ions in the same solution in order to obtain simultaneous deposition of Fe and Cd? (b) What would have to be the activity of Ni^{++} ions to obtain Ni and Fe simultaneously?

64. The overvoltage of O_2 on a Pt anode is 0.72 volt. Which ion will be deposited first from a solution of pH $= 7$ containing Cl^- at 0.1 activity?

65. A solution containing Zn^{++}, Na^+, and H^+ ions at unit activity is electrolyzed with a Hg cathode. Explain what will occur at the cathode as the potential is gradually increased.

66. A solution is made originally 0.1 molar in Ag^+ ions and 0.25 molar in KCN. If the dissociation constant of $Ag(CN)_2^-$, $K = [Ag^+][CN^-]^2/[Ag(CN)_2^-]$, is 3.8×10^{-19}, what will be the concentration of Ag ions in this solution, and what will be the deposition potential of Ag? For this calculation assume activities equal to concentrations. $Ans. \ 1.52 \times 10^{-17}; \ -0.196$ volt.

67. A solution is 0.1 molar in Au^+ and 0.1 molar in Ag^+ ions. The standard reduction potential of Au^+ is 1.68 volts, while the dissociation constant of $Au(CN)_2^-$, $K = [Au^+][CN^-]^2/[Au(CN)_2^-]$, is 5×10^{-39}. What concentration of NaCN will have to be added to the given solution in order to deposit Au and Ag simultaneously?

13
Kinetics of Homogeneous Reactions

CHEMICAL KINETICS is the branch of physical chemistry which concerns itself with the study of the velocity of chemical reactions and with the elucidation of the mechanisms by which they proceed.

A question of importance which thermodynamics does not touch upon is: *How rapidly and by what mechanism does a reaction take place?* Thermodynamics considers only the energy relations between reactants and the products of a reaction. It makes no attempt to indicate the stages through which the reactants may have to pass to reach the final products, nor is it concerned with the rate at which equilibrium is attained. Chemical kinetics complements thermodynamics by supplying information about the rate of approach to equilibrium and also, whenever possible, about the mechanism responsible for the conversion of reactants to products.

Not all reactions lend themselves readily to kinetic study. Many ionic reactions are so rapid that to all appearances they are instantaneous. Explosions and certain other reactions, of which $N_2O_4 \longrightarrow 2\,NO_2$ is an example, also proceed so quickly that either it is impossible to determine their rate, or special methods have to be devised to do so. On the other hand, some reactions are so slow that months and even years are necessary to observe any perceptible transformation at ordinary temperatures. Between these extremes lie the reactions whose velocities are conveniently measurable. In this category fall many gaseous reactions, as well as many reactions in solution involving both organic and inorganic substances.

The rates of reactions depend on the nature of the reacting substances, the temperature, and the concentrations of the reactants. An increase in temperature leads almost invariably to an increase in velocity; in fact, for many reactions an increase of 10°C in temperature may increase the rate of reaction twofold or more. Again, with the exception of certain (zero order) reactions, upon which concentration is without effect, an increase in the initial concentration of reactants also results in an acceleration of the rate. However, the rate for any given initial concentration does not remain constant during the course of the reaction, but is highest at the beginning and decreases with time as the reactants are consumed. Theoretically, infinite time would be required for the rate to fall to zero. Practically, however, the rate becomes so low after a time that to all intents and purposes the reaction may be considered completed within a finite time interval.

Further, many reactions are influenced by the presence of substances which possess the ability of accelerating or decelerating the rates of such reactions. Substances possessing this property of affecting the rates of chemical reactions are called *catalysts*, while the reactions thus affected are designated as *catalyzed reactions*. Again, certain reactions are uninfluenced by light, while others, called *photochemical reactions*, are greatly stimulated when light of appropriate frequency is permitted to pass through the reacting mixture.

Reactions may be classified kinetically as being either *homogeneous* or *heterogeneous*. A reaction is said to be kinetically homogeneous if it takes place in one phase only. If two or more phases are involved in the process, as in a gaseous reaction proceeding on the surface of a solid catalyst or on the walls of the container, the reaction is said to be heterogeneous. The principles governing the kinetics of some heterogeneous reactions will be discussed in Chapter 20. Here attention will be devoted to three types of homogeneous reactions, namely, (a) uncatalyzed reactions, (b) chain reactions, and (c) catalyzed reactions. Photochemical reactions will be discussed in Chapter 19.

MEASUREMENT OF REACTION RATE

Kinetic studies of reactions are generally carried out at constant temperature. A reaction mixture of known composition is prepared and thermostated, and the decrease in concentration of reacting substances, or the appearance of reaction products, is measured as a function of time by some suitable means. From such concentration-time data the kinetic behavior of the process can be deduced with the aid of certain principles which will be developed in the following. Again, the dependence of the rate on temperature is obtained by repeating this procedure at a number of suitable temperatures.

The most obvious way of following the concentration changes occurring in a reaction is to remove samples from the system at various time intervals, stop the reaction, and then analyze the samples to determine the concentration of either a reactant or a product. However, if at all possible, chemists prefer to follow the concentration changes in a reacting system by employing some physical property which changes with time, and from which the requisite concentrations can be deduced. In gaseous reactions involving a volume change, the property most commonly observed is the variation in pressure at constant volume, or the change in volume at constant pressure. The latter method, which also can be employed sometimes in liquid systems, is known as dilatometry. Other physical means which have been used under suitable conditions are conductivity, refractive index variation, gas evolution, spectroscopy and colorimetry, viscosity, light scattering, polarography, magnetic susceptibility, and mass spectrometry.

ORDER AND MOLECULARITY OF REACTIONS

The rate of a chemical reaction is the rate at which the concentrations of reacting substances vary with time, i.e., $-dC/dt$, where C is the concentration of reactant and t the time. The minus sign is used to denote that the concentration decreases with time. The dependence of this rate on the concentrations of reacting substances is given by the *law of mass action*. This law states that *the rate of any reaction is at each instant proportional to the concentrations of the reactants, with each concentration raised to a power equal to the number of molecules of each species participating in the process.* Thus, for the reaction

$$A \longrightarrow \text{Products} \tag{a}$$

the rate should be proportional to C_A, for

$$2\,A \longrightarrow \text{Products} \tag{b}$$
and
$$A + B \longrightarrow \text{Products} \tag{c}$$

to C_A^2 and $C_A C_B$, respectively, while for

$$A + 2\,B \longrightarrow \text{Products} \tag{d}$$
or
$$2\,A + B \longrightarrow \text{Products} \tag{e}$$

to $C_A C_B^2$ and $C_A^2 C_B$. These proportionalities indicate that different rate dependencies on concentration are obtained at constant temperature for various reactions, and to distinguish these the term *order of reaction* is employed. By the order of a chemical reaction is meant *the sum of all the exponents to which the concentrations in the rate equation are raised.* Thus when the rate of a reaction is given by

$$\frac{-dC}{dt} = kC_1^{n_1}C_2^{n_2}C_3^{n_3} \cdots \tag{1}$$

where k is a constant, the reaction orders of the individual constituents are n_1, n_2, n_3, etc., and the order of the reaction as a whole, n, is

$$n = n_1 + n_2 + n_3 + \cdots \tag{2}$$

In examples (a) to (e) the order of reaction is identical with the number of molecules of reactants participating in the reaction. Thus, the first order reaction (a) is also *unimolecular*, the second order reactions (b) and (c) are *bimolecular*, while the third order reactions (d) and (e) are *termolecular*. Although this identity of reaction order and the number of molecules reacting as given by the stoichiometric equation of the process is observed in many reactions, it does not hold in all cases. For the reactions

$$CH_3CHO \longrightarrow CH_4 + CO$$
$$3\,KClO \longrightarrow KClO_3 + 2\,KCl$$

we would anticipate *a priori* that the first of these would be first order, and the second, third order. Actually both reactions are found to be second order. Since the reaction order *may* differ from the number of molecules participating in the reaction, a distinction must be made between the *molecularity*, i.e., the number of molecules involved in the step leading to reaction, and the order of a reaction. According to the former, reactions are unimolecular, bimolecular, etc., depending on whether one, two, or more molecules are involved in the rate determining step. On the other hand, the term reaction order is confined to the dependence of the observed rate on the concentrations of the reactants.

It is evident that the order, or for that matter the molecularity, of a reaction cannot always be predicted from the stoichiometric equations of a reaction but that each and every reaction must be investigated kinetically to ascertain both order and molecularity. Further, care must be exercised in interpreting an observed reaction order and in inferring from it the molecularity and the mechanism of the process. When the order of a reaction agrees with the rate equation obtained by applying the law of mass action to the stoichiometric equation of the reaction, it is fairly safe to assume that the reaction proceeds as given by the chemical equation. When the two are not the same, however, a mechanism has to be devised which is in accord both with the observed reaction order and with the over-all equation of the process.

In developing the rate equations for reactions of various orders we shall proceed on the supposition that molecularity and order are identical. Whenever necessary attention will be directed to any discrepancy between the two.

FIRST ORDER REACTIONS

According to the law of mass action the rate of any unimolecular reaction

$$A \longrightarrow \text{Products} \tag{3}$$

should at any time t be directly proportional to the concentration of A, C_A, present in the system at that instant, i.e.,

$$\frac{-dC_A}{dt} = k_1 C_A \tag{4}$$

The proportionality factor k_1 is called the *specific rate* or the *specific velocity constant* of the first order reaction. On setting $C_A = 1$ in Eq. (4), the significance of this constant may be seen to be the speed of the reaction when the concentration of A is constant and equal to unity. Again, the units of k_1 follow from $k_1 = (-1/C_A)(dC_A/dt)$ as a reciprocal time or frequency. For any first order reaction k_1 should be a constant characteristic of the reaction, independent of the concentration, and a function of the temperature only.

Before integrating this differential equation, it is desirable to transform it into an alternate form. To do this let a be the initial concentration of A, and x the decrease in concentration of A due to reaction up to time t. Then $C_A = a - x$ at time t,

$$\frac{-dC_A}{dt} = \frac{-d(a - x)}{dt} = \frac{dx}{dt}$$

and Eq. (4) becomes

$$\frac{dx}{dt} = k_1(a - x) \tag{5}$$

This equation gives the rate for a first order reaction in terms of the initial concentration and the amount of substance reacted. Integrating Eq. (5) and remembering that at the start of the reaction $t = 0$ and $x = 0$, and at time t $x = x$, we obtain

$$\int_0^x \frac{dx}{a-x} = \int_0^t k_1 dt$$

$$\left[-\ln(a-x)\right]_0^x = \left[k_1 t\right]_0^t$$

$$\ln \frac{a}{a-x} = k_1 t \tag{6}$$

Any first order reaction must satisfy Eq. (6). To ascertain whether a particular reaction obeys this equation, several methods are available. First, knowing the initial concentration and the concentrations of reactant at various elapsed times, a, $(a-x)$, and t corresponding to the latter may be substituted into the equation and k_1 solved for. If the reaction is first order, a series of k_1's are thus obtained which are constant within the accuracy of the experiment. However, if the k_1 values exhibit an appreciable drift the reaction is not first order, and higher order equations must then be tried to find one that will satisfy the observed data.

Equation (6) may also be tested graphically. On rearrangement this equation becomes

$$\ln(a-x) = -k_1 t + \ln a$$

or

$$\log_{10}(a-x) = \left(\frac{-k_1}{2.303}\right) t + \log_{10} a \tag{7}$$

Since for any one experiment a is constant, a plot of log $(a-x)$ vs. t should yield a straight line in which the y intercept will be $\log_{10} a$ and the slope $(-k_1/2.303)$. Consequently, when such a plot constructed from the experimental data is found to be linear, the reaction is first order; and, by taking the slope of the line, k_1 follows as

$$k_1 = -2.303 \text{ (slope)} \tag{8}$$

A third means of testing first order reactions is known as the *fractional life method*. In this method the time necessary to decompose a definite fraction of the reactant, usually one-half, is determined for a number of different values of a. When one-half of the reactant has undergone decomposition, $a - x = a/2$, and the time, $t_{1/2}$, necessary for this to occur follows from Eq. (6) as

$$t_{1/2} = \frac{1}{k_1} \ln \frac{a}{a/2}$$

$$= \frac{\ln 2}{k_1} \tag{9}$$

The quantity $t_{1/2}$ is known as the *half-life period* of the reaction. According to Eq. (9) *the half-life period of any first order reaction is independent of the initial concentration*. In other words, it takes a first order reaction just as much time to go halfway to completion when the initial concentration is high as when it is low. That this requirement is met by first order reactions is shown by the data given in Table 13–1 for the thermal decomposition of gaseous acetone at 601°C. Although the initial pressure of the acetone is seen to vary about fourfold, the half-life period nevertheless remains constant within the experimental accuracy. Taking as an average $t_{1/2} = 81$ sec, we find for the rate constant of this reaction

$$k_1 = \frac{\ln 2}{t_{1/2}} = \frac{0.693}{81}$$
$$= 0.0086 \text{ sec}^{-1}$$

In calculating rate constants the common practice is to express the time in minutes or seconds. Occasionally, with very slow reactions time may be expressed in hours or even days. Concentrations, on the other hand, should invariably be expressed in moles per liter for reactions in

TABLE 13–1.   HALF-LIFE PERIODS
FOR THERMAL DECOMPOSITION
OF ACETONE AT 601°C*

a (mm Hg)	$t_{1/2}$ (sec)
98	86
192	78
230	85
296	80
362	77

* Hinshelwood and Hutchison, *Proc. Roy. Soc.*, 111A, 345 (1926).

solution, and in these or some suitable pressure units for gaseous reactions. In any case, for complete clarity the units of time and concentration for which k is given should be stated, as well as the temperature. However, for first order reactions, *and only for these*, the units in which the concentrations are expressed do not matter, since k_1 depends on the ratio of two concentrations. Consequently, as long as a and $a - x$ are expressed in the same units the value of the ratio is unaffected, and any convenient units may be employed.

EXAMPLES OF FIRST ORDER REACTIONS

Examples of homogeneous gaseous first order reactions are the thermal dissociations of nitrous oxide, nitrogen pentoxide, acetone, propionic aldehyde, various aliphatic ethers, azo compounds, amines, and ethyl bromide, and the isomerization of d-pinene to dipentene. As typical of these may be taken the thermal decomposition of azoisopropane to hexane and nitrogen

$$(CH_3)_2CHN{=}NCH(CH_3)_2 \longrightarrow N_2 + C_6H_{14}$$

which was investigated by Ramsperger[1] over the pressure range 0.0025 to 46 mm Hg and between 250 and 290°C. The rate of reaction was followed by pressure measurements with a McLeod gage. The only data necessary were the initial pressure of the reactant and the total pressures of the system at various stages of decomposition. From these pressures the rate constant of the process may be calculated as follows: If we let P_i be the initial pressure of azoisopropane, P the total pressure, P_A the pressure of azoisopropane, and x the decrease in pressure of azoisopropane, all at time t, then at each stage of the reaction $P_A = P_i - x$, $P_{N_2} = P_{C_6H_{14}} = x$, and the total pressure of the system is given by

$$\begin{aligned} P &= P_A + P_{N_2} + P_{C_6H_{14}} \\ &= (P_i - x) + x + x \\ &= P_i + x \end{aligned}$$

From this equation we find $x = P - P_i$, and on substitution of this value for x into $P_A = P_i - x$, we get

$$\begin{aligned} P_A &= P_i - x \\ &= P_i - (P - P_i) \\ &= 2P_i - P \end{aligned}$$

Since a in Eq. (6) is proportional to P_i, and $(a - x)$ to P_A, the expression for k_1 is

$$\begin{aligned} k_1 &= \frac{2.303}{t} \log_{10} \frac{P_i}{P_A} \\ &= \frac{2.303}{t} \log_{10} \frac{P_i}{2P_i - P} \end{aligned}$$

The constants thus calculated from experimental data for a typical run at 270°C are shown in Table 13–2. Column 1 gives the time in seconds, column 2 the total pressure P, and column 3 the values of k_1. P_i is the pressure of the system at $t = 0$, which for the experiment in question was 35.15 mm Hg. Since k_1 exhibits satisfactory constancy, the decomposition of azoisopropane must be a first order reaction.

[1] Ramsperger, *J. Am. Chem. Soc.*, **50**, 714 (1928).

An example of a first order reaction in solution is the decomposition of benzene diazonium chloride, $C_6H_5N{=}NCl$, in water. This compound, known only in solution, dissociates readily on heating to liberate nitrogen. As the nitrogen all comes from the compound, the volume of it liberated may be used as a measure of the decrease in concentration of the diazo salt. Cain and Nicoll[2] utilized this fact in their kinetic study of this reaction. A freshly prepared solution of the salt was quickly heated to the desired temperature, and the reaction was then permitted to proceed

TABLE 13–2.   DECOMPOSITION OF AZOISOPROPANE
AT 270°C

t (sec)	P (mm Hg)	k_1 (sec^{-1})
0	35.15	—
180	46.30	2.12×10^{-3}
360	53.90	2.11
540	58.85	2.07
720	62.20	2.03
1020	65.55	1.96
		Mean $= 2.06 \times 10^{-3}$

in a thermostat. The course of the reaction was followed by measuring in a gas burette the volumes V of nitrogen evolved at various time intervals. To complete the required data, the initial concentration, expressed as cubic centimeters of nitrogen V_0, was calculated from the amount of diazo compound originally present. In terms of V_0 and V, the velocity constant of the reaction follows as

$$k_1 = \frac{2.303}{t} \log_{10}\left(\frac{a}{a-x}\right) = \frac{2.303}{t} \log_{10}\left(\frac{V_0}{V_0 - V}\right)$$

The results obtained in an experiment at 50°C with an amount of benzene diazonium chloride equivalent to 58.3 cc nitrogen are shown in Figure 13–1, where a plot of $\log_{10}(V_0 - V)$ vs. t is given. The plot is linear with slope $= -0.0308$. Consequently,

$$\begin{aligned} k_1 &= -2.303 \text{ (slope)} \\ &= -2.303 (-0.0308) \\ &= 0.0709 \text{ min}^{-1} \end{aligned}$$

Other first order reactions in solution are the decomposition of nitrogen pentoxide and the isomerization of d-pinene in various organic solvents,

---

[2] Cain and Nicoll, *J. Chem. Soc.*, **81**, 1412 (1902).

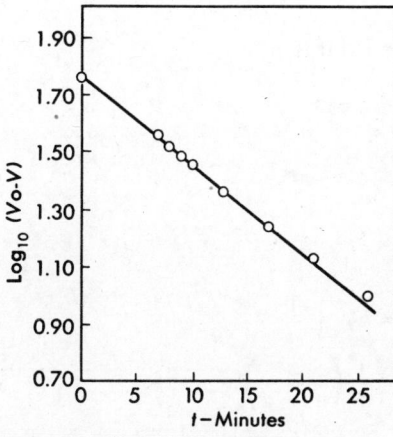

Figure 13–1. Plot of $\log_{10}$ $(V_0 - V)$ vs. t for decomposition of benzene diazonium chloride at 50°C.

the decomposition of malonic, trichloracetic, and acetonedicarboxylic acids in water, and the dissociation of various diazonium salts in the same solvent.

SECOND ORDER REACTIONS

Any bimolecular reaction can in general be represented by the equation

$$A + B \longrightarrow \text{Products} \tag{10}$$

If we let a and b be the initial concentrations of A and B respectively, and x the decrease in concentration of each at time t, then the concentrations of A and B will be $(a - x)$ and $(b - x)$, and the rate equation for the second order reaction becomes

$$\frac{dx}{dt} = k_2(a - x)(b - x) \tag{11}$$

Here k_2 is the specific rate constant for such a reaction and is equal to the rate when both A and B are unity. Its units are $C^{-1}t^{-1}$, and its magnitude depends on the nature of the reaction, the temperature, and the units in which t and C are expressed. Integration of this equation, again remembering that $x = 0$ when $t = 0$, yields

$$k_2 = \frac{2.303}{t(a - b)} \log_{10} \frac{b(a - x)}{a(b - x)} \tag{12}$$

This is the equation which any second order reaction must obey.

Equation (12) is considerably simplified when A and B are present in equal initial concentrations, or when A and B are the same, as in the reaction

$$2\,A \longrightarrow \text{Products} \tag{13}$$

For these conditions Eq. (11) reduces to

$$\frac{dx}{dt} = k_2(a - x)^2 \tag{14}$$

and on integration becomes either

$$\left(\frac{1}{a - x}\right) = k_2 t + \frac{1}{a} \tag{15}$$

or

$$k_2 = \frac{1}{at}\left(\frac{x}{a - x}\right) \tag{16}$$

These expressions for second order reactions may be tested by the methods described for first order reactions. Values of a, b, and x, or a and x may be substituted along with t and k_2 solved for, or k_2 may be evaluated graphically. For the latter procedure Eq. (12) is best written in the form

$$\frac{2.303}{(a - b)} \log_{10}\left(\frac{a - x}{b - x}\right) = k_2 t - \frac{2.303}{(a - b)} \log_{10}\left(\frac{b}{a}\right) \tag{17}$$

from which we see that, since k_2, a, and b are constant for any given experiment, a plot of the left-hand side of this equation against t should yield a straight line with slope equal to k_2. On the other hand, to test the rate law when the concentrations are the same, Eq. (15) is used. According to this equation a plot of $1/(a - x)$ vs. t should be linear, with slope equal to k_2 and y intercept equal to $1/a$.

The half-life method cannot be used with reactions where the concentrations of A and B are different, since A and B will have different times for half reaction. However, it may be used where the concentrations are the same, or where the two reacting molecules are identical. In these instances $x = a/2$ at the half-life point, and the period of half-life, $t_{1/2}$, follows from Eq. (16) as

$$t_{1/2} = \frac{1}{k_2} \cdot \frac{(a/2)}{a(a/2)}$$

$$= \frac{1}{k_2 a} \tag{18}$$

i.e., *for a second order reaction the period of half-life is inversely proportional to the first power of the initial concentration.* Knowing $t_{1/2}$ and a, k_2 is readily evaluated from Eq. (18).

It may be stated as a general rule that the period of half-life for any n order reaction is inversely proportional to the $(n - 1)$ power of the initial concentration, i.e.,

$$t_{1/2} \propto \frac{1}{a^{n-1}} \tag{19}$$

The proportionality constant for Eq. (19) depends on the reaction order and is $(\ln 2)/k_1$ for first order, $1/k_2$ for second order, and $\frac{3}{2} k_3$ for third order reactions.

EXAMPLES OF SECOND ORDER REACTIONS

Homogeneous gas reactions of second order are very common. In this category are included various thermal dissociations, such as those of hydrogen iodide, nitrogen dioxide, ozone, chlorine monoxide, nitrosyl chloride, formaldehyde, and acetaldehyde; the combination of hydrogen and iodine to form hydrogen iodide; the polymerization of ethylene; the hydrogenation of ethylene; and others. The behavior of all such reactions is exemplified by the thermal decomposition of acetaldehyde. This reaction, investigated by Hinshelwood and Hutchison,[3] was found to be almost totally homogeneous and to proceed according to the equation

$$2 \, CH_3CHO \longrightarrow 2 \, CH_4 + 2 \, CO$$

Since in this reaction there is an increase in pressure at constant volume on dissociation, the change in pressure observed on a manometer attached to the system may be employed to follow the reaction course. From these pressure measurements k_2 is calculated as follows.

If we let P_i be the initial pressure of acetaldehyde and x the decrease in pressure after time t, then the pressure of the reactant at time t is $(P_i - x)$. Again, when the pressure of acetaldehyde drops by x, the pressures of methane and carbon monoxide must each increase by x. We obtain thus for the total pressure of the system, P,

$$
\begin{aligned}
P &= P_{CH_3CHO} + P_{CH_4} + P_{CO} \\
&= (P_i - x) + x + x \\
&= P_i + x
\end{aligned}
$$

from which x follows as $x = P - P_i$. Further, since $P_i - x$ is proportional to $(a - x)$ and P_i to a, substitution of these into Eq. (16) yields for k_2

$$
\begin{aligned}
k_2 &= \frac{1}{at}\left(\frac{x}{a - x}\right) \\
&= \frac{1}{P_i t}\left(\frac{x}{P_i - x}\right)
\end{aligned}
$$

---

[3] Hinshelwood and Hutchison, *Proc. Roy. Soc.*, **111A**, 380 (1926).

In Table 13–3 are given data obtained by Hinshelwood and Hutchison for this reaction during an experiment at 518°C with an initial pressure of acetaldehyde equal to 363 mm Hg. The good constancy of k_2 confirms both the reaction order and the correctness of the formulation.

TABLE 13–3.   THERMAL DECOMPOSITION OF ACETALDEHYDE
AT 518°C
$(P_i = 363$ mm Hg$)$

t (sec)	$x = P - P_i$ (mm Hg)	$k_2 = \dfrac{1}{P_i t}\left(\dfrac{x}{P_i - x}\right)$
42	34	6.79×10^{-6}
73	54	6.59
105	74	6.71
190	114	6.64
242	134	6.66
310	154	6.55
384	174	6.61
480	194	6.59
665	224	6.68
840	244	6.72
1070	264	6.86
1440	284	6.88

Average $= 6.69 \times 10^{-6}$ mm^{-1} sec^{-1}

Among the many reactions in solution which are of second order may be mentioned the saponifications of various esters by bases, the conversion of nitroparaffins to aci-nitroparaffins by bases, the reactions of alkyl halides with amines, the hydrolyses of esters, amides, and acetals, esterifications of organic acids, the combination of NH_4^+ and CNO^- ions to form urea, and the reactions of bromacetates with thiosulfates and thiocyanates.

As a prototype of these reactions may be taken the saponification of ethyl butyrate by hydroxyl ions in water solution, namely,

$$CH_3CH_2CH_2COOC_2H_5 + OH^- \longrightarrow CH_3CH_2CH_2COO^- + C_2H_5OH$$

which was studied by Williams and Sudborough[4] at 20°C. The reaction mixture was prepared by bringing together solutions of ethyl butyrate and barium hydroxide so as to yield an initial concentration a of the ester and b of the base. The course of the reaction was followed by removing periodically samples of solution and titrating these with standard acid to

[4] Williams and Sudborough, *J. Chem. Soc.*, **101**, 415 (1912).

ascertain the concentration of unreacted barium hydroxide. The concentrations of base thus determined give directly $(b - x)$, while a minus the amount of reacted base yields $(a - x)$. Table 13–4 shows some of the data

TABLE 13–4.   SAPONIFICATION OF ETHYL BUTYRATE
BY BARIUM HYDROXIDE AT 20°C

t (min)	$a - x$	$b - x$	k_2 [Eq. (12)]
	(arbitrary units)		
0	19.75	20.85	—
5	14.75	15.85	0.0032
17	9.40	10.50	0.0030
36	5.93	7.03	0.0029
65	3.57	4.67	0.0030

obtained in this manner, and the values of k_2 calculated from these by Eq. (12). As an alternate test a plot called for by Eq. (17) may be prepared to see whether a straight line results. Such a plot, given in Figure 13–2, is linear.

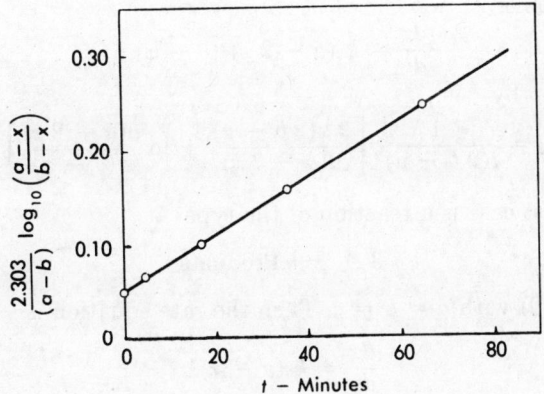

Figure 13–2.   Plot of $\dfrac{2.303}{(a - b)} \log_{10} \dfrac{(a - x)}{(b - x)}$ vs. t for the saponification of ethyl butyrate at 20°C.

THIRD ORDER REACTIONS

The most general type of termolecular reaction is one where three different molecules react, namely,

$$A + B + C \longrightarrow \text{Products} \tag{20}$$

If the reactants are all present at different initial concentrations a, b, c, and x is the decrease in the concentration of each at time t, then the third order rate equation is

$$\frac{dx}{dt} = k_3(a - x)(b - x)(c - x) \tag{21}$$

The integrated form of Eq. (21) is rather complex. For a simpler case where two of the initial concentrations are equal, say $a = b$, Eq. (21) reduces to

$$\frac{dx}{dt} = k_3(a - x)^2(c - x) \tag{22}$$

which on integration yields for k_3

$$k_3 = \frac{1}{t(c - a)^2}\left[\frac{x(c - a)}{a(a - x)} + \ln\frac{c(a - x)}{a(c - x)}\right] \tag{23}$$

On the other hand, when two of the molecules are identical, as in

$$2\,A + B \longrightarrow \text{Products} \tag{24}$$

the concentrations at any time are $(a - 2\,x)$ and $(b - x)$. Then the rate equation becomes

$$\frac{dx}{dt} = k_3(a - 2\,x)^2(b - x) \tag{25}$$

and k_3 is given by

$$k_3 = \frac{1}{t(2\,b - a)^2}\left[\frac{2\,x(2\,b - a)}{a(a - 2\,x)} + \ln\frac{b(a - 2\,x)}{a(b - x)}\right] \tag{26}$$

The simplest case is a reaction of the type

$$3\,A \longrightarrow \text{Products} \tag{27}$$

or reaction (20) with $a = b = c$. Then the rate equation is

$$\frac{dx}{dt} = k_3(a - x)^3 \tag{28}$$

and on integration

$$k_3 = \frac{1}{2\,t}\left[\frac{1}{(a - x)^2} - \frac{1}{a^2}\right] \tag{29}$$

$$= \frac{1}{2\,ta^2}\left[\frac{x(2\,a - x)}{(a - x)^2}\right] \tag{30}$$

From (30) the period of half-life of a third order reaction follows as

$$t_{1/2} = \frac{3}{2\,k_3a^2} \tag{31}$$

and hence for such a reaction $t_{1/2}$ is inversely proportional to a^2.

Only five homogeneous gas reactions are definitely known to be third order, and every one of these involves the interaction of nitric oxide with either chlorine, bromine, oxygen, hydrogen, or heavy hydrogen. As an example of one of these may be taken the work of Hinshelwood and Green[5] on the reaction

$$2\ NO + H_2 \longrightarrow N_2O + H_2O$$

For hydrogen present in excess, k_3 for this reaction should be given by Eq. (26). Now, at a time t' when the nitric oxide is half consumed $2\ x = a/2$, and this equation becomes

$$k_3 = \frac{1}{t'(2\ b - a)^2}\left[\frac{2\ b - a}{a} + \ln\frac{2\ b}{4\ b - a}\right] \qquad (32)$$

This is the relation Hinshelwood and Green used to evaluate k_3 from various initial pressures of nitric oxide ($a = P_{NO}$) and hydrogen ($b = P_{H_2}$), and the observed times, t', of half decomposition of nitric oxide. Their results, given in Table 13–5, yield fairly satisfactory constants, and indi-

TABLE 13–5.   KINETICS OF THE REACTION
$2\ NO + H_2 \longrightarrow N_2O + H_2O$ at 826°C

P_{NO} (mm Hg)	P_{H_2} (mm Hg)	t' (sec)	k_3
110	316	270	1.19×10^{-7}
152	404	204	0.91
359	400	89	1.12
144	323.5	227	1.10
131	209.5	264	1.39
370	376	92	1.17
300	404	100	1.09
232	313	152	1.19

cate that the reduction of nitric oxide by hydrogen is a third order reaction.

Reactions in solution which appear to be third order are the oxidation of ferrous sulfate in water, the reaction between iodide and ferric ions in aqueous solution, and the action between benzoyl chloride and alcohols in ether solution. The decomposition of hypobromous acid at constant pH in the pH range of 6.4 to 7.8 has also been found to be third order with respect to the acid.[6]

[5] Hinshelwood and Green, *J. Chem. Soc.*, **128**, 730 (1926).
[6] Prutton and Maron, *J. Am. Chem. Soc.*, **57**, 1652 (1935).

PSEUDO-MOLECULAR REACTIONS

There are many reactions which obey a first order rate equation although in reality they are bi- or ter-molecular. As an example of these may be taken the decomposition of carbonyl sulfide in water, namely,

$$COS + H_2O \longrightarrow CO_2 + H_2S$$

According to the law of mass action this reaction should be second order, with the rate dependent on the concentration of both the carbonyl sulfide and the water. Actually, however, the rate is found to be first order with respect to the carbonyl sulfide and independent of the water. Reactions exhibiting such behavior are said to be *pseudo-molecular*.

The pseudo-unimolecular nature of this reaction is explainable by the fact that water is present in such excess that its concentration remains practically constant during the course of the reaction. Under these conditions $b - x = b$, and the rate equation becomes

$$\frac{dx}{dt} = k_2(a - x)b \tag{33}$$

On integration this leads to

$$k = bk_2 = \frac{2.303}{t} \log_{10} \frac{a}{a - x} \tag{34}$$

which is the equation for a first order reaction. It is evident, however, that the new constant k is not independent of the concentration, as is the case with true first order constants, but may vary with b if the latter is changed appreciably. When such is the case, the true constant k_2 can be obtained from k by dividing the latter by b.

Pseudo-molecular reactions are encountered whenever one or more of the reactants remain constant during the course of an experiment. This is the case with reactions conducted in solvents which are themselves one of the reactants, as in the decomposition of carbonyl sulfide in water, or in the esterification of acetic anhydride in alcohol

$$(CH_3CO)_2O + 2 C_2H_5OH \longrightarrow 2 CH_3COOC_2H_5 + H_2O$$

Again, this is also true of reactions subject to catalysis, in which case the concentration of the catalyst does not change. The decomposition of diacetone alcohol to acetone in aqueous solution is catalyzed by hydroxyl ions, with the rate proportional to the concentration of the alcohol and that of the base. Since the concentration of the base does not change within any one experiment, however, the rate equation reduces to one of first order with respect to the alcohol. But the rate constants k obtained for various concentrations of base are not identical, as may be seen from Table 13–6. To obtain from these the true second order velocity constant, the k's must be divided by the hydroxyl ion concentration. When this is done excellent k_2 values result, as column 3 indicates.

TABLE 13–6.   DECOMPOSITION OF DIACETONE
ALCOHOL IN WATER AT 25°C*
(Catalyst: NaOH)

C Conc. of NaOH	k (min^{-1})	$k_2 = k/C$
0.0205	0.0455	2.22
0.0292	0.0651	2.23
0.0518	0.1160	2.24
0.0710	0.1576	2.22
0.1045	0.2309	2.21

* LaMer and Miller, *J. Am. Chem. Soc.*, **57**, 2674 (1935).

REVERSIBLE OR OPPOSING REACTIONS

The above formulation of the rate equations involves the tacit assumption that the processes proceed in the direction indicated without any tendency for the reactions to reverse themselves. Such an assumption is justifiable only with reactions whose point of equilibrium lies way over on the side of the products. When this is not the case, the products formed initiate a reaction counter to the forward process whose rate increases as the products accumulate, and eventually becomes equal to that of the forward reaction. At this point the over-all rate becomes zero, and the system is in equilibrium.

Reactions exhibiting this tendency to reverse themselves are called *reversible* or *opposing reactions*. As an illustration of one of these and of the method of dealing with them kinetically may be taken the oxidation of nitric oxide to nitrogen dioxide

$$2\,NO + O_2 \underset{k_2}{\overset{k_1}{\rightleftharpoons}} 2\,NO_2$$

investigated by Bodenstein and Lindner.[7] They found that below 290°C the rate of the forward reaction is third order and proceeds as written without complications. However, above this temperature the rate of dissociation becomes noticeable and leads to a decrease in the rate of disappearance of nitric oxide and oxygen. To correct for this back reaction we proceed as follows. If we let a and b be the initial concentrations

[7] Bodenstein and Lindner, *Z. physik. Chem.*, **100**, 87 (1922).

of nitric oxide and oxygen respectively, and x the amount of oxygen reacted in time t, then at t the concentration of oxygen is $(b - x)$, of nitric oxide $(a - 2\,x)$, while that of nitrogen dioxide formed $2\,x$. The rate of the forward reaction is hence given by

$$\left(\frac{dx}{dt}\right)_f = k_3(a - 2\,x)^2(b - x) \tag{35}$$

and that of the reverse reaction by

$$\left(\frac{dx}{dt}\right)_r = k_2(2\,x)^2 \tag{36}$$

Since these two rates oppose each other, the net rate of the forward reaction, dx/dt, must be the difference between them, or

$$\frac{dx}{dt} = \left(\frac{dx}{dt}\right)_f - \left(\frac{dx}{dt}\right)_r$$
$$= k_3(a - 2\,x)^2(b - x) - k_2(2\,x)^2 \tag{37}$$

This equation involves two constants, and hence another relation between k_3 and k_2 is required before the constants can be evaluated. For obtaining this relation we utilize the fact that when the reaction reaches equilibrium $dx/dt = 0$, and, therefore,

$$k_2 = \frac{k_3(a - 2\,x_e)^2(b - x_e)}{(2\,x_e)^2} \tag{38}$$

where x_e is the value of x at equilibrium. On substitution of this value of k_2 into the integrated form of Eq. (37), an expression is obtained involving only k_3 which may be used to test the kinetics of the reaction. It will be observed that here not only the initial concentrations and x at various times are required, but also the value of x at equilibrium, x_e. To find the latter the reaction is permitted to proceed until equilibrium is attained, and x_e is measured.

In Table 13-7 are shown some results obtained by Bodenstein and Lindner during an experiment at 339°C. Column 2 gives the values of k_3 calculated by using only Eq. (35), while column 3 gives the values of k_3 obtained after correction for the back reaction. Whereas the constants in column 2 diminish regularly with time, those in column 3 are constant and confirm the applicability of Eq. (37) to this reaction.

Other reactions which are reversible, and which can be treated in a manner analogous to the one described, are the mutarotation of glucose, the combination of hydrogen and iodine to form hydrogen iodide, the hydrogenation of ethylene, and the hydrolyses of various esters.

TABLE 13–7. KINETICS OF THE REACTION
2 NO + O$_2$ $\longrightarrow$ 2 NO$_2$ at 339°C

t (min)	k_3 [Eq. (35)]	k_3 [Eq. (37)]
4	0.58 × 10^{-5}	0.58 × 10^{-5}
5	0.58	0.59
6	0.57	0.59
8	0.54	0.58
10	0.53	0.59
12	0.50	0.59
17	0.46	0.61
22	0.37	0.63

CONSECUTIVE REACTIONS

Chemical reactions such as

$$A \xrightarrow{k_1} B \xrightarrow{k_2} C \qquad \qquad (39)$$

which proceed from reactants to products through one or more inter-mediate stages are called *consecutive reactions*. In these reactions each stage has its own rate and its own velocity constant. Further, whether one or another of these particular rates, or some combination of all, is measured experimentally depends on the relative magnitudes of the velocity constants of the various stages. It is a well-established fact that in any sequence of reactions of varying speed the one that is slowest will determine the rate of the over-all reaction. This follows of necessity because any succeeding stage has to wait for any preceding before it can take place. Consequently, if, in a reaction such as that represented by Eq. (39), $k_1 \gg k_2$, the conversion of B to C will determine the rate of formation of the product. On the other hand, if $k_2 \gg k_1$, the formation of B from A will control the rate, and C will result from B as soon as the latter appears. However, when k_1 and k_2 are comparable in magnitude, the rate of the over-all reaction depends on both constants, and the situation becomes more complex.

An instance of a consecutive reaction proceeding in two stages with appreciably different velocity constants is the decomposition of sodium hypochlorite in alkaline solutions. Although the stoichiometric equation of the process

$$3 \text{ NaClO} \longrightarrow 2 \text{ NaCl} + \text{NaClO}_3 \qquad \qquad (40)$$

suggests a third order reaction, the process is actually second order. To explain this result the suggestion has been advanced that the reaction proceeds in two steps, the first of which is

$$2 \, \text{NaClO} \longrightarrow \text{NaCl} + \text{NaClO}_2 \tag{40a}$$

and the second

$$\text{NaClO}_2 + \text{NaClO} \longrightarrow \text{NaCl} + \text{NaClO}_3 \tag{40b}$$

The sum of Eqs. (40a) and (40b) is Eq. (40). To decide which of these two reactions is rate determining, Förster and Dolch[8] investigated not only the decomposition of sodium hypochlorite but also the reaction between sodium chlorite and hypochlorite to form the chlorate. They found that the rate of the latter reaction is about 25 times that of the former, and consequently the rate of Eq. (40a) must control the rate of the over-all reaction given by Eq. (40).

The exact mathematical analysis of consecutive reactions with comparable velocity constants is as a rule very difficult unless the reactions are of the simplest kind, such as the one given in Eq. (39). To show the method of attack, let a be the initial concentration of A, x the amount of it decomposed in time t, and y the concentration of C formed at the same instant. Then at time t we have $C_A = (a - x)$, $C_C = y$, and $C_B = x - y$, since the total amount of B formed is x, and y of it has decomposed to form C. From these the rate of disappearance of A follows as

$$\frac{dx}{dt} = k_1(a - x) \tag{41}$$

while the rate of decomposition of B, which is the same as the rate of formation of C, is

$$\frac{dy}{dt} = k_2(x - y) \tag{42}$$

To find the dependence of the concentrations of A, B, and C on time these rate equations must be solved. This is accomplished by integrating Eq. (41) first, finding x, substituting it into Eq. (42), and then integrating the latter. As a result we obtain the following expressions for the concentrations:

$$C_A = (a - x) = ae^{-k_1 t} \tag{43a}$$

$$C_B = (x - y) = \frac{k_1 a}{k_2 - k_1}(e^{-k_1 t} - e^{-k_2 t}) \tag{43b}$$

$$C_C = \frac{a}{k_2 - k_1}[(k_2 - k_2 e^{-k_1 t}) - (k_1 - k_1 e^{-k_2 t})] \tag{43c}$$

[8] Förster and Dolch, Z. *Elektrochem.*, **23**, 137 (1917).

Figure 13–3 shows C_A, C_B, and C_C as a function of time for $a = 1$, $k_1 = 0.01$, and $k_2 = 0.02$. It will be observed that, whereas the concentration of A falls and that of C increases continuously, the concentration of B rises to a maximum and then decreases with time. This behavior of the concentration of the intermediate product is characteristic of consecutive reactions with comparable values of velocity constants and may be used to identify such reactions.

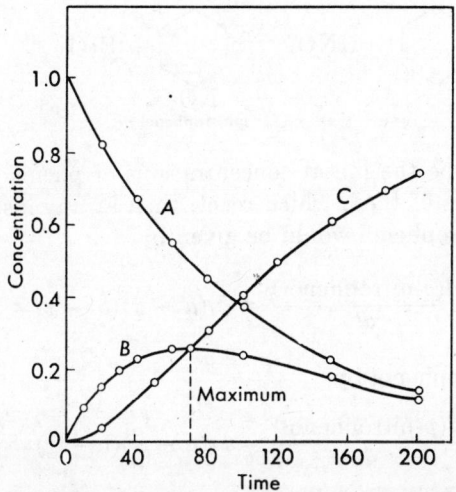

Figure 13–3. Dependence of concentration on time in consecutive reactions.

In practice the picture is not quite so simple, because usually the over-all reaction is known and the intermediate steps have to be deduced from it. Under these conditions it may be necessary to follow not only the rate of disappearance of A, but also the rates of formation of B and C before analysis of the reaction becomes possible. Further, with reactions of higher order the mathematics become more formidable, and various artifices have to be relied upon to arrive at a solution of the kinetic processes. Nevertheless, a number of consecutive reactions have been unraveled, and these substantiate the general validity of the attack outlined above.

PARALLEL REACTIONS

In *parallel reactions* the reacting substances, instead of proceeding along one path to yield a given set of products, also follow one or more other paths to give different products. Thus, in the nitration of phenol, *p*- and *o*-nitrophenols are formed simultaneously from phenol and nitric acid by the parallel reactions:

$$\text{OH} + HNO_3 \xrightarrow{k_1} \text{OH} NO_2 + H_2O \tag{44a}$$

o-nitrophenol

$$\text{OH} + HNO_3 \xrightarrow{k_2} \text{OH} + H_2O \tag{44b}$$

$$NO_2$$
p-nitrophenol

If we let a and b be the initial concentrations of phenol and nitric acid, and x the amounts of these which react, then at any instant the rate of formation of *o*-nitrophenol would be given by

$$\frac{d(\text{o-nitrophenol})}{dt} = k_1(a - x)(b - x) \tag{45}$$

and that of *p*-nitrophenol by

$$\frac{d(\text{p-nitrophenol})}{dt} = k_2(a - x)(b - x) \tag{46}$$

From these the rate of disappearance of the reactants, dx/dt, follows as the *sum* of Eqs. (45) and (46), or

$$\frac{dx}{dt} = (k_1 + k_2)(a - x)(b - x) \tag{47}$$

Again, the ratio of Eq. (45) to Eq. (46) yields

$$\frac{d(\text{o-nitrophenol})}{d(\text{p-nitrophenol})} = \frac{k_1(a - x)(b - x)}{k_2(a - x)(b - x)}$$

$$= \frac{k_1}{k_2} \tag{48}$$

Consequently, by following the disappearance of the reactants the sum of the rate constants can be found, while from the rates of formation of the individual products the ratio k_1/k_2 can be obtained. From this sum and ratio are evaluated then the individual constants.

Parallel reactions are very common, particularly in organic chemistry. The reaction yielding the largest amount of a product is generally referred to as the main reaction, while the others, yielding smaller amounts of products, are called the side reactions.

EFFECT OF TEMPERATURE ON REACTION VELOCITY

Increase in temperature leads to a tremendous increase in reaction velocity and hence in rate constants. The only known exception to the generality of this statement is the reaction $2 NO + O_2 \longrightarrow 2 NO_2$, which exhibits a small negative temperature coefficient. This rapid acceleration of reaction rate is observed both in gaseous and liquid phase reactions, as may be seen from the typical data cited in Table 13–8. In the

TABLE 13–8.   VARIATION OF REACTION RATE CONSTANTS WITH TEMPERATURE

Decomposition of CH_3CHO in Gas Phase*		Decomposition of $CO(CH_2COOH)_2$ in Aqueous Solution†		
$T°K$	k_2 [sec^{-1} (moles/liter)$^{-1}$]	$T°K$	k_1 (sec^{-1})	$t_{1/2}$ (sec)
703	0.011	273	2.46×10^{-5}	28,200.
733	0.035	283	10.8	6,410.
759	0.105	293	47.5	1,460.
791	0.343	303	163.	425.
811	0.79	313	576.	120.
836	2.14	323	1850.	37.4
865	4.95	333	5480.	12.6

* Hinshelwood and Hutchison, *Proc. Roy. Soc.*, **111A**, 380 (1926).
† Wiig, *J. Phys. Chem.*, **34**, 596 (1930).

second order decomposition of gaseous acetaldehyde an increase of 162°C in temperature results in a 450-fold increase in the velocity constant. On the other hand, in the first order decomposition of acetonedicarboxylic acid

$$CO(CH_2COOH)_2 \longrightarrow CO(CH_3)_2 + 2 CO_2$$

in aqueous solution the constants increase 2200-fold for only a 60°C change in temperature. For the latter reaction are tabulated also the half-life periods at various temperatures. These exhibit the tremendous range of 28,200 to 12 sec on passing from 0° to 60°C.

Arrhenius first pointed out that the variation of rate constants with temperature can be represented by an equation similar to that used for equilibrium constants, namely,

$$\frac{d \ln k}{dT} = \frac{E_a}{RT^2} \tag{49}$$

In this *Arrhenius equation* k is the reaction rate constant, T the absolute temperature, R the gas constant in calories, and E_a a quantity characteristic of the reaction with the dimensions of an energy. E_a is known as the *energy of activation* and plays a very important role in chemical kinetics. Its significance will be discussed presently.

When Eq. (49) is integrated on the supposition that E_a is a constant, we obtain

$$\ln k = -\frac{E_a}{RT} + C'$$

or

$$\log_{10} k = \left(\frac{-E_a}{2.303\,R}\right)\frac{1}{T} + C \tag{50}$$

where C' and C are constants of integration. However, if we integrate between the limits $k = k_1$ at $T = T_1$ and $k = k_2$ at $T = T_2$, then

$$\log_{10}\frac{k_2}{k_1} = \frac{E_a}{2.303\,R}\left(\frac{T_2 - T_1}{T_1 T_2}\right) \tag{51}$$

From Eq. (51) it is evident that as soon as two values of k are available at two different temperatures E_a may be evaluated; or, when E_a and a value of k at some one temperature are known, k at another temperature may be calculated.

To test the validity of the Arrhenius equation we resort to Eq. (50). According to this equation a plot of $\log_{10} k$ against $1/T$ should be a straight line with

$$\text{slope} = \frac{-E_a}{2.303\,R} \tag{52}$$

and y intercept equal to C. Consequently, if such a plot is found to be linear the equation is confirmed. Further, by taking the slope of the line, E_a may be calculated readily through Eq. (52).

In Figure 13–4 plots are shown of $\log_{10} k$ vs. $1/T$ for the data given in Table 13–8. As predicted by the Arrhenius equation, both reactions yield satisfactory straight lines. From these we find the slope for the acetaldehyde decomposition curve to be -9920, while that for the acetonedicarboxylic acid decomposition -5070. Hence the energy of activation for the first of these reactions is

$$E_a = -4.58(-9920)$$
$$= 45,500 \text{ cal mole}^{-1}$$

and for the second

$$E_a = -4.58(-5070)$$
$$= 23,200 \text{ cal mole}^{-1}$$

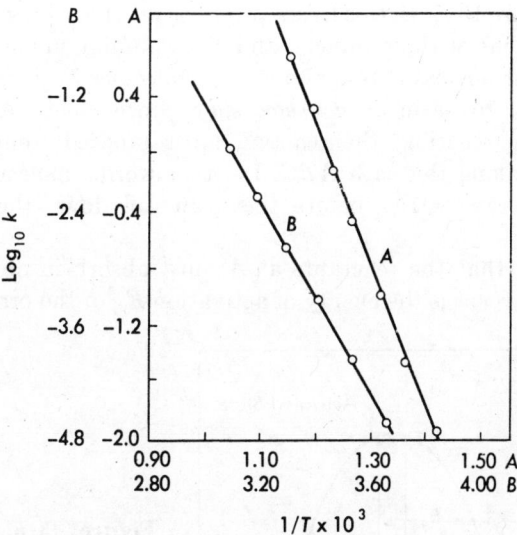

Figure 13–4. Variation of reaction rate with temperature: *A*, decomposition of acetaldehyde in gas phase; *B*, decomposition of acetonedicarboxylic acid in aqueous solution.

THE ENERGY OF ACTIVATION

In reactions involving two or more molecules it is logical to presuppose that before reaction can take place the molecules must come into contact with each other; in other words, they must collide. If collision is a sufficient cause for reaction, the rate of reaction should equal the rate of collision. However, when the number of molecules actually reacting in a gaseous reaction, as obtained from the observed velocity constants, is compared with the total number of molecules colliding, calculated from kinetic theory, the latter is found to exceed the number undergoing transformation by many powers of ten. This discrepancy can be explained only by the assumption that molecules in order to react must be in some special configuration at collision, or that they must be in some exceptionally high energy state, or both. Although configuration does play a part in certain reactions, the appearance of the E_a term in the Arrhenius equation and other considerations definitely favor an exceptional energy state as a prime requisite for reaction; i.e., *molecules must be activated before they can react on collision.*

According to the concept of activation, reactants do not pass directly to products but must first acquire sufficient energy to pass over an activation energy barrier. The ideas involved can be made clear with the aid of Figure 13–5. In this figure A represents the average energy of the reactants, C that of the products, and B the minimum energy which the reactants must possess in order to react. Molecules in state B are said to be *activated* or to be in an *activated state*. Since molecules must be activated before reaction, the reaction must proceed from A to C not directly, but along the path ABC. In other words, molecules must first climb the energy barrier before they can roll down the hill to form products.

The energy that the reactants at A must absorb in order to become activated and react is the energy of activation E_{a_1} of the process $A \longrightarrow C$.

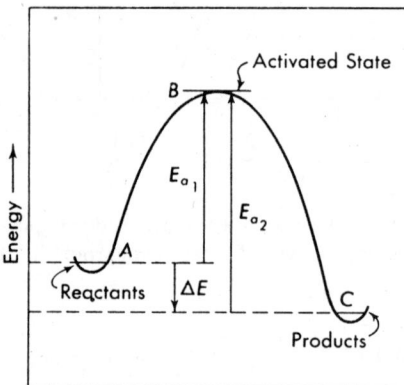

Figure 13–5. Energy of activation.

This energy is obviously $E_{a_1} = E_B - E_A$, i.e., the difference in energy between the activated state and that corresponding to the average energy of the reactants. By the same argument the energy of activation E_{a_2} of the reverse process $C \longrightarrow A$ must be $E_{a_2} = E_B - E_C$. Consequently, the difference, ΔE, between E_{a_1} and E_{a_2} follows as

$$\begin{aligned} \Delta E &= E_{a_1} - E_{a_2} \\ &= (E_B - E_A) - (E_B - E_C) \\ &= E_C - E_A \end{aligned} \tag{53}$$

But ΔE is the difference in energy between products and reactants, or the *heat of reaction at constant volume*. This must mean that in going from the activated state to the products the molecules give up, not only the energy absorbed on activation from A to B, but also the energy ΔE corresponding to the difference in the energy levels of C and A. Hence it may be concluded that the concept of an activation energy in no way violates the thermodynamic relations between the energies of products and reactants. All it does is to introduce two thermal quantities whose difference always leads to the heat of reaction.

THE COLLISION THEORY OF BIMOLECULAR REACTIONS

The collision theory of reaction velocity attempts to account for the observed kinetics of reactions in terms of the molecular behavior of the reacting systems. We shall consider this theory first for bimolecular and then for unimolecular and termolecular reactions.

According to the collision theory of bimolecular reactions two molecules in order to react must (a) collide and (b) possess on collision sufficient energy for them to be activated. This means that out of all the collisions occurring only those will be fruitful which involve active molecules. Consequently, if Z is the number of molecules colliding per cubic centimeter per second in a reacting system containing 1 mole of reactant per liter, and q the fraction of these that are activated, then the specific rate of the bimolecular reaction in molecules per cubic centimeter per second, k, should be given by

$$k = Zq \qquad (54)$$

For bimolecular *gaseous* reactions Z and q can be calculated with the aid of the kinetic theory of gases. We have seen in Chapter 1 Eq. (46), that for a gas containing only *one kind* of molecule the number of colliding molecules per cubic centimeter per second is

$$Z_{11} = \sqrt{2}\,\pi\sigma^2 v(n^*)^2 \qquad (55)$$

where σ is the molecular diameter, v the average molecular velocity in centimeters per second, and n^* the number of molecules per cubic centimeter. Substituting $v = 0.921\sqrt{(3\,RT/M)}$ from Eq. (43) of the same chapter, and collecting constants, Z_{11} becomes

$$Z_{11} = 0.921\sqrt{2}\,\pi\sigma^2(n^*)^2\sqrt{\frac{3\,RT}{M}}$$

$$= 6.51 \times 10^4\,\sigma^2(n^*)^2\sqrt{\frac{T}{M}} \qquad (56)$$

However, when two *different* molecules are involved, $Z = Z_{12}$ and is given by

$$Z_{12} = n_1^* n_2^* \left(\frac{\sigma_1 + \sigma_2}{2}\right)^2 \sqrt{\frac{8\,\pi RT(M_2 + M_1)}{M_1 M_2}}$$

$$= 1.14 \times 10^4 (\sigma_1 + \sigma_2)^2 n_1^* n_2^* \sqrt{\frac{T(M_2 + M_1)}{M_1 M_2}} \qquad (57)$$

Here σ_1 and σ_2 are the molecular diameters of the respective molecules, M_1 and M_2 their molecular weights, and n_1^* and n_2^* the numbers of molecules of each type per cubic centimeter at temperature T. $Z = Z_{11}$ is used when all the reacting molecules are the same, as in 2 HI $\longrightarrow$ H$_2$ + I$_2$; on the other hand, $Z = Z_{12}$ is utilized when the two molecules are unlike, as in the reaction H$_2$ + I$_2$ $\longrightarrow$ 2 HI.

The fraction of activated molecules q can also be ascertained from kinetic theory. The molecules in a gas are in constant motion during which they enter into repeated collisions with each other. These collisions cause a redistribution of translational energy, resulting in a small fraction of the molecules acquiring an energy considerably above the average translational energy of all the molecules. From the Maxwell distribution law for molecular velocities it can be shown that, in any gas containing n^* molecules per cubic centimeter at a temperature T, the number of these, n', which will possess an energy E or higher is

$$n' = n^* e^{-E/RT} \tag{58}$$

If E is identified with E_a, the energy of activation, the *fraction* of molecules activated follows from Eq. (58) as

$$\frac{n'}{n^*} = e^{-E_a/RT} = \frac{\text{number of activated molecules}}{\text{total number of molecules}} \tag{59}$$

and this is also q. The expression for k becomes, therefore,

$$k = Z e^{-E_a/RT} \tag{60}$$

where Z is given by Eq. (56) for like molecules or by Eq. (57) for unlike molecules.

We are in a position now to test this theory with an actual calculation. For this purpose we shall take the reaction 2 HI $\longrightarrow$ H$_2$ + I$_2$ at 556°K, for which the observed value of k is 3.5×10^{-7} sec^{-1} (moles/liter)$^{-1}$, and $E_a = 44,000$ cal. For hydrogen iodide $\sigma = 3.5 \times 10^{-8}$ cm, $M = 127.9$, and since 1 mole of gas per liter contains 6.0×10^{23} molecules, the number per cubic centimeter is $n^* = 6.0 \times 10^{20}$. Inserting these values of σ, n^*, T, and M into Eq. (56), Z_{11} follows as

$$Z_{11} = 6.51 \times 10^4 (3.5 \times 10^{-8})^2 (6.0 \times 10^{20})^2 \sqrt{\frac{556}{127.9}}$$

$$= 6.0 \times 10^{31} \text{ molecules}$$

Again

$$q = e^{-E_a/RT}$$
$$= e^{-44,000/1.99 \times 556}$$
$$= 5.2 \times 10^{-18}$$

Therefore,

$$k = Zq = 6.0 \times 10^{31} \times 5.2 \times 10^{-18}$$
$$= 3.1 \times 10^{14} \text{ molecules reacting per cc per sec}$$

Multiplying this number by 1000 and dividing by Avogadro's number to obtain the number of moles reacting per liter, we find

$$k = \frac{3.1 \times 10^{14} \times 1000}{6.0 \times 10^{23}} = 5.2 \times 10^{-7} \text{ mole per liter per sec}$$

as against the experimentally observed value of $k = 3.5 \times 10^{-7}$. Keeping in mind the absolute nature of the calculation and the uncertainties involved, the agreement between theory and experiment must be considered very good.

Similar concordance between the collision theory and experiment has been obtained with several other bimolecular gas reactions, and also with a number of bimolecular reactions in solution by extending to the latter the ideas outlined above. However, there are many reactions, both in the gas phase and in solution, for which the theory yields results that are too high by factors ranging up to 10^9. For such reactions it is customary to write Eq. (60) in the form

$$k = PZe^{-E_a/RT} \tag{61}$$

where P, called the *probability factor*, is inserted to allow for the disparity between calculated and observed rate constants. P may have values ranging from unity for reactions obeying the collision theory to about 10^{-9}. It has been suggested that such "slow" reactions require not only collision between molecules which are active, but also a preferred critical orientation of the reacting molecules. If this is the case, not all collisions between activated molecules will result in reaction, and thus the calculated rate will be higher than the observed. On this basis it may be possible to account for probability factors as low as $10^{-3} - 10^{-4}$, but it is very questionable whether such an explanation can account for P's lower than these. In fact, it will be shown later that the collision theory omits from consideration an *entropy of activation* which cannot be disregarded.

THE COLLISION THEORY OF UNIMOLECULAR REACTIONS

Offhand it is difficult to see how the collision theory could possibly be employed to explain the mechanism of unimolecular reactions. In unimolecular processes only one molecule participates in the reaction, and consequently the question arises: How do molecules in unimolecular reactions attain their energy of activation? The answer to this question was first suggested by Lindemann[9] in 1922. Lindemann pointed out that the

[9] Lindemann, *Trans. Faraday Soc.*, **17**, 598 (1922).

behavior of unimolecular reactions can be explained on the basis of bimolecular collisions provided we postulate that a *time lag exists between
activation and reaction* during which activated molecules may either react
or be deactivated to ordinary molecules. Consequently, the rate of reaction will not be proportional to all the molecules activated, but only to
those which remain active. The Lindemann hypothesis can be formulated
by the scheme

$$A + A \underset{k_2}{\overset{k_1}{\rightleftarrows}} A + A^* \tag{62a}$$

$$A^* \overset{k_3}{\longrightarrow} \text{Products} \tag{62b}$$

where A represents inactive and A^* activated molecules, and the various
k's the rate constants for the respective rate processes. Further, the rate
of disappearance of A to form products will be proportional to the concentration of A^*, namely,

$$-\frac{d[A]}{dt} = k_3[A^*] \tag{63}$$

where the brackets indicate the concentrations of the various species.

Since $[A^*]$ is not known, it is necessary to obtain a relation for the
concentration of A^* in terms of A. To do this we resort to a postulate
known as the *stationary* or *steady state principle*. This principle states
that when a short-lived reaction intermediate exists at low concentration
in a system, the rate of formation of the intermediate can be considered
to be equal to its rate of disappearance. Applying this principle to A^*, we
see that it is formed in Eq. (62a) with a rate $k_1[A]^2$, and it disappears with
a rate $k_2[A][A^*]$ in Eq. (62a) and a rate $k_3[A^*]$ in Eq. (63). Consequently,

$$k_1[A]^2 = k_2[A][A^*] + k_3[A^*]$$

and

$$[A^*] = \frac{k_1[A]^2}{k_3 + k_2[A]} \tag{64}$$

Substituting Eq. (64) into Eq. (63) we obtain

$$-\frac{d[A]}{dt} = \frac{k_1 k_3[A]^2}{k_3 + k_2[A]} \tag{65}$$

Equation (65) predicts two limiting possibilities. When $k_2[A] \gg k_3$, the
equation reduces to

$$-\frac{d[A]}{dt} = \frac{k_1 k_3[A]^2}{k_2[A]}$$

$$= \frac{k_1 k_3}{k_2}[A]$$

$$= k[A] \tag{66}$$

where $k = k_1k_3/k_2$. This case corresponds to a situation in the reaction where the concentration of A is high enough to produce appreciable deactivation of A^* by collision with inactive molecules. Under such conditions, Eq. (66) predicts a first order reaction. On the other hand, when $k_3 \gg k_2[A]$, Eq. (65) gives

$$-\frac{d[A]}{dt} = \frac{k_1k_3}{k_3}[A]^2$$
$$= k_1[A]^2 \tag{67}$$

i.e., the reaction should be second order. This situation should obtain at low concentrations of A where the rate of activation becomes so slow as to be rate-controlling.

Such changes from first order kinetics in gaseous reactions at higher pressures to second order at lower pressures actually have been observed in many gaseous reactions. Depending on the reaction and the temperature, the transition from first to second order generally starts at pressures between 10 and 200 mm Hg, and it is usually not completed until the pressure gets fairly low. To show what happens, it is convenient to define a parameter k' by the relation

$$k' = -\frac{1}{[A]}\frac{d[A]}{dt} \tag{68}$$

In terms of k' Eq. (65) becomes

$$k' = \frac{k_1k_3[A]}{k_3 + k_2[A]} \tag{69}$$

and hence k' should be a function of $[A]$. For low pressures this equation should reduce to $k' = k_1[A]$, which would correspond to a second order reaction, while for higher pressures we should have $k' = k_1k_3/k_2$, a constant. That such is actually the case may be seen from Figure 13–6, where a plot of k' vs. P_i, the initial pressure, is given for the decomposition of azomethane[10] at 330°C, namely,

$$(CH_3)_2N_2 \longrightarrow C_2H_6 + N_2$$

As expected, k' is independent of P_i at the higher pressures where the reaction is first order, and linearly dependent on P_i at the low pressures where the reaction is second order. In between the rate is a combination of first and second order reactions, and is given by Eq. (69).

When Eq. (60) is used to estimate for unimolecular reactions the number of molecules becoming activated due to bimolecular collisions,

[10] H. C. Ramsperger, *J. Am. Chem. Soc.*, **49**, 912 (1927).

it is found that the number of molecules calculated is considerably *smaller* than the number actually entering reaction. To allow for this fact, the suggestion has been advanced that in unimolecular reactions not only are translational energies involved in activation but also vibrational. With the latter included, the fraction of the molecules becoming activated is much larger, and the number of molecules activated can be reconciled with the number observed to react in many unimolecular reactions both in the gas phase and in solution.

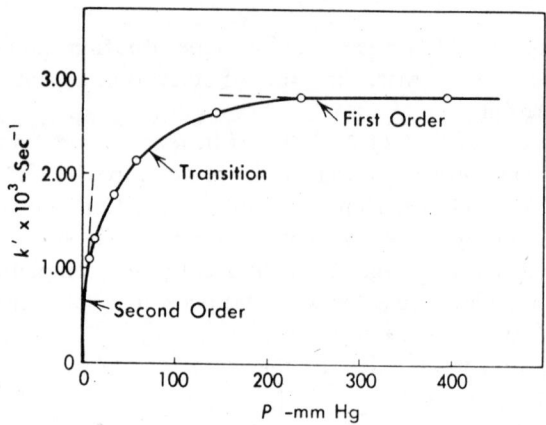

Figure 13-6.   Plot of k' vs. P_i for decomposition of azomethane at 330°C.

According to Kassel,[11] the energy of activation of a unimolecular reaction may be defined as the minimum energy per mole, translational, vibrational, and rotational, which a molecule must possess in order to react. Furthermore, an appreciable part of this energy must be localized in the weakest bond in the molecule. When such a bond receives the requisite energy it is loosened or broken, and decomposition results.

KINETIC THEORY OF TERMOLECULAR REACTIONS

Kinetic theory considerations indicate that simultaneous three-body collisions are not very probable, and hence the mechanism of third order reactions cannot be explained on this basis. Trautz actually showed that the third order reactions observed in gases can be accounted for in terms of bimolecular reactions. Thus, for the interaction between NO and Cl_2, namely,

$$2 \, NO + Cl_2 \longrightarrow 2 \, NOCl \tag{a}$$

[11] Kassel, *J. Phys. Chem.*, **32**, 225 (1928).

he postulated the sequence

$$NO + Cl_2 \underset{}{\rightleftharpoons} NOCl_2 \qquad\qquad \textbf{(b)}$$

$$NOCl_2 + NO \xrightarrow{k'} 2\,NOCl \qquad\qquad \textbf{(c)}$$

the sum of which is reaction (a). Step (b) is taken to be an equilibrium, for which the constant K is

$$K = \frac{[NOCl_2]}{[NO][Cl_2]}$$

and hence

$$[NOCl_2] = K[NO][Cl_2] \qquad\qquad \textbf{(70)}$$

Further, the rate of formation of product is given by step (c), namely,

$$\frac{d[NOCl]}{dt} = k'[NO][NOCl_2] \qquad\qquad \textbf{(71)}$$

Substituting Eq. (70) into Eq. (71) we get

$$\frac{d[NOCl]}{dt} = k'K[NO]^2[Cl_2] \qquad\qquad \textbf{(72)}$$

which is a third order reaction with $k_3 = k'K$. In a similar manner can be explained the other third order gas reactions observed with NO.

CHAIN REACTIONS

In 1906 Bodenstein and Lind, in studying the kinetics of combination of hydrogen and bromine to form hydrogen bromide, found that the rate of the process could not be represented by any equation deduced on a simple basis, but had to be expressed between 200° and 300°C by the relation

$$\frac{d[HBr]}{dt} = \frac{k[H_2][Br_2]^{1/2}}{1 + k'\dfrac{[HBr]}{[Br_2]}} \qquad\qquad \textbf{(73)}$$

A satisfactory explanation of Eq. (73) was finally proposed by Christiansen, Herzfeld, and Polanyi in 1919–1920. These men postulated that the first stage in the reaction is a dissociation of bromine molecules to bromine atoms

$$(1) \qquad Br_2 \xrightarrow{k_1} 2\,Br$$

followed next by the reactions

$$(2) \quad Br + H_2 \xrightarrow{k_2} HBr + H$$

$$(3) \quad H + Br_2 \xrightarrow{k_3} HBr + Br$$

$$(4) \quad H + HBr \xrightarrow{k_4} H_2 + Br$$

$$(5) \quad Br + Br \xrightarrow{k_5} Br_2$$

According to this mechanism, HBr is formed in steps (2) and (3), and disappears in step (4). Consequently, the rate of formation of HBr is given by

$$\frac{d[\text{HBr}]}{dt} = k_2[\text{Br}][\text{H}_2] + k_3[\text{H}][\text{Br}_2] - k_4[\text{H}][\text{HBr}] \qquad \text{(a)}$$

Since the atoms H and Br are short-lived intermediates, we can apply to them the steady state principle. We thus obtain

$$\frac{d[\text{H}]}{dt} = k_2[\text{Br}][\text{H}_2] - k_3[\text{H}][\text{Br}_2] - k_4[\text{H}][\text{HBr}] = 0 \qquad \text{(b)}$$

$$\frac{d[\text{Br}]}{dt} = k_1[\text{Br}_2] + k_3[\text{H}][\text{Br}_2] + k_4[\text{H}][\text{HBr}] - k_2[\text{Br}][\text{H}_2] - k_5[\text{Br}]^2 = 0 \qquad \text{(c)}$$

Simultaneous solution of Eqs. (b) and (c) yields

$$[\text{Br}] = \left[\frac{k_1[\text{Br}_2]}{k_5}\right]^{1/2} \qquad \text{(d)}$$

and

$$[\text{H}] = \frac{k_2(k_1/k_5)^{1/2}[\text{H}_2][\text{Br}_2]^{1/2}}{k_3[\text{Br}_2] + k_4[\text{HBr}]} \qquad \text{(e)}$$

which, on substitution into equation (a), finally gives

$$\frac{d[\text{HBr}]}{dt} = \frac{2k_2(k_1/k_5)^{1/2}[\text{H}_2][\text{Br}_2]^{1/2}}{1 + \left(\dfrac{k_4}{k_3}\right)\dfrac{[\text{HBr}]}{[\text{Br}_2]}} \qquad \text{(74)}$$

Equations (73) and (74) are identical in form. They show that $k = 2k_2(k_1/k_5)^{1/2}$ and $k' = k_4/k_3$; i.e., k and k' are composites of the various constants for the individual reaction steps.

A reaction like the above, proceeding in a series of successive stages initiated by a suitable primary process, is called a *chain reaction*. In the hydrogen-bromine and also in the hydrogen-chlorine reaction, which is similar in behavior, the initial step is the appearance of bromine or chlorine atoms. These atoms may result from thermal or photochemical dissociation of the molecules, or they may be produced by the introduction of metallic vapors, such as sodium, which yield atoms through a reaction of the type

$$\text{Na} + \text{Cl}_2 \longrightarrow \text{NaCl} + \text{Cl}$$

In other reactions the chain may be initiated by the appearance of *free radicals*, such as CH_3, C_2H_5, or CH_3CO, which then react with molecules but are eventually regenerated to propagate the chain. Finally, a chain may start at the wall of the reacting vessel. This appears to be the case in the reactions $H_2 + O_2$, $CS_2 + O_2$, and $CO + O_2$, which have been shown to be chain processes.

An example of a reaction proceeding by a free radical mechanism is the gaseous decomposition of ethane, namely,

$$C_2H_6 \longrightarrow C_2H_4 + H_2$$

This reaction is found experimentally to be first order, i.e.,

$$-\frac{d[C_2H_6]}{dt} = k[C_2H_6] \tag{75}$$

Since free radicals have been detected in this reaction, the following mechanism has been proposed to account for the observed results:

$$
\begin{aligned}
(1) & \quad C_2H_6 \xrightarrow{k_1} 2\ CH_3 \\
(2) & \quad CH_3 + C_2H_6 \xrightarrow{k_2} CH_4 + C_2H_5 \\
(3) & \quad C_2H_5 \xrightarrow{k_3} C_2H_4 + H \\
(4) & \quad H + C_2H_6 \xrightarrow{k_4} H_2 + C_2H_5 \\
(5) & \quad H + C_2H_5 \xrightarrow{k_5} C_2H_6
\end{aligned}
$$

This reaction scheme, with the aid of the usual steady state approximations, leads to Eq. (75) with

$$k = \left(\frac{k_1 k_3 k_4}{2\ k_5}\right)^{1/2} \tag{76}$$

Inspection of steps (1) to (5) for the hydrogen-bromine reaction will reveal that not all the stages of a chain reaction operate to continue a chain once started, but that some of them lead to a termination or "breaking" of chains. Thus, reactions (1) to (4) favor the propagation of the chain by formation of hydrogen or bromine atoms, while reaction (5) results in removal of bromine atoms, and hence in chain stoppage. The same would be true of combination of two hydrogen atoms to form a hydrogen molecule. Another important factor in terminating chains is collision of chain propagators with the walls of the reacting vessel. These collisions may result in either deactivation or reaction of the active agent with the walls. The consequence in either case is a broken chain.

Chain reactions are encountered in the oxidations of various gaseous hydrocarbons, phosphine, and methyl alcohol; in the decomposition of ozone; and in photochemical reactions. Again, indications are that some reactions in solution, such as oxidation of sodium sulfite by oxygen and the decomposition of certain acids by sulfuric acid, also proceed by a chain mechanism.

THE THEORY OF ABSOLUTE REACTION RATES

The theory of absolute reaction rates,[12] frequently also called the transition-state theory, is based on statistical mechanics and represents an alternate approach to reaction kinetics. This theory postulates that molecules before undergoing reaction must form an activated complex in equilibrium with the reactants, and that the rate of any reaction is given by the rate of decomposition of the complex to form the reaction products. For a reaction between a molecule of A and one of B, the postulated steps can be represented by the scheme

$$A + B \rightleftharpoons \underset{\substack{\text{Activated} \\ \text{Complex}}}{[A \cdot B]^*} \overset{k}{\longrightarrow} \underset{}{\text{Products}} \qquad (77)$$

$$\underset{\text{Reactants}}{A + B}$$

The activated complex is assumed to be endowed with certain properties of an ordinary molecule and to possess some, although temporary, stability. On the basis of these ideas Eyring was able to show that the rate constant k of *any* reaction, no matter what the molecularity or order may be, is given by the expression

$$k = \frac{RT}{Nh} K^* \qquad (78)$$

In this equation R is the gas constant in ergs mole^{-1} degree^{-1}; N, Avogadro's number; h, Planck's constant, equal to 6.625×10^{-27} erg-sec; T, the absolute temperature; and K^*, the equilibrium constant for the formation of the activated complex from the reactants. We may resort to thermodynamics and write for K^*

$$\ln K^* = -\frac{\Delta F^*}{RT}$$
$$= -\frac{(\Delta H^* - T\Delta S^*)}{RT} \qquad (79)$$

where ΔF^*, ΔH^*, and ΔS^* are, respectively, the free energy, enthalpy, and entropy of activation. Introducing Eq. (79) into Eq. (78), we obtain for k

$$k = \frac{RT}{Nh} e^{\Delta S^*/R} e^{-\Delta H^*/RT} \qquad (80a)$$

or

$$\ln k = \ln \left(\frac{RT}{Nh}\right) + \frac{\Delta S^*}{R} - \frac{\Delta H^*}{RT} \qquad (80b)$$

[12] Eyring, *J. Chem. Physics*, **3**, 107 (1935); Wynne-Jones and Eyring, *J. Chem. Physics*, **3**, 492 (1935).

Consequently, when k and ΔH^* of a reaction are known at a given temperature, ΔS^* may be found.

Equation (80) is the fundamental relation of the transition-state theory. The ΔH^* required for use of this equation is obtained from the experimentally observed activation energy E_a as follows: For reactions in solution, for gaseous reactions where the rate constants are expressed in pressure units, and for first order gas reactions with k_1 in concentration units, ΔH^* is related to E_a by the expression

$$\Delta H^* = E_a - RT \tag{81}$$

However, for second order gas reactions with k in concentration units

$$\Delta H^* = E_a - 2\,RT \tag{82}$$

while for third order reactions with k in such units

$$\Delta H^* = E_a - 3\,RT \tag{83}$$

Further, for use in Eqs. (80a) or (80b) the time in the k's must be expressed in seconds. Under such conditions the ΔS^* obtained will be for a standard state corresponding to the concentration units employed in expressing k. Thus, if k is expressed in moles per liter the standard state will be 1 mole per liter; for gases in atmospheres, each gas at 1 atm pressure; for number of molecules per cc, 1 molecule per cc, etc.

Comparison of Eq. (80a) with Eq. (61) of the collision theory shows that essentially

$$PZ = \frac{RT}{Nh}\,e^{\Delta S^*/R} \tag{84}$$

and consequently the difficulty occasioned in the latter theory by the appearance of the probability factor P can be reconciled by the introduction of the entropy of activation. Although at present it is not possible to estimate theoretically ΔS^* for any but the simplest reactions, the theory of absolute reaction rates does represent a step forward in elucidating chemical kinetics. Further, its use has also proved fruitful in a number of other directions.[13]

PRIMARY SALT EFFECT IN IONIC REACTIONS

The rate constants of reactions in solution involving either nonelectrolytes or nonelectrolytes and ions are essentially unaffected by the presence of electrolytes. On the other hand, the velocity constants of reactions

[13] See Glasstone, Laidler, and Eyring, *The Theory of Rate Processes*, McGraw-Hill Book Company, Inc., New York, 1941.

between ions are sensitive to variation of the ionic strength of the solution and change with the latter in a manner dependent on the charges of the reacting ions.

To explain the existence of this *primary salt effect*, Brönsted[14] in 1922 suggested a theory of ionic reactions which is in essence a special case of the theory of absolute reaction rates. We shall follow here a treatment due to Bjerrum because it is simpler than Brönsted's. Bjerrum postulated that reacting ions go to form an activated complex in equilibrium with the reactants and that the rate of reaction is proportional to the *concentration* of the complex. Thus, any ionic reaction between A^{z_A} and B^{z_B} can be represented by the scheme

$$A^{z_A} + B^{z_B} \underset{\text{reactants}}{\rightleftharpoons} \underset{\text{activated complex}}{[(A \cdot B)^{(z_A+z_B)}]*} \xrightarrow{k'} \text{Products} \qquad (85)$$

where z_A and z_B are the charges of the two ions, while $(z_A + z_B)$ is the charge on the activated complex. If we let C_X be the concentration of the complex at any instant, then the rate at any time t is given by

$$\frac{dx}{dt} = k'C_X \qquad (86)$$

But, since the complex is in equilibrium with the reactants, the thermodynamic equilibrium constant K must be

$$K = \frac{a_X}{a_A a_B}$$

$$= \frac{C_X f_X}{(C_A f_A)(C_B f_B)} \qquad (87)$$

The a's, C's, and f's represent, respectively, the activities, concentrations, and activity coefficients of the respective species. From Eq. (87) C_X follows as

$$C_X = (KC_A C_B)\frac{f_A f_B}{f_X}$$

which on substitution into Eq. (86) yields for the rate

$$\frac{dx}{dt} = k'KC_A C_B \cdot \frac{f_A f_B}{f_X}$$

$$= \left(k_0 \frac{f_A f_B}{f_X}\right) C_A C_B \qquad (88)$$

where $k_0 = k'K$.

Equation (88) indicates that the experimentally evaluated constant k, as obtained from the ordinary rate equation for the process depicted in

---

[14] See Brönsted, *Chem. Rev.*, **5**, 231 (1928), for a very exhaustive treatment of this subject.

Eq. (85), namely,

$$\frac{dx}{dt} = kC_A C_B$$

is not the true reaction constant k_0 but involves also the activity coefficients of the reactants and the complex. In fact,

$$k = k_0 \frac{f_A f_B}{f_X} \tag{89}$$

Since the activity coefficients of the charged reactants and complex depend on the ionic strength of the solution, the velocity constant k will also be a function of the ionic strength. By applying the Debye-Hückel theory of activity coefficients, it can be shown that for *dilute* solutions k should vary with ionic strength μ according to the equation

$$\log_{10} \frac{k}{k_0} = 2\, A z_A z_B \sqrt{\mu} \tag{90}$$

where A is the Debye-Hückel constant.[15] According to Eq. (90), a plot of $\log_{10} k$ or of $\log_{10} k/k_0$ against $\sqrt{\mu}$ should be linear with a slope equal

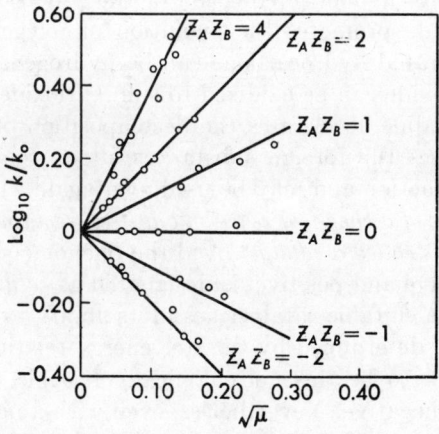

Figure 13-7. Primary salt effect in ionic reactions.

to $(2\, A z_A z_B)$, i.e., *dependent only on the product $z_A z_B$ of the charges of the reacting ions.* That this requirement is in line with observation is indicated by Figure 13-7, where a plot of $\log_{10} k/k_0$ vs. $\sqrt{\mu}$ is given for reactions of various $z_A z_B$ values. The solid lines are those predicted by Eq. (90), while the points are experimental results.

It should be noticed that when $z_A z_B = 0$, as for a reaction between a

[15] See Table 11-11, Chapter 11.

nonelectrolyte and an ion, the primary salt effect is essentially zero. This will be true for a reaction of the type

$$CH_2ICOOH + CNS^- \longrightarrow CH_2(CNS)COOH + I^-$$

Again, when $z_A z_B$ is positive, as for the reaction

$$CH_2BrCOO^- + S_2O_3^{--} \longrightarrow CH_2(S_2O_3)COO^- + Br^- \quad (z_A z_B = +2)$$

the salt effect will be positive, and k will increase with ionic strength. Lastly, when $z_A z_B$ is negative, as for

$$[Co(NH_3)_5Br]^{++} + OH^- \longrightarrow [Co(NH_3)_5OH]^{++} + Br^- \quad (z_A z_B = -2)$$

the salt effect is negative, and k will fall with increasing ionic strength.

CATALYSIS

It is a well-known fact that the velocity of many reactions can be changed by introducing certain substances other than the reactants appearing in the stoichiometric equation of the process. Thus, a pinch of manganese dioxide promotes the evolution of oxygen from potassium chlorate, unsaturated hydrocarbons can be hydrogenated in presence of nickel, sulfur dioxide can be oxidized to sulfur trioxide in the presence of platinum, and iodine accelerates the decomposition of nitrous oxide. In all these processes the foreign substances added remain unchanged at the end of the reaction and may be used over again. *Any substance which influences the rate of a chemical reaction but itself remains unchanged chemically at the end is called a catalyst.* And the phenomenon of rate acceleration, both negative and positive, is designated as *catalysis*.

The power of a suitable catalyst lies in its ability to change the rate at which a reaction determined by the free energy relations can take place. For any reaction to be thermodynamically feasible the change in free energy must be negative. Nevertheless, even with a negative ΔF the rate of transformation may be so slow as to make the reactants appear to be inert. In such cases the purpose of a catalyst is to speed the reaction and to permit a more rapid approach to equilibrium. From a thermodynamic standpoint, therefore, catalysis introduces no complications into the energy relations of the system. Free energy, heat of reaction, and entropy calculations for a reaction remain the same whether the process takes place *per se* or is accelerated by a catalyst.

Since a catalyst cannot change the ΔF^0 of a reaction, neither can it modify the equilibrium constant. Consequently, if the forward rate is affected so must also be the reverse rate. In other words, a catalyst must accelerate equally both the forward and reverse reactions. This is found to be the case. Occasionally a catalyst may appear to shift the point of equilibrium, but in such instances study will reveal that either the catalyst participates actively as a reactant, or some other complications exist which have not been allowed for.

The activity of a catalyst increases generally with its concentration, although this is not invariably true. Further, the concentration of the catalyst appears in the rate equation. This suggests that the catalyst participates as a reactant but that it is regenerated on completion of the sequence of reaction steps. However, this function of a catalyst as modifier of the mechanism of a reaction is not its only effect. Use of a catalyst also leads usually to a decrease of activation energy; and this lowering of the energy barrier between reactants and products results in a higher rate of reaction.

Catalysis may be homogeneous or heterogeneous, depending on whether the catalyst forms a single phase with the reactants or constitutes a separate phase. Here we shall consider homogeneous catalysis in gases and in solution, while heterogeneous catalysis will be discussed in Chapter 20.

HOMOGENEOUS CATALYSIS IN GASES

A familiar example of catalysis of a gas reaction by another gas is the oxidation of sulfur dioxide to sulfur trioxide in presence of nitric oxide in the chamber sulfuric acid process. Nitric oxide catalyzes also the oxidation of carbon monoxide to carbon dioxide and the decomposition of nitrous oxide to nitrogen and oxygen. The latter reaction is catalyzed as well by iodine, chlorine, and bromine vapors. Iodine vapor is also effective in accelerating the decomposition of methyl ethyl, diethyl, and diisopropyl ethers. Without the catalyst these decompositions proceed unimolecularly with activation energies of 54,500, 53,000, and 61,000 cal for the ethers in the order mentioned. However, in presence of iodine vapor the decompositions take place by a bimolecular mechanism involving a molecule of the ether and of iodine, and with activation energies of 38,000, 34,300, and 28,500 cal respectively.

For more extended discussion we may choose the decomposition of acetaldehyde. This decomposition proceeds of itself as a second order reaction with an energy of activation of 45,500 cal. But when iodine vapors are introduced into the system, it is found that the rate is given by $dx/dt = kC_{\text{acetaldehyde}}C_{\text{I}_2}$, where k is about 10,000 times larger at 518°C than the constant for the reaction in absence of iodine. Further, the energy of activation is reduced to 32,500 cal. To account for these observations it has been suggested that, instead of decomposing directly to methane and carbon monoxide, the reaction in presence of iodine vapor takes place in the two steps:

$$CH_3CHO + I_2 \longrightarrow CH_3I + HI + CO$$
$$CH_3I + HI \longrightarrow CH_4 + I_2$$

The first of these is considered to be the slower and therefore rate determining. As this alternate mechanism involves a lower energy of activation, the reaction can occur at a faster rate.

HOMOGENEOUS CATALYSIS IN SOLUTION

Homogeneously catalyzed reactions in solution are very common, as may be judged from Table 13–9. Of particular interest are the reactions catalyzed by hydrogen or hydroxyl ions, and by acids and bases in the generalized sense. Reactions catalyzed by hydrogen ions only are said to be subject to *specific hydrogen ion catalysis*. For such reactions the rate is proportional to the concentration of the substrate, i.e., the reacting molecule or ion, and to the concentration of the catalyzing hydrogen ions. Again, in *specific hydroxyl ion catalysis* the rate is proportional to the concentrations of substrate and hydroxyl ions. The latter type of catalysis is illustrated by the data given in Table 13–6 for the hydroxyl ion catalyzed decomposition of diacetone alcohol in water.

In distinction to specific hydrogen ion and hydroxyl ion catalysis, we have *generalized acid* or *generalized base catalysis*. In generalized acid catalysis every species present in solution which is an acid in the Brönsted sense acts as a catalyst for the reaction. Thus, if a substrate S, subject to generalized acid catalysis, is allowed to react in a solution of acetic acid and sodium acetate in water, we have in this solution the acids H_2O, H_3O^+, and $CH_3COOH(HA)$, and hence in such a solution the rate of reaction of the substrate will be given by

$$\frac{dx}{dt} = k_{H_2O}C_{H_2O}C_S + k_{H_3O^+}C_{H_3O^+}C_S + k_{HA}C_{HA}C_S$$
$$= C_S(k_{H_2O}C_{H_2O} + k_{H_3O^+}C_{H_3O^+} + k_{HA}C_{HA}) \tag{91}$$

From a knowledge of the sundry concentrations, and by choosing suitable experimental conditions, the various k's, called *catalytic coefficients*, can be evaluated, and the over-all reaction rate can thus be ascertained. On the other hand, in generalized basic catalysis the substrate is catalyzed by bases in the generalized sense. If a substrate S, subject to this type of catalysis, is permitted to react in the acetate-acetic acid mixture mentioned above, the bases present are H_2O, OH^-, and acetate ions (A^-), and the rate equation becomes now

$$\frac{dx}{dt} = k_{H_2O}C_{H_2O}C_S + k_{OH^-}C_{OH^-}C_S + k_A\text{-}C_A\text{-}C_S$$
$$= C_S(k_{H_2O}C_{H_2O} + k_{OH^-}C_{OH^-} + k_A\text{-}C_A\text{-}) \tag{92}$$

By utilizing again suitable methods the cataly'ic coefficients of various bases may be found, and from these the total rate of the reaction.

TABLE 13–9. HOMOGENEOUSLY CATALYZED REACTIONS IN SOLUTION

Reaction	Catalyst	Solvent
Decomposition of H_2O_2	HBr, HCl, HI	H_2O
Oxidation of persulfates	Ag^+	H_2O
$Ce^{++++} + S_2O_3^{--}$	I^-	H_2O
Inversion of menthone	$C_2H_5O^-$	C_2H_5OH
Inversion of sucrose	H^+	H_2O
Hydrolysis of esters, amides, acetals	H^+	H_2O
Decomposition of diazoacetic acid	H^+	H_2O
Hydrolysis of sulfamic acid	H^+	H_2O
Hydrolysis of pyrophosphates	H^+	H_2O
Decomposition of nitroso-triacetone-amine	OH^-	H_2O
Decomposition of triacetone alcohol	OH^-	H_2O
Conversion of acetone to triacetone alcohol	OH^-	H_2O
Hydrolysis of ethyl orthocarbonate, ortho-acetate and orthopropionate	Generalized acid	H_2O
Oxidation of phosphorous and hypophosphorous acids by iodine	Generalized acid	H_2O
Rearrangement of N-bromacetanilide	Generalized acid	C_6H_6, C_6H_5Cl
Decomposition of nitramide	Generalized base	H_2O, isoamyl alcohol, m-cresol
Isomerization of nitromethane	Generalized base	H_2O
Mutarotation of glucose	Generalized acid-base	H_2O
Enolization of acetone	Generalized acid-base	H_2O
Acetylation of β-naphthol	Generalized acid-base	CH_3COOH

Finally, there are some reactions, such as the enolization of acetone or the mutarotation of glucose, which are subject to *generalized acid-base catalysis*. In such reactions the over-all rate is the sum of the products of the catalytic coefficient, the concentration of substrate, and the concentration of all acid or basic species present; i.e., the total rate equation is the sum of expressions of the type given in Eqs. (91) and (92).

In generalized acid, base, or acid-base catalysis the catalytic coefficients of various acids and bases vary greatly in magnitude. For generalized acid catalysis the coefficients increase as the acid strength of the catalyzing acid goes up, while in basic catalysis the same parallelism is found with the basic strength of the catalyzing base.

NEGATIVE CATALYSIS

The rates of some reactions are readily inhibited by the presence of small quantities of various substances. As examples of this *negative catalysis* may be mentioned the inhibiting action of bromine on the methane-chlorine reaction, and the action of a number of alcohols and aldehydes on the oxidation of sodium sulfite in solution. The readiness with which such substances inhibit the reaction when present in even small quantities suggests that we are dealing here with chain reactions, and that the action of the negative catalysts results from their ability to react with chain propagators to bring about termination of chains. This idea has actually been confirmed by Bäckström for inhibition of the sodium sulfite oxidation, and it is quite probable that a similar explanation may be valid in other cases as well.

REFERENCES

1. E. S. Amis, *Kinetics of Chemical Change in Solution*, The Macmillan Company, New York, 1949.
2. R. P. Bell, *Acid-Base Catalysis*, Oxford University Press, Oxford, 1941.
3. S. W. Benson, *The Foundations of Chemical Kinetics*, McGraw-Hill Book Company, Inc., New York, 1960.
4. Friess, Lewis, and Weissberger, *Investigation of Rates and Mechanisms of Reactions*, Interscience Publishers, Inc., New York, 1961.
5. Frost and Pearson, *Kinetics and Mechanism*, John Wiley & Sons, Inc., New York, 1961.
6. Glasstone, Laidler, and Eyring, *The Theory of Rate Processes*, McGraw-Hill Book Company, Inc., New York, 1941.
7. K. J. Laidler, *Chemical Kinetics*, McGraw-Hill Book Company, Inc., New York, 1950.
8. E. A. Moelwyn-Hughes, *Kinetics of Reactions in Solution*, Oxford University Press, New York, 1947.
9. E. W. R. Steacie, *Atomic and Free Radical Reactions*, Reinhold Publishing Corporation, New York, 1954.
10. C. Walling, *Free Radicals in Solution*, John Wiley & Sons, Inc., New York, 1957.

PROBLEMS

1. At 100°C the gaseous reaction $A \rightarrow 2B + C$ is observed to be first order. On starting with pure A it is found that at the end of 10.0 minutes the total pres-

sure of the system is 176.0 mm Hg, and after a long time 270.0 mm Hg. From these data find (a) the initial pressure of A, (b) the pressure of A at the end of 10.0 minutes, (c) the rate constant of the reaction, and (d) the half-life period of the reaction. $Ans.$ (d) 10.7 min.

2. An artificially produced radioactive isotope decomposes according to the first order law with a half-life period of 15 min. In what time will 80% of the sample be decomposed?

3. From the mean rate constant for the decomposition of azoisopropane at 270°C given in Table 13–2 calculate (a) the percentage of the original sample decomposed after 25 sec, and (b) the time required for the reaction to go 95% to completion.

4. Heppert and Mack [$J. Am. Chem. Soc.$, **51**, 2706 (1929)] gave the following data for the vapor phase decomposition of ethylene oxide into methane and carbon monoxide at 414.5°C:

Time (min)	0	5	7	9	12	18
Pressure (mm)	116.51	122.56	125.72	128.74	133.23	141.37

Show that the decomposition follows a first order reaction, and calculate the mean specific rate constant. $Ans.$ $k_1 = 0.0123$ min^{-1}.

5. At 25°C the half-life period for the decomposition of N_2O_5 is 5.7 hr and is independent of the initial pressure of N_2O_5. Calculate (a) the specific rate constant and (b) the time required for the reaction to go 90% to completion.

6. The catalyzed decomposition of H_2O_2 in aqueous solution is followed by removing equal volume samples at various time intervals and titrating them with $KMnO_4$ to determine the undecomposed H_2O_2. The results thus obtained are:

Time (min)	5	10	20	30	50
Ml of KMnO$_4$	37.1	29.8	19.6	12.3	5.0

Show graphically that the reaction is first order. Ascertain from the plot (a) the value of the rate constant, and (b) the ml of $KMnO_4$ required for the titration of the sample removed at $t = 0$. $Ans.$ (a) 0.0437 min^{-1}.

7. The rate of saponification of methyl acetate at 25°C was studied by making up a solution 0.01 molar in both base and ester, and titrating the mixture at various times with standard acid. The following data were thus obtained:

Time (min)	3	5	7	10	15	21	25
Concentration of base found	0.00740	0.00634	0.00550	0.00464	0.00363	0.00288	0.00254

Show by a graphical method that the reaction is second order, and determine the specific rate constant. $Ans.$ 11.7 (mole/liter)$^{-1}$ min^{-1}.

8. From the data of the preceding problem, calculate the time required for the reaction to proceed 95% to completion when the initial concentrations of base and ester are both 0.004 mole/liter. What will be the half-life period in this case?

9. The following reaction [Maron and LaMer, *J. Am. Chem. Soc.*, **60,** 2588 (1938)]

$$CH_3CH_2NO_2 + OH^- \longrightarrow H_2O + CH_3CH{=}NO_2^-$$

is second order, with the rate proportional to the concentrations of nitroethane and OH^- ions. The specific rate constant at 0°C is 39.1 (moles/liter)$^{-1}$ min^{-1}. After what time will the base be 99% neutralized in a solution containing 0.005 mole/liter of nitroethane and 0.003 mole/liter of NaOH at 0°C?

10. The following table gives kinetic data [Slator, *J. Chem. Soc.*, **85,** 1286 (1904)] for the reaction between $Na_2S_2O_3$ and CH_3I at 25°C, the concentrations being expressed in arbitrary units:

Time (min)	0	4.75	10	20	35	55	∞
$Na_2S_2O_3$	35.35	30.5	27.0	23.2	20.3	18.6	17.1
CH_3I	18.25	13.4	9.9	6.1	3.2	1.5	0

Show that the reaction is second order, and calculate the mean specific rate constant.   *Ans.* 1.97×10^{-3}.

11. The conversion of hydroxyvaleric acid into valerolactone at 25°C in 0.025 N HCl solution was followed by titration with standard base. The data obtained were:

Time (min)	0	48	124	289	∞
Cc of base	19.04	17.60	15.80	13.37	10.71

What is the reaction order and mean specific rate constant?

12. At 25°C the specific rate constant for the hydrolysis of ethyl acetate by NaOH is 6.36 (moles/liter)$^{-1}$ min^{-1}. Starting with concentrations of base and ester of 0.02 mole/liter, what proportion of ester will be hydrolyzed in 10 min?

13. For the reduction $2\,FeCl_3 + SnCl_2 \longrightarrow 2\,FeCl_2 + SnCl_4$ in aqueous solution the following data were obtained at 25°C

t (min)	1	3	7	11	40
y	0.01434	0.02664	0.03612	0.04102	0.05058

where y is the amount of $FeCl_3$ reacted in moles per liter. The initial concentrations of $SnCl_2$ and $FeCl_3$ were respectively 0.03125 and 0.0625 mole/liter. Show that the reaction is third order, and calculate the average specific rate constant.   *Ans.* 86 (mole/liter)$^{-2}$ min^{-1}.

14. A substance decomposes according to a zero order reaction with a rate constant k. (a) What will be the expression for the period of half-life when the initial concentration is a? (b) How long will it take the reaction to go to completion?

15. Show by differentiation that Eq. (12) actually is the integral solution of Eq. (11).

16. Suppose that a single substance A reacts with an n order rate to give products, where n may be an integer or a fraction. Taking C_0 to be the initial concentration of A, and C the concentration at any time t, deduce the following:

   (a) The expression for the rate of reaction
   (b) The integrated form of the rate equation
   (c) Whether the rate equation obtained in (b) applies to all values of n.

   17. Letting t_f be the life period corresponding to any fraction f of A reacted in time t_f, deduce from the equation obtained in problem 16(b) the relation between f, C_0, and t_f. Show how this equation can be used to find both the reaction order and the value of the specific rate constant.

   18. Consider the reversible reaction $H_2(g) + I_2(g) \underset{k_2}{\overset{k_1}{\rightleftharpoons}} 2\,HI(g)$. Assuming that we start with equal initial concentrations of the two reacting gases, (a) set up the rate expression for the reaction, (b) integrate it, and (c) show how the two rate constants can be evaluated when the equilibrium constant K is known.

   19. The racemization of pinene is a first order reaction. In the gas phase the specific rate constant was found to be 2.2×10^{-5} min^{-1} at $457.6°K$ and 3.07×10^{-3} at $510.1°K$. From these data estimate the energy of activation, and the specific rate constant at $480°K$. *Ans.* $E_a = 43,600$ cal mole^{-1}.

   20. In the following table are listed specific rate constants k for the decomposition N_2O_5 at various temperatures:

$t°C$	k (sec^{-1})
0	7.87×10^{-7}
25	3.46×10^{-5}
35	1.35×10^{-4}
45	4.98×10^{-4}
55	1.50×10^{-3}
65	4.87×10^{-3}

Determine graphically the energy of activation, and find the specific rate constant at 50°C.

   21. At 378.5°C the half-life period for the first order thermal decomposition of ethylene oxide is 363 min, and the energy of activation of the reaction is 52,000 cal/mole. From these data estimate the time required for ethylene oxide to be 75% decomposed at 450°C. *Ans.* 13.5 min.

   22. The specific rate constants for the second order neutralization of 2-nitropropane by base in aqueous solution are given by the expression [Maron and LaMer, *J. Am. Chem. Soc.*, **60**, 2588 (1938)]:

$$\log_{10} k = \frac{-3163.0}{T} + 11.899$$

The time is in minutes, while the concentrations are in moles per liter. Calculate the energy of activation, and the half-life period at 10°C when the initial concentrations of base and acid are each 0.008 mole/liter.

23. The molecular diameters of O_2 and H_2 respectively are 3.39×10^{-8} cm and 2.47×10^{-8} cm. When 1 g of O_2 and 0.1 g of H_2 are mixed in a 1-liter flask at $27°C$, what will be the number of collisions per cubic centimeter per second?
Ans. 2.76×10^{29}.

24. The energy of activation for a bimolecular gaseous decomposition is 20,000 cal/mole. Calculate the fraction of the molecules having sufficient energy to decompose at $27°C$ and at $227°C$.

25. The gas phase reaction $2\,A \longrightarrow B + C$ is bimolecular with an activation energy of 24,000 cal/mole. The molecular weight and diameter of A are, respectively, 60 and 3.5 Å. Calculate from kinetic theory the specific rate constant for the decomposition at $27°C$.   *Ans.* 3.8×10^{-7} (mole/liter)$^{-1}$ sec^{-1}.

26. Show that Eq. (76) follows from the mechanism postulated for the first order decomposition of ethane given on page 583.

27. The decomposition of gaseous ozone, $2\,O_3 \longrightarrow 3\,O_2$, exhibits a rate law given by

$$-\frac{d[O_3]}{dt} = k\,\frac{[O_3]^2}{[O_2]}$$

Show that this rate law follows from the mechanism

$$(1) \quad O_3 \underset{k_2}{\overset{k_1}{\rightleftharpoons}} O_2 \cdot O$$

$$(2) \quad O_3 + O \overset{k_3}{\longrightarrow} 2\,O_2$$

on the supposition that $k_2 \gg k_3$. What is the resulting expression for k?

28. Suppose that in the preceding problem step (1) is assumed to involve an equilibrium with a constant K. (a) What will be the resulting equation for the rate of reaction? (b) What will be the expression for the rate constant?

29. Suppose that the experimentally observed energy of activation for the decomposition of ozone in problem 27 is E_a. (a) How is E_a related to the energies of activation for the individual reaction steps? (b) What is the significance of $(E_{a_1} - E_{a_2})$?   *Ans.* (a) $E_a = E_{a_1} + E_{a_3} - E_{a_2}$.

30. What will be the relation of E_a for the ozone decomposition to the energies involved in the mechanism postulated in problem 28?

31. For the hydrolysis of sulfamic acid [Maron and Berens, *J. Am. Chem. Soc.*, **72**, 3571 (1950)] $k = 1.16 \times 10^{-3}$ (mole/liter)$^{-1}$ sec^{-1} at $90°C$, while $E_a = 30,500$ cal mole^{-1}. From these data find (a) ΔF^*, (b) ΔH^*, and (c) ΔS^* of the reaction.
Ans. (a) 26,400 cal mole^{-1}; (b) 29,800 cal mole^{-1}; (c) 9.4 eu mole^{-1}.

32. From the results of the preceding problem find the equilibrium constant at $90°C$ for the activation process

$$NH_2SO_2O^- + H_3O^+ = [NH_2SO_2O^- \cdot H_3O^+]^*$$

Using this constant, calculate the concentration of the activated complex when each ion is present at a concentration of 0.001 mole per liter. Assume concentrations to be identical with the activities of the ions.

33. According to the Brönsted theory, will the specific rate constants of the following reactions remain unchanged, increase, or diminish as the ionic strength is increased:

(a) $NH_4^+ + CNO^- \longrightarrow CO(NH_2)_2$

(b) The saponification of an ester.

(c) $S_2O_8^{--} + I^- \longrightarrow$ products.

34. For a certain reaction catalyzed by hydrogen ions, the following specific rate constants were observed at 25°C in solutions containing formic acid (HA) and sodium formate (NaA):

HA (mole/liter)	NaA (mole/liter)	k (min^{-1})
0.250	0	2.55
0.250	0.025	0.682
0.250	0.100	0.172
0.250	0.250	0.0679

At this temperature the ionization constant of formic acid is 1.77×10^{-4}. Find the catalytic coefficient of the H$^+$ ions.

35. For the reaction

$$NH_2SO_2OH + H_2O = NH_4HSO_4$$

in aqueous solution at 80.35°C, the following constants were observed at the indicated ionic strengths:

$\mu \times 10^3$	k (mole/liter)$^{-1}$ hr^{-1}
5.06	1.07
11.21	1.02
15.85	0.976
22.94	0.886

From these data ascertain (a) the valencies of the reacting ions, and (b) the rate constant corrected for primary salt effects. At this temperature the Debye-Hückel constant $A = 0.574$.

14
Atomic
Structure

THE FIRST modern attempt at an explanation of the reactivity of chemical elements is Dalton's atomic theory formulated in 1808. Dalton postulated that every element consists of indivisible particles called atoms, each of which has the same mass for a given element, and that these atoms react with each other to form compounds. However, in setting up the atom as the unit of elementary chemical combination, he made no effort to establish any relation between the atoms of various elements. This was first done in 1815 by Prout who, reasoning from the proximity of the atomic weights of elements to whole numbers on the basis of hydrogen as unity, suggested that all elements were composed of multiples of hydrogen atoms. But further atomic weight determinations, particularly of such elements as chlorine and copper, showed marked differences from whole numbers, so this principle fell into disrepute. Only when the existence of isotopes was established at a much later date, and the true significance of deviations from close proximity to whole numbers was appreciated, did Prout's hypothesis become re-established as an important principle in considerations of atomic constitution.

Another important principle which indicated the existence of a regular relationship among the elements was the periodic law discovered independently by Mendeléeff (1869) in Russia and Lothar Meyer (1870) in Germany. This law, that the chemical and physical properties of the elements are periodic functions of their atomic weights, pointed to regularities in the structure of atoms and to the fact that certain structures repeated themselves periodically to yield similarity of chemical and physical properties. Nevertheless, it was not until much later that the reason for this behavior was found, and it was observed that the regularity does not follow atomic weight but atomic number, i.e., the numerical sequence of an element in the periodic table.

Finally, the electrical researches of Nicholson and Carlisle, Davy, Berzelius, and Faraday showed that matter and electricity are intimately associated and that electricity itself is corpuscular in nature. These findings, followed by studies of electric discharges through rarefied gases and the discovery of radioactivity, eventually established that atoms are divisible and that on subdivision all atoms consist of the same structural units. Primarily these are the electron, the proton, and the neutron. A description of these units, their properties, and the manner in which they enter into the makeup of atoms, as far as is known, will be the concern of this chapter.

THE ELECTRON

Gases are as a rule poor conductors of electricity. However, when a tube filled with gas is evacuated to pressures of 0.01 mm or lower and an electric potential is applied across a pair of electrodes sealed into the tube, a discharge takes place between the electrodes during which a stream of *cathode rays* is emitted *from the cathode*. These rays travel in straight lines perpendicular to the cathode surface, they produce a temperature rise in any object they strike, they pass through thin films of metals interposed in their path but are stopped by thicker foils, and they can make any opaque object placed in their path cast a sharp shadow. When they strike the wall of the tube, a fluorescence is produced. The fact that these rays can cast a sharp shadow suggests that they consist of material particles and not electromagnetic radiation. Again, these particles can be deflected by electric and magnetic fields, indicating that they are electrically charged; and, further, the direction of deflection is always such as to indicate that these particles always bear a *negative charge*. Finally, no matter what the nature of the cathode or the gas in the tube may be, the particles are always the same.

These particles constituting the cathode rays have been named *electrons*. The fact that electrons are independent of the nature of the source from which they come suggests that they are constituents of all matter. To study more precisely the properties of electrons, Thomson utilized a combination of electric and magnetic fields placed across the path of these through the discharge tube. His apparatus, shown in Figure 14–1, consisted of a tube T, i1.to which were sealed the cathode C, the anode A, a slit system S for defining the electron beam, and a pair of electrodes D and D' for applying an electric field. The electromagnet for supplying the magnetic field (not shown) was placed outside the tube. A beam of electrons emanating from C passes through a hole in the anode A, is collimated as a fine pencil by the slit S, and in absence of any applied electric or magnetic fields strikes the phosphorescent tube wall at point P. However, when a magnetic field is applied across the electron path, the beam is deflected upward or downward from P according to the direction and strength of the field. If now an electric field of correct direction is superimposed over the magnetic field, it is possible to adjust the electric field

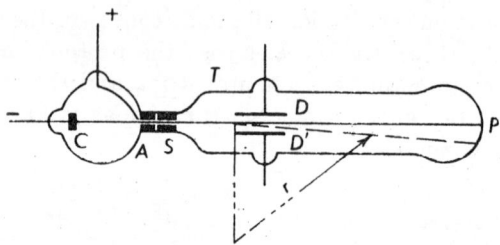

Figure 14-1.   J. J. Thomson's apparatus for determination of e/m of electron.

strength until the displaced beam is returned to its original position at P. From the relations of the forces involved, the ratio of the electron charge e to its mass m may then be evaluated, and Thomson found thus e/m to be 1.79×10^7 electromagnetic units per gram. More refined measurements have modified this value to

$$e/m = 1.75880 \times 10^7 \text{ emu g}^{-1}$$
$$= 1.75880 \times 10^8 \text{ abs. coulombs g}^{-1} \qquad (1)$$

In order to resolve this ratio, the charge on the electron must be ascertained independently. This determination was first carried out with a high degree of precision in 1913 by R. A. Millikan through his famous oil-drop experiments.

A schematic diagram of Millikan's apparatus is shown in Figure 14-2. It consisted of a chamber B, immersed in a thermostat (not shown) and filled with air whose pressure could be controlled by a vacuum pump and read on a manometer M. At the lower end of the chamber were installed two condenser plates C and C', across which could be applied a potential from the battery S. The space between the plates was illuminated by a light source located in front of a window in the chamber, W_1, and could be observed through a low-power telescope mounted in front of the apparatus and fitted with cross hairs a definite distance apart.

By means of an atomizer A, a spray of fine droplets of oil was introduced into B. In due course one of these droplets worked its way down toward C and passed through an opening O in this plate into the space between C and C'. As soon as this happened the opening O was closed, and the rate of fall of the droplet through the air under gravity was determined by measuring the time necessary for it to pass between the cross hairs of the telescope. At this stage a beam of X rays from the source R was passed into the space between the plates through the window W_2. This beam caused the air molecules to become ionized, forming ions which would

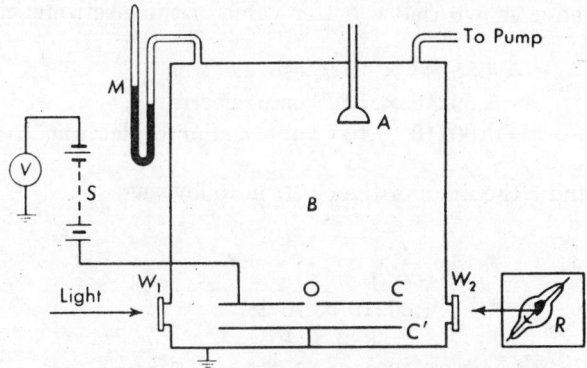

Figure 14–2.   Millikan's oil-drop apparatus.

from time to time be captured by the oil droplet and impart to it a charge. When the electrostatic field was now applied from S across the condenser plates, an oil droplet with a charge upon it could be made to fall faster or actually rise against gravity with a velocity dependent on the direction of the field, its magnitude, and the charge on the drop. The rate of rise or fall could again be observed by timing the passage between the cross hairs. In fact, once a particle was trapped between the plates, it could be made to fall or rise at will by either shutting off or turning on the electrostatic field, and hence the measurements on any particle could be repeated a number of times.

If we let v_1 be the rate of fall of the particle under gravity g, v_2 the rate of rise of the particle against gravity when an electrostatic field strength X is applied, and m' and e' the mass of and charge on the oil droplet, then in the oil drop experiments these quantities are connected by the relation

$$\frac{v_1}{v_2} = \frac{m'g}{Xe' - m'g} \tag{2}$$

The values of m' required in Eq. (2) in order to find e' were obtained by use of a modified form of Stokes's law. From his many experiments with all types of drops and various field strengths Millikan found that e' was not constant. However, he did find that there was a *common factor* for all the values of e' which made the various charges observed *whole number multiples of the common factor*. This is an excellent confirmation of the atomicity of electricity and indicates that the various oil droplets had captured and retained one, two, or more gaseous ions on their surfaces. The value of this least common factor deduced by Millikan was 4.774×10^{-10} electrostatic units, and this must be, therefore, the charge of the unit of electricity, or the electron.

Since this work a redetermination of the viscosity of air and other corrections have shown that a better value of the electronic charge, e, is

$$
\begin{aligned}
e &= 4.80298 \times 10^{-10} \text{ esu/electron} \\
&= 1.60210 \times 10^{-20} \text{ emu/electron} \\
&= 1.60210 \times 10^{-19} \text{ abs. coulomb/electron}
\end{aligned} \tag{3}
$$

From e/m and e the mass of the electron follows as

$$
\begin{aligned}
m &= \frac{e}{(e/m)} \\
&= \frac{1.60210 \times 10^{-19}}{1.75880 \times 10^8} \\
&= 9.1091 \times 10^{-28} \text{ g per electron}
\end{aligned} \tag{4}
$$

This mass of the electron may be compared with the mass of a hydrogen atom, which is, of course, the atomic weight of hydrogen divided by Avogadro's number, namely,

$$
m_\text{H} = \frac{1.00797}{6.0225 \times 10^{23}} = 1.673 \times 10^{-24} \text{ g}
$$

From m_H and m it is seen that the mass of the hydrogen atom is 1836 times greater than the mass of the electron.

The mass of the electron given in Eq. (4) is the *rest mass*, i.e., the mass when the electron is either at rest or moving with velocities that are low compared with that of light. However, when the electron is moving at very high speeds, then according to the theory of relativity the mass of the electron is increased in line with the equation

$$
m = \frac{m_0}{\sqrt{1 - (v/c)^2}} \tag{5}
$$

where m_0 is the mass of the electron at rest, m the mass when the electron is moving with a velocity v, and c the velocity of light. Equation (5) shows that m increases with v, until at $v = c$ the mass of the electron becomes infinite. This equation is applicable also to any other body of rest mass m_0 moving with extremely high velocities.

WAVE NATURE OF THE ELECTRON

Heretofore the electron has been treated as a particle. However, it was pointed out in Chapter 2 that electrons possess also wave properties. From this it must be concluded that, depending on the manner of observation, electrons have the faculty of behaving both as particles and waves. Thus in a discharge tube the electrons exhibit their attributes as particles, whereas in electron diffraction experiments they act as electromagnetic waves and yield diffraction patterns on reflection from crystalline surfaces.

This duality of behavior, exhibited also by light and even atoms, is an illustration of the *Heisenberg uncertainty principle* enunciated in 1927, which states that *it is impossible to define simultaneously the exact momentum and position of a body.* Precisely, Heisenberg showed that the product of the uncertainty in the position of a body Δx, and the uncertainty in the momentum Δp, and hence the velocity, is even in the perfect experiment related to Planck's constant h by the expression

$$(\Delta x)(\Delta p) \geq \frac{h}{4\pi} \tag{6}$$

Consequently, as soon as any attempt is made to define exactly the position of a body, i.e., make Δx very small, Δp becomes large. Similarly, any attempt to define Δp exactly leads to large uncertainties in x. Now, in studying the electron in discharge experiments we concentrate on the exact definition of its velocity and momentum, and arrive thus at the conclusion that the electron is a particle. On the other hand, when electron diffraction experiments are performed, the emphasis is on the position of the electron. Hence its momentum becomes an ambiguous quantity, and the electron behaves like a wave. From these considerations it may be deduced that the corpuscular nature of the electron becomes manifest when the particular experiment performed involves definition of its velocity, momentum, or energy, whereas the wave properties come to the fore when its position is being fixed.

THE PROTON

Since the atoms composing matter are generally electrically neutral, and since the mass of even the lightest atom, hydrogen, is very much greater than that of the electron, there must evidently be present in atoms positive electricity with which, in all probability, most of the mass is associated. The problem is, therefore, to find the unit of this positive charge and to ascertain its mass.

As a preliminary indication of the answer to this problem, let us consider the simplest and lightest of atoms, hydrogen. Under ordinary conditions, this atom is electrically neutral and has a mass of 1.67×10^{-24} g, which is 1836 times as great as that of the electron. When this atom is converted into an ion, it is found that its charge is exactly equal to that of the electron but is opposite in sign. Further, the mass of the positively charged hydrogen ion, or *proton*, is very nearly the same as that of the hydrogen atom. These facts suggest that the unit of positive electricity is the proton with a mass 1835 times that of the electron and that the hydrogen atom is composed of one proton and one electron.

It is not necessary to rely completely on this type of conjectural reasoning, for there are direct ways of producing positive rays and measuring their e/m ratio. In 1886 Goldstein discovered that there are present in a discharge tube not only cathode rays composed of electrons moving toward the anode, but also positive rays moving in the opposite direction. By perforating the cathode these *positive* or *canal rays* can be made to pass through the holes to the rear, and there be studied by the effect of electric and magnetic fields upon them. By such methods it was established that these rays consist of positively charged particles which, unlike the electrons, have associated with them practically all of the mass of the atoms from which they come. Further, these particles are not emitted by the anode, but originate between the electrodes from ionization of the gas atoms through electron bombardment. By appropriate means these rays can be generated also by emission of positive ions from anodes, in which instance the rays consist of ions of the metal used. The significant fact revealed by all such studies is that, although unit positive charge may be associated with ions of various masses, no particle found is of a mass lower than that of the proton. Of added importance is the observation that the masses of all ions heavier than that of hydrogen are within close limits whole number multiples of the mass of the proton. These facts indicate that the proton is the unit of positive electricity and that like electrons the protons enter into the constitution of all atoms.

POSITIVE RAY ANALYSIS AND ISOTOPES

A method of investigating positive rays is the Aston *mass-spectrograph*, with which masses of positively charged particles can be determined with an accuracy greater than one part in a million. The principle of the mass spectrograph can be understood with the aid of the schematic diagram in Figure 14–3. A beam of positive rays, narrowed by means of the slits S_1 and S_2, is passed between two charged plates P_1 and P_2. Under the influence of the electrostatic field between the plates the beam spreads out by downward deflection of the particles, the extent of deflection being determined by the charge and velocity of each particle. From the spread-out beam only those ions are selected which can pass through the narrow slit S_3, and these are sent into an electromagnetic field M directed to deflect the ions in a direction opposite to that of the electric field. From the relations obtaining for the apparatus it can be shown that, for any given electric and magnetic field strength, all particles of a given charge to mass ratio will be brought to a focus at a point such as F. Furthermore, the focus of all particles of various charge to mass ratios will lie on a straight line situated along GH. Consequently, a photographic plate placed along GH will record by a narrow exposed band the point of focus of all the ions of a fixed charge to mass ratio, while a series of these dark

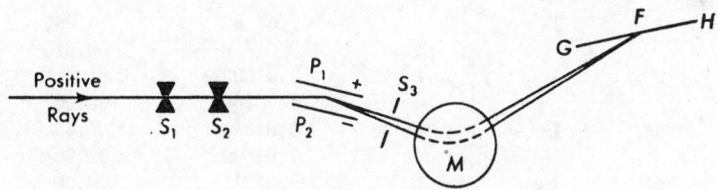

Figure 14–3. Schematic diagram of Aston's mass spectrograph.

bands will give the loci of all ions of various charge to mass ratios in the beam. The striking similarity of such an exposed plate to an ordinary spectrum plate accounts for the name given to this apparatus.

More recent designs of mass spectrographs dispense with the photographic plate by substituting for it detecting and recording devices which yield directly plots of current intensity vs. atomic mass. This is done by focusing all ions of a given charge to mass ratio onto a slit, and then allowing the positive current to pass into the detecting and recording instruments.

From a study of the mass spectra it is possible to deduce very precise values of the atomic weights. The mass spectrograph also reveals that many elements, which from ordinary atomic weight determinations were considered to consist of atoms of the same mass, are in reality mixtures of atoms of different masses although of the same atomic number. Such atoms of different mass but same atomic number are called *isotopes*. Thus chlorine, with a chemical atomic weight of 35.453, was shown to consist of two isotopes of atomic weights 35 and 37 mixed in such proportion as to yield this over-all atomic weight. Again, magnesium, atomic weight 24.312, was shown to consist of isotopes of relative masses 24, 25, and 26. In every instance the abundance of the isotopes in the mixture is such as to yield the chemically observed atomic weight.

In Table 14–1 are given the atomic weights of a number of the lighter elements as deduced from the mass spectrograph. Inspection of this table reveals that the masses of the various elements are very nearly whole numbers and multiples of the atomic weight of the hydrogen atom. This fact lends credence to the Prout hypothesis. Again, the existence of isotopes accounts for fractional atomic weights deduced by chemical methods and shows why no satisfactory regularity could be observed in these to substantiate Prout's suggestion.

TABLE 14–1.   ATOMIC MASSES DEDUCED FROM MASS SPECTROMETRY

Atomic No.	Atom	Rounded Mass	Atomic Weight (C^{12} basis)	Abundance (%)
1	H	1	1.00782	99.985
		2	2.01410	0.015
2	He	3	3.01603	10^{-5}
		4	4.00260	100.00
3	Li	6	6.01511	7.4
		7	7.01599	92.6
4	Be	9	9.01217	100.00
5	B	10	10.01295	18.83
		11	11.00942	81.17
6	C	12	12.00000	98.892
		13	13.00333	1.108
7	N	14	14.00307	99.64
		15	15.00009	0.36
8	O	16	15.99491	99.76
		17	16.99913	0.04
		18	17.99916	0.20

HYDROGEN AND DEUTERIUM

Among the methods which have been employed to separate or enrich isotopes are the mass spectrograph, diffusion, centrifuging, thermal diffusion, electrolysis, fractional distillation, and chemical exchange. It is not possible to describe here either these methods or the results obtained. Nevertheless, it is of interest to point out the differences found between ordinary hydrogen and its heavier isotope *deuterium*, symbol D. Deuterium was discovered spectroscopically in 1932 by Urey and his co-workers at Columbia University. In ordinary hydrogen gas deuterium is present as about one part in 6600 parts of light hydrogen. The heavier isotope is usually concentrated by electrolysis of aqueous alkali solutions, in which instance the lighter hydrogen atoms escape more readily than the heavier deuterium, and the solution becomes more concentrated in D_2O, *heavy water.* By repeating the electrolysis a number of times, pure D_2O can be prepared and studied.

In Table 14–2 are contrasted some physical properties of hydrogen and deuterium, while in Table 14–3 are compared the properties of ordinary and heavy water. These data show appreciable differences in the properties of H_2 and D_2, and of H_2O and D_2O. The differences in the two forms of

TABLE 14–2. COMPARISON OF PROPERTIES OF HYDROGEN ISOTOPES

Property	H_2	D_2
Molar volume of solid	26.15 cc	23.17 cc
Triple point	13.92°K	18.58°K
Heat of fusion	28 cal/mole	47 cal/mole
Boiling point	20.38°K	23.50°K
Heat of vaporization at triple point	217.7 cal/mole	303.1 cal/mole

TABLE 14–3. COMPARISON OF PROPERTIES H_2O AND D_2O

Property	H_2O	D_2O
Relative density, 25°/25°C	1.0000	1.1079
Melting point	0.00°C	3.82°C
Boiling point	100.00°C	101.42°C
Surface tension, 20°C	72.75 dynes/cm	67.8 dynes/cm
Temperature of maximum density	4.0°C	11.6°C
Dielectric constant, 25°C	78.54	78.25
Viscosity, 20°C	10.09 millipoises	12.6 millipoises
Heat of fusion	1436 cal/mole	1510 cal/mole
Heat of vaporization	10,480 cal/mole	10,740 cal/mole

water affect significantly the solubility, conductance, equilibrium relations, and the rates of various reactions conducted in these media as solvents. However, the differences between hydrogen and deuterium are much more pronounced than are those between isotopes of other elements, where the relative difference in mass is much less than it is here.

THE NEUTRON

In 1920 Rutherford suggested that there may also be present in atoms a particle of essentially the same mass as the proton but with no charge. The existence of such uncharged particles, called *neutrons*, was actually established in 1932 by Chadwick. These particles have been found to be independent of their source in their properties, to be unaffected by electric and magnetic fields, indicating thus absence of charge, and to have a mass of 1.00866 atomic weight units, which is almost identical with that of the proton, 1.00728.

RUTHERFORD'S THEORY OF THE ATOM

Although the preceding evidence points to protons, neutrons, and electrons as the constituents of all matter, it does not indicate how these are arranged to yield the various atoms. An early attempt to solve this problem was Rutherford's proposal of a nuclear atom. Rutherford's theory is based on observations made by Geiger, Marsden, and himself on the scattering of α particles by metallic foils. These particles are emission products of certain radioactive disintegrations. They have a mass approximately four times that of the proton and bear two positive charges. They are in fact helium nuclei, i.e., helium atoms from which two electrons have been removed. When Geiger and Marsden bombarded thin metallic foils with these α particles, they found that whereas most of them went through the foils unaffected, about one particle in 20,000 suffered a violent deflection through angles of 90° or greater. In seeking an explanation of these results it must be realized that electrons, because of their low mass, would not be able to deflect α particles. Again, to cause the sharp deflection of the particles through the large observed angles would require strong forces which can arise only from interaction of particles of the same charge as those used in the bombardment. Finally, since most of the α particles go through unaffected, these sources of deflection cannot be continuous throughout the foil, but must be localized in spaces that are relatively small compared to the space occupied by the foil as a whole. Basing his argument on these points, Rutherford suggested that atoms consist of miniature solar systems in which all of the positive charge is located in a nucleus at the center of the atom, while the electrons required for electrical balance revolve about the nucleus at some distance from it, much as the planets revolve about the sun. As in terms of this picture the atom consists mostly of empty space, α particles directed against a thin metallic foil should be able to pass through it without deflection as long as they do not approach too near to a positive nucleus. However, when an α particle is so directed as to come near the nucleus, the strong repulsive force developed between the nucleus and the positive α particle causes the latter to be deflected sharply.

From a mathematical analysis of the scattering Rutherford was able to deduce that the charge on the nucleus responsible for the observed deflections was equal approximately to one-half the atomic weight of the metal constituting the foil. Further, he was able to estimate that the dimensions of the nuclei were of the order of 10^{-12}–10^{-13} cm, dimensions which along with the volume of the electrons involved are only 10^{-12}–10^{-15} of the volumes actually occupied by the atoms. From these figures it follows that only $1/10^{12}$ to $1/10^{15}$ of the volume of an atom is occupied by protons, electrons, and neutrons, with the rest being nothing but empty space.

THE CHARGE ON THE NUCLEUS

Rutherford's earlier researches on α-particle scattering did not lead him to an exact definition of the nuclear charge. The first precise statement that the *nuclear charge of an atom is identical with its atomic number* was made by van der Broeck in 1913. Nevertheless, the first direct evidence on this point is due to Moseley's studies (1913, 1914) of the X radiations emitted by various elements.

X rays are emitted when a metallic target is bombarded with rapidly moving electrons. No X radiation from a given target is observed until the electrons acquire a certain threshold speed which is characteristic for each metal. Once the required speed of bombarding electrons is attained or exceeded, each metal emits X radiation which consists of series of lines possessing fixed frequencies characteristic of the target substance. The various sets of lines thus obtained can be divided into series designated in turn as the K, L, M, etc., series, while the lines within each series are referred to in sequence as the α, β, γ, etc., lines. Thus, the K_α line is the first line in the K series, the L_β is the second line in the second series, etc.

Moseley undertook the task of studying the X-ray spectra of elements lying between aluminum and gold in the periodic table to ascertain whether any regularities exist among them. As a result of this exhaustive investigation he found that the spectra observed followed the same sequence as do the elements in the periodic table. And, what is more important, he also found that if we take the frequencies of any particular line in all the elements, such as the K_α or K_β lines, then the frequencies are related to each other by the equation

$$\sqrt{\nu} = a(Z - k) \tag{7}$$

where ν is the frequency of any particular K_α or K_β, etc., line, Z is the *atomic number* of the element in the periodic table, and a and k are constants for any particular type of line. The regularity thus observed can be obtained only when the *atomic number* is used, but not with the atomic weight. This suggests that the fundamental quantity involved in X radiation is not the atomic weight but the atomic number. Furthermore, since X radiation arises from energy shifts taking place in extranuclear electrons, and since X radiation of the various elements exhibits regularity with atomic number, Moseley concluded that the change in atomic number from element to element represents the regular variation in the number of extranuclear electrons in the neutral atom; i.e., *the number of extranuclear electrons is identical with the atomic number of an element.* Moreover, since the number of extranuclear electrons in an atom must be equal to the charge of the nucleus for over-all electroneutrality, it must follow also that the atomic number gives the number of units of positive charge on the nucleus. The correctness of this conclusion has subsequently been verified by Chadwick's work on scattering of α particles, and by certain observations on radioactive disintegration.

DISTRIBUTION OF ELECTRONS, PROTONS, AND NEUTRONS IN AN ATOM

We are prepared now to consider the number of electrons, protons, and neutrons in an atom and their distribution. On the atomic weight scale the mass of the proton or neutron is essentially unity. Further, because of the relatively low mass of electrons compared to that of protons or neutrons, the mass of atoms will be due primarily to the latter two. Therefore, if we let A be the atomic weight of any atom, this will also be the sum of the protons and neutrons in it. Now, Moseley showed that the number of extranuclear electrons is Z, where Z is the atomic number, and that the nuclear charge is also Z. Since only the protons carry a charge, this must mean that there are present in the nucleus Z protons and $(A - Z)$ neutrons.

The following then is the picture of an atom. Each atom consists of a nucleus composed of Z protons and $(A - Z)$ neutrons. To balance the nuclear charge Z electrons are distributed outside the nucleus and revolve about it as a center. Thus, in the hydrogen atom with $A = 1$ and $Z = 1$ there must be only one proton in the nucleus and one electron revolving about it. Again, in beryllium with $A = 9$ and $Z = 4$ there must be a nucleus composed of four protons and five neutrons, and four external electrons. Finally, in an atom such as uranium, with $A = 238$ and $Z = 92$, there must be present 92 protons and 146 neutrons in the nucleus, and 92 satellite electrons.

Rutherford believed that the external electrons may occupy any and all positions outside the nucleus and possess thereby energies which can vary continuously. However, this concept of continuous variation in energy is contradicted by atomic spectra, which are not continuous but discontinuous. Again, since a revolving electron is a charged body in motion, then according to classical electrodynamics such a body should radiate energy continuously. As a result of this loss of energy by radiation the orbit of revolution of an electron should get smaller and smaller, until eventually the electron should fall into the nucleus and be retained there. Such behavior on the part of electrons has never been observed. To overcome these difficulties inherent in the Rutherford atom, Niels Bohr advanced in 1913 his now famous theory of atomic structure. Before this theory can be presented, however, two other subjects must be introduced first, namely, the quantum theory of radiation and the emission of line spectra by excited atoms.

THE QUANTUM THEORY OF RADIATION

When radiation strikes any surface of a body, part of the radiant energy is generally reflected, part is absorbed, and part is transmitted. The reason for the incomplete absorption is that ordinary bodies are as a rule imperfect absorbers of radiation. In contrast to these we have the *black body*, which by definition is the perfect absorber of energy and retains any radiant energy that strikes it. Although a blackened metallic surface or carbon black approximates fairly closely a black body, a hollow sphere, blackened on the inside and with a small opening, meets the defined condition much more satisfactorily. Any radiation that enters through the small opening is reflected repeatedly from the walls of the enclosure until all of the energy eventually becomes absorbed.

A black body is not only a perfect absorber of radiant energy, but also a perfect radiator. In fact, of all bodies heated to a given temperature, it is the black body which will radiate the maximum amount of energy possible for the given temperature. Furthermore, a black body is in thermal equilibrium with its surroundings and radiates in any given time per unit area the same amount of energy as it absorbs.

The total amount of energy E radiated by a black body per unit area and time is given by the *Stefan-Boltzmann fourth power law*, namely,

$$E = \sigma T^4 \tag{8}$$

where T is the absolute temperature, while σ is a universal constant equal to 5.6697×10^{-5} for energy in ergs, time in seconds, and area in square centimeters. This energy is not emitted with a single frequency, nor is it uniformly distributed along the spectrum. Lummer and Pringsheim showed that the energy emitted depends on the temperature and wave length in the manner indicated in Figure 14–4. For each temperature there is a wave length at which the energy radiated is a maximum. Again, the position of this maximum shifts with increase in temperature toward lower wave lengths, and the maximum itself is the more pronounced the higher the temperature.

Wien in 1896 and Lord Rayleigh in 1900 made attempts to account for this distribution of black body radiation on the basis of classical concepts of continuous emission of radiation. Although Wien's equation proved fairly satisfactory at low wave lengths and Rayleigh's at high ones, neither theory could account completely for the observed phenomena. This failure of classical theories of radiation led Max Planck in 1900 to discard these and to come forward with the bold hypothesis that *black*

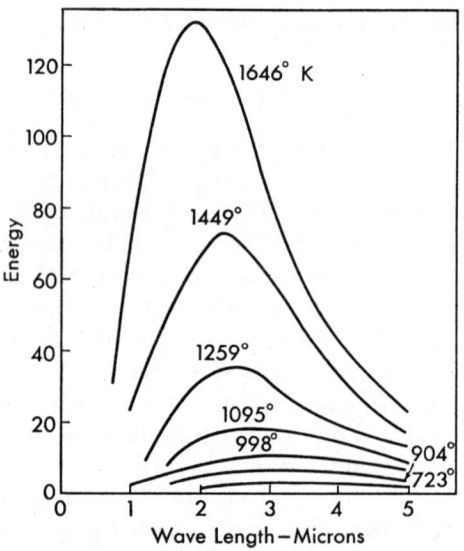

Figure 14-4.   Distribution of energy in black body radiation.

bodies radiate energy not continuously, but discontinuously in energy packets called quanta, given by the relation

$$E = h\nu \tag{9}$$

E is the quantum of energy radiated, ν is the frequency, and h, called *Planck's constant* is a universal constant equal to 6.6256×10^{-27} erg-sec. Equation (9) is the fundamental relation of the *quantum theory of radiation*. Planck considered the black body to consist of oscillators of molecular dimensions, each with a fundamental vibration frequency ν, and that each oscillator could emit energy either in the unit quantum $E_1 = h\nu$, or in *whole number multiples* n thereof, $E_n = nh\nu$. On this basis he was able to deduce for the energy E_λ radiated by a black body at any wave length λ the relation

$$E_\lambda = \frac{8\pi hc}{\lambda^5} \frac{1}{e^{hc/kT\lambda} - 1} \tag{10}$$

where c is the volocity of light, T the absolute temperature, and k the gas constant per molecule, i.e., R/N. This equation not only reproduced in excellent fashion the spectral distribution of energy, but reduced as well for low wave lengths to Wien's equation and for high wave lengths to the Rayleigh equation.

Planck's success with the quantization of black body radiation led Albert Einstein in 1905 to a generalization of the quantum theory. Einstein postulated that *all radiant energy must be absorbed or emitted by a body in quanta whose magnitude depends on the frequency according to Eq. (9), or multiples thereof.* Furthermore, Einstein argued that radiation is not only emitted or absorbed in quanta, but it is also propagated through space in these units, called *photons* of light. This theory ascribes to light a corpuscular character similar to that established for electricity and matter.

Einstein's theory of the corpuscular nature of all light has been amply substantiated by observations on electron emission produced by light, by the inverse photoelectric effect, i.e., the emission of radiation by electron bombardment, as in the generation of X rays, by the Compton effect arising from collisions between photons and electrons, and other phenomena. Details may be found in books on the quantum theory.

THE LINE SPECTRA OF ATOMS

Bodies on being heated to high temperatures emit radiation which can be passed through a spectrograph, resolved there into its component wave lengths, and recorded photographically. When such spectrograms are analyzed, it is found that solids generally give spectra that are continuous. On the other hand, gases and vapors yield a series of lines, called *line spectra*, or bands, called *band spectra*, which consist of many lines close together. The *band spectra* are radiations emitted by *molecules*, whereas the *line spectra* are due to *atoms*. At present we are interested only in the latter, and so we shall turn to a discussion of these.

Of all the elements the simplest line spectrum is that exhibited by atomic hydrogen. The spectrum of this element consists of a number of lines which can be classified into groups or series. Each series of lines is related by a formula, which for the *Balmer* series, appearing in the visible spectral range, takes the form

$$\frac{1}{\lambda} = \bar{\nu} = R_H \left(\frac{1}{2^2} - \frac{1}{n^2} \right) \tag{11}$$

Here λ is the wave length of the line, whose reciprocal, $\bar{\nu}$, is called the *wave number*, i.e., the number of waves per centimeter, R_H is a constant called the *Rydberg constant* and equal to 109,677.58 cm^{-1}, and n is a running number taking on for the various lines of this series values of 3, 4, 5, etc. Similarly, the lines in the various other series found in hydrogen can be expressed by formulas analogous to Eq. (11). In fact, the lines of all series can be represented by the general expression

$$\bar{\nu} = R_{\text{H}} \left(\frac{1}{n_1^2} - \frac{1}{n_2^2} \right) \qquad (12)$$

where the values of n_1 and n_2 for the various series are summarized in Table 14–4. The significant fact to be observed here is that every line in

TABLE 14–4. SPECTRAL SERIES OBSERVED
IN ATOMIC HYDROGEN

Series	n_1	n_2	Spectral Region
Lyman	1	2, 3, 4, . . .	Ultraviolet
Balmer	2	3, 4, 5, . . .	Visible
Paschen	3	4, 5, 6, . . .	Infrared
Brackett	4	5, 6, 7, . . .	Infrared
Pfund	5	6, 7, . . .	Infrared

any given series can be represented as a difference of two terms, one R_{H}/n_1^2, where n_1 has a fixed value for a given series, and a second R_{H}/n_2^2, where n_2 can take on a series of integral consecutive values beginning with $n_2 = n_1 + 1$. The significance of this important regularity will appear as soon as the Bohr theory is presented.

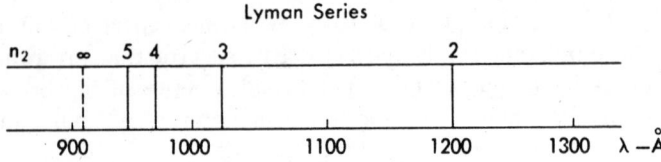

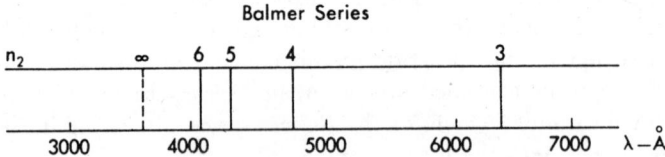

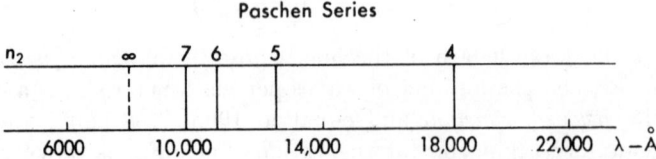

Figure 14–5. Series in spectra of atomic hydrogen.

In Figure 14–5 are shown the spectral lines given by Eq. (12) for three of the series and the values of n_2 which correspond to them. The dotted lines are for the wave lengths which result when $n_2 = \infty$, and they represent the wave length limits for each series. As may be seen, the spacing of the lines is large for the small values of n_2, and narrows rapidly as n_2 is increased.

The spectra of other elements are more complicated. The spectra of ionized gaseous atoms such as singly ionized helium, He^+, doubly ionized lithium, Li^{++}, or trebly ionized beryllium, Be^{+++}, bear a striking resemblance to that of hydrogen, and these are said to be hydrogenic in character. With heavier atoms, however, and with unionized atoms of the elements mentioned, the situation is more complex, and a different method of classification is necessary. Nevertheless, the spectra of all these elements can be arranged into series; and every series can again be represented as a difference of two terms, one fixed and characteristic of the series, another integrally variable with each line.

BOHR'S THEORY OF ATOMIC STRUCTURE

To account for the line spectra of elements and to circumvent the objections leveled against the Rutherford atom, Niels Bohr advanced in 1913 a theory of atomic structure radically different from any that preceded it. Along with Rutherford, Bohr considered the atom to consist of a nucleus with electrons revolving about it. However, whereas Rutherford placed no restrictions on the electron orbits, Bohr postulated that the only possible orbits for electron revolution are those for which the angular momentum is a *whole number* multiple n of the quantity $(h/2 \pi)$, h being Planck's constant. This assumption constitutes a *quantization of the angular momentum* of the electron. Bohr further postulated that as long as any electron stays in a given orbit it does not radiate energy, despite the demands of classical electrodynamics, and hence the energy of an electron remains constant as long as it does not change orbits. This second postulate introduces the concept that there are definite *energy levels* or *stationary states* within the atom in which an electron possesses a definite and invariable energy content. Finally, Bohr assumed as his third postulate that each line observed in the spectrum of an element results from the passage of an electron from an orbit in which the energy is E_2 to one of lower energy E_1 and that this difference in energy is emitted as a quantum of radiation of frequency ν in line with the equation

$$\Delta E = E_2 - E_1 = h\nu \tag{13}$$

Bohr applied these ideas to an atom consisting of a nucleus and one external electron. Consider in general a nucleus of charge $+Ze$, A in

Figure 14–6, around which revolves in an orbit or radius r an electron B of charge e, mass m, and tangential velocity v. The force of electrostatic attraction between the nucleus and the electron is given by Coulomb's law as Ze^2/r^2. This attractive force is opposed by the centrifugal force

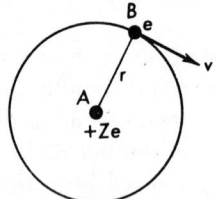

Figure 14–6.   Bohr's model of one-electron atom.

of the electron, mv^2/r, which tends to make it escape from its orbit. For a stable situation it is necessary that these forces be equal, namely, that

$$\frac{Ze^2}{r^2} = \frac{mv^2}{r}$$

from which we get

$$r = \frac{Ze^2}{mv^2} \tag{14}$$

Now, the angular momentum of the electron is mvr, and by Bohr's postulate it should equal

$$mvr = \frac{nh}{2\pi}$$

Consequently,

$$v = \frac{nh}{2\pi mr} \tag{15}$$

where n is an integer. Substituting Eq. (15) into (14) we obtain for r

$$r = \frac{n^2h^2}{4\pi^2 mZe^2} \tag{16}$$

The total energy E of the electron in its orbit is the sum of its kinetic and potential energies. The kinetic energy is given by $1/2\ mv^2$, while the potential energy by $-Ze^2/r$. Hence

$$E = \frac{1}{2}\ mv^2 - \frac{Ze^2}{r}$$

which, on introduction of mv^2 from Eq. (14), becomes

$$E = \frac{Ze^2}{2\ r} - \frac{Ze^2}{r}$$

$$= -\frac{Ze^2}{2\ r} \tag{17}$$

Substituting Eq. (16) for r into Eq. (17), E follows as

$$E = - \frac{2\,\pi^2 m Z^2 e^4}{n^2 h^2} \qquad (18)$$

Equation (18) gives the total energy of a single electron revolving about a nucleus of atomic number Z. This equation involves known constants and the integer n which may have only integral values 1, 2, 3, etc. For any given value of n, which is called the *principal quantum number*, the energy of the electron will be fixed, and this will be the constant energy of the particular energy level or stationary state. The energies of the various levels, since they are proportional to $1/n^2$, will decrease according to the square of the quantum number, so that the level for which $n = 2$ will have only 1/4 the energy of the level for which $n = 1$, the one with $n = 3$, 1/9 the energy, etc. In other words, the energy levels which the electron can occupy do not change continuously, but involve abrupt changes in energy on passage from one level to another.

According to Bohr's third postulate, spectral lines are supposed to result when an electron jumps from one energy level to another, and thereby the atom emits a quantum of energy whose frequency is given by Eq. (13). This point can be tested very readily. Suppose the electron is originally in an orbit for which $n = n_2$ and that it changes into an orbit with $n = n_1$. In its initial state the electron will have, according to Eq. (17), the energy

$$E_{n_2} = - \frac{2\,\pi^2 m Z^2 e^4}{h^2 n_2^2}$$

while in the final state

$$E_{n_1} = - \frac{2\,\pi^2 m Z^2 e^4}{h^2 n_1^2}$$

The difference in energy, $\Delta E = E_{n_2} - E_{n_1}$ should be emitted then as a quantum of energy $h\nu$. Hence the frequency ν of the emitted line should be given by

$$\nu = - \frac{2\,\pi^2 m Z^2 e^4}{h^3 n_2^2} + \frac{2\,\pi^2 m Z^2 e^4}{h^3 n_1^2}$$

$$= \frac{2\,\pi^2 m Z^2 e^4}{h^3} \left(\frac{1}{n_1^2} - \frac{1}{n_2^2} \right) \qquad (19)$$

and the wave number $\bar{\nu} = \nu/c$ by

$$\bar{\nu} = \frac{2\,\pi^2 m Z^2 e^4}{h^3 c} \left(\frac{1}{n_1^2} - \frac{1}{n_2^2} \right)$$

$$= R_\infty Z^2 \left(\frac{1}{n_1^2} - \frac{1}{n_2^2} \right) \qquad (20)$$

where

$$R_\infty = \frac{2\,\pi^2 m e^4}{h^3 c} \tag{21}$$

Equation (20) has been derived on the supposition that the nucleus was sufficiently heavy to be stationary, i.e., that its mass was infinite. When such is not the case, then Eq. (20) has to be modified to

$$\bar{\nu} = \left(\frac{m'}{m + m'}\right) R_\infty Z^2 \left(\frac{1}{n_1^2} - \frac{1}{n_2^2}\right) \tag{22}$$

where, as before, m is the mass of the electron and m' is the mass of the nucleus of atomic number Z.

Applied to hydrogen, for which $Z = 1$, Eq. (22) should take the form

$$\bar{\nu} = \left(\frac{m'}{m + m'}\right) R_\infty \left(\frac{1}{n_1^2} - \frac{1}{n_2^2}\right) \tag{23}$$

This equation is strikingly similar to Eq. (12) for the spectral lines of hydrogen. Further, if the theory is valid, then we should also have

$$\begin{aligned} R_H &= \left(\frac{m'}{m + m'}\right) R_\infty \\ &= \left(\frac{m'}{m + m'}\right)\left(\frac{2\,\pi^2 m e^4}{h^3 c}\right) \end{aligned} \tag{24}$$

Substitution of the values of the constants into Eq. (24) gives $R_H = 109{,}677.61$ cm^{-1} as against the observed value of $109{,}677.58$ cm^{-1}. This remarkably close agreement constitutes an excellent confirmation of Bohr's postulates and his explanation of the origin of the hydrogen spectra.

Further comparison of Eqs. (14) and (23) supplies the significance of the various series observed in the spectral lines emitted by the hydrogen atom. In the Lyman series $n_1 = 1$, and hence the lines observed result from passage of electrons from the second, third, fourth, etc., orbits into the first. Similarly, the Balmer series, for which $n_1 = 2$, arise from passage of electrons from the third, fourth, etc., orbits into the second orbit. Likewise the Paschen, Brackett, and Pfund series arise from electron transitions from higher orbits into the third, fourth, and fifth orbits respectively. These various transitions for the sundry series are illustrated schematically in the energy level diagram for the hydrogen atom shown in Figure 14–7. Each horizontal line in this figure shows the energy content of a particular level corresponding to the indicated quantum number, while the vertical lines represent the various electron transitions from which arise the spectral lines observed in each series.

In a similar manner Eq. (23) has been used to account for other hydrogenic spectra, i.e., spectra arising from a nucleus around which revolves only one electron.

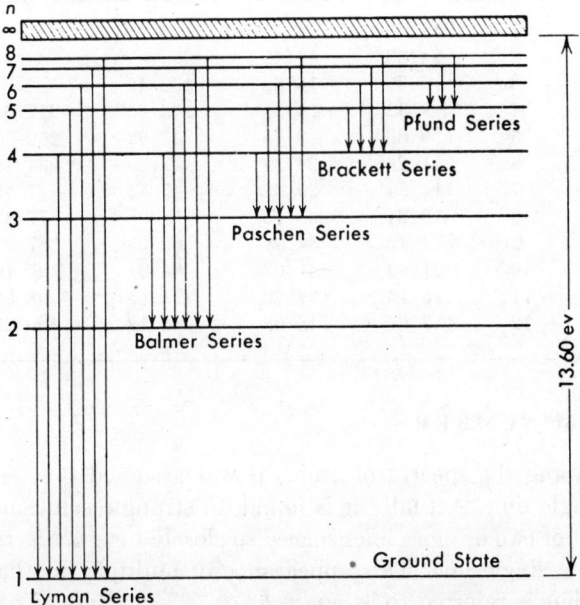

Figure 14–7.   Origin of spectral series in atomic hydrogen (schematic).

IONIZATION POTENTIALS

In Figure 14–7 the lowest energy level of the electron is the one for which $n = 1$. This level is called the *ground state* of the electron. Again, the highest energy level is the one for which $n = \infty$, and this state corresponds to the removal of the electron from the influence of the nucleus, i.e., ionization of the atom. The energy, I, required to produce ionization from the normal or ground state of the electron is called the *ionization potential*, and it is usually expressed in electron volts, eV.[1] In the case of the hydrogen atom only one ionization potential is present, namely, $I = 13.60$ eV. However, in atoms containing several external electrons there will be an ionization potential for each electron, and the magnitude of this potential will increase with each removed electron.

The ionization potentials of some atoms are given in Table 14–5.

---

[1] 1 eV = 1.60210 $\times$ 10^{-12} erg per particle = 23,061 cal mole^{-1}.

TABLE 14-5. IONIZATION POTENTIALS OF ATOMS

Element	Atomic No.	I (eV)				
		1	2	3	4	5
H	1	13.60				
He	2	24.58	54.40			
Li	3	5.39	75.62	122.42		
Be	4	9.32	18.21	153.85	217.66	
B	5	8.30	25.15	37.92	259.30	338.5
C	6	11.26	24.38	47.86	64.48	390.1
N	7	14.54	29.61	47.43	77.45	97.4
O	8	13.61	35.15	54.93	77.39	113.0
F	9	17.42	34.98	62.65	87.23	113.7
Ne	10	21.56	41.07	64.0	97.16	
Na	11	5.14	47.29	71.65	98.88	
Mg	12	7.64	15.03	80.12	109.29	

QUANTUM NUMBERS

In describing the spectra of atoms it was assumed that each emitted line is a single unit. Actually it is found on strong resolution that many lines consist of two or more lines spaced so closely together as to constitute apparently a single line. This appearance of multiple fine lines in place of a single line is referred to as *multiplet* or *fine structure* of spectral lines.

Now, although the Bohr theory does account for the various lines in hydrogen or hydrogenic spectra, it does not explain their fine structure. The appearance of this fine structure suggests that any energy level corresponding to a given value of the principal quantum number n may in fact be composed of a series of sublevels, each with its own quantum number, but differing only slightly in total energy from every other sublevel. On this basis a transition from an upper to a lower level would lead to a number of closely spaced lines depending on the sublevel into which a given electron enters.

Detailed spectroscopic study has in fact shown that four quantum numbers are required to characterize each electron in a given atom. These are: (a) the principal quantum number, n; (b) the azimuthal quantum number, l; (c) the magnetic quantum number, m; and (d) the spin quantum number, s.

The principal quantum number n of any electron in an atom determines the major energy level to which an electron belongs. This number may take on only the integer values, 1, 2, 3, etc., depending on which level the electron is in. Corresponding to each value of n there are n possible values of the azimuthal quantum number l which determine the angular momentum of the electron, and which may be looked upon as sublevels of the major level. The values of l range from $l = 0$ up to $l = (n - 1)$ in integral steps. Thus, when $n = 1$ there is only one value of l possible, $l = 0$. Again, for $n = 2$ we may have $l = 0$ and $l = 1$, whereas for $n = 4$ we have $l = 0, 1, 2$, and 3.

The magnetic quantum number m determines the possible quantized space orientations of the orbital angular momentum. It may be looked upon as indicating the manner in which each sublevel can be split into still finer levels differing slightly from each other in energy, but still belonging to a particular sublevel. The values which this quantum number can have are determined by the values of l, and range from $-l \rightarrow 0 \rightarrow +l$, i.e., a total of $(2l + 1)$ values of m for each value of l. For example, for $l = 0$ there is only one value of m, namely, $m = 0$. Again, for $l = 1$ we may have $m = -1, 0$, and $+1$, a total of three values. On the other hand, if $l = 3$, we have seven values of m, these being $m = -3, -2, -1, 0, +1, +2, +3$.

Finally the *spin quantum number* s arises from a suggestion made by Uhlenbeck and Goudsmit in 1925 that an electron in its motion in an orbit may rotate or spin about its own axis. Such a spin would contribute to the angular momentum of the electron and would modify thus the energy relations. Assuming that this spin is also quantized, there are only two possible values which s may have, $s = +\frac{1}{2}$ or $s = -\frac{1}{2}$, depending on whether the electron spins in one direction or in another.

The four quantum numbers and the values they may take on may be summarized as follows:

1. *Principal quantum number*, n. This number can only have the integral values $n = 1, 2, 3$, etc.
2. *Azimuthal quantum number*, l. For each value of n there may be n values of l, namely, $l = 0, 1, 2, \cdots (n - 1)$.
3. *Magnetic quantum number*, m. For each value of l there may be $(2l + 1)$ values of m. These range from $m = -l$ through $m = 0$ to $m = +l$ in integral values; i.e., $m = 0, \pm 1, \pm 2, \cdots \pm l$.
4. *Spin quantum number*, s. There are only two possible values of s, $s = \frac{1}{2}$ and $s = -\frac{1}{2}$ for each value of l.

By specifying its four quantum numbers, the "address" of any electron in a given atom is completely defined; i.e., the four quantum numbers locate each electron in the major energy level (n), the particular sublevel (l), the sub-sublevel (m), and the direction of its spin (s).

Nothing said thus far would preclude several electrons in an atom from occupying the same energy level, and having, therefore, four identical quantum numbers. However, study of spectra reveals that this is not possible, and this fact led Pauli to enunciate in 1923 his *exclusion principle*. According to this principle *no two electrons in an atom may have all four quantum numbers the same.* To exist in the same atom electrons may have three quantum numbers identical, but the fourth must be different. We shall see presently that this simple principle is of fundamental value, and aids greatly in deducing the possible distribution of electrons in the atoms of various elements.

ELECTRON SHELLS AND SUBSHELLS

Before turning to the question of electron distribution in atoms, it is necessary to develop the concept of *shells and subshells*. Just as the energy levels of electrons in atoms may be considered to be divided into levels and sublevels, so may the electrons be considered to exist in groups and subgroups, referred to respectively as shells and subshells. All electrons possessing the same principal quantum number n are said to be present in the same shell. Again, all the electrons in a given shell which occupy the sublevels with the same l value are said to be present in the same subshell. Thus all the electrons for which $n = 1$ and $l = 0$ occupy not only the same shell, but the same subshell. Likewise all electrons for which $n = 2$ and $l = 1$ are present in the second shell and in the $l = 1$ subshell.

This concept finds its justification in the explanation of the origin of X-ray spectra. The ordinary line spectra of elements are due to electrons present in the outermost shell, i.e., the *optical electrons*. Because the optical electrons are most weakly bound to the nucleus, they are exicted more readily than electrons embedded deeper in the atom, and they yield, therefore, the more easily obtainable line spectra. On the other hand, the X-ray spectra are much more difficult to excite, as they involve much higher emitted energies, and hence X-rays must arise from electrons closer to the nucleus than the optical electrons. For these reasons the generally accepted theory of the origin of X-ray spectra is this. On bombardment of an atom by high-speed electrons, collisions occur between bombarding electrons and those in the atom. When the energy of the missile particles is great enough, these may dislodge electrons deep within the atom, leaving vacant spaces. To reoccupy the vacancies, electrons from higher levels will then pass into these, causing the emission of a spectrum. When this passage is from levels of $n = 2$ or higher into the $n = 1$ level, the result is the K series of X-ray spectra. Again, when the jumps are from $n = 3$ levels or above into the $n = 2$ level, the emission is the L series. Similarly, all transitions into the $n = 3$ level yield the M series, those into the $n = 4$ level the N series, etc., provided the atom contains enough electrons to yield all these series. When this is not the case, as with the lighter atoms, only the K series, or the K and part of the L series are observed. It will be seen that this theory of the origin of X-ray spectra ascribes their appearance not to electron jumps from sublevel to sublevel, but to electron passage from one major level to another, i.e., from shell to shell. The particular sublevel in a shell from which an electron comes determines only the fine structure of the line, not its position. The latter is determined primarily by the change in the principal quantum number n.

With the aid of this theory the X-ray spectra of elements can readily be accounted for, and this constitutes, therefore, a confirmation of the concept of electron shells. By association with the series of spectra resulting from vacancies in various shells, it is customary to refer to the shell for which $n = 1$ as the K shell, the one for which $n = 2$ as the L shell, the one with $n = 3$ as the M shell, etc.

THE PERIODIC TABLE AND ATOMIC STRUCTURE

If theories of atomic structure are to be of any value, they must be able to explain the differences in the chemical reactivity of various elements, and why the chemical and physical properties of elements repeat themselves in the manner represented by the periodic table. The periodicity of the elements definitely rules out mass as the determining factor in chemical reactivity, and hence the nuclei of atoms cannot possibly be responsible for chemical behavior. We must seek our explanation, therefore, in the configurational architecture of the external electrons in order to ascertain why elements act as they do. It is one of the crowning achievements of modern theories of atomic structure that they have been able to shed a great deal of light on this point and to account for the observed periodic repetition of chemical properties.

In passing from one element to another in the periodic table, the atomic number increases by one and so does the nuclear charge. To preserve the electroneutrality of the atoms, this progressive increase in nuclear charge must be accompanied by a simultaneous increase in the number of orbital electrons. Consequently, in passing from hydrogen with $Z = 1$ to lawrencium with $Z = 103$, the number of electrons about the nucleus must increase progressively by one from a single electron for hydrogen to 103 for lawrencium. As these electrons enter the outer structure of the atom, they must become arranged in shells and subshells, each containing the number of electrons commensurate with the number present and the number that may be crowded into each. This much about the electronic arrangement appears to be clear. But the problems which still remain are (a) how many electrons may be located in each shell and subshell, and (b) where does each successive electron go?

The first satisfactory suggestion about the number of electrons in various shells and their possible arrangement was advanced in 1921 independently by Bury and Bohr. These men proposed that the maximum number of electrons in each shell should be given by $2n^2$, where n is the principal quantum number. This would give for the maximum number of electrons in successive shells 2, 8, 18, 32, and 50. Further, Bury and Bohr suggested that *there can be present no more than eight electrons in the outermost shell of an atom before the next shell is started.* In other words, in shells where more than eight electrons can be present, only eight of these enter, then a new shell is started, and the incompleted shell is left to be filled in later.

That the Bury-Bohr assignment of electrons to the various shells is correct can be proved with the aid of the Pauli exclusion principle and the rules given for the possible values of quantum numbers. For the K shell $n = 1$, and hence the values which l, m, and s may have are $l = 0$, $m = 0$, and $s = +\frac{1}{2}$ or $s = -\frac{1}{2}$. Since no two electrons may have the same four quantum numbers, the following are the only possibilities for four different quantum numbers in the K shell:

$$n = 1 \qquad l = 0 \qquad m = 0 \qquad s = +\frac{1}{2}$$

$$n = 1 \qquad l = 0 \qquad m = 0 \qquad s = -\frac{1}{2}$$

This means that in the K shell there is only one subshell, $l = 0$, in which only two electrons can be accommodated. On the other hand, for $n = 2$ we may have $l = 0$ and $l = 1$, $m = 0$, $m = -1$, and $m = +1$, and $s = \frac{1}{2}$ or $s = -\frac{1}{2}$. These possibilities lead to the following combinations of four different quantum numbers:

$$n = 2 \qquad l = 0 \qquad m = 0 \qquad s = +\frac{1}{2}$$

$$n = 2 \qquad l = 0 \qquad m = 0 \qquad s = -\frac{1}{2}$$

$$n = 2 \qquad l = 1 \qquad m = 0 \qquad s = +\frac{1}{2}$$

$$n = 2 \qquad l = 1 \qquad m = 0 \qquad s = -\frac{1}{2}$$

$$n = 2 \qquad l = 1 \qquad m = -1 \qquad s = +\frac{1}{2}$$

$$n = 2 \qquad l = 1 \qquad m = -1 \qquad s = -\frac{1}{2}$$

$$n = 2 \qquad l = 1 \qquad m = 1 \qquad s = +\frac{1}{2}$$

$$n = 2 \qquad l = 1 \qquad m = 1 \qquad s = -\frac{1}{2}$$

On this basis the L shell should consist of eight electrons, with two of these in the $l = 0$ subshell, and six in the $l = 1$ subshell. In a like manner it can be shown that the M shell can contain 18 electrons, with 2 of these in the $l = 0$ subshell, 6 in the $l = 1$ subshell, and 10 in the $l = 2$ subshell; the N shell, 32 electrons with the subshell distribution 2, 6, 10, and 14; while the O shell has a possible maximum of 50 electrons. These numbers are exactly those predicted by Bohr and Bury.

ARRANGEMENT OF ELECTRONS IN ATOMS

The arrangement of electrons in atoms has been deduced from a combined use of the Pauli exclusion principle, the Bohr-Bury postulates, and spectroscopic study. In order to appreciate the significance of the results, reference must first be made to the periodic classification of the elements shown in Table 14–6. Here the elements are ordered according to chemical similarity in vertical *groups*, and horizontally into repetitive *periods*. The first period consists of only 2 elements, hydrogen and helium. The second and third, again, contain 8 elements each, with each period terminating with a rare gas, i.e., neon and argon. On the other hand, before the next rare gas, krypton, is reached in the fourth period 18 elements are traversed. Of these the 8 elements potassium, calcium, gallium, germanium, arsenic, selenium, bromine, and krypton behave more or less normally and exhibit fairly close similarity to preceding members of their own group, while the 10 elements starting with scandium and ending with zinc are somewhat more unusual in their behavior. The latter are generally referred to as the *transition elements*. In the fifth period the situation is exactly the same as in the fourth, the period consisting again of 18 elements, 8 of which are more or less "normal," while 10 elements, beginning with yttrium and ending with cadmium, are transition elements. However, when we come to the sixth period, we find that 32 elements are involved here. A study of these shows that they may be divided into three groups: (a) 8 more or less "normal" elements, namely, cesium, barium, thallium, lead, bismuth, polonium, astatine, and radon; (b) 10 transition elements ranging from lanthanum to mercury; and (c) the rare earths falling between $Z = 58$ and $Z = 71$. The latter elements exhibit such striking similarity in chemical properties that the only way to accommodate them in the table is to list all of them as a group in the space between barium and hafnium. Finally, the seventh period contains francium, radium, and the elements of the actinide series, among which are included all the transuranium elements discovered thus far.

The explanation of the above-described behavior lies in the arrangement of the electrons shown in Table 14–7. In this table the common practice is followed of designating the major shells by the letters K, L, M, etc., corresponding to $n = 1, 2, 3$, and so on, and the subshells by symbols such as $1s$, $2p$, $3d$, $4f$, etc. In the latter the numerals represent the number

TABLE 14-6. PERIODIC ARRANGEMENT OF THE ELEMENTS

Period	I A	II A	III B	IV B	V B	VI B	VII B	VIII			I B	II B	III A	IV A	V A	VI A	VII A	Rare Gases
1	1 H																	2 He
2	3 Li	4 Be											5 B	6 C	7 N	8 O	9 F	10 Ne
3	11 Na	12 Mg											13 Al	14 Si	15 P	16 S	17 Cl	18 Ar
4	19 K	20 Ca	21 Sc	22 Ti	23 V	24 Cr	25 Mn	26 Fe	27 Co	28 Ni	29 Cu	30 Zn	31 Ga	32 Ge	33 As	34 Se	35 Br	36 Kr
5	37 Rb	38 Sr	39 Y	40 Zr	41 Nb	42 Mo	43 Tc	44 Ru	45 Rh	46 Pd	47 Ag	48 Cd	49 In	50 Sn	51 Sb	52 Te	53 I	54 Xe
6	55 Cs	56 Ba	57–71 *	72 Hf	73 Ta	74 W	75 Re	76 Os	77 Ir	78 Pt	79 Au	80 Hg	81 Tl	82 Pb	83 Bi	84 Po	85 At	86 Rn
7	87 Fr	88 Ra	89–103 **															

Rare Earths

*	57 La	58 Ce	59 Pr	60 Nd	61 Pm	62 Sm	63 Eu	64 Gd	65 Tb	66 Dy	67 Ho	68 Er	69 Tm	70 Yb	71 Lu

Actinide Series

**	89 Ac	90 Th	91 Pa	92 U	93 Np	94 Pu	95 Am	96 Cm	97 Bk	98 Cf	99 Es	100 Fm	101 Mv	102 No	103 Lw

TABLE 14–7. ARRANGEMENT OF ELECTRONS IN VARIOUS ELEMENTS

Shell		K	L		M			N				O				P				Q	
Subshell		1s	2s	2p	3s	3p	3d	4s	4p	4d	4f	5s	5p	5d	5f	6s	6p	6d	6f	7s	7p
At. No.	Elem.																				
1	H	1																			
2	He	2																			
3	Li	2	1																		
4	Be	2	2																		
5	B	2	2	1																	
6	C	2	2	2																	
7	N	2	2	3																	
8	O	2	2	4																	
9	F	2	2	5																	
10	Ne	2	2	6																	
11	Na	2	2	6	1																
12	Mg	2	2	6	2																
13	Al	2	2	6	2	1															
14	Si	2	2	6	2	2															
15	P	2	2	6	2	3															
16	S	2	2	6	2	4															
17	Cl	2	2	6	2	5															
18	Ar	2	2	6	2	6															
19	K	2	2	6	2	6		1													
20	Ca	2	2	6	2	6		2													
21	Sc	2	2	6	2	6	1	2													
22	Ti	2	2	6	2	6	2	2													
23	V	2	2	6	2	6	3	2													
24	Cr	2	2	6	2	6	5	1													
25	Mn	2	2	6	2	6	5	2													
26	Fe	2	2	6	2	6	6	2													
27	Co	2	2	6	2	6	7	2													
28	Ni	2	2	6	2	6	8	2													
29	Cu	2	2	6	2	6	10	1													
30	Zn	2	2	6	2	6	10	2													
31	Ga	2	2	6	2	6	10	2	1												
32	Ge	2	2	6	2	6	10	2	2												
33	As	2	2	6	2	6	10	2	3												
34	Se	2	2	6	2	6	10	2	4												
35	Br	2	2	6	2	6	10	2	5												
36	Kr	2	2	6	2	6	10	2	6												

TABLE 14–7 (*Continued*)

Shell		K	L		M			N				O				P				Q	
Subshell		1s	2s	2p	3s	3p	3d	4s	4p	4d	4f	5s	5p	5d	5f	6s	6p	6d	6f	7s	7p
At. No.	Elem.																				
37	Rb	2	2	6	2	6	10	2	6			1									
38	Sr	2	2	6	2	6	10	2	6			2									
39	Y	2	2	6	2	6	10	2	6	1		2									
40	Zr	2	2	6	2	6	10	2	6	2		2									
41	Cb	2	2	6	2	6	10	2	6	4		1									
42	Mo	2	2	6	2	6	10	2	6	5		1									
43	Tc	2	2	6	2	6	10	2	6	6		1									
44	Ru	2	2	6	2	6	10	2	6	7		1									
45	Rh	2	2	6	2	6	10	2	6	8		1									
46	Pd	2	2	6	2	6	10	2	6	10											
47	Ag	2	2	6	2	6	10	2	6	10		1									
48	Cd	2	2	6	2	6	10	2	6	10		2									
49	In	2	2	6	2	6	10	2	6	10		2	1								
50	Sn	2	2	6	2	6	10	2	6	10		2	2								
51	Sb	2	2	6	2	6	10	2	6	10		2	3								
52	Te	2	2	6	2	6	10	2	6	10		2	4								
53	I	2	2	6	2	6	10	2	6	10		2	5								
54	Xe	2	2	6	2	6	10	2	6	10		2	6								
55	Cs	2	2	6	2	6	10	2	6	10		2	6			1					
56	Ba	2	2	6	2	6	10	2	6	10		2	6			2					
57	La	2	2	6	2	6	10	2	6	10		2	6	1		2					
58	Ce	2	2	6	2	6	10	2	6	10	2	2	6			2					
59	Pr	2	2	6	2	6	10	2	6	10	3	2	6			2					
60	Nd	2	2	6	2	6	10	2	6	10	4	2	6			2					
61	Pm	2	2	6	2	6	10	2	6	10	5	2	6			2					
62	Sm	2	2	6	2	6	10	2	6	10	6	2	6			2					
63	Eu	2	2	6	2	6	10	2	6	10	7	2	6			2					
64	Gd	2	2	6	2	6	10	2	6	10	7	2	6	1		2					
65	Tb	2	2	6	2	6	10	2	6	10	8	2	6	1		2					
66	Dy	2	2	6	2	6	10	2	6	10	9	2	6	1		2					
67	Ho	2	2	6	2	6	10	2	6	10	10	2	6	1		2					
68	Er	2	2	6	2	6	10	2	6	10	11	2	6	1		2					
69	Tm	2	2	6	2	6	10	2	6	10	13	2	6			2					
70	Yb	2	2	6	2	6	10	2	6	10	14	2	6			2					
71	Lu	2	2	6	2	6	10	2	6	10	14	2	6	1		2					
72	Hf	2	2	6	2	6	10	2	6	10	14	2	6	2		2					
73	Ta	2	2	6	2	6	10	2	6	10	14	2	6	3		2					
74	W	2	2	6	2	6	10	2	6	10	14	2	6	4		2					
75	Re	2	2	6	2	6	10	2	6	10	14	2	6	5		2					
76	Os	2	2	6	2	6	10	2	6	10	14	2	6	6		2					

TABLE 14–7 (Continued)

Shell		K	L		M			N				O				P				Q	
Subshell		1s	2s	2p	3s	3p	3d	4s	4p	4d	4f	5s	5p	5d	5f	6s	6p	6d	6f	7s	7p
At. No.	Elem.																				
77	Ir	2	2	6	2	6	10	2	6	10	14	2	6	9							
78	Pt	2	2	6	2	6	10	2	6	10	14	2	6	9		1					
79	Au	2	2	6	2	6	10	2	6	10	14	2	6	10		1					
80	Hg	2	2	6	2	6	10	2	6	10	14	2	6	10		2					
81	Tl	2	2	6	2	6	10	2	6	10	14	2	6	10		2	1				
82	Pb	2	2	6	2	6	10	2	6	10	14	2	6	10		2	2				
83	Bi	2	2	6	2	6	10	2	6	10	14	2	6	10		2	3				
84	Po	2	2	6	2	6	10	2	6	10	14	2	6	10		2	4				
85	At	2	2	6	2	6	10	2	6	10	14	2	6	10		2	5				
86	Rn	2	2	6	2	6	10	2	6	10	14	2	6	10		2	6				
87	Fr	2	2	6	2	6	10	2	6	10	14	2	6	10		2	6			1	
88	Ra	2	2	6	2	6	10	2	6	10	14	2	6	10		2	6			2	
89	Ac	2	2	6	2	6	10	2	6	10	14	2	6	10		2	6	1		2	
90	Th	2	2	6	2	6	10	2	6	10	14	2	6	10		2	6	2		2	
91	Pa	2	2	6	2	6	10	2	6	10	14	2	6	10	2	2	6	1		2	
92	U	2	2	6	2	6	10	2	6	10	14	2	6	10	3	2	6	1		2	
93	Np	2	2	6	2	6	10	2	6	10	14	2	6	10	5	2	6			2	
94	Pu	2	2	6	2	6	10	2	6	10	14	2	6	10	6	2	6			2	
95	Am	2	2	6	2	6	10	2	6	10	14	2	6	10	7	2	6			2	
96	Cm	2	2	6	2	6	10	2	6	10	14	2	6	10	7	2	6	1		2	
97	Bk	2	2	6	2	6	10	2	6	10	14	2	6	10	8	2	6	1		2	
98	Cf	2	2	6	2	6	10	2	6	10	14	2	6	10	9	2	6	1		2	
99	Es	2	2	6	2	6	10	2	6	10	14	2	6	10	10	2	6	1		2	
100	Fm	2	2	6	2	6	10	2	6	10	14	2	6	10	11	2	6	1		2	
101	Mv	2	2	6	2	6	10	2	6	10	14	2	6	10	12	2	6	1		2	
102	No	2	2	6	2	6	10	2	6	10	14	2	6	10	13	2	6	1		2	
103	Lw	2	2	6	2	6	10	2	6	10	14	2	6	10	14	2	6	1		2	

of the major shell, while the letters, s, p, d, and f stand for $l = 0, 1, 2,$ and 3 respectively. Starting with hydrogen, we see that there is only one electron present in the K shell. In helium a second electron is added to this shell, and since only two electrons can be present in any energy level with $n = 1$, this shell becomes completely occupied and closed with this element. The next element, lithium, must start, therefore, a new shell, which is the L level, and electrons in the succeeding elements continue to occupy this shell until it also becomes completely filled when neon is reached. The only thing the added electron in sodium can do now is to start the

M shell, and the following elements up to argon continue the filling-in process in normal sequence until the first two subshells of the *M* level are occupied.

But when we come to potassium, the electron instead of filling in the 3*d* level starts a new shell. This behavior is in line with the Bohr-Bury theory that no external shell can contain more than eight electrons. Calcium follows potassium, but scandium, instead of continuing the process, starts to fill in the 3*d* subshell. This entrance of electrons into an inner shell continues through gallium until this subshell becomes completely occupied. Once this happens the tendency initiated by potassium is continued, and electrons again enter the outer shell in sequence until eight are present.

In the electron behavior just described lies the difference between the "normal" and the transition elements. Whereas in the "normal" elements electrons are added in sequence to the outermost electronic shell, in the transition elements the outermost shell remains essentially stationary, while succeeding electrons enter the shell immediately below the surface. Because of the filling-in process the length of the period is extended. Again, since the distribution of electrons in the outer shell is primarily responsible for the chemical properties of elements, we may anticipate that "normal" elements will act differently from the transition elements, and such is in fact the case.

The situation in the fourth period starting with rubidium and ending with xenon is essentially the same as in the third period. Like potassium, rubidium initiates a new shell, and strontium follows suit. But, starting with yttrium and continuing through cadmium, succeeding electrons enter the 4*d* level until it is filled, then the building up of the 5*p* subshell is continued to xenon. Here again the transition elements arise from entrance of electrons into the shell below the surface.

The sixth period starts just like the preceding two, with cesium and barium entering the 6*s* subshell and lanthanum the 5*d* sublevel. However, the electrons in cerium, instead of following lanthanum, begin to fill in the unoccupied 4*f* sublevel. Once initiated, this filling in of the second shell from the surface persists until it is terminated with lutecium. Hafnium continues then the process started by lanthanum, and when this is ended with mercury, thallium and the succeeding elements proceed to build up the *P* shell. Reference to Table 14–6 will show that the elements which fill the gap in the *N* shell are exactly the ones for which no room could be found in the periodic classification, namely, the rare earths. Since in these elements the external electronic configuration remains essentially unchanged from element to element, they all exhibit great similarity of chemical properties and act as if they were a single element in the periodic sequence.

Finally, in the seventh period a new shell is started into which enter the electrons for francium and radium. With actinium begins a new group of elements, the actinide series, where the succeeding electrons start first to fill the $6d$ level, and then enter the open $5f$ shell.

Thus far the emphasis has been on the reason for the periods and their length. However, the similarity of chemical properties within groups also follows from the table. Inspection of the *outermost* electronic configurations of hydrogen, lithium, sodium, potassium, rubidium, cesium, and francium shows that they are the same, and consist of a single electron in a new shell. Again, the outer electronic configurations of the elements in group IIA involve two electrons, group IIIA three electrons, etc. The electronic configuration of the rare gases is always such as to necessitate the inception of a new shell by a succeeding element. This fact indicates that a rare gas configuration must be a highly stable arrangement of electrons and accounts for the relative reluctance these elements exhibit to enter combination with other atoms.

In referring to the electronic arrangement of an atom, it is customary to give the information in the form $1s^2 2s^2 p^6 3s^2 p^6 d^{10} 4s^1$. Here the digits 1, 2, 3, 4, etc., refer to the principal quantum number and represent, respectively, the K, L, M, N, etc., shells. Again, the letters s, p, d, etc., refer to the subshells, while the superscripts give the number of electrons present in each of the latter. Thus the combination given above involves the presence of two s electrons in the K shell; two s and six p electrons in the L shell; two s, six p, and 10 d electrons in the M shell; and one s electron in the N shell. The element corresponding to this arrangement is copper.

SPECTRAL TERMS

The energy levels of an atom possessing a single orbital electron are described in terms of the four quantum numbers n, l, m, and s. When more than one electron is present, a different scheme has to be employed to designate the energy levels. The system commonly used is that based on Russell-Saunders coupling, and takes the form

$$n^{(2S'+1)}L_J$$

We shall explain now the significance of this symbology.

The symbol n refers to the major energy level of the atom. For the lowest or ground state $n = 1$, for the next higher one 2, etc. Again, the azimuthal quantum numbers of the various electrons in the atom, l_i, can be combined vectorially to give a sum $L = \Sigma l_i$. Like the l_i's, L can have only integral values such as 0, 1, 2, 3, etc., and these are denoted, respectively, by the letters S, P, D, F, etc. In making such summations closed shells can be ignored, as they do not contribute to L. Hence only the external electrons need be considered.

The spins s_i of the various electrons in an atom can be combined in an algebraic sum $S' = \Sigma s_i$ to yield a resultant spin quantum number S' which can have values of 0, ½, 1, ¾, etc. These combined spins can be coupled vectorially with L to give resultant angular momenta whose values are $\sqrt{J(J+1)}\ (h/2\pi)$, where J may equal $J = L + S', L + S' - 1, L + S' - 2, \cdots L - S'$, i.e., a total of $(2\,S' + 1)$ values of J for each value of L. If S' is greater than L, only positive values of $L - S'$ or zero are significant.

On the basis of the above combinations of electron quantum numbers an atomic energy level, or spectral term as it is called, may have designations such as $1^2S_{1/2}$, 2^3P_0, or $4^2F_{7/2}$. The first of these means that $n = 1$, $L = 0$, $J = ½$, and $2\,S' + 1 = 2$. Hence $S' = ½$. The second, in turn, gives $n = 2$, $L = 1$, $J = 0$, and $S' = 1$. Finally, $4^2F_{7/2}$ represents $n = 4$, $L = 3$, $J = \frac{7}{2}$, and $S' = ½$. The sum $(2\,S' + 1)$ is called the *multiplicity* of the energy state. These terms are read in sequence "one doublet S a half," "two triplet P zero," and "four doublet F seven halves."

Not all transitions between energy levels in an atom are possible. Selection rules limit allowed transitions to changes in L of $\Delta L = \pm 1$,

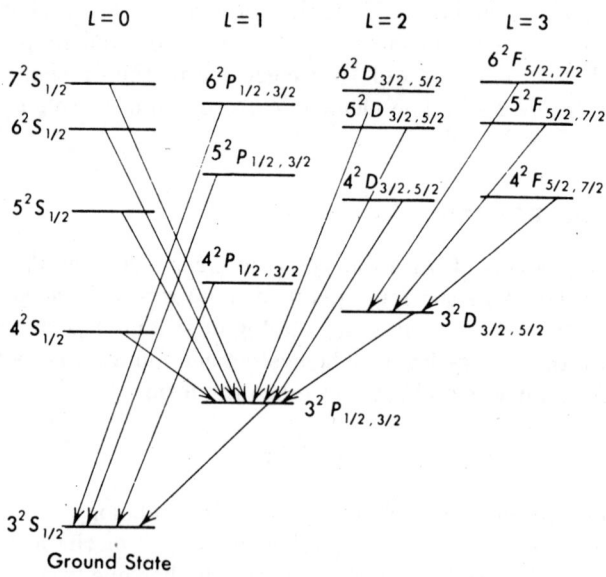

Figure 14–8. Schematic energy level diagram for neutral sodium atom with fine structure omitted. States with two J values consist of two levels closely spaced.

and to changes in J of $\Delta J = 0$ and ± 1. The relative positioning of various energy levels, and the operation of the selection rules, may be judged from the energy level diagram for sodium shown in Figure 14–8.

QUANTUM MECHANICS

Until 1925 all attempts to deal with atomic and molecular structural problems were based on classical mechanics, in which the structural units of the atom were treated as particles with determined momenta and positions upon which were superimposed postulated quantized conditions such as those of Bohr. In 1925–1926 Werner Heisenberg and Erwin Schrödinger independently proposed a new *quantum or wave mechanics* which takes cognizance of the wave and particle duality of matter and which deals with atomic and molecular scale problems in a totally different manner. Although formally different, the matrix mechanics of Heisenberg has been shown to be equivalent to the wave mechanics of Schrödinger. Since the latter approach lends itself somewhat more readily to a physical interpretation than the former, we shall employ here Schrödinger's wave mechanical treatment.

The basis for handling all problems by wave mechanics is the Schrödinger equation. This equation cannot be derived from first principles or basic postulates. Rather, it is arrived at in a manner to be described below, and the assumption is then made that this equation is valid for the treatment of all types of problems which involve atomic and molecular behavior.

Consider a system, such as a string, vibrating with a frequency ν and producing a wave traveling in the x direction with a velocity u_x and amplitude (height of wave) α. For such a system classical mechanics tells us that

$$\frac{\partial^2 \alpha}{\partial x^2} = \frac{1}{u_x^2} \frac{\partial^2 \alpha}{\partial t^2} \tag{25}$$

where t is the time necessary for the wave to reach a position x. The solution of this equation can be obtained in the form

$$\alpha = f_1(x) f_2(t) \tag{26}$$

where $f_1(x)$ is a function only of x, and $f_2(t)$ can be written as

$$f_2(t) = A \sin 2 \pi \nu t \tag{27}$$

where A is the maximum amplitude of the wave. Consequently, Eq. (26) becomes

$$\alpha = f_1(x) A \sin 2 \pi \nu t \tag{28}$$

If we differentiate Eq. (26) with respect to x at constant t, we obtain

$$\frac{\partial^2 \alpha}{\partial x^2} = f_2(t) \frac{\partial^2 f_1(x)}{\partial x^2} \tag{29}$$

Again, a similar differentiation of Eq. (28) with respect to t at constant x gives

$$\frac{\partial^2 \alpha}{\partial t^2} = -4\pi^2 \nu^2 f_1(x)(A \sin 2\pi\nu t)$$

$$= -4\pi^2 \nu^2 f_1(x) f_2(t) \tag{30}$$

and hence on substitution of Eqs. (29) and (30) into Eq. (25) we get

$$\frac{\partial^2 f_1(x)}{\partial x^2} = -\frac{4\pi^2 \nu^2}{u_x^2} f_1(x) \tag{31}$$

But wave length $\lambda = u_x/\nu$. Making this substitution, Eq. (31) becomes

$$\frac{\partial^2 f_1(x)}{\partial x^2} = -\frac{4\pi^2}{\lambda^2} f_1(x) \tag{32}$$

Equation (32) applies to a wave moving in the x direction only. If we consider a wave moving in three dimensions, then $f_1(x)$ can be replaced by ψ, where $\psi = \psi(x,y,z)$, and Eq. (32) is transformed to

$$\frac{\partial^2 \psi}{\partial x^2} + \frac{\partial^2 \psi}{\partial y^2} + \frac{\partial^2 \psi}{\partial z^2} = -\frac{4\pi^2}{\lambda^2} \psi \tag{33}$$

or

$$\nabla^2 \psi = -\frac{4\pi^2}{\lambda^2} \psi \tag{34}$$

where the symbol ∇^2 (del squared) represents

$$\nabla^2 = \frac{\partial^2}{\partial x^2} + \frac{\partial^2}{\partial y^2} + \frac{\partial^2}{\partial z^2} \tag{35}$$

Again, according to de Broglie's hypothesis a wave of length λ can be considered to be a particle of mass m moving with a velocity v in line with Eq. (5) of Chapter 2, namely, $\lambda = h/mv$. Substituting the latter into Eq. (34), we obtain

$$\nabla^2 \psi = -\frac{4\pi^2 m^2 v^2}{h^2} \psi \tag{36}$$

Finally, the total energy of the particle, E, is the sum of its potential energy, U, and its kinetic energy $E_k = \frac{1}{2} mv^2$. Consequently,

$$E_k = (E - U) = \frac{1}{2} mv^2$$

and

$$mv^2 = 2(E - U)$$

Substitution of the latter expression into Eq. (36) thus yields

$$\nabla^2\psi + \frac{8\,\pi^2 m}{h^2}\,(E - U)\psi = 0 \tag{37}$$

which is the *Schrödinger wave equation,* and the postulated basis of all quantum mechanical behavior of matter.

NATURE AND SIGNIFICANCE OF ψ

Equation (37) implies in essence that a moving body of mass m, velocity v, potential energy U, and total energy E has a wave associated with it of an amplitude given by the wave function ψ. For any physical situation ψ must be finite, single-valued, and continuous. The particular values of ψ which yield satisfactory solutions of Eq. (37) are called *eigenfunctions* or *characteristic wave functions,* while the values of the energy which correspond to the eigenfunctions are called the *eigenvalues* of the system.

The physical significance of ψ is nebulous. However, ψ^2, or $\psi\psi^*$, where ψ^* is the complex conjugate of ψ, can be interpreted as the probability of finding a particle within a certain domain in space. For a one-dimensional problem this domain is the interval between x and $x + dx$, for a two-dimensional problem an element of area, and for a three-dimensional problem an element of volume.

In general it is not difficult to set up the differential wave equation for a particular situation. However, the solution of the equation may be very complicated and laborious, and various approximations or assumptions may have to be used to arrive at some idea of the answer. We proceed now to illustrate the use of the Schrödinger equation, and the nature of the solutions it gives for some simple problems.

THE FREE PARTICLE

Consider a particle of mass m moving with a velocity v under conditions where $U = 0$. Then Eq. (37) becomes

$$\nabla^2\psi + \frac{8\,\pi^2 m}{h^2}\,E\psi = 0$$

or, on division by ψ,

$$\frac{1}{\psi}\,\nabla^2\psi + \frac{8\,\pi^2 m}{h^2}\,E = 0 \tag{38}$$

In a Cartesian coordinate system the total energy of the particle can be considered to consist of components E_x, E_y, and E_z along the three axes such that

$$E = E_x + E_y + E_z \tag{39}$$

Further, we can take

$$\psi = \psi_x \psi_y \psi_z \tag{40}$$

where ψ_x is a function of x only, ψ_y a function of y only, and ψ_z a function of z only. Substituting Eqs. (39) and (40) into Eq. (38), along with the expression for ∇^2 given by Eq. (35), we get

$$\left(\frac{1}{\psi_x} \frac{\partial^2 \psi_x}{\partial x^2} + \frac{8\,\pi^2 m E_x}{h^2} \right) + \left(\frac{1}{\psi_y} \frac{\partial^2 \psi_y}{\partial y^2} + \frac{8\,\pi^2 m E_y}{h^2} \right) + \left(\frac{1}{\psi_z} \frac{\partial^2 \psi_z}{\partial z^2} + \frac{8\,\pi^2 m E_z}{h^2} \right) = 0$$

$$\tag{41}$$

where each term in parentheses is a function of one variable only. Since each variable can be varied independently, each of these terms must be equal to zero, and we thus obtain three differential equations with separated variables to be solved.

The solutions of these equations are

$$\psi_x = A_x \sin\left(\frac{2\,\pi x}{h} \sqrt{2\,mE_x} \right) \tag{42a}$$

$$\psi_y = A_y \sin\left(\frac{2\,\pi y}{h} \sqrt{2\,mE_y} \right) \tag{42b}$$

$$\psi_z = A_z \sin\left(\frac{2\,\pi z}{h} \sqrt{2\,mE_z} \right) \tag{42c}$$

where A_x, A_y, and A_z are constants. Further, to meet the boundary conditions for ψ, we must also have

$$E_x = \frac{1}{2}\,mv_x^2 \tag{43a}$$

$$E_y = \frac{1}{2}\,mv_y^2 \tag{43b}$$

$$E_z = \frac{1}{2}\,mv_z^2 \tag{43c}$$

where v_x, v_y, and v_z are the components of the velocity v along the three coordinate axes. From Eq. (40) we see that the eigenfunction of ψ is the product of Eqs. (42a) to (42c), while Eq. (39) leads to

$$E = \frac{1}{2}\,m(v_x^2 + v_y^2 + v_z^2)$$

$$= \frac{1}{2}\,mv^2 \tag{44}$$

This result may have been anticipated, for, when $U = 0$, the total energy of the system is merely the kinetic energy $E_k = \frac{1}{2}\,mv^2$.

THE PARTICLE IN A BOX

Consider now the same particle as in the preceding section when it is constrained to move in a rectangular box of dimensions a, b, and c in length. Within the box, i.e., between $x = 0$ and a, $y = 0$ and b, and $z = 0$ and c, the potential energy is zero, but at the boundaries it increases suddenly to infinity and remains everywhere outside the box at this value.

If we represent U now as the sum of its components along the three coordinate axes, namely,

$$U = U_x + U_y + U_z \tag{45}$$

then, by proceeding in the same manner as above, it is possible to separate the Schrödinger equation into three differential equations of the form

$$\frac{1}{\psi_x} \frac{\partial^2 \psi_x}{\partial x^2} + \frac{8\,\pi^2 m}{h^2} (E_x - U_x) = 0 \tag{46}$$

for each of the variables. Solution of Eq. (46) yields between $x = 0$ and $x = a$ the eigenfunctions

$$\psi_x = \sqrt{\frac{2}{a}} \sin\left(\frac{n_x \pi x}{a}\right) \tag{47}$$

and the eigenvalues for the energy

$$E_x = \frac{n_x^2 h^2}{8\,ma^2} \tag{48}$$

where n_x is a quantum number taking on the values $n_x = 1, 2, 3$, etc. Similarly, the other differential equations give

$$\psi_y = \sqrt{\frac{2}{b}} \sin\left(\frac{n_y \pi y}{b}\right) \tag{49}$$

$$\psi_z = \sqrt{\frac{2}{c}} \sin\left(\frac{n_z \pi z}{c}\right) \tag{50}$$

and

$$E_y = \frac{n_y^2 h^2}{8\,mb^2} \tag{51}$$

$$E_z = \frac{n_z^2 h^2}{8\,mc^2} \tag{52}$$

where n_y and n_z are again quantum numbers equal to 1, 2, 3, etc. From Eqs. (47), (49), and (50), ψ follows as

$$\psi = \psi_x \psi_y \psi_z$$

$$= \sqrt{\frac{8}{abc}} \sin\left(\frac{n_x \pi x}{a}\right) \sin\left(\frac{n_y \pi y}{b}\right) \sin\left(\frac{n_z \pi z}{c}\right) \tag{53}$$

and the total energy from Eqs. (48), (51), and (52) as

$$E = E_x + E_y + E_z$$
$$= \frac{h^2}{8\,m}\left(\frac{n_x^2}{a^2} + \frac{n_y^2}{b^2} + \frac{n_z^2}{c^2}\right) \tag{54}$$

It is of interest to observe that, whereas the energies of a free particle can vary continuously, those for a particle in a box are quantized. However, the differences in energy between various levels are small due to the presence of h^2 in the numerators of the energy equations.

Figure 14–9 shows plots of ψ_x and ψ_x^2 vs. x for several values of n_x and $a = 2$. From Figure 14–9(a) it may be seen that the number of half-

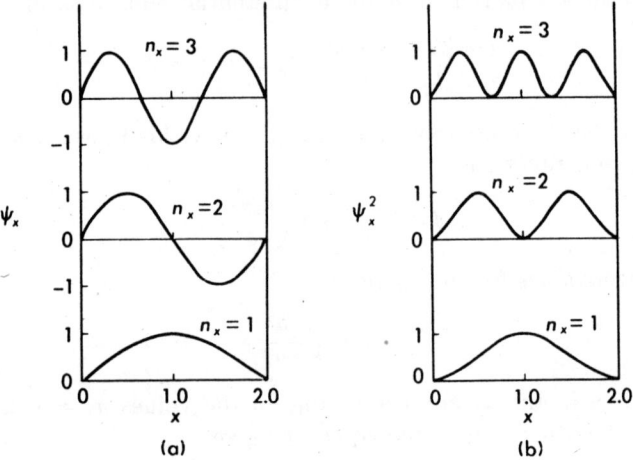

Figure 14–9. Plots of ψ_x and $\psi_x{}^2$ vs. x for $a = 2$.

waves given by ψ_x is identical with the value of n_x. Similarly, Figure 14–9(b) shows that n_x is also the number of maxima exhibited by ψ_x^2.

QUANTUM MECHANICS OF THE HYDROGEN ATOM

Consider an atom consisting of a nucleus of mass m' and charge Ze around which revolves one electron of mass m. In mechanics such a combination can be looked upon as a particle of *reduced mass* μ given by

$$\frac{1}{\mu} = \frac{1}{m} + \frac{1}{m'} \tag{55}$$

revolving about the center of mass at a distance r from the origin. Again, the potential energy of such a particle is $U = -Ze^2/r$. Inserting these

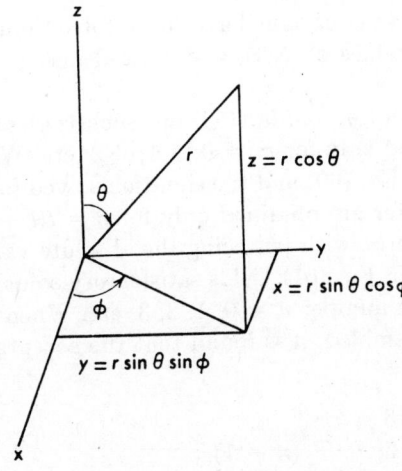

$z = r \cos \theta$

$x = r \sin \theta \cos \phi$

$y = r \sin \theta \sin \phi$

Figure 14–10. Spherical polar coordinates.

values of μ and U into the Schrodinger equation we get

$$\nabla^2\psi + \frac{8\,\pi^2\mu}{h^2}\left(E + \frac{Ze^2}{r}\right)\psi = 0 \qquad (56)$$

Equation (56) is best handled in spherical polar coordinates, whose relation to Cartesian coordinates is shown in Figure 14–10. In terms of r, θ, and φ, Eq. (56) becomes

$$\frac{1}{r^2}\frac{\partial}{\partial r}\left(r^2\frac{\partial\psi}{\partial r}\right) + \frac{1}{r^2\sin^2\theta}\frac{\partial^2\psi}{\partial\varphi^2} + \frac{1}{r^2\sin\theta}\frac{\partial}{\partial\theta}\left(\sin\theta\frac{\partial\psi}{\partial\theta}\right)$$
$$+ \frac{8\,\pi^2\mu}{h^2}\left(E + \frac{Ze^2}{r}\right)\psi = 0 \quad (57)$$

If we write now

$$\psi = R \cdot \Theta \cdot \Phi \qquad (58)$$

where R is a function of r only, Θ a function of θ only, and Φ a function of only φ, then Eq. (57) can be separated into the three differential equations[2]

$$\frac{\partial^2\Phi}{\partial\varphi^2} + m^2\Phi = 0 \qquad (59)$$

$$\frac{1}{\sin\Theta}\frac{\partial}{\partial\Theta}\left(\sin\Theta\frac{\partial\Theta}{\partial\Theta}\right) - \frac{m^2}{\sin^2\theta}\Theta + \beta\Theta = 0 \qquad (60)$$

and

$$\frac{1}{r^2}\frac{\partial}{\partial r}\left(r^2\frac{\partial R}{\partial r}\right) - \frac{\beta}{r^2}R + \frac{8\,\pi^2\mu}{h^2}\left(E + \frac{Ze^2}{r}\right)R = 0 \qquad (61)$$

[2] For details see Pauling and Wilson, *Introduction to Quantum Mechanics*, McGraw-Hill Book Company, Inc., New York, 1935.

where m and β are constants. These equations have to be solved now in sequence to obtain the allowed values of Φ, Θ, and R, and also of the energy E.

The boundary conditions for solution of Eq. (59) are such that satisfactory values of Φ can be obtained only for $m = 0, \pm 1, \pm 2$, etc. When these values of m are inserted into Eq. (60) and the equation solved for Θ, then permissible values of the latter are obtained only for $\beta = l(l + 1)$, where $l = |m|, |m| + 1, |m| + 2$, etc., with $|m|$ being the absolute values of m. Finally, with these values of β Eq. (61) yields satisfactory solutions only for values of a total quantum number $n = 0, 1, 2, 3$, etc. When the various quantum numbers are assembled, it is found that the acceptable solutions of Eq. (57) demand that

$$n = 0, 1, 2, 3, \cdots$$
$$l = 0, 1, 2, 3, \cdots (n - 1)$$
and
$$m = -l \rightarrow 0 \rightarrow +l$$

for each value of l. These are exactly the values postulated before for, respectively, the principal, azimuthal, and magnetic quantum numbers of an electron. However, whereas these quantum numbers were introduced before on a postulational basis, now they follow as requirements for the solution of the Schrödinger equation for a hydrogenic atom.

Solution of Eq. (57) does not yield the spin quantum numbers. To obtain these it is necessary to introduce into the equation relatavistic effects. These give then the $s = +\frac{1}{2}$ or $s = -\frac{1}{2}$ obtained by Uhlenbeck and Goudsmit for postulated electron spin.

Reference to the above discussion shows that the values of the Φ portion of the total wave function ψ depend only on the magnetic quantum number m. On the other hand, Θ depends on l and m, and R on n and l. Consequently ψ, given by Eq. (58), must be determined by the values of all three quantum numbers.

Finally, solution of Eq. (57) yields also the total energy E of the electron in the atom. The expression obtained is

$$E = -\frac{2\pi^2 \mu Z^2 e^4}{n^2 h^2} \tag{62}$$

If we take m' to be infinite in Eq. (55), then $\mu = m$, and Eq. (62) becomes identical with Eq. (18) for E obtained by Bohr for a hydrogenic atom.

THE HYDROGENLIKE WAVE FUNCTIONS

The wave function for an electron in an atom is called an *atomic orbital*. For any electron the orbital is determined by the values of the three quantum numbers n, l, and m. Further, for a hydrogenlike atom the total wave function is the product of the radial contribution to ψ, namely R, and the two angular portions Θ and Φ. The expressions for R and $\Theta\Phi$ deduced from solution of Eq. (57) for values of $n = 1$ to 3 are summarized in Table 14–8, where $q = Zr/a_0$, and

$$a_0 = \frac{h^2}{4\,\pi^2\mu e^2} = 0.529 \text{ Å}$$

The latter may be seen from Eq. (16) to be the first Bohr orbit for the

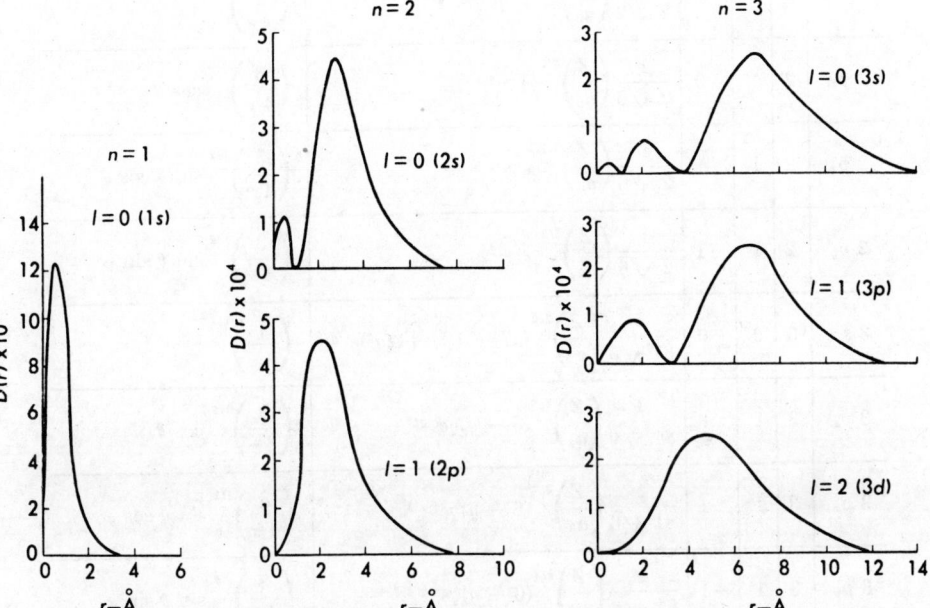

Figure 14–11.   Radial distribution functions for hydrogen atom.

hydrogen atom. The orbitals for the hydrogen atom follow from Table 14–8 on setting $Z = 1$.

Consider first the radial contributions to ψ for the hydrogen atom. In each instance R^2 represents the probability of finding the electron at a point between r and $r + dr$ in space. However, it is more illuminating to consider the probability of finding the electron in a spherical shell of thickness dr at a distance r from the center of mass, which is practically identical with the position of the nucleus. This probability $D(r)$, called the *radial distribution function*, is given by the expression

$$D(r) = 4\,\pi r^2 R^2 \tag{63}$$

Plots of $D(r)$ vs. r for various values of n and l are shown in Figure 14–11.

TABLE 14–8. THE HYDROGENLIKE WAVE FUNCTIONS*

Orbital Designation	n	l	m	R	$\Theta\Phi$
$1\,s$	1	0	0	$2\left(\dfrac{Z}{a_0}\right)^{3/2}e^{-q}$	$\left(\dfrac{1}{4\pi}\right)^{1/2}$
$2\,s$	2	0	0	$\dfrac{1}{2\sqrt{2}}\left(\dfrac{Z}{a_0}\right)^{3/2}(2-q)e^{-q/2}$	$\left(\dfrac{1}{4\pi}\right)^{1/2}$
$2\,p_z$	2	1	0	$\dfrac{1}{2\sqrt{6}}\left(\dfrac{Z}{a_0}\right)^{3/2}qe^{-q/2}$	$\left(\dfrac{3}{4\pi}\right)^{1/2}\cos\theta$
$2\,p_x$	2	1	1	$\dfrac{1}{2\sqrt{6}}\left(\dfrac{Z}{a_0}\right)^{3/2}qe^{-q/2}$	$\left(\dfrac{3}{4\pi}\right)^{1/2}\sin\theta\cos\varphi$
$2\,p_y$	2	1	-1	$\dfrac{1}{2\sqrt{6}}\left(\dfrac{Z}{a_0}\right)^{3/2}qe^{-q/2}$	$\left(\dfrac{3}{4\pi}\right)^{1/2}\sin\theta\sin\varphi$
$3\,s$	3	0	0	$\dfrac{2}{81\sqrt{3}}\left(\dfrac{Z}{a_0}\right)^{3/2}(27-18q+2q^2)e^{-q/3}$	$\left(\dfrac{1}{4\pi}\right)^{1/2}$
$3\,p_z$	3	1	0	$\dfrac{4}{81\sqrt{6}}\left(\dfrac{Z}{a_0}\right)^{3/2}(6-q)qe^{-q/3}$	$\left(\dfrac{3}{4\pi}\right)^{1/2}\cos\theta$
$3\,p_x$	3	1	1	$\dfrac{4}{81\sqrt{6}}\left(\dfrac{Z}{a_0}\right)^{3/2}(6-q)qe^{-q/3}$	$\left(\dfrac{3}{4\pi}\right)^{1/2}\sin\theta\cos\varphi$
$3\,p_y$	3	1	-1	$\dfrac{4}{81\sqrt{6}}\left(\dfrac{Z}{a_0}\right)^{3/2}(6-q)qe^{-q/3}$	$\left(\dfrac{3}{4\pi}\right)^{1/2}\sin\theta\sin\varphi$
$3\,d_{z^2}$	3	2	0	$\dfrac{4}{81\sqrt{30}}\left(\dfrac{Z}{a_0}\right)^{3/2}q^2e^{-q/3}$	$\left(\dfrac{5}{16\pi}\right)^{1/2}(3\cos^2\theta-1)$
$3\,d_{xz}$	3	2	1	$\dfrac{4}{81\sqrt{30}}\left(\dfrac{Z}{a_0}\right)^{3/2}q^2e^{-q/3}$	$\left(\dfrac{15}{4\pi}\right)^{1/2}\sin\theta\cos\theta\cos\varphi$
$3\,d_{yz}$	3	2	-1	$\dfrac{4}{81\sqrt{30}}\left(\dfrac{Z}{a_0}\right)^{3/2}q^2e^{-q/3}$	$\left(\dfrac{15}{4\pi}\right)^{1/2}\sin\theta\cos\theta\sin\varphi$
$3\,d_{x^2-y^2}$	3	2	2	$\dfrac{4}{81\sqrt{30}}\left(\dfrac{Z}{a_0}\right)^{3/2}q^2e^{-q/3}$	$\left(\dfrac{15}{16\pi}\right)^{3/2}\sin^2\theta\cos 2\varphi$
$3\,d_{xy}$	3	2	-2	$\dfrac{4}{81\sqrt{30}}\left(\dfrac{Z}{a_0}\right)^{3/2}q^2e^{-q/3}$	$\left(\dfrac{15}{16\pi}\right)^{3/2}\sin^2\theta\sin 2\varphi$

* $q = Zr/a_0$

For the ground state of the hydrogen atom, i.e., $n = 1$ and $l = 0$, $D(r)$ shows a maximum at $r = a_0 = 0.529$ Å, and hence the most probable position for the electron is identical with the radius predicted by Bohr for the first electronic orbit. But, whereas Bohr predicted that the electron will be only at this distance, the plot shows definite probabilities for finding the electron at distances both shorter and longer than a_0.

The radial distribution functions for $n = 2$ and 3 are less sharp than the one for $n = 1$, and they extend over wider values of r. These facts indicate a lesser localization of the electron with increase in distance from the nucleus and decrease in attraction between the latter and the electron. Further, for the electron in $2s$, $3s$, and $3p$ states we observe a relatively large maximum in $D(r)$ accompanied by smaller maxima at lower values of r. Consequently, in these states the electron has definite probabilities for occupying more than one preferred distance from the nucleus, although this preference decreases as the nucleus is approached.

Inspection of Table 14–8 reveals that all orbitals with identical values of l and m have the same dependence of $\Theta\Phi$ on the two angles irrespective of the values of n and R. Consequently, all s orbitals will have the same shape, and so will all p_z orbitals, p_x orbitals, etc. For s orbitals $\Theta\Phi$ is a constant, and hence all such orbitals will have spherical symmetry. However, since all other types give dependence of $\Theta\Phi$ on the angles, all orbitals other than s will have shapes which will not be spherically symmetrical.

Three-dimensional plots of $\Theta\Phi$ are shown in Figure 14–12 on the next page. As expected, the s orbitals are spherical. The p orbitals consist of two portions which together resemble a dumbbell; and, although equivalent, they differ from each other in their orientation along the axes. However, the various d orbitals differ both in shape and space orientation.

QUANTUM MECHANICAL TREATMENT OF MORE COMPLEX ATOMS

We have seen above that for hydrogenlike atoms the variables in the Schrödinger equation are separable, and the equation can be solved exactly by analytic means. However, this is no longer true when more than one electron is present. To illustrate the nature of the complexities, consider the case of an atom consisting of two electrons and a nucleus of charge Ze. For such an atom the potential energy is given by

$$U = -\frac{Ze^2}{r_1} - \frac{Ze^2}{r_2} + \frac{e^2}{r_{12}} \tag{64}$$

where r_1 is the distance of electron (1) from the nucleus, r_2 that of electron (2), and r_{12} is the distance of separation of the two electrons. If we let ψ

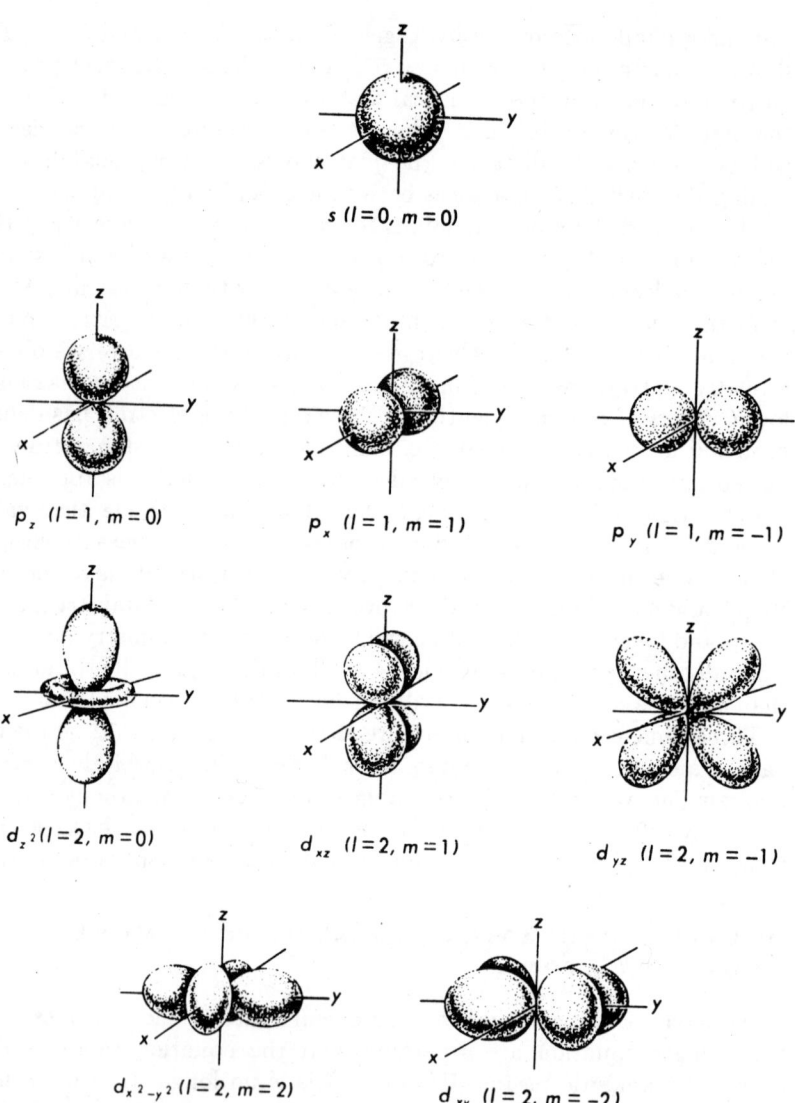

$s \; (l = 0, \; m = 0)$

$p_z \; (l = 1, \; m = 0)$ $p_x \; (l = 1, \; m = 1)$ $p_y \; (l = 1, \; m = -1)$

$d_{z^2} (l = 2, \; m = 0)$ $d_{xz} \; (l = 2, \; m = 1)$ $d_{yz} \; (l = 2, \; m = -1)$

$d_{x^2 - y^2} \; (l = 2, \; m = 2)$ $d_{xy} \; (l = 2, \; m = -2)$

Figure 14–12.   Plots of the angular parts of the hydrogen atom wave functions.

be the total wave function and m the mass of the electron, then Eq. (37) yields

$$\nabla_1^2 \psi + \nabla_2^2 \psi + \frac{8 \pi^2 m}{h^2} \left(E + \frac{Ze^2}{r_1} + \frac{Ze^2}{r_2} - \frac{e^2}{r_{12}} \right) \psi = 0 \qquad (65)$$

Here E is the total energy of the system, and ∇_1^2 and ∇_2^2 refer to the co-ordinates of the two electrons. The variables in Eq. (65) are not separable, and the equation cannot be solved directly by analytic methods. Similar problems are encountered in other cases involving the presence of more than one electron.

To handle such equations, various approximation procedures for arriving at solutions have been developed. Of these, the one used most frequently is the *variation method*, whose basis is the variation principle. Suppose that we start by assuming suitable functions for ψ and then use these to find E from the appropriate Schrödinger equation. The variation principle states that all such calculated values of E will be larger than the true E, and will give the latter only when the correct function for ψ is chosen, i.e., we shall get $E_{\text{calc}} \geqslant E$. Consequently, of all the functions chosen those giving the lowest energy will be best, and in the limit the correct function for ψ will yield the correct value for E. This principle supplies a criterion for the direction of the calculations and the choice of a suitable function for ψ.

The manner in which such trial calculations are made can be gathered from the following considerations. Equation (37) may be rearranged to

$$\frac{-h^2}{8\,\pi^2 m}\,\nabla^2\psi + U\psi = E\psi$$

or

$$\left(\frac{-h^2}{8\,\pi^2 m}\,\nabla^2 + U\right)\psi = E\psi \tag{66}$$

Let us define now the *Hamiltonian operator*, or simply the *Hamiltonian*, by the expression

$$H = \frac{-h^2}{8\,\pi^2 m}\,\nabla^2 + U \tag{67}$$

Then Eq. (66) can be written simply as

$$H\psi = E\psi \tag{68}$$

H is a noncommuting operator, i.e., $H\psi \neq \psi H$. However, since in any given situation E is a constant, $E\psi = \psi E$. In the expression for the Hamiltonian ∇^2 represents the sum of the ∇^2 terms for all electrons present, namely,

$$\nabla^2 = \nabla_1^2 + \nabla_2^2 + \nabla_3^2 + \cdots$$
$$= \Sigma \nabla_1^2 \tag{69}$$

Thus in the case of the hydrogen atom, where one electron is present, $\nabla^2 = \nabla_1^2$. However, for a two-electron atom $\nabla^2 = \nabla_1^2 + \nabla_2^2$, as may be seen from Eq. (65).

The general solution of Eq. (68) for E takes the form

$$E = \frac{\int \psi H \psi \, d\tau}{\int \psi^2 \, d\tau} \tag{70}$$

.v⟨...⟩ is an element of volume, and the integration has to be carried out over all space. Hence, once a function is assumed for ψ, it may be substituted into Eq. (70), and the corresponding energy evaluated. The quantity $\psi^2 d\tau$ in the denominator of Eq. (70) represents the probability of finding the electron or electrons involved in the volume element $d\tau$. Consequently, the integral $\int \psi^2 d\tau$ should equal unity when all space is considered. This will be true only if the ψ function chosen includes the correct proportionality coefficients in the ψ expression selected. We say then that the ψ function is normalized. However, when the appropriate coefficients are not included in the ψ function, then $\int \psi^2 d\tau \neq 1$, and both integrals in Eq. (70) must be evaluated to find E.

Such trial calculations are usually very laborious and prohibitively time-consuming when carried out by ordinary means. However, the advent of digital computers has reduced greatly both of these difficulties, and has made feasible calculations on systems which would have been practically impossible to handle before.

THE HELIUM ATOM

A good example of the application of the variation method is the calculation of the energy of the helium atom in the ground state. This energy is known experimentally to be -78.98 eV, i.e., the sum of the ionization potentials for both electrons as given in Table 14–5. For this atom, consisting of a nucleus and two electrons, the potential energy is given by Eq. (64) with $Z = 2$. Again, the expression for H is

$$
\begin{aligned}
H &= \frac{-h^2}{8\pi^2 m}(\nabla_1^2 + \nabla_2^2) + U \\
&= \frac{-h^2}{8\pi^2 m}(\nabla_1^2 + \nabla_2^2) + \left(-\frac{2e^2}{r_1} - \frac{2e^2}{r_2} + \frac{e^2}{r_{12}}\right)
\end{aligned}
\tag{71}
$$

Using this equation for H, and assuming the unnormalized wave function ψ to be the product of two hydrogenlike wave functions, namely,

$$
\psi = e^{-Zr_1/a_0} e^{-Zr_2/a_0}
\tag{72}
$$

Unsold obtained in 1927 a value of $E = -74.81$ eV. In the next approximation Kellner took ψ to be the same as given by Eq. (72), but assumed Z to be an adjustable parameter Z'. With $Z' = 1.6875$ he was able to get -77.47 eV, a much better result. The value of $Z' = 1.69$ rather than 2 is interpreted to mean that one electron in the helium atom screens the other from the nucleus, and thereby reduces the net positive charge of the latter. Finally, after a number of other attempts, Hylleraas obtained in 1930 $E = -78.99$ eV by using for ψ a polynomial with 14 terms.

The ground state of the helium atom involves the presence of two $1s$ electrons for which the total wave function is ψ. To occupy the same orbital without violating the Pauli exclusion principle, it is necessary that the two electrons have opposite spins. Hence the spin quantum number for one of the electrons must be $\frac{1}{2}$ and for the other $-\frac{1}{2}$. Without these opposed spins no stable orbital can be formed.

OTHER MULTI-ELECTRON ATOMS

The above treatment of the helium atom is an example of the manner in which quantum mechanics approaches the problem of dealing with multi-electron atoms. In this approach each electron in the atom is represented by a wave function ψ_e called an atomic orbital. The wave function for the whole atom, ψ, is then taken to be some suitable combination of the atomic orbitals of the electrons present. Again, the location of an electron in an atom is determined by its quantum numbers, and the latter determine also the nature of ψ_e. Thus, as in the hydrogen atom, ψ_e depends on the quantum numbers n, l, and m. Since for each combination of these three quantum numbers we may have electrons with spins of $\pm\frac{1}{2}$, each atomic orbital can refer to two electrons, namely, one for which the spin is $\frac{1}{2}$ and another for which the spin is $-\frac{1}{2}$.

Atomic orbitals are designated in the same manner as the electrons to which they correspond. For instance, an $1s$ electron is said to be in a $1s$ atomic orbital, a $2p$ electron in a $2p$ orbital, a $3d$ electron in a $3d$ orbital, etc. If we suppose now that the orbitals for electrons in various atoms are similar to those for the hydrogen atom, then either Table 14–8 or Figure 14–11 shows that only one type of s orbital is possible for a given value of n, three types of p orbitals for $n = 2$ or higher, and five types of d orbitals for $n = 3$ or higher. Hence the maximum number of electrons which can be accommodated in an s orbital is two On the other hand, the p electrons may be distributed among p_x, p_y, and p_z orbitals, and so the maximum number of electrons which these can hold is $3 \times 2 = 6$. Similarly, d electrons may exist in the five d orbitals, with the maximum number of d electrons possible being $5 \times 2 = 10$. These maximum numbers for s, p, and d electrons are identical with those deduced before from application of the Pauli exclusion principle, and given in Table 14–7.

These considerations make it possible to assign electrons to various orbitals in an atom by means of the *aufbau or building up* principle. According to this principle we imagine an atom to be built up by starting with the nucleus and adding to it progressively the number of electrons required to form the atom in its ground state. Each electron will enter in sequence the available orbital of lowest energy, then the next higher one, and so on. By this means we obtain as before for hydrogen the electron arrangement $1s^1$, for helium $1s^2$, for oxygen $1s^2 2s^2 2p^4$, etc. However, to ascertain the distribution of p electrons among the three types of p orbitals, or d electrons among the five types of d orbitals, supplementary information generally is needed. Thus it is found that in nitrogen, with the electron arrangement $1s^2 2s^2 2p^3$, one p electron occupies each of the three p orbitals. Hence here none of the p orbitals is filled. As a rule all p orbitals have to be occupied by at least one electron before orbital filling takes place. The reason for this behavior is the fact that spin uncoupled electrons lead to higher binding energies in the atom than do coupled ones. As a consequence the atom is stabler with the former arrangement than with the latter.

REFERENCES

1. Fano and Fano, *Basic Physics of Atoms and Molecules*, John Wiley & Sons, Inc., New York, 1959.
2. W. Heitler, *Elementary Wave Mechanics*, Oxford University Press, New York, 1956.
3. G. Herzberg, *Atomic Spectra and Atomic Structure*, Dover Publications, New York, 1944.
4. W. Kauzmann, *Quantum Chemistry*, Academic Press, New York, 1957.
5. H. G. Kuhn, *Atomic Spectra*, Academic Press, New York, 1962.
6. E. Merzbacher, *Quantum Mechanics*, John Wiley & Sons, Inc., New York, 1961.
7. Pauling and Wilson, *Introduction to Quantum Mechanics*, McGraw-Hill Book Company, Inc., New York, 1935.
8. K. S. Pitzer, *Quantum Chemistry*, Prentice-Hall, Inc., Englewood Cliffs, N. J., 1953.
9. R. S. Shankland, *Atomic and Nuclear Physics*, The Macmillan Company, New York, 1960.

PROBLEMS

1. A spherical particle of unit density falls freely in air with a constant velocity of 0.500 cm sec^{-1}. The viscosity coefficient of the air is 0.000180 poise. Using Stokes's law, find the radius of the particle. *Ans.* 6.43×10^{-4} cm.

2. A spherical particle of unit density and radius of 1.00×10^{-4} cm is trapped between two parallel charged plates 2.40 cm apart. To keep the particle at rest, a potential difference of 2052 volts has to be applied across the plates. From these data find (a) the applied electric field in volts cm^{-1}, and (b) the number of unit charges of electricity on the particle. One electrostatic unit of potential is equal to 300 volts. *Ans.* (a) 855 volts cm^{-1}; (b) 3.

3. What applied electric field in volts cm^{-1} will be required to make the particle described in problem 2 rise against gravity with a velocity $v_2 = v_1/5$?

4. Suppose that the polarity of the plates in problem 2 is reversed. What applied electric field in volts cm^{-1} will be required to make the particle fall with a velocity $v_2 = 1.20\, v_1$?

5. What will be the mass of an electron moving with a speed of (a) 5.00×10^9 cm sec^{-1}, and (b) 2.90×10^{10} cm sec^{-1}?   *Ans.* (a) 9.24×10^{-28} g.

6. (a) What is the mass of the proton at rest? (b) What will be its mass when it is moving with a speed of 2.00×10^{10} cm sec^{-1}?

7. According to the Heisenberg uncertainty principle, what will be the minimum uncertainty in the position of an electron when its momentum is known within Δp equal to (a) 1×10^{-2}, (b) 1×10^{10}, and (c) zero g-cm sec^{-1}?

8. Silver possesses two stable isotopes whose atomic weights are 106.911 and 108.909 on the carbon 12 scale. Determine the abundance of each of these isotopes in ordinary silver.   *Ans.* Ag^{107}: 52.00%.

9. Magnesium possesses three stable isotopes. For two of these the atomic weights are 23.985 and 24.986, while the abundances are, respectively, 78.60 and 10.11%. Find the atomic weight of the third isotope.

10. What is the number of electrons, protons, and neutrons in an atom of gold of atomic mass 197?

11. The atomic weight of one of the isotopes of Pt is 193.977. (a) Calculate the masses of the electrons and of the nuclei in one gram atomic weight of this isotope. (b) What would be the mass of the nuclei calculated from the rest masses of the neutrons and protons present in the atom?   *Ans.* (b) 195.570 g.

12. (a) Calculate the amount of energy in calories which would be emitted per second by a perfect black body of 1 mm^2 in area kept at 2000°K. (b) How much energy will be absorbed by this body per second?

13. What will be the energy radiated by the black body described in problem 12 when its temperature is raised to 10,000°K?

14. Assuming that the exponential is large compared with unity in the denominator of Eq. (10), derive the expression for the wave length λ_m at which the distribution of energy in black body radiation is a maximum at any constant temperature T.

15. Using the relation derived in problem 14, calculate λ_m at 1000°K.
        Ans. $\lambda_m = 2.88$ microns.

16. The wave length of a certain line in the Balmer series is observed to be 4341 Å. To what value of n_2 does this line correspond?

17. Calculate the wave number, wave length, and frequency of the first line in the Lyman series.   *Ans.* $\nu = 2.466 \times 10^{15}$ sec^{-1}.

18. Calculate the wave number, wave length, and frequency of the third line of the Paschen series.

19. Using the Bohr theory, calculate the radii of the first and tenth orbits in the hydrogen atom.

20. Calculate the velocities of the electron in the first and tenth orbits of the hydrogen atom.   *Ans.* $v_1 = 21.9 \times 10^7$ cm sec^{-1}.

21. Calculate the kinetic, potential, and total energies of the electron in the first Bohr orbit of the hydrogen atom.

22. Calculate the kinetic, potential, and total energies of the electron in the tenth Bohr orbit of the hydrogen atom.

23. (a) What total amount of energy in calories would be required to shift all the electrons from the first Bohr orbit to the sixth Bohr orbit in a gram atom of hydrogen? (b) Through what distance would each electron have to move? (c) What frequency of radiation would be emitted if the electrons returned to their initial state?   *Ans.* (a) 304,800 cal.

24. What is the frequency and wave length of radiation emitted by the transition of the electron from the second to the first Bohr orbit in singly ionized helium? Assume that the Rydberg constant is the same as R_H for hydrogen.

25. Calculate the energy in calories required to ionize one gram atomic weight of hydrogen.

26. Calculate the energy in calories required to produce from neutral He atoms one gram atomic weight of (a) He^+ ions and (b) He^{++} ions.

Ans. (b) 1,821,000 cal.

27. What possible values may the other quantum numbers assume when the principal quantum number is 5?

28. According to the Pauli principle, what maximum number of electrons may be found in each of the l subshells in problem 27?

29. Explain the fundamental difference between the electronic configurations of the group IA elements (Li, Na, K, etc.) and those of the group IB elements (Cu, Ag, Au).

30. Give the electronic arrangements for the following: (a) Li, (b) Be$^+$, (c) O, and (d) Ca.   *Ans.* (a) $1s^2 2s^1$.

31. To which neutral atom corresponds the electronic arrangement $1s^2 2s^2 p^6 3s^2 p^6 d^5 4s^1$?

32. Give the values of L, S', and J for the following spectral terms: (a) $^1S_0$, (b) $^2P_{3/2}$, (c) $^3F_2$, (d) $^4F_{3/2}$.   *Ans.* (a) $L = 0$, $S' = 0$, $J = 0$.

33. List the possible values of J for each of the following spectral terms: (a) ^1S, (b) ^2S, (c) ^2P, (d) ^3F.

34. Figure 14–7 shows that the 3^2P state for sodium consists of a doublet level with spectral terms given by $3^2P_{1/2}$ and $3^2P_{3/2}$. The transitions from these levels to the ground state result in the well-known sodium D line, which consists of two closely spaced lines. For the transition $3^2P_{3/2} \rightarrow 3^2S_{1/2}$, $\bar{\nu} = 16,978.0$ cm^{-1}, while for the transition $3^2P_{1/2} \rightarrow 3^2S_{1/2}$, $\bar{\nu} = 16,960.9$ cm^{-1}. (a) Which of these levels lies closer to the ground state? (b) What is the difference in energy between the two 3^2P states?

35. The 3^2D state of sodium consists of two levels with spectral terms $3^2D_{3/2}$ and $3^2D_{5/2}$. Similarly, the 3^2P state of sodium consists of two levels with spectral terms $3^2P_{1/2}$ and $3^2P_{3/2}$. (a) What transitions may be expected to occur between the two states? (b) What should be the multiplicity of the expected line?

Ans. (b) Quadruplet.

36. The spectral line observed for the transitions described in problem 35 is actually a triplet. Which of the possible transitions is forbidden by the selection rules?

37. Consider a particle of mass m constrained to move in a cubic box of side a. (a) What is the quantum mechanical expression for the energy of the particle? (b) How does the energy depend on the mass of the particle? (c) How does the energy depend on the size of the box?

38. An electron moves in a cubic box 1 cm on edge. How much energy, in ergs, will be required to raise the electron from its lowest level to the state where (a) $n_x = 2$, $n_y = 1$, $n_z = 1$; (b) $n_x = 1$, $n_y = 2$, $n_z = 1$; (c) $n_x = 1$, $n_y = 1$, $n_z = 2$?   *Ans.* (a) $\Delta E = 1.81 \times 10^{-26}$ erg.

39. A proton moves in a cubic box 1 cm on edge. How much energy will be required to raise the proton from its lowest energy level to the state where $n_x = n_y = n_z = 2$?

40. The average kinetic energy of a gas molecule is $\frac{3}{2} kT$, where k is the Boltzmann constant. What will have to be the average value of $(n_x^2 + n_y^2 + n_z^2)$ in order that a molecule of He may possess a temperature of 300°K when confined in a cubical box 100 Å on edge?

41. Determine for a hydrogenlike atom the radial distribution function for a 2s orbital relative to that for a 2p orbital.

42. Using the result obtained in problem 41, calculate for the hydrogen atom the probabilities of finding the electron in the 2s orbital relative to those in the 2p orbital at distances from the nucleus of (a) $r/a_0 = 1$, (b) $r/a_0 = 2$, and (c) $r/a_0 = 3$.   *Ans.* (a) 3.

43. Using the result obtained in problem 41, calculate for singly ionized helium the probabilities of finding the electron in the 2s orbital relative to those in the 2p orbital at distances from the nucleus of (a) $r/a_0 = \frac{1}{2}$, (b) $r/a_0 = 1$, and (c) $r/a_0 = \frac{3}{2}$.

15
Nature of
Chemical
Bonding

THE PERIODIC recurrence of chemical properties of the elements on increase of atomic number, and hence mass, rules out the nucleus and points to the extranuclear electrons as the seat of chemical reactivity. Again, we have seen that, independently of the total number of electrons present, the electronic configuration in the external shells of all atoms belonging to a given group is essentially identical. The latter fact strongly suggests that chemical reactivity is associated primarily with the outermost electrons in an atom, and does not depend to any degree on electrons in shells much below the surface. Therefore, any explanation for the laws of chemical combination and valency must be sought in the behavior of the outermost electrons, and in the interaction of these with the similar electrons of other atoms.

In dealing with atomic combination it is convenient to classify the bonds formed into three basic types, namely, (a) ionic or electrovalent bonds, (b) covalent bonds, and (c) metallic bonds. Each of these types represents certain basic behavior which can be defined and described. However, not all bonding found in molecules falls clearly into one of these categories. In many instances the bonds encountered reflect properties which indicate a character intermediate between the basic types. In such cases the bond is treated as a combination of the basic types involved, with each type making a definite contribution to the nature of the bond in question.

THE IONIC BOND

The first attempt to explain valence in terms of electrons was made by W. Kossel in 1916. The starting point of Kossel's theory was the observation that the stability of the rare gases is attained when the external shell of the atom contains two electrons in helium and eight electrons in every other case. Kossel suggested, therefore, that all atoms tend to reach rare gas configurations either by taking on or losing electrons. For example, an atom of sodium contains two closed shells with one electron outside of these. Again, chlorine contains two closed shells and seven external electrons. Now, to attain a rare gas configuration sodium can either divest itself of the one external electron it possesses and become the positive ion Na^+, or it may acquire seven electrons and become Na^{-7}. Obviously the first of these processes should be easier to accomplish, and so the tendency of sodium will be to lose an electron and become Na^+. On the other hand the chlorine atom may acquire a rare gas configuration either by gaining one electron or by losing all seven. Here evidently the first process is the more probable, and hence chlorine would tend to form the ion Cl^-. These natural tendencies come into play when sodium and chlorine are brought together. Writing a dot to represent the electron, we obtain thus the reaction

$$\text{Na} + \cdot \overset{\cdot\cdot}{\underset{\cdot\cdot}{\text{Cl}}} : \; = \text{Na}^+ + [\; : \overset{\cdot\cdot}{\underset{\cdot\cdot}{\text{Cl}}} : \;]^-$$

in which an electron is transferred from the sodium to the chlorine to yield a sodium and a chloride ion, each with a rare gas configuration. Similarly, the reaction of barium and sulfur can be represented by the electron transfers

$$\text{Ba} + \overset{\cdot\cdot}{\underset{\cdot\cdot}{\text{S}}} : \; = \text{Ba}^{++} + [\; : \overset{\cdot\cdot}{\underset{\cdot\cdot}{\text{S}}} : \;]^{--}$$

This type of atomic interaction, involving the *outright transfer* of one or more electrons from one atom to another, leads to the formation of ions which are held together by electrostatic attraction. Because of the electrostatic nature of the binding force, the bond between the atoms is said to be *ionic* or *electrovalent,* and the valence exhibited is said to be an *electrovalence.*

Ionic binding is found in gaseous molecules such as the alkali halides, and also in ionic crystals. Consider first a single molecule consisting of a pair of ions of charges Z_1e and Z_2e. The electrostatic potential energy of attraction between these ions will be given by $-Z_1Z_2e^2/r$, where r is the distance of separation of the two ions. Besides this attraction, such a pair of ions will also experience a repulsion, particularly at small values of r, due to overlapping of electron clouds and interaction of the like-charged nuclei. The potential energy due to repulsion can be represented by a term of the form $be^{-r/\rho}$, where b and ρ are constants. Consequently, the total potential energy of such a molecule can be written as

$$U = -\frac{Z_1 Z_2 e^2}{r} + be^{-r/\rho} \tag{1}$$

Equation (1) gives $U = 0$ when $r = \infty$, i.e., the potential energy of the molecule is zero when the ions are separated from each other. As

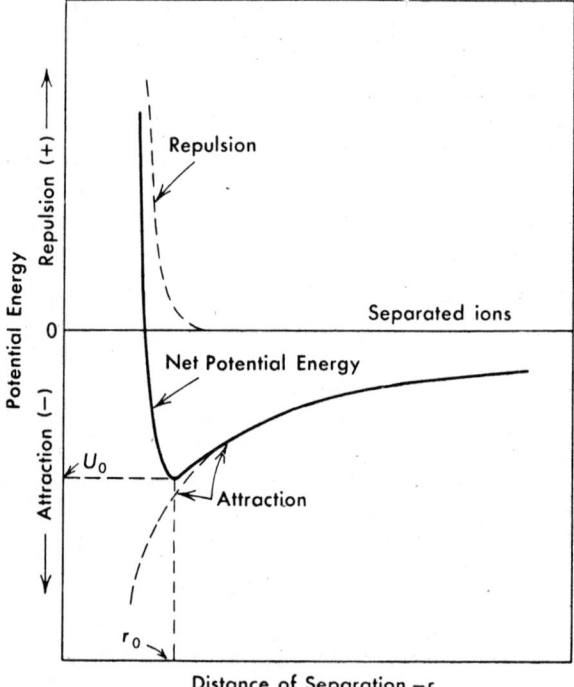

Figure 15–1.   Potential energy diagram for ionic bonding.

the ions are brought closer to each other the attraction and repulsion potentials increase in the manner shown schematically by the dotted curves in Figure 15–1. The net potential energy U is the sum of the repulsion and attraction potentials, and is the solid curve in the figure. The latter curve shows a minimum at $r = r_0$ and $U = U_0$, which correspond to the equilibrium separation and binding energy of the molecule.

For this minimum we must have

$$\left(\frac{dU}{dr}\right)_{r=r_0} = 0$$

and hence differentiation of Eq. (1) with respect to r gives

$$\frac{Z_1 Z_2 e^2}{r_0^2} - \frac{be^{-r_0/\rho}}{\rho} = 0$$

or
$$be^{-r_0/\rho} = \frac{\rho Z_1 Z_2 e^2}{r_0^2} \qquad (2)$$

On substitution of Eq. (2) into Eq. (1) we obtain

$$U_0 = -\frac{Z_1 Z_2 e^2}{r_0} + \frac{\rho Z_1 Z_2 e^2}{r_0^2}$$

$$= -\frac{Z_1 Z_2 e^2}{r_0}\left(1 - \frac{\rho}{r_0}\right) \qquad (3)$$

The term ρ/r_0 is equal generally to 0.10–0.15, and consequently the repulsion operates to reduce the electrostatic attraction energy between the ions by about 10–15% at the equilibrium position.

LATTICE ENERGY OF IONIC CRYSTALS

The lattice energy of an ionic crystal is the enthalpy change, ΔH_c, required to decompose 1 mole of the crystal into its constituent gaseous ions at any temperature T. Born and Mayer showed in 1932 that this energy can be calculated theoretically on the basis of ideas analogous to those developed in the preceding section. In what follows we shall limit ourselves to a discussion of crystals of the alkali halides, for which $Z_1 = Z_2 = 1$.

Born and Mayer took the equilibrium potential energy of 1 mole of an alkali halide crystal at 0°K to be

$$U_0 = -\frac{A'Ne^2}{r_0}\left(1 - \frac{\rho}{r_0}\right) - \frac{C}{r_0^6} + E_0 \qquad (4)$$

where A', ρ, C, and E_0 are constants for any given crystal, N is Avogadro's number, and r_0 is the distance between ions. In Eq. (4) the first term is Eq. (3) modified by the introduction of the *Madelung constant* A', which takes into account the arrangement and spacing of the ions in the crystal lattice. Again, the second term represents the van der Waals attractions among the ions, while the last term is the zero-point energy of the crystal. The latter is the residual vibrational energy present in the crystal at absolute zero.

The parameters A', ρ, C, and E_0 can be obtained from various properties of the solid, while r_0 follows from the X-ray analysis of the crystal. Hence U_0 can be calculated. Further, since $\Delta H_c = - U_0$ at 0°K, the lattice energy can be obtained at this temperature and then corrected to any desired temperature. Values of ΔH_c obtained in this manner for 25°C are given in the last column of Table 15–1.

TABLE 15–1. LATTICE ENERGIES OF ALKALI HALIDES AT 25°C

Salt	$-\Delta H_f$	$\frac{1}{2} \Delta H_d$	ΔH_s	I	A	ΔH_c (kcal mole^{-1}) Born-Haber Cycle	Born-Mayer Theory
NaF	137	18	26	118	84	215	213
KF	135	18	22	100	84	191	190
RbF	133	18	20	96	84	183	182
NaCl	98	29	26	118	87	184	183
KCl	104	29	22	100	87	168	164
RbCl	105	29	20	96	87	163	161
NaBr	86	27	26	118	82	175	175
KBr	94	27	22	100	82	161	159
RbBr	96	27	20	96	82	157	154

The values of ΔH_c obtained above can be checked by means of a cycle devised by Born and Haber in 1919. The steps in this *Born-Haber cycle* for any temperature T are

$$
\begin{array}{ccc}
\text{MX(s)} & \xrightarrow[\;(2)\;]{\Delta H_c} & \text{M}^+(g) + \text{X}^-(g) \\
{\scriptstyle(1)}\uparrow{\scriptstyle \Delta H_f} & & {\scriptstyle(4)}\uparrow I \qquad\qquad \uparrow -A \\
\text{M(s)} + \tfrac{1}{2}\text{X}_2(g) & \xrightarrow[\;(3)\;]{\Delta H_s + \frac{1}{2}\Delta H_d} & \text{M}(g) + \text{X}(g)
\end{array}
$$

The first step involves the formation of 1 mole of the alkali halide MX from the elements, for which ΔH_f is the heat of formation, while the second step is the dissociation of the lattice to form the gaseous ions $\text{M}^+(g)$ and $\text{X}^-(g)$, for which the thermal effect is the lattice energy ΔH_c. These same gaseous ions can be obtained also by proceeding along steps (3) and (4). Step (3) involves, first, the sublimation of M(s) to M(g) for which ΔH_s is the heat of sublimation of the metal, and also the dissociation of the molecule $\text{X}_2(g)$ into 2X(g). For one atom required here the thermal effect is $\frac{1}{2} \Delta H_d$, where ΔH_d is the heat of dissociation of the molecule $\text{X}_2(g)$. Consequently, the total thermal effect for step (3) is $\Delta H_s + \frac{1}{2} \Delta H_d$. Finally, M(g) can be converted to $\text{M}^+(g)$ plus an electron by supplying the ionization energy I, while $X(g)$ can be converted to $X^-(g)$ by having it react with an electron. The thermal effect for the latter process is defined as

$$X(g) + e = X^-(g) \qquad \Delta H = -A \tag{5}$$

where A is the *electron affinity* of the atom for the electron. Since the over-all ΔH of a particular reaction sequence is independent of the path, then it must follow that

$$\Delta H_f + \Delta H_c = \Delta H_s + \frac{1}{2}\Delta H_d + I - A$$

or $\qquad \Delta H_c = \Delta H_s + \frac{1}{2}\Delta H_d + I - A - \Delta H_f \tag{6}$

Equation (6) allows the calculation of ΔH_c from available thermochemical data for ΔH_s, ΔH_d, I, A, and ΔH_f. Such data for a number of alkali halides at 25°C are given in Table 15–1. The values of ΔH_c calculated from these, and shown in the second last column of the table, agree well with the lattice energies obtained from the Born-Mayer theory.

THE COVALENT BOND

Although the above explanation is satisfactory for ionic substances, it can hardly account for the formation of molecules such as CH_4, NO, H_2, N_2, and others of this category. In these no ions can be detected, and no reason is immediately apparent why, for instance, two atoms of hydrogen should combine into a molecule. To get around this difficulty, G. N. Lewis proposed in 1916 that union between atoms can be attained also from *sharing of electrons in pairs*. For example, if one hydrogen atom with a single external electron were to share this electron with another hydrogen atom, the result would be a hydrogen molecule, H:H, in which each hydrogen atom could claim the pair and possess thereby the helium structure. Again, the formation of a molecule such as CH_4 would arise from the sharing of electrons between hydrogen and carbon as follows:

$$4\,H\cdot + \cdot \overset{\displaystyle .}{\underset{\displaystyle .}{C}}\cdot = H : \overset{\displaystyle ..}{\underset{\displaystyle ..}{C}} : H$$

Here also each hydrogen attains by sharing electrons the rare gas configuration of helium, while the carbon, by being able to share its own as well as the four acquired electrons, has the rare gas configuration of neon. In all such sharings *every electron pair corresponds to a single valence bond.* To differentiate the sharing type of atomic combination from electrovalent union, the terms *covalence* and *covalent bond* are employed.

Covalent binding can be used as well to account for the formation of double and triple bonds in molecules. Thus the formation of an oxygen molecule can be represented by

$$\overset{..}{\underset{..}{O}} : + : \overset{..}{\underset{..}{O}} = \overset{..}{\underset{..}{O}} :: \overset{..}{\underset{..}{O}}$$

in which instance four electrons, corresponding to two valence bonds, are shared by the two atoms. Similarly, the electronic structure of ethylene can be written as

$$H : C :: C : H$$
$$\overset{..}{H} \quad \overset{..}{H}$$

where the four electrons between the two carbons indicate the double bond. Finally, in triple bonds three electron pairs are shared as in N_2, $:N:::N:$, or C_2H_2, $H:C:::C:H$. In N_2 each nitrogen contributes three electrons to the formation of the covalent triple bond, whereas in C_2H_2 each carbon supplies three electrons.

A covalent bond can arise also from the sharing of electrons supplied by one atom only. The bond then is said to be coordinate covalent. An example of this is found in the formation of the NH_4^+ ion from ammonia and the hydrogen ion, namely,

$$H \qquad\qquad\qquad \left[\quad H \quad \right]^+$$
$$\overset{..}{H : N :} + H^+ = \left[H : \overset{..}{N} : H \right]$$
$$H \qquad\qquad\qquad \left[\quad H \quad \right]$$

In the ammonia molecule four electron pairs are present, of which only three are shared. When a hydrogen ion comes near this molecule the nitrogen allows the proton to share with it the free pair of electrons, and the result is thus the formation of the ammonium ion. Once established, the bond between the acceptor hydrogen and the donor nitrogen is indistinguishable from the other N—H bonds, and the positive charge becomes the property of the whole group rather than of any one hydrogen. As other examples of coordinate covalent binding may be cited the formation of $BF_3 \cdot NH_3$ from BF_3 and NH_3, and of BF_4^- from BF_3 and F^-. In both these instances the acceptor atom, boron, possesses only six electrons about it. In order to reach the stable octet of electrons, it accepts an electron pair from the donors NH_3 and F^- to form the respective complexes. By this means both the donor and acceptor atoms attain rare gas configurations.

NATURE OF BOND AND PHYSICAL PROPERTIES

The manner in which a chemical bond arises reflects significantly in the properties of the resulting compounds. Because electrovalency leads to the formation of ions held together by electrostatic attraction, compounds possessing this type of binding are ionized in the solid as well as in the molten states. In the latter state they are good conductors of electricity, and, on solution in such solvents as water, they yield strong electrolytes which also conduct a current. Further, because of the strong electrostatic forces between the electrically charged ions the compounds exhibit relatively high melting and boiling points. In contrast to the electrovalent compounds those resulting from covalency are unionized and hence nonconducting, and they exhibit much lower melting and boiling points. Again, whereas electrovalent compounds are generally more soluble in polar solvents such as water or ethyl alcohol, covalent compounds are much more readily soluble in typical organic solvents such as benzene, ether, or carbon tetrachloride. These differences in the two types of binding are reflected by the data given in Table 15–2 for several electrovalent and covalent compounds.

TABLE 15–2. COMPARISON OF PROPERTIES OF ELECTROVALENT
AND COVALENT COMPOUNDS

	Electrovalent		Covalent	
Property	NaCl	$BaCl_2$	CCl_4	CH_4
Melting point	801°C	962°C	−22.8°C	−182.6°C
Boiling point	1413°C	1560°C	76.8°C	−161.4°C
Equiv. conductance at m.p.	133	64.6	0	0
Solubility in H_2O at 20°C				
(g/100 g H_2O)	36.0	35.7	Insoluble	9 cc at 20°C
Solubility in ether	Insoluble	Insoluble	∞	91 cc ⸱t 20°C

The products resulting from coordinate covalent binding may, as we have seen, be both uncharged molecules, as $BF_3 \cdot NH_3$, or ions, as NH_4^+ or BF_4^-. In this category belong also such compounds as the hydrates, ammoniates, alcoholates, etherates, etc. For this variety of compounds it is very difficult to state with clarity any property differentiations. As a matter of fact, all gradations of behavior ranging from pure covalency on the one hand to almost electrovalency on the other may be found. In general the evidence seems to indicate that coordinate covalent binding tends to increase the melting and boiling points of compounds containing only covalent bonds.

EXCEPTIONS TO THE OCTET RULE

The Lewis-Kossel theory can account for the formation of many compounds by electron transfer and sharing. In doing this the theory relies on the concept that for stability there can be present no more than eight electrons about an atom. However, many compounds are known for which this electron octet rule is apparently exceeded, namely, PCl_5, SF_6, OsF_8, etc. In terms of an electron pair per bond, the phosphorus atom in PCl_5 would have to be surrounded by 10 electrons, the sulfur in SF_6 by 12, and the osmium in OsF_8 by 16. Again, in a molecule such as B_2H_6 there are present insufficient electrons to bind by sharing both the two borons and the six hydrogens. Since each boron possesses three electrons, the total of six for the two would be just sufficient to bind the hydrogens without leaving any for the bond between the borons. Still, although not very stable, B_2H_6 exists.

To account for the existence, and in some instances high stability, of compounds apparently violating the octet rule, the suggestion has been made that more than eight electrons may be coordinated about an atom. However, Sugden pointed out that the octet rule can be preserved provided it is granted that in some cases a bond can be formed by the sharing of a single electron rather than a pair. Thus for phosphorus pentachloride the structure may be written as a combination of three electron pairs and two singlet electron bonds, namely,

$$
\begin{array}{c}
\text{Cl} \\
\quad\;\; \overset{..}{} \quad \text{Cl} \\
\text{Cl} : \text{P} : \\
\quad\;\; \overset{..}{} \quad \text{Cl} \\
\text{Cl}
\end{array}
$$

for SF_6 as a combination of two pairs and four singlets, and for OsF_8 as a combination of eight singlet bonds. Likewise the formation of B_2H_6 may be accounted for by the structure

$$
\begin{array}{c}
\text{H} \quad \text{H} \\
\overset{..}{} \quad \overset{..}{} \\
\text{H} : \text{B} : \text{B} : \text{H} \\
\overset{.}{} \quad \overset{.}{} \\
\text{H} \quad \text{H}
\end{array}
$$

We shall see below that single electron bonds are possible, but not in the form postulated above. Quantum mechanics shows that the stability of single electron bonds arises from a phenomenon called resonance, in which the single electron, instead of being localized on a particular atom, actually oscillates between the atoms involved in the bond. Such resonance leads to a binding energy which makes possible the formation of the bond and the stabilization of the molecule in which it is present.

QUANTUM MECHANICS AND THE COVALENT BOND

Although the Lewis theory specifies that a covalent bond involves the sharing of electrons in pairs, it does not explain how such sharing leads to bonding. Possibly the most valuable contribution that quantum mechanics has made to chemistry has been its elucidation of the nature of covalence, and of the factors that are involved in the formation of the covalent bond.

There is only one molecule for which an exact solution of the wave equation has been obtained analytically, namely, the hydrogen molecule ion H_2^+. For other relatively simple molecules only approximate solutions can be arrived at, but these are sufficient to give an insight into the influences which lead to bond formation, and to the behavior of the electrons in such bonds. Complex molecules still offer many handicaps to their elucidation, but even here the results obtained for the simpler molecules have been of great help.

THE HYDROGEN MOLECULE ION

Before proceeding to normal molecules, it is very instructive to consider first the nature of the bonding forces involved in the formation of the hydrogen molecule ion. This ion, produced by bombardment of hydrogen gas with electrons, is essentially a hydrogen molecule with one electron removed by ionization. It consists, therefore, of two protons and one electron, and it may be viewed as a molecule formed from a hydrogen atom and a hydrogen ion, namely,

$$H + H^+ = H_2^+$$

Spectroscopic measurements show that the binding energy of this ion in the ground state is $E' = -2.791$ eV relative to $H + H^+$ at infinite separation, and that the internuclear equilibrium distance is $r_0 = 1.058$ Å.

In terms of the distances defined in Figure 15–2, the potential energy

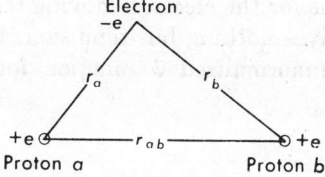

Figure 15–2. Definition of distances for hydrogen molecule ion.

for this ion is given by

$$U = -\frac{e^2}{r_a} - \frac{e^2}{r_b} + \frac{e^2}{r_{ab}} \qquad (7)$$

Again, assuming the nuclei to be stationary, the Hamiltonian for the system is

$$H = -\frac{h^2}{8\pi^2 m}\nabla^2 + U$$

$$= -\frac{h^2}{8\pi^2 m}\nabla^2 - \frac{e^2}{r_a} - \frac{e^2}{r_b} + \frac{e^2}{r_{ab}} \qquad (8)$$

where m is the mass of the electron and ∇^2 refers to its coordinates. Consequently, the Schrodinger equation for the hydrogen molecule ion follows as

$$H\psi = E\psi$$

or

$$\left(-\frac{h^2}{8\pi^2 m}\nabla^2 - \frac{e^2}{r_a} - \frac{e^2}{r_b} + \frac{e^2}{r_{ab}}\right)\psi = E\psi \qquad (9)$$

where ψ is the wave function for the electron and E is the energy of the system.

The variables in Eq. (9) are separable in elliptical coordinates, and the expressions obtained can be solved for ψ and for E as a function of r_{ab}. A plot of the latter is similar in appearance to the solid curve in Figure 15–1, and yields for the ground state of H_2^+ a minimum at $E = -16.389$ eV and $r_{ab} = 1.058$ Å. This value of E corresponds to separation of the ion into two protons and an electron. To convert E to a reference state of $H + H^+$, the ionization energy of the hydrogen atom must be added to E, and we obtain thus $E' = E + 13.598 = -16.389 + 13.598 = -2.791$ eV at $r_{ab} = r_0 = 1.058$ Å. These values for E' and r_0 agree exactly with the observed ones.

Although the exact solution gives agreement between theory and experiment, the equations resulting from it are too complicated to give much insight into the nature of the bonding forces. For this reason it is instructive to examine some approximate trial solutions more amenable to interpretation. An initial attempt to view the H_2^+ ion as a combination of a hydrogen atom and a hydrogen ion leads to no binding at all distances of separation. Similarly, the assumption that $\psi = \psi_a - \psi_b$, where ψ_a and ψ_b are the $1s$ hydrogen atom wave functions for the electron moving respectively about nucleus a and nucleus b, gives nothing but repulsion. However, bonding is obtained when the unnormalized ψ function for the electron in H_2^+ is taken to be

$$\psi = \psi_a + \psi_b \qquad (10)$$

Substitution of Eq. (10) for ψ and Eq. (8) for H into Eq. (70) of the last chapter yields for E as a function of r_{ab} a plot similar to the solid curve in Figure 15–1, with a minimum at $r_0 = 1.32$ Å and $E = -15.36$ eV. Hence $E' = -1.76$ eV.

These values of E' and r_0 do not agree too well with the experimental ones, indicating that Eq. (10) for ψ is too simple. However, this is not the point of interest at the moment. The important facts are that this equation leads to bonding of the two protons and the electron in H_2^+, and also to information regarding the nature of the bonding. The expression for the energy of the ground state given by Eq. (10) for ψ is

$$E = -\frac{e^2}{2\,a_0} + \frac{e^2}{r_{ab}} + \frac{J + K}{1 + \Delta} \tag{11}$$

where a_0 is the radius of the first Bohr orbit for the hydrogen atom, and J, K, and Δ are functions of r_{ab}. The function Δ is always positive and it need not concern us further. Again, J is a function which measures the Coulomb interaction of the electron in a $1s$ orbital on nucleus a with nucleus b, while K, the *resonance integral*, gives the energy resulting from exchange of the electron between the two nuclei. Evaluation of the terms in Eq. (11) shows that the biggest contribution to bonding is made here by K. Consequently, the stability of the hydrogen molecule ion is due primarily to the resonance energy which results from the fact that the electron is not localized on either nucleus, but oscillates so as to be part of the time on one or the other of the nuclei.

It is of interest to point out that other functions for ψ have been found which give a much closer approximation to experiment than Eq. (10). In fact, H. M. James proposed a function for ψ which yields $r_0 = 1.058$ Å and $E' = -2.786$ eV. This value of E' is only 0.005 eV larger than the experimental.

Examination of the electron density distribution in the H_2^+ ion by means of ψ obtained in the exact solution shows contour lines which not only embrace both nuclei, but also give a high concentration of electronic charge in the space between them. The latter concentration of negative charge attracts the positive nuclei, and acts thus as a means for holding them together.

THE HYDROGEN MOLECULE

This molecule, composed of two protons and two electrons, has been found spectroscopically to have in its lowest state a binding energy of $E' = -4.745$ eV and an internuclear distance of $r_0 = 0.740$ Å when the energy of the two separated atoms is taken as zero. In terms of the distances shown in Figure 15–3, the potential energy of the molecule follows as

$$U = -\frac{e^2}{r_{a1}} - \frac{e^2}{r_{a2}} - \frac{e^2}{r_{b1}} - \frac{e^2}{r_{b2}} + \frac{e^2}{r_{12}} + \frac{e^2}{r_{ab}} \tag{12}$$

and the Hamiltonian as

$$H = -\frac{h^2}{8\pi^2 m}(\nabla_1^2 + \nabla_2^2) + U \tag{13}$$

where m is the mass of the electron, and ∇_1^2 and ∇_2^2 refer to the coordinates of the two electrons. Consequently, the wave equation for the hydrogen molecule is

$$H\psi = E\psi$$

$$\left[-\frac{h^2}{8\pi^2 m}(\nabla_1^2 + \nabla_2^2) + U\right]\psi = E\psi \tag{14}$$

In Eq. (14) ψ is the total wave function for the two electrons in the molecule, and E is the energy referred to the completely separated protons

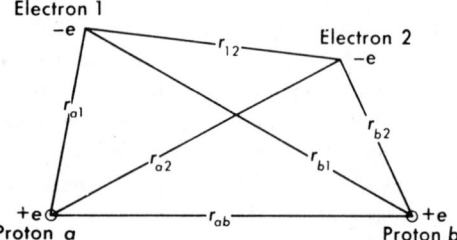

Figure 15–3.   Definition of distances for hydrogen molecule.

and electrons. The problem now is to solve Eq. (14) for the ground state of the hydrogen molecule on the assumption that a function for ψ can be constructed from $1s$ orbitals for the hydrogen atom.

The first approach to this problem was made by Heitler and London in 1927 on the basis of a formulation called the *valence bond method.* Consider the hydrogen molecule at large values of r_{ab}. The system will consist then of two normal hydrogen atoms, each involving a proton and an electron. Suppose that under these conditions nucleus a has associated with it electron (1), and nucleus b electron (2). For the first of these atoms we can designate the $1s$ orbital wave function by $\psi_a(1)$, for the second by $\psi_b(2)$, and the wave function of the two atoms by $\psi_a(1)\psi_b(2)$. However, it is also possible that on separation nucleus a may have associated with it electron (2) and nucleus b electron (1). The wave function for the two atoms will be then $\psi_a(2)\psi_b(1)$. Heitler and London postulated that on bringing the two atoms together the wave function

ψ_{VB} of the molecule may be taken to be a linear combination of the two products, or

$$\psi_{VB} = \psi_a(1)\psi_b(2) + \psi_a(2)\psi_b(1) \tag{15}$$

Use of Eq. (15) for ψ in the solution of the Schrödinger equation shows that this function does lead to bonding between the hydrogen atoms. As in the case of the hydrogen molecule ion, the bonding results largely from the energy arising from the exchange of electrons between the two atoms, and the concentration of electron charge in the space between them. However, the bonding energy and equilibrium distance obtained for the ground state of the molecule are considerably higher than the observed values, being $E' = -3.14$ eV and $r_0 = 0.869$ Å.

In the above formulation of ψ each electron is associated with a particular nucleus, and the bonding results when the electrons are brought near each other. In the sense that the association of the two electrons leads to the formation of a bond, the Heitler-London approach represents the quantum mechanical equivalent of the electron pair covalent bond postulated by G. N. Lewis. Further, since both electrons in the hydrogen molecule ground state must occupy the same $1s$ orbital, the Pauli exclusion principle demands that their spins must be opposite, i.e., for one the spin must be $\frac{1}{2}$ and for the other $-\frac{1}{2}$. Consequently, electron pairing to form a covalent bond can occur only between electrons of opposite spins.

An alternate procedure for setting up the expressions for ψ is the *molecular orbital method* developed by Hund, Mulliken, Hückel, and others. In this method the electron in a molecule is not considered to move in the sphere of a particular nucleus, but rather in the sphere of all nuclei present. Hence the wave function for each electron will depend on all the nuclei present, and the total wave function will be a product of the wave functions for the various electrons involved in the bonding.

Consider first the two protons in the hydrogen molecule and only one of the electrons. This situation corresponds to the case of the hydrogen molecule ion, for which Eq. (10) gives an approximate acceptable wave function. By analogy with this equation we can write for the wave function of electron (1)

$$\psi_1 = \psi_a(1) + \psi_b(1) \tag{16}$$

Such a function for a single electron in a molecule is called a *molecular orbital*. In this instance the molecular orbital is constructed from a linear combination of atomic orbitals. Similarly, if we now consider electron (2) in the field of the two nuclei, then by linear combination of atomic orbitals we can write the molecular orbital function for this electron as

$$\psi_2 = \psi_a(2) + \psi_b(2) \tag{17}$$

Hence ψ follows from Eqs. (16) and (17) as

$$\psi_{MO} = \psi_1\psi_2$$
$$= [\psi_a(1) + \psi_b(1)][\psi_a(2) + \psi_b(2)] \tag{18}$$

This function for ψ yields $E' = -2.68$ eV and $r_0 = 0.850$ Å for the ground state of the hydrogen molecule. The result is somewhat worse here than that obtained by the valence bond method.

Both the valence bond and molecular orbital approaches can be modified and ramified to yield wave functions which give better concordance between theory and experiment. Thus James and Coolidge developed a wave function for the hydrogen molecule which gives the correct value for r_0 and $E' = -4.720$ eV from evaluation of 13 terms in the expression. By extending the calculation to 50 terms, Kolos and Roothaan were able to get for E' a value practically identical with the observed $E' = -4.745$ eV.

DIFFERENCE BETWEEN VALENCE BOND AND MOLECULAR ORBITAL METHODS

From the results given above the impression may be gathered that the valence bond method is superior to the molecular orbital approach. This may be true for the hydrogen molecule, but it is not always true when the methods are applied to more complex molecules. Depending on the situation encountered, either of the methods may give the better results. However, whether better or worse, both methods represent approximations which lead either to qualitative or, at best, semiquantitative conclusions regarding the bonding involved in the formation of molecules. Since the molecular orbital method is usually less complex mathematically than the valence bond formulation, it is employed more frequently for the examination of problems involving both molecular structure and interaction between molecules.

In view of the wide use of these methods, it is of interest to ascertain the basic difference between them. This can be done by comparing Eqs. (15) and (18) for ψ of the hydrogen molecule. If we multiply out Eq. (18), we obtain

$$\psi_{MO} = \psi_a(1)\psi_b(2) + \psi_a(2)\psi_b(1) + \psi_a(1)\psi_a(2) + \psi_b(1)\psi_b(2) \tag{19}$$

The first two terms in this equation are identical with ψ_{VB} as given by Eq. (15). Consequently,

$$\psi_{MO} = \psi_{VB} + \psi_a(1)\psi_a(2) + \psi_b(1)\psi_b(2) \tag{20}$$

and we see that the molecular orbital method adds to ψ of the valence bond method the last two terms in Eq. (20). These terms correspond to situations where both electrons are either on nucleus a or nucleus b, i.e., the structures $H_a^- H_b^+$ and $H_a^+ H_b^-$. These terms therefore are called ionic contributions to the wave function. Since for the hydrogen molecule ψ_{VB} gives a better result for E' than ψ_{MO}, the indications are that such ionic forms are not very probable in H_2, and hence they should not be included in the wave function on an equal basis with the first two terms. However, in other molecules, particularly when the atoms involved are different, such ionic contributions may be very important, and their inclusion in the wave function may have a significant effect on the results obtained.

BONDING AND ANTIBONDING MOLECULAR ORBITALS

Consider the combination of two atoms A and B to form a diatomic molecule AB. Let ψ_A be the atomic orbital of an electron in atom A which can enter combination, and ψ_B the atomic orbital of a similar electron in atom B. When these two atoms are brought together, the two atomic orbitals may combine to form a molecular orbital ψ containing the two electrons. If we assume ψ to be a linear combination of the atomic orbitals, then we can write

$$\psi = \psi_A + \lambda \psi_B \tag{21}$$

where λ is a factor which weights the relative contributions of ψ_A **and** ψ_B to ψ. Mathematical argument shows that λ here may equal $\pm c$, **where** c is a constant dependent on the nature of A and B. On introduction of $\lambda = \pm c$ into Eq. (21), we obtain two separate expressions for ψ, namely,

$$\psi_b = \psi_A + c\psi_B \tag{22}$$
and
$$\psi_a = \psi_A - c\psi_B \tag{23}$$

The function ψ_b, representing the sum of the atomic orbitals, leads to bonding of the two atoms. ψ_b is thus a *bonding molecular orbital*, and the electrons in it are called bonding electrons. On the other hand, the function ψ_a, corresponding to the difference between the two atomic orbitals, leads to repulsion between A and B. Hence ψ_a is an *antibonding molecular orbital*, and the electrons in it are said to be antibonding.

For a stable molecule to be formed from the two atoms, it is necessary that the number of bonding electrons present in the molecule be larger than the number of antibonding ones. Whether a given electron is bonding or antibonding depends on the interaction of the atomic orbitals.

If these orbitals overlap appreciably and lead to a concentration of electronic charge between the nuclei, then bonding results. However, when the atomic orbitals repel each other, no overlap or concentration of charge between A and B is possible. We then get a strong repulsion between the two nuclei, and no bonding can take place.

BONDING IN DIATOMIC MOLECULES

A diatomic molecule in which A and B are identical is said to be *homonuclear*, while one with A and B different is said to be *heteronuclear*. Examples of the first category are H_2, O_2, and N_2; of the second, HCl, HBr, or HF.

For homonuclear diatomic molecules $c = 1$ in Eqs. (22) and (23), and hence ψ_A and ψ_B contribute equally to the formation of both bonding and antibonding molecular orbitals. In such molecules molecular orbitals are formed from similar atomic orbitals, i.e., s orbitals of one atom combine with s orbitals of the other, p orbitals with p orbitals, etc. The types of molecular orbitals formed from atomic s and p orbitals are shown in Figure 15-4. From this figure it may be seen that combination of s orbitals leads to both bonding and antibonding molecular orbitals which are symmetrical with respect to an axis joining the two atoms. Symmetrical molecular orbitals are designated as σ orbitals for bonding and σ^* for antibonding. To indicate the particular s atomic orbitals involved in the formation of the molecular orbitals, symbols such as $\sigma 1s$, $\sigma 2s$, or $\sigma^* 1s$, $\sigma^* 2s$ are employed, which show that $1s$, $2s$, etc., atomic orbitals are involved in the formation of the σ or σ^* orbitals.

If we take the x axis as the one joining the two atoms in p orbital interaction, then the combination of two p_x orbitals leads again to symmetrical or σ molecular orbitals. However, under such conditions $p_y - p_y$ or $p_z - p_z$ combinations lead to molecular orbitals which are not symmetrical about the line joining A and B, as may be seen from part (3) of the figure. These unsymmetrical combinations are called π orbitals for bonding and π^* orbitals for antibonding. Here, again, symbology such as $\pi 2p$, $\pi 3p$, or $\pi^* 2p$, $\pi^* 3p$ is used to indicate the source of the p orbitals involved in the formation of the π or π^* bonds.

These ideas on molecular orbital formation, along with the concept of bonding and antibonding electrons, can be used qualitatively to predict what happens when two atoms are brought together to form homonuclear diatomic molecules. Consider first the combination of two hydrogen atoms, each with a $1s^1$ electron configuration. These two atomic orbitals form a ground state $\sigma(1s)^2$ molecular orbital containing two bonding electrons. Since a pair of bonding electrons can be taken to form a single covalent bond, the molecule formed can be represented as H—H.

The helium atom has the configuration $1s^2$. When two such atoms are brought together, the resulting orbitals give for the combination $\sigma(1s)^2\sigma^*(1s)^2$, i.e., two bonding and two antibonding electrons. Since there is no excess of bonding electrons present, no stable He_2 molecule can be formed.

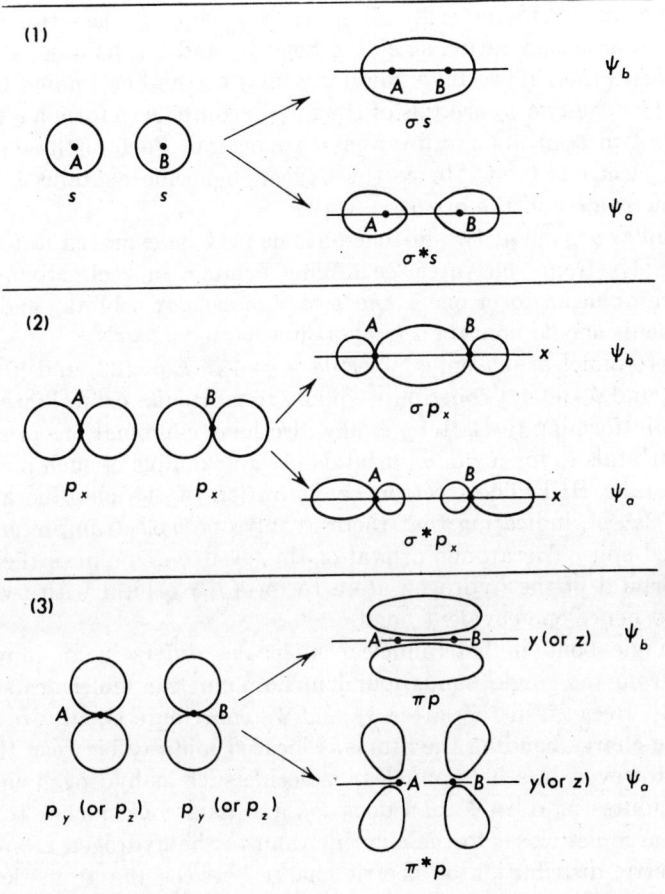

Figure 15–4.   Molecular orbitals formed from s and p atomic orbitals.

As a more complicated case of single bond formation may be taken the Na_2 molecule. The sodium atom contains 11 electrons in the configuration $1s^2 2s^2 2p^6 3s^1$. When two such atoms are combined, the resulting molecular configuration is found to be $[\sigma(1s)^2\sigma^*(1s)^2\sigma(2s)^2\sigma^*(2s)^2$ $\sigma(2p_x)^2\sigma^*(2p_x)^2\pi(2p_y)^2\pi^*(2p_y)^2\pi(2p_z)^2\pi^*(2p_z)^2]\sigma(3s)^2$. It will be observed that for the closed shells the bonding electrons are balanced off by antibonding ones, and hence we need consider only the outer or valence shell electrons. Since there are two of these in the $\sigma(3s)$ orbital, a single bond is formed between the two sodium atoms.

The oxygen molecule represents a case of double bond formation. The electron configuration of the oxygen atom is $1s^2 2s^2 2p^4$. Of the four p electrons, two are in a p_z orbital, and one each in the p_x and p_y orbitals. For formation of the molecule all electrons except the last two may be omitted from consideration since the bonding and antibonding electrons cancel each other. If we take now the x-axis as the line joining the two atoms, then the two p_x orbitals of the atoms combine to form a σ orbital, while the two p_y orbitals unite to give a π orbital. Each of these orbitals has two electrons in it. Hence the oxygen molecule contains a double bond; one is a σ and the other a π type.

By similar argument two nitrogen atoms may be expected to form the molecule N_2 from the three $2p$ atomic orbitals in each atom. These should combine to form one σ and two π molecular orbitals, and hence the molecule should contain a triple bond, namely, N≡N.

For heteronuclear diatomic molecules $c \neq 1$ in Eqs. (22) and (23), and hence ψ_A and ψ_B do not contribute equally to molecular orbital formation. Further, in forming the latter we may also have combinations of s and p atomic orbitals to form σ or σ^* orbitals. As an example of such molecules we may take HCl. The electron configuration of the chlorine atom is $1s^2 2s^2 2p^6 3s^2 3p^5$, indicating that there is only one p electron present with uncoupled spin. The atomic orbital of this electron combines then with the $1s$ orbital of the hydrogen atom to form a σ orbital with two electrons and hence one covalent bond.

The single bond in heteronuclear molecules differs in an important respect from the single bonds found in homonuclear molecules such as H_2 or Cl_2. Because in the latter ψ_A and ψ_B contribute equally to ψ, the electronic charge bonding the atoms is located midway between the two atoms. However, in a heteronuclear molecule such as hydrogen chloride, ψ_{Cl} contributes more to ψ than does ψ_H, and as a consequence the electronic charge lies closer to the chlorine than to the hydrogen atom. This unsymmetric distribution of electric charge between the atoms leads to the presence of a permanent dipole moment in the molecule, as we shall see in the next chapter, and it also makes the chlorine atom negative relative to the hydrogen atom. We thus get a bond which, besides having covalent properties, possesses some partial ionic character as well.

THE ELECTRONEGATIVITIES OF ATOMS

Chemists have long been aware of the fact that some atoms in a molecule have a greater tendency to attract electrons to themselves than others. The tendency of an atom to attract electrons to itself is called the *electronegativity* of the atom.

Pauling was the first to show that an electronegativity scale for atoms can be developed from bond energies as follows. Consider two atoms A and B which form in a molecule a bond A—B of bond energy ΔH_{A-B}. If we let now ΔH_{A-A} and ΔH_{B-B} be the bond energies of the bonds A—A and B—B, then any difference, Δ, between ΔH_{A-B} and $[(\Delta H_{A-A})(\Delta H_{B-B})]^{1/2}$ can be ascribed to ionic energy in the bond A—B. This ionic energy may be assumed to be related to the electronegativities of the two atoms, x_A and x_B, by the expression $\Delta = 30(x_A - x_B)^2$ when all energies are taken in kcal mole^{-1}. We get thus

$$(x_A - x_B)^2 = \frac{\Delta H_{A-B} - [(\Delta H_{A-A})(\Delta H_{B-B})]^{1/2}}{30} \tag{24}$$

from which $(x_A - x_B)$ can be found on introduction of the bond energies. Finally, this difference can be resolved by assuming a value for the electronegativity of one of the elements. When x of fluorine is taken to be 4.0, then the electronegativities which follow for some other elements are those shown in Table 15–3.

TABLE 15–3. ELECTRONEGATIVITIES OF SOME ELEMENTS

H 2.1						
Li 1.0	Be 1.5	B 2.0	C 2.5	N 3.0	O 3.5	F 4.0
Na 0.9	Mg 1.2	Al 1.5	Si 1.8	P 2.1	S 2.5	Cl 3.0
K 0.8	Ca 1.0	Sc 1.3	Ge 1.8	As 2.0	Se 2.4	Br 2.8
Rb 0.8	Sr 1.0	Y 1.2	Sn 1.8	Sb 1.9	Te 2.1	I 2.5

Two other methods for calculating electronegativities are the following. A suggestion by Mulliken leads to the relation

$$x = \frac{I + A}{125} \tag{25}$$

where I is the first ionization potential of the atom and A its electron affinity, both expressed in kcal per mole. Again, Allred and Rochow[1] give for x the equation

$$x = 0.359 \frac{Z'}{r^2} + 0.744 \tag{26}$$

where Z' is the effective nuclear charge of the atom attracting the electrons and r its covalent radius. For an atom A, r is one-half of the internuclear distance in the molecule A_2. These equations give values of x not significantly different from those listed in Table 15–3.

Since difference in electronegativity of the atoms constituting a bond $A—B$ leads to unsymmetric distribution of electric charge between them, a relation may be expected to exist between the percentage of ionic character, y, of the bond and $(x_A - x_B)$. Pauling gives for this relation the equation

$$\log (100 - y) = 2.00 - 0.11(x_A - x_B)^2 \tag{27}$$

An alternate expression proposed for y is

$$y = 16(x_A - x_B) + 3.5(x_A - x_B)^2 \tag{28}$$

The values of y obtained from Eqs. (27) and (28) are frequently quite different.

BONDING IN SOME POLYATOMIC MOLECULES

Diatomic molecules are of necessity linear. However, polyatomic molecules may be linear, as N_2O and CO_2, or nonlinear, as H_2O, H_2S, and NH_3. For the former molecules the angle between two bonds is 180°, while for the latter it is less than 180° but still definite for any given molecule. The formation of these molecules can again be explained in terms similar to those used for diatomic molecules.

Consider first the formation of the water molecule. The electronic configuration of the oxygen atom is $1s^2 2s^2 2p^4$, where the distribution of the p electrons is $2p_z^2 2p_x 2p_y$. Neglecting all filled atomic orbitals, we have available for bonding the two atomic orbitals $2p_x$ and $2p_y$. As shown in Figure 15–5, each of these atomic orbitals can interact with a $1s$ orbital of a hydrogen atom to form a σ molecular orbital containing two electrons. The result is thus the appearance of two O—H bonds with an angle of 90° between them. The bond angle actually observed in H_2O is 104.45°. Part of this spread in angle is accounted for by the fact that the electronegativity of oxygen is greater than that of hydrogen. As a result the hydrogen atoms develop a positive charge which leads to a repulsion between them and to an increase in the bond angle.

---

[1] Allred and Rochow, *J. Inorg. and Nuclear Chem.*, **5**, 264, 269 (1958).

The situation in H_2S is analogous to that in H_2O, except that the sulfur orbitals involved in the formation of the two S—H bonds are the $3p_x$ and the $3p_y$ orbitals. The observed angle between the bonds is 92.2°, which is quite close to the expected 90°.

Similar reasoning explains as well the formation of the NH_3 molecule. Atomic nitrogen has three external $2p$ electrons, all uncoupled, giving thus the configuration $2p_x2p_y2p_z$. The orbitals of these electrons can interact with the $1s$ orbitals in three hydrogen atoms to form three σ orbitals corresponding to three N—H bonds. Since the various p orbitals

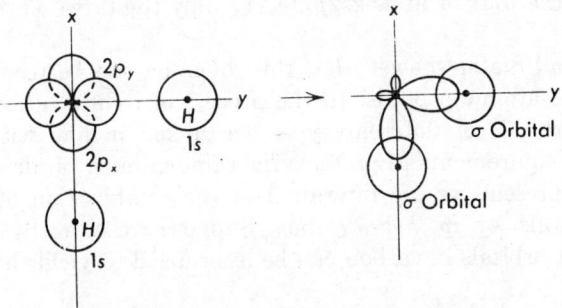

Figure 15-5.   Formation of H_2O molecule.

are oriented at 90° to each other, this should also be the angle between the N—H bonds. The bond angles observed in NH_3 are 107.3°. Here, again, part of the increase in angle can be accounted for by the repulsion between hydrogen atoms resulting from the difference in electronegativity of nitrogen and hydrogen.

The formation of N_2O can be visualized as addition of an oxygen atom to N≡N. In N_2 each nitrogen still has two unreacted $2s$ electrons. If in one of these atoms the spin of the $2s^2$ electrons is uncoupled and one of the electrons is moved to a p orbital, then the resulting s and p orbitals can interact with the two p orbitals in oxygen to form a double bond. We thus get N≡N=O. Similarly, uncoupling of the $2s$ electrons in carbon gives one s and three p orbitals. Interaction of these four orbitals with the four p orbitals available in two oxygen atoms leads to the formation of a linear CO_2 molecule with the structure O=C=O. However, as we shall see later, such static structures for these molecules do not account for all their properties.

HYBRIDIZATION OF ATOMIC ORBITALS

Spectroscopic evidence indicates that atomic carbon, with a ground state electron configuration $1s^22s^22p^2$, has two unpaired electrons. Carbon therefore should be divalent. However, in $COCl_2$, CCl_4, and in organic compounds carbon is tetravalent. Further, organic chemists have established beyond doubt that in saturated compounds the four valence bonds of carbon are identical, and that they are tetrahedrally oriented in space with bond angles of 109.47°.

Carbon can be made tetravalent by uncoupling the two $2s$ electrons and promoting one of them into the unoccupied p orbital. This process would require an input of about 65 kcal per gram atom of energy, but this energy should be more than recoverable when the atom undergoes reaction. Still, such unpairing and promotion do not solve the problem because, of the four orbitals $2s2p_x2p_y2p_z$, only the three p orbitals would be identical.

Pauling and Slater showed that this dilemma can be resolved by *hybridization* of atomic orbitals. In the process of hybridization a number of atomic orbitals of different types are mixed in line with quantum mechanical requirements to produce the same number of identical *hybrid orbitals*. At present we are interested in the combination of one s and three p orbitals, or sp^3 *hybridization*. Suppose we start by postulating that the four orbitals in carbon can be hybridized to yield the four wave functions

$$\psi_1 = k_1(a_1s + b_1p_x + c_1p_y + d_1p_z) \tag{29a}$$
$$\psi_2 = k_2(a_2s + b_2p_x + c_2p_y + d_2p_z) \tag{29b}$$
$$\psi_3 = k_3(a_3s + b_3p_x + c_3p_y + d_3p_z) \tag{29c}$$
$$\psi_4 = k_4(a_4s + b_4p_x + c_4p_y + d_4p_z) \tag{29d}$$

where s, p_x, p_y, and p_z represent the wave functions for the respective electrons. These equations offer an infinite number of possible hybrids. However, if we assume that $a_1 = b_1 = c_1 = d_1$, then mathematical argument shows that all the k's are equal to $\frac{1}{2}$, and all the coefficients except c_2, d_2, b_3, d_3, b_4, and c_4 are equal to 1 while the latter are equal to -1. We obtain thus

$$\psi_1 = \frac{1}{2}(s + p_x + p_y + p_z) \tag{30a}$$
$$\psi_2 = \frac{1}{2}(s + p_x - p_y - p_z) \tag{30b}$$
$$\psi_3 = \frac{1}{2}(s - p_x + p_y - p_z) \tag{30c}$$
$$\psi_4 = \frac{1}{2}(s - p_x - p_y + p_z) \tag{30d}$$

These equations yield four identical orbitals pointing toward the corners of a tetrahedron. Consequently sp^3 hybridization of the four carbon orbitals confirms the identity of the four carbon bonds in compounds such as CH_4, CCl_4, or C_2H_6, and also their tetrahedral disposition in space.

sp^2 HYBRIDIZATION

sp^3 hybridization does not explain the formation or behavior of double bonds in olefinic hydrocarbons such as $H_2C{=}CH_2$. For these we have to postulate sp^2 hybridization of the one s orbital with the p_x and p_y orbitals in carbon. The p_z orbital is left unhybridized. By following an argument similar to the above, three identical hybrid orbitals can be obtained whose equations are

$$\psi_1 = (1/\sqrt{3})[s + \sqrt{2}\,p_x] \tag{31a}$$
$$\psi_2 = (1/\sqrt{3})[s - (1/\sqrt{2})p_x + (\sqrt{\tfrac{3}{2}})p_y] \tag{31b}$$
$$\psi_3 = (1/\sqrt{3})[s - (1/\sqrt{2})p_x - (\sqrt{\tfrac{3}{2}})p_y] \tag{31c}$$

These orbitals lie in the xy plane at angles 120° apart. The loops of the p_z orbital project above and below the plane, as shown in Figure 15–6(a). When two such carbon atoms are combined along with four hydrogen atoms to form ethylene, the hydrogens attach themselves to the b and c orbitals in the plane, while the a orbitals of the two carbon atoms combine to form a σ bond along the C—C line in Figure 15–6(b). The second C—C

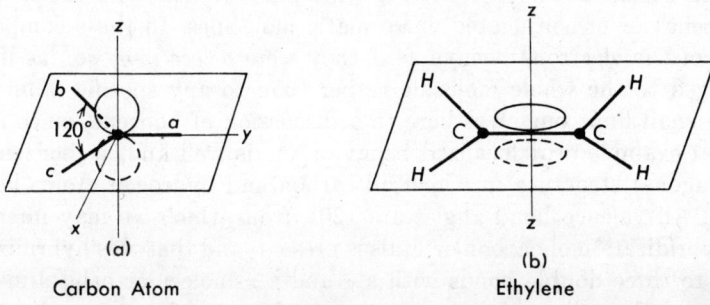

(a)

Carbon Atom

(b)

Ethylene

Figure 15–6. sp^2 hybridization of orbitals in carbon.

bond is formed from the p_z orbitals of the two atoms and it is thus a π bond, as shown by the "balloons" which lie above and below the xy plane.

Since the energy of the σ bond is higher than that of the π bond, it is believed that the latter is the one which opens when olefinic compounds undergo reaction at the double bond.

sp HYBRIDIZATION

The acetylene molecule, C_2H_2, is linear and contains a triple bond. Its formation can be explained by hybridizing the s and p_x orbitals in carbon to obtain two hybrids whose wave functions are

$$\psi_1 = (1/\sqrt{2})(s + p_x) \tag{32a}$$
$$\psi_2 = (1/\sqrt{2})(s - p_x) \tag{32b}$$

These hybrid orbitals are identical and linear but they face in opposite directions, i.e., the angle between them is 180°. When two carbon atoms with such hybridization are combined along with two hydrogens, we get a σ bond by union of a hybrid on each carbon atom. In turn, the hydrogens attach themselves to the two outward-facing hybrid orbitals. The two remaining bonds between the carbons result from formation of two π orbitals by combination of the unhybridized p_y and p_z orbitals on each carbon. We obtain thus a linear molecule containing a triple bond composed of one σ and two π molecular orbitals.

THE STRUCTURE OF BENZENE

In all discussions of bonding up to now electrons were assigned to definite atomic orbitals, and these interacted with the orbitals of similar electrons in other atoms to form molecular orbitals of definite location in the molecule. However, this concept of *localized* atomic and molecular orbitals breaks down when attempts are made to deal with the structure and behavior of conjugated or aromatic molecules. In these compounds some of the electrons behave as if they were *delocalized*, i.e., as if they belonged to the whole molecule rather than to any specific atom in it.

We shall limit ourselves here to a discussion of benzene, since it is a typical example of delocalized behavior. As is well known, benzene has a hexagonal structure in which all carbon and hydrogen atoms lie in a plane. All valence bond angles are 120°, from which we may infer that sp^2 hybridization of carbon orbitals is present, and that this hybridization leads to three double bonds with a σ and a π molecular orbital in each. We arrive thus at the Kekule structure for benzene. However, some other facts do not square with this conclusion. Typical reactions of double bonds are addition, whereas the common reactions of benzene are substitution. Again, the C—C bond length in benzene is 1.39 Å, whereas the length for C$=$C is 1.33 Å. Finally, the bonding energy of C_6H_6 is greater by 37 kcal than the energy expected for a molecule containing three C$=$C, three C—C, and six C—H bonds.

Molecular orbital theory approaches the benzene problem by delocalizing the six $2p_z$ electrons which are supposed to be involved in the formation of the three π orbitals of the double bonds. By doing this it makes these electrons a property of not only the carbon atoms from which they come, but of all the six carbons present. This procedure is analogous to the hybridization of s and p atomic orbitals, except that now six $2p_z$ orbitals are hybridized by writing

$$\psi = k_1\psi_1 + k_2\psi_2 + k_3\psi_3 + k_4\psi_4 + k_5\psi_5 + k_6\psi_6 \qquad (33)$$

In Eq. (33) ψ is the total hybridized orbital function, the k's are constants, and the ψ_i's are the atomic orbital wave functions for the various $2p_z$

electrons. Solution of Eq. (33) leads to six possible molecular orbitals of different energy for the six electrons involved. To obtain the greatest stability in the normal C_6H_6 molecule, these electrons have to be placed in the three molecular orbitals of lowest energy. Further, each molecular orbital must contain two electrons of opposite spins. We thus obtain three delocalized molecular orbitals to take the place of the three π orbitals involved in the localized double bonds. Since these three molecular orbitals are delocalized, they cannot be expected to act as normal double bonds of definite position in the molecule.

The energy of the delocalized electrons is found from Eq. (33) to be

$$E = 6\alpha + 8\beta$$

where α and β are the values of certain integrals. This value of E may be compared with

$$E' = 6\alpha + 6\beta$$

which is the energy of the six $2p_z$ electrons when located in the π orbitals of three double bonds. Hence the difference between E and E' is

$$\Delta E = 2\beta$$

As β is negative, we see that delocalization of the p electrons leads to stronger bonding and a more stable benzene molecule.

The valence-bond method treats the benzene problem differently. Kekule and Dewar proposed for benzene the structures shown in Figure 15–7, which differ from each other only in the location of the double

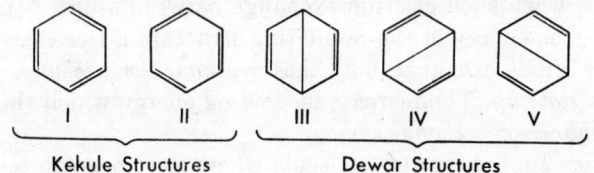

I   II   III   IV   V

Kekule Structures   Dewar Structures

Figure 15–7.   Kekule and Dewar structures for benzene.

bonds. Suppose now that the wave function for benzene, ψ, is a linear combination of the wave functions for the structures I to V. We obtain thus

$$\psi = a_1\psi_I + a_2\psi_{II} + a_3\psi_{III} + a_4\psi_{IV} + a_5\psi_V \tag{34}$$

where the a's are constants. Solution of this equation for the benzene ground state yields $a_1 = a_2$ and $a_3 = a_4 = a_5$. Hence the Kekule structures contribute equally to ψ, and so do the three Dewar structures. Further, since it was also found that $a_3 = 0.4341a_1$, the two Kekule structures constitute 61% of ψ and the three Dewar structures 39%.

From Eq. (34) the total energy of a hybridized benzene molecule follows as

$$E_t = Q + 2.61\,J$$

where Q and J are values of integrals appearing in the calculation. Again, the total energy of either structure I or II can be shown to be

$$E_t' = Q + 1.50\,J$$

Consequently,

$$\Delta E_t = 1.11\,J$$

which, with J negative, shows the hybrid combination to be more stable than the Kekule structure. Toward this value of ΔE_t the two Kekule structures contribute 81% and the three Dewar structures 19%. The Kekule structures are thus much more important.

RESONANCE

A fundamental contribution of wave mechanics to the theory of molecular structure is the concept of *resonance* or *exchange energy*. We have already seen in the case of H_2^+ that stability was attained from exchange of the electron between the two protons. Again, in the hydrogen molecule greater stability was gained when the electrons attached to the individual atoms exchanged places. The two configurations resulting from the exchange in both instances are not static, but involve continuous oscillation of the molecule from one of these to the other. Quantum mechanics shows that, when such electron exchange between atoms is possible and does occur, the energy of the resonating structure is *less* than that of the molecule without exchange, i.e., that *resonance or exchange leads to an increase in stability*. The increase in binding energy which thus results is called *resonance* or *exchange energy*.

Resonance appears in any molecule where electrons can oscillate from one position to another. As a result of this oscillation the molecule may take on a number of resonance structures, each of which does not differ significantly in total energy from every other. Thus we have seen above that benzene is a resonance hybrid of the five structures shown in Figure 15–7, and this hybridization leads to a resonance energy of 37 kcal per mole. Similarly, CO_2 can be represented as a resonance hybrid of the three structures

$$\text{O:C:::O} \qquad \text{O::C::O} \qquad \text{O:::C:O}$$
$$\text{I} \qquad\qquad \text{II} \qquad\qquad \text{III}$$

which yield a molecule stabler by 36 kcal per mole than the assumption that CO_2 consists only of structure II, namely, $O{=}C{=}O$.

It should be emphasized that *a resonance hybrid is not a mixture* of the postulated structures. Rather, it is a structure whose nature, within our limited means of description, we can characterize only in terms of composites of structures which we can visualize and write down.

BEHAVIOR OF TRANSITION ELEMENTS

The discussion thus far has dealt with elements involving only s and p electrons and falling within the first three periods of the periodic table. When we come to the transition elements of the fourth and higher periods, d electrons as well must be used to account for the observed valencies. Thus, whereas the formation of $CoCl_2$ involves use of the two $4s$ electrons of cobalt, for formation of $CoCl_3$ we must employ not only these but also one $3d$ electron.

Another striking characteristic of transition elements is the fact that they readily form coordination compounds in which appear ions such as $Co(NH_3)_6^{+3}$, $Co(CN)_6^{-3}$, $Fe(CN)_6^{-3}$, and $Fe(CN)_6^{-4}$, or $Ni(CN)_4^{-2}$, $AuBr_4^{-1}$, and $CuCl_4^{-2}$. In these ions the cobalt and iron atoms have coordinated about them six groups or *ligands*, while the nickel, gold, and copper have four ligands. To explain the formation of such ions, the valence-bond theory postulates the hybridization of d, s, and p orbitals in the central atoms.

Consider the ion $Co(CN)_6^{-3}$, a typical example of a central atom with a coordination number of six. The formation of this ion can be represented by the reaction

$$Co^{+3} + 6[:C \equiv N]^- = Co(CN)_6^{-3} \tag{35}$$

Neutral cobalt has an outer electron configuration of $3d^7 4s^2$, and hence for Co^{+3} the configuration must be $3d^6$. These six d electrons can be distributed among the five possible d orbitals in the manner shown in Figure 15–8(a). Suppose that we now borrow six electrons from the carbon atoms in the six CN^- groups and add them to Co^{+3}. This process will make the CN groups neutral, and the cobalt will become Co^{-3} with the electron distribution given in the lower part of Figure 15–8(a). Here it is evident that the six unfilled atomic orbitals are not alike. However, by d^2sp^3 *hybridization* it is possible to obtain six identical orbitals which point to the corners of an octahedron. These hybrid orbitals then can react with the orbitals of the six odd electrons on the carbon atoms in the neutral CN groups to form the complex $Co(CN)_6^{-3}$, which thus has a structure consisting of the central ion, Co^{+3}, and six groups disposed octahedrally about it as shown in Figure 15–9. Similar explanations apply to other complexes of coordination number six, where the B ligands all may be the same or some may be different.

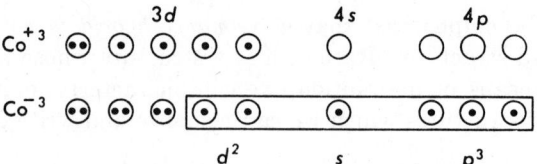

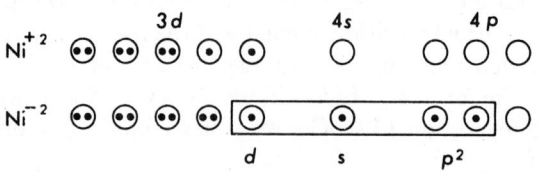

(a) d^2sp^3 hybridization

(b) dsp^2 hybridization

Figure 15–8.   Hybridization of d, s, and p orbitals.

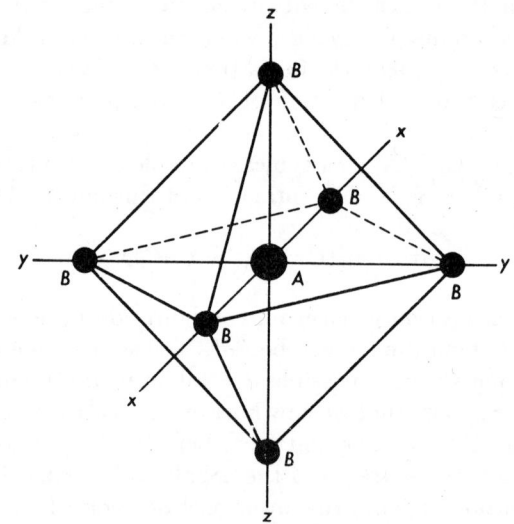

Figure 15–9.   Octahedral orientation of groups about central atom with coordination number of six.

$Ni(CN)_4^{-2}$ may be taken as a typical case of a central ion with coordination number four. The formation of this ion can be represented by the equation

$$Ni^{+2} + 4\ CN^- = Ni(CN)_4^{-2} \qquad (36)$$

The external electronic configuration of Ni^{+2} is $3d^8$, with the electron distribution shown in Figure 15–8(b). If we borrow four electrons from the CN^- groups and add them to the Ni^{+2}, then we obtain Ni^{-2} and the electron distribution given in the lower part of the figure. Here also the orbitals are different. To make them the same, dsp^2 *hybridization* can be resorted to in this case. The result of such hybridization is four identical orbitals directed toward the corners of a square plane. These orbitals then can react with the four unfilled orbitals in the carbons of the four CN groups to yield the arrangement given in Figure 15–10. Similar argu-

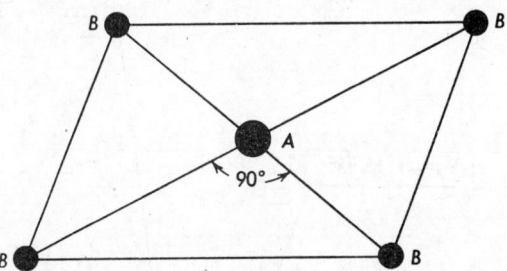

Figure 15–10.   Square planar orientation resulting from dsp^2 hybridization of central atom orbitals.

ments apply to other complexes with coordination number four, and the B ligands again may be the same or different.

Besides those mentioned, there are other types of *dsp* hybridization possible which result in various other geometric configurations for complex ions and molecules. A discussion of these may be found in the books by Pauling, Coulson, and Cotton and Wilkinson listed at the end of the chapter.

CRYSTAL AND LIGAND FIELD THEORIES

The valence-bond treatment ascribes complex ion formation to co-valence arising from hybridization of d, s, and p orbitals in the central atom, and interaction of these hybrid orbitals with orbitals in the ligands. On the other hand, the *crystal field theory* goes to the opposite extreme, and deals with the formation of such ions in terms of purely electrostatic interactions between the ligands and the d orbitals in the central ion.

Consider first a complex ion with a coordination number of six, such as $Co(CN)_6^-$, which has the octahedral orientation shown in Figure 15–9. In this complex the ligands are disposed along the x, y, and z axes, and hence their negative charge is also concentrated along these directions. If we examine the shapes of the five possible d orbitals as given in Figure 14–11 of the preceding chapter, we see that the d_{z^2} and $d_{x^2-y^2}$ orbitals also have their negative charge concentrated in these directions, while the remaining three orbitals have their charge concentration lying between the axes. We may expect, therefore, that the d_{z^2} and $d_{x^2-y^2}$ orbitals of a d electron on the central ion will experience a much greater electrostatic repulsion than will the other three orbitals coming within the field of the ligands. Consequently, these orbitals will contribute less to binding of the ligands to the central ion, i.e., the energy of the d_{z^2} and $d_{x^2-y^2}$ orbitals will be higher than that of the other three.

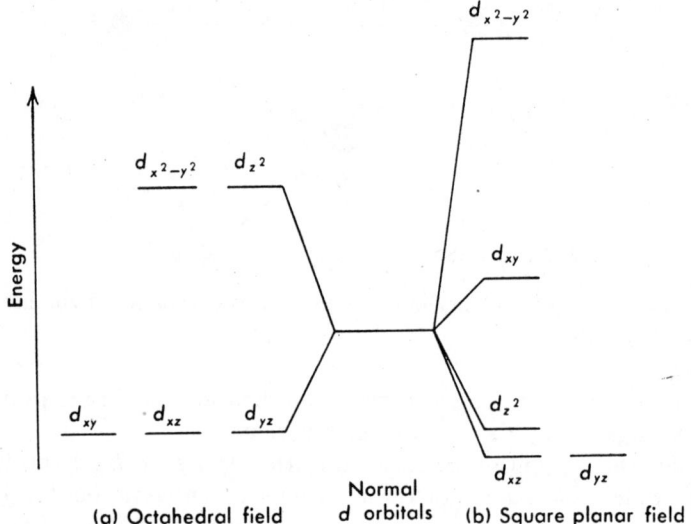

Figure 15–11.   Splitting of d orbitals in ligand fields.

The undisturbed five orbitals of a d electron all have the same energy. Further, the total energy of the five orbitals must remain constant. Hence, when the energy of the d_{z^2} and $d_{x^2-y^2}$ orbitals is raised by electrostatic repulsion, the energy of the other three orbitals must decrease. We thus obtain a splitting of the d orbitals of a d electron into the two groups of levels shown in Figure 15–11(a). Here three of the orbitals have identical energies lower than those of the original d orbitals, while two have the same but higher energies. This type of splitting of d orbital levels should occur in all instances of octahedral orientation of six coordinate groups.

In the case of square planar coordination of ligand groups, as in $Ni(CN)_4^{-2}$, the splitting of d orbital levels expected is that shown in Figure 15–11(b). This type of splitting may be deduced again from the shape of the d orbitals and Figure 15–10, where the x and y axes are taken to lie along the square diagonals. Here we have four sets of levels, only one of which contains two identical orbitals.

The extent of level splitting which occurs in any instance depends on the nature of the ligands and of the central ion. For a weak ligand field the extent of splitting is small, while for a strong ligand field it is large, as shown in Figure 15–12. Further, the strength of the ligand field determines the manner in which the d electrons in the central ion distribute themselves among the available d orbitals. As an example, consider the six d electrons in Co^{+3}. In a weak field, where the upper levels are not too far removed from the lower ones, the electron distribution is that given in Figure 15–12(a). Here four of the d orbitals are occupied by single unpaired electrons and one by a pair with opposed spins. However,

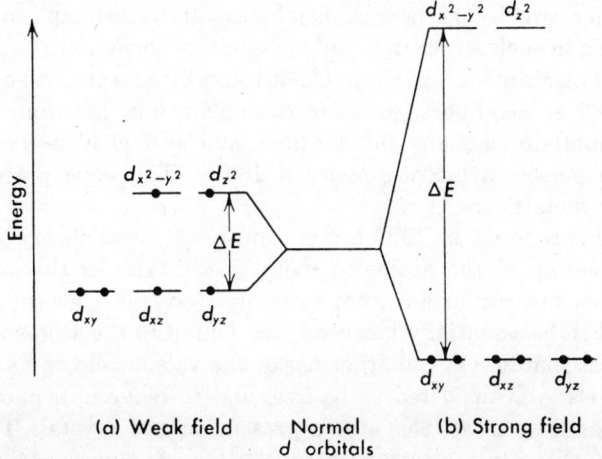

Figure 15–12. Distribution of the d^6 electrons of Co^{+3} in weak and strong octahedral ligand fields.

in a strong ligand field the upper orbital levels are considerably removed from the lower ones, and consequently all electrons concentrate in the latter, Figure 15–12(b). We thus obtain three completely occupied d orbitals with paired electrons and two empty ones. Since the magnetic properties of the Co^{+3} ion complexes should depend on the number of unpaired electrons present, we may expect differences in the magnetic behavior of Co^{+3} complexes in weak and strong ligand fields. This is actually the case. The ability of the crystal field theory to account for the magnetic properties of coordination compounds is one of the important advantages which this theory has over the valence-bond approach.

However, it is not possible to explain all properties of coordination complexes in terms of electrostatic effects only. Evidence indicates that such compounds frequently possess covalent character as well. To take care of this situation, the *ligand field theory* combines the valence-bond method and crystal field theory, and treats coordination complexes as a composite of the two types of orbital interaction. By suitable combination of the two approaches, the ligand field theory is able to deal more effectively with the binding in such compounds and with the explanation of their behavior.

THE METALLIC BOND

Most metals crystallize in either body-centered cubic, face-centered cubic, or close-packed hexagonal lattices. In the first of these arrangements each atom of the metal is surrounded by 14 neighboring atoms, in the other two by 12 nearest neighbors. If we attempt to deal with the bonding in such structures, we immediately encounter the problem of insufficient electrons. Thus, if we take lithium with its one valence electron and 14 nearest neighbors, we have to explain how an atom of lithium can be bound to so many other atoms and still yield a crystal stable enough to possess a melting point of 186°C. The same problem arises with other metals.

F. Bloch proposed in 1928 a quantum-mechanical theory to account for the bonding of the atoms in metallic crystals. In this *band theory* all electrons present in an atom in completely filled energy levels are considered to be essentially localized, i.e., bound to the atoms with which they are associated. On the other hand, the valence electrons in unfilled energy levels are considered to be free, and to move in a potential field which extends over all the atoms present in the crystal. The atomic orbitals of these free electrons in one atom can overlap those of the electrons in other atoms to yield delocalized molecular orbitals which produce bonding among all the atoms present. The molecular orbitals of these free electrons are called the *conduction orbitals* of the metal.

The energy levels of electrons in isolated atoms are discrete and generally well spaced. However, the presence of other atoms in the crystal affects these discrete levels, and changes *each level* into a band of closely spaced levels equal in number to the number of atoms present in the entire metallic structure. If the total number of atoms present is large, each isolated level becomes practically a continuous band. Further, when the spacing of the original levels and of the atoms in the metal is large, then the bands arising from the original electron levels are distinct and separated from each other by appreciable energy gaps. However, when the levels and distances are close, then the bands may cross and overlap each other.

The picture of the electronic structure of a metal given by this theory is then the following. A solid metal can be considered to possess bands of electrons which are separated from each other by energy gaps. Further, these bands can be completely filled with localized electrons, or partially filled with free electrons whose molecular orbitals extend over all the atoms in the crystal. For a metal such as sodium, these ideas can be represented by the diagram shown in Figure 15–13(a). In this figure the shaded areas represent the portions of bands filled by electrons, and the spaces between bands the energy gaps which must be surmounted to pass from one band to another. In sodium all the bands are filled except that containing the $3s$ electrons, and hence these electrons are free to participate in delocalized interaction and bonding. Further, since the $3s$ electrons are mobile, application of an electric potential across solid sodium

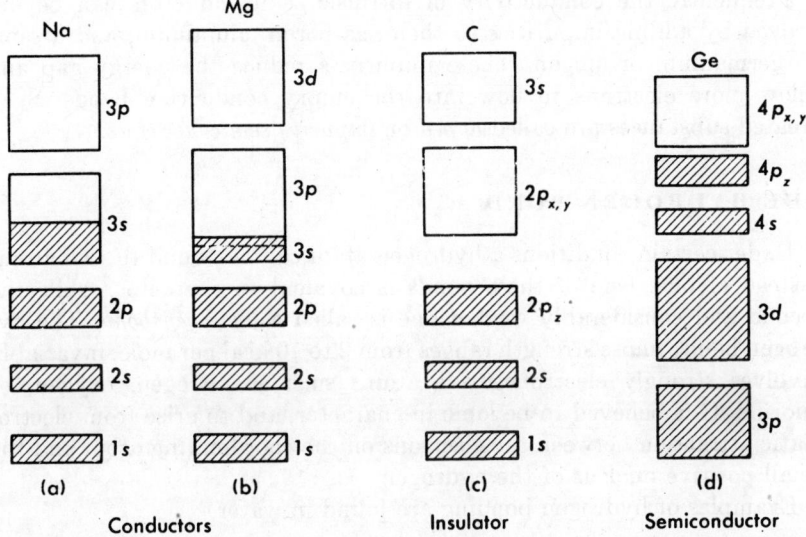

Figure 15–13.   Band models of conductors, insulators, and semiconductors.

will readily set these electrons in motion. Hence sodium is a good *conductor* of electricity.

The band structure for diamond shown in Figure 15–13(c) indicates that all bands are completely filled, and there are thus no free electrons to make the diamond conducting. This substance is, therefore, a poor conductor of electricity, or an *insulator*. To get conduction in an insulator it is necessary to apply a sufficiently high potential difference to raise electrons from the upper filled band, through the energy gap, and into the lowest empty band. When this is done a breakdown of the dielectric takes place, and conduction becomes possible.

Magnesium, with its external $3s^2$ configuration, may be expected to be an insulator. However, in this metal the $3s$ and $3p$ bands are sufficiently close to overlap, Figure 15–13(b), and we thus obtain an unfilled band with free electrons which make this substance a good metallic conductor.

Germanium and silicon are intrinsic *semiconductors*, i.e., their conductivities are intermediate between those of insulators and good electronic conductors. The explanation for this behavior can be found in the band diagram for germanium shown in Figure 15–13(d). Here the $4p_x$ band is filled, but the empty band immediately above it is sufficiently close to make the gap between the two small. Through thermal activation electrons in the upper closed band can acquire sufficient energy to span the small energy gap and move into the empty band. Once there, these electrons become mobile, and impart a degree of conductivity to the metal.

Frequently the conductivity of intrinsic semiconductors can be improved by adding impurities to them, as boron, aluminum, and arsenic to germanium or silicon. These impurities reduce the energy gap and allow more electrons to flow into the empty conduction bands. Such treated substances are called *doped* or *impurity semiconductors*.

THE HYDROGEN BOND

Under certain conditions a hydrogen atom can be bound to two atoms instead of one. One of these bonds is covalent in character, while the second and considerably weaker one is called a *hydrogen bond*. The hydrogen bond, whose strength ranges from 2 to 10 kcal per mole, invariably involves strongly electronegative atoms such as nitrogen, oxygen, or fluorine. It is believed to be ionic in character, and to arise from electrostatic attraction between the electrons on the coordinating atom and the small positive nucleus of the hydrogen.

Examples of hydrogen bonding are found in water,

$$
\begin{array}{c}
\qquad\qquad\qquad \text{H} \\
\qquad\qquad\quad\ \diagup \\
\text{H—O—H}\text{----}\text{O} \\
\qquad\qquad\quad\ \diagdown \\
\qquad\qquad\qquad \text{H}
\end{array}
$$

in alcohols

$$
\begin{array}{c}
\text{R} \\
\ \diagdown \\
\quad \text{O}\text{----}\text{HOR} \\
\ \diagup \\
\text{H}
\end{array}
$$

in carboxylic acids

$$
\begin{array}{c}
\qquad \text{O}\text{----}\text{H—O} \\
\qquad \| \qquad\qquad\ \diagdown \\
\text{R—C} \qquad\qquad \text{C—R} \\
\qquad \diagdown \qquad\qquad\ \| \\
\qquad \text{O —H}\text{----}\text{O}
\end{array}
$$

in nitrophenols

$$
O\text{—}H
$$

N=O

O

O----H—N—H

and in amides   R—C   C—R

H—N—H----O

In the nitrophenols hydrogen bonding results in the formation of a second ring in which the properties of both the nitro and hydroxyl groups are modified. However, with carboxylic acids and amides hydrogen bonding leads to an association of two molecules, while in water and the alcohols the bonding may well extend throughout the entire mass. In terms of hydrogen bonding it is understandable why water, acids, alcohols, and amides are associated liquids, whereas hydrocarbons, carbon tetrachloride, and chloroform are not. The same explanation applies to liquid hydrogen fluoride, where as a result of hydrogen bond formation between the fluorine of one molecule and the hydrogen of another the dimer H_2F_2 is formed.

Hydrogen bonding is not confined to the liquid state. X-ray studies of ice and solid carboxylic acids, and vapor density measurements on the latter, reveal that hydrogen bonding extends also into the solid and vapor phases. The presence of hydrogen bonds in the liquid state is indicated by higher boiling points, heats of vaporization, and viscosities than shown by corresponding normal liquids. Hydrogen bonding can be detected also by infrared studies. Thus the spectrum of aceto-acetic acid ester gives no bands associated with the hydroxyl group, while in other molecules the hydroxyl bands are displaced from their normal positions.

REFERENCES

1. Cartmell and Fowles, *Valency and Molecular Structure*, Butterworths Scientific Publications, London, 1956.
2. Cotton and Wilkinson, *Advanced Inorganic Chemistry*, Interscience Publishers, Inc., New York, 1962.
3. C. A. Coulson, *Valence*, Oxford University Press, New York, 1960.
4. W. Kauzmann, *Quantum Chemistry*, Academic Press Inc., New York, 1957.
5. C. Kittel, *Elementary Solid State Physics*, John Wiley & Sons, Inc., New York, 1962.

6. L. Pauling, *Nature of the Chemical Bond,* Cornell University Press, Ithaca, N. Y., 1960.

7. Pauling and Wilson, *Introduction to Quantum Mechanics,* McGraw-Hill Book Company, Inc., New York, 1935.

8. Pimentel and McClellan, *The Hydrogen Bond,* W. H. Freeman and Company, San Francisco, 1960.

9. K. S. Pitzer, *Quantum Chemistry,* Prentice-Hall, Inc., Englewood Cliffs, N. J., 1953.

10. A. Streitwieser, *Molecular Orbital Theory for Organic Chemists,* John Wiley & Sons, Inc., New York, 1961.

PROBLEMS

1. On the basis of the octet rule show the external electronic configurations of the following substances: (a) H_2O_2, (b) CH_3COOH, (c) H_2SO_4, and (d) $HClO_4$.

2. Show the external electronic configurations of the following complex ions: (a) $[Cu(NH_3)_4]^{++}$, (b) $[(CH_3)_4N]^+$, and (c) $[Ag(CN)_2]^-$.

3. If nitrogen can have no more than eight electrons in its valence shell, what is the most reasonable electronic formula for R_3NO, where R is an alkyl group?

4. (a) Neglecting repulsion, calculate the potential energy between a K^+ and a Cl^- ion separated by a distance of 10 Å. (b) What is the potential energy in kcal per mole of KCl?   *Ans.* (b) -33.1 kcal mole^{-1}.

5. Repeat problem 4 for a separation distance of 3.14 Å.

6. For KCl $\rho = 0.396 \times 10^{-8}$ cm and $r_0 = 3.14$ Å. Using Eq. (3), find the potential binding energy of KCl in kcal per mole.

7. For crystalline KCl at 25°C the Madelung constant is 1.75, $\rho = 0.396 \times 10^{-8}$ cm, $r_0 = 3.14$ Å, $E_0 = 1.4$ kcal mole^{-1}, and the van der Waals attraction energy is 3.9 kcal mole^{-1}. Find the lattice energy of the crystal in kcal mole^{-1}.
$$Ans. \ \Delta H_c = 164.0 \text{ kcal mole}^{-1}.$$

8. The lattice energy of CsCl at 25°C is 151.3 kcal mole^{-1}. For this substance the Madelung constant is 1.763, $r_0 = 3.560$ Å, $\rho = 0.424 \times 10^{-8}$ cm, and $E_0 = 1.0$ kcal mole^{-1}. From these data find (a) the van der Waals attraction energy per mole, and (b) the constant C in Eq. (4).

9. For LiBr(s) $\Delta H_f = -84$ kcal mole^{-1}. Further, for Li $\Delta H_s = 34$ and $I = 124$ kcal mole^{-1}. Using these data and any other needed information from Table 15-1, calculate the crystal energy of LiBr by means of the Born-Haber cycle.
$$Ans. \ \Delta H_c = 187 \text{ kcal mole}^{-1}.$$

10. Derive by means of the Born-Haber cycle the expression for ΔH_c of ionic crystals such as MX, where M is a divalent metal, and X is oxygen, sulfur, or selenium.

11. For MgO(s) $\Delta H_c = 931$ and $\Delta H_f = -145$ kcal mole^{-1}. Further, for magnesium $\Delta H_s = 34$ kcal mole^{-1}, $I_1 = 7.64$ eV and $I_2 = 15.03$ eV, while for

$O_2(g)$ $\Delta H_d = 118$ kcal mole^{-1}. Calculate from these data the electron affinity of oxygen.   *Ans. A* $= -170$ kcal mole^{-1}.

12. For $MgS(s)$ $\Delta H_c = 794$ and $\Delta H_f = -81$ kcal mole^{-1}, while for $S_2(g)$ $\Delta H_d = 132$ kcal mole^{-1}. Using these data and those given in the preceding problem, calculate the electron affinity of sulfur.

13. Suggest an explanation why the electron affinities of O^{-2} and S^{-2} are negative.

14. (a) Write the molecular orbital configuration for a molecule formed from two Li atoms. (b) How many bonds can be present between the two atoms?

15. (a) Write the molecular orbital configuration for the combination of two Be atoms. (b) How many bonds can be formed between the two atoms?

16. (a) How many bonds can be formed in the molecule C_2? (b) What is the nature of these bonds?

17. The energies of the following bonds are

F—F	36.6 kcal mole^{-1}
Cl—Cl	58.0 '' ''
Cl—F	60.6 '' ''

Assuming the electronegativity of fluorine to be 4.0, calculate the electronegativity of chlorine.   *Ans.* $x_{Cl} = 3.3$.

18. Using the electronegativities given in Table 15–3 and the requisite data from the preceding problem, calculate the bond energy for Cl—F.

19. The ionization potential of Li is 5.39 eV. Assuming the electron affinity of the element to be zero, calculate its electronegativity.

20. Taking the electronegativity of magnesium to be that given in Table 15–3, and its first ionization potential to be 7.64 eV, calculate the electron affinity of the element.

21. (a) Calculate the percentage of ionic character in the bond Br—F by means of Eq. (27). (b) Repeat the calculation using Eq. (28).   *Ans.* (a) 30%.

22. (a) Calculate the percentage of ionic character in the bond Sr—Cl by means of Eq. (27). (b) Repeat the calculation using Eq. (28).

23. (a) Explain the formation of Cl_2O. (b) What bond angle is predicted for this molecule. The observed angle is 110.8°.

24. (a) Explain the formation of PH_3. (b) What bond angles are predicted for this molecule? The observed angles are 93.3°.

25. (a) Explain the formation of $SiCl_4$ on the basis of sp^3 hybridization. (b) What are the predicted bond angles for this molecule?

26. In the CO molecule the binding energy is larger than may be expected for a C=O bond. How can this fact be explained?

27. (a) Explain the formation of the $Fe(CN)_6^{-3}$ complex ion on the basis of d, s, and p hybridization. (b) What type of hybridization is required?

28. (a) Explain the formation of the $Fe(CN)_6^{-4}$ complex ion on the basis of d, s, and p hybridization. (b) What type of hybridization is required?

29. (a) Explain the formation of the $AuBr_4^-$ ion on the basis of d, s, and p hybridization. (b) What type of hybridization is required? (c) What is the orientation of the Br^- ions about the central atom?

30. For octahedral coordination, what will be the distribution of the d electrons in Fe^{+2} when the ligand field is (a) weak and (b) strong?

31. For octahedral coordination, what will be the distribution of d electrons in Fe^{+3} when the ligand field is (a) weak and (b) strong?

32. Experimental evidence indicates that liquid HCN and liquid NH_3 are hydrogen bonded. Show how this bonding may take place.

16

Investigation of Molecular Structure

$\mathbf{T}$HE COMPOSITION of molecules can be determined by chemical analysis, and their molecular weight by vapor density or colligative property measurements. Again, by study of their reactions, a chemist can ascertain or infer the sequence of atoms in the molecule, and some-times also the molecular geometry. However, these observations do not give direct information on bond lengths, bond angles, polarity of bonds, internal molecular and atomic motions, or other quantitative details about the molecule. To obtain such information, means have to be employed which are more refined in character, and which measure more directly the properties sought. A discussion of some of the methods used for this purpose, and of some of the results obtained, is the concern of this chapter.

THE MOLAR REFRACTION

Lorenz and Lorentz showed in 1880 that the expression

$$R_s = \left(\frac{n^2 - 1}{n^2 + 2}\right)\frac{1}{\rho} \tag{1}$$

should be a constant independent of temperature for any given sub-stance. In this equation R_s is the *specific refraction* of the substance, n the index of refraction, and ρ the density measured at the same tem-perature as n. From Eq. (1) the molar refraction R_m is obtained on multiplying R_s by the molecular weight of the substance M, namely,

$$R_m = R_s M = \left(\frac{n^2 - 1}{n^2 + 2}\right)\frac{M}{\rho} \tag{2}$$

The index of refraction of any medium is the ratio of the velocity of light in a vacuum to that in the given medium. It is measured by means of such optical instruments as the Pulfrich, Abbé, or immersion refractometers. Measured in this manner the index of refraction is not constant but increases as the wave length of light is decreased. To obtain comparable results it is necessary, therefore, to use light of a fixed frequency. For this purpose the common practice is to employ the yellow light of the sodium-D line from, say, a sodium vapor lamp. However, white light may be employed in the Abbé or immerison refractometers through the use of an incorporated prism which eliminates color fringes and reduces the results to sodium light.

The molar refraction R_m defined by Eq. (2) has been demonstrated to be an additive property for any given substance provided all measurements are referred to a given wave length of light. Some atomic and bond contributions evaluated for sodium-D light are given in Table 16-1. The molar refraction may be seen to depend on the number and

TABLE 16-1.   MOLAR REFRACTION CONTRIBUTIONS
(For Sodium-D Light)

Carbon	2.418	Oxygen (in OH group, O—)	1.525
Hydrogen	1.100	Oxygen (in CO group, O=)	2.211
Chlorine	5.967	Oxygen (in ethers, O—)	1.643
Bromine	8.865	3-membered ring	0.71
Iodine	13.900	4-membered ring	0.48
Double bond	1.733		
Triple bond	2.398		

nature of the atoms present, and also on the character of the binding. These values may be used to compare calculated with observed molar refractions, and thus to check the structure of molecules.

To illustrate the calculation of molar refractions we may take acetic acid, CH_3COOH. For this substance at 22.9°C the density is 1.046 g cc^{-1}, the index of refraction for sodium light 1.3715, and the molecular weight is 60.05 g mole^{-1}. The observed value of the molar refraction is, therefore,

$$R_m = \left(\frac{n^2 - 1}{n^2 + 2}\right)\frac{M}{\rho}$$

$$= \left[\frac{(1.3715)^2 - 1}{(1.3715)^2 + 2}\right]\frac{60.05}{1.046}$$

$$= 13.303 \text{ cc mole}^{-1}$$

To compare with this value we find from Table 16–1 for acetic acid

$$
\begin{aligned}
2\ \text{C} &= 2 \times 2.418 = & 4.836 \\
4\ \text{H} &= 4 \times 1.100 = & 4.400 \\
1\ \text{O—} &= 1 \times 1.525 = & 1.525 \\
1\ \text{O}= &= 1 \times 2.211 = & \underline{2.211} \\
& & 12.972\ \text{cc mole}^{-1}
\end{aligned}
$$

The two values are seen to agree within 0.06 unit. Since the index of refraction is dimensionless, the units of the molar refraction are the units of M/ρ, i.e., volume. R_m is expressed thereby in cubic centimeters per m...

Calculated and observed molar refractions agree as a rule very clos...

Exceptions arise only in open chain molecules with conjugated double bonds and in certain ring systems, where the phenomenon of *optical exaltation* is observed, i.e., the observed value is generally higher than the calculated. Further, the molar refraction principle can be applied to gases and liquids, as well as to solids. The molar refraction of gases is usually found to be identical with that of the corresponding liquids. Solids are studied best by dissolving them first in a solvent and measuring the refractive index n and density ρ of the solution. The molar refraction of the solution, $R_{1,2}$, follows then as

$$
R_{1,2} = \left(\frac{n^2 - 1}{n^2 + 2}\right)\left(\frac{N_1 M_1 + N_2 M_2}{\rho}\right) \tag{3}
$$

where M_1 and M_2 are the molecular weights of the solvent and solute, while N_1 and N_2 are their mol fractions in solution. $R_{1,2}$ is in turn related to the individual molar refractions R_1 and R_2 by the equation

$$
R_{1,2} = N_1 R_1 + N_2 R_2 \tag{4}
$$

and hence when R_1 of the solvent is known, R_2 may be found from Eq. (4). This value of R_2 may then be compared with values calculated from the molar refraction contributions given in Table 16–1. Such application of molar refraction to mixtures works out very satisfactorily.

ELECTRICAL POLARIZATION OF MOLECULES

Any molecule is composed of positively charged nuclei and negatively charged electrons. When such a molecule is introduced into an electric field between two charged plates, the field will act to attract the positive nuclei toward the negative plate and the electrons toward the positive plate, as shown in Figure 16–1(a). The result is thus an electrical distortion or *polarization* of the molecule to form an *electric dipole*, Figure 16–1(b), with the positive charge at one end and the negative charge at the other. This polarization lasts only as long as the field is applied. When the field is removed the distortion disappears, and the molecule reverts to its original condition. For this reason this type of electrical distortion of a molecule is called *induced polarization*, and the electric dipole formed is called an *induced dipole*.

Unlike the original molecule, where the centers of positive and negative electricity were coincident, in the induced dipole these charges are separated by a distance l. Again, since the molecule as a whole is neutral, the positive charge at one end, z_+, must be equal but be opposite in sign to the negative charge, z_-, at the other end. To such an induced dipole

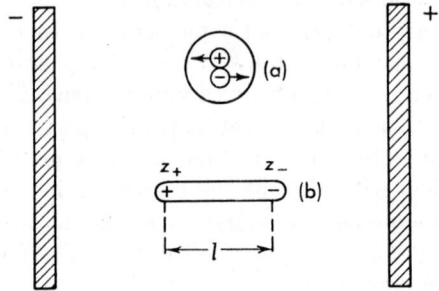

Figure 16–1.   Polarization of a molecule in an electric field: (a) original state; (b) polarized.

may be ascribed an electric moment μ_i, which by definition is the value of the charge at *one* end of the dipole multiplied by the distance between the charges, namely,

$$\mu_i = zl \tag{5}$$

The subscript i is used to denote that the dipole moment is induced only in an electric field.

The magnitude of μ_i is determined by the electric field strength X acting on the molecule in accordance with the relation

$$\mu_i = \alpha X \tag{6}$$

Here α is a constant called the *polarizability* of the given molecule. Clausius and Mosotti have shown from electromagnetic theory that the constant α is in turn related to the *dielectric constant D* of the medium between the plates, i.e., of the molecules, by the equation

$$\left(\frac{D-1}{D+2}\right)\frac{M}{\rho} = \frac{4}{3}\pi N\alpha = P_i \tag{7}$$

In this equation M and ρ are the molecular weight and density of the molecules, while N is Avogadro's number. Since both N and α are constants independent of temperature, the quantity P_i should also be a constant independent of the temperature and should be determined only by the nature of the molecules. P_i is called the *induced molar polarization* of the given substance, and it gives the electrical distortion produced in 1 mole of a substance when the electric field strength is unity. As D is unitless, the units of P_i are those of M/ρ, namely volume, and are expressed in cubic centimeters per mole.

The dielectric constant is a property of any given medium. For a vacuum $D = 1$, but for any other medium D is greater than unity. The dielectric constant of a substance can be determined by measuring first the capacity of a condenser with a vacuum between the plates, C_0, and next the capacity of the same condenser when filled with the given substance, C. Then the dielectric constant follows from electrical theory as

$$D = \frac{C}{C_0} \tag{8}$$

To measure the capacity, various electronic circuits are available which use alternating currents with frequencies of 10^6 to 10^7 cycles per second.

With the dielectric constant and density of a substance measured at a given temperature, the molar polarization can be calculated by Eq. (7). When this is done, it is found that the molar polarizations of such substances as oxygen, carbon dioxide, nitrogen, and methane are constant and independent of the temperature. On the other hand, for substances such as hydrogen chloride, chloroform, nitrobenzene, and methyl chloride, the molar polarization is not constant but decreases as the temperature is raised. The explanation of this anomalous behavior has proved of great import to studies of molecular structure.

PERMANENT DIPOLE MOMENTS

To account for the variation of the molar polarization of certain substances with temperature, Peter Debye made in 1912 the significant suggestion that such molecules possess a *permanent* dipole moment of their own. The permanent dipole moment would arise from the fact that in some molecules the centers of positive and negative electricity may not coincide. As a result there would be present in the molecule even outside of an electric field positive and negative centers of electrical gravity a distance l apart; and, consequently, such a molecule will be a permanent dipole and possess a permanent dipole moment μ of magnitude zl. Outside of an electric field the permanent dipoles of various molecules in an aggregation would, because of thermal agitation, be oriented more or less randomly in space. However, when placed in an electric field two disturbing effects will arise. First, the electric field will tend to rotate and orient the permanent dipoles in the direction of the field; and second, the field will tend to polarize the molecules. Were the molecules perfectly stationary, the orienting effect of the electric field would result in an alignment of the dipoles at an angle of 180° to the direction of the field and 90° to the condenser plates. But the perfect alignment of the dipoles is disturbed by thermal agitation of the molecules, and hence the resulting orientation will be some point intermediate between the original position of the dipole in space and the final completely oriented position. This situation in an electric field is illustrated in Figure 16–2, where (a) gives the initial

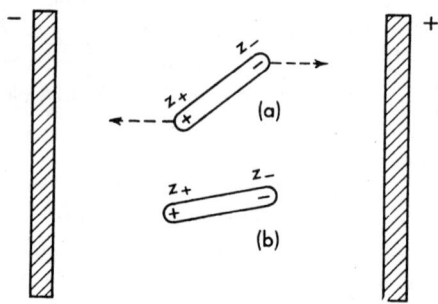

Figure 16–2. Polarization of a molecule with permanent dipole: (a) original state; (b) polarized and oriented position.

orientation of the molecule with the permanent dipole, and (b) the orientation resulting from the effect of the applied field.

In the absence of a permanent dipole moment in a molecule the molar polarization, as calculated from the measured dielectric constant by means of Eq. (7), gives only the induced molar polarization P_i. On the other hand, when a dipole moment is present, this quantity measures not only the induced polarization, but also the *molar orientation polarization* P_o. The *total molar polarization* P_t is, therefore,

$$P_t = \left(\frac{D-1}{D+2}\right)\frac{M}{\rho} = P_i + P_o \tag{9}$$

As before, the induced molar polarization is given by the Clausius-Mosotti equation, namely, $P_i = \frac{4}{3}\pi N\alpha$. Again, Debye showed that the orientation polarization P_o should be equal to

$$P_o = \frac{4}{3}\pi N\left(\frac{\mu^2}{3\,kT}\right) \tag{10}$$

where μ is the *permanent dipole moment* of the molecule, k is the gas constant per molecule, i.e., R/N, and T is the temperature. Substituting these values of P_i and P_o into Eq. (9), the total molar polarization, as given by the measured dielectric constant, becomes

$$P_t = \left(\frac{D-1}{D+2}\right)\frac{M}{\rho} = \frac{4}{3}\pi N\alpha + \frac{4}{3}\pi N\left(\frac{\mu^2}{3\,kT}\right) \tag{11}$$

In Eq. (11) the first term on the right is a constant which may be represented by A. In the second term, again, all quantities are constant except T, and hence this term may be written as B/T. With these substitutions Eq. (11) takes the form

$$P_t = \left(\frac{D-1}{D+2}\right)\frac{M}{\rho} = A + \frac{B}{T} \tag{12}$$

which shows that *for molecules with permanent dipoles the total molar*

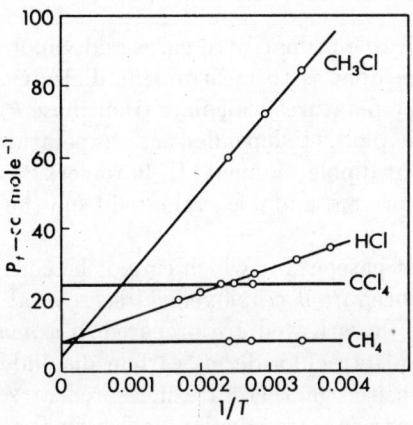

Figure 16–3. Variation of total polarization with temperature for polar and nonpolar compounds.

polarization P_t should vary linearly with $1/T$. Further, the slope of a plot of P_t vs. $1/T$ should be equal to

$$B = \frac{4\,\pi N\mu^2}{9\,k} \tag{13}$$

from which, on insertion of the values of N and k, μ should follow as

$$\mu = 0.0128\,\sqrt{B}\times 10^{-18} \tag{14}$$

Therefore, by plotting P_t against $1/T$ and taking the slope B of the linear plot, μ may be calculated from Eq. (14), and thereby the permanent dipole moment of the molecule in question may be found.

Debye's theory of permanent dipole moments is confirmed by experimental data in every respect. In Figure 16–3 are given plots of P_t vs. $1/T$ for hydrogen chloride and methyl chloride which show that for these substances P_t varies linearly with $1/T$, and hence these substances must possess permanent dipole moments. However, since the molar polariza-

tions of methane and carbon tetrachloride are independent of temperature, no permanent dipole moments are present in these substances, and
all electrical polarization is due, therefore, to induction.

Molecules possessing permanent dipole moments are said to be *polar*
and those without permanent dipoles *nonpolar*. The units of the dipole
moment are those of the electronic charge multiplied by a distance. As
the electronic charge is of the order of 10^{-10} electrostatic units and molecular distances are of the order of 10^{-8} cm, the dipole moments of molecules
will always have a magnitude of $10^{-10} \times 10^{-8} = 10^{-18}$. The quantity
10^{-18} is called a *Debye unit*, symbol D. A dipole moment may be given
thus as 1.6×10^{-18} or $1.6\ D$.

DETERMINATION OF DIPOLE MOMENTS

Equation (11) should be applicable primarily to gases and vapors. For
such substances all that need be done is to measure the dielectric constants and densities at several temperatures, calculate from these P_t, and
plot the latter against $1/T$. If the plot is independent of temperature, the
substance possesses no permanent dipole moment. If, however, P_t varies
with $1/T$, a permanent dipole is present and μ is evaluated from the slope
by Eq. (14).

With substances which are not gaseous, or which cannot be converted
readily to a gas, an alternate procedure is employed. First, several *dilute*
solutions of the substance to be investigated are prepared in a *nonpolar*
solvent, such as benzene, carbon tetrachloride, or carbon disulfide, and
the dielectric constants and densities of these are measured at several
temperatures. Next, at each temperature are calculated the molar polarizations of the solutions, $P_{1,2}$, from the relation

$$P_{1,2} = \left(\frac{D-1}{D+2}\right)\left(\frac{N_1 M_1 + N_2 M_2}{\rho}\right) \tag{15}$$

where N_1 and N_2 are the mol fractions of solvent and solute in the solution, M_1 and M_2 the respective molecular weights, and ρ the density of
the solution. Third, from $P_{1,2}$, defined also as

$$P_{1,2} = N_1 P_1 + N_2 P_2 \tag{16}$$

where P_1 and P_2 are the molar polarizations of the solvent and solute,
P_2 of the solute is calculated from the known values of N_1, N_2 and P_1.
If the values of P_2 thus determined are constant at each temperature for a
number of concentrations, P_2 is the molar polarization of the pure solute.
However, if P_2 is found to vary with concentration due to effect of the
solvent, then P_2 is plotted against N_2, and the curve is extrapolated to
$N_2 = 0$ to yield P_2 freed of solvent influence. Finally, the values of P_2

thus obtained at each temperature are plotted against $1/T$, and μ is found from the slope in the same manner as for gases.

Other procedures are also available, but these need not be described here. Some results for the permanent dipole moments of various substances obtained either in the gas phase or in solution are shown in Table 16–2.

TABLE 16–2.   DIPOLE MOMENTS OF VARIOUS MOLECULES
(In Debye Units)

Inorganic Molecules	μ	Organic Molecules	μ
H_2, Cl_2, Br_2, I_2, N_2	0	CH_4, C_2H_6, C_2H_4, C_2H_2	0
CO_2, CS_2, $SnCl_4$, SnI_4	0	CCl_4, CBr_4	0
HCl	1.03	C_6H_6, naphthalene, diphenyl	0
HBr	0.78	CH_3Cl	1.86
HI	0.38	CH_3Br	1.80
H_2O	1.84	C_2H_5Br	2.03
HCN	2.93	CH_3OH	1.70
NH_3	1.46	CH_3NH_2	1.24
SO_2	1.63	CH_3COOH	1.74
N_2O	0.17	p-dichlorbenzene	0
CO	0.12	m-dichlorbenzene	1.72
PH_3	0.55	o-dichlorbenzene	2.50
PCl_3	0.78	p-chlornitrobenzene	2.83
$AsCl_3$	1.59	m-chlornitrobenzene	3.73
$AgClO_4$	4.7	o-chlornitrobenzene	4.64

MOLECULAR STRUCTURE AND DIPOLE MOMENTS

The absence of a dipole moment in molecules such as hydrogen, chlorine, and nitrogen indicates that the electron pairs binding the atoms are situated equidistant between the constituents of the molecule. Were this not true a dipole moment would be present. Again, in linear molecules such as carbon dioxide and carbon disulfide, even if the electron pairs are not equidistant between the atoms, the electric moment on one side of the molecule balances the moment on the other side, and the net dipole moment is therefore zero. The same considerations apply also to such molecules as stannic chloride, carbon tetrachloride, and the paraffin hydrocarbons, both saturated and unsaturated. In aromatic molecules such as benzene, naphthalene, and diphenyl the absence of a net dipole moment can be accounted for most readily on the basis of a planar structure for these. On the other hand, the presence of a dipole moment in the hydrogen halides and carbon monoxide definitely points to a nonequidistant distribution of the electron pairs between the two atoms. In all probability this pair lies closer to the halogens and oxygen than to the hydrogen or carbon, and this leads to the formation of a dipole. In a similar manner may be explained the dipole moments appearing in the alkyl halides, nitriles, amines, and alcohols.

The fact that water and sulfur dioxide have appreciable dipole moments rules out linear structures for these molecules. If we postulate that the two hydrogens in water and the two oxygens in sulfur dioxide lie on the same side of the central atom, there will be no tendency for the moments to cancel each other, and an appreciable net moment will result. Quantitative calculations indicate that this is the only way in which the dipole moments of these molecules can be accounted for. In a like manner the dipole moment of ammonia can be explained on the basis of a triangular pyramidal structure in which the nitrogen is situated at the apex of the pyramid, and the hydrogens lie at the other three corners.

It is of interest to observe that in p-dichlorbenzene, where the chlorines are at opposite ends of the benzene ring, the dipole moment is zero, and hence electrical symmetry is present. However, when substitution of the second chlorine takes place in a meta or ortho position, the symmetry is destroyed and a dipole appears. Furthermore, the dipole is larger the greater the asymmetry, as may be seen from a comparison of the μ values of m- and o-dichlorbenzenes. On the other hand, when the two substituents in the benzene ring are different, as in chlornitrobenzene, dipole moments are present in all three forms. Nevertheless the dipole moment of the para form is least and increases successively with meta and ortho substitution.

MOLECULAR SPECTRA

Just as study of atomic spectra has proved helpful in elucidating the structure of atoms, so the spectra emitted by gaseous molecules have yielded valuable information on the structure of molecules. Unlike the *line* spectra emitted by atoms, molecules of a gas when excited emit a *band* spectrum which strong resolution reveals to be composed of many lines very closely spaced. In general the sorting and classification of lines in a band spectrum is a difficult and tedious task. Nevertheless, the spectra of most diatomic and many polyatomic molecules have been unraveled, and from these a deep insight has been obtained into what occurs within the molecule.

Three types of molecular spectra are recognized, namely, (a) *rotation spectra*, (b) *vibration-rotation spectra*, and (c) *electronic band spectra*. To appreciate the origin of these, an understanding is required of the energy levels in a molecule. For this purpose consider a diatomic molecule AB.

This molecule consists of two nuclei corresponding to the atoms A and B and their accompanying electrons. As in the constituent atoms, the various electrons in the molecule may exist in a number of energy levels such that on excitation an electron may shift from one to another. Once thus excited, the electron on return will emit a spectral line of frequency determined by the energy difference of the two levels and the quantized Bohr relation. Were this the only manner in which the molecule could absorb energy, the electronic spectrum of a molecule would be a line spectrum

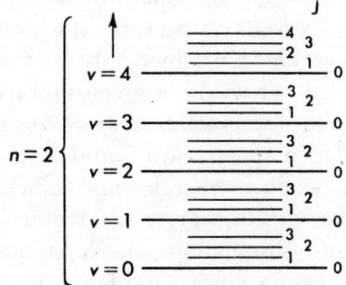

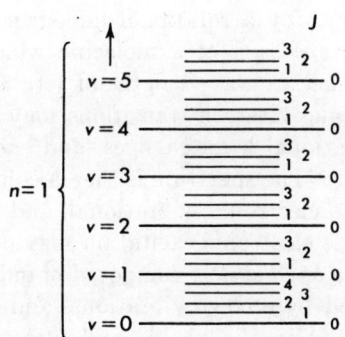

Figure 16–4. Molecular energy levels (schematic).

akin to that of atoms. But a molecule may also absorb energy to cause a vibration of the nuclei with respect to each other, and to set the molecule as a whole rotating. These vibrational and rotational motions are quantized and are superimposed upon the kinetic motion of the molecule. From the standpoint of band spectra the latter motion may be disregarded, but the other two cannot be. The incidence of quantized vibration and rotation coupled with the electronic levels leads to the diagrammatic picture of the energy levels of a molecule shown in Figure 16–4. The group of levels represented by $n = 1$ corresponds to the lowest electronic energy level of the molecule. Similarly, the group given by $n = 2$ is the energy structure of the next higher, excited, electronic level. Each of these electronic levels is subdivided into the *vibrational* sublevels designated by the vibrational quantum numbers v on the left. In turn, each of the vibrational sublevels has associated with it a number of *rotational* quantum levels, the energy of which is determined by the rotational quantum numbers J shown on the right. In a like manner every higher electronic level of a molecule will consist of vibrational and rotational levels of the form shown in Figure 16–4, and will lie in sequence above those for $n = 2$.

From this description it is evident that the energy required for excitation of various emissions will be least for rotation, and will be increasingly greater for vibration and then electron transfer. If the exciting energies are kept suitably small, it is possible to produce only transitions from one rotational quantum level to another within a given vibrational level, and the result is then the appearance of the *rotational spectrum alone*. Since the energies involved are small these spectra are found in the far infrared. On the other hand, when the exciting energies are sufficiently large to cause vibration level transitions within a given electronic level, emissions are observed corresponding to the changes in the vibration quantum numbers. Further, since a change in vibrational levels involves also changes in rotational levels, the over-all result is a vibrational spectrum wherein each line is accompanied by a rotational fine-structure. We obtain thus the *vibration-rotation spectrum* of a molecule which is generally located in the near infrared. i.e., at wave lengths of 1 to 50 μ.[1] Finally, with still higher exciting energies electron transitions may take place which are accompanied by vibrational level changes, and each of these in turn by rotational fine-structure. The spectrum is then a complex *electronic band* consisting of lines due to electronic, vibrational, and rotational transitions. As more than a single electronic excitation may occur, the complete spectrum may consist of a *band system* composed of individual electronic bands, each accompanied by its own vibrational and rotational lines. Electronic bands are found in the visible and ultraviolet spectral ranges.

ROTATION SPECTRA

Pure rotation spectra are exhibited only by molecules which possess a permanent dipole moment to interact with the exciting electromagnetic radiation. This means that only molecules such as those with $\mu \neq 0$ in Table 16–2 will give rotation spectra in the far infrared, while those with $\mu = 0$ will not.

The rotational spectrum of a *diatomic* molecule can be accounted for by

[1] $1 \mu = 10^{-4}$ cm $= 10,000$ Å.

considering the molecule as a first approximation to be a rigid rotator, i.e., a rigid dumbbell joined along its line of centers by a bond equal in length to the distance r_0 between the two nuclei. Wave mechanics shows that the energy E_J of such a rigid rotator in any given rotational quantum level J is given by

$$E_J = \left(\frac{h^2}{8\pi^2 I}\right) J(J+1) \tag{17}$$

In this equation h is Planck's constant, while I is the *moment of inertia* of the molecule given in terms of the two atomic masses, m_1 and m_2, and the internuclear distance r_0 by the relation

$$I = \left(\frac{m_1 m_2}{m_1 + m_2}\right) r_0^2 \tag{18}$$

J may have integral values of 0, 1, 2, 3, etc. For a transition from a rotational level of quantum number J' to one of higher quantum number J, the energy difference would correspond to

$$\Delta E_r = E_J - E_{J'} = \left(\frac{h^2}{8\pi^2 I}\right)[J(J+1) - J'(J'+1)] \tag{19}$$

Rotational transitions are restricted to changes in J corresponding to $\Delta J = J - J' = 1$. Inserting this condition into Eq (19), we obtain

$$\Delta E_r = h\nu = \left(\frac{2h^2}{8\pi^2 I}\right) J \tag{20}$$

and hence

$$\nu = \left(\frac{h}{4\pi^2 I}\right) J \tag{21}$$

Here the values of J start with $J = 1$.

Since the quantity in parentheses is a constant for a given molecule, Eq. (21) shows that the rotational spectrum should consist of a number of equally spaced lines the frequencies of which are determined by J. Hence, by determining the frequencies of lines corresponding to various J values in the pure rotation spectrum of a molecule, the moment of inertia of the molecule may be obtained through Eq. (21) and from it the internuclear distance of the atoms r_0 by means of Eq. (18).

Linear polyatomic molecules like $H-C\equiv N$ or $H-C\equiv C-Cl$ also have only one moment of inertia. For these, Eqs. (17), (20), and (21) are valid and may be used to find I. However, since I involves now several interatomic distances, the latter cannot be recovered from the one measured moment of inertia. Isotopic substitution may be resorted to for such compounds. Thus, from a study of HCN and DCN, two moments of inertia can be obtained. Assuming that such substitution does not affect the bond distances, the latter can be recovered from the two observed moments.

Nonlinear molecules have three moments of inertia, two of which may be identical in certain instances. These three moments correspond to the respective rotations of the molecule as a whole about a set of x, y, and z axes passed through the center of gravity of the molecule. Such molecules again require special and more elaborate analysis in order to obtain the moments and bond lengths from their rotational spectra.

Although pure rotation spectra offer the simplest means of deducing moments of inertia of molecules, technical difficulties in far infrared studies impair high accuracy. For this reason this quantity is deduced more frequently from the other types of spectra which can be measured more conveniently and precisely, or by microwave spectroscopy.

MICROWAVE SPECTROSCOPY

In infrared work the energy source is a heated bar or filament which supplies nonmonochromatic radiation. The latter is passed through a suitable prism which disperses the light and allows selection of a narrow band of desired wave length to be transmitted through the substance under test. The intensity of the transmitted light is measured then by means of a thermopile and compared with that of the incident beam. On the other hand, in microwave spectroscopy an electronically controlled oscillator is used to generate monochromatic electromagnetic energy of frequency corresponding to a wave length range of 0.1 to several centimeters. Since the energies at such wave lengths are of the same order of magnitude as those involved in rotational transitions in molecules, the microwaves may be used to measure these. The procedure employed is to pass the monochromatic energy through a gaseous sample of the substance under investigation, and to measure the intensity of the transmitted radiation by means of an electronic receiver and cathode-ray oscillograph. By varying the frequency of the oscillator and observing the intensity of the transmitted beams, it is possible to obtain data from which moments of inertia and internuclear distances of the molecules involved can be calculated.

The microwave method is again applicable only to molecules which possess permanent dipole moments. Compared with infrared measurements, its accuracy is very high and corresponds to about ± 0.001 Å in r_0. The method may also be used for identification and analysis of substances which can be obtained in gaseous form for observation. For details see the book by Gordy, Smith, and Trambarulo.[2]

[2] Gordy, Smith, and Trambarulo, *Microwave Spectroscopy*, John Wiley & Sons, Inc., New York, 1953.

VIBRATION-ROTATION SPECTRA

Vibration-rotation spectra are exhibited by diatomic molecules with permanent dipole moments, i.e., heteronuclear diatomic molecules, and polyatomic molecules with and without permanent dipole moments. Homonuclear diatomic molecules, such as O_2, N_2, or Cl_2, have no permanent dipoles and do not show vibration-rotation spectra. On the other hand, the exciting radiation can induce an oscillating dipole moment in polyatomic molecules which have no permanent dipole moment of their own, and hence these can undergo vibrational-rotational transitions. A vibration-rotation spectrum for HBr is shown in Figure 16-5.

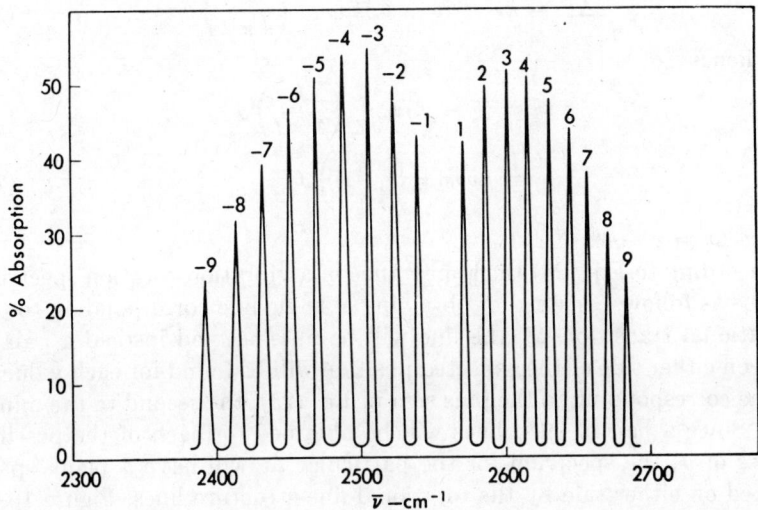

Figure 16-5.   Vibration-rotation spectrum of HBr for transition from $v' = 0$ to $v = 1$.

To a first approximation the vibratory motion of the nuclei of a diatomic molecule can be represented as the vibration of a *simple harmonic oscillator;* i.e., an oscillator in which the restoring force is proportional to the displacement in accordance with Hooke's law. For such an oscillator wave mechanics shows that the vibrational energy E_v is related to the fundamental vibrational frequency v_0 by the relation

$$E_v = \left(v + \frac{1}{2}\right)hv_0 \tag{22}$$

Here v is the *vibrational quantum number,* which may take on the values $v = 0, 1, 2,$ etc. Equation (22) reveals that such an oscillator retains in its lowest vibrational level $v = 0$ the energy $E_0 = \frac{1}{2} hv_0$. This residual energy, called the *zero-point energy* of the oscillator, cannot be removed from the molecule even by cooling it to 0°K.

For a vibrational transition from any level of quantum number v' to another of quantum number v, the absorbed energy is given by

$$\Delta E_v = E_v - E_{v'} = \left(v + \frac{1}{2}\right)h\nu_0 - \left(v' + \frac{1}{2}\right)h\nu_0 = (v - v')h\nu_0 \quad (23)$$

Upon these vibrational transitions are superposed accompanying rotational changes which *add* and *subtract* from Eq. (23) the energy change given by Eq. (20). The energy change resulting is thus

$$\Delta E = \Delta E_v \pm \Delta E_r$$

$$\Delta E = h\nu = (v - v')h\nu_0 \pm \left(\frac{h^2}{4\,\pi^2 I}\right)J \quad (24)$$

and hence

$$\nu = (v - v')\nu_0 \pm \left(\frac{h}{4\,\pi^2 I}\right)J$$

$$= \Delta v \nu_0 \pm \left(\frac{h}{4\,\pi^2 I}\right)J \quad (25)$$

where $\Delta v = v - v'$.

According to Eq. (25) the appearance of a vibration-rotation spectrum will be as follows: Instead of obtaining a single line corresponding to the vibrational transition Δv, the line will be missing, and instead a pair of lines on either side of the expected position will be found for each value of J, one corresponding to the plus sign in Eq. (25), the second to the minus sign. Since a pair of such lines will be obtained for each of the possible values of J, the spectrum for the particular Δv will have a blank space flanked on either side by the rotational fine-structure lines, Figure 16–6.

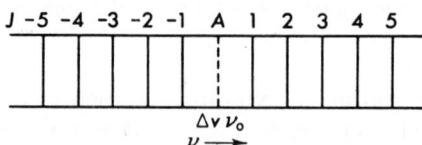

Figure 16–6.   Diagram of vibration-rotation spectrum (schematic).

As one unit such as Figure 16–6 is obtained for each possible change in vibrational quantum number, the complete vibration-rotation spectrum will be composed of a number of such units.

The bands corresponding to $v' = 0$ and $v = 1, 2$, etc., are called respectively the *fundamental* band, the *first harmonic* band, the *second harmonic* band, etc. By ascertaining in the fundamental band the frequency corresponding to the missing vibration line, i.e., position A in Figure 16–6, ν_0

is obtained directly. Similarly, the frequencies at the positions corresponding to A in the first and second harmonics gives $2\nu_0$ and $3\nu_0$. These may be used, then, to check the value of ν_0, the fundamental vibration frequency of the molecule. With ν_0 known, measurement of the line frequencies for various values of J permits calculation of the moment of inertia through Eq. (25), and from it, in turn, the internuclear distance r_0. In this manner it is possible to ascertain from vibration-rotation spectra the fundamental vibration frequency of the molecule, the moment of inertia, and the distance of separation of the nuclei.

Equation (25) predicts a constant spacing of vibration-rotation lines. Actually the spacings observed are variable due to complications to be mentioned later. In Table 16–3 are given measured frequencies in wave

TABLE 16–3.   VIBRATION-ROTATION SPECTRUM OF HCL FOR TRANSITION
FROM $v' = 0$ TO $v = 1$
$(\bar{\nu}_0 = 2890 \text{ cm}^{-1})$

J	$\bar{\nu}_{obs.}$	$\bar{\nu}_{obs.} - \bar{\nu}_{calc.}$	J	$\bar{\nu}_{obs.}$	$\bar{\nu}_{obs.} - \bar{\nu}_{calc.}$
1	2910	-1	-1	2869	0
2	2930	-2	-2	2847	0
3	2949	-4	-3	2825	-1
4	2967	-7	-4	2803	-2
5	2985	-11	-5	2779	-5
6	3002	-15	-6	2755	-8
7	3017	-19	-7	2731	-10

numbers, $\bar{\nu}_{obs.}$, for some lines in the fundamental vibration-rotation band of HCl, and a comparison of these with the frequencies calculated for this substance by means of Eq. (25), $\bar{\nu}_{calc.}$. The agreement between the two sets of $\bar{\nu}$ values is fairly good when J is low, but becomes worse as J increases. Further, the spacing between the lines decreases with increase in J, and is not the same for positive and negative values of J.

Under strong resolution the individual lines in the vibration-rotation spectrum of HCl are found to be doublets, with the weaker members possessing wave lengths about 27 Å longer than the stronger ones. This splitting is due to the presence in HCl of the two chlorine isotopes Cl^{35} and Cl^{37}. These different masses give different ν_0 and I values for HCl^{35} and HCl^{37}, and produce thus splitting of the lines.

Although a diatomic molecule can have only one fundamental frequency of vibration, the number of such frequencies increases with the number of atoms. For a linear polyatomic molecule the number of possible modes of vibration the molecule *may* have is $(3n - 5)$, where n is the number of atoms in the molecule, while for a nonlinear molecule it is $(3n - 6)$. Thus, in passing from a diatomic to a triatomic linear molecule the possible vibration frequencies increase from one to four, and for a nonlinear molecule from one to three. Not all the possible vibrations are necessarily active, nor are all of the frequencies necessarily different. Nevertheless, the added modes of motion complicate the situation, and consequently the resulting spectra are much more complex and difficult to interpret.

ELECTRONIC SPECTRA

Electronic spectra arise from excitation of electrons in a molecule from one energy level to a higher one. As a result of this excitation the molecule will absorb radiation corresponding to the energy difference between the two electronic levels involved. Since during the excitation a change takes place also in vibrational and rotational levels, the electronic emission is always accompanied by vibrational and rotational transitions. Thus, if we let E_e', E_v', and E_r' be, respectively, the electronic, vibrational, and rotational energies of the molecule before the transition, the E_e, E_v, and E_r the corresponding quantities after the change, then the total energy of the molecule in the initial state is

$$E' = E_e' + E_v' + E_r'$$

and, in the final state,

$$E = E_e + E_v + E_r$$

The energy change involved in the electronic transition is, consequently,

$$\Delta E = (E_e - E_e') + (E_v - E_v') + (E_r - E_r')$$
$$= \Delta E_e + \Delta E_v + \Delta E_r \tag{26}$$

and the resulting frequency of radiation follows as

$$\nu = \frac{\Delta E}{h} = \frac{\Delta E_e + \Delta E_v + \Delta E_r}{h} \tag{27}$$

As ΔE_v and ΔE_r may have different values depending on the rotational and vibrational quantum numbers involved, the number of lines possible for a given change in E_e is large and leads to a complex electronic band. Further, with different values of ΔE_e possible as well, the complete band spectrum will consist of a band system composed of a series of individual bands.

Despite their complexity electronic band spectra are frequently studied in preference to the other types. This is because these spectra occur in the visible or ultraviolet ranges and hence may be photographed and inspected easily. In general the information deduced from electronic bands is the same as that obtained from vibration-rotation spectra. Besides, it is possible to ascertain from electronic band spectra information about the excited states of the molecule as well as the energy of dissociation of the molecule into its component atoms.

When the electronic band spectrum of a molecule is inspected, it is found that at one end of the spectrum the distance of separation of the lines becomes smaller and smaller, until at a given position the line spectrum terminates, and a space of continuous emission starts. The position of the limit where the lines end gives the frequency of radiation necessary for *dissociation* of the molecule. When the dissociation is into normal atoms, the spectroscopic dissociation energy D_s can be calculated directly from the observed frequency. However, when the dissociation results in one or more excited atoms, a correction has to be applied for the energy of excitation.

The above discussion of the molecular spectra of diatomic molecules is a simplification of the actual situation. The molecule as a whole is not quite a rigid rotator nor a simple harmonic oscillator. Because of these defects in ideal behavior corrections have to be introduced which tend to complicate Eqs. (21), (23), and (25). For the purpose at hand, however, the treatment given should suffice.

TABLE 16–4. MOLECULAR CONSTANTS OF DIATOMIC MOLECULES

Molecule	$m^* \times 10^{24}$ (g)	$I \times 10^{40}$ (g-cm²)	$r_0 \times 10^8$ (cm)	$\nu_0 \times 10^{-12}$ (sec⁻¹)	D_s kcal/mole
H_2	0.8371	0.460	0.742	131.8	103.2
N_2	11.63	13.94	1.095	70.75	170.3
O_2	13.28	19.35	1.207	47.38	117.1
Cl_2	29.04	114.8	1.988	16.94	57.1
Br_2	66.35	345.8	2.283	9.689	45.5
I_2	105.16	748.0	2.667	6.427	35.6
HCl	1.627	2.645	1.275	86.63	102.2
CO	11.39	14.49	1.128	65.05	256.2
NO	12.40	16.43	1.151	57.08	149.6

Table 16–4 summarizes the molecular constants of some diatomic molecules deduced from band spectra in the manner described. In this table column 2 gives the reduced mass of the molecule, i.e., $m^* = (m_1 m_2)/(m_1 + m_2)$, columns 3, 4, and 5 the moments of inertia, the interatomic distances, and the fundamental frequencies of vibration, respectively, while column 6 gives the spectroscopically deduced dissociation energies. These constants are for molecules which contain in each instance the most abundant isotope of the element involved. Again, Table 16–5 gives bond lengths and bond angles for some polyatomic molecules. These results were obtained by either band or microwave spectroscopy.

TABLE 16-5. BOND LENGTHS AND BOND ANGLES
OF POLYATOMIC MOLECULES

Molecule	Bond	Bond Length, Å	Bond Angle	
Linear				
N_2O	NN	1.126		
	NO	1.186		
COS	CO	1.161		
	CS	1.558		
HCN	CH	1.064		
	CN	1.156		
HCCCl	CH	1.052		
	CC	1.211		
	CCl	1.632		
Nonlinear				
NH_3	NH	1.008	HNH	107.3°
H_2O	OH	0.958	HOH	104.5°
SO_2	SO	1.432	OSO	119.5°
CH_4	CH	1.091	HCH	109.5°
CH_2Cl_2	CH	1.068	HCH	112°
	CCl	1.772	ClCCl	111.8°
C_2H_6	CH	1.107	HCH	109.3°
	CC	1.536		
CH_3OH	CH	1.096	HCH	109.3°
	CO	1.427	COH	108.9°
	OH	0.956		

RAMAN SPECTRA

An alternate and simpler method than band spectroscopy for obtaining vibrational and rotational frequencies of molecules is through observations of the *Raman effect*. In 1928 Raman found that when light of a definite frequency ν was passed through a gas, liquid, or solid, and the scattered radiation was observed at right angles to the direction of the incident beam, the scattered radiation contained lines not only of the original light, but also some of lower, and occasionally higher, frequency. The lines shifted toward the lower frequencies are called *Stokes lines*, those toward the higher frequencies *anti-Stokes lines*. Raman further found that the difference $\Delta\nu$ between the incident and any given scattered line was *constant and characteristic of the substance irradiated*, and *completely independent of the frequency ν of the incident radiation*. Such differences $\Delta\nu$ between incident and scattered lines are called *characteristic Raman frequencies*.

The Raman effect arises from absorption by a molecule from the incident radiation of sufficient quanta of energy to cause transitions from lower to higher vibrational levels, or from lower to higher rotational levels. In vibrational excitation a *single* quantum of energy is removed in magnitude equal to the energy necessary for passage from one vibrational quantum level to the next. In rotations, however, *two* quanta are extracted to cause a change in rotational level from a value J' to $J = (J' + 2)$. As a result of such removal of energy from the original beam the energy of the latter must decrease, and so must the frequency. We thus obtain the Stokes lines. On the other hand, the molecule, instead of absorbing energy, may emit some by rotational or vibrational transitions from higher levels to lower. When this happens the emitted energy is added to that of the incident light, the frequency of the latter becomes larger, and anti-Stokes lines are obtained.

The Raman frequencies observed for various rotational and vibrational transitions are identical with the same transitions observed in band spectra. Since this is true, vibrational and rotational frequencies may be ascertained more conveniently from the Raman effect, which may be studied in the visible spectral range rather than in the more difficult infrared. Still another advantage is the fact that molecules without permanent dipole moments, such as O_2, N_2, and Cl_2, also exhibit the Raman effect. Consequently, the rotational and vibrational changes in these can be observed as well by Raman but not by infrared spectroscopy.

ELECTRON DIFFRACTION OF GASES

Electron diffraction of gases offers still another powerful tool for arriving at the geometric configuration of molecules and the distance between atoms in these. The experimental arrangement involves generation of electrons by a hot filament, acceleration of these by a constant electric field of about 40,000 volts cm^{-1}, passage of the electrons through a sample of gas or vaporized liquid or solid kept at a low pressure, and, finally, impingement of the electrons upon a photographic plate placed in their path. The image formed on the latter consists of a series of concentric rings similar to those obtained in the powder method of X-ray analysis.

The principle of the procedure is also similar to that of the powder method. The electrons, acting as waves of length ca. 0.05 Å, are diffracted by the atoms in the gas molecules. Since these molecules are randomly oriented with respect to the electron beam, some of the diffracted rays experience interference, while others are reinforced. The result is a series of rings whose positions and intensities depend on the geometry of the apparatus, the configurational structure of the molecules, and the atomic radii. From such rings and their intensities can be ascertained whether the molecules are linear or nonlinear, whether they are planar or nonplanar, the bond distances, and also the angles between bonds in the case of nonlinear molecules.

POTENTIAL ENERGY DIAGRAMS OF DIATOMIC MOLECULES

Spectroscopic molecular constants can be used to construct the potential energy diagrams of diatomic molecules. For the *lowest* or *ground electronic state* of a molecule the potential energy diagram will have the general form given in Figure 16–7. To any excited state an analogous

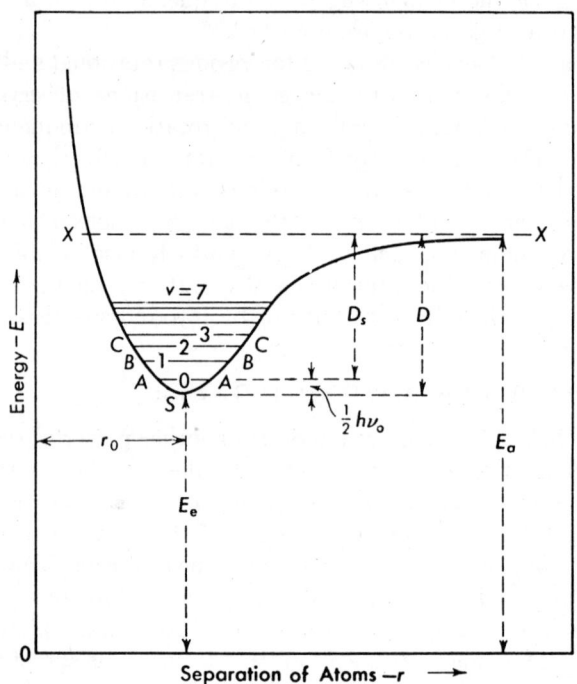

Figure 16–7.   Potential energy diagram of diatomic molecule.

potential energy curve will correspond, except that because of the higher energy content it will lie above the one for the ground state.

The total energy E of any molecule consists of the electronic energy E_e, the vibrational energy E_v, the rotational energy E_r, and the energy of thermal motion E_t; i.e.,

$$E = E_e + E_v + E_r + E_t \qquad (28)$$

At the absolute zero of temperature E_t and E_r are both zero, and E_v

reduces to the zero-point energy of the molecule, namely, $E_0 = \frac{1}{2} h\nu_0$. At the absolute zero, therefore, the energy is given by

$$E_{T=0} = E_e + \frac{1}{2} h\nu_0 \tag{29}$$

For such a stable molecule in its ground state point S in Figure 16–7, corresponding to the normal distance of separation r_0, gives the electronic energy content E_e. When the zero-point energy is added to this quantity, the energy level is raised to the line A–A, and we arrive thus at the energy of the molecule at the absolute zero. As the vibrational frequencies of the molecule are excited, the total energy levels are raised to the lines B–B, C–C, etc., in accordance with the increase in the vibrational quantum number v. The spacings of the lines within the potential energy diagram depend on the differences in energy between vibrational quantum levels. For an ideal harmonic oscillator the spacings between the lines would be constant, but actually they get smaller as v increases. When eventually line X–X is reached, the molecule dissociates into the two neutral atoms. The difference in energy between X–X and $E = 0$, namely E_a, is the energy content of the two resulting atoms. If from this energy we subtract the electronic energy E_e of the original molecule, the result is the energy of dissociation D of the molecule. However, if instead of subtracting E_e we deduct $E_e + \frac{1}{2} h\nu_0$, we obtain the spectroscopic energy of dissociation D_s. Hence

$$D = D_s + \frac{1}{2} h\nu_0 \tag{30}$$

A diagram such as Figure 16–7 indicates, therefore, the energy of a molecule in its ground electronic state and in various vibrational states as a function of the distance of separation. To obtain the total energy of the molecule at any given temperature, it would be necessary to add to $(E_e + E_v)$ the rotational and thermal energy contributions.

In dealing with specific heats and other thermodynamic properties of molecules, the consideration is always with energy differences, and hence in absence of electronic excitation E_e cancels out. Under such conditions the energy reference point may be transposed to point S; i.e., the energy at S may be taken as zero. On this basis the potential energy E_P of any diatomic molecule can be derived from an empirical equation due to Morse, namely,

$$E_P = D[1 - e^{-a(r-r_0)}]^2 \tag{31}$$

Here again D is the energy of dissociation, r_0 the equilibrium separation distance of the molecule, r the distance of separation at any other point, and a is a constant characteristic of a particular molecule. The constant a

may be evaluated from the relation

$$a = \pi \nu_0 \sqrt{\left(\frac{2\,m_1 m_2}{m_1 + m_2}\right)\frac{1}{D}} \qquad (32)$$

where ν_0 is the fundamental vibration frequency of the molecule, while m_1 and m_2 are the masses of the two atoms involved. In using Eq. (32) D must be expressed in ergs per molecule, but in Eq. (31) it may have any desired units, as kilogram calories per mole. Finally, to establish the vibrational levels, i.e., the lines A–A, B–B, etc., in Figure 16–7, recourse is had to Eq. (22). By substituting into this relation $v = 0$ the zero-point energy follows, $v = 1$ the level B–B, $v = 2$ the level C–C, etc. The lines thus obtained will be idealized and equidistant. For their true positions information on the anharmonicity of the oscillator is required.

ABSORPTION OF LIGHT BY LIQUIDS AND SOLUTIONS

Because of molecular interaction in liquids and solutions, rotational fine structure disappears in the infrared and only continuous vibrational bands are observed. Again, in the visible and ultraviolet regions, where electronic transitions occur, the vibrational fine structure is also suppressed to a large degree, and hence the electronic bands as well are generally continuous.

Maximum vibrational absorptions take place in the infrared region. On the other hand, the location of electronic bands depends on the energy involved in the electronic transition responsible for the absorption. If the energy of the latter is large, the absorption will occur primarily in the ultraviolet. With smaller energies, however, the absorption band will appear in the near ultraviolet or in the visible light regions.

The color exhibited by a substance is determined by the light the substance transmits in the visible range, and this, in turn, is related to the spectral range in which the substance absorbs light. Any substance that absorbs only in the ultraviolet will transmit all white light, and as a consequence the substance will appear colorless. If, however, the substance shows a color in white light, the color indicates the light which was *not* absorbed. For instance, carbon black has the particular color because it absorbs nonselectively the white light that reaches it. Again, sodium chloride solutions are colorless because they do not absorb selectively any light in the visible region. On the other hand, a copper sulfate solution is blue because it absorbs out of the white light the yellow and red, leaving the blue to be transmitted to the eye. It is evident that the characteristic of the absorbing substance is not the light it transmits but rather the light it absorbs. Consequently, it is by studying the latter that information about the absorbing substance can be obtained. Further, it also follows that only colored substances can be studied in the visible region. Substances which are colorless do not absorb white light appreciably, and hence these must be investigated in the ultraviolet.

SPECTROPHOTOMETRY

In the spectral range where absorption is appreciable the intensity of the initial light is reduced greatly on passing through the absorbing medium. On the contrary, to either side of the absorption band the intensity is reduced considerably less or not at all. The position of an absorption band can be ascertained, therefore, by measuring the decrease in intensity of the original light after passage through an absorbing medium as a function of the wave length. Then, on plotting the fraction of the original light absorbed against the wave length, the position of the absorption band will be given by a maximum in the curve. Conversely, if we plot instead the fraction of light transmitted vs. the wave length, a minimum in the plot will indicate the position of maximum absorption.

The decrease in intensity of incident light of any given wave length on passage through an absorbing substance is given by *Lambert's law*. This law states that the rate of decrease of intensity with thickness of absorbing material, $-dI_t/dl$, where I_t is the intensity at thickness l, is proportional to the intensity of the light at point l, namely, that

$$-\frac{dI_t}{dl} = kI_t \qquad (33)$$

In Eq. (33) the proportionality constant k is called the *absorption coefficient* and is characteristic of the absorbing medium. Since at $l = 0$ we have the original intensity I_0, the intensity I_t at any point l can be found from Eq. (33) by integration between these limits. We obtain thus

$$\int_{I_0}^{I_t} \frac{dI_t}{I_t} = \int_{l=0}^{l=l} - kdl$$

$$\ln \frac{I_t}{I_0} = -kl \qquad (34)$$

or

$$\frac{I_t}{I_0} = e^{-kl} \qquad (35)$$

The last equation may be written also as

$$\frac{I_t}{I_0} = 10^{-k'l} \qquad (36)$$

in which case $k' = k/2.303$ is the *extinction coefficient* of the substance.

Consequently, $\ln I_t$ falls off linearly, and I_t exponentially, with the distance l the light travels through the absorbing medium.

For absorbing solutes the decrease in intensity with l is proportional not only to I_t, but also to the concentration of the solution C. For solutions, therefore,

$$- \frac{dI_t}{dl} = \varepsilon I_t C \tag{37}$$

where ε, called the *molar absorption coefficient*, is a proportionality constant determined by the nature of the absorbing solute and the wave length of light used. Integration of Eq. (37) between the same limits as before, and at constant C, yields

$$\int_{I_0}^{I_t} \frac{dI_t}{I_t} = - \int_{l=0}^{l=l} \varepsilon C dl$$

$$\ln \frac{I_t}{I_0} = - \varepsilon C l \tag{38}$$

and

$$\frac{I_t}{I_0} = e^{-\varepsilon C l} \tag{39}$$

Equations (37), (38), and (39) are expressions of *Beer's law* for absorption of light by solutions. These show that the intensity of the transmitted light falls off exponentially with both C and l, or that the logarithm of the transmitted intensity I_t is proportional to both the concentration and the length of the absorbing path.

As before the natural logarithm can be changed in Eq. (39) to base 10, in which case

$$\frac{I_t}{I_0} = 10^{-\varepsilon' C l} \tag{40}$$

The constant $\varepsilon' = \varepsilon/2.303$ is called the *molar extinction coefficient* of the absorbing solution.

The ratio I_t/I_0 is determined with some form of spectrophotometer. At present spectrophotometers utilizing photoelectric cells are available which give I_t/I_0 directly. Once I_t/I_0 for a pure liquid is determined and the thickness of the cell used is known, the absorption or extinction coefficient of the substance for the given wave length can readily be calculated. Similarly, from the measured I_t/I_0 and the known values of l and C may be found the molar absorption or extinction coefficients. Next, to ascertain the wave length at which maximum absorption takes place, we may plot these coefficients, their logarithms, or log (I_0/I_t) vs. λ, in which case the maximum absorptions will be given by maxima in the curves. On the other hand, when I_t/I_0 is plotted vs. λ, then the minima in the curves will correspond to the maxima in absorption. An absorption curve for benzene in solution is shown in Figure 16–8.

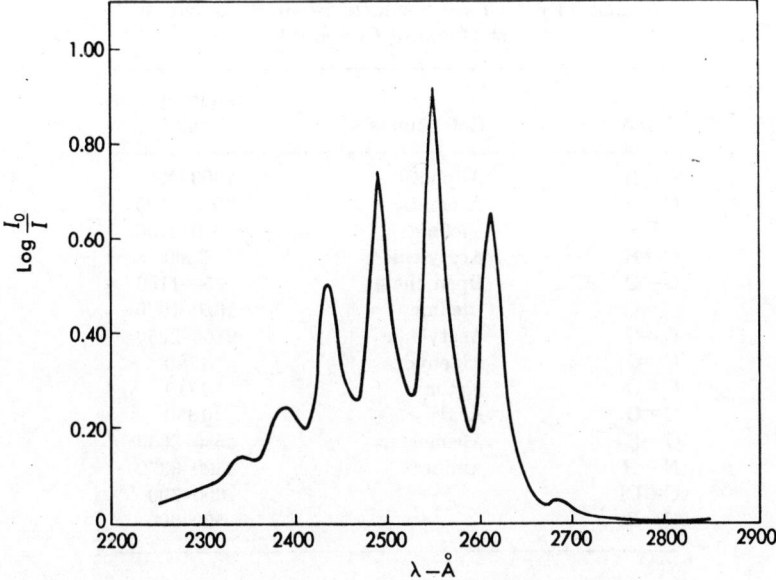

Figure 16-8.   Absorption curve for benzene in solution.

SPECTROPHOTOMETRY AND MOLECULAR STRUCTURE

The absorption spectrum of a compound is very characteristic, and so comparison of spectra of known and unknown compounds can be used to identify the latter. Further, extensive study of the spectra of organic compounds has shown that certain frequencies can be associated with specific bonds in molecules. Hence the presence or absence of such bonds in a compound can be ascertained by seeing whether the frequencies for the anticipated bonds do or do not appear. By this means it is possible to determine the nature of bonds in various types of compounds. Generally, characteristic bond frequencies are not fixed in position, but depend to some degree on the nature of the groups attached to the linkages in question. Consequently the frequencies associated with particular bonds can usually be specified only in terms of a range of values, as may be seen from the characteristic bond frequencies given in Table 16-6.

Absorption bands can also be used to determine the concentration of compounds in solution. By using a wave length at or near an absorption maximum, and by measuring I_t/I_0 for a solution of known concentration in a cell of known length, the absorption or extinction coefficients of the compound can be calculated from Eqs. (39) or (40). Then, by using ε or ε', and finding I_t/I_0 for a solution of unknown concentration, C can be calculated from the same equations.

TABLE 16–6.   CHARACTERISTIC BOND FREQUENCIES
FOR ORGANIC COMPOUNDS

Bond	Compounds	Frequency – $\bar{\nu}$ cm^{-1}
C—H	Aliphatic	2800–3000
C—H	Aromatic	3000–3100
C—H	Olefinic	3000–3100
C—H	Acetylenic	3300
C—C	Open chain	750–1100
C=C	Olefinic	1620–1670
C≡C	Acetylenic	2100–2250
C=O	Aldehydes	1780
C=O	Ketones	1710
C=O	Acids	1650
O—H	Alcohols	3580–3650
N—H	Amines	3300–3370
C—Cl	—	600–700
C—Br	—	500–600

MAGNETIC PROPERTIES OF SUBSTANCES

The force f acting between two magnetic poles of strength p_1 and p_2 separated by a distance r is given by

$$f = \frac{p_1 p_2}{q r^2} \tag{41}$$

Here q is the *magnetic permeability* of the medium between the poles, and it represents the tendency of magnetic lines of force to pass through the medium relative to the tendency for a vacuum. For the latter $q = 1$, while for other media q may be greater or less than unity. When $q < 1$, the substance composing the medium is said to be *diamagnetic*, and the lines of force prefer to pass through a vacuum rather than through the substance. On the other hand, when $q > 1$, the substance is said to be *paramagnetic*, and the tendency for lines of force to pass through the substance is greater than through a vacuum. Finally, substances with very large q are said to be *ferromagnetic*. These include Fe, Co, Ni, and their alloys.

The *specific magnetic susceptibility* χ of a substance is related to q by the relation

$$\chi = \frac{q - 1}{4 \pi \rho} \tag{42}$$

where ρ is the density of the substance in g cc^{-1}. From Eq. (42) it may be seen that χ is negative for diamagnetic materials and positive for paramagnetic ones. Its units are cc g^{-1}. Again, the *molar magnetic susceptibility*, χ_M, is given by

$$\chi_M = M\chi \tag{43}$$

where M is the molecular weight of the substance involved. The units of χ_M are thus cc mole^{-1}.

The specific magnetic susceptibility of a substances can be readily measured by Gouy's method. In this method a cylinder containing the sample is suspended by wire from one arm of a balance until its lower part is centered between the poles of an electromagnet. The cylinder is counterbalanced in this position by weights placed on the other balance pan. When the current is turned on the cylinder will be drawn downward into the electromagnet if the substance in it is paramagnetic, or it will be pushed upward if the substance is diamagnetic. In the first instance weights have to be added to the pan to restore balance, while in the second weights have to be removed. In either case the change in mass, Δm, is related to χ of the substance by the relation

$$\Delta mg = \frac{1}{2}(\rho\chi - \rho_a\chi_a)AH^2 \tag{44}$$

where g is the acceleration of gravity, ρ the density and A the cross-sectional area of the specimen, H the field strength of the magnet in gausses, χ_a the specific magnetic susceptibility of air, and ρ_a its density. At room temperature $\rho_a\chi_a = 0.0300 \times 10^{-6}$. The calculation of χ can be simplified by measuring in the apparatus at the same value of H a substance of known magnetic susceptibility, χ_s, such as water. Then

$$\frac{\Delta m}{\Delta m_s} = \frac{\rho\chi - \rho_a\chi_a}{\rho_s\chi_s - \rho_a\chi_a} \tag{45}$$

The molar magnetic susceptibility is calculated from χ by means of Eq. (43).

Table 16–7 shows χ_M values for some elements and compounds at room temperature.

DIAMAGNETIC SUBSTANCES

The magnetic susceptibility of diamagnetic substances is independent of temperature, and is about the same for a substance in the gas and liquid states. It arises from the action of the magnetic field upon the electrons moving about their nuclei. Application of an electric field changes the velocity of the electrons, and, as a consequence, a magnetic field is induced which opposes the applied magnetic field. Hence both χ and χ_M are negative.

TABLE 16-7. MOLAR MAGNETIC SUSCEPTIBILITIES AT ROOM TEMPERATURE
(Cgs units)

Substance	$\chi_M \times 10^6$	Substance	$\chi_M \times 10^6$
Diamagnetic		*Paramagnetic*	
Sb(s)	−99.0	Al(s)	16.5
C(s)	− 6.0	Mg(s)	13.5
Br_2(l)	−56.4	$CuCl_2$(s)	1,180.0
H_2(g)	− 4.0	$TiCl_3$(s)	1,110.0
Cl_2(g)	−40.5	NO(g)	1,461.0
N_2(g)	−12.0	MnS(s)	3,850.0
$CaCl_2$(s)	−54.7	UCl_3(s)	3,460.0
C_6H_5COOH(s)	−70.3	O_2(g)	3,449.0
H_2O(l)	−13.0	$NiCl_2$(s)	6,145.0
CCl_4(l)	−66.6	CoI_2(s)	10,760.0
CO_2(g)	−21.0	$CoSO_4$(s)	10,000.0
H_2S(g)	−25.5	$FeSO_4$(s)	10,200.0
NH_3(g)	−18.0	$FeCl_2$(s)	14,750.0
C_2H_2(g)	−12.5	$MnSO_4 \cdot 5\,H_2O$(s)	14,700.0

P. Pascal has shown that, as in the case of molar refraction, χ_M can be calculated from atomic and bond contributions which are supposed to apply to all molecules. Values of some of these contributions are given in Table 16-8. These are used in exactly the same manner as those for the molar refractions. For instance, for C_6H_5COOH we have

$$
\begin{aligned}
7\ C &= 7(-6.00 \times 10^{-6}) = -42.00 \times 10^{-6} \quad '' \\
6\ H &= 6(-2.93 \times 10^{-6}) = -17.58 \quad '' \\
1\ O\text{—} &= 1(-4.61 \times 10^{-6}) = - 4.61 \quad '' \\
1\ O\text{=} &= 1(-3.36 \times 10^{-6}) = - 3.36 \quad '' \\
1\ \text{benzene ring} &= 1(-1.4\ \times 10^{-6}) = - 1.4 \quad '' \\
\hline
\chi_M &= -69.0\ \times 10^{-6}
\end{aligned}
$$

The observed value is $\chi_M = -70.3 \times 10^{-6}$ cc mole^{-1}.

TABLE 16-8. ATOMIC AND BOND CONTRIBUTIONS TO χ_M

H	− 2.93 × 10⁻⁶	Cl	−20.1 × 10⁻⁶
C	− 6.00	Br	−30.6
O(in alcohols and ethers)	− 4.61	C=C bond	+ 5.5
O(in ketones)	+ 1.73	C=N bond	+ 8.2
O(in C=O)	− 3.36	N=N bond	+ 1.9
N(in open chains)	− 5.57	C≡C bond	+ 0.8
N(in ring)	− 4.61	C=C—C=C group	+10.6
N(in amines)	− 1.54	Benzene ring	− 1.4
F	−11.5		

PARAMAGNETIC SUBSTANCES

The magnetic susceptibility of paramagnetic substances varies with temperature in line with the Curie law, namely,

$$\chi_M = \alpha_M + \frac{C_M}{T} \tag{46}$$

where α_M and C_M are constants for a given substance. In this equation α_M represents the diamagnetic molar susceptibility of a substance induced by the applied magnetic field as discussed in the preceding section. Again, to explain C_M, Langevin suggested in 1905 that each molecule, atom, or ion of a paramagnetic substance is a tiny magnet which possesses a definite inherent magnetic moment of magnitude μ_m. In a magnetic field each of these tiny magnets will tend to arrange itself parallel to the applied field. However, this tendency will be opposed by the thermal motion of the particles, and hence the arrangement actually taken will be dependent on the temperature. On this basis Langevin was able to show that

$$\frac{C_M}{T} = \frac{\mu_m^2 N}{3\,kT} \tag{47}$$

where k is the Boltzmann constant. Inserting Eq. (47) into Eq. (46), we obtain thus

$$\chi_M = \alpha_M + \frac{\mu_m^2 N}{3\,kT} \tag{48}$$

When χ_M data are available at several temperatures, then μ_m can be evaluated by plotting χ_M vs. $1/T$ and taking the slope of the resulting straight line. This slope is obviously $\mu_m^2 N/3\,k$, from which μ_m can be found. The more common practice is to evaluate μ_m from χ_M measured at one temperature. For this purpose α_M is first calculated by the additivity rule for atomic and bond contributions, and this value of α_M is then used in Eq. (48) to find μ_m.

The units of μ_m are ergs gauss^{-1}. Frequently, however, μ_m is expressed in Bohr magnetons. It can be shown theoretically that the magnetic moment of the electron in the hydrogen atom is a whole number multiple of a quantity μ_B given by

$$\mu_B = \frac{eh}{4\,\pi mc} \tag{49}$$

where e and m are, respectively, the charge and mass of the electron and c the velocity of light. μ_B is called a *Bohr magneton*, and it is equal to 9.2732×10^{-21} erg gauss^{-1}. Therefore, to obtain μ_m in Bohr magnetons, it has to be divided by this value of μ_B.

PARAMAGNETISM AND UNPAIRED ELECTRONS

We have seen before (page 632) that the quantum number J is a vectorial combination of the azimuthal quantum number L and the spin quantum number S'. Quantum theory shows that μ_m is related to J by the expression

$$\mu_m = g'\mu_B \sqrt{J(J + 1)} \tag{50}$$

where g' is called the Landé splitting factor. However, in pure liquids, solutions, and solids the surrounding molecules wipe out the effective contribution of L to J, and hence the latter becomes equal to S'. Under these conditions $g' = 2$, and Eq. (50) becomes thus

$$\mu_m = 2\,\mu_B \sqrt{S'(S' + 1)} \tag{51}$$

and Eq. (48)

$$\chi_M - \alpha_M = \frac{4\,\mu_B^2 N}{3\,kT}\,[S'(S' + 1)] \tag{52}$$

S' depends on the number of unpaired electrons in a molecule, with each unpaired electron contributing $\frac{1}{2}$ to S'. Consequently, Eq. (52) can be used to find the number of unpaired electrons present in a molecule. Table 16-9 gives the values of $\chi_M - \alpha_M$ at 25°C predicted by Eq. (52) for different numbers of unpaired electrons. Since here α_M is practically negligible compared with χ_M, the last column of the table gives essentially the latter. Comparison of Tables 16-7 and 16-9 shows that there must be present one unpaired electron in, say, $CuCl_2$, two in UCl_3, three in $NiCl_2$, etc.

TABLE 16-9.   RELATION OF NUMBER OF UNPAIRED ELECTRONS TO χ_M

No. of Unpaired Electrons	S'	$\chi_M - \alpha_M$ at 25°C
0	0	0
1	0.5	$1{,}260 \times 10^{-6}$
2	1.0	3,350
3	1.5	6,290
4	2.0	10,060
5	2.5	14,670
6	3.0	20,120
7	3.5	26,410

NUCLEAR SPIN AND MAGNETIC MOMENT

Nuclei of atoms are composed of protons and neutrons. Like electrons, these particles spin on their own axes, and they possess thus angular momenta equal to $\frac{1}{2}\left(\frac{h}{2\pi}\right)$ for each particle. The resultant angular momentum for all the protons and neutrons present in a nucleus is called *nuclear spin*. This nuclear spin is given by $\sqrt{I(I + 1)}\left(\frac{h}{2\pi}\right)$, where I is the spin quantum number for a nucleus. I may have values of 0, $\frac{1}{2}$, 1, $1\frac{1}{2}$, etc. For nuclei with even sums of protons plus neutrons I is integral, i.e., $I = 0, 1, 2, \cdots$. On the other hand, when the sums are odd, then $I = \frac{1}{2}, \frac{3}{2}, \frac{5}{2}, \cdots$. Thus, for the nuclei of helium and oxygen, $I = 0$, for hydrogen and fluorine $\frac{1}{2}$, for deuterium and nitrogen 1, for sodium $\frac{3}{2}$, and for chlorine $\frac{5}{2}$.

Protons and neutrons in a nucleus possess magnetic properties which lead to a magnetic moment μ'_n for the nucleus. This magnetic moment is related to the nuclear spin quantum number I by the equation

$$\mu'_n = g_n \mu_N \sqrt{I(I + 1)} \tag{53}$$

where g_n is a quantity analogous to the Landé splitting factor and μ_N, the *nuclear magneton*, is given by

$$\mu_N = \frac{eh}{4\pi m_p c} \tag{54}$$

where m_p is the mass of the proton. Equation (54) defines μ_N, but it does *not* give the magnetic moment of the proton. The value of μ_N is 5.0505×10^{-24} erg gauss^{-1}, and hence μ_N is considerably smaller than the Bohr magneton μ_B.

ORTHO AND PARA STATES OF MOLECULES

The hydrogen atom has a nuclear spin of $\frac{1}{2}$. When two such atoms are combined to form an H_2 molecule, two possible combinations of nuclear spin can arise. In one of these the two spins can be parallel, in which case the resultant spin quantum number is $I = \frac{1}{2} + \frac{1}{2} = 1$, and we obtain thus *ortho-hydrogen*. When, however, the two spins are antiparallel, then $I = \frac{1}{2} - \frac{1}{2} = 0$, and we get then *para-hydrogen*. These two forms of hydrogen differ from each other in their low temperature heat capacities and also in their electronic band spectra. Ortho-hydrogen yields a spectrum with fairly intense rotational lines corresponding to *odd* values of the rotational quantum number J. On the other hand, para-hydrogen shows a spectrum with less intense lines corresponding to *even* values of J. Electronic band spectra of ordinary hydrogen gas exhibit alternate intense and less intense rotational lines. This must mean that ordinary hydrogen is a mixture of ortho- and para-hydrogen. From a comparison of the intensities of the rotational lines in ordinary hydrogen it has been established that in this gas the ratio of ortho- to para-hydrogen is fixed, and is equal to three parts ortho to one part para in equilibrium with each other.

Para-hydrogen has been prepared and studied. Some of its properties, such as vapor pressure, boiling point, triple point, specific heat, and electrical conductivity, have been found to be different from those of ordinary hydrogen. It has been established, in fact, that the above properties of ordinary hydrogen are the averages of the properties of the ortho- and para-hydrogen composing the mixture.

Ortho and para states are observed only when the two atoms composing the molecule are identical and when they possess nuclear spin. When the nuclei of the atoms do not possess spin, or when the two atoms are different, no ortho-para states can exist. Thus D_2 exists in ortho and para states, but O_2 and HCl do not. In O_2 the nuclei do not have spin. Again, although nuclear spin is present in both H and Cl, the two atoms are different in hydrogen chloride and hence no ortho-para states are possible.

NUCLEAR MAGNETIC RESONANCE

The magnetic moment of a nucleus, as given by Eq. (53), is a vector quantity with components in various allowable directions. The maximum observable component μ_n in any particular allowed direction is

$$\mu_n = g_n \mu_N I' \tag{55}$$

where I' may be $I, I - 1, I - 2, \cdots -I$, a total of $(2I + 1)$ values. In the absence of an external magnetic field all these orientations have the same energy. However, when a uniform magnetic field of strength H is applied, then these various allowed orientations of μ_n split into levels of different potential energies. Relative to the undisturbed nucleus, these potential energies are given by

$$E = -\mu_n H \tag{56}$$

or, on substitution of Eq. (55) for μ_n,

$$E = -g_n \mu_N H I' \tag{57}$$

Consequently, E depends on both I' and H. For a given value of H the levels resulting from Eq. (57) will be the following. When $I = I' = 0$, then $E = 0$. Hence nuclei with zero nuclear spin will not show splitting in a magnetic field. Again, when $I = \frac{1}{2}$, then $I' = \frac{1}{2}$ and $-\frac{1}{2}$, and the energies of the levels will be $-\frac{1}{2} g_n \mu_N H$ and $+\frac{1}{2} g_n \mu_N H$. Similarly, for $I = \frac{3}{2}$ we shall have $I' = \frac{3}{2}, \frac{1}{2}, -\frac{1}{2}$, and $-\frac{3}{2}$, and thus the energies will be minus these multiples of $g_n \mu_N H$. The nuclear energy levels in a magnetic field for several values of I are shown in Figure 16–9.

The difference in energy, ΔE, between two levels I_1' and I_2' follows from Eq. (57) as

$$\Delta E = E_2 - E_1 = -g_n\mu_N H(I_2' - I_1')$$

However, selection rules limit transitions only to adjacent levels, i.e., $\Delta I' = I_2' - I_1' = \pm 1$. Hence

$$\Delta E = \pm g_n\mu_N H \tag{58}$$

where the plus sign refers to absorption and the minus sign to emission

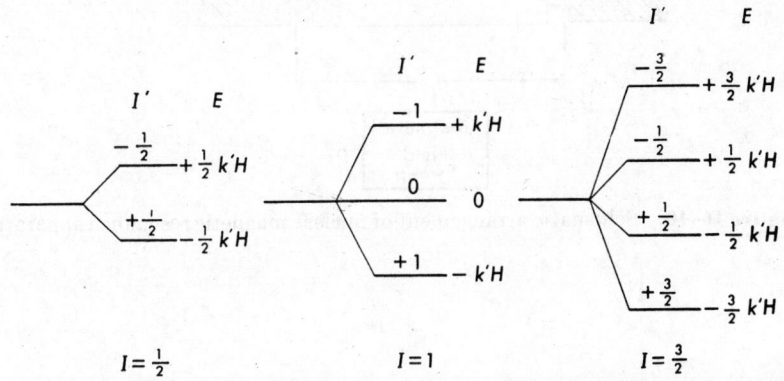

Figure 16–9. Splitting of nuclear energy levels in a magnetic field ($k' = g_n\mu_N$).

of energy. Since $\Delta E = h\nu$, the frequency which corresponds to a transition is

$$\nu = \frac{\Delta E}{h} = \pm \frac{g_n\mu_N H}{h} \tag{59}$$

The frequencies involved are in megacycles per second and they lie, therefore, in the radio-frequency range.

Purcell and Block showed independently in 1946 how these transition frequencies can be measured by means of a nuclear magnetic resonance (nmr) method. A schematic diagram of the apparatus is shown in Figure 16–10. The radio-frequency oscillator, A, is adjusted to deliver some preset frequency ν in the coil located between the magnet poles. Surrounding the sample S, placed as shown, is a receiver coil to pick up the broadcast signal. This signal is carried to a receiver and amplifier B, and then to a cathode-ray oscillograph indicator or recorder. D is a variable magnetic field sweep which controls the field strength between the poles. When this field strength is varied no absorption of energy by the sample takes place until the right-hand side of Eq. (59) becomes equal to the preset frequency. When this frequency is reached the sample is in resonance with the applied frequency, absorption occurs, and stops again

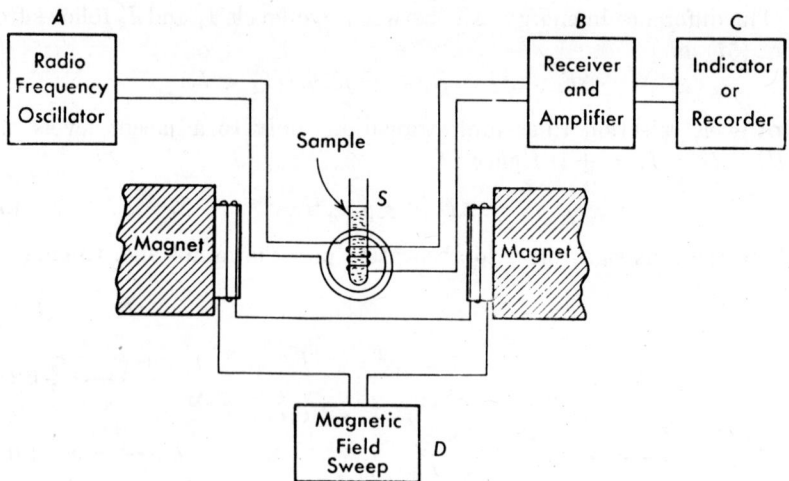

Figure 16–10. Schematic arrangement of nuclear magnetic resonance apparatus.

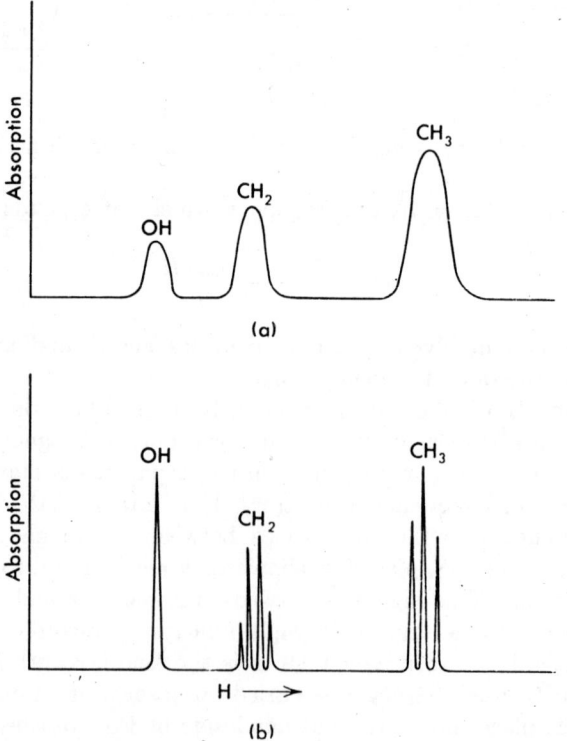

Figure 16–11. Nmr spectrum of ethyl alcohol. (a) low resolution; (b) high resolution.

when H is raised further. The result is then a trace such as Figure 16–11(a), where the peaks represent the values of H at which the sample is in resonance with the applied frequency. A plot such as Figure 16–11 is called an nmr spectrum.

INTERPRETATION OF NMR SPECTRA

To indicate the manner in which nmr spectra are interpreted and used to obtain information about molecular structure, we shall take as a specific example the behavior of ethyl alcohol, CH_3CH_2OH. In this molecule the carbon and oxygen atoms have no nuclear spin and hence no magnetic moments. On the other hand, the hydrogen atom has a spin quantum number of $\frac{1}{2}$, and hence we would expect for this atom a single absorption frequency given by Eq. (59). However, as Figure 16–11(a) indicates, this molecule actually yields three absorption peaks. The reason for this behavior lies in the fact that CH_3CH_2OH actually contains protons in three different environments, namely, those in the CH_3 group, those in the CH_2 group, and the one attached to the oxygen. Each of these types of hydrogen is screened differently by surrounding electrons, and hence the electrical environment of each is different. As a consequence each type of hydrogen atom interacts differently with the applied magnetic field, and shows thus a different absorption frequency. We obtain thereby three absorption peaks instead of one.

This environmental effect is expressed quantitatively in terms of the *chemical shift*, δ, defined as

$$\delta = \frac{H_s - H_r}{H_r} \cdot 10^6 \qquad (60)$$

In this equation H_s is the magnetic field strength at which a maximum appears for a given type of proton, while H_r is the field strength for some reference substance, such as water. Thus for the protons in the methyl group $\delta = 4.1$, while for the proton in the hydroxyl group $\delta = -0.1$.

When the nmr spectrum of CH_3CH_2OH is studied at high resolution, it reveals the fine structure shown in Figure 16–11(b). Here the CH_2 and CH_3 peaks have split into four and three peaks respectively. This fine structure arises from magnetic interactions among the various protons, which lead to slight shifts in magnetic moments and hence absorption frequencies.

From the above discussion it is evident that nmr spectra allow a study of the electrical environments in which nuclei with spin find themselves, as well as of the interactions among such nuclei. Hence nmr offers a means of investigating the finer details of molecular structure not revealed by the other methods.

INTERMOLECULAR FORCES

Throughout the book frequent reference has been made to intermolecular or van der Waals forces and the important role these play. However, very little has been said about the nature of these forces or their cause.

Intermolecular forces leading to attraction are due to three actions, namely, (a) the orientation, (b) the induction, and (c) the dispersion effects. There is also a repulsion which comes into play between the molecules on close approach. The orientation effect, present only in molecules with permanent electric dipoles, arises from the attractions and orientations such dipoles exercise upon each other. For molecules with dipole moment μ, this force of attraction between a pair of molecules has been shown to be

$$f_c = \left(\frac{4 \mu^4}{kT}\right) \frac{1}{r^7} \tag{61}$$

where k is the Boltzmann constant, T the absolute temperature, and r the distance of separation of the molecules. Again, the presence of permanent dipoles causes polarization in neighboring molecules, and thereby induced dipoles. Interaction of the induced and permanent dipoles leads to an attractive force, f_i, between a pair of molecules given by

$$f_i = (12 \ \alpha\mu^2) \frac{1}{r^7} \tag{62}$$

In this expression for the force resulting from the induction effect r and μ have the same significance as before, while α is the polarizability of the molecules.

The existence of the dispersion effect was first pointed out by F. London. London showed that, due to vibrations of electron clouds with respect to the nuclei of atoms in a molecule, tiny instantaneous dipoles of a specific orientation are formed. These dipoles induce dipole moments in neighboring atoms, and the latter interact with the original dipoles to produce an attraction between the molecules. For a pair of molecules the attractive force, f_d, due to this dispersion effect is given by

$$f_d = \left(\frac{9 \ h\nu_0\alpha^2}{2}\right) \frac{1}{r^7} \tag{63}$$

where h is Planck's constant and ν_0 is a characteristic frequency for the oscillation of the charge distribution.

Finally, the repulsion between molecules arises from the interaction on close approach of the electron atmospheres and nuclei of the atoms in one molecule with those in another. As a result a repulsive force, f_r, is developed whose magnitude can be represented by

$$f_0 = -\frac{B}{r^n} \tag{64}$$

where B is a constant for a given substance and n may range from 10 to 13. On adding Eqs. (61)–(64), we obtain for the total force f acting between a pair of molecules

$$
\begin{aligned}
f &= f_o + f_i + f_d + f_r \\
&= \left(\frac{4\,\mu^4}{kT}\right)\frac{1}{r^7} + (12\,\alpha\mu^2)\frac{1}{r^7} + \left(\frac{9\,h\nu_0\alpha^2}{2}\right)\frac{1}{r^7} - \frac{B}{r^n} \\
&= \frac{A}{r^7} - \frac{B}{r^n}
\end{aligned}
\tag{65}
$$

where
$$A = \left(\frac{4\,\mu^4}{kT}\right) + (12\,\alpha\mu^2) + \left(\frac{9\,h\nu_0\alpha^2}{2}\right) \tag{66}$$

From Eq. (65) it may be seen that the attractive forces between mole cules are inversely proportional to the seventh power of the distance between them, while the force of repulsion is proportional to a higher inverse power of the distance. Further, Eq. (66) shows that in molecules possessing no permanent dipole moments, i.e., $\mu = 0$, only the dispersion forces are operative, and hence these are responsible for the van der Waals forces in such substances as the rare gases, H_2, Cl_2, and the straight chain hydrocarbons. On the other hand, for molecules which have a permanent dipole moment all the effects contribute to the observed van der Waals interactions.

TABLE 16–10.   COMPARISON OF ENERGIES AND DISTANCES INVOLVED IN VARIOUS INTERACTIONS

Interaction	Bond Energy (Kcal mole^{-1})	Bond Distance (Å)
Primary valence	50–200+	1–2
Hydrogen bonding	4–10	2–3
Intermolecular attraction	0.5–5	3–5

Finally, Table 16–10 gives a comparison of the energies per mole and distances involved in primary valence, hydrogen bonding, and van der Waals attractions. From this table it may be seen that primary valence involves the highest energies, van der Waals interactions the smallest, with the hydrogen bonds being intermediate between the two.

REFERENCES

1. E. P. Andrew, *Nuclear Magnetic Resonance*, Cambridge University Press, Cambridge, 1955.
2. P. P. Bauman, *Absorption Spectroscopy*, John Wiley & Sons, Inc., New York, 1962.
3. A. G. Gaydon, *Dissociation Energies and Spectra of Diatomic Molecules*, Chapman and Hall, Ltd., London, 1953.
4. J. B. Goodenough, *Magnetism and the Chemical Bond*, John Wiley & Sons, Inc., New York, 1963.
5. Gordy, Smith, and Trambarulo, *Microwave Spectroscopy*, John Wiley & Sons, Inc., New York, 1953.
6. G. Herzberg, *Infrared and Raman Spectra of Polyatomic Molecules*, D. Van Nostrand Company, Inc., Princeton, N.J., 1945.
7. G. Herzberg, *Spectra of Diatomic Molecules*, D. Van Nostrand Company, Inc., Princeton, N.J., 1950.
8. K. Nakamoto, *Infrared Spectra of Inorganic and Coordination Compounds*, John Wiley & Sons, Inc., New York, 1963.
9. Pople, Schneider, and Bernstein, *High-Resolution Nuclear Magnetic Resonance*, McGraw-Hill Book Company, Inc., New York, 1959.
10. P. W. Selwood, *Magnetochemistry*, Interscience Publishers, Inc., New York, 1956.
11. R. S. Shankland, *Atomic and Nuclear Physics*, The Macmillan Company, N.Y., 1960.
12. A. Weissberger, *Physical Methods of Organic Chemistry*, Interscience Publishers, Inc., New York, 1960. Chapters 18, 26, 28, 29, 30, 39, 40, 41, 42, and 43.
13. W. West, *Chemical Applications of Spectroscopy*, Interscience Publishers, Inc., New York, 1956.

PROBLEMS

1. Calculate the molar refractions of the following compounds

Compound	Density	Refractive Index
(a) Methyl chloroformate	1.223	1.3868
(b) Tertiary butyl chloride	0.843	1.3869
(c) Secondary butyl alcohol	0.806	1.3924

and compare your results with those obtained from the additivity rule.

Ans. (a) Obs. 18.18, calc. 17.84.

2. The density of allyl chloride ($CH_2{=}CH{-}CH_2Cl$) is 0.938 g cc^{-1} at 20°C. Calculate the index of refraction, and compare your answer with the observed value of 1.4154.

3. The refractive index of naphthalene decahydride ($C_{10}H_{18}$) is 1.4804 at 18°C. Calculate the density and compare with the observed value of 0.895 g cc^{-1}.

4. The density of ethyl alcohol at 20°C is 0.789 g cc^{-1} while that of methyl alcohol is 0.792 g cc^{-1}. Assuming that these substances form an ideal solution, calculate by using Table 16-1 the refractive index of a solution containing 50% by weight of each constituent. *Ans.* 1.348.

5. The dielectric constant of $CH_4(g)$ at 0°C and 1 atm pressure is 1.00094. Assuming that methane behaves as an ideal gas, calculate (a) the induced molar polarization and (b) the polarizability of this substance.

Ans. (a) 7.02 cc; (b) 2.79×10^{-24} cc.

6. For $SO_2(g)$ at 0°C and 1 atm pressure the dielectric constant is 1.00993. This gas has a permanent dipole moment of 1.63 *D*. Assuming that SO_2 behaves as an ideal gas, calculate per mole (a) the total, (b) the orientation, and (c) the induced polarizations.

7. From the following data giving the dielectric constant *D* at various temperatures, determine the dipole moment of gaseous HCl by a graphical method:

$t°C$	D
-75	1.0076
0	1.0046
100	1.0026
200	1.0016

The pressure is 1 atm in every case. Assume the ideal gas law in calculating the density of HCl. *Ans.* 1.18 *D*.

8. From the dipole moment calculate the mean distance between the centers of positive and negative electricity in the HBr molecule.

9. The dielectric constant of cyclohexane at 20°C is 2.033 while its density is 0.7784 g cc^{-1}. In turn, for a solution of ethyl ether in cyclohexane, for which the mol fraction of solute is 0.04720 and density is 0.7751 g cc^{-1}, the dielectric constant is 2.109. Assuming the solution to be ideal, find the total molar polarization of the ethyl ether.

10. From the internuclear distance of 1.151 Å for NO, calculate the moment of inertia of the molecule. *Ans.* $I = 1.643 \times 10^{-39}$ g cm^2.

11. Calculate, in calories per mole, the rotational energies of NO in states with $J = 1, 2,$ and 3.

12. Calculate the rotational frequencies in wave numbers which will result when NO passes from states of (a) $J = 1 \rightarrow 0$, (b) $J = 2 \rightarrow 1$, and (c) $J = 3 \rightarrow 2$.

13. Calculate the energy in calories per mole required to raise NO from its lowest rotational level to the level with $J = 4$. *Ans.* 97.42 cal $mole^{-1}$.

14. From data given in Table 16-4 calculate the zero point energy of HCl in calories per mole.

15. Calculate the energy difference in calories per mole between I_2 in its lowest energy level and I_2 in its first vibrational and rotational state.

16. Assuming that I_2 could absorb in the infrared, calculate the frequencies in wave numbers which would result for the fundamental vibration-rotation lines of I_2 when $J = \pm 1$. *Ans.* $\bar{\nu} = 214.45$ and 214.31 cm^{-1}.

17. Repeat problem 16 for $J = \pm 2$ and $J = \pm 3$.

18. Assuming that I_2 could absorb in the infrared, calculate the frequencies in wave numbers which would result for the first harmonic vibration-rotation lines of the I_2 molecule when $J = \pm 2$.

19. How many principal moments of inertia and fundamental modes of vibration may be present in methyl fluoride?

20. The band spectrum of Br_2 terminates at a wave length of 5107 Å. The energy of the excited atoms formed in the dissociation is 10,400 cal in excess of that possessed by the normal atoms. What is the spectroscopic dissociation energy of Br_2 in calories per mole?   *Ans.* 45,500 cal mole^{-1}.

21. Given the following bond distances

$$
\begin{array}{ll}
\text{C—C} & 1.54 \text{ Å} \\
\text{C—H} & 1.07 \text{ Å} \\
\text{C—Cl} & 1.77 \text{ Å}
\end{array}
$$

calculate the atomic radii of H, C, and Cl on the supposition that the atoms are spheres which contact each other.

22. Starting with Eq. (17), deduce the expressions for the Raman frequencies of Stokes and anti-Stokes lines on transition from an initial rotational state of quantum number J' to a final state of quantum number J. Express the results in terms of J only.

23. Calculate, in units of $(h/8 \pi^2 I)$, the Stokes and anti-Stokes Raman frequencies which will result from rotational transitions into a final state of quantum number $J = 4$.   *Ans.* Stokes: $+14$; anti-Stokes: -22.

24. Using data given in Table 16–4, set up the Morse potential energy function for the Cl_2 molecule.

25. From the equation obtained in the preceding problem, find the work required, in calories per mole, to separate the two chlorine atoms in Cl_2 a distance of 0.10 Å from their equilibrium position in the molecule.

26. A certain glass filter, 0.35 cm in thickness, transmits 45.3% of light of $\lambda = 6000$ Å. What are the absorption and extinction coefficients of the glass?   *Ans.* $k = 2.26$ cm^{-1}; $k' = 0.982$ cm^{-1}.

27. Suppose it is desired to use the filter mentioned in the preceding problem to reduce the transmission of light of $\lambda = 6000$ Å to 0.1%. What thickness of filter will be required?

28. Using a cell 1 cm in thickness, the following data for K_2CrO_4 solutions were obtained at $\lambda = 3660$ Å in a photoelectric colorimeter:

I_t/I_0	Concentration (moles/liter $\times$ 10^4)
0.420	0.80
0.275	1.20
0.175	1.60
0.110	2.00

Plot these data to test Beer's law, and from the plot determine the molar absorption and extinction coefficients of K_2CrO_4 at this wave length.

29. Using the results of the preceding problem, calculate what will be the per cent of light of $\lambda = 3660$ Å transmitted by a 1.00×10^{-4} molar solution of K_2CrO_4 when placed in a cell 5 cm in thickness.

30. The molar magnetic susceptibility of $H_2O(l)$ at 25°C is -13.0×10^{-6} cc per mole, while the density is 0.9970 g per cc. Find (a) the specific magnetic susceptibility, and (b) the magnetic permeability of liquid water at this temperature.   *Ans.* (a) $\chi = -0.722 \times 10^{-6}$ cc g^{-1}; (b) $q = 0.99999$.

31. For La(s) at room temperature $\chi_M = 118.0 \times 10^{-6}$ cc per gram atom, while the density is 6.15 g per cc. Calculate (a) the specific magnetic susceptibility, and (b) the magnetic permeability of this element.

32. Using Table 16–8, calculate χ_M for benzene and compare the result with the observed value of -54.8×10^{-6} cc per mole.

33. For $O_2(g)$ $\chi_M = 3449.0 \times 10^{-6}$ cc per mole at 25°C. Assuming α_M to be negligible compared with χ_M, calculate the inherent magnetic moment of this molecule.   *Ans.* $\mu_m = 2.66 \times 10^{-20}$ erg gauss^{-1}.

34. For $Tb_2(SO_4)_3$ at 20°C $\chi_M = 7.820 \times 10^{-2}$ cc per mole. Assuming α_M to be negligible, calculate (a) the electron spin quantum number for the molecule, and (b) the number of unpaired electrons in the molecule.

35. Using the results of the preceding problem, find the magnetic moment of the $Tb_2(SO_4)_3$ molecule in Bohr magnetons.   *Ans.* $\mu_m = 13.54\ \mu_B$.

36. For Li^7 $g_n = 2.167$ and $I = 3/2$. Calculate the magnetic moment of this atomic nucleus in (a) nuclear magnetons, and (b) ergs per gauss.
$$Ans.\ (a)\ \mu'_n = 4.195\ \mu_N.$$

37. For H^1 $g_n = 5.57$ and $I = 1/2$. What will be the energies of the nuclear levels which will result when this atom is placed in a magnetic field of 5000 gauss? Express E in nuclear magnetons.   *Ans.* $E_1 = -13,900\ \mu_N$, $E_2 = 13,900\ \mu_N$.

38. For deuterium $g_n = 0.86$ and $I = 1$. Find the energies, in nuclear magnetons, of the levels which will result when this atom is placed in a magnetic field of 5000 gauss.

39. Using the data given in problem 37, find (a) the energy difference between nuclear levels, and (b) the resonance frequency for H^1 in a field of 5000 gauss.

40. Using the data given in problem 38, find (a) the energy difference between nuclear levels, and (b) the resonance frequency for D in a field of 5000 gauss.

41. For argon $\mu = 0$ and $h\nu_0 = 2.47 \times 10^{-11}$ erg. Calculate the total attractive force between two argon molecules separated by a distance of 4 Å.
$$Ans.\ 1.80 \times 10^{-6}\ dyne.$$

42. For water $\mu = 1.84 \times 10^{-18}$ esu cm, $\alpha = 1.48 \times 10^{-24}$ cc, and $h\nu_0 = 2.88 \times 10^{-11}$ erg. Calculate (a) the total force of attraction at 300°K between a pair of water molecules 5 Å apart, and (b) the percentages of the total force contributed by the orientation, induction, and dispersion effects.

43. Assuming that for a certain substance $A = B = 15.0 \times 10^{-2}$ for force in dynes and distance in Å, construct a plot of the total force operating between two molecules of the substance as a function of distance from $r = 0$ to $r = 5$ Å. Take $n = 10$.

17

Nuclear Chemistry

IN OUR DISCUSSION of atomic and molecular structure emphasis has been placed on the external electronic arrangement of the orbital electrons in atoms and their interactions, as these are the factors responsible for valency and chemical reactivity. The problem which requires attention now is the nature of the atomic nucleus. We have seen that the nuclei of atoms consist of protons in number equal to the atomic number of the element, and of neutrons in number equal to the difference between the atomic mass A and the nuclear charge Z, i.e., $(A - Z)$. No further insight can be obtained from pursuing mass spectrographic or spectroscopic studies, since these leave the nuclei unaffected. Neither can ordinary chemical investigation be of help, for chemical reactivity is associated only with the external electrons. As a matter of fact, the only phenomenon which does involve changes in the atomic nucleus is radioactivity, both natural and artificial, so we shall turn to a brief exposition of this subject.

NATURAL RADIOACTIVITY

In 1895 Henri Becquerel discovered that uranium salts emit radiation which can cause fogging of photographic plates. Subsequently it was shown that this radiation can ionize air, is emitted from the element as well as its salts, and is not at all affected by temperature or the source of the uranium. This spontaneous emission of radiation by an element is called *radioactivity*, while the elements which exhibit this behavior are said to be *radioactive*.

In 1898 Marie and Pierre Curie found that the mineral pitchblende (mainly U_3O_8) exhibited radioactivity more pronounced than uranium itself. This suggested the presence of elements more active than uranium, and Madame Curie actually succeeded in isolating from this mineral two new radioactive elements, polonium and radium. At about the same time thorium and actinium were also shown to be radioactive. Since that time over 30 elements have been proved to be naturally radioactive, and many others have been induced to become radioactive by artificial means to be described later.

The rays emitted by radioactive elements can be detected and measured by their ability to ionize gases. Another means is to employ a Wilson *cloud chamber*. When an ionizing particle passes through supersaturated water vapor, droplets of water tend to condense along the path of motion. To take advantage of this fact Wilson designed an apparatus which consists of a chamber filled with dust-free air saturated with water vapor, and fitted with a piston. As the particles enter through a window the air is made to expand, the temperature drops, and the vapor becomes supersaturated. As a result a fog track deposits along the path of the particles which can be photographed. From such fog tracks the velocity and energy of the particles can be ascertained.

A similar arrangement is the *bubble chamber* invented by Glaser in 1952. In the bubble chamber the ionizing particles pass through a superheated liquid such as hydrogen, and cause gas bubbles to be formed along their path of motion. These bubbles make the tracks visible, and they can thus be photographed and studied in the same manner as those in the Wilson cloud chamber.

The rays emitted by radioactive elements and their salts have been shown to consist of three types of radiation, α, β, and γ rays. Study of the α rays in a magnetic field has revealed that these consist of material particles possessing a mass four times and a positive charge twice that of the proton. This would make the charge and mass of the α particles identical with that of the helium nucleus, and this identity has actually been established. These particles have the ability to ionize gases and penetrate matter, as Rutherford showed in his scattering experiments. The initial velocity of the α particles on emission is very high, ranging from $1-2 \times 10^9$ cm sec^{-1} as against 3×10^{10} cm sec^{-1} for the velocity of light.

β rays have also been shown to consist of particles, but with a much lower mass and a negative charge. Their e/m ratio is the same as that of the electron, and this, in fact, is what they are—electrons. These rays again can ionize gases, affect a photographic plate, and penetrate matter.

But whereas their penetrating power is about 100 times that of α particles, their ionizing power is only about $\frac{1}{100}$ as great because of their much lower mass. Finally, the velocity of the emitted electrons depends on their source and is very large, approaching in some instances very close to the velocity of light.

The γ rays are quite different from either the α or β rays. They are totally unaffected by electric and magnetic fields and behave in every respect as electromagnetic radiation of the same nature as X rays. Measurements of their wave length indicate that they are of the order of 10^{-8}–10^{-11} cm in length and hence are shorter than X radiation. Like the other rays they affect photographic plates, ionize gases, and penetrate matter. In penetrating power they are about 10 to 100 times as effective as β rays, but in ionizing power they are proportionately weaker.

THEORY OF RADIOACTIVE DISINTEGRATION

The fact that radioactive elements emit α particles indicates that the portion of the atom involved in the emission must be the nucleus. To account for this nuclear radiation Rutherford and Soddy advanced in 1903 their theory of *radioactive disintegration*. Rutherford and Soddy proposed that the nuclei of radioactive elements are unstable, and decompose spontaneously by emission of an α or β particle to form a *new element* of different chemical and physical properties. Thus, when uranium I with $A = 238$ and $Z = 92$ emits an α particle to yield uranium X_1, the mass should decrease by four units to $A = 234$, and the nuclear charge by two units to $Z = 90$. Since the nuclear charge has decreased, two external electrons which are no longer necessary will also be lost, and we obtain a new element of $Z = 90$ and $A = 234$, which is an isotope of thorium. Again, when uranium X_1 emits a β particle the nuclear mass remains unchanged, but the nuclear charge is increased to $Z = 91$. The new product, uranium X_2, has now a mass of 234 and an atomic number of 91 and hence is an isotope of protactinium. This idea, that as a result of radioactive spontaneous disintegration new elements are formed, has been amply verified, and is a well-established principle in the field of radioactivity.

From the examples cited it may be observed that the product of a radioactive decay may itself be radioactive and undergo further decomposition. This is generally the case. For example, uranium I yields a whole series of consecutive disintegration products until eventually radium G, a stable isotope of lead, is reached. Analogously thorium goes through a sequence of disintegrations until the stable thorium D, again an isotope of lead, is attained. Each of the intermediate steps involves the emission of either an α or a β particle, and only rarely both. The γ radiation arises from energy rearrangements within the nucleus after an α or β

particle ejection. When the rearrangement leads to a lower potential energy, the excess is evolved as a quantum of energy of the frequency of the γ ray. Not all nuclear transformations are accompanied by emission of γ rays.

The transformation brought about by the emission of an α or β particle from the nucleus is summarized in the Fajans-Soddy-Russell *displacement law*. This law states that whenever a parent nucleus emits an α particle its atomic number is decreased by *two units*, and the new element is shifted two positions *to the left* in the periodic table from that of the parent. On the other hand, when the parent nucleus emits a β particle, the atomic number is increased by *one*, and hence the product is shifted one place *to the right* in the periodic table. This law is merely an epitome of the changes described in the above examples.

RATE OF RADIOACTIVE DECAY

Radioactive disintegrations proceed at rates which cannot be modified by any known chemical or physical means. For a given element this rate at any instant of time is proportional only to the number N of the nuclei present, namely,

$$-\frac{dN}{dt} = kN \tag{1}$$

where k is a rate constant characteristic of the element. Integration of Eq. (1) between the limits $N = N_0$ at $t = 0$ and $N = N$ at $t = t$ yields

$$t = \frac{1}{k} \ln \frac{N_0}{N} \tag{2}$$

from which the half-life period, $t_{1/2}$, of the radioactive element follows as

$$t_{1/2} = \frac{1}{k} \ln \frac{N_0}{(N_0/2)}$$
$$= \frac{\ln 2}{k} \tag{3}$$

Equation (3) shows that the time necessary for one-half of a quantity of radioactive substance to be disintegrated is a constant independent of the initial amount present and is characteristic only of the element in question. For the naturally radioactive elements $t_{1/2}$ may range from 10^{-11} sec to 1.3×10^{10} years.

RADIOACTIVE SERIES

Naturally, radioactive elements belong to either the uranium, thorium, or actinium series. In all series the end products of final disintegration are isotopes of lead. The uranium series is shown in Table 17–1. The symbol gives the name of the element, its rounded mass A as a superscript, and the atomic number Z as a subscript on the left. In each instance a particular element is formed from the one above it through emission of the indicated radiation. Elements designated as (a) and (b) both result from the substance above them. Thus, $_{84}Po^{218}$ disintegrates to give $_{82}Pb^{214}$ through emission of an α particle and $_{85}At^{218}$ through emission of an electron. Both are then transformed into $_{83}Bi^{214}$.

TABLE 17–1. THE URANIUM RADIOACTIVE SERIES

Element	Symbol	Radiation	Half-Life
Uranium	$_{92}U^{238}$	α,γ	4.50×10^9 yrs
Thorium	$_{90}Th^{234}$	β,γ	24.1 days
Protactinium	$_{91}Pa^{234}$	β,γ	1.14 min
Uranium	$_{92}U^{234}$	α,γ	2.48×10^5 yrs
Thorium	$_{90}Th^{230}$	α,γ	8.22×10^4 yrs
Radium	$_{88}Ra^{226}$	α,γ	1.620×10^3 yrs
Radon	$_{86}Rn^{222}$	α	3.83 days
Polonium	$_{84}Po^{218}$	α,β	3.05 min
(a) Lead (99.96%)	$_{82}Pb^{214}$	β,γ	26.8 min
(b) Astatine (0.04%)	$_{85}At^{218}$	α	2.0 sec
Bismuth	$_{83}Bi^{214}$	α,β,γ	19.7 min
(a) Polonium (99.96%)	$_{84}Po^{214}$	α	1.5×10^{-4} sec
(b) Thallium (0.04%)	$_{81}Tl^{210}$	β,γ	1.32 min
Lead	$_{82}Pb^{210}$	β,γ	22 yrs
Bismuth	$_{83}Bi^{210}$	α,β,γ	4.85 days
(a) Polonium (ca. 100%)	$_{84}Po^{210}$	α,γ	138 days
(b) Thallium (ca. 10^{-5}%)	$_{81}Tl^{206}$	β	4.23 min
Lead	$_{82}Pb^{206}$	—	—

In the thorium series all elements have rounded mass numbers A which are whole number multiples n of 4, i.e., $A = 4n$. For the uranium series $A = 4n + 2$, while for the actinium series $A = 4n + 3$. Another series with $A = 4n + 1$, called the neptunium series, is observed only with certain artificially radioactive elements.

MASS AND NUCLEAR BINDING ENERGY

Mass is not conserved in nuclear reactions but mass plus energy is. This means that for any loss in mass an equivalent amount of energy must appear, and vice versa. Invariably the actual isotopic atomic mass M of an element is smaller than the sum of the masses of the protons, neutrons, and electrons present in it. The *mass defect*, ΔM, for an isotope is defined as

$$\Delta M = ZM_H + (A - Z)M_n - M \qquad\qquad (4)$$

where A and Z have the same significance as before, M_H is the atomic weight of the hydrogen atom, and M_n that of the neutron. On the C^{12} scale $M_H = 1.00782$ and $M_n = 1.00866$. The quantity ΔM measures the binding energy of a nucleus relative to its constituents once it is converted to energy units. Einstein has shown that the interconversion of mass and energy is given by the relation

$$E = mc^2 \qquad\qquad (5)$$

where E is the energy in ergs, m the mass in grams, and c the velocity of light in cm sec^{-1}. Consequently, for a mass defect ΔM the energy liberated must be

$$E = \Delta Mc^2 \qquad\qquad (6)$$

and this is the binding energy of the nucleus.

For a given ΔM the energy evolved is very large. For instance, $_2\text{He}^4$ has an atomic weight of 4.00260 and it is equivalent to the combination of two hydrogen atoms and two neutrons. Using Eq. (4) we obtain $\Delta M = 0.03036$ g, and hence

$$\begin{aligned} E &= 0.03036(2.9979 \times 10^{10})^2 \\ &= 2.732 \times 10^{19} \text{ ergs} \\ &= 6.530 \times 10^{11} \text{ cal} \\ &= 28.32 \text{ MeV} \end{aligned}$$

An MeV is one million electron volts, and it is equal to 2.3061×10^{10} cal per mole. Still another way of expressing this energy is in terms of MeV per *nucleon*, i.e., single nuclear particle, which is the binding energy divided by the mass number A. For the example cited above the binding energy per nucleon is $28.32/4 = 7.08$ MeV.

When $\Delta M = 1$ g, Eq. (6) gives 931.48 MeV. Hence the binding energy per gram atomic weight of an isotope can be calculated simply from the relation $E = 931.48 \Delta M$ MeV.

Figure 17–1 shows a plot of binding energies per nucleon vs. mass number. Except for fluctuations at the lower mass numbers, the binding energies increase with A, go through a maximum near $A = 60$, and then decrease.

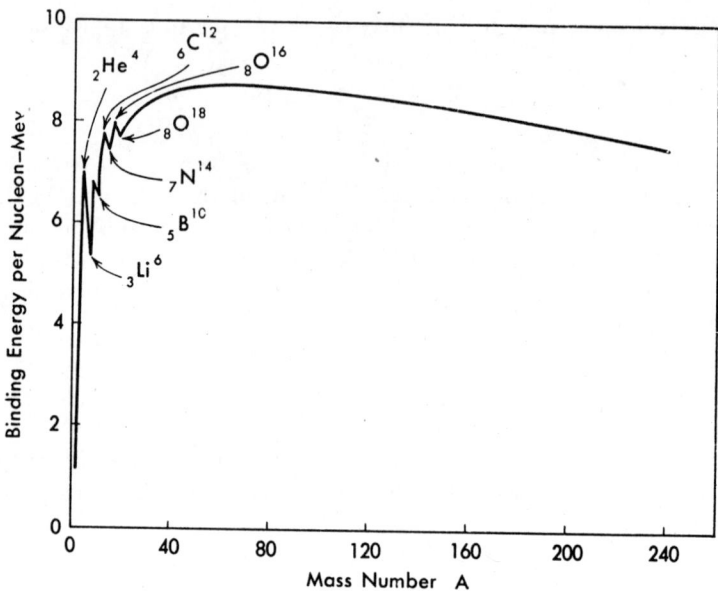

Figure 17-1.   Variation of nuclear binding energy with mass number.

TRANSMUTATION OF ELEMENTS

In the preceding discussion examples were given of the transmutation of one element into another. Until about 1919 the only instances of such transformations were the natural radioactive disintegrations in which a more complex element changed into simpler elements by emission of either α or β particles. All these processes were spontaneous and could not be controlled in any known manner. However, since 1919 various means have been developed by which hundreds of transmutations of elements have been brought about *artificially*. These all entail the bombardment of elements with relatively high-speed atomic particles such as α rays, neutrons, protons, deuterons, and electrons as the transmuting projectiles. For certain processes high-energy γ rays are also suitable. Until relatively recently the only sources of such high-speed, and therefore high-energy, missiles were the α and β particles emitted by naturally radioactive elements. At present apparatus is available, such as powerful transformers, electrostatic generators, high voltage discharge tubes, betatrons, and the cyclotron developed by Lawrence at the University of California, by means of which charged particles can be formed and accelerated to terrific velocities. Another powerful source of missiles is the nuclear reactor or pile. The neutrons and other particles generated in such piles are very effective in accomplishing the transmutations of various elements.

THE POSITRON

Dirac predicted in 1928 that there should be present in matter a particle similar to the electron but of positive charge, i.e., a *positron*. In 1932 Carl Anderson, while studying cosmic rays, detected in the Wilson cloud chamber a fog track which could only be produced by a positively charged particle of very low mass. Anderson ascribed this track to the passage of a positron. This conclusion was confirmed quickly by other workers, and the existence of positrons was thus definitely established.

The positron has been shown to have a mass identical with that of the electron, and a charge equal in magnitude but opposite in sign. However, whereas the electron is stable, the positron is very short-lived. On release from a nucleus a positron loses its kinetic energy very rapidly, in 10^{-10} sec or so, and combines then with an electron to cause the annihilation of both particles. To compensate for the destroyed mass 2 quanta of X radiation are formed whose total energy is equal to the mass destroyed in accordance with the Einstein relation $E = mc^2$.

Various methods are available at present for the production of positrons. When beryllium, magnesium, or aluminum are bombarded by α particles, these elements become artificially radioactive and emit positrons. In each instance the process involves two consecutive *nuclear* reactions, the first of which is the formation of an artificially radioactive element, the second the emission of the positron. For aluminum these reactions are

$$_{13}\mathrm{Al}^{27} + {}_2\alpha^4 \longrightarrow {}_{15}\mathrm{P}^{30} + {}_0n^1 \tag{7}$$
$$_{15}\mathrm{P}^{30} \longrightarrow {}_{14}\mathrm{Si}^{30} + e^+ \tag{8}$$

In these equations e^+ is the symbol for the positron while n is the symbol for the neutron. Thus Eq. (7) states that aluminum, with mass 27 and nuclear charge 13, reacts with an α particle, mass 4 and atomic number 2, to form phosphorus, of mass 30 and charge 15, and a neutron of unit mass and zero charge. Again, Eq. (8) states that the phosphorus is radioactive and decomposes to yield a positron and silicon of mass 30 and atomic number 14.

Similar reactions result also from bombardment of nitrogen, fluorine, sodium, phosphorus, and potassium with α particles, from the bombardment of various elements with protons or deuterons, and from the interaction of lead with γ rays.

NEUTRONS

Equation (7) shows a means of producing neutrons, namely, by bombardment of a suitable metal with α particles. The reaction given involves the capture of an α particle by the aluminum nucleus with the resultant formation of phosphorus and the emission of a neutron. Other important methods of neutron generation are the bombardment of deuterium and lithium atoms with deuterons (d)

$$_1H^2 + _1d^2 \longrightarrow _2He^3 + _0n^1 \tag{9}$$
$$_3Li^7 + _1d^2 \longrightarrow _4Be^8 + _0n^1 \tag{10}$$

and the interaction of deuterium with γ radiation

$$_1H^2 + \gamma \text{ rays} \longrightarrow _1H^1 + _0n^1 \tag{11}$$

An important source of neutrons is the atomic reactor, where neutrons are formed as a result of nuclear fission. Such neutrons are fast, i.e., their energy is between 1 and 10 MeV. These fast neutrons can be slowed down by passage through matter. When the energy drops below 1 eV we obtain slow or thermal neutrons. The latter are particularly subject to capture by various nuclei, and they are thus very effective in initiating nuclear reactions.

Neutrons are radioactive, and they decay with a half-life period of 12.8 min to protons and electrons, namely,

$$_0n^1 = _1p^1 + e \tag{12}$$

where p is the proton. During this process 0.78 MeV of energy is evolved per gram atom. However, neutrons generally undergo capture and reaction before they can disintegrate, and hence their decay can be observed only with the isolated particles.

ARTIFICIAL RADIOACTIVITY

In 1934 I. and F. Joliot discovered that some products of nuclear reactions are unstable and tend to change to more stable nuclei by *spontaneous* positron emission. Since that time many such elements have been observed, and it has been found that the emissions may involve not only positrons, but also electrons and γ radiation. Since this type of radioactivity does not arise spontaneously but must be brought about by preliminary appropriate bombardment, the phenomenon is referred to as *artificial or induced radioactivity*. In each instance the elements exhibiting induced radioactivity are unstable isotopes of the known elements. They do not appear among the stable isotopes because their life period is usually short, and hence it is hardly to be expected that they will be found in nature.

Artificially radioactive elements behave very much like those which are naturally radioactive. They disintegrate in each instance according to the same rate law, Eq. (1), and exhibit thus constant half-life periods characteristic of the element in question. Further, although some artificially radioactive elements may be produced by several different methods, once formed each element decomposes at the same rate and has thus the same life period. There is, however, one respect in which the two do differ. Whereas natural radioactivity is confined only to the very massive nuclei, artificial radioactivity can be induced in light nuclei as well. Thus not only Bi^{210}, Pb^{209}, Pb^{205}, Tl^{206}, and Au^{198} are artificially radioactive, but also such light nuclei as H^3, He^6, Be^7, and C^{10}.

In Table 17-2 are given some examples of artificially radioactive elements, the methods by which they are produced, their half-life periods, and the nature of the emitted radiation.

TABLE 17-2. SOME ARTIFICIALLY RADIOACTIVE ELEMENTS

Element	Method of Production	Half-Life Period	Emitted Radiation
$_1H^3$	$_1H^2 + _1d^2$	12 years	e
$_2He^6$	$_4Be^9 + _0n^1$	0.87 sec	e
$_5B^{12}$	$_5B^{11} + _1d^2$	0.022 sec	e
$_6C^{11}$	$_5B^{10} + _1d^2$	20.5 min	e^+
$_6C^{14}$	$_6C^{13} + _1d^2$	6000 years	e
$_7N^{13}$	$_6C^{12} + _1p^1$	10.1 min	e^+
$_8O^{15}$	$_7N^{14} + _1d^2$	2.1 min	e^+
$_{11}Na^{24}$	$_{11}Na^{23} + _1d^2$	14.8 hours	e
$_{15}P^{30}$	$_{13}Al^{27} + _2\alpha^4$	2.55 min	e^+
$_{18}Ar^{41}$	$_{18}Ar^{40} + _0n^1$	110 min	e
$_{19}K^{38}$	$_{19}K^{39} + _0n^1$	7.5 min	e^+
$_{35}Br^{80}$	$_{35}Br^{79} + _0n^1$	18 min	e
$_{53}I^{124}$	$_{51}Sb^{121} + _2\alpha^4$	4.0 days	e^+
$_{79}Au^{198}$	$_{79}Au^{197} + _0n^1$	2.7 days	e

TYPES OF NUCLEAR REACTIONS

Many types of nuclear reactions are possible and have been observed. Such reactions are generally classified by giving first the nature of the bombarding particle and then the nature of the lightest particle in the products. Thus a (p,n) reaction is one resulting from attack of a proton on some nucleus to produce another nucleus and a neutron. Similarly a (d,p) reaction involves use of a deuteron to form a proton.

Table 17-3 gives a number of observed nuclear reaction types, examples of these, and the simplified notation by which they are represented. In all these reactions the sum of the mass numbers on the left-hand side of the equation must equal the same sum on the right-hand side, and a similar balance must apply also to the sums of the atomic numbers.

Reactions similar to the above have been used to prepare eleven transuranium elements which do not appear in nature. These are neptunium

TABLE 17-3. TYPES OF NUCLEAR REACTIONS

Type	Example	Representation
γ,n	$_1\mathrm{H}_2^2 + \gamma = {}_1\mathrm{H}^1 + {}_0n^1$	$_1\mathrm{H}_2^2(\gamma,n)_1\mathrm{H}^1$
p,γ	$_3\mathrm{Li}^7 + {}_1p^1 = {}_4\mathrm{Be}^8 + \gamma$	$_3\mathrm{Li}^7(p,\gamma)_4\mathrm{Be}^8$
p,n	$_3\mathrm{Li}^7 + {}_1p^1 = {}_4\mathrm{Be}^7 + {}_0n^1$	$_3\mathrm{Li}^7(p,n)_4\mathrm{Be}^7$
p,α	$_3\mathrm{Li}^7 + {}_1p^1 = {}_2\mathrm{He}^4 + {}_2\alpha^4$	$_3\mathrm{Li}^7(p,\alpha)_2\mathrm{He}^4$
d,p	$_{11}\mathrm{Na}^{23} + {}_1d^2 = {}_{11}\mathrm{Na}^{24} + {}_1p^1$	$_{11}\mathrm{Na}^{23}(d,p)_{11}\mathrm{Na}^{24}$
d,n	$_1\mathrm{H}^2 + {}_1d^2 = {}_2\mathrm{He}^3 + {}_0n^1$	$_1\mathrm{H}^2(d,n)_2\mathrm{He}^3$
$d,2\,n$	$_{92}\mathrm{U}^{238} + {}_1d^2 = {}_{93}\mathrm{Np}^{238} + 2\,{}_0n^1$	$_{92}\mathrm{U}^{238}(d,2\,n)_{93}\mathrm{Np}^{238}$
n,γ	$_{48}\mathrm{Cd}^{113} + {}_0n^1 = {}_{48}\mathrm{Cd}^{114} + \gamma$	$_{48}\mathrm{Cd}^{113}(n,\gamma)_{48}\mathrm{Cd}^{114}$
n,p	$_7\mathrm{N}^{14} + {}_0n^1 = {}_6\mathrm{C}^{14} + {}_1p^1$	$_7\mathrm{N}^{14}(n,p)_6\mathrm{C}^{14}$
$n,2\,n$	$_{92}\mathrm{U}^{238} + {}_0n^1 = {}_{92}\mathrm{U}^{237} + 2\,{}_0n^1$	$_{92}\mathrm{U}^{238}(n,2\,n)_{92}\mathrm{U}^{237}$
n,α	$_{27}\mathrm{Co}^{59} + {}_0n^1 = {}_{25}\mathrm{Mn}^{56} + {}_2\alpha^4$	$_{27}\mathrm{Co}^{59}(n,\alpha)_{25}\mathrm{Mn}^{56}$
α,p	$_7\mathrm{N}^{14} + {}_2\alpha^4 = {}_8\mathrm{O}^{17} + {}_1p^1$	$_7\mathrm{N}^{14}(\alpha,p)_8\mathrm{O}^{17}$
α,n	$_{13}\mathrm{Al}^{27} + {}_2\alpha^4 = {}_{15}\mathrm{P}^{30} + {}_0n^1$	$_{13}\mathrm{Al}^{27}(\alpha,n)_{15}\mathrm{P}^{30}$
$\alpha,2\,n$	$_{95}\mathrm{Am}^{241} + {}_2\alpha^4 = {}_{97}\mathrm{Bk}^{243} + 2\,{}_0n^1$	$_{95}\mathrm{Am}(\alpha,2\,n)_{97}\mathrm{Bk}^{243}$

(93), plutonium (94), americium (95), curium (96), berkelium (97), californium (98), einsteinium (99), fermium (100), mendelevium (101), nobelium (102), and lawrencium (103).

THE NEUTRINO AND ANTINEUTRINO

In nuclear disintegrations involving the emission of β particles, the ejected electrons have a distribution of energies E which range up to a maximum value E_{max}. When a mass-energy balance is made using E_{max} for the electrons, the principle of conservation of mass-energy is obeyed. However, it is not obeyed for all electron energies which are less than E_{max}. Further, it is found that the laws of conservation of nuclear spin and angular momentum are also violated. To get around these difficulties, Pauli suggested in 1927 that during such disintegrations there is emitted from the nucleus another particle, called the *neutrino*, whose properties would preserve the principles of conservation of mass-energy, spin, and angular momentum. These properties would have to be no charge, zero mass, a nuclear spin of $\dfrac{1}{2}\left(\dfrac{h}{2\,\pi}\right)$, energies corresponding to $E_{max} - E$, and the particles would have to move with the speed of light.

In 1934 Fermi indicated that not only may a neutrino exist, but also an *antineutrino*, i.e., a particle identical to the neutrino but of opposite spin. He further proposed that β decay takes place according to the reaction

$$_0n^1 \longrightarrow {}_1p^1 + e + \bar{\nu} \tag{13}$$

and positron emission according to

$$_1p^1 \longrightarrow {}_0n^1 + e^+ + \nu \tag{14}$$

where ν is the neutrino and $\bar{\nu}$ the antineutrino. These decay schemes agree well with experiment. Further, Reines and Cowan obtained in 1953 direct proof of the existence of the antineutrino, and they deduced from it the properties of the neutrino which are in very good accord with the theoretical requirements.

ANTIPROTONS AND ANTINEUTRONS

Antiparticles such as the electron and the positron, the neutrino and the antineutrino, suggest the possible existence of antiprotons and anti-neutrons. In 1955 Segré and his associates at the University of California produced *antiprotons* by the bombardment of a copper target with protons accelerated to energies of over 6 billion electron volts (BeV). The mass and spin of the antiprotons was found to be identical with that of protons, but the charge was opposite, i.e., -1. Through similar experiments Cork and his associates, in 1956, obtained *antineutrons*. These again are identical with neutrons, except that their magnetic moment is of opposite sign. Like electrons and positrons, protons and antiprotons and neutrons and antineutrons annihilate each other on collision to yield a variety of decomposition products.

MESONS AND HYPERONS

In 1935 H. Yukawa predicted the existence of nuclear particles with a mass intermediate between that of the electron and the proton. Since then three kinds of such particles, called *mesons*, have been found in both cosmic rays and in the laboratory. One variety, the μ meson, has been shown to have a mass 206 times that of the electron and a charge which may be positive or negative. A second variety, the π meson, has a mass about 270 times that of the electron, and charges which may be positive, negative, or zero. Finally, K mesons have a mass ca. 970 times that of the electron, and again their charge may be positive, negative, or zero.

·Particles have also been obtained from nuclei whose masses are larger than those of the proton. The masses of such *hyperons* have been found to be about 2180, 2330, and 2580 times that of the electron at rest. Some of these particles are electrically neutral, while others are positively or negatively charged.

Mesons and hyperons are very short-lived. Depending on their nature, mesons decompose into other mesons, electrons, positrons, neutrinos, antineutrinos, and γ rays. Hyperons, in turn, break down to other hyperons, mesons, neutrons, and protons.

The discovery of these particles further complicates an already complex situation. We have seen that the nucleus can be considered to be composed of protons and neutrons. At the same time the nucleus also can yield α particles, electrons, positrons, neutrinos, antineutrinos, antiprotons, antineutrons, mesons, and hyperons. How all these nuclear particles arise is unknown, as no theory has as yet been developed which can account satisfactorily for the structure of the atomic nucleus and the forces which bind its constituents together.

NUCLEAR FISSION

For the nuclear reactions described above the change in mass produced by bombardment of a nucleus is always small. In each case the nucleus either increases in mass by capture of a particle or loses some mass by emission of a particle, but never does it rupture to yield fragments of appreciably smaller masses. However, in 1934 Enrico Fermi discovered that uranium when bombarded with thermal neutrons yielded electrons and some strange products of decay. In 1939 Hahn and Strassmann showed that what happens here is a *rupture* of the uranium nucleus to form fragments which have a mass and atomic number considerably lower than that of uranium. Thus they showed that two of the major fragments resulting from uranium ($Z = 92$) fission were barium ($Z = 56$) and lanthanum ($Z = 57$). This process is, therefore, not a relatively minor nuclear transformation, but a drastic neutron bombardment initiated smashing of a nucleus into considerably smaller fragments.

Further research has established that of the various uranium isotopes the one primarily responsible for the fission is U^{235}. Accompanying the breakup of this nucleus is the emission of a tremendous amount of energy which is associated with the various fragments. Measurements indicate that the total energy thus released is as high as 4×10^{12} cal per gram atom ruptured. The liberation of such high energies immediately raises the prospect of the utilization of atomic energy as a source of power. But the realization of this prospect is faced with a difficulty. Ordinary uranium consists of 99.28 per cent U^{238}, 0.71 per cent U^{235}, and about 0.01 per cent U^{234}. To obtain from this isotopic mixture the desired U^{235}, methods of separation are required which can yield appreciable quantities of this isotope. Such methods have been developed, and they are at present under the control of the United States Atomic Energy Commission.

U^{238} can be made fissionable by converting it into Pu^{239} through neutron bombardment. The following series of reactions then sets in:

$$_{92}U^{238} + _0n^1 \longrightarrow _{92}U^{239}$$
$$_{92}U^{239} \longrightarrow _{93}Np^{239} + e$$
$$_{93}Np^{239} \longrightarrow _{94}Pu^{239} + e$$

i.e., the U^{238} on capture of a neutron is transformed into U^{239}, which is radioactive, and passes over with emission of an electron into the transuranic element neptunium. The latter, in turn, is also radioactive, and emits an electron to form plutonium of atomic number 94. Plutonium like U^{235} undergoes fission. Another fissionable isotope is U^{233}, which can be obtained from Th^{232} by irradiation with slow neutrons. The reactions involved are

$$_{90}Th^{232} + _0n^1 \longrightarrow _{90}Th^{233}$$
$$_{90}Th^{233} \longrightarrow _{91}Pa^{233} + e$$
$$_{91}Pa^{233} \longrightarrow _{92}U^{233} + e$$

NUCLEAR FUSION

From the calculation given on page 739, it is evident that tremendous quantities of energy can be obtained by formation of heavier nuclei from those of light elements such as hydrogen, deuterium, and helium, i.e., by the *fusion* of lighter nuclei into heavier ones. Such fusion processes are responsible for the radiant energy of the sun and for the energy released by the hydrogen bomb.

The energy of the sun is believed to arise from the thermonuclear reactions

$$_1H^1 + _1H^1 \longrightarrow _1H^2 + e^+ + \bar{\nu}$$
$$_1H^2 + _1H^1 \longrightarrow _2He^3 + \gamma$$
$$_2He^3 + _2He^3 \longrightarrow _2He^4 + 2 _1H^1$$

for which the total energy change is 26 MeV, and the net effect is the conversion of four protons into one He^4 nucleus. In hotter stars the same net reaction occurs by a different mechanism involving intermediary use of carbon, nitrogen, and oxygen. In absence of protons He^4 nuclei can fuse to form Be^8, which can then react further to form still heavier elements.

APPLICATIONS

Many of the isotopes and nuclear particles described in this chapter find extensive application in the fields of physics, chemistry, biology, and medicine. We have already seen how α particles, protons, neutrons, and deutrons can serve as projectiles for the transmutation of elements, for the production of artificial radioactivity, and for nuclear fission. γ rays are employed in medicine for the treatment of cancer as well as in nuclear studies. In the former application radon resulting from radium decomposition is collected, compressed, and sealed into tiny ampoules which are then used to treat the infected areas.

Radioactive elements, both natural and artificial, have been employed to determine the solubility of difficultly soluble salts and to estimate the surface area of adsorbents. Both of these applications depend on the fact that a radioactive element added to the same stable element will persist in a definite ratio irrespective of the state the element may be in. Consequently, when, say, a trace of radioactive lead is added to a solution of ordinary lead, and the lead is precipitated as either the sulfate or chromate, a determination of the radioactivity of the filtrate readily yields the concentration of lead still dissolved. Similarly, from the activity of a material adsorbed from a mixture of stable and radioactive isotopes it is possible to estimate the extent of adsorption, the number of adsorbed molecules, and the area covered by these.

Isotopes, stable and radioactive, have been used to trace the course of chemical and physiological reactions. Used in this manner these substances are called *isotopic tracers*. Deuterium, and the isotopes of carbon, oxygen, and nitrogen, have been used to elucidate the mechanism of reactions and to follow the course these elements take in passing through an animal body. In the first instance the rates of reactions with the ordinary isotopic species are measured, then the heavier or lighter isotopes are incorporated, and the rates are measured again. From such data it is frequently possible to infer the path followed by the reaction, and hence the mechanism. In physiologic work the uncommon isotopes are incorporated in various foods, as D and O in water or the carbohydrates, N in proteins, C in carbohydrates or fats, fed to animals, and the concentration of these in various organs estimated. From the accumulation of any of these tracers in an organ it is possible to decide just where each food goes. For these purposes radioactive isotopes are also of value, for they permit the tracing of a food by the emission intensity. In this manner have been used radioactive phosphorus and iodine.

Another interesting application of isotopes is in *exchange reactions*. When a compound containing hydroxyl or carboxyl groups is placed in heavy water, it is found that deuterium exchanges itself for ordinary hydrogen atoms in these compounds. This fact indicates that the hydrogen atoms in these compounds are not static, but ionize to be replaced by the more plentiful deuterium atoms. Similar exchange has been found in the hydrogens of acetylene, acetone, and chloroform, but not in some of the paraffin hydrocarbons. In like manner various halogens have been found to exchange with their ions in solution, indicating a mobile equilibrium in action, but not with alkyl halides. A particularly interesting fact revealed by exchange studies is that sulfur atoms in sulfates do not exchange while those in thiosulfates do, and that, whereas no exchange can take place between manganous and permanganate ions, interchange does occur rapidly between manganous and manganic ions.

REFERENCES

1. Calvin *et al.*, *Isotopic Carbon*, John Wiley & Sons, Inc., New York, 1949.
2. Friedlander and Kennedy, *Introduction to Radiochemistry*, John Wiley & Sons, Inc., New York, 1955.
3. A. Glassner, *Introduction to Nuclear Science*, D. Van Nostrand Company, Inc., Princeton, N.J., 1961.
4. I. Kaplan, *Nuclear Physics*, Addison-Wesley Publishing Company, Reading, Mass., 1955.
5. Pollard and Davidson, *Applied Nuclear Physics*, John Wiley & Sons, Inc., New York, 1951.
6. R. S. Shankland, *Atomic and Nuclear Physics*, The Macmillan Company, New York, 1960.
7. Wahl and Bonner, *Radioactivity Applied to Chemistry*, John Wiley & Sons, Inc., New York, 1951.
8. A. Weissberger, *Physical Methods of Organic Chemistry*, Interscience Publishers, Inc., New York, 1960. Chapters 50 and 51.
9. Whitehouse and Putnam, *Radioactive Isotopes*, Oxford University Press, London, 1953.

PROBLEMS

1. How many protons and neutrons are present in a Pb nucleus of atomic mass 206?

2. Assuming that the nucleus in problem 1 is composed of α particles and neutrons, how many of each of these particles would be present?

3. Uranium of mass number 235 undergoes natural radioactive disintegration with the emission of an α particle. (a) What are the mass and atomic number of the product? (b) What is the element formed?   *Ans.* (b) Thorium.

4. Pu^{241} is radioactive and decomposes with emission of a β particle.
(a) What are the mass and atomic number of the product?
(b) What is the element formed?

5. Rn^{222} has a half-life period of 3.83 days. What fraction of the sample will remain undecomposed at the end of 10 days?   *Ans.* 0.164.

6. For Na^{24} $t_{1/2} = 14.8$ hrs. In what period of time will a sample of this substance lose 90% of its radioactive intensity?

7. A sample of radioactive I^{133} gave with a Geiger counter 3150 counts min^{-1} at a certain time, and 3055 counts min^{-1} exactly 1 hr later. Find the half-life period of I^{133}.

8. One gram of Ra is placed in an evacuated tube whose volume is 5.0 cc. Assuming that each Ra nucleus yields 4 He atoms which are retained in the tube, what will be the partial pressure at 27°C of the He produced at the end of a year? For Ra the period of half-life is 1590 years. *Ans.* 28.5 mm Hg.

9. The rate of radioactive decomposition corresponding to 3.7×10^{10} disintegrations per second is called a *curie*. What weight of Ra^{226}, whose half-life period is 1620 years, will be required to yield 1 millicurie of radiation?

Ans. 1.0×10^{-3} g.

10. What weight of Ra^{225}, for which $t_{1/2} = 14.8$ days, is required to yield 1 millicurie of radiation?

11. A sample of radioactive substance shows an intensity of 2.30 millicuries at a time t, and an intensity of 1.62 millicuries 600 sec later. What is the half-life period of the radioactive material?

12. $_{89}Ac^{228}$ is an isotope which appears in one of the naturally radioactive series. To which series does this isotope belong?

13. To which radioactive series does $_{89}Ac^{227}$ belong?

14. Calculate in MeV the energy equivalent to one gram atomic weight of electrons at rest. *Ans.* 0.511 MeV.

15. Taking into consideration the variation of electron mass with velocity, calculate in MeV the total energy equivalent to one gram atomic weight of electrons moving with a velocity equal to ½ that of light. What fraction of the total energy is the kinetic energy of the electrons?

16. Repeat problem 12 for electrons moving with a velocity equal to 0.90 that of light.

17. The atomic weight of $_{36}Kr^{84}$ is 83.9115. Calculate (a) the mass defect, (b) the binding energy in MeV, and (c) the binding energy in MeV per nucleon for this isotope. *Ans.* (c) 8.71 MeV.

18. For $_{92}U^{238}$ the binding energy per nucleon is 7.576 MeV. What is the atomic weight of this isotope?

19. $_{3}Li^{6}$ when bombarded with neutrons emits α particles. Write a balanced equation for the nuclear reaction involved.

20. On bombardment of $_{36}Kr^{82}$ with deuterons, protons are emitted. Write the equation for the nuclear reaction.

21. $_{79}Au^{197}$ on bombardment with neutrons yields $_{79}Au^{198}$. The latter substance is radioactive and emits electrons. Write the equations for the process.

22. $_{13}Al^{27}$ on bombardment with α particles yields $_{15}P^{30}$, which is radioactive and emits a positron. Write the equations for the nuclear reactions involved.

23. $_{94}Pu^{239}$ captures a neutron to form $_{94}Pu^{240}$. The latter captures another neutron to form the radioactive $_{94}Pu^{241}$, which decomposes then with emission of an electron. Write the equations for the nuclear changes.

24. For the thermonuclear reaction $4_{1}H^{1} \rightarrow {}_{2}He^{4}$ the energy liberated is 26 MeV per gram atomic weight of He formed. In turn, the heat of combustion of a gram of coal is about 7200 cal per gram. How many tons of coal would be required to produce the same amount of energy?

18
Statistical
Mechanics

A MACROSCOPIC system is composed of many microscopic constituents, such as atoms, molecules, or ions. Thermodynamics defines the equilibrium properties of a macroscopic system in terms of the variables upon which it depends. Thus the energy of a gas depends upon the temperature, volume, and the number of moles of gas present. For fixed values of the variables the system properties are determined, and they can change only when the variables are altered. Further, these properties can be evaluated when suitable macroscopic measurements are made. However, at no time does thermodynamics exhibit any curiosity about the nature, structure, or energy states of the microscopic constituents, nor about the manner in which these microscopic factors combine to give the observed macroscopic properties of the system.

Quite evidently the thermodynamic properties of a system cannot be independent of the nature and state of the microscopic constituents which compose it. It should be possible, therefore, to calculate the macroscopic properties of systems from suitable summation of the microscopic properties. *Statistical mechanics* is the area of science which concerns itself with this problem, both for systems in equilibrium and systems away from equilibrium. Here we shall be involved only with the equilibrium aspects of the subject.

THE THERMODYNAMIC PROBABILITY OF A SYSTEM

Consider a system composed of N identical and distinguishable particles, such as gas molecules, at temperature T, volume V, and total energy E. In this system not all the particles will have the same average energy E/N. Rather, if each particle has n allowable energy levels, then there will be a distribution of particles among the various energy levels such that n_0 particles may be present in the level with energy ϵ_0, n_1 in level ϵ_1, n_2 in level ϵ_2, $\cdots$, and n_n in level ϵ_n. Still, no matter what the distribution may be, it will have to be true that the sum of the particles present is N, i.e.,

$$\Sigma n_i = N \tag{1}$$

and that the total energy of the system is E, namely,

$$\Sigma n_i \epsilon_i = E \tag{2}$$

It is evident that there are many ways in which the particles can distribute themselves among the various levels and still meet the conditions specified by Eqs. (1) and (2). We are interested at present in the number of such possible distributions. The problem here is the same as that involved in the distribution of N numbered balls among n boxes such that one box contains n_0 balls, the second n_1 balls, etc. The mathematical answer to this problem is

$$W' = \frac{N!}{n_0! n_1! n_2! \cdots n_n!} \tag{3}$$

where W' is the possible number of arrangements, $N!$ is N factorial and represents the product $(1 \times 2 \times 3 \times \cdots N)$, and the same applies to the terms in the denominator. Thus, if $N = 6$ and $n_0 = 1, n_1 = 2, n_2 = 3$, then

$$W' = \frac{1 \times 2 \times 3 \times 4 \times 5 \times 6}{1(1 \times 2)(1 \times 2 \times 3)} = 60$$

The quantity W' is called the *thermodynamic probability* for a system of distinguishable particles. On taking logarithms of both sides, Eq. (3) becomes

$$\ln W' = \ln N! - [\ln n_0! + \ln n_1! + \ln n_2! + \cdots \ln n_n!]$$
$$= \ln N! - \Sigma \ln n_i! \tag{4}$$

When N is large, as, say, Avogadro's number of particles, then $\ln N!$ can be approximated by Stirling's formula, namely,

$$\ln N! = N \ln N - N \tag{5}$$

Under these conditions n_i is also large, and hence

$$\Sigma \ln n_i! = \Sigma n_i \ln n_i - \Sigma n_i$$
$$= \Sigma n_i \ln n_i - N \tag{6}$$

On substitution of Eqs. (5) and (6) into Eq. (4), the latter becomes

$$\ln W' = N \ln N - N - \Sigma n_i \ln n_i + N$$
$$= N \ln N - \Sigma n_i \ln n_i \tag{7}$$

THE MOST PROBABLE DISTRIBUTION

It is a basic tenet of statistical mechanics that the most probable distribution of the particles in a system will be such as to make W' a maximum. For a maximum $\delta W'$, and hence also $\delta \ln W'$, will have to be zero. Applying this condition to Eq. (7) we get

$$\delta \ln W' = \delta \Sigma n_i \ln n_i = 0$$
and so
$$\Sigma(1 + \ln n_i)\delta n_i = 0 \tag{8}$$

At the same time we shall also get from Eqs. (1) and (2), since N and E are constant,

$$\Sigma \delta n_i = 0 \tag{9}$$
and
$$\Sigma \epsilon_i \delta n_i = 0 \tag{10}$$

If the last two equations be multiplied respectively by the undetermined coefficients α' and β, and the equations added to Eq. (8), we obtain

$$\Sigma(\ln n_i + 1 + \alpha' + \beta\epsilon_i)\delta n_i = 0$$
or
$$\Sigma(\ln n_i + \alpha + \beta\epsilon_i)\delta n_i = 0 \tag{11}$$

where $\alpha = \alpha' + 1$. Now, within the constraints imposed by the constancy of E and N, the variations in n_i can occur independently and need not be zero. Consequently, to satisfy Eq. (11) each term in the summation must be zero. Hence

$$\ln n_i + \alpha + \beta\epsilon_i = 0$$
and
$$n_i = e^{-\alpha}e^{-\beta\epsilon_i} \tag{12}$$

The value of the term $e^{-\alpha}$ can be found by use of Eq. (1). Since

$$N = \Sigma n_i = e^{-\alpha}\Sigma e^{-\beta\epsilon_i}$$
$$e^{-\alpha} = \frac{N}{\Sigma e^{-\beta\epsilon_i}} \tag{13}$$

Again, from application of Eq. (2) to the energy of, say, an ideal gas, it can be shown that

$$\beta = \frac{1}{kT} \tag{14}$$

where k is the Boltzmann constant. Inserting Eqs. (13) and (14) into Eq. (12), we obtain

$$n_i = \frac{Ne^{-\epsilon_i/kT}}{\Sigma e^{-\epsilon_i/kT}}$$

or

$$\frac{n_i}{N} = \frac{e^{-\epsilon_i/kT}}{\Sigma e^{-\epsilon_i/kT}} \tag{15}$$

For complete generality Eq. (15) requires a slight modification. In the above discussion we tacitly assumed that each energy level is nondegenerate, i.e., it is composed only of a single level. However, when such is not the case, then each level must be assigned a statistical weight, g_i, and Eq. (15) becomes

$$\frac{n_i}{N} = \frac{g_i e^{-\epsilon_i/kT}}{\Sigma g_i e^{-\epsilon_i/kT}} \tag{16a}$$

$$= \frac{g_i e^{-\epsilon_i/kT}}{Q} \tag{16b}$$

where

$$Q = \Sigma g_i e^{-\epsilon_i/kT} \tag{17}$$

is called the *partition function* for the system.

Equation (16) is known as the *Boltzmann distribution law*, and it gives at any temperature T the fraction of the total number of particles in a system which in the most probable or equilibrium state will possess the energy ϵ_i. Hence this equation gives the most probable distribution of the particles in a system among all the allowable energy levels.

THE PARTITION FUNCTION

The ratio of the number of particles in any state of energy ϵ_i relative to that in state of energy ϵ_0 follows from Eq. (16a) as

$$\frac{n_i}{n_0} = \frac{g_i e^{-\epsilon_i/kT}}{g_0 e^{-\epsilon_0/kT}}$$

$$= \frac{g_i e^{-(\epsilon_i - \epsilon_0)/kT}}{g_0} \tag{18}$$

For computational purposes it is convenient to consider $\epsilon_0 = 0$, and to take all ϵ_i values relative to this ground state. On this basis Eq. (18) becomes

$$n_i = \frac{n_0}{g_0} g_i e^{-\epsilon_i/kT} \tag{19}$$

and on substitution of this equation into Eq. (1) we get for N

$$
\begin{aligned}
N &= \frac{n_0}{g_0} g_0 + \frac{n_0}{g_0} g_1 e^{-\epsilon_1/kT} + \frac{n_0}{g_0} g_2 e^{-\epsilon_2/kT} + \cdots \\
&= \frac{n_0}{g_0} (g_0 + g_1 e^{-\epsilon_1/kT} + g_2 e^{-\epsilon_2/kT} + \cdots) \\
&= \frac{n_0}{g_0} \sum g_i e^{-\epsilon_i/kT} \\
&= \frac{n_0}{g_0} Q
\end{aligned}
\tag{20}
$$

As we shall see in the following, it is possible to express the thermodynamic properties of a system in terms of Q and the derivatives $(\partial \ln Q/\partial T)_V$ and $(\partial^2 \ln Q/\partial T^2)_V$. From differentiation of Eq. (17) it can be shown that

$$
\left(\frac{\partial \ln Q}{\partial T}\right)_V = \frac{1}{T}\left(\frac{Q'}{Q}\right)
\tag{21}
$$

and

$$
\left(\frac{\partial^2 \ln Q}{\partial T^2}\right)_V = \frac{1}{T^2}\left[\left(\frac{Q''}{Q}\right) - \left(\frac{Q'}{Q}\right)^2 - 2\left(\frac{Q'}{Q}\right)\right]
\tag{22}
$$

where

$$
Q' = \sum g_i \left(\frac{\epsilon_i}{kT}\right) e^{-\epsilon_i/kT}
\tag{23}
$$

and

$$
Q'' = \sum g_i \left(\frac{\epsilon_i}{kT}\right)^2 e^{-\epsilon_i/kT}
\tag{24}
$$

Consequently, when necessary both Q and its derivatives can be obtained from the summations indicated in Eqs. (17), (23), and (24).

SYSTEMS OF INDEPENDENT PARTICLES

In the discussion which follows we shall confine ourselves to systems of independent particles. These are systems in which no interactions take place among particles, and hence the latter can behave as independent entities with a total energy which is merely the sum of the energies of the individual particles. The systems which meet this condition are ideal gases, and hence in essence the relations to be deduced apply only to these. In this connection it should be remembered that the energy of ideal gases is independent of pressure and volume, and depends only on the temperature and quantity of gas considered.

Further, in what follows we shall take N to be Avogadro's number of particles, so that the expressions given will correspond in all instances to 1 mole of gas.

THE ENERGY OF A SYSTEM

Let E be the total energy of a system and E_0 its energy in the ground state. Then

$$E - E_0 = \Sigma n_i \epsilon_i \tag{25}$$

where n_i is the number of particles present in the level of energy ϵ_i relative to the ground state, and the summation is taken over all energy states present. Elimination of n_0/g_0 between Eqs. (19) and (20) yields

$$n_i = \frac{N}{Q} g_i e^{-\epsilon_i/kT}$$

and hence Eq. (25) can be written as

$$\begin{aligned}
E - E_0 &= \frac{N}{Q} \sum g_i \epsilon_i e^{-\epsilon_i/kT} \\
&= \frac{NkT}{Q} \sum g_i \left(\frac{\epsilon_i}{kT}\right) e^{-\epsilon_i/kT} \\
&= NkT\left(\frac{Q'}{Q}\right)
\end{aligned} \tag{26}$$

But Eq. (21) shows that $(Q'/Q) = T\left(\dfrac{\partial \ln Q}{\partial T}\right)_V$. Consequently,

$$E - E_0 = RT^2 \left(\frac{\partial \ln Q}{\partial T}\right)_V \tag{27}$$

where the gas constant R has been substituted for kN.

SEPARATION OF THE PARTITION FUNCTION

For systems involving noninteracting particles the total energy of the system can be written as the sum of the translational, rotational, vibrational, and electronic energies, i.e.,

$$E = E_t + E_r + E_v + E_e \tag{28}$$

If we now use Eq. (27) to express each separate form of energy, then we obtain

$$E_t - E_{0(t)} = RT^2 \left(\frac{\partial \ln Q_t}{\partial T}\right)_V \tag{29}$$

$$E_r - E_{0(r)} = RT^2 \left(\frac{\partial \ln Q_r}{\partial T}\right)_V \tag{30}$$

$$E_v - E_{0(v)} = RT^2 \left(\frac{\partial \ln Q_v}{\partial T}\right)_V \tag{31}$$

$$E_e - E_{0(e)} = RT^2 \left(\frac{\partial \ln Q_e}{\partial T}\right)_V \tag{32}$$

where Q_t, Q_r, Q_v, and Q_e are the partition functions, respectively, for translation, rotation, vibration, and electronic energy. Substituting Eq. (27) for E and the above expressions for the various forms of energy into Eq. (28), we get

$$E_0 + RT^2 \left(\frac{\partial \ln Q}{\partial T}\right)_V = [E_{0(t)} + E_{0(r)} + E_{0(v)} + E_{0(e)}]$$
$$+ RT^2 \left[\left(\frac{\partial \ln Q_t}{\partial T}\right)_V + \left(\frac{\partial \ln Q_r}{\partial T}\right)_V \right.$$
$$+ \left(\frac{\partial \ln Q_v}{\partial T}\right)_V + \left.\left(\frac{\partial \ln Q_e}{\partial T}\right)_V\right]$$
$$= E_0 + RT^2 \left[\frac{\partial \ln (Q_t Q_r Q_v Q_e)}{\partial T}\right]_V$$

and hence

$$Q = Q_t Q_r Q_v Q_e \tag{33}$$

Equation (33) makes it possible to consider separately each Q_i, and the separate contribution which it makes to the total value of a thermodynamic property. The latter is obtained then as the sum of the individual contributions.

THE PARTITION FUNCTION FOR TRANSLATION

We have seen in Chapter 14 that the energies of a particle moving in a rectangular box of dimensions a, b, and c are given by Eq. (54) of that chapter, namely,

$$\epsilon_t = \frac{h^2}{8\,m} \left(\frac{n_x^2}{a^2} + \frac{n_y^2}{b^2} + \frac{n_z^2}{c^2}\right) \tag{34}$$

Here ϵ_t was substituted for E to keep the symbols consistent. For translation $g_i = 1$. Hence Q_t is given by

$$Q_t = \Sigma e^{-\epsilon_t/kT}$$
$$= \Sigma e^{-\frac{h^2}{8mkT}\left(\frac{n_x^2}{a^2}\right)} e^{-\frac{h^2}{8mkT}\left(\frac{b_y^2}{b^2}\right)} e^{-\frac{h^2}{8mkT}\left(\frac{n_z^2}{c^2}\right)} \tag{35}$$

Consider now motion in the x direction only. The partition function Q_x for this motion will be

$$Q_x = \sum_{n_x = 0}^{n_x = \infty} e^{-\frac{h^2}{8mkT}\left(\frac{n_x^2}{a^2}\right)} \tag{36}$$

Since the energy levels are very closely spaced, the summation in

Eq. (36) can be replaced by the integral

$$Q_x = \int_0^\infty e^{-\frac{h^2}{8mkT}\left(\frac{n_x^2}{a^2}\right)} dn_x$$

$$= \frac{(2\pi mkT)^{1/2}}{h} a \tag{37}$$

Obviously, similar expressions follow for motion in the y and z directions. Hence we obtain from Eq. (35)

$$Q_t = Q_x Q_y Q_z$$

$$= \frac{(2\pi mkT)^{3/2}}{h^3} (abc)$$

$$= \frac{(2\pi mkT)^{3/2}}{h^3} V \tag{38}$$

where V is the volume of the container occupied by the particle.

THE THERMODYNAMIC FUNCTIONS FOR TRANSLATION

On taking logarithms of Eq. (38) we get

$$\ln Q_t = \ln \frac{(2\pi mk)^{3/2} V}{h^3} + \tfrac{3}{2} \ln T \tag{39}$$

and hence

$$\left(\frac{\partial \ln Q_t}{\partial T}\right)_V = \tfrac{3}{2}\left(\frac{1}{T}\right) \tag{40}$$

Since $E_{0(t)} = 0$, substitution of the latter equation into Eq. (29) yields the well-known results

$$E_t = \tfrac{3}{2} RT \tag{41}$$

and

$$H_t = E_t + RT$$

$$= \tfrac{5}{2} RT \tag{42}$$

From these we get

$$C_{v(t)} = \left(\frac{\partial E_t}{\partial T}\right)_V = \tfrac{3}{2} R \tag{43}$$

and

$$C_{p(t)} = \left(\frac{\partial H_t}{\partial T}\right)_P = \tfrac{5}{2} R \tag{44}$$

The translational entropy of the system is given by

$$S_t = S_{0(t)} + \int_0^T \frac{C_{v(t)}}{T} dT \tag{45}$$

where $S_{0(t)}$ is the translational entropy of the gas at $T = 0$. Again, differentiation of Eq. (29) with respect to T at constant V yields

$$C_{v(t)} = \frac{\partial}{\partial T}\left[RT^2\left(\frac{\partial \ln Q_t}{\partial T}\right)_V\right] \qquad (46)$$

which may be inserted into Eq. (45) to give

$$S_t = S_{0(t)} + \int_0^T \frac{1}{T}\frac{\partial}{\partial T}\left[RT^2\left(\frac{\partial \ln Q_t}{\partial T}\right)_V\right]dT$$

This equation can be integrated by parts. The result is

$$S_t = S_{0(t)} + RT\left(\frac{\partial \ln Q_t}{\partial T}\right)_V + R\ln Q_t - R\ln Q_{0(t)} \qquad (47)$$

where $Q_{0(t)}$ is the value of Q_t at $T = 0$. Further, it can be shown that

$$S_{0(t)} = R\ln Q_{0(t)} - R\ln N + R$$

and hence Eq. (47) becomes

$$S_t = RT\left(\frac{\partial \ln Q_t}{\partial T}\right)_V + R\ln\left(\frac{Q_t}{N}\right) + R \qquad (48)$$

Finally, substitution of Eq. (39) for $\ln Q_t$ and Eq. (40) for $\left(\frac{\partial \ln Q_t}{\partial T}\right)_V$ leads to

$$S_t = R[\tfrac{3}{2}\ln M + \tfrac{3}{2}\ln T + \ln V] + C_1 \qquad (49)$$

where

$$C_1 = R\left[\tfrac{5}{2} + \tfrac{3}{2}\ln\left(\frac{2\pi k}{h^2}\right) - \tfrac{5}{2}\ln N\right] \qquad (50a)$$

$$= -11.073 \text{ cal degree}^{-1} \qquad (50b)$$

and $M = mN$ is the molecular weight of the gas.

An alternate form of Eq. (49) is obtained on setting $V = RT/P$. Then

$$S_t = R[\tfrac{3}{2}\ln M + \tfrac{5}{2}\ln T - \ln P] + C_2 \qquad (51)$$

where

$$C_2 = C_1 + R\ln R \qquad (52a)$$

$$= -2.315 \text{ cal degree}^{-1} \qquad (52b)$$

With E_t, H_t, and S_t known, both the Helmholtz and Gibbs free energies can readily be obtained. Since $A_t = E_t - TS_t$, we get from Eqs. (41) and (49)

$$\frac{A_t}{T} = -R[\tfrac{3}{2}\ln M + \tfrac{3}{2}\ln T + \ln V] + (\tfrac{3}{2}R - C_1) \qquad (53)$$

Similarly, for $F_t = H_t - TS_t$ we obtain from Eqs. (42) and (51)

$$\frac{F_t}{T} = -R[\tfrac{3}{2}\ln M + \tfrac{5}{2}\ln T - \ln P] + (\tfrac{5}{2}R - C_2) \qquad (54)$$

In using the above equations V is expressed in cc, P in atmospheres, and R under the logarithm in cc-atmospheres mole^{-1} degree^{-1}. However, R outside the logarithms can be taken in calories mole^{-1} degree^{-1} in order to obtain the thermodynamic quantities in calories per mole.

The standard state for an ideal gas is the gas at 1 atm pressure. Since E_t, H_t, $C_{p(t)}$, and $C_{v(t)}$ for an ideal gas are independent of pressure, Eqs. (41) to (44) give directly these quantities for the standard state. However, to obtain S_t^0, A_t^0, and F_t^0, P must be set equal to unity in Eqs. (51) and (54), and $V = RT/P = RT$ in Eqs. (49) and (53).

MONATOMIC GASES

Monatomic gases at moderate temperatures are systems which exhibit only translation. At high temperatures these gases undergo electronic excitation which must be considered in calculating the energy, entropy, etc., of these substances. Table 18–1 compares the entropies of several of the rare gases obtained by means of the third law and calculated from Eq. (51). The agreement between the two sets of values is within the uncertainty involved in the third law measurements.

TABLE 18–1.   ENTROPIES OF MONATOMIC
GASES AT 25°C

Gas	S^0 — calories mole^{-1} deg^{-1}	
	Eq. (51)	Third Law
Neon	34.95	35.01
Argon	36.99	36.95
Krypton	39.20	39.17

THERMODYNAMIC FUNCTIONS FOR ROTATION, VIBRATION, AND ELECTRONIC EXCITATION

If we let Q_i be the partition function for rotation, vibration, or electronic excitation, then the energy corresponding to Q_i will be according to Eq. (27),

$$E_i - E_{0(i)} = RT^2 \left(\frac{\partial \ln Q_i}{\partial T} \right)_V \tag{55}$$

and hence

$$C_{v(i)} = \frac{\partial}{\partial T} \left[RT^2 \left(\frac{\partial \ln Q_i}{\partial T} \right)_V \right]$$

$$= RT \left[2 \left(\frac{\partial \ln Q_i}{\partial T} \right)_V + T \left(\frac{\partial^2 \ln Q_i}{\partial T^2} \right)_V \right] \tag{56}$$

Further, since only translation contributes to pressure, $PV = 0$ for internal motions. Consequently, for these

$$H_i = E_i \qquad (57)$$

and $$C_{p(i)} = C_{v(i)} \qquad (58)$$

By employing the same argument as used in connection with the translational entropy, we can obtain for S_i an expression similar to Eq. (47), namely,

$$S_i = S_{0(i)} + RT \left(\frac{\partial \ln Q_i}{\partial T} \right)_V + R \ln Q_i - R \ln Q_{0(i)}$$

However, here $S_{0(i)} = R \ln Q_{0(i)}$. Hence

$$S_i = R \ln Q_i + RT \left(\frac{\partial \ln Q_i}{\partial T} \right)_V \qquad (59)$$

Further, with $PV = 0$, $F_i = A_i = E_i - TS_i$. Utilizing Eqs. (56) and (59) we thus get

$$F_i - E_{0(i)} = -RT \ln Q_i \qquad (60)$$

Since the thermodynamic functions for rotation, vibration, and electronic excitation do not depend on pressure, the above equations give directly the standard state values for these. Again, inspection of the above equations shows that to find the various functions we need $\ln Q_i$, and its first and second derivatives with respect to T. To obtain these we need Q_i as a function of T, or spectroscopic data to calculate Q_i, Q_i', and Q_i''.

ROTATION

The rotational energy of a molecule, ϵ_r, depends on the rotational quantum number J. Further, for rotation $g_i = (2J + 1)$. Hence the general partition function for rotation is given by

$$Q_r = \Sigma(2J + 1)e^{-\epsilon_r/kT} \qquad (61)$$

Without any simplifications or assumptions Q_r can be obtained from the rotational spectra of a molecule by performing the summation called for in Eq. (61). Again, Q_r' and Q_r'' can be obtained from the summations indicated in Eqs. (23) and (24). From these then may be calculated the rotational contributions to the various thermodynamic quantities.

Such a procedure, although exact, is very laborious. Thus, to find Q_r, Q_r', and Q_r'' for $CO(g)$ at 25°C, about 40 terms have to be calculated and summed for each quantity. For this reason it is desirable to simplify the calculation by obtaining expressions for Q_r from models of the rotational behavior of a molecule.

For a diatomic molecule the simplest model is the rigid rotator, for which Eq. (17) of Chapter 16 gives

$$\epsilon_r = \left(\frac{h^2}{8\pi^2 I}\right) J(J+1) \tag{62}$$

Substitution of Eq. (62) into Eq. (61) yields

$$Q_r = \Sigma(2J+1)e^{-\alpha J(J+1)} \tag{63}$$

where

$$\alpha = \frac{h^2}{8\pi^2 I k T} \tag{64}$$

If we assume that the energy levels are closely spaced, which will be true when $\alpha \leq 0.01$, then the summation in Eq. (64) can be replaced by the integral

$$Q_r = \frac{1}{\sigma}\int_0^\infty (2J+1)e^{-\alpha J(J+1)}\, dJ \tag{65}$$

where the symmetry factor σ is 2 for symmetrical molecules such as $O^{16}O^{16}$ or $Cl^{35}Cl^{35}$, and 1 for unsymmetrical molecules such as HCl, CO, or $O^{16}O^{18}$. On integration Eq. (65) gives

$$Q_r = \frac{1}{\sigma\alpha}$$
$$= \frac{8\pi^2 I k T}{\sigma h^2} \tag{66}$$

from which follow

$$\ln Q_r = \ln\left(\frac{8\pi^2 I k}{\sigma h^2}\right) + \ln T \tag{67}$$

$$\left(\frac{\partial \ln Q_r}{\partial T}\right)_V = \frac{1}{T} \tag{68}$$

and

$$\left(\frac{\partial^2 \ln Q_r}{\partial T^2}\right)_V = -\frac{1}{T^2} \tag{69}$$

These expressions can be substituted into Eqs. (55) to (60) to obtain the rotational contributions to the various thermodynamic quantities.

When $\alpha > 0.01$, Eq. (66) can be modified to

$$Q_r = \frac{1}{\sigma\alpha}\left(1 + \frac{\alpha}{3} + \frac{\alpha^2}{15}\right) \tag{70}$$

With the last equation satisfactory results can be obtained up to $\alpha = 0.30$.

Equation (66) or (70) can also be applied to polyatomic linear molecules. For the latter $\sigma = 1$ if the molecule has no plane of symmetry, as in NNO, and $\sigma = 2$ when a plane of symmetry is present, as in OSO.

For rigid nonlinear molecules Q_r is given by

$$Q_r = \frac{\pi^{1/2}}{\sigma h^3} (8 \pi^2 k)^{3/2} (I_1 I_2 I_3)^{1/2} T^{3/2} \qquad (71)$$

where I_1, I_2, and I_3 are the principal moments of inertia of the molecules for respective rotation about the x, y, and z axes. Here σ is the number of identical appearances the molecule can adopt on rotation about the coordinate axes. Thus, for H_2S or H_2O $\sigma = 2$, for NH_3 or AsH_3 $\sigma = 3$, for CH_4 $\sigma = 12$, while for SF_6 $\sigma = 24$.

The use of the above equations can be illustrated with the calculation of the standard entropy of $CO(g)$ at 25°C. At this temperature CO is still in its lowest electronic and vibrational levels, and hence $S^0 = S_t^0 + S_r^0$. From Eq. (51) we find $S_t^0 = 35.93$ eu mole^{-1}. Again, direct summation of the spectroscopically observed rotational levels gives $Q_r = 112.89$ and $Q_r' = 112.66$. On substitution of Eq. (21) into Eq. (59), the expression for S_r^0 becomes

$$
\begin{aligned}
S_r^0 &= R \left[\ln Q_r + \frac{Q'}{Q} \right] \\
&= 1.9872 \left[\ln 112.89 + \frac{112.66}{112.89} \right] \\
&= 11.38 \text{ eu mole}^{-1}
\end{aligned}
$$

Therefore $S^0 = 35.93 + 11.38 = 47.31$ eu mole^{-1} at 25°C. If instead of direct summation we employ Eq. (66) for Q_r of the rigid rotator with $I = 14.49 \times 10^{-40}$ g cm^2 and $\sigma = 1$, then we get $Q_r = 107.27$. Hence from combination of Eqs. (68) and (59) we get

$$
\begin{aligned}
S_r^0 &= R[\ln Q_r + 1] \\
&= 1.9872[\ln 107.27 + 1] \\
&= 11.28 \text{ eu mole}^{-1}
\end{aligned}
$$

and $S^0 = 35.93 + 11.28 = 47.21$ eu mole^{-1}. Use here of Eq. (70) for Q_r does not improve the result significantly.

VIBRATION

A diatomic molecule has only one mode of vibration and a single fundamental vibration frequency ν_0. Considered as a harmonic oscillator, the vibrational energy of such a molecule is given by Eq. (22) of Chapter 16, namely,

$$
\begin{aligned}
\epsilon_v' &= (v + \tfrac{1}{2})h\nu_0 \\
&= \tfrac{1}{2} h\nu_0 + vh\nu_0 \\
&= \epsilon_{0(v)} + \epsilon_v
\end{aligned}
$$

where $\epsilon_{0(v)} = \frac{1}{2} h\nu_0$ is the zero point energy of the oscillator, and

$$\epsilon_v = vh\nu_0 \tag{72}$$

is the energy of the oscillator relative to the ground state. Since for vibration $g_i = 1$, the vibrational partition function for such a molecule follows as

$$Q_v = \Sigma e^{-vh\nu_0/kT} \tag{73}$$

If we set now

$$u = \frac{h\nu_0}{kT} \tag{74}$$

then the summation shown in Eq. (73) can be evaluated to yield

$$Q_v = \frac{1}{1 - e^{-u}} \tag{75}$$

Differentiation of the latter with respect to T at constant V results in

$$\left(\frac{\partial \ln Q_v}{\partial T}\right)_V = \frac{1}{T}\left(\frac{u}{e^u - 1}\right) \tag{76}$$

and

$$\left(\frac{\partial^2 \ln Q_v}{\partial T^2}\right)_V = \frac{1}{T^2}\left[\frac{u^2 e^u}{(e^u - 1)^2} - \frac{2u}{e^u - 1}\right] \tag{77}$$

When Eqs. (73)–(75) are substituted into Eqs. (55)–(60), analytic expressions are obtained for the vibrational contributions to the various thermodynamic properties.

The labor involved in using these equations for calculations is obviated by available tables for the harmonic oscillator which give for values of u or $\bar{\nu}/T$ the corresponding values of $(E - E_0)_v/T$, $(F - E_0)_v/T$, and $C_{v(v)}$. From the former two S_v can be found by use of the relation

$$S_v = \frac{(E - E_0)_v}{T} - \frac{(F - E_0)_v}{T}$$

since here again $PV = 0$. Hence for vibration $H_v = E_v$, $A_v = F_v$, and $C_{p(v)} = C_{v(v)}$.

Linear polyatomic molecules containing n atoms have $3n - 5$ modes of vibration, while nonlinear ones have $3n - 6$ modes. Consequently, such molecules also possess these numbers of fundamental vibration frequencies. In dealing with these, the common practice is to apply the above equations to each vibrational mode separately, and then to add the separate contributions in order to obtain the total contribution made by vibration to a given thermodynamic property. This procedure can be illustrated with the following example. Suppose that we wish to find S_v^0 of $CO_2(g)$ at 25°C. CO_2 is a linear molecule and hence it possesses $(3 \times 3 - 5) = 4$ fundamental vibration frequencies. In wave numbers

TABLE 18-2. VIBRATIONAL ENTROPY OF $CO_2(g)$ AT 25°C

$\bar{\nu}$	$\dfrac{\bar{\nu}}{T}$	$\dfrac{(E^0 - E_0^0)_v}{T}$	$\dfrac{(F^0 - E_0^0)_v}{T}$	S_v^0
672.3	2.255	0.2618	−0.0791	0.341
672.3	2.255	0.2618	−0.0791	0.341
1351.2	4.532	0.0192	−0.0029	0.022
2396.4	8.038	0.0002	−0.0000	0.000
			Total	0.704

these are $\bar{\nu}_1 = \bar{\nu}_2 = 672.3$, $\bar{\nu}_3 = 1351.2$, and $\bar{\nu}_4 = 2396.4$. These frequencies, along with the above-mentioned tabulations, lead to the results shown in Table 18-2. Hence the vibrational contribution to the entropy of $CO_2(g)$ at 25°C is 0.70 eu mole^{-1} out of a total $S_{25°C}^0 = 51.05$ eu mole^{-1}.

THE ELECTRONIC PARTITION FUNCTION

The electronic partition function of an atom or molecule is given by

$$Q_e = \Sigma g_i e^{-\epsilon_e/kT} \tag{78}$$

where ϵ_e is the electronic energy relative to the ground state. At ordinary temperatures most atoms and molecules are still in the ground state and hence $\epsilon_e = 0$ and $Q_e = g_0$. However, such is not always the case. In any event, summation is always employed to find Q_e, Q_e', and Q_e''.

Table 18-3 gives data and illustrates the calculation of Q_e, Q_e', and Q_e'' for Ni(g) at 25°C. Here $g_i = (2J + 1)$, where J is the subscript on the spectral term, and $\bar{\nu}$ is the frequency in cm^{-1} of the energy levels relative to the ground state. From the values of Q_e, Q_e', and Q_e'' then may be calculated all the electronic contributions to the thermodynamic functions by means of Eqs. (55) to (60). Here again, $PV = 0$, and hence all the quantities obtained correspond to standard states.

TABLE 18-3. CALCULATION OF Q_e, Q_e', AND Q_e'' FOR Ni(g) AT 25°C

Spectral Term	g_i	$\bar{\nu}$ − cm^{-1}	$\epsilon_e \times 10^{14}$	$g_i e^{-\epsilon_e/kT}$	$g_i\left(\dfrac{\epsilon_e}{kT}\right)e^{-\epsilon_e/kT}$	$g_i\left(\dfrac{\epsilon_e}{kT}\right)^2 e^{-\epsilon_e/kT}$
$^3F_4$	9	0	0	9.0000	0	0
$^3D_3$	7	204.82	4.0672	2.6061	2.5750	2.5445
$^3D_2$	5	879.82	17.471	0.0717	0.3044	1.2921
$^3F_3$	7	1322.15	26.453	0.0113	0.0729	0.4685
$^3D_1$	3	1713.11	34.018	0.0008	0.0063	0.0521
$^3F_2$	5	2216.55	44.014	0.0001	0.0012	0.0128
			Sum	11.6900	2.9598	4.3700

RESULTS OF STATISTICAL CALCULATIONS

Table 18–4 gives a comparison of S^0 values at 25°C for ideal gases obtained by third law measurements and by statistical calculations. For all gases listed except the last four the agreement is good. In the case of CO and N_2O the third law results are lower than the statistical ones.

TABLE 18–4. COMPARISON OF THIRD LAW AND STATISTICAL
ENTROPIES OF IDEAL GASES AT 25°C

Substance	S^0 — eu mole^{-1}	
	Third Law	Statistical
AsF_3	69.05	69.08
CO_2	51.11	51.05
CH_4	44.46	44.47
Cl_2	53.31	53.31
HCl	44.47	44.66
H_2S	49.10	49.10
N_2	45.93	45.77
NH_3	45.96	45.98
O_2	49.10	49.01
SO_2	59.24	59.29
CO	46.22	47.31
N_2O	51.43	52.55
CH_3—Cd—CH_3	72.40	69.42
CH_3—CCl_3	76.22	74.05

The explanation advanced for this discrepancy is that at $T = 0$ the crystals of these substances are not perfect. Hence the assumption that $S_0^0 = 0$ is not valid, and the third law results are the ones believed to be in error. On the other hand, for the last two molecules the statistical values are too low. The explanation here lies in the ideas of free and hindered rotation of groups. In calculating the rotational contribution of a molecule, cognizance was taken only of the rotation of the molecule as a whole. However, in these and other such molecules groups of atoms can rotate as a unit about other groups to which they are joined by single bonds. Thus, in CH_3—Cd—CH_3 the methyl groups can rotate about the bonds linking them to Cd, and similarly in CH_3—CCl_3 the CCl_3 group can rotate about the bond joining it to CH_3. Such rotations can be free, i.e., unhindered by potential energy barriers, or hindered. In any case, whether free or hindered, such rotation contributes to the thermodynamic functions and should be allowed for in the calculations.

The theory of free and hindered rotation of groups in molecules is discussed in detail by Pitzer and Brewer,[1] and by Aston.[2] For CH_3—Cd—CH_3 the theory shows that the internal rotation of the methyl groups is free, and that this rotation contributes 2.93 eu mole^{-1} to S^0 at 25°C. When this correction is made we get $S^0 = 69.42 + 2.93 = 72.35$ eu mole^{-1}, a result in good agreement with the third law value. On the other hand, the rotation of the groups in CH_3—CCl_3 proves to be hindered, and to have a potential energy barrier of $V_0 = 2967$ cal mole^{-1}. From this barrier is calculated a contribution of 2.16 eu mole to S^0. Hence $S^0 = 74.05 + 2.16 = 76.21$ eu mole^{-1}, which is in excellent agreement with the third law value of 76.22 eu.

Very few molecules show free internal rotation. In most instances the rotation is hindered, with potential energy barriers which may range from several hundred to several thousand calories per mole.

STATISTICAL CALCULATION OF EQUILIBRIUM CONSTANTS

For translation

$$F_t^0 = H_t^0 - TS_t^0$$
$$= E_t^0 + RT - TS_t^0$$

If we insert into this expression Eq. (29) for E_t^0 and Eq. (48) for S_t^0 we get

$$F_t^0 = RT^2 \left(\frac{\partial \ln Q_t^0}{\partial T}\right)_V + RT - RT^2 \left(\frac{\partial \ln Q_t^0}{\partial T}\right)_V - RT \ln \left(\frac{Q_t^0}{N}\right) - RT$$

$$= -RT \ln \left(\frac{Q_t^0}{N}\right) \tag{79}$$

where Q_t^0 is the partition function for the standard state, i.e., $P = 1$ atm. Using Eq. (79) for translation, and Eq. (60) for the standard free energies of rotation, vibration, and electronic excitation, we obtain for the total F^0 of an ideal gas

$$F^0 = F_t^0 + F_r^0 + F_v^0 + F_e^0$$

$$= -RT \ln \left(\frac{Q_t^0}{N}\right) + E_{0(r)}^0 - RT \ln Q_r^0 + E_{0(v)}^0$$

$$- RT \ln Q_v^0 + E_{0(e)}^0 - RT \ln Q_e^0$$

$$= (E_{0(r)}^0 + E_{0(v)}^0 + E_{0(e)}^0) - RT \ln \left(\frac{Q_t^0 Q_r^0 Q_v^0 Q_e^0}{N}\right)$$

$$= E_0^0 - RT \ln \left(\frac{Q^0}{N}\right) \tag{80}$$

[1] Pitzer and Brewer, *Thermodynamics*, McGraw-Hill Book Company, Inc., New York, 1961.
[2] J. G. Aston in Taylor and Glasstone's *Treatise on Physical Chemistry*, D. Van Nostrand Company, Inc., Princeton, N.J., 1942.

where E_0^0 is the total standard ground state energy of the gas and Q^0 the total partition function.

Equation (80) can be used to derive the relation between the equilibrium constant of a reaction and the Q^0's of the substances involved. Consider in general a reaction involving ideal gases such as

$$\alpha A(g) + bB(g) = cC(g) + dD(g)$$

In terms of Eq. (80) for F^0 per mole, the equation for ΔF^0 of the reaction will be

$$\Delta F^0 = [(cE_{0_C}^0 + dE_{0_D}^0) - (\alpha E_{0_A}^0 + bE_{0_B}^0)] - RT \ln \frac{(Q_C^0/N)^c (Q_D^0/N)^d}{(Q_A^0/N)^\alpha (Q_B^0/N)^b}$$

$$= \Delta E_0^0 - RT \ln \frac{(Q_C^0/N)^c (Q_D^0/N)^d}{(Q_A^0/N)^\alpha (Q_B^0/N)^b} \tag{81}$$

where ΔE_0^0 is the standard energy change of the reaction at $T = 0$. But $\Delta F^0 = -RT \ln Kp$. Therefore,

$$\ln Kp = -\frac{\Delta E_0^0}{RT} + \ln \frac{(Q_C^0/N)^c (Q_D^0/N)^d}{(Q_A^0/N)^\alpha (Q_B^0/N)^b} \tag{82}$$

This equation shows that Kp of a reaction can be calculated from the total partition functions of reactants and products when ΔE_0^0 is available. The latter can be obtained for some reactions from spectroscopic data, while for others thermal data are required to obtain ΔE_0^0.

An alternate and more common method of getting Kp from spectroscopic data is the following. The partition function Q^0 of each substance is used to calculate $(F^0 - E_0^0)/T$ at various temperatures. These then are combined to yield at each temperature $(\Delta F^0 - \Delta E_0^0)/T$, which, on addition of $\Delta E_0^0/T$, becomes $\Delta F^0/T$. Then $\ln Kp$ follows from the relation $\ln Kp = -\Delta F^0/RT$. This method of calculating Kp can be illustrated with the following example. For the reaction

$$S_2(g) = 2 S(g)$$

the partition functions of $S_2(g)$ and $S(g)$ give at 4000°K

$$S_2(g): \left(\frac{F^0 - E_0^0}{T}\right)_{S_2} = -68.274 \text{ cal mole}^{-1}$$

$$S(g): \left(\frac{F^0 - E_0^0}{T}\right)_{S} = -48.498 \text{ cal mole}^{-1}$$

Hence

$$\frac{\Delta F^0 - \Delta E_0^0}{T} = 2\left(\frac{F^0 - E_0^0}{T}\right)_{S} - \left(\frac{F^0 - E_0^0}{T}\right)_{S_2}$$

$$= 2(-48.498) - (-68.274)$$

$$= -28.722 \text{ cal}$$

Again, for $S_2(g)$ the spectroscopic dissociation energy is $D_s = 102,650$

cal mole^{-1}, and this is also ΔE_0^0 for the reaction in question. Therefore,

$$\frac{\Delta F^0}{T} = \frac{\Delta E_0^0}{T} - 28.722 = \frac{102,650}{4000} - 28.722$$

$$= 25.662 - 28.722$$

$$= -3.060 \text{ cal}$$

and so

$$\ln Kp = -\frac{\Delta F^0}{RT} = \frac{3.060}{1.987}$$

$$= 1.540$$

$$Kp = 4.664 \text{ at } 4000°K$$

The value of Kp thus obtained is for pressures expressed in atmospheres. To illustrate another method of getting ΔE_0^0, consider the reaction

$$S_2(g) + 2 \, O_2(g) = 2 \, SO_2(g)$$

For this reaction at 25°C $(\Delta F^0 - \Delta E_0^0)/T = 29.50$, $\Delta H^0 = -172,900$ cal, and $\Delta S^0 = -33.67$ eu. Hence

$$\Delta F^0 = \Delta H^0 - T\Delta S^0$$

$$= -172,900 - 298.15(-33.67)$$

$$= -162,860 \text{ cal at } 25°C$$

Insertion of this value of ΔF^0 at 25°C into the expression for $(\Delta F^0 - \Delta E_0^0)/T$ at this temperature gives then

$$-162,860 - \Delta E_0^0 = 298.15(29.50)$$

and hence

$$\Delta E_0^0 = -162,860 - 298.15(29.50)$$

$$= -171,660 \text{ cal}$$

This value of ΔE_0^0 can be used at other temperatures to find ΔF^0 and thereby also Kp.

ENTROPY AND PROBABILITY

At constant energy and volume a system is in equilibrium when its entropy is a maximum. Again, the thermodynamic probability is also a maximum at equilibrium. From these facts it may be expected that a relationship exists between the entropy S and the probability W, namely, that $S = f(W)$. To get an idea of the nature of this function, consider two systems of identical particles for which the entropies are S_1 and S_2 and the probabilities W_1 and W_2. When these systems are combined the total entropy is $S = S_1 + S_2$, while the total probability becomes $W_1 W_2$ and thus $S = f(W_1 W_2)$. Consequently,

$$S = f(W_1) + f(W_2) = f(W_1 W_2)$$

The function which satisfies this relation must be logarithmic, namely,

$$S = k_1 \ln W + k_2 \tag{83}$$

where k_1 and k_2 are constants to be evaluated.

If Eq. (83) has general validity, it should apply among other things to a perfect crystal at 0°K. For such a crystal $S = 0$ by the third law. Again, since the crystal is perfect, there is only one arrangement for it and $W = 1$. On inserting these values into Eq. (83) we see that $k_2 = 0$, and hence

$$S = k_1 \ln W \tag{84}$$

The nature of k_1 can be found by using Eq. (84) to calculate the translational entropy of an ideal gas. However, before we can do this, an expression for W must be obtained from Eq. (3) for W' through the introduction of two modifications. These involve the insertion of statistical weights at this stage, and the quantum statistical recognition that identical particles are actually indistinguishable.[3] With these modifications the expression for W becomes

$$W = \frac{g_0^{n_0} g_1^{n_1} g_2^{n_2} \cdots g_n^{n}}{n_0! n_1! n_2! \cdots n_n!} \tag{85}$$

and

$$\ln W = \Sigma n_i \ln g_i - \Sigma n_i \ln n_i + N \tag{86}$$

when logarithms are taken and the factorials are expanded by Stirling's approximation formula. Now, Eq. (16b) applied to translation gives

$$n_i = \frac{N g_i e^{-\epsilon_i/kT}}{Q_t}$$

and so Eq. (86) can also be written as

$$\ln W = \sum n_i \ln g_i - \sum n_i \ln \left[\left(\frac{N}{Q_t} \right) g_i e^{-\epsilon_i/kT} \right] + N \tag{87}$$

Equation (87) can be simplified as follows. The term

$$-\sum n_i \ln \left[\left(\frac{N}{Q_t} \right) g_i e^{-\epsilon_i/kT} \right] = -\sum n_i \ln \left(\frac{N}{Q_t} \right) - \sum n_i \ln g_i + \sum \frac{n_i \epsilon_i}{kT}$$

$$= N \ln \left(\frac{Q_t}{N} \right) - \sum n_i \ln g_i + \frac{1}{kT} \sum n_i \epsilon_i$$

$$= N \ln \left(\frac{Q_t}{N} \right) - \sum n_i \ln g_i$$

$$+ \frac{RT^2}{kT} \left(\frac{\partial \ln Q_t}{\partial T} \right)_V \tag{88}$$

---

[3] In connection with the discussion of S_t this correction was made by including in $S_{0(t)}$ on page 759 the term $-(R \ln N - R) = -k(N \ln N - N) = -k \ln N!$

since $\Sigma n_i = N$ and

$$\sum n_i \epsilon_t = E_t - E_{0(t)} = RT^2 \left(\frac{\partial \ln Q_t}{\partial T} \right)_V$$

Therefore, on substitution of Eq. (88) into Eq. (87) we get

$$\ln W = \sum n_i \ln g_i + N \ln \left(\frac{Q_t}{N} \right) - \sum n_i \ln g_i + \frac{RT^2}{kT} \left(\frac{\partial \ln Q_t}{\partial T} \right)_V + N$$

$$= N \ln \left(\frac{Q_t}{N} \right) + \frac{RT}{k} \left(\frac{\partial \ln Q_t}{\partial T} \right)_V + N$$

and on multiplication of both sides of this equation by the Boltzmann constant k

$$k \ln W = Nk \ln \left(\frac{Q_t}{N} \right) + RT \left(\frac{\partial \ln Q_t}{\partial T} \right)_V + Nk$$

$$= R \ln \left(\frac{Q_t}{N} \right) + RT \left(\frac{\partial \ln Q_t}{\partial T} \right)_V + R \qquad \textbf{(89)}$$

Finally, comparison of Eqs. (48) and (89) shows that the right-hand side of the latter is equal to S_t. Consequently, $S_t = k \ln W$, and hence $k_1 = k$. Equation (84) thus becomes

$$S = k \ln W \qquad \textbf{(90)}$$

and this is the general relation between entropy and probability.

BOSE-EINSTEIN AND FERMI-DIRAC STATISTICS

The Boltzmann statistics discussed in this chapter are based on the Boltzmann distribution law, which, in turn, is deduced from the concept of distinguishable particles. As we have seen, this concept requires modification in dealing with translational entropy, since actually the particles are indistinguishable. Further, although the modified Boltzmann statistical approach yields satisfactory results for large assemblies of atoms and molecules at ordinary and high temperatures, it breaks down when applied to H_2 and He at temperatures below several degrees Kelvin, and it also does not apply to photons, electrons, and other elementary particles.

To rectify this situation, two quantum statistical approaches have been developed which are based on quantum mechanics and on the concept of indistinguishable particles. *Bose-Einstein statistics* apply in general to systems with symmetric wave functions, i.e., wave functions whose sign does not change on interchange of the coordinates of any two particles in the system. Systems possessing such wave functions are photons, and atoms or molecules containing nuclei with an *even* number of protons plus neutrons in them. For these the Bose-Einstein distribution law is

$$n_i = \frac{g_i}{Be^{\epsilon_i/kT} - 1} \tag{91}$$

where B is a constant at any given temperature.

Fermi-Dirac statistics are used for systems with antisymmetric wave functions, i.e., wave functions which change sign when the coordinates of any two particles are interchanged. Systems with such wave functions are electrons, protons, and atoms or molecules with *odd* numbers of fundamental particles in their nuclei. For these the distribution law takes the form

$$n_i = \frac{g_i}{Be^{\epsilon_i/kT} + 1} \tag{92}$$

When $Be^{\epsilon_i/kT}$ is large compared with unity, Eqs. (91) and (92) both reduce to the Boltzmann distribution law. Under such conditions $B = Q/N$, and hence

$$n_i = \frac{g_i}{Be^{\epsilon_i/kT}}$$
$$= \frac{Ng_ie^{-\epsilon_i/kT}}{Q}$$

REFERENCES

1. Aston and Fritz, *Thermodynamics and Statistical Mechanics*, John Wiley & Sons, Inc., New York, 1959.
2. F. H. Crawford, *Heat, Thermodynamics, and Statistical Physics*, Harcourt, Brace & World, Inc., New York, 1963.
3. N. Davidson, *Statistical Mechanics*, McGraw-Hill Book Company, Inc., New York, 1962.
4. T. H. Hill, *Introduction to Statistical Mechanics*, Addison-Wesley Publishing Company, Reading, Mass., 1960.
5. Mayer and Mayer, *Statistical Mechanics*, John Wiley & Sons, Inc., New York, 1940.
6. G. S. Rushbrooke, *Introduction to Statistical Mechanics*, Oxford University Press, Oxford, 1949.

PROBLEMS

1. Four identical, but distinguishable, particles are to be distributed between two identical containers so that each container has two particles in it. What is the number of possible arrangements for the particles?   *Ans.* Six.

2. Suppose that the particles in problem 1 are indistinguishable. What is the number of possible arrangements for the particles?

3. A very large number of identical, but distinguishable, particles, N, is to be distributed between two containers, and each of these is to have $N/2$ particles in it. What is the number of ways in which the particles can be arranged?

Ans. 2^N.

4. Explain why the answer to problem 1 is 6 and not $2^4 = 16$.

5. Suppose that the particles in problem 3 are indistinguishable. In how many ways can the particles be distributed between the two containers?

6. A certain system of N particles possesses, among others, two energy levels for which $g_1 = 2$ and $E_1 = 10,000$ cal mole^{-1}, and $g_2 = 3$ and $E_2 = 14,000$ cal mole^{-1}. What will be the ratio of the number of particles in the two states at (a) 300°K and (b) 1000°K? *Ans.* (b) 0.200.

7. For a system composed of Avogadro's number of molecules $g_0 = 1$ and the total partition function at 500°K is 6.31×10^{34}. (a) What fraction of the total number of molecules will be in the ground state? (b) What will be the total number of molecules in this state?

8. Derive Eq. (21) of this chapter.

9. Derive Eq. (22) of this chapter.

10. Calculate the translational partition function for an ideal gas contained in a volume of 30 liters at 500°K. The molecular weight of the gas is 100.

Ans. 6.31×10^{34}.

11. Calculate the standard translational entropy of $O_2(g)$ at 25°C.

Ans. 36.32 eu mole^{-1}.

12. Find for $O_2(g)$ at 25°C the values of (a) E_t^0, (b) H_t^0, (c) $C_{v(t)}^0$, (d) $C_{p(t)}^0$, (e) A_t^0, and (f) F_t^0.

13. Calculate for $CO(g)$ at 25°C the values of (a) E_t^0, (b) H_t^0, (c) $C_{v(t)}^0$, (d) $C_{p(t)}^0$, (e) S_t^0, (f) A_t^0, and (g) F_t^0. *Ans.* (e) $S_t^0 = 35.93$ eu mole^{-1}.

14. Calculate the total entropy of $Xe(g)$ at 298.15°K and 0.10 atm. pressure. At this temperature Xe is in its lowest electronic state.

15. Deduce expressions for the rotational, vibrational, or electronic contributions to the thermodynamic functions in terms of Q_i, Q_i', and Q_i''.

16. Assuming linear molecules to be rigid rotators, set up the expressions for the rotational contributions made by such molecules to the various thermodynamic quantities.

17. The moment of inertia for the nitrogen molecule is 13.94×10^{-40} g cm^2. Calculate the rotational partition function for $N_2(g)$ at 25°C. *Ans.* 51.60.

18. Using Q_r obtained in problem 17 and the equations deduced in problem 16, calculate the rotational contributions to the thermodynamic functions of $N_2(g)$ at 25°C.

19. $N_2(g)$ is in its lowest vibrational and electronic states at 25°C. Utilizing the result obtained in problem 17, calculate the total standard entropy of $N_2(g)$ at this temperature. *Ans.* 45.75 eu mole^{-1}.

20. Calculate the total values of C_v and C_p for nitrogen at 25°C. What is their dependence on temperature?

21. For the linear molecule $N_2O(g)$ $I = 66.4 \times 10^{-40}$ g cm². Calculate $S_t^0 + S_r^0$ for $N_2O(g)$ at 25°C. *Ans.* 52.61 eu mole⁻¹.

22. For $NH_3(g)$ $I_1 = I_2 = 2.814 \times 10^{-40}$, $I_3 = 4.452 \times 10^{-40}$ g cm², and $\sigma = 3$. Calculate $S_t^0 + S_r^0$ per mole of NH_3 at 25°C.

23. Set up the equations for the thermodynamic properties of Avogadro's number of simple harmonic oscillators.

24. When T gets large $u \to 0$. What limiting forms do the equations obtained in problem 23 approach under these conditions? (*Hint:* For small values of x, $e^x = 1 + x$.)

25. For $I_2(g)$ $\nu_0 = 6.427 \times 10^{12}$ sec⁻¹. Using a table of e^x and e^{-x} values given in mathematical handbooks, calculate for this gas the vibrational partition function at 300°K. *Ans.* 1.557.

26. Using the equations obtained in problem 23 and the data given in problem 25, calculate the contributions made by vibration to the thermodynamic properties of $I_2(g)$ at 300°K.

27. With the aid of the data given in Table 18–3, calculate (a) $(E^0 - E_0^0)_e$, (b) S_e^0, and (c) $(F^0 - E_0^0)_e$ for $Ni(g)$ at 25°C. *Ans.* (a) 150 cal mole⁻¹.

28. Using the results of the preceding problem, calculate the total values of (a) $E^0 - E_0^0$, (b) $H^0 - E_0^0$, and (c) S^0 for $Ni(g)$ at 25°C.

29. For $Br_2(g)$ at 2000°K $C_p = 9.03$ cal mole⁻¹ degree⁻¹. Does electronic excitation contribute to the thermodynamic properties of $Br_2(g)$ at this temperature? Explain your answer.

30. The values of $(F^0 - E_0^0)/T$ in calories mole⁻¹ degree⁻¹ for the following gases at 298.15°K and 1000°K are

	298.15°K	1000°K
$H_2(g)$	−24.436	−32.752
$S_2(g)$	−47.242	−56.48
$H_2S(g)$	−41.174	−51.24

Again, for the reaction $2\,H_2(g) + S_2(g) = 2\,H_2S(g)$, $\dfrac{\Delta F^0}{T} = -117.86$ cal degree⁻¹ at 25°C. From these data find (a) ΔE_0^0 and (b) log K_p of the reaction at 1000°K. *Ans.* (b) 4.316.

19
Photochemistry

$\mathbf{P}$HOTOCHEMISTRY concerns itself with the study of the effect of radiant energy on chemical reactions and with the rates and mechanisms by which reactions initiated by light proceed.

Ordinary or *thermal* reactions are initiated by activation brought about through molecular collisions. It is characteristic of all such reactions that they can occur only when the reactions are accompanied by a free energy decrease. If a free energy increase is involved, no reaction is possible. However, thermal activation is not the only means by which the energy of atoms and molecules can be raised sufficiently to cause reaction. We have seen that atoms and molecules can absorb radiation. In fact, with absorption of a sufficiently large quantum of radiant energy a molecule may be ruptured. Such absorption of light by an atom or molecule leads to its excitation; and if the activation is sufficiently great, chemical reaction may result. It is in this manner that absorbed light can affect the rate of a chemical reaction and frequently bring about chemical changes under conditions where thermal activation alone would not be effective.

The rate of thermal uncatalyzed reactions at any fixed concentration can be varied only by change of temperature. With photochemical reactions, however, the rate can be controlled also by varying the intensity of the light used for irradiation. In the latter reactions the number of molecules activated depends on the intensity of the light, and hence the concentration of activated molecules will be proportional to the light intensity to which the reactant is exposed. With sufficiently intense light sources it is thus possible to attain reaction rates at ordinary temperatures which would not result thermally except at considerably elevated temperatures. Again, since photochemical activation does not depend to any degree on temperature, the rate of activation is usually temperature-independent. Any increase in the rate of a photochemical reaction with temperature is due primarily to thermal reactions which follow the activation process.

Further, not only spontaneous reactions can be made to proceed photochemically, but also many reactions attended by a free energy increase. In the spontaneous reactions the light acts to speed up the thermal reactions, i.e., it acts essentially as a catalyst. In the nonspontaneous reactions, on the other hand, the radiant energy supplied to the system may increase the free energy of the reactants sufficiently to make the ΔF negative. An outstanding example of such a process is *photosynthesis*. Under the action of sunlight and promoted by chlorophyll (the green coloring matter of vegetation), carbon dioxide and water are combined in plants into complex carbohydrate materials and oxygen. On removal of the light the products oxidize slowly back to carbon dioxide and water, releasing at the same time the energy accumulated from the sun's radiation. Other examples of such reversals of thermodynamically spontaneous processes are the photochemical conversion of oxygen to ozone and the decomposition of HCl to hydrogen and chlorine.

THE GROTTHUS-DRAPER LAW

The Grotthus-Draper law, referred to also as the first law of photochemistry, states that *only light which is absorbed can be effective in producing chemical change.* However, it does not necessarily follow that absorbed light will always cause reaction. Atoms and molecules can absorb radiation only to re-emit it either as a line or band spectrum. Under such conditions the absorption does not result in reaction. The circumstances under which absorption of light does lead to reaction will be described later.

For application of the Grotthus-Draper law to processes where absorption of light does lead to chemical change it is necessary to know the amount of light absorbed. If we let I_0 be the intensity of the light that enters a medium and I_t the intensity of the transmitted light, then the intensity of the light absorbed, I_a, must be

$$I_a = I_0 - I_t \tag{1}$$

Now, for substances other than solutions or gases the intensity of the transmitted light is given by Lambert's law, namely,

$$I_t = I_0 e^{-kl}$$

Again, for solutions or gases I_t is given by Beer's law

$$I_t = I_0 e^{-\epsilon Cl}$$

From these relations the light absorbed by media other than solutions or gases follows then as

$$\begin{aligned} I_a &= I_0 - I_0 e^{-kl} \\ &= I_0(1 - e^{-kl}) \end{aligned} \tag{2}$$

while that absorbed by solutions or gases is

$$\begin{aligned} I_a &= I_0 - I_0 e^{-\epsilon Cl} \\ &= I_0(1 - e^{-\epsilon Cl}) \end{aligned} \tag{3}$$

THE EINSTEIN LAW OF PHOTOCHEMICAL EQUIVALENCE

Another principle of importance is the *law of photochemical equivalence* proposed by Einstein. This law, sometimes called the second law of photochemistry, states that *any molecule or atom activated by light absorbs only one quantum of the light which causes the activation.* The energy acquired by a single atom or molecule in absorbing the one quantum depends on the frequency of the irradiating light and is given by the Planck relation $\Delta E = h\nu$. Per mole or gram atomic weight the energy absorbed will be Avogadro's number times $h\nu$, or

$$\Delta E = Nh\nu \text{ per mole} \tag{4}$$

The quantity of energy defined by Eq. (4) is called an *einstein*, and its magnitude for wave length λ in Å is given by

$$\begin{aligned} \Delta E &= \frac{1.196 \times 10^{16}}{\lambda} \text{ ergs mole}^{-1} \\ &= \frac{2.859 \times 10^{8}}{\lambda} \text{ cal mole}^{-1} \end{aligned} \tag{5}$$

The energies contained in an einstein for various wave lengths of light are shown in Table 19-1. From this table it is evident that the energy absorbed per mole by a system varies widely with the wave length of the light used. Ordinarily activating radiation is confined to wave lengths

TABLE 19-1. ENERGIES PER EINSTEIN AT VARIOUS
WAVE LENGTHS OF LIGHT

Wave Length (Å)	Spectral Range	Energy per Einstein (cal)
1	X rays	2.86×10^8
1000	Ultraviolet	285,900
2000	Ultraviolet	142,950
3000	Ultraviolet	95,300
4000–4500	Visible (violet)	71,470–63,530
4500–5000	Visible (blue)	63,530–57,180
5000–5750	Visible (green)	57,180–49,720
5750–5900	Visible (yellow)	49,720–48,460
5900–6500	Visible (orange)	48,460–43,980
6500–7500	Visible (red)	43,980–38,120
8000	Infrared	35,740
9000	Infrared	31,770
10,000	Infrared	28,590

of 2000–10,000 Å, and hence activating energies may range from 28,000 to 143,000 cal per mole.

The photochemical equivalence law applies only to the absorption or primary photochemical process. When as a result of the primary absorption only one molecule decomposes, and the products enter no further reaction, the number of molecules reacting will be equal to the number of energy quanta absorbed. More frequently, however, a molecule activated photochemically initiates a sequence of thermal reactions as a result of which several or many reactant molecules may undergo chemical change. Under such conditions there will be no 1:1 relation between reacting molecules and the number of energy quanta absorbed. Again, in certain processes where deactivation occurs, less than one molecule may react per quantum. To express the relation between the number of molecules entering reaction and the number of quanta absorbed, the *quantum yield* or efficiency of a process ϕ is introduced. This quantity is defined as

$$\phi = \frac{\text{Number of molecules reacting in a given time}}{\text{Number of quanta of light absorbed in the same time}}$$

$$= \frac{\text{Number of moles reacting in a given time}}{\text{Number of einsteins of light absorbed in the same time}} \quad (6)$$

and gives the number of molecules observed to undergo chemical transformation per quantum of absorbed energy. The quantum efficiency of a reaction may vary from almost zero to about 10^6 observed for the combination of hydrogen and chlorine. Nevertheless, no matter how large or small ϕ may be, it is generally accepted that the Einstein equivalence law is valid and that the primary step initiating a photochemical change is the absorption of a single quantum of energy by a molecule or atom. Whatever happens thereafter is beyond the concern of this law.

CONSEQUENCES OF LIGHT ABSORPTION BY ATOMS

When an atom absorbs radiant energy, the energy, depending on its amount, may cause the atom to ionize, or it may go to produce electronic excitation. Here we are primarily interested in the latter possibility. An electronically excited atom has a life period of about 10^{-7} to 10^{-8} sec. If during this brief period the atom suffers no collision with another particle to which some or all of the excess energy can be passed on, the atom will re-emit part or all of this energy as radiation. Such an emission of radiation by an excited atom or molecule is called *fluorescence*. When the excited electron returns to its initial state and emits thereby radiation of exactly the same frequency as that absorbed, we obtain *resonance fluorescence*. However, the excited electron may pass also to an energy level intermediate between the excited and ground states. The frequency of the emitted fluorescence ν' will then be less than that of the absorbed light ν. In rare instances ν' may actually be greater than ν, i.e., more radiant energy is emitted than has been absorbed. When this happens it is usually an indication that some kinetic energy has been converted into energy of internal motion and has been radiated along with the excitation energy of the atom.

Ordinarily fluorescent emission ceases as soon as the incident radiation is removed. However, in some instances the fluorescence may persist for some time after the light has been shut off, and we have then *phosphorescence*. Phosphorescence and fluorescence arise in the same way, except that in the former the emission of the radiation takes place much more slowly.

When a photochemically excited atom undergoes collision with another atom or molecule before it has a chance to fluoresce, the fluorescence may be *quenched*, i.e., the intensity of the fluorescent emission may be diminished or stopped. The quenching of fluorescence is due to transfer of energy from the excited atom to the particle with which it collides. As a result of this energy transfer the following changes may occur:

1. An excited atom may activate another atom with which it collides. An example is the activation of thallium atoms by excited mercury vapor, namely,

$$Hg^* + Tl = Hg + Tl^* \tag{7}$$

The asterisk denotes the activated or excited particle.

2. The excited atom may collide with a molecule and activate it. An instance is the activation of hydrogen by excited cadmium atoms

$$Cd^* + H_2 = Cd + H_2^* \tag{8}$$

3. An excited atom may react with the colliding molecule. Such a process is the reaction

$$Hg^* + O_2 = HgO + O \tag{9}$$

4. Finally, an excited atom may collide with a molecule and by energy exchange cause the molecule to dissociate. A dissociation thus brought about by an excited atom is called *photosensitization*. It is exemplified by the photosensitized dissociation of hydrogen gas by excited mercury vapor, namely,

$$Hg^* + H_2 = Hg + 2 H \tag{10}$$

Whether or not fluorescence will be quenched depends greatly on the concentration of the fluorescent atoms and of the quenching substance. In a gas at low pressure the time interval between collisions is usually greater than the life of the excited atoms, and hence very little quenching will occur. At higher pressures the time interval between collisions is shorter, and consequently appreciable quenching will take place. Since in a liquid medium collisions are very frequent, the fluorescence emitted by liquid media will generally be appreciably quenched. Further, the quenching of fluorescence is to a large degree specific. Thus in the quenching of mercury fluorescence oxygen gas is most effective, followed by hydrogen and carbon monoxide, while helium and argon are very inefficient quenchers at the same pressure.

The photochemical importance of the quenching of fluorescence lies in the fact that excited atoms or molecules resulting from the quenching process may react further to continue the photochemically initiated change. Thus in reactions such as (7) and (8) the activated products may combine with other reactants to yield new products. Similarly, the atoms formed in reactions (9) and (10) may enter further reaction. Nevertheless, the absorbed light is involved only in the initial activation of some atom. Any succeeding reactions are generally thermal in character.

CONSEQUENCES OF LIGHT ABSORPTION BY MOLECULES

When a molecule absorbs incident radiation the latter may go either to activate the molecule or to dissociate it. Whether one or the other of these possibilities occurs depends on the amount of energy absorbed, and this, in turn, depends on the frequency of the light. A molecule activated photochemically will emit the energy as fluorescence unless it manages to collide with another molecule or atom to which the excitation energy may be passed on. In the latter eventuality the product of the collision may be again an activated particle, a new molecule, or a photo-sensitized reaction in which the quenching molecule is dissociated. The products of the quenching process may then continue the sequence of changes by further combination with atoms or molecules. As in atomic fluorescence, the extent of quenching varies with the pressure or concentration and depends also on the nature of the substances involved.

Absorption of radiant energy may lead also to dissociation of the absorbing molecule. In fact, in most photochemical reactions involving molecules the primary step is usually dissociation of some molecule into atoms, simpler molecules, or free radicals, which by further interaction either with each other or with different molecules continue the reaction sequence. As in atomic systems, the primary photochemical stage is the dissociation. The secondary reactions proceed by thermal means.

PHOTOCHEMICAL KINETICS

The rate laws which photochemical reactions follow are generally more complex than those for thermal reactions. In the first place, the rate of the primary activating process is controlled by the intensity of the activating radiation used, and is proportional to it. The primary process is then followed by one or more reactions whose nature must be known before a rate law for the over-all process can be deduced. Further, in many instances even the nature of the activating step is unknown. To obtain information about all the possible changes that may occur it is necessary to measure the rates at which various reactants disappear, the rates at which products are formed, and also the effect of the intensity of light as well as its frequency on these. From the data thus collected it is frequently possible to postulate a mechanism to account for the observed facts. Of considerable help in this connection are atomic and molecular spectra of the reactants involved. For some reactions these spectra allow the definite establishment of the nature of the primary step. In others the processes that cannot occur are eliminated, and thereby the number of mechanisms to be considered is reduced.

It is neither expedient nor necessary to formulate here the various rate expressions which photochemical reactions follow. We shall confine ourselves instead to an illustration of the procedure by setting up the relations for several relatively simple cases.

Consider first a hypothetical reaction

$$A_2 \longrightarrow 2A$$

which proceeds by photochemical activation. Assume further that the reaction follows the mechanism:

(a) $A_2 + h\nu \longrightarrow A_2^*$ (Activation) k_1
(b) $A_2^* \longrightarrow 2\,A$ (Dissociation) k_2
(c) $A_2^* + A_2 \longrightarrow 2\,A_2$ (Deactivation) k_3

The first stage in this sequence is the absorption of a quantum of light by A_2 with the formation of an activated molecule. This activated molecule may undergo now dissociation according to reaction (b), or it may be deactivated by collision with an inactive molecule of A_2 according to reaction (c). The rate constants with which these reactions occur are indicated to the right of each step.

The final product A is formed only in reaction (b). Consequently the rate of formation of A must be given by

$$\frac{dC_A}{dt} = k_2 C_{A_2^*} \tag{11}$$

To eliminate from this rate expression the unavailable concentration of active molecules, we resort to the concept of the stationary state (see page 578). If we apply this concept of the stationary state to the intermediate A_2^*, we observe that A_2^* is formed by reaction (a) and that it disappears by reactions (b) and (c). For reaction (a) the rate is determined only by the rate at which the light is absorbed, i.e., the rate is directly proportioned to the intensity of the absorbed light I_a. We obtain thus

$$\frac{dC_{A_2^*}}{dt} = k_1 I_a \tag{12}$$

The rate of disappearance of A_2^* is given in turn by the sum of the rates of reactions (b) and (c), namely,

$$-\frac{dC_{A_2^*}}{dt} = k_2 C_{A_2^*} + k_3 C_{A_2^*} C_{A_2} \tag{13}$$

Equating (12) and (13) to obtain the condition for the stationary state, we find that

$$k_1 I_a = k_2 C_{A_2^*} + k_3 C_{A_2^*} C_{A_2}$$
$$C_{A_2^*} = \frac{k_1 I_a}{k_2 + k_3 C_{A_2}} \tag{14}$$

Inserting $C_{A_2^*}$ from Eq. (14) into Eq. (11), the rate of formation of A is seen to be

$$\frac{dC_A}{dt} = k_2 C_{A_2^*} = \frac{k_2 k_1 I_a}{k_2 + k_3 C_{A_2}} \tag{15}$$

Finally, since for every two molecules of A formed one molecule of A_2

reacts, the photochemical efficiency of the process ϕ will be

$$\phi = \frac{1}{2\,I_a} \cdot \frac{dC_A}{dt} = \frac{1}{2}\left[\frac{k_2 k_1}{k_2 + k_3 C_{A_2}}\right] \tag{16}$$

As a second example may be taken the photochemical chlorination of chloroform in the gas phase

$$Cl_2 + CHCl_3 + h\nu = CCl_4 + HCl \tag{17}$$

For this reaction Schumacher and Wolff[1] found that the rate of formation of carbon tetrachloride is given by the expression

$$\frac{dC_{CCl_4}}{dt} = kC_{Cl_2}^{1/2}I_a^{1/2} \tag{18}$$

To account for this rate, the following mechanism has been proposed:

$$
\begin{array}{llll}
\text{(a)} & Cl_2 + h\nu \longrightarrow 2\,Cl & & k_1 I_a \\
\text{(b)} & Cl + CHCl_3 \longrightarrow CCl_3 + HCl & & k_2 C_{Cl} C_{CHCl_3} \\
\text{(c)} & CCl_3 + Cl_2 \longrightarrow CCl_4 + Cl & & k_3 C_{CCl_3} C_{Cl_2} \\
\text{(d)} & 2\,CCl_3 + Cl_2 \longrightarrow 2\,CCl_4 & & k_4 C_{CCl_3}^2 C_{Cl_2}
\end{array}
$$

To the right of each reaction is given the rate at which it proceeds. In these reactions carbon tetrachloride is formed only in steps (c) and (d), and hence the rate of formation of this substance must be

$$\frac{dC_{CCl_4}}{dt} = k_3 C_{CCl_3} C_{Cl_2} + k_4 C_{CCl_3}^2 C_{Cl_2} \tag{19}$$

To eliminate from this expression C_{CCl_3} we assume first a stationary state for this species. Since CCl_3 is formed only in step (b) and it is removed in steps (c) and (d), the rate of the former must be equal to the sum of the rates of the latter two, namely,

$$k_2 C_{Cl} C_{CHCl_3} = k_3 C_{CCl_3} C_{Cl_2} + k_4 C_{CCl_3}^2 C_{Cl_2} \tag{20}$$

Further, since Cl is also a short-lived intermediate, a stationary state may be assumed for it as well. This means that the rate of steps (a) and (c) equals the rate of step (b), or

$$k_1 I_a + k_3 C_{CCl_3} C_{Cl_2} = k_2 C_{Cl} C_{CHCl_3} \tag{21}$$

Adding Eqs. (20) and (21), we obtain

$$k_1 I_a = k_4 C_{CCl_3}^2 C_{Cl_2}$$

and
$$C_{CCl_3} = \left[\frac{k_1 I_a}{k_4 C_{Cl_2}}\right]^{1/2} \tag{22}$$

[1] Schumacher and Wolff, Z. physik. Chem., **25B**, 161 (1934).

Substituting C_{CCl_3} from Eq. (22) into Eq. (19), the rate of formation of CCl_4 follows as

$$\frac{dC_{CCl_4}}{dt} = k_3 \left[\frac{k_1 I_a}{k_4 C_{Cl_2}} \right]^{1/2} C_{Cl_2} + k_4 \left[\frac{k_1 I_a}{k_4 C_{Cl_2}} \right] C_{Cl_2}$$

$$= k I_a^{1/2} C_{Cl_2}^{1/2} + k_1 I_a \tag{23}$$

where $k = k_3 (k_1/k_4)^{1/2}$. If it is assumed now that the second term on the right in Eq. (23) is negligible compared with the first term, Eq. (23) reduces to

$$\frac{dC_{CCl_4}}{dt} = k I_a^{1/2} C_{Cl_2}^{1/2} \tag{24}$$

which is identical with the observed rate, Eq. (18).

In photochemical rate equations time is generally expressed in seconds, concentrations either in molecules or moles per cc. For the first of these concentration units I_a must be expressed in quanta of light absorbed per cc per second, i.e., the total number of quanta absorbed in 1 second divided by the volume of the absorbing medium. For the second concentration unit, on the other hand, I_a must be taken in einsteins absorbed per cc per second, namely, the total number of einsteins absorbed in 1 second divided by the volume in cc.

EXPERIMENTAL STUDY OF PHOTOCHEMICAL REACTIONS

To measure the rate of a photochemical reaction it is necessary to irradiate a reaction mixture with light of a selected wave length and to observe the manner in which the concentration of reactants or products varies with time. For this purpose some arrangement such as that shown in Figure 19–1 is required. In this diagram A is a light source emitting radiation of suitable intensity in the desired spectral range. To select radiation of only a single wave length or to confine the radiation to a narrow band, the light is passed through the lens B into a monochromator or filter at C. From C the light enters the cell D immersed in a thermostat and containing the reaction mixture. Finally, the light transmitted through D falls on some suitable recorder E, where its intensity is determined.

The light sources used depend on the spectral range in which radiation is desired and include filament lamps, carbon and metal arcs, and various gas discharge tubes. The reaction cells may be glass or quartz vessels, usually with optically plane windows for entrance and exit of light. In some instances metal cells have been used with windows cemented to the ends. Glass can be used only in the visible spectral range. Below 3500 Å not only the cells but also any other optical parts through which the light passes must be of quartz. With gases no provision need be made for stirring; solutions, however, must be agitated.

The radiant energy is measured with some form of thermopile or actinometer. The thermopile is essentially a multijunction thermocouple consisting usually of silver and bismuth soldered to metal strips blackened with lamp, platinum, or bismuth black. The radiation falling on the blackened strips is absorbed almost completely and nonselectively, and is converted into heat. The heat thus generated raises the temperature of the hot junctions above the cold, and the current generated thereby is measured. The current produced is proportional to the energy absorbed,

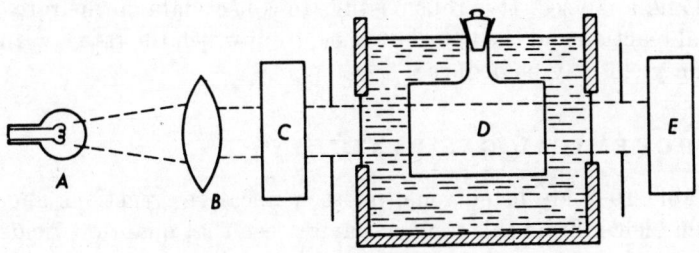

Figure 19–1.   Apparatus for study of photochemical reactions (schematic).

and this, in turn, depends on the intensity of the incident light and the area exposed to it. Thermopiles are calibrated with standard light sources.

Photocells may be used in place of thermopiles. Still other devices are chemical actinometers, which are merely gas mixtures or solutions sensitive to light. When radiation impinges upon these, a chemical reaction ensues whose extent is determined by the amount of energy absorbed. The most common of these is the uranyl oxalate actinometer, consisting of 0.05 molar oxalic acid and 0.01 molar uranyl sulfate (UO_2SO_4) in water. Under the action of light the reaction

$$H_2C_2O_4 \longrightarrow H_2O(l) + CO_2(g) + CO(g)$$

takes place, whose extent can be ascertained by titrating the remaining oxalic acid with permanganate solution. For best results this actinometer should be calibrated against a thermopile in the spectral range in which it is to be employed. When this is not possible, the quantum yields recorded in the literature may be utilized.[2] This actinometer is applicable to radiations lying between 2000 and 5000 Å.

To obtain the total radiant energy absorbed in unit time the procedure

[2] See Noyes and Leighton, *The Photochemistry of Gases*, Reinhold Publishing Corporation, New York, 1941, Table 13, p. 82.

is as follows. First the empty cell, or the cell filled with solvent alone in the case of solutions, is interposed in the light beam, and a reading is taken. This gives the total energy incident in a given time upon the system. Next the cell with the reactants is substituted, and the reading is again taken, which gives now the total energy transmitted. The difference between these readings is the total energy absorbed by the reacting mixture in the given time. If this time is 1 second, the total energy absorbed divided by the volume of the reacting mixture is I_a, the intensity of the radiation absorbed.

The rate of the chemical reaction taking place in the system is ascertained in the usual manner. For this purpose the change in some physical property can be followed, or samples can be removed periodically from the cell and analyzed. It is thus possible to collect data on the rate of the chemical reaction and the light intensity, from which the rate law and the quantum yield may be deduced.

PHOTOCHEMICAL GAS REACTIONS

In Table 19–2 are listed some photochemical gas reactions and their quantum yields. From this table it may be seen that quantum yields vary

TABLE 19–2.   SOME NONSENSITIZED PHOTOCHEMICAL GAS REACTIONS

Reaction	Wave Length (Å)	Quantum Yield	Remarks
$2 NH_3 = N_2 + 3 H_2$	2100	0.25	Depends on pressure
$SO_2 + Cl_2 = SO_2Cl_2$	4200	1	
$HCHO = H_2 + CO$	2500–3100	1–100	At 100–400°C
$CH_3CHO = CH_4 + CO$	2500–3100	1–138	At 100–400°C
$2 HI = H_2 + I_2$	2070–2820	2	Constant over wide range of pressures and temperatures
$2 HBr = H_2 + Br_2$	2070–2530	2	
$H_2 + Br_2 = 2 HBr$	Below 6000	2	Near 200°C. Very small at 25°C
$CH_3N:NCH_3 = C_2H_6 + N_2$	3660	2	Up to 300°C
$3 O_2 = 2 O_3$	1700–2530	1–3	
$nC_2H_2 = (C_2H_2)_n$	2150	9	Near room temperature
$CO + Cl_2 = COCl_2$	4000–4360	ca. 10^3	Falls with temperature. Depends also on pressure of reactants
$H_2 + Cl_2 = 2 HCl$	4000–4360	Up to 10^6	Varies with P_{H_2} and impurities

within very wide limits, as from 0.25 for ammonia decomposition to 10^6 for the combination of chlorine and hydrogen. Several of these reactions will be discussed now to illustrate some of the mechanisms by which they proceed.

THE PHOTOLYSIS OF AMMONIA

The *photolysis*, or photochemical decomposition, of ammonia was studied by Wiig,[3] who found that the reaction proceeds quantitatively according to

$$2 \text{ NH}_3(g) \longrightarrow \text{N}_2(g) + 3 \text{ H}_2(g)$$

with an average quantum yield of 0.25 up to 500 mm pressure of NH_3. The following mechanism has been proposed to explain these results:

(a) $\text{NH}_3 + h\nu \longrightarrow \text{NH}_2 + \text{H}$
(b) $\text{NH}_2 + \text{H} \longrightarrow \text{NH}_3$
(c) $\text{H} + \text{H} \longrightarrow \text{H}_2$
(d) $\text{NH}_2 + \text{NH}_2 \longrightarrow \text{N}_2\text{H}_4$
(e) $\text{N}_2\text{H}_4 + \text{H} \longrightarrow \text{NH}_3 + \text{NH}_2$
(f) $\text{NH}_2 + \text{NH}_2 \longrightarrow \text{N}_2 + 2 \text{ H}_2$

From this mechanism it can be shown that the quantum yield should be small and that it should vary with pressure. These conclusions are in accord with the observed results.

THE COMBINATION OF HYDROGEN AND BROMINE

The photochemical combination of hydrogen and bromine was studied by Bodenstein and Lütkemeyer[4] between 160° and 218°C. They found that at any given temperature the rate of formation of hydrogen bromide is given by

$$\frac{dC_{\text{HBr}}}{dt} = \frac{kC_{\text{H}_2}I_a^{1/2}}{1 + k'\left[C_{\text{HBr}}/C_{\text{Br}_2}\right]} \tag{25}$$

and that the quantum yield ϕ, as given by

$$\phi = \frac{1}{I_a} \cdot \frac{dC_{\text{HBr}}}{dt} = \frac{kC_{\text{H}_2}I_a^{-1/2}}{1 + k'\left[C_{\text{HBr}}/C_{\text{Br}_2}\right]} \tag{26}$$

increased with temperature up to 2 near 200°C, and was greater above it. These results are explainable by a series of reaction steps identical with those given on page 581 for the thermal chain reaction between these two substances, except that the dissociation indicated by step (a) takes

[3] Wiig, *J. Am. Chem. Soc.*, 57, 1559 (1935); 59, 827 (1937).
[4] Bodenstein and Lütkemeyer, *Z. physik. Chem.*, 114, 208 (1925).

place photochemically. The variation of the quantum yield with temperature is accountable by changes in the magnitudes of k and k' as the temperature is raised.

THE HYDROGEN-CHLORINE REACTION

In the photolysis of ammonia the sequence of reactions leading to the final products is initiated by formation of NH_2 and H. Once formed, these particles are removed in every subsequent reaction, and neither of these intermediates is regenerated except in the secondary step (e). Consequently, a reaction cycle is started primarily by the first step, while practically every subsequent one operates to remove the active intermediates. On the other hand, in the hydrogen-bromine reaction the cycle initiated by the appearance of bromine atoms in reaction (a) is kept going by regeneration of these atoms in reactions (c) and (d), and by formation of hydrogen atoms in (b). Every bromine or hydrogen atom thus formed is just as capable of starting a reaction as step (a), and hence each is able to start a new chain. Such reactions are called, therefore, *photochemical chain reactions.* Nevertheless the chains here do not acquire great length because of their rapid termination, and the quantum yield is thus low.

The situation is different, however, in the hydrogen-chlorine reaction. Here the chains acquire great length, and as many as 10^6 molecules of hydrogen chloride have been observed to be formed per absorbed quantum of light. As a rule the yield, although still high, is less than the figure quoted, particularly when impurities are present in the reaction mixture. By reacting with atoms of hydrogen and chlorine, these impurities operate to terminate chains, thus reducing the quantum efficiency of the process. The chains are terminated also by the walls of the reaction vessel.

At moderate gas pressures, and with comparable proportions of chlorine to hydrogen, Bodenstein and Unger,[5] and others, found that the rate of formation of HCl is given by

$$\frac{dC_{HCl}}{dt} = kI_aC_{H_2} \tag{27}$$

This rate equation can be accounted by the following mechanism:

(a) $Cl_2 + h\nu \longrightarrow 2\ Cl$ k_1
(b) $Cl + H_2 \longrightarrow HCl + H$ k_2
(c) $H + Cl_2 \longrightarrow HCl + Cl$ k_3
(d) Cl (at walls) $\longrightarrow \frac{1}{2}\ Cl_2$ k_4

Here every reaction continues the chain except the last one.

[5] Bodenstein and Unger, *Z. physik. Chem.*, **11B**, 253 (1930).

In all combinations of atoms, free radicals, or atoms and free radicals to form stable molecules, considerable energy is evolved. Unless this energy is removed, the resulting molecule cannot be stable. Consequently, in all such reactions it is necessary to assume either that a third body is involved in the collision which carries off the energy, or that the reaction occurs at the walls. The third body may be a molecule of the reactants, the products, or some other neutral impurity present in the system.

PHOTOSENSITIZED GAS REACTIONS

When a reaction mixture is exposed to light to which the reactants are insensitive, no reaction will take place. However, it is possible to introduce into the reaction mixture molecules or atoms which will absorb the light, become excited, and then pass on this energy to one of the reactants and thereby activate it for reaction. A substance acting in this manner is called a *photosensitizer*, while the reaction resulting is said to be *photosensitized*.

An example of such a reaction is the combination of carbon monoxide and hydrogen sensitized by mercury vapor. The products are primarily formaldehyde and some glyoxal. The steps involved may be written as:

$$
\begin{aligned}
&\text{(a)} \quad && \text{Hg} + h\nu \longrightarrow \text{Hg}^* \\
&\text{(b)} \quad && \text{Hg}^* + \text{H}_2 \longrightarrow 2\,\text{H} + \text{Hg} \\
&\text{(c)} \quad && \text{H} + \text{CO} \longrightarrow \text{HCO} \\
&\text{(d)} \quad && \text{HCO} + \text{H}_2 \longrightarrow \text{HCHO} + \text{H} \\
&\text{(e)} \quad && 2\,\text{HCO} \longrightarrow \text{HCHO} + \text{CO} \\
&\text{(f)} \quad && 2\,\text{HCO} \longrightarrow \text{CHO—CHO (glyoxal)}
\end{aligned}
$$

Step (a) indicates the activation of the mercury vapor by light absorption, step (b) a transfer of the activation energy to a hydrogen molecule. The energy absorbed by the latter causes it to dissociate into atoms which initiate then the remainder of the reaction cycle.

Many mercury-photosensitized reactions have been investigated. These include decompositions of H_2, NH_3, H_2O, PH_3, AsH_3, various hydrocarbons, alcohols, ethers, acids, and amines; hydrogenations such as that of ethylene, propylene, and butylene; combinations, such as that of oxygen and hydrogen to water, oxygen to ozone, and hydrogen and nitrogen to ammonia; and polymerizations, such as that of ethylene. The dissociation of hydrogen is sensitized also by xenon, the polymerization of ethylene by cadmium vapor. The latter is effective as well in the decomposition of ethane and propane into hydrogen, methane, and higher hydrocarbons. The halogens have also been employed frequently as sensitizers. Thus chlorine, bromine, and iodine sensitize the decomposition of ozone; bromine that of Cl_2O; while chlorine promotes the combination of oxygen and hydrogen to water and of carbon monoxide and oxygen to carbon dioxide.

In Table 19–3 are given several photosensitized gas reactions along with their quantum yields and the wave lengths at which these were obtained.

TABLE 19–3.   SOME PHOTOSENSITIZED GAS REACTIONS

Reaction	Sensitizer	Wave Length (Å)	Quantum Yield	Remarks
$2 O_3 = 3 O_2$	Cl_2	4300	2–30	Increases with O_3 concentration
$2 H_2 + O_2 = 2 H_2O$	Cl_2	4300	2	
$H_2 + CO = HCHO$	Hg	2537	2	
$2 CO + O_2 = 2 CO_2$	Cl_2	4050–4360	1000	Depends on temperature and concentration

PHOTOCHEMICAL REACTIONS IN THE LIQUID PHASE

Many substances undergo photochemical reaction when liquefied or dissolved in a solvent. Again such reactions may be initiated by direct light absorption on the part of a reactant, or they may be photosensitized. A *priori* it may be anticipated that the quantum efficiency of a photochemical process will be less in the liquid phase than for the same reaction in the gas phase. The reason for this is that in the liquid phase an active molecule or atom may readily be deactivated by frequent collisions with other molecules, or by reaction with the solvent. Furthermore, because of the very short mean free path in the liquid phase, free radicals or atoms when formed photochemically will tend to recombine before they have a chance to get very far from each other. The net effect of these processes will be to keep the quantum yield relatively low. In fact, only those reactions may be expected to proceed to any extent for which the primary products of the photochemical act are relatively stable particles. Otherwise the active intermediates will tend to recombine or react with the solvent and thereby keep the yield low.

That these considerations apply in many instances is shown by the data in Table 19–4. In every case except the last the yield is lower in the liquid phase than in the gas for the same wave length of irradiating light. However, in a number of reactions, of which the photolysis of $Ni(CO)_4$ is an example, just the reverse is true. This reversal may be due to formation of intermediates of kinetic energy sufficiently high to separate them rapidly, thereby preventing their combination, or to the specific effect of solvent which makes it inefficient as a deactivator. That solvents may exert specific effects is borne out by the observation that the same reaction conducted in different media may have different quantum efficiencies under the same conditions. Thus for the photolysis of ClO_2 at 4358 Å $\phi = 1.0$ in CCl_4 and only 0.20 in water. Similarly the photolysis of $N_2CH_2COOC_2H_5$ at 2600 Å gives $\phi = 1.1$ in heptane, 1.42 in methyl alcohol, and 2.8 in water.

TABLE 19–4. COMPARISON OF QUANTUM YIELDS IN THE GAS AND LIQUID PHASES

		Quantum Yield	
Reaction	Wave Length (Å)	Gas	Liquid Phase
$2\,NH_3 = N_2 + 3\,H_2$	2100	0.14–0.32	0 (In liquid NH_3)
$CH_3COOH = CH_4 + CO_2$	<2300	1	0.45 (In H_2O)
$Cl_2O = Cl_2 + \tfrac{1}{2}\,O_2$	4358	3.2	1.8 (In CCl_4)
$NO_2 = NO + \tfrac{1}{2}\,O_2$	4050	0.50	0.03 (In CCl_4)
$Pb(CH_3)_4 = Pb + 2\,C_2H_6$	2357	1.1	0.4 (In hexane)
$Ni(CO)_4 = Ni + 4\,CO$	3010–3135	0	2.8 (In CCl_4)

In Table 19–5 are given results for several other sensitized and non-sensitized reactions in solution. The mechanisms for some of these may be indicated. The quantum efficiency of unity for the chloracetic acid reaction may be accounted by the simple sequence

$$ClCH_2COOH + h\nu \longrightarrow ClCH_2COOH^*$$
$$ClCH_2COOH^* + H_2O \longrightarrow HOCH_2COOH + HCl$$

For the cis-butene iodination Forbes and Nelson[6] found the rate to be

$$\frac{dC_{C_4H_8I_2}}{dt} = kC_{C_4H_8}C_{I_2}I_a \tag{28}$$

To deduce this rate law it is necessary to assume that the mechanism involved is

(a) $\quad I_2 + h\nu \longrightarrow 2\,I$

(b) $\quad I + C_4H_8 \rightleftharpoons C_4H_8I$

(c) $\quad C_4H_8I + I_2 \longrightarrow C_4H_8I_2 + I$

(d) $\quad C_4H_8I_2 + I \longrightarrow C_4H_8I + I_2$

(e) $\quad I \longrightarrow \tfrac{1}{2}\,I_2$

The reaction is initiated by formation of iodine atoms which subsequently become involved in reactions (b), (d), and (e). Step (c), however, operates to regenerate these. Further, it is necessary to postulate that step (b)

[6] Forbes and Nelson, *J. Am. Chem. Soc.*, **59**, 693 (1937).

TABLE 19-5.   FURTHER EXAMPLES OF PHOTOCHEMICAL REACTIONS IN SOLUTION

Reaction	Solvent	Wave Length (Å)	Quantum Yield	Remarks
Nonsensitized				
$2 HI = H_2 + I_2$	H_2O	2070	0.34	Depends on concentration and wave length
cis-$C_4H_8 + I_2 = C_4H_8I_2$	$CHCl_3$	4360	2.48	Each reactant 0.01 molar at $-55°C$
$2 Fe^{+++} + I_2 = 2 Fe^{++} + 2 I^-$	H_2O	5790	1	
$ClCH_2COOH + H_2O =$				
$HOCH_2COOH + HCl$	H_2O	2537	1	0.3–0.5 molar
$H_2C_2O_4 = H_2O + CO + CO_2$	H_2O	2650	0.01	
$2 H_2O_2 = 2 H_2O + O_2$	H_2O	3100	7–80	
Sensitized				
$H_2C_2O_4 = H_2O + CO + CO_2$	H_2O	2540–4350	0.49–0.60	0.05 molar acid 0.01 molar UO_2^{++} as sensitizer
$2 CCl_3Br + O_2 = 2 COCl_2 + Br_2$	CCl_3Br	4070–4360	0.9	Br_2 as sensitizer
Maleic ester = Fumaric ester	CCl_4	4360	295	Br_2 as sensitizer

proceeds in both directions. From these equations an expression equivalent to the observed relation (28) may be derived.

The exact mechanism by which the UO_2^{++}-sensitized dissociation of oxalic acid occurs is uncertain. Indications are that a complex is formed between the acid and UO_2^{++}. In the reactions involving bromine the sensitization is probably caused by dissociation of the molecule into bromine atoms, which continue then the reaction.

PHOTOCHEMICAL EFFECTS IN SOLIDS

Solid substances may also be affected by light. Thus silver salts exposed to light tend to darken, and sodium chloride crystals in ultraviolet light develop a yellowish tinge which can be removed by heating, and dyes tend to fade in sunlight.

The outstanding application of photolysis in the solid state is in photography. A photographic emulsion is a dispersion of finely divided silver chloride or bromide in a gelatin base. When such an emulsion is exposed to light, the halides are acted upon to form an invisible latent image of density proportional to the light intensity. To make the image visible, it must be developed by treatment with a reducing agent which acts primarily on the exposed portions of the emulsion to yield a finely divided deposit of silver. A photographic plate thus developed is treated with a fixing agent, such as sodium thiosulfate, to dissolve the undeveloped silver halide, but not the part exposed. On the resulting negative the portions that were lightest on the object photographed appear darkest, and vice versa. To obtain from the negative a print in which the light densities are rectified, the above process is repeated by photographing the negative. The exposure thus obtained will have, when developed, the light and dark portions of the negative reversed, and these will correspond, therefore, to the relative light and dark shadings of the original object.

Photographic emulsions consisting of silver halides alone are sensitive only to light near the ultraviolet. To extend the range of sensitivity to longer wave lengths, various dyes which absorb in the yellow and red are incorporated into the emulsion. Photographic emulsions have also been developed which are sensitive to near-infrared radiation and thus make possible night photography.

FLASH PHOTOLYSIS

In ordinary photochemical studies the light enters the system continuously but at a relatively low intensity level. As a result the concentration of short-lived intermediates is very low, and their observation is difficult. In *flash photolysis*, on the other hand, the system is exposed to a very powerful beam of light for a very short period of time, say 10^{-3} to 10^{-4} second. This powerful light flash produces a much higher concentration of intermediates, such as atoms or free radicals, whose presence, concentration, and state can be ascertained by continuous observation of their absorption spectra. From such data it is possible to identify these intermediates, to obtain their concentration, and even to evaluate in certain instances the rate constants of photochemical reaction steps involving these species.

EFFECT OF TEMPERATURE ON PHOTOCHEMICAL REACTIONS

The effect of temperature on photochemical reactions is quite different from that on thermal ones. For the latter an increase of 10°C in temperature generally leads to a two- or threefold increase in the rate. In photochemical reactions, however, the same increase in temperature results as a rule in only a very small increase in rate. For most such reactions the temperature coefficient is not far from unity. Occasionally it does approach a value close to that for a thermal reaction, as in the potassium oxalate-iodine reaction, but such behavior is exceptional. Further, is some photochemical reactions, of which the chlorination of benzene is an example, the unusual phenomenon of decrease in rate with temperature is encountered.

To explain the observed temperature coefficients of photochemical reactions it is necessary to consider separately the effect of this variable on the primary and secondary reactions in the complete process. The primary light absorption process should be practically temperature-independent. Again, since secondary reactions in photochemical processes are thermal in character, these should have temperature coefficients akin to those of ordinary reactions. However, most secondary reactions in photochemical processes involve interaction between atoms or free radicals, or of these with molecules. For such reactions the energy of activation is usually small or even zero. As the temperature coefficient of a reaction is determined by the magnitude of the activation energy, we may conclude that even for the secondary reactions the temperature coefficient should be smaller than for thermal ones involving only molecules. The net result should be, therefore, a small temperature coefficient for the over-all process. This is generally the case.

When a large temperature coefficient is observed in a photochemical reaction it is usually an indication that one or more of the intermediate steps have a high activation energy. Another possibility is that some step in the reaction sequence is an equilibrium whose variation with temperature involves an appreciable positive heat of reaction. Thus suppose that an observed velocity constant k is composed of the product of a rate constant k_1 and an equilibrium constant K; i.e.,

$$k = k_1 K \tag{29}$$

Taking logarithms and differentiating with respect to temperature, Eq. (29) yields

$$\frac{d \ln k}{dT} = \frac{d \ln k_1}{dT} + \frac{d \ln K}{dT} \tag{30}$$

The first term on the right is E_a/RT^2, where E_a is the energy of activation of the step involving k_1. Again, the second term is $\Delta H/RT^2$, where ΔH is the heat of the reaction for which K is the equilibrium constant. Substitution of these into Eq. (30) gives

$$\frac{d \ln k}{dT} = \frac{E_a}{RT^2} + \frac{\Delta H}{RT^2}$$
$$= \frac{E_a + \Delta H}{RT^2} \tag{31}$$

Even if E_a is small, a large positive ΔH will give a large value for the quantity on the right in Eq. (31), and hence the temperature coefficient will be large. On the other hand, if ΔH is negative and numerically greater than E_a, then $(E_a + \Delta H)$ will be negative, and so will $d \ln k/dT$. Under these conditions the reaction will have a negative temperature coefficient. This may well be the explanation for the decrease in rate with temperature observed in the photochemical chlorination of benzene and some other reactions of this type.

PHOTOCHEMICAL EQUILIBRIUM

Consider a reaction

$$A + B \xrightarrow{\text{light}} C + D \tag{32}$$

proceeding as indicated under the influence of absorbed radiation. Suppose, further, that the products are not photosensitive, but that they combine in a thermal reaction to reform the reactants, namely,

$$C + D \xrightarrow{\text{thermal}} A + B \tag{33}$$

As these reactions proceed, a state will eventually be reached in which the rates of the two become equal, and the equilibrium

$$A + B \underset{\text{thermal}}{\overset{\text{light}}{\rightleftharpoons}} C + D \tag{34}$$

is established. Similarly, both forward and reverse reactions may be light-sensitive, in which case an equilibrium of the type

$$A + B \underset{\text{light}}{\overset{\text{light}}{\rightleftharpoons}} C + D \tag{35}$$

is attained. Equilibria such as (34) or (35), resulting from opposing processes in which one or both reactions are photosensitive, are called *photochemical equilibria*. An example of the first of these is the dimerization of anthracene

$$2\,C_{14}H_{10} \underset{\text{thermal}}{\overset{\text{light}}{\rightleftharpoons}} C_{28}H_{20} \tag{36}$$

and of the second the decomposition of sulfur trioxide

$$2\,SO_3 \underset{\text{light}}{\overset{\text{light}}{\rightleftharpoons}} 2\,SO_2 + O_2 \tag{37}$$

The equilibrium constants for photochemical equilibria are *constant only for a given light intensity*, and vary as the latter is changed. Further, they do *not* correspond to the equilibrium constants obtained for the reactions under purely thermal conditions. For instance, thermal equilibrium calculations show that to obtain 30 per cent dissociation of SO_3 at 1 atm pressure this substance must be heated to 630°C. On the other hand, Coehn and Becker[7] found that photochemically SO_3 may be dissociated to the extent of about 35 per cent at 45°C. Again, whereas the thermal equilibrium constant varies very markedly with temperature, the photochemical equilibrium constant was found at constant light intensity to be temperature-independent between 50 and 800°C. From these observations, and others, it is evident that the usual equilibrium considerations do not apply to photochemical reactions. A moment's reflection will indicate the reason. In thermal equilibrium we are concerned only with the normal free energy relations of products and reactants as they apply to the conditions of these at the temperature and pressure in question. In photochemical equilibria, on the other hand, these free energy relations are modified by the free energy supplied by the light. Addition of the latter changes the ΔF° of the reaction, and thereby also the equilibrium constant. In other words, by supplying free energy in the form of light it is possible to make a reaction go in a direction which from ordinary thermodynamic considerations alone would appear to be impossible.

CHEMILUMINESCENCE

In photochemical reactions chemical change results from absorption of light. The converse of this process is the emission of light by a system at ordinary temperatures as a result of a chemical reaction. Such emission of "cold light," called *chemiluminescence*, is a well-known phenomenon.

When yellow phosphorus is oxidized in oxygen or air at low pressures and at temperatures between -10 and 40°C, the phosphorus is converted to P_2O_5 with emission of a visible greenish-white luminescence. Similarly, the oxidation of certain Grignard reagents, silicon compounds, and 3-amino-phthalic acid hydrazine in alkaline solution is accompanied by light emission. In all these instances part or all of the energy change of the reaction, instead of appearing as heat, goes to activate electronically some molecule. This activated molecule emits then the excitation energy as radiation and reverts to a normal state.

The chemiluminescence which is possibly best understood is that emitted in the reaction of alkali metal vapors with halogens. When streams of alkali metal vapors and halogens at pressures of 10^{-2} to 10^{-3} mm Hg are mixed, alkali halides are formed with the emission of luminescence. Polanyi and his co-workers showed that in such reactions a chain sets in as a result of which an activated alkali atom is obtained. This atom

---

[7] Coehn and Becker, *Z. Elektrochem.*, **13**, 545 (1907); *Z. physik. Chem.*, **70**, 88 (1909).

on return to its ground state emits the observed radiation. In the sodium-chlorine reaction the amount of light obtained is only about 1 quantum per 1000 sodium chloride molecules formed, while in the sodium-iodine reaction it is only 1 quantum per 2000 sodium iodide molecules. The significant fact established in all these reactions is that at no time is radiation obtained whose energy exceeds the heat of the respective reaction. Usually the light emitted is lower in energy than the latter, with the balance being dissipated as translational kinetic energy of the atom.

Chemiluminescence phenomena are not confined to the laboratory. Every student is familiar with the cold light emitted by fireflies. This light arises from the oxidation of a protein substance called luciferin. Certain marine species are also luminescent. Finally, chemiluminescent glow, due to oxidation, may be observed above marshes where wood and vegetation undergo decay. In all these instances part of the energy change accompanying a given process is emitted as radiation instead of heat.

RADIATION CHEMISTRY

Chemical activation can be obtained not only by the use of ultraviolet, visible, and near infrared light, but also by α, β, and γ rays, protons, neutrons, deuterons, and high voltage electric discharges. The energies involved in these initiators are very high and may amount to 50 or more million electron volts (MeV). The primary effect of the passage of such high energy radiation through matter is the production of extensive ionization, both by direct impact and by the secondary radiation resulting from the initial ionization. The heavier particles, i.e., α particles, protons, neutrons, and deuterons, are more effective in producing ionization than are the lighter and faster electrons and γ rays. The reason for this is that the heavier particles, by moving slower, have a greater opportunity to collide with molecules and cause thereby more extensive ionization.

The ions produced in this manner may be discharged by interaction with electrons, or, more importantly, they may initiate reactions which may lead to the formation of new products. The latter are frequently substances of lower molecular weight than the initial molecules, but such is not always the case. Thus irradiation of water yields not only hydrogen but also hydrogen peroxide. Again, irradiation of a monomer mixture can produce polymerization. Further, a polymer exposed to a high energy source is found to exhibit properties quite different from those of the original substance, and more in line with what is to be expected of a product of higher molecular weight. Such modification of polymer properties by high energy radiation is of great interest at present, and considerable research is being conducted in this direction.

REFERENCES

1. G. F. J. Garlick, *Luminescent Materials*, Oxford University Press, New York, 1949.
2. Masson, Boekelheide, and Noyes in *Catalytic, Photochemical, and Electrolytic Reactions*, Interscience Publishers, Inc., New York, 1956.
3. Noyes and Leighton, *The Photochemistry of Gases*, Reinhold Publishing Corporation, New York, 1941.
4. P. Pringsheim, *Fluorescence and Phosphorescence*, Interscience Publishers, Inc., New York, 1949.
5. Pringsheim and Vogel, *Luminescence of Liquids and Solids*, Interscience Publishers, Inc., New York, 1946.
6. E. I. Rabinowitch, *Photosynthesis and Related Processes*, Interscience Publishers, Inc., New York, 2 vols., 1945–1954.
7. Rollefson and Burton, *Photochemistry and the Mechanism of Chemical Reactions*, Prentice-Hall, Inc., Englewood Cliffs, N.J., 1939.

PROBLEMS

1. A 2-mm layer of Pyrex glass will transmit 10% of incident radiation of 3000 Å. What percentage of light of the same wave length will be absorbed by a 1-mm layer of the glass?   *Ans.* 68.4%.

2. A certain substance in a cell of length l absorbs 10% of the incident light. What fraction of the incident light will be absorbed in a cell five times as long?

3. In a cell of a certain length and at a pressure of 100 mm Hg gaseous acetone transmits 25.1% of the incident light of wave length 2650 Å. Assuming Beer's law to apply, calculate the pressure at which 98% of the incident light will be absorbed by acetone in the same cell and at the same temperature.
                              Ans. 283 mm Hg.

4. From the data given in the preceding problem find what percentage of the incident radiation will be absorbed by acetone at a pressure of 200 mm Hg in a cell one-third as long as used before.

5. Calculate the energy per mole of light having wave lengths of (a) 850 Å, (b) 2500 Å, and (c) 5 μ. In what spectral region does each wave length fall?
                              Ans. (a) 3.36×10^5 cal., ultraviolet.

6. The energy of activation for the thermal decomposition of $N_2O(g)$ is 53,000 cal mole^{-1}. Calculate the frequency and wave length of the radiation corresponding to this energy, and tell in what region of the spectrum it is to be found.

7. From the energy of activation given in the preceding problem calculate (a) the fraction of the total number of molecules, and (b) the actual number of molecules, which can be activated thermally in 1 liter of N_2O at a pressure of 50 mm Hg and 27°C.

8. In the photochemical combination of $H_2(g)$ and $Cl_2(g)$ a quantum efficiency of about 1×10^6 is obtained with a wave length of 4800 Å. How many moles of HCl(g) would be produced under these conditions per calorie of radiant energy absorbed? *Ans.* 33.6 moles.

9. A certain system absorbs 2.0×10^{16} quanta of light per second. At the end of 10 min it is observed that 0.001 mole of the irradiated substance has reacted. What is the quantum efficiency of the process?

10. The incident radiation from a monochromatic source of 2537 Å on a slit is 80 ergs/sec/sq mm. The radiation is passed through a slit whose area is 5 sq mm and onto a cell whose face transmits 30% of the incident radiation. If the solution in the cell transmits 10% of the radiation, how much energy in cal is absorbed by the solution per hour? *Ans.* 9.3×10^{-3} cal.

11. If the quantum efficiency in the preceding problem is unity for the substance decomposing and if all the radiation absorbed by the solution is used in the photochemical decomposition, how many moles of substance would be decomposed per hour?

12. The dissociation energy of H_2 is 103,200 cal mole^{-1}. If H_2 is illuminated with radiation whose wave length is 2537 Å and is thereby dissociated, how much of the original radiant energy per mole will be converted into kinetic energy? *Ans.* 10,100 cal.

13. When gaseous HI is illuminated with radiation of 2530 Å, it is observed that 1.85×10^{-2} mole decomposes per 1000 cal of radiant energy absorbed. Calculate the quantum efficiency of the reaction.

14. An uranyl oxalate actinometer is irradiated for 15 min with light of 4350 Å. At the end of this time it is found that oxalic acid equivalent to 12.0 cc of 0.001 molar $KMnO_4$ has been decomposed by the light. At this wave length the quantum efficiency of the actinometer is 0.58. Find the average intensity of the light used in (a) ergs per sec and (b) in quanta per sec. *Ans.* (a) 1.58×10^5 ergs sec^{-1}.

15. Assuming stationary states for Cl and H atoms, show that the mechanism proposed for the photochemical combination of $H_2(g)$ and $Cl_2(g)$ leads to the observed rate equation (27). What is the expression for k in this equation in terms of the rate constants for reactions (a) to (d)?

16. Assuming stationary states for Br and H atoms, deduce Eq. (25) for the rate of the photochemical combination of $H_2(g)$ and $Br_2(g)$ on the basis of the mechanism given on page 581 for the thermal reaction. What are the expressions in this case for k and k' in terms of the individual reaction rate constants?

17. Suppose that in a certain over-all reaction the observed velocity constant k, in terms of the velocity constants of the individual reactions involved, k_1, k_2, and k_3, is given by

$$k = \frac{k_1 k_2}{k_3^{1/2}}$$

What will be the relation between the observed activation energy and the activation energies of the individual reaction steps?

18. In a certain reaction $A + B = 2D$ the forward reaction proceeds photochemically with a rate given by

$$\frac{dD_f}{dt} = kI_aC_AC_B$$

while the reverse reaction proceeds thermally and bimolecularly with a velocity constant k'. Deduce the expression for the concentration of D at photochemical equilibrium.   *Ans.* $C_D = (kI_aC_AC_B/k')^{1/2}$.

19. Assuming that the rate of thermal dissociation of HBr(g) molecules is negligible compared to the rate of the photochemical combination of $H_2(g)$ and $Br_2(g)$, derive the expression for the concentration of HBr(g) at photochemical equilibrium.

20. For the photochemical dimerization of anthracene

$$2 C_{14}H_{10} \longrightarrow C_{28}H_{20}$$

the following mechanism has been proposed:

(a) $C_{14}H_{10} + h\nu \longrightarrow (C_{14}H_{10})^*$ k_1
(b) $(C_{14}H_{10})^* + C_{14}H_{10} \longrightarrow C_{28}H_{20}$ k_2
(c) $C_{28}H_{20} \longrightarrow 2 C_{14}H_{10}$ k_3
(d) $(C_{14}H_{10})^* \longrightarrow C_{14}H_{10} + h\nu'$ k_4

(a) Derive the expression for the rate of formation of the dimer.
(b) Deduce from the latter the expression for the concentration of dimer at photochemical equilibrium.
(c) How does this concentration depend upon the intensity of absorbed light?

21. What will be the wave length of γ rays possessing energies of (a) 1 MeV and (b) 10 MeV?

22. Calculate the velocities of 100 eV electrons and protons.

23. Calculate the velocities of 100 eV neutrons, deuterons, and α particles.

20

Surface Phenomena and Catalysis

SOLIDS, liquids, and solutions exhibit many properties which can be explained only in terms of the action of their surfaces. These include surface and interfacial tensions, adsorption, the spreading cf liquids on surfaces, insoluble surface films, and the catalytic activity of various solid surfaces for many types of chemical reactions. The present chapter is devoted to a discussion of these phenomena, and their theoretical and practical implications.

THE SURFACE TENSION OF LIQUIDS

Within the body of a liquid a molecule is acted upon by molecular attractions which are distributed more or less symmetrically about the molecule. At the surface, however, a molecule is only partially surrounded by other molecules, and as a consequence it experiences only an attraction toward the body of the liquid. This latter attraction tends to draw the surface molecules inward, and in doing so makes the liquid behave as if it were surrounded by an invisible membrane. This behavior of the surface, called *surface tension,* is the effect responsible for the resistance a liquid exhibits to surface penetration, the nearly spherical shape of falling water droplets, the spherical shape of mercury particles on a flat surface, the rise of liquids in capillary tubes, and the flotation of metal foils on liquid surfaces. From a purely thermodynamic point of view surface tension may be thought of as due to the tendency of a liquid to reduce its surface to a point of minimum potential surface energy, a condition requisite for stable surface equilibrium. Since a sphere has the smallest area for a given volume, the tendency of a liquid particle should be to draw itself into a sphere due to the action of surface tension, as is actually the case.

Since the natural tendency of a liquid is to decrease its surface, any increase in surface can only be accomplished with the expenditure of work. Consider a liquid film contained within a rectangular wire frame, $ABCD$, as shown in Figure 20–1. The side $CD = l$ is movable. If a force f

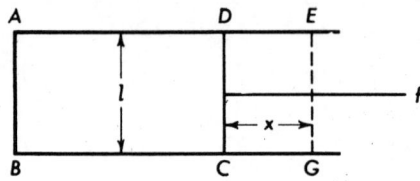

Figure 20–1.

is required to move the wire CD against the force of surface tension acting in the film along CD, the work done, w, in moving the wire from CD to EG is

$$w = fx \tag{1}$$

The force acting, f, must, of course, be balanced by the force of surface tension along CD. If we designate by γ the force *per centimeter* along CD, and since there are two surfaces to the film,

$$f = 2\,\gamma l \tag{2}$$

and

$$w = fx = 2\,\gamma lx \tag{3}$$

From Eq. (2) γ may be defined as $\gamma = f/2\,l$, or the *force in dynes acting along 1 cm length of surface*. However, $2\,lx$ is the area of new surface of liquid generated by CD, ΔA. Hence Eq. (3) becomes also

$$w = \gamma(2\,lx) = \gamma\Delta A$$

and

$$\gamma = \frac{w}{\Delta A} \tag{4}$$

Consequently, γ may be considered also as the *work in ergs necessary to generate a square centimeter of surface area*, and is, therefore, referred to frequently as the free surface energy of a liquid per square centimeter of area.

The surface tension is a characteristic property of each liquid and differs greatly in magnitude for different liquids. Of the various methods available for measuring surface tension, such as the tensiometer, drop weight, bubble pressure, or capillary rise methods, the last is by far the most important and is considered the standard. The capillary rise method for the estimation of surface tension is based on the fact that most liquids when brought into contact with a fine glass capillary tube will rise in the tube to a level above that of the liquid outside the tube. This will occur only when the liquid wets glass, i.e., adheres to it. If the liquid does not wet glass, as mercury for instance, the level inside the capillary will fall below that outside, and the mercury will exhibit a convex surface, as against the concave surface in the first instance.

To understand the theory of the capillary rise method, consider a fine capillary tube of uniform radius r immersed in a vessel containing a liquid that wets glass, Figure 20-2. By wetting the inner wall of the capillary

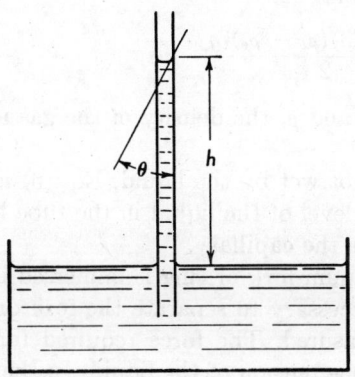

Figure 20-2. Capillary rise method for determination of surface tension.

the surface of the liquid is increased. To decrease its free surface the liquid must rise within the capillary. As soon as this happens, however, the glass is again wet, and again the liquid draws itself upward. This process does not continue indefinitely, but stops when the force of surface tension acting upward becomes equal to the force due to the column of liquid acting downward. If we call γ the surface tension in dynes per centimeter of inner circumference, and consider the force to be acting at an angle θ, called the contact angle, with the vertical, then the force due to surface tension is

$$f_1 = 2\pi r\gamma \cos \theta$$

This force is balanced by that due to the column of liquid of height h, or

$$f_2 = \pi r^2 h \rho g$$

where ρ is the density of the liquid and g is the acceleration of gravity in centimeters per second per second. Therefore, since at equilibrium $f_2 = f_1$,

$$2\pi r\gamma \cos\theta = \pi r^2 h\rho g$$

and

$$\gamma = \frac{rh\rho g}{2\cos\theta} \tag{5}$$

For most liquids which wet glass θ is essentially zero, and $\cos\theta = 1$. Then

$$\gamma = \frac{rh\rho g}{2} \tag{6}$$

and γ may be calculated provided we know the radius of the capillary the density of the liquid, and the height to which the liquid will rise in the capillary.

For precise work two corrections must be applied, one for the volume of the meniscus and another for the density of the gas above the liquid. Under these conditions Eq. (6) becomes

$$\gamma = \frac{(h + r/3)(\rho_l - \rho_v)rg}{2} \tag{7}$$

where ρ_l is the density of the liquid and ρ_v the density of the gas above the liquid.

When the surface of the glass is not wet by the liquid, Eq. (6) is still applicable, but h is now the drop in level of the liquid in the tube below that of the level of the liquid outside the capillary.

In the tensiometer method a platinum fork or ring is immersed in the liquid to be tested, and the force necessary to separate the fork or ring from the liquid surface is then measured. The force required for this operation can be related to the surface tension of the liquid. In the drop weight method, on the other hand, the liquid whose surface tension is to be determined is allowed to pass very slowly through a calibrated capillary tip so as to form an approximately spherical drop. When the drop has attained a definite weight, depending on the size of the capillary tip and the surface tension of the liquid, a portion of it will detach itself and fall. Harkins and Brown[1] have shown that the surface tension of the liquid may be determined from the weight of the falling drop W, and the radius of the capillary tip by the relation

$$\gamma = \frac{Wg}{2\pi r\phi} \tag{8}$$

where ϕ is a complicated function of $(r/V^{1/3})$, V being the volume of the drop. For the nature of the function ϕ and details of the method the student is referred to Harkins's papers on the subject.

---

[1] Harkins and Brown, *J. Am. Chem. Soc.*, **41**, 499 (1919).

As may be seen from Table 20–1, the surface tension of liquids decreases as the temperature is increased, and becomes zero at the critical temperature. The variation of surface tension with temperature may be represented by the Ramsay-Shields equation

$$\gamma \left(\frac{M}{\rho_l}\right)^{2/3} = k(t_c - t - 6) \qquad (9)$$

where γ is the surface tension at temperature t, and M, ρ_l, and t_c are the molecular weight, density, and critical temperature of the liquid, respectively. Since M/ρ_l is the molar volume of the liquid, $(M/\rho_l)^{2/3}$ is a quantity proportional to the molar surface of the liquid, and hence the product

TABLE 20–1. SURFACE TENSION OF LIQUIDS AT VARIOUS TEMPERATURES
(Dynes cm^{-1})

Liquid	0°C	20°C	40°C	60°C	80°C	100°C
Water	75.64	72.75	69.56	66.18	62.61	58.85
Ethyl alcohol	24.05	22.27	20.60	19.01	—	—
Methyl alcohol	24.5	22.6	20.9	—	—	15.7
Carbon tetrachloride	—	26.8	24.3	21.9	—	—
Acetone	26.2	23.7	21.2	18.6	16.2	—
Toluene	30.74	28.43	26.13	23.81	21.53	19.39
Benzene	31.6	28.9	26.3	23.7	21.3	—

on the left of Eq. (9) is frequently referred to as the *molar surface energy*. k is a constant which is supposed to be independent of temperature.

The Ramsay-Shields equation has been found to be valid for many liquids up to within 30° to 50° of the critical temperature. However, the equation predicts that γ will be zero at $t = (t_c - 6)$ and that it will become negative at the critical temperature. To obviate this difficulty Katayama suggested the modified equation

$$\gamma \left(\frac{M}{\rho_l - \rho_v}\right)^{2/3} = k'(t_c - t) \qquad (10)$$

where ρ_v is the density of the vapor at temperature t. In this equation $\gamma = 0$ when $t = t_c$.

INTERFACIAL TENSION AND SPREADING OF LIQUIDS

When two immiscible or partially miscible liquids A and B are contacted, it is found that an *interfacial tension* exists at the boundary between the two layers. This interfacial tension, γ_{AB}, can be measured by methods very similar to those employed to determine the surface tension of pure liquids. Its value is generally intermediate between the surface tensions of the two liquids, γ_A and γ_B, but sometimes it is lower than both, as may be seen from the data given in Table 20-2.

If a column of pure liquid 1 cm² in cross section is pulled apart, two surfaces are created, each of which is 1 cm² in area. Since the surface tension is also the work necessary to create unit surface area, the work done in pulling the liquid apart is then

$$w_c = 2\,\gamma \tag{11}$$

where w_c is the *work of cohesion* of the liquid. In a like manner, if we consider a column composed of two immiscible or partially miscible liquids,

TABLE 20-2.   INTERFACIAL TENSION OF LIQUIDS AT 20°C
(Dynes cm⁻¹)

A	B	γ_A	γ_B	γ_{AB}
Water	Benzene	72.75	28.88	35.0
"	Carbon tetrachloride	"	26.8	45.0
"	n-Octane	"	21.8	50.8
"	n-Hexane	"	18.4	51.1
"	Mercury	"	470.	375.
"	n-Octyl alcohol	"	27.5	8.5
"	Ethyl ether	"	17.0	10.7

then the work required to separate one liquid from the other against the acting interfacial tension is given by

$$w_a = \gamma_A + \gamma_B - \gamma_{AB} \tag{12}$$

where w_a is now the *work of adhesion*. These two types of work lead to a very important quantity, called the *spreading coefficient*, and given by

$$S_{BA} = w_a - w_{c_B} \tag{13a}$$
$$= \gamma_A - \gamma_B - \gamma_{AB} \tag{13b}$$

S_{BA} is the coefficient for the spreading of liquid B on the surface of A, while w_{c_B} is the work of cohesion of liquid B. A positive value of S_{BA} indicates that when a small quantity of liquid B is placed on the surface of A, it will spread across the surface like oil on water. On the other hand, when S_{BA} is negative no spreading will take place, and the added liquid will remain as a drop on the surface.

THE SURFACE TENSION OF SOLUTIONS

The effect of dissolved substances on the surface tension of the solvent is exemplified by the three types of curves shown in Figure 20-3. In solutions of Type I addition of solute leads to an increase in surface tension, but the increase is generally not large. Such behavior is exhibited by strong electrolytes, sucrose, and aminobenzoic acid in water, or aniline in cyclohexane. On the other hand, with nonelectrolytes or weak electrolytes in water the behavior most often encountered is that given by

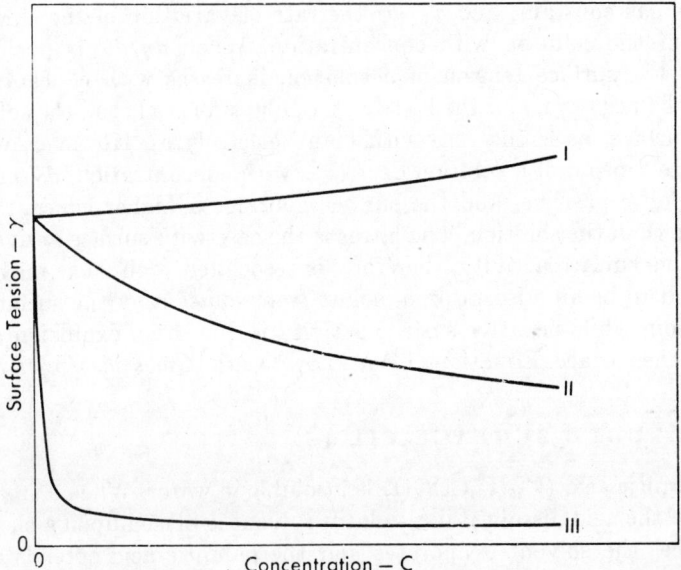

Figure 20-3.   Dependence of surface tension of solutions on solute concentration.

curves of Type II. Here the solutions exhibit surface tensions which decrease regularly and more or less gradually with increase in solute concentration. Finally, Type III curves are given by aqueous solutions of soaps, certain sulfonic acids and sulfonates, and other types of organic compounds. These substances, called *surface active agents*, possess the ability to lower the surface tension of water to a low value even at very low concentrations. Thus, a concentration of 0.0035 mole of sodium oleate per liter of water is sufficient to lower the surface tension of the latter from 72 to about 30 dynes cm^{-1} at 25°C.

Solutes which give a decrease of surface tension with increase in concentration are said to exhibit positive surface activity, while those which lead to an increase in surface tension are said to have a negative surface activity.

In 1878 J. Willard Gibbs showed surface activity to be due to unequal distribution of a solute between the surface and the body of a solution. By a purely thermodynamic argument he deduced that, if a solute distributes itself so that the surface contains q moles of solute per 1 cm² of area in excess of that present in the body of the solution, then for dilute solutions q at equilibrium should be given by

$$q = -\frac{C}{RT}\frac{d\gamma}{dC} \tag{14}$$

where C is the concentration of the solution, T the absolute temperature, R the gas constant, and $d\gamma/dC$ the rate of variation of the surface tension of the solution with concentration. When $d\gamma/dC$ is positive, i.e., when the surface tension of a solution increases with concentration, q must be negative, and the body of the solution is richer in the solute than the surface, as is the case with many electrolytes. However, when the surface tension of a solution decreases with concentration, $d\gamma/dC$ is negative, q is positive, and the surface contains a higher concentration of solute than the solution. The latter is the case with surface active agents. Positive surface activity, therefore, is associated with what may be considered to be an adsorption of solute from solution by the surface of the solution, while negative surface activity is due to an expulsion of solute from the surface. Equation (14) is known as the *Gibbs adsorption equation*.

INSOLUBLE SURFACE FILMS

Palmitic acid ($C_{15}H_{31}COOH$) is insoluble in water. When a small quantity of the acid dissolved in a volatile solvent is placed upon a clean water surface, the solvent evaporates and the palmitic acid spreads over the surface until a film only one molecule thick results. The surface tension, γ, of the surface covered by this unimolecular film is less than that of pure water, γ_0. The difference between the two, namely,

$$f = \gamma_0 - \gamma \tag{15}$$

is defined as the *surface pressure*, f, which is the force acting per unit length of the film.

Unimolecular films analogous to those of palmitic acid can be obtained with many substances which are insoluble or only slightly soluble in water. The surface pressure of such films depends on the area occupied per molecule. In 1917 Irving Langmuir showed how the surface pressure can be determined as a function of molecular area by means of a *surface balance*. This apparatus consists essentially of a shallow trough containing water. Near one end of the trough is a fixed barrier extending across the trough and floating on the water. The barrier is equipped with a mechanism for determining any horizontal pressure exerted upon it. Near the other end of the trough is a movable bar also touching the water and capable of compressing the film against the fixed barrier. If the number of molecules placed on the surface is known, the area they occupy is given by the area between the fixed and movable barriers. Further, the total

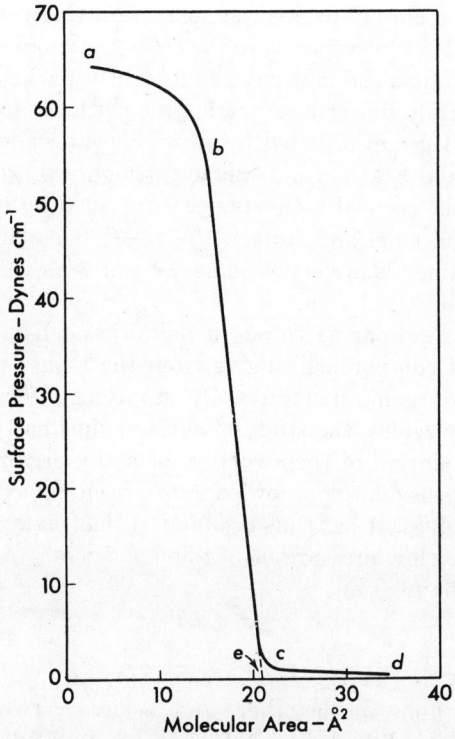

Figure 20-4. Force-area plot for palmitic acid on water at 16°C.

force exerted on the fixed bar divided by its length gives the surface pressure. By varying the positions of the movable bar and observing the corresponding pressures, it is possible to obtain f values for various areas, σ, occupied by a molecule in the unimolecular surface film.

An f–σ plot for palmitic acid films on water at 16°C is shown in Figure 20-4. At large molecular areas, as along cd, the surface pressure is low. Between c and b the pressure climbs rapidly on compression and then tapers off again along ab. The reasons for the observed behavior are as follows: For areas between d and c the molecules on the surface are sufficiently separated from each other to allow considerable compression without appreciable increase in surface pressure. However, as point c is approached the unimolecular film becomes tightly packed, and hence further reduction in area requires considerable force. As a consequence the pressure rises sharply along cb. Finally, between b and a the molecules in the film are so tightly packed that any attempt at still greater reduction of the area causes the film to buckle. The extension of the line bc to $f = 0$, point e, is taken, therefore, to be the area occupied by a molecule of palmitic acid in a closely packed monolayer on the water surface.

This area of 21 Å² corresponds to the cross section of the linear palmitic acid molecule. Hence the molecules in a tightly packed monolayer must be aligned vertically like straws in a bundle, with the hydrophilic (water-loving) —COOH group oriented towards the water and the hydrophobic (water-hating) hydrocarbon end sticking straight up. That this is the correct picture of the packed monolayer is confirmed by calculations of the length of the molecule, and from it the C—C bond distance. The bond length thus obtained is in very good agreement with values deduced from X-ray diffraction.

Measurements similar to those on palmitic acid have been made on various types of compounds ranging from the higher paraffin hydrocarbons to complex organic structures. By supplying information on the size and shape of molecules, the study of surface films has proved very valuable in the elucidation of the structure of many organic compounds. At the same time considerable knowledge has been gained on the nature of the films themselves. It has thus been found that many substances on the surface behave at low pressures as if the films were two-dimensional ideal gases obeying the relation

$$f\sigma = kT \tag{16}$$

where $k = R/N$ is the Boltzmann constant, i.e., the gas constant per molecule. Other films, on the other hand, behave as two-dimensional non-ideal gases, or even liquids or solids. In fact, detailed f-σ plots for some substances resemble strikingly the isothermals for carbon dioxide shown in Figure 1–10.

ADSORPTION

Attention has already been directed to the fact that the molecular forces at the surface of a liquid are in a state of unbalance or unsaturation. The same is true of the surface of a solid, where the molecules or ions in the surface of a crystal do not have all their forces satisfied by union with other particles. As a result of this unsaturation, solid and liquid surfaces tend to satisfy their residual forces by attracting onto and retaining on their surfaces gases or dissolved substances with which they come in contact. This phenomenon of concentration of a substance *on the* surface of a solid or liquid is called *adsorption*. The substance thus attracted to a surface is said to be the *adsorbed phase*, while the substance to which it is attached is the *adsorbent*.

*Ad*sorption should be carefully distinguished from *ab*sorption. In the latter process a substance is not only retained on the surface, but passes through the surface to become distributed throughout the body of a solid or liquid. Thus water is *ab*sorbed by a sponge, or water vapor is *ab*sorbed by anhydrous calcium chloride to form a hydrate; but acetic acid in solution and various gases are *ad*sorbed by charcoal. Where doubt exists as to whether a process is true adsorption or absorption, the noncommittal term *sorption* is sometimes employed.

ADSORPTION OF GASES BY SOLIDS

Although it is probable that all solids adsorb gases to some extent, adsorption as a rule is not very pronounced unless an adsorbent possesses a large surface for a given mass. For this reason silica gel and charcoals obtained from various sources, such as wood, bone, coconut shells, and lignite, are particularly effective as adsorbing agents. These substances have a very porous structure and with their large exposed surfaces can take up appreciable volumes of various gases. The extent of adsorption can further be increased by "activating" the adsorbents in various ways. Thus wood charcoal can be "activated" by heating between 350 and 1000°C in a vacuum or in air, steam, and certain other gases to a point where the adsorption of carbon tetrachloride at 24°C can be increased from 0.011 g per gram of charcoal to 1.48 g. The activation involves apparently a distilling out of hydrocarbon impurities from a charcoal and leads thereby to exposure of a larger free surface for possible adsorption.

The amount of gas adsorbed by a solid depends on the natures of the adsorbent and gas being adsorbed, the area of the adsorbent, the temperature, and the pressure of the gas. The specificity with which certain gases are adsorbed by a given solid can be seen from Table 20–3, where the volumes of various gases adsorbed by a gram of charcoal at 15°C are given. The volumes of gas have all been reduced to 0°C and 1 atm pressure. A comparison of the relative volumes of various gases adsorbed by a solid reveals that in general the extent of adsorption parallels the increase in the critical temperature of the gases. This parallelism suggests that gases which liquefy easily are more readily adsorbed, but it does not necessarily indicate that the gases exist as liquids on the surface. A similar correlation is obtained with boiling points.

As may be expected, an increase in the surface area of adsorbent increases the total amount of gas adsorbed. Since the surface area of adsorbing agents cannot always be determined readily, common practice is to employ the mass of adsorbent as a measure of the surface available and to express the amount of adsorption per unit mass of adsorbing agent used.

In adsorption a true equilibrium is established between the gas in contact with a solid and the gas on the surface; i.e., for a given gas and adsorbent the extent of adsorption under any condition of temperature and pressure is definite and reproducible. Like all equilibria, the adsorption process is strongly affected by temperature. Invariably an increase in temperature leads to a decrease in the amount adsorbed, and vice

TABLE 20–3.   ADSORPTION OF GASES BY CHARCOAL AT 15°C
(1 g of adsorbent)

Gas	Volume Adsorbed (cc)	Critical Temperature (°K)
H_2	4.7	33
N_2	8.0	126
CO	9.3	134
CH_4	16.2	190
CO_2	48	304
HCl	72	324
H_2S	99	373
NH_3	181	406
Cl_2	235	417
SO_2	380	430

versa. Thus, at 600 mm pressure a gram of charcoal adsorbs about 10 cc of nitrogen at 0°C, 20 cc at −29°C, and 45 cc at −78°C. These data indicate that adsorption of gas by a solid is accompanied by evolution of heat which is called the *heat of adsorption*.

TYPES OF ADSORPTION

Study of adsorption of various gases on solid surfaces has revealed that the forces operative in adsorption are not the same in all cases. Two types of adsorption are generally recognized, namely, *physical* or van der Waals adsorption and *chemical* or activated adsorption. Physical adsorption is characterized by low heats of adsorption, of the order of 10,000 cal or less per mole of adsorbate, and by the fact that the adsorption equilibrium is reversible and is established rapidly. This is the type observed in the adsorption of various gases on charcoal. The forces responsible for the adsorption here are of the same kind as are involved in deviations of gases from ideal behavior and in liquefaction, i.e., van der Waals forces. On the other hand, activated or chemical adsorption is accompanied by much higher heat changes, ranging from 20,000 to as high as 100,000 cal, and leads to a much firmer attachment of the gas to the surface. As these heats are of the same order of magnitude as those involved in chemical reactions, it is quite certain that the adsorption consists of a combination of gas molecules with the surface to form a surface compound. In the adsorption of oxygen on tungsten it has actually been found that tungsten trioxide distills from the surface at about 1200°K; however, even above this temperature oxygen remains on the surface apparently as WO. As other examples of chemical adsorption may be mentioned that of carbon monoxide on tungsten, oxygen on silver, gold, platinum, and carbon, and hydrogen on nickel.

Many cases of adsorption are neither one type nor the other, but a combination of both. Again, some systems show physical adsorption at low temperatures and chemisorption as the temperature is raised. This is true with the adsorption of hydrogen on nickel. In general chemical adsorption is more specific in nature than physical and is to be found only where there is tendency toward compound formation between a gas and adsorbent. However, since van der Waals forces are not specific in nature, physical adsorption may be found in all instances, although it may possibly be masked by the stronger chemical type.

ADSORPTION ISOTHERMS

The relation between the amount of substance adsorbed by an adsorbent and the equilibrium pressure or concentration at constant temperature is called an *adsorption isotherm*. Five general types of isotherms have been observed in the adsorption of gases on solids. These are shown in Figure 20–5. In cases of chemisorption only isotherms of type I are encountered, while in physical adsorption all five types occur.

In isotherms of type I the amount of gas adsorbed per given quantity of adsorbent increases relatively rapidly with pressure and then much more slowly as the surface becomes covered with gas molecules. To represent the variation of the amount of adsorption per unit area or unit mass with pressure, Freundlich proposed the equation

$$y = kP^{1/n} \tag{17}$$

Here y is the weight or volume of gas adsorbed per unit area or unit mass of adsorbent, P is the equilibrium pressure, and k and n are empirical constants dependent on the nature of solid and gas and on the temperature. This equation may be tested as follows. Taking logarithms of both sides, Eq. (17) becomes

$$\log_{10} y = \log_{10} k + \frac{1}{n} \log_{10} P \tag{18}$$

If $\log_{10} y$ is plotted now against $\log_{10} P$, a straight line should result with slope equal to $1/n$ and ordinate intercept equal to $\log_{10} k$. Figure 20–6 shows such a plot for the adsorption of nitrogen on mica at 90°K. In this plot y is in mg/cm², while P is in dynes/cm². Although the requirements of

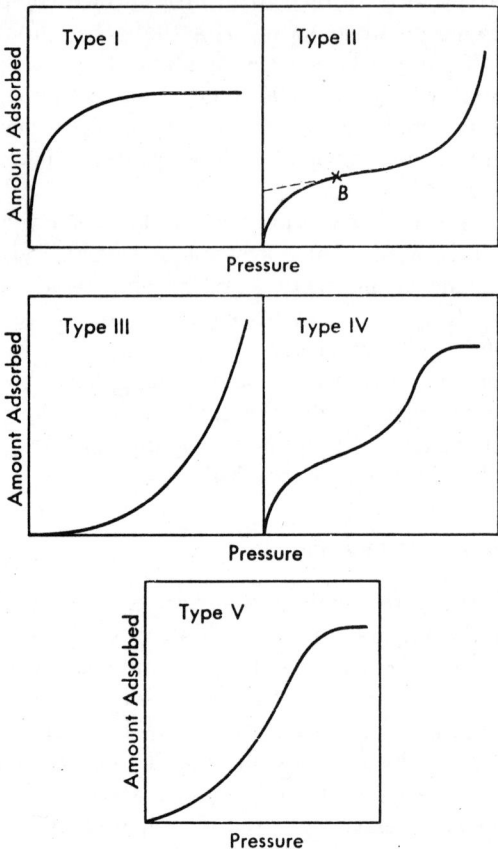

Figure 20–5.   Types of adsorption isotherms.

the equation are met satisfactorily at the lower pressures, at higher pressures the experimental points curve away from the straight line, indicating that this equation does not have general applicability in reproducing adsorption of gases by solids.

THE LANGMUIR ADSORPTION EQUATION

A much better equation for type I isotherms was deduced by Irving Langmuir[2] from theoretical considerations. Langmuir postulated that

[2] I. Langmuir, *J. Am. Chem. Soc.*, **38**, 2221 (1916); **40**, 1316 (1918).

gases in being adsorbed by a solid surface cannot form a layer more than a *single molecule* in depth. Further, he visualized the adsorption process as consisting of two opposing actions, a condensation of molecules from the gas phase onto the surface and an evaporation of molecules from the surface back into the body of the gas. When adsorption first starts, every molecule colliding with the surface may condense on it. However, as adsorption proceeds, only those molecules may be expected to be adsorbed which strike a part of the surface not already covered by adsorbed molecules. The result is that the initial *rate* of condensation of molecules on a

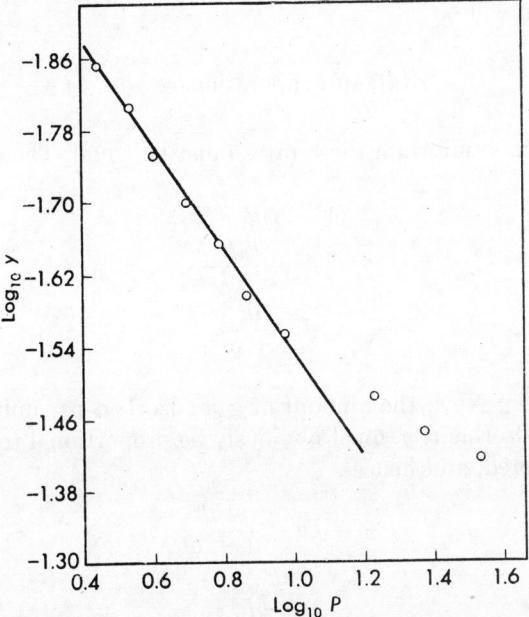

Figure 20–6. Application of Freundlich equation to adsorption of N_2 on mica at 90°K.

surface is highest and falls off as the area of surface available for adsorption is decreased. On the other hand, a molecule adsorbed on a surface may, by thermal agitation, become detached from the surface and escape into the gas. The rate at which desorption will occur will depend, in turn, on the amount of surface covered by molecules and will increase as the surface becomes more fully saturated. These two rates, condensation and desorption, will eventually become equal, and when this happens an adsorption equilibrium will be established.

These ideas can be formulated mathematically. If we let θ be the fraction of the total surface covered by adsorbed molecules at any instant, then the fraction of surface bare and available for adsorption is $(1 - \theta)$. Since, according to kinetic theory, the rate at which molecules strike unit area of a surface is proportional to the pressure of the gas, the rate of condensation of molecules should be determined both by the pressure and the fraction of surface bare, or

$$\text{Rate of condensation} = k_1(1 - \theta)P$$

where k_1 is a constant of proportionality. On the other hand, if we let k_2 be the rate at which molecules evaporate from unit surface when the surface is fully covered, then for a fraction θ of surface covered the rate of evaporation will be

$$\text{Rate of evaporation} = k_2\theta$$

For adsorption equilibrium these rates must be equal. Therefore,

$$k_1(1 - \theta)P = k_2\theta$$
$$\theta = \frac{k_1P}{k_2 + k_1P}$$
$$= \frac{bP}{1 + bP} \tag{19}$$

where $b = k_1/k_2$. Now, the amount of gas adsorbed per unit area or per unit mass of adsorbent, y, must obviously be proportional to the fraction of surface covered, and hence

$$y = k\theta = \frac{kbP}{1 + bP}$$
$$= \frac{aP}{1 + bP} \tag{20}$$

where the constant a has been written for the product kb.

Equation (20) is the *Langmuir adsorption isotherm*. The constants a and b are characteristic of the system under consideration and are evaluated from experimental data. Their magnitude depends also on the temperature. At any one temperature the validity of the Langmuir adsorption equation can be verified most conveniently by first dividing both sides of Eq. (20) by P and then taking reciprocals. The result is

$$\frac{P}{y} = \frac{1}{a} + \left(\frac{b}{a}\right)P \tag{21}$$

Since a and b are constants, a plot of P/y vs. P should yield a straight line with slope equal to b/a and an ordinate intercept equal to $1/a$. Figure

20–7 gives such a plot of the same data as used for Figure 20–6. This excellent straight line confirms the Langmuir adsorption equation and indicates its superiority to the Freundlich isotherm. Furthermore, this test, and many similar ones, lend support to the correctness of Langmuir's mechanism of the adsorption process and to his assumption that the adsorbent is covered only by a unimolecular layer of gas molecules.

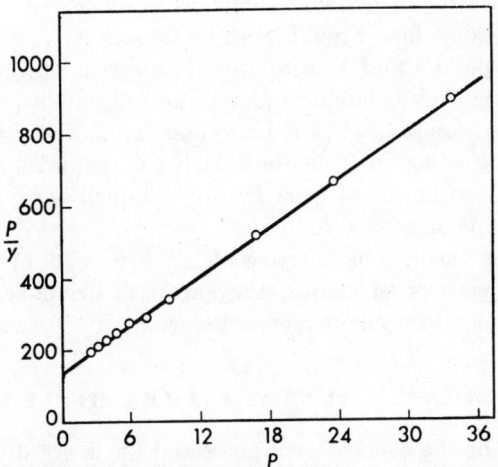

Figure 20–7.   Application of Langmuir equation to adsorption of N_2 on mica at 90°K.

TYPES II–V ISOTHERMS

The explanation proposed for type II and type III isotherms is that in these the adsorption is *multimolecular*, i.e., the adsorption involves the formation of many molecular layers on the surface rather than a single one. On this postulate Brunauer, Emmett, and Teller[3] derived for these two types of isotherms the relation

$$\frac{P}{v(P^0 - P)} = \frac{1}{v_m c} + \left(\frac{c - 1}{v_m c}\right)\frac{P}{P^0} \tag{22}$$

In this equation v is the volume, reduced to standard conditions, of gas adsorbed at pressure P and temperature T, P^0 the saturated vapor pressure of the adsorbate at temperature T, v_m the volume of gas, reduced to standard conditions, adsorbed when the surface is covered with a unimolecular layer, and c is a constant at any given temperature equal

---

[3] Brunauer, Emmett, and Teller, *J. Am. Chem. Soc.*, **60**, 309 (1938); **62**, 1723 (1940); S. Brunauer, *The Adsorption of Gases and Vapors*, Princeton University Press, Princeton, 1943.

approximately to

$$c = e^{(E_1 - E_L)/RT} \tag{23}$$

Here E_1 is the heat of adsorption of the first layer, and E_L is the heat of liquefaction of the gas. The isotherms of type II follow when $E_1 > E_L$ and those of type III when $E_1 < E_L$.

Equation (22) can be tested by plotting $P/v(P^0 - P)$ vs. P/P^0. The plot should be a straight line with slope given by $(c - 1)/v_m c$ and intercept by $1/v_m c$. From these v_m and c can be found.

To explain types IV and V isotherms it has been suggested that substances exhibiting such behavior undergo not only multilayer adsorption but also condensation of gas in the pores and capillaries of the adsorbent. The two types arise again from the relative magnitudes of E_1 and E_L. When $E_1 > E_L$, isotherms of type IV are obtained here, whereas when $E_1 < E_L$ isotherms of type V follow.

Although these theories have proved fairly successful in explaining the more complex types of isotherms, they are still insufficient to account quantitatively for all the phenomena observed.

DETERMINATION OF SURFACE AREA OF ADSORBENTS

The question of the surface area possessed by finely divided solids is important not only in adsorption, but also in contact catalysis and many other fields. Brunauer, Emmett, and Teller[4] showed how it is possible to use the adsorption of gases by such materials for determination of their surface areas, and thus furnished a very powerful tool being very widely used at present.

The B-E-T method is based on the postulate, for which there is considerable justification, that in adsorption of gases exhibiting type II isotherms point B in Figure 20–5 corresponds to the adsorbed volume necessary to yield a monolayer of gas on the surface. This point is equal to v_m in Eq. (22). If such be the case, then the area of the solid per definite weight of adsorbent is given by

$$\Sigma = \left(\frac{P_0 v_B}{RT_0}\right) NS \tag{24}$$

In this equation Σ is the area in $\mathring{A}^2$, $P_0 = 1$ atm, $T_0 = 273.2°K$, R is the gas constant, v_B the volume corresponding to point B, N Avogadro's number, and S the area occupied on the surface by a single gas molecule. The general practice is to convert Σ to square meters per gram or to acres per pound.

---

[4] Brunauer, Emmett and Teller, *J. Am. Chem. Soc.*, **57**, 1754 (1935); **59**, 1553 (1937); **59**, 2682 (1937).

For area determinations the gas most commonly used is nitrogen at its normal boiling point ($-195.8°C$) or at liquid air temperature ($-183°C$). At these temperatures the area of the nitrogen molecule is generally taken to be 16.2 Å^2.

Another method for determining the areas of solids was proposed by Harkins and Jura.[5] This method involves plotting $\log_{10} P/P^0$ vs. $1/v^2$, where all the symbols have the same significance as before, and taking the slope of the linear portion of the curve. The area follows then from the slope, A, as

$$\Sigma = k \sqrt{-A} \qquad (25)$$

where k is a constant for a given gas and temperature. For areas in square meters per gram of adsorbent k is equal to 4.06 for N_2 at $-195.8°C$, 13.6 for n-butane at $0°C$, 16.9 for n-heptane at $25°C$, and 3.83 for water vapor at the same temperature. This method gives generally very good agreement with the results obtained by the B-E-T method.

More recently, Maron, Bobalek, and Fok[6] described a procedure for the determination of the surface area of carbon blacks by adsorption of soap from aqueous solution. The method involves the ascertaining of the amount of soap required to cover the surface of a unit mass of the black with a monolayer of soap. From a knowledge of this quantity and the area occupied on the surface by a soap molecule, the surface area of the carbon black can be calculated.

ADSORPTION OF SOLUTES BY SOLIDS

Solid surfaces also can adsorb dissolved substances from solution. When a solution of acetic acid in water is shaken with activated carbon, part of the acid is removed by the carbon and the concentration of the solution is decreased. Similarly, activated carbon can be used to remove ammonia from solutions of ammonium hydroxide, phenolphthalein from solutions of acids or bases, etc. Again, freshly precipitated silver chloride tends to adsorb either silver or chloride ions, depending on which are in excess, while arsenic trisulfide tends to adsorb sulfide ions from solutions in which they are precipitated.

As a rule activated carbon is much more effective in adsorbing nonelectrolytes from a solution than electrolytes, and the extent of adsorption is usually the greater the higher the molecular weight of the adsorbate. Conversely, inorganic solids tend to adsorb electrolytes more readily than nonelectrolytes. This tendency of adsorbents to attract certain substances

[5] Harkins and Jura, *J. Am. Chem. Soc.*, **66**, 1366 (1944).
[6] Maron, Bobalek, and Fok, *J. Colloid. Sci.*, 11, 21 (1956).

in preference to others occasionally leads to the phenomenon of *negative adsorption*, i.e., the concentration of a solute is actually increased after treatment with the adsorbing agent. Dilute solutions of potassium chloride agitated with blood charcoal offer a case in point. The explanation suggested for negative adsorption is that the solvent, in this case water, is adsorbed in preference to the electrolyte, and as a consequence the concentration of the solute is raised. However, at high concentrations potassium chloride exhibits positive adsorption, with the salt being adsorbed rather than the water.

Adsorption from solution follows generally the principles laid down for adsorption of gases and is subject to the same factors. As was pointed out, some adsorbents are specifically more effective in attracting certain substances to their surface than others. Again, an increase in temperature operates to decrease the extent of adsorption, while an increase in surface area increases it. Adsorption of solutes, like that of gases, involves the establishment of an equilibrium between the amount adsorbed on the surface and the concentration of the substance in solution. The variation of extent of adsorption with concentration of solute is usually represented by the Freundlich equation, which seems to work better with this type of adsorption than with gases. For this purpose, Eq. (17) is written in the form

$$y = kC^{1/n} \qquad (24)$$

where y is the mass of substance adsorbed per unit mass of absorbent, C is the equilibrium concentration of the solute being adsorbed, while k

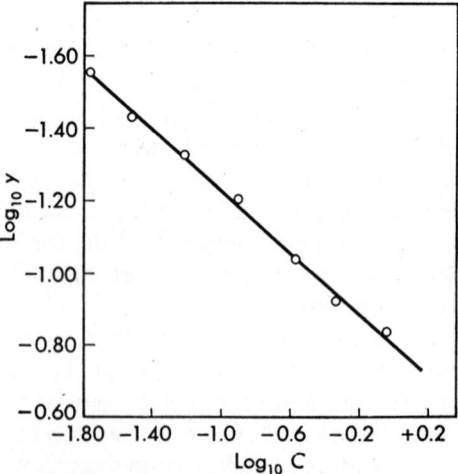

Figure 20–8.   Application of Freundlich equation to adsorption of acetic acid on charcoal at 25°C.

and n are again empirical constants. By taking as before logarithms of both sides of Eq. (26), we obtain

$$\log_{10} y = \log_{10} k + \frac{1}{n} \log_{10} C \tag{27}$$

and hence a plot of $\log_{10} y$ vs. $\log_{10} C$ should be linear with slope equal to $1/n$ and intercept equal to $\log_{10} k$. Figure 20–8 shows such a plot of data on the adsorption at 25°C of acetic acid from aqueous solution by blood charcoal. In the figure y is expressed in grams of acid adsorbed per gram of charcoal, while C is in moles per liter. From the slope and intercept of the satisfactory straight line we find $n = 2.32$, $k = 0.160$, and hence these data may be represented at 25°C by the equation $y = 0.160\ C^{1/2.32}$.

APPLICATIONS OF ADSORPTION

Adsorption finds very extensive application both in the research laboratory and in industry. Adsorption of gases on solids is utilized in the laboratory to preserve the vacuum between walls of Dewar containers designed for the storage of liquid air or liquid hydrogen. Activated charcoal, placed between the walls, tends to adsorb any gases which appear due to glass imperfections or diffusion through the glass. Again, adsorption plays a very important part in various aspects of catalysis of gaseous reactions by solid surfaces. This subject will be developed in more detail below. Further, all gas masks are merely devices containing an adsorbent or a series of adsorbents which, by preferentially removing poisonous gases from the atmosphere, purify the air for breathing. In a similar manner various adsorbents are used in industry to recover solvent vapors from air, or particular solvents from mixtures of other gases.

As applications of adsorption from solution may be mentioned the clarification of sugar liquors by charcoal, the removal of coloring matter from various other types of solutions, and the recovery of dyes from dilute solutions in a number of solvents. Adsorption has also been used for the recovery and concentration of vitamins and other biologic substances and finds utility now in a method called *chromatographic analysis*. This scheme for estimation of small quantities of substance depends on progressive and selective adsorption of a number of constituents present simultaneously in a solution or gas.

Surface active agents find extensive application in detergents, paints, waterproofing, lubrication, and a host of other fields. All surfaces are covered with layers of either gaseous, liquid, or solid films, and the problem of displacing these frequently becomes very important. Substances that can displace these adhering materials are called *wetting agents*. Wetting agents lower the interfacial tensions by preferential adsorption and permit thereby the wetting of a surface by a liquid. When the wetting involves also a dispersion of the displaced film, we have a *detergent*, i.e., a substance suitable for removal of dirt and grime. Typical among the wetting agents used commercially are aliphatic alcohols, sulfonated higher alcohols, sulfonated alkyl naphthalenes, and soaps of various kinds.

CATALYSIS OF GASEOUS REACTIONS BY SOLID SURFACES

Many gaseous reactions are catalyzed by solids. The feature characterizing all such reactions is the fact that they proceed not in the gas phase but at the surface of the solid catalyst. For this reason this type of process is frequently referred to as *contact catalysis*. The catalytic reaction involves at least four distinct steps, namely: (a) adsorption of the reacting gases on the surface of the catalyst; (b) activation of the adsorbed reactants; (c) reaction of the activated gas in the adsorbed phase; and (d) diffusion of the products of the reaction from the surface into the gas phase. Although any one of these processes may be slowest and therefore rate-determining, it has been found possible to approach the kinetics of such reactions from the standpoint of adsorption and to formulate rate equations in terms of principles based on the Langmuir adsorption isotherm.

The theory of contact catalysis starts with the Langmuir concept that a gas is adsorbed on the surface of a solid in a layer *one molecule deep*. Further, the entire surface need not be covered with adsorbed gas. In fact, parts may be covered and parts may be bare, but it is only the covered portions which will lead to reaction and hence will determine the rate. If, then, we are to employ the law of mass action to set up the rate equations of such reactions, we must consider not the concentration or pressure of the reactants in the gas phase, but their concentration in the adsorbed film. The function of the gas phase is merely to control the concentration of molecules in the adsorbed layer. Since the latter concentration is at any instant proportional to the *fraction of surface covered*, the rate of reaction must also be proportional to this fraction, and hence

$$\text{Rate} = k'\theta \tag{28}$$

where θ is the fraction of surface covered by reactant, while k' is a proportionality constant.

We have seen that for any adsorbed gas θ is given by

$$\theta = \frac{bP}{1 + bP} \tag{29}$$

where b is a constant and P is the pressure of the gas in contact with the adsorbent. This is the expression for θ to be inserted into Eq. (28). However, in two limiting cases Eq. (29) can be simplified as follows. When a gas is only weakly adsorbed b is small, and bP may be disregarded compared with unity. Under these conditions θ becomes

$$\theta = bP \tag{30}$$

i.e., *the fraction of the surface covered when a gas is only slightly adsorbed is directly proportional to the pressure.* Again, when a gas is strongly adsorbed, as when either b or P or both are large, then $bP \gg 1$, and Eq. (29) reduces to

$$\theta = \frac{bP}{bP} = 1 \tag{31}$$

Therefore, *when a gas is strongly adsorbed the surface is practically completely covered with a film of gas at all times, and $\theta = 1$.*

Another deduction which can be made from Eq. (29) is also of importance for our purposes. Since θ is the fraction of surface covered, $(1 - \theta)$, the fraction of surface bare, is

$$(1 - \theta) = 1 - \left(\frac{bP}{1 + bP}\right)$$
$$= \frac{1}{1 + bP} \tag{32}$$

For strong adsorption $bP \gg 1$, and hence

$$(1 - \theta) = \frac{1}{bP} = \frac{b'}{P} \tag{33}$$

From Eq. (33) we see that, when a gas is strongly adsorbed on a surface, *the fraction of the latter still free and available for adsorption of other gases is inversely proportional to the pressure of the adsorbed gas.*

With the aid of these equations we are prepared now to illustrate the kinetics of heterogeneous reactions in a number of simple cases. The examples will be confined to reactions in which only a single gas participates as a reactant. For more complex situations the student may refer to the book by Schwab, Taylor, and Spence listed at the end of the chapter.

ONE REACTANT GAS SLIGHTLY ADSORBED

Consider a gas A decomposing on the surface of a solid catalyst according to the reaction

$$A \longrightarrow \text{Products} \tag{34}$$

Let θ_A be the fraction of the surface covered by A at time t and pressure P. Then according to Eq. (28) the rate of the reaction should be

$$\text{Rate} = k'\theta_A$$

When A is only slightly adsorbed, θ_A is given by Eq. (30). Hence

$$-\frac{dP}{dt} = k'bP = kP \tag{35}$$

where $k = k'b$ is the rate constant of the reaction. On integration Eq. (35) becomes

$$k = \frac{2.303}{t} \log_{10}\left(\frac{P_i}{P}\right) \tag{36}$$

Here P_i is the initial pressure of A, and P its pressure at any time t. This equation is identical in form with the equation for a first order homogeneous reaction. For this reason Eqs. (35) or (36) are said to represent *first order* heterogeneous reactions.

TABLE 20–4.   DECOMPOSITION OF N_2O
ON GOLD AT 900°C

t (sec)	P (mm Hg)	k (sec^{-1})
0	200	—
900	167	2.01×10^{-4}
1800	136	2.15
3180	100	2.18
3900	86	2.17
4800	70	2.19
6000	54	2.18
7200	44	2.10

Quite a number of reactions have been found to follow Eq. (36). Among these may be mentioned the decomposition of phosphine on glass, porcelain, and silica; the decomposition of formic acid vapors on glass, platinum, and rhodium; the decomposition of nitrous oxide on gold; and the dissociation of hydrogen iodide on platinum. In Table 20–4 are shown some data for the dissociation of nitrous oxide on gold[7] which may be considered typical of all such reactions.

[7] Hinshelwood and Pritchard, *Proc. Roy. Soc.*, 108A, 211 (1925).

ONE REACTANT GAS STRONGLY ADSORBED

Since for a strongly adsorbed reactant $\theta = 1$, the rate of the reaction is given by

$$-\frac{dP}{dt} = k \tag{37}$$

i.e., the rate of the reaction is *constant and independent of pressure.* A reaction obeying Eq. (37) is said to be of *zero order.* On integration Eq. (37) becomes

$$k = \frac{P_i - P}{t} \tag{38}$$

Hinshelwood and Burk[8] found the decomposition of ammonia on tungsten to be practically independent of the initial pressure and to proceed in accordance with equation (37). This may be seen from the results quoted in Table 20-5. The same behavior has been observed in the decomposition of ammonia on the surfaces of molybdenum and osmium, and in the dissociation of hydrogen iodide on gold.

TABLE 20-5.   DECOMPOSITION OF NH_3
ON TUNGSTEN AT 856°C
$(P_i = 200$ mm Hg$)$

t (sec)	$P_i - P$ (mm Hg)	k (mm Hg sec^{-1})
100	14	0.14
200	27	0.14
300	38	0.13
400	48	0.12
500	59	0.12
1000	112	0.11

ONE REACTANT GAS MODERATELY ADSORBED

When the reactant is moderately adsorbed, the rate of reaction is proportional to the fraction of surface covered as given by Eq. (29), namely,

$$-\frac{dP}{dt} = k'\left(\frac{bP}{1 + bP}\right) = \frac{kP}{1 + bP} \tag{39}$$

This equation can frequently be approximated by the expression

$$-\frac{dP}{dt} = kP^n \tag{40}$$

[8] Hinshelwood and Burk, *J. Chem. Soc.*, **127**, 1116 (1925).

which makes the rate proportional to the amount adsorbed as given by the Freundlich isotherm. Stock and Bodenstein[9] used Eq. (40) to represent the decomposition of stibine on antimony at 25°C, while Eq. (39) was found to hold in the decomposition of nitrous oxide on indium oxide.[10]

RETARDED REACTIONS

In some reactions the catalytic surface adsorbs not only the reactant but also a product of the reaction. In fact, the product may be adsorbed more strongly than the reactant. Since a strongly adsorbed product diminishes the surface available for condensation of reactant, the fraction of the surface covered by the latter must decrease, and thereby also the rate of reaction. Hence *the effect of a strongly adsorbed product is to retard the reaction producing it.*

Consider specifically the reaction

$$A \longrightarrow B + C \tag{41}$$

where the reactant A is weakly adsorbed, while the product B is strongly adsorbed. Let P_A and P_B be the pressures of A and B respectively at time t, and let $(1 - \theta_B)$ be the fraction of the surface not covered by B. The rate of reaction of A will then be proportional to the rate at which molecules of A can become attached to the surface, namely, to the fraction of surface exposed and the pressure of A. Hence

$$-\frac{dP_A}{dt} = k'(1 - \theta_B)P_A \tag{42}$$

But when a gas is strongly adsorbed, $(1 - \theta)$ is given by Eq. (33), which in this case takes the form $(1 - \theta_B) = b'/P_B$. Substituting this value of $(1 - \theta_B)$ into Eq. (42), we obtain for the rate of reaction

$$-\frac{dP_A}{dt} = \frac{k'b'P_A}{P_B}$$
$$= \frac{kP_A}{P_B} \tag{43}$$

i.e., the rate is directly proportional to the pressure of the reactant and *inversely* to the pressure of the product responsible for the retardation.

The decomposition of ammonia on platinum at 1138°C was found by Hinshelwood and Burk[11] to conform essentially to Eq. (43). Nitrogen appears to have no effect on the reaction, but the hydrogen formed

[9] Stock and Bodenstein, *Berichte*, **40**, 570 (1907)
[10] Schwab, Staeger, and v. Baumbach, *Z. physik. Chem.*, **21B**, 65 (1933).
[11] Hinshelwood and Burk, *J. Chem. Soc.*, **127**, 1114 (1925).

exhibits a strong retardation, as may be seen from Table 20-6. In this table column 1 gives the decrease in the pressure of ammonia observed in 120 seconds on starting in each case with 100 mm, while column 2 gives the pressures of added hydrogen present initially. For a threefold increase in the pressure of the retarding gas the decomposition is decreased in an equal period from 33 to 10 per cent of the total.

TABLE 20-6. HYDROGEN RETARDATION OF NH_3
DECOMPOSITION ON Pt AT 1138°C
(P_i = 100 mm Hg)

ΔP_{NH_3} in 120 sec (mm)	P_{H_2} initially added (mm)
33	50
27	75
16	100
10	150

When the product responsible for the retardation is moderately rather than strongly adsorbed, Eq. (42) is still applicable, but $(1 - \theta_B)$ is given by Eq. (32). For this case we obtain, therefore,

$$-\frac{dP_A}{dt} = \frac{kP_A}{1 + bP_B} \qquad (44)$$

This equation has been applied to the decomposition of nitrous oxide on platinum, cadmium oxide, cupric oxide, and nickel oxide. The retarding gas is oxygen.

THE ORDER OF HETEROGENEOUS REACTIONS

The order of a heterogeneous reaction is defined as the total power to which the pressures appear in the rate equation. A more suitable criterion of the order n of a hererogeneous reaction is the period of half-life, which is inversely proportional to the $(n - 1)$ power of the initial pressure. By studying the period of half-life of a reaction as a function of the initial pressure, it is possible to evalute $n - 1$, and hence the reaction order.

The value of n so obtained is not necessarily the true order of the reaction as it takes place on the surface. Rather, it is the apparent order ascertained by following the surface reaction indirectly through its influence on the pressure of the gas phase. As we have seen in the examples cited, n may or may not be a whole number. When a single reacting gas is weakly adsorbed, the reaction is first order and $n = 1$. Again, when a single reactant is strongly adsorbed, $n = 0$, and the reaction proceeds independently of the pressure. On the other hand, for the intermediate adsorption involved in the decomposition of stibine on antimony, $n = 0.6$. The last two results would be hard to understand unless it is remembered that in heterogeneous reactions only apparent orders are obtained and that changes may proceed on the surface with orders which may be quite different.

For a single reacting gas only slightly adsorbed, and where products exhibit no retardation, the apparent and true orders may be inferred to be identical. For fractional and zero order reactions the true and apparent orders cannot possibly be the same. In such cases the true order must be deduced from the nature of the reaction and from any other data available. This correlation of reaction orders is further complicated by adsorption of and retardation due to products. In general, the effect of a strongly adsorbed product, i.e., to the extent that the free surface is inversely proportional to its pressure, is to yield an apparent order *one less than the true order*. This can readily be shown by integrating the rate equation of such a process and solving for the time of half-life. Consequently, when an apparent zero order reaction is found to be retarded by a strongly adsorbed product, it may be assumed that the reaction has a true first order.

EFFECT OF TEMPERATURE ON HETEROGENEOUS REACTIONS

As for homogeneous reactions, the influence of temperature on the rate constant k of heterogeneous reactions is given by the Arrhenius equation

$$\frac{d \ln k}{dT} = \frac{E_a}{RT^2} \tag{45}$$

E_a is called here the *apparent energy of activation*. The apparent energy of activation, evaluated from the observed velocity constants, is not necessarily the energy required to activate the reactants on the *surface*, which is the true activation energy. The latter may be modified by the heats of adsorption of reactants, or reactants and products, to yield an apparent activation energy which may be quite different from the true. A discussion of the relation between these two activation energies is given by Hinshelwood.[12]

Comparison of the energies of activation of gaseous reactions at solid surfaces with those of the same reactions proceeding homogeneously shows almost invariably the former to be considerably lower than the latter. Consequently the presence of the catalyst leads to a lowering of the

[12] C. N. Hinshelwood, *Kinetics of Chemical Change*, Oxford University Press, Oxford, 1940.

net energy required for the activation of a given reaction. With this lower energy barrier more molecules can enter reaction in a given time than is possible with the higher energy threshold present in homogeneous reactions. This lowering of the activation energy is brought about by change in the mechanism of the reaction caused by presence and participation of the solid surface in the reaction sequence.

CATALYTIC POISONS

Small quantities of foreign substances added to a reacting system are frequently sufficient to impair seriously the catalytic acitvity of a surface. Such substances are termed *catalytic poisons*. Poisons may be of two kinds, temporary and permanent. In temporary poisoning the surface decreases in activity, or loses it entirely, only for the period that the poison is in contact with it. As soon as the poison is removed from the presence of the catalyst, the activity is restored. The diminution in activity is due in most instances to a strong preferential adsorption of the poison on the catalyst surface. With sufficiently strong adsorption the reactant may be displaced completely, and the entire surface may be covered with an inactive blanket of the poison.

An illustration of temporary poisoning is the retarding effect exercised by carbon monoxide on the hydrogenation of ethylene in presence of copper.[13] Into this class fall also the reactions in which retardation by products takes place. In expressing the rate of such reactions the pressure of the poison appears in the denominator of the rate equation whether the poison takes part in the reaction or not.

Permanent poisoning, on the other hand, involves a chemical interaction between the surface and the poison to form a new surface which is catalytically inert. Activity can be re-established only by chemical rejuvenation. Among the permanent poisons volatile silicon and sulfur compounds are particularly fatal to the life of many catalysts. So are arsenic compounds, especially to platinum.

The very small quantities sufficient to poison a catalytic surface have led H. S. Taylor to suggest that only a fraction of the total exposed surface is catalytically active. He, as well as others, showed that the quantity of poison effective in stopping activity is sometimes too small to cover the whole surface with a film one molecule deep. In his theory of *active centers* Taylor accounts for this by considering the surface as irregular, with high catalytic activity localized in certain spots that are more elevated than others. As soon as those spots become covered with poison, the activity of the surface diminishes to practically zero. Further-

[13] Pease, *J. Am. Chem. Soc.*, **45**, 1196, 2235 (1923); Pease and Stewart, *ibid.*, **47**, 1235 (1925).

more, since those spots are only a small part of the total surface, the amount of substance necessary to poison a catalyst is correspondingly small.

Catalytic poisoning is occasionally very specific and may be taken advantage of to control the products of a reaction. For instance, the decomposition of alcohol in presence of copper proceeds in two stages,

$$CH_3CH_2OH \longrightarrow CH_3CHO + H_2$$
$$CH_3CHO \longrightarrow CH_4 + CO$$

the first of which is uninfluenced by water vapor, while the second is considerably retarded. Therefore, by using alcohol with some water in it, Armstrong and Hilditch[14] were able to increase the yield of acetaldehyde and to prevent materially the formation of undesired methane.

REFERENCES

1. N. K. Adam, *The Physics and Chemistry of Surfaces*, Oxford University Press, New York, 1949.
2. A. W. Adamson, *The Physical Chemistry of Surfaces*, Interscience Publishers, Inc., New York, 1960.
3. J. J. Bikerman, *Surface Chemistry*, Academic Press, New York, 1958.
4. S. Brunauer, *The Adsorption of Gases and Vapors*, Princeton University Press, Princeton, 1943.
5. W. E. Garner, *Chemisorption*, Butterworths Scientific Publications, London, 1957
6. R. H. Griffith, *The Mechanism of Contact Catalysis*, Oxford University Press, New York, 1946.
7. W. D. Harkins, *The Physical Chemistry of Surface Films*, Reinhold Publishing Corporation, New York, 1952.
8. C. N. Hinshelwood, *Kinetics of Chemical Change*, Oxford University Press, New York, 1940.
9. Komarewsky and Riesz, in *Catalytic, Photochemical and Electrolytic Reactions*, Interscience Publishers, Inc., New York, 1956.
10. Schwab, Taylor, and Spence, *Catalysis*, D. Van Nostrand Company, Inc., Princeton, N.J., 1937.
11. A. Weissberger, *Physical Methods of Organic Chemistry*, Interscience Publishers, Inc., New York, 1959, Chaps. 13 and 14.

PROBLEMS

1. The radius of a given capillary is 0.105 mm. A liquid whose density is 0.800 g/cc rises in this capillary to a height of 6.25 cm. Calculate the surface tension of the liquid.

[14] Armstrong and Hilditch, *Proc. Roy. Soc.*, **97A**, 262 (1920).

2. The surface tension of mercury at 0°C is 480.3 dynes/cm while the density is 13.595 g/cc. If it is desired to obtain a *fall* in height of 10.0 cm, what radius of glass capillary tube will have to be used? *Ans.* 0.00720 cm.

3. In measuring the surface tension of a liquid by the drop-weight method, 12 drops of the liquid falling from a tip whose diameter is 0.800 cm are found to weigh 0.971 g. If $\phi = 0.600$ under these conditions, what is the surface tension of the liquid? *Ans.* 52.60 dynes/cm.

4. From the following data calculate the critical temperature of CO_2:

t (°C)	ρ_l (g/cc)	γ (dynes/cm)
0	0.927	4.50
20	0.772	1.16

5. From the following data for CH_3Cl

t (°C)	0	10	20
γ (dynes/cm)	19.5	17.8	16.2
ρ_l (g/cc)	0.955	0.937	0.918
ρ_v (g/cc)	0.00599	0.00820	0.0110

determine the constant k in the Ramsay-Shields equation and the critical temperature of the liquid. *Ans.* $k = 2.05$; $t_c = 140$°C.

6. Using the data given in the preceding problem, determine the constant k' in the Katayama equation and t_c of CH_3Cl.

7. At 20°C the surface tensions of water and mercury are, respectively, 72.8 and 483 dynes cm^{-1}, while the interfacial tension between the two is 375 dynes cm^{-1}. Calculate (a) the work of cohesion of mercury, (b) the work of adhesion, and (c) the spreading coefficient of mercury on water. Will mercury spread on water? *Ans.* No.

8. Using the data given in problem 7, find (a) the work of cohesion of water, and (b) the spreading coefficient of water on mercury. Will water spread over a mercury surface?

9. At 19°C the surface tensions of solutions of butyric acid in water, γ, can be represented accurately by the equation

$$\gamma = \gamma_0 - a \ln (1 + bC)$$

where γ_0 is the surface tension of water, while a and b are constants. Set up the expression for the excess surface concentration q as a function of C.

10. For butyric acid the constants in the preceding problem are $a = 13.1$ and $b = 19.62$. Calculate q at a concentration of 0.20 mole per liter.

Ans. 4.32×10^{-10} mole cm^{-2}.

11. From the data given in problems 9 and 10 calculate the limiting value of q as C becomes large. (*Hint:* Assume $bC \gg 1$.)

12. Assuming that the only molecules present in the surface are those corresponding to the excess, calculate from the result of problem 11 the area in Å^2 occupied by a molecule of butyric acid in the solution surface. *Ans.* 30.5 Å^2.

13. In a surface balance experiment involving the spreading of myristic acid on water, 1.53×10^{-7} mole of the acid was spread over an area 12.00 cm wide and 24.40 cm long. The force exerted on the fixed barrier, whose effective length was 11.20 cm, was found to be 122.0 dynes. Find (a) the surface pressure, and (b) the area in Å^2 occupied by a molecule of the acid.

Ans. (a) 10.9 dynes cm^{-1}; (b) 31.8 Å^2.

14. The density of the straight-chain paraffin hydrocarbon *n*-dodecane ($C_{12}H_{26}$) is 0.751 g cc^{-1} at 20°C. If the cross-sectional area of the molecule is 20.7 Å^2, find (a) the length of the molecule, and (b) the average distance between two adjacent carbon atoms, both in Å.

15. At 25°C and a surface pressure of 0.10 dyne cm^{-1}, lauric acid occupies an area of 3100 Å^2 per molecule on a water surface. Assuming the film to be a two-dimensional ideal gas, calculate the gas constant in ergs mole^{-1} degree^{-1}, and compare the result with the accepted value.

16. The following are data for the adsorption of CO on wood charcoal at 0°C. The pressure P is in mm Hg, while x is the volume of gas in cc, measured at standard conditions, adsorbed by 2.964 g of charcoal.

P	x	P	x
73	7.5	540	38.1
180	16.5	882	52.3
309	25.1		

Find graphically the constants k and n of the Freundlich equation.

Ans. $k = 0.088$ cc/g; $n = 1.26$.

17. Find graphically the constants a and b in the Langmuir equation which fit the data given in problem 16.

18. From the results of the preceding problem calculate the volume of CO at standard conditions adsorbed by 1 g of wood charcoal at 0°C when the partial pressure of CO is 400 mm. *Ans.* 10.5 cc.

19. In a laboratory experiment 50 cc of a solution originally 0.2 molar in CH_3COOH was permitted to come to equilibrium with a 5 g sample of charcoal. At equilibrium 25 cc of the CH_3COOH solution required 30.0 cc of 0.1 N NaOH for neutralization. Calculate the weight of CH_3COOH adsorbed per gram of charcoal and the equilibrium concentration of the acid. *Ans.* $c = 0.120$ molar.

20. The following data were obtained for the adsorption of acetone on charcoal from an aqueous solution at 18°C:

y (millimoles/g)	C (millimoles/liter)
0.208	2.34
0.618	14.65
1.075	41.03
1.50	88.62
2.08	177.69
2.88	268.97

Evaluate the constants k and n of the Freundlich equation.

21. For the adsorption of a substance A from aqueous solution by charcoal at 25°C the Freundlich constants are $n = 3.0$ and $k = 0.50$ for y in grams per gram and C in grams per liter. What weight of A will be adsorbed by 2 g of charcoal from 1 liter of a solution containing originally 2 g of the substance?   *Ans.* 1 g.

22. In the adsorption of N_2 at 90.1°K on a certain solid, the following volumes of gas, reduced to standard conditions, were found to be adsorbed per gram of solid at the indicated relative pressures:

P/P^0	v (in cc)
0.05	51.3
0.10	58.8
0.15	64.0
0.20	68.9
0.25	74.2

Show that these data follow a type II isotherm, and evaluate the constants v_m and c in Eq. (22).   *Ans.* $v_m = 57.3$ cc/g; $c = 109$.

23. From the results of the preceding problem find the area of the solid in square meters per gram, and the value of $E_1 - E_L$ in calories per mole.

24. Using the data given in problem 22, find the area of the solid in square meters per gram by the Harkins-Jura method.

25. Langmuir measured the volumes, reduced to 1 atm pressure, of various gases adsorbed at 20°C by cover glasses having a total surface area of 1966.0 cm². His results were

	Volume (mm³)	Molecular Diameter (Å)
H_2O	354	2.20
CO_2	64.0	4.18
N_2	49.0	3.75

Assuming that each molecule occupies an area equal to the square of its diameter, calculate in each case the molecular thickness of the surface layer.

26. Calculate the pressure of N_2O in contact with a gold surface at 900°C after $2\frac{1}{2}$ hr when the initial pressure is 350 mm. After what time will the decomposition be 95% complete? Use the data given in Table 20–4.   *Ans.* 51 mm; 233 min.

27. Integrate Eq. (40) to obtain k in terms of P, t, n, and the initial pressure P_i.

28. Using the result obtained in problem 27, deduce the expression for the period of half-life in terms of n, k, and P_i.

29. Stock and Bodenstein give the following data for the decomposition of SbH_3 on Sb at 25°C:

Time (min)	0	5	10	15	20	25
$P_{Stibine}$ (atm)	1.000	0.731	0.509	0.327	0.189	0.093

Show that the decomposition may be represented satisfactorily by the rate expression $-dP/dt = kP^n$, where $n = 0.60$, and determine the rate constant.

30. Using the data in problem 29, determine graphically the half-life period of the reaction and compare your result with that given by the equation derived in problem 28.

31. Kunsman [*J. Am. Chem. Soc.*, **50**, 2100 (1928)] reported the following data for the decomposition of NH_3 on W at 1100°K:

Initial pressure of NH_3 (mm)	265	130	58	16
Half-life period (min)	7.6	3.7	1.7	1.0

Show that the reaction is approximately zero order, and calculate the mean specific rate constant.

32. From the data given in Table 20–5 estimate the percentage of NH_3 decomposed after 1 hr, and the time required for the decomposition of 75% of the original sample.   *Ans.* 100%; 1154 sec.

33. According to Hinshelwood the decomposition of N_2O on Pt follows the rate equation

$$\frac{-d(a - x)}{dt} = \frac{k(a - x)}{1 + bx}$$

where a is the initial pressure of N_2O in mm, x is the decrease in pressure of N_2O in time t, while k and b are constants. On integration this rate equation yields

$$k = \frac{1 + ab}{t} \ln \frac{a}{a - x} - \frac{bx}{t}$$

From the following data obtained at 741°C with a = 95 mm:

t (sec)	315	750	1400	2250	3450
x (mm)	10	20	30	40	50

determine the constants k and b by a graphical method.

Ans. $k = 3.36 \times 10^{-4}$; $b = 0.0254$.

34. Using the constants obtained in problem 33, calculate the half-life period for the decomposition of N_2O on Pt at 741°C when the initial pressure of the gas is 200 mm.

35. Derive the rate expression for the kinetics of a heterogeneous reaction in which two reactants are moderately adsorbed and the products do not retard the reaction.

36. Formic acid decomposes into CO_2 and H_2 on a gold surface, following a first order law. Hinshelwood and Topley [*J. Chem. Soc.*, **123**, 1014 (1923)] observed a rate constant of 5.5×10^{-4} at 140.0°C and 9.2×10^{-3} at 185.0°C. Calculate the apparent energy of activation.   *Ans.* 23,450 cal mole^{-1}.

37. The decomposition of phosphine on fused silica follows the first order law. At 828°K the half-life period is 580 sec, while at 956°K it is 22 sec. Determine (a) the apparent energy of activation of the reaction, and (b) the half-life period of the reaction at 800°K.

21
Colloids

T HE SIZE of solute particles in ordinary solutions is generally between 1 and 10 Å, or 0.1–1 mμ.[1] There are many systems, however, where
the particles are considerably larger and may in fact range from the upper
limit for ordinary solutions up to several microns in size. Systems which
consist of media with dissolved or dispersed particles ranging approximately from 1 mμ up to several microns in size are called *colloids*. Dispersions with particles larger than several microns are considered to be
coarse mixtures. Colloids occupy, then, an intermediate position between
solutions of relatively low molecular weight substances on the one hand,
and coarse mixtures on the other. The dividing line between ordinary
solutions and colloids, or between the latter and coarse mixtures, is not
sharp, since many of the characteristics of the systems shade into each
other without discontinuity. Consequently, classification is frequently
difficult, and the designation used becomes a matter of arbitrary choice.

Colloids may be grouped into three general classes, each of which depends on the manner in which the large size of the particles arises. These
are (a) colloidal dispersions, (b) solutions of macromolecules, and (c)
association colloids. The dispersions consist of suspensions in a medium of

[1] 1 μ (micron) $= 10^{-6}$ meter $= 10^{-4}$ cm $= 10,000$ Å
   1 mμ (millimicron) $= 10^{-3}$ micron $= 10^{-7}$ cm $= 10$ Å.

insoluble substances in masses containing many individual molecules. Examples are colloidal dispersions of gold, As_2S_3, or oil in water. The macromolecular solutions, on the other hand, are true solutions of molecules so large that they fall into the colloidal range. Examples of these are aqueous solutions of proteins and polyvinyl alcohol, or solutions of rubber and various other high-polymeric materials in organic solvents. Finally, the association colloids consist of solutions of soluble and relatively low molecular weight substances which, at a particular concentration in each instance, associate to form aggregates of colloidal size. Soap solutions are an outstanding example of this category.

In the following discussion these three classes of colloids and their properties will be considered in turn.

COLLOIDAL DISPERSIONS

Dispersions of all kinds of substances in various media can be prepared, be these substances crystalline or noncrystalline, and be they electrolytes or nonelectrolytes. The particles in dispersion are usually beyond the range of the ordinary microscope, and as a rule they will not be retained by ordinary filters. Unlike true solutions, which are homogeneous, colloidal dispersions are considered to be heterogeneous, with the medium as one phase and the dispersed substance as the other. The medium in which the dispersion takes place is called the *disperse medium*, the substance dispersed the *disperse phase*, while the complete colloidal solution is referred to as a *disperse system*. These terms are analogous to solvent, solute and solution in ordinary solutions. Furthermore, dispersions are thermodynamically unstable and tend to coagulate and precipitate on

TABLE 21-1. TYPES OF COLLOIDAL DISPERSIONS

Disperse Phase	Disperse Medium	Name	Examples
Solid	Gas	Aerosol	Smokes
Solid	Liquid	Sol or Suspensoid	AgCl, Au, As_2S_3, or S in H_2O
Solid	Solid	—	Glasses colored with dispersed metals (ruby glass)
Liquid	Gas	Aerosol	Fogs, mists, clouds
Liquid	Liquid	Emulsion or Emulsoid	Dispersions of H_2O in oil or oils in H_2O
Liquid	Solid	Gels	Jellies, minerals with liquid inclusions (opal)
Gas	Gas	—	Unknown
Gas	Liquid	Foam	Whipped cream
Gas	Solid	—	Pumice stone

standing unless suitable precautions are taken. Once a disperse phase precipitates, it will not redisperse spontaneously; i.e., the precipitation of the disperse phase is an irreversible process. Since both the disperse phase and disperse medium may be solid, liquid, or gaseous, dispersions may be classified, like solutions, according to the state of the two to yield the nine types of disperse systems given in Table 21–1. Actually these reduce to eight, since no colloidal dispersion of one gas in another has ever been observed. The names generally ascribed to some of these types as well as examples of each category are included.

Of the eight types observed, those of particular importance and interest are the sols or suspensoids, emulsions, gels, aerosols, and foams. However, as an extended discussion of all these would take us too far afield, only sols, emulsions, and gels will be considered here.

SOLS AND THEIR PREPARATION

Dispersions of solids in liquids can be subdivided broadly into *lyophobic* and *lyophilic* sols. Lyophobic (solvent-hating) sols are those dispersions in which there is very little attraction between disperse phase and medium, as in dispersions of various metals and salts in water. Lyophilic (solvent-loving or -attracting) sols, on the other hand, are dispersions in which the disperse phase exhibits a definite affinity for the medium, and as a result extensive solvation of the colloidal particles takes place. Examples of lyophilic sols are dispersions of the lyophobic type to which have been added, say, gelatin, glue, or casein. The latter substances are adsorbed by the disperse phase and impart to it lyophilic character. For dispersions in water the terms *hydrophobic* and *hydrophilic* are generally substituted for the more inclusive designations.

As dispersions involve a state of subdivision of matter intermediate between true solutions and coarse particles, they may be prepared either by *condensation* of particles in true solution or by *dispersion* of coarse particles into smaller aggregates. Within each of these general methods of preparation are a number of specific means of bringing about colloid formation. Some of these are chemical in nature and involve reactions like double exchange, oxidation, and reduction. Others in turn are physical, like change of solvent, use of an electric arc, or of a colloid mill for disintegration of bulk solids. A detailed discussion of the preparation of sols is given by McBain.[2]

In preparing suspensoids, sols frequently result which contain besides the colloidal particles appreciable quantities of electrolyte. To obtain the pure colloid this electrolyte has to be removed. For this purification of a

[2] J. W. McBain, *Colloid Science*, D. C. Heath and Company, Boston, 1950, Chap. 5.

sol three methods are available, namely, (a) dialysis, (b) electrodialysis, and (c) ultrafiltration. In *dialysis* the sol is purified by permitting the electrolyte to diffuse through a porous membrane, such as parchment, cellophane, or collodion, which is permeable to molecules of solvent and low molecular weight solutes, but not to colloidal particles. The usual procedure for dialyzing a sol is to enclose it in a sack made of the material chosen for a dialyzing membrane and then to immerse the membrane in distilled water. On prolonged contact with the water the ions in solution pass through the membrane into the water; but the colloidal particles, not able to pass through the membrane, remain behind to yield a pure sol. Sometimes a continuous stream of water is kept passing about the membrane to assure a maximum concentration gradient for diffusion.

Dialysis is, at best, a very slow process, requiring days, and occasionally weeks, for its completion. It is usually not carried to a point where all of the electrolyte is removed, since highly dialyzed sols, by losing too much electrolyte, become unstable and tend to precipitate readily.

In *electrodialysis* the dialyzing process is accelerated by applying a potential difference across the membrane. Under the influence of this applied field ions migrate faster and are thereby removed from a sol much more readily.

Ultrafiltration is a process similar to filtration of an ordinary precipitate, except that a membrane is used which will permit passage of electrolytes and medium but not of colloid. For this purpose the diaphragms mentioned previously, or ordinary filter paper impregnated with collodion, may be employed. Other filter media are unglazed porcelain and finely sintered glass. Since as a rule ultrafiltration proceeds very slowly, pressure or suction is used to speed the process. By these means it is possible to separate, in the form of a slime, colloid particles from media containing electrolytes. These slimes may be suspended, then, in pure media.

PROPERTIES OF SUSPENSOIDS

The properties of suspensoids are best considered under a number of specific heads, namely, (a) physical, (b) colligative, (c) optical, (d) kinetic, and (e) electrical properties.

The *physical properties* of sols depend on whether the sols are lyophobic or lyophilic. For dilute lyophobic sols such properties as density, surface tension, and viscosity are not very different from those of the medium. This is to be anticipated, since such sols, by showing very slight interaction between suspended matter and medium, affect the properties of the latter little. On the other hand, lyophilic sols show a high degree of solvation of the suspensoid, and as a consequence the physical properties of the medium are modified. This is particularly noticeable in the viscosity, which is much higher for the sol than for the medium. Furthermore, the surface tension of the sol is frequently lower than that of the pure medium.

Sols exhibit *colligative properties*, but the effects observed are *very much smaller* than for ordinary solutions. In fact, with the exception of osmotic pressure, the effects are practically negligible. The reason for this lies in the difference in particle size of the two types of dispersions. When, say, 0.01 mole of solute is dissolved in 1000 g of solvent to form a true solution, the number of particles resulting is 0.01 N, where N is Avogadro's number. However, if the same amount of substance is colloidally dispersed in the same amount of medium to form aggregates containing 1000 molecules per particle, the number of particles present will be only $\frac{1}{1000}$ of that present in the true solution, and hence the colligative effect will be reduced by the same amount. In freezing point lowering, boiling point elevation, and vapor pressure lowering such a decrease in number of particles leads to practically no observable difference between the properties of the sol and the pure medium. But the osmotic pressure, being larger in magnitude than the others, still manifests itself, although again considerably reduced below that of a true solution of the same concentration.

OPTICAL PROPERTIES OF SOLS

An outstanding characteristic of colloidal dispersions is their pronounced scattering of light. Light passing through a vacuum has all of its intensity transmitted in the direction of travel. However, when the light travels through a medium in which discrete particles are present, such particles interfere with the motion of the light, and cause part of its energy to be scattered in all directions in space. Such scattering, called the *Tyndall effect*, does not involve a change in the wave length of the light, and it is produced by all types of systems, be they gases, liquids, solutions, or colloids. As a result of this scattering light is observed not only in the direction in which the light is traveling, but also at various angles to the direction of the primary beam.

The amount of light scattered by pure gases and liquids is very small. Again, in very dilute liquid solutions or dispersions the extent of scattering depends on the size of the solute or colloid particles present, and on the refractive index ratio $m = n/n_0$, where n is the refractive index of the dissolved or suspended particles and n_0 that of the medium. The scattering is the more pronounced the larger the size of the particles present and the higher the value of m. Since in ordinary solutions the solute particles are small, such solutions again are relatively weak scatterers. However, colloidal dispersions with their large particles are good scatterers of light, and they exhibit Tyndall effects on passage of light which can readily be observed and measured.

The general theory of light scattering by isotropic, noninteracting, spherical particles of any uniform size was developed by Gustav Mie[3] in 1908. Although the mathematical results of the theory are very complex, they can nevertheless be summarized as follows. Suppose that unpolarized incident light of wave length λ_0 in vacuo and initial intensity I_0 is passed through a dispersion, and suppose, further, that the intensity of light scattered at any angle θ to the direction of the primary beam is I'_θ. Mie theory predicts, then, that the ratio $I_\theta = I'_\theta/I_0$ will be a function of θ, m, and the parameter $\alpha = \pi D/\lambda_m$, where D is the diameter of the scattering particles and λ_m is the wave length of the light in the medium. The two wave lengths λ_m and λ_0 are related by the equation

$$\lambda_m = \frac{\lambda_0}{n_0} \tag{1}$$

The same conclusion applies to the vertical component of I_θ, namely, $I_{\theta v}$. Since both I_θ and $I_{\theta v}$ depend on D, they can both be employed to determine particle size of colloidal dispersions by means of light scattering. Besides, the total scattered light also depends on m and α, and hence this quantity can be used as well to find D.

Of the various means by which D can be obtained,[4] we shall discuss only two methods: (a) the minimum intensity method, and (b) the transmission method.

MINIMUM INTENSITY METHOD

Dispersions with particles larger than about 1800 Å in diameter exhibit minima, or minima and maxima, in the intensity of the vertical component of the scattered light, i.e., in plots of $I_{\theta v}$ vs. θ. Recently Maron and Elder[5] showed, on the basis of Mie theory and experimentally, that the angular positions of the *first minimum*, θ_1, are related to D, λ_m, and values of m between 1.00 and 1.55 by the equation

$$\frac{D}{\lambda_m} \sin \frac{\theta_1}{2} = 1.062 - 0.347\, m \tag{2}$$

Consequently, when m and λ_m are known and θ_1 is measured for a dispersion, D can then be calculated by means of Eq. (2).

[3] G. Mie, *Ann. Physik*, **25**, 377 (1908).
[4] For a discussion and references see Maron and Elder, *J. Colloid Sci.*, **18**, 107, 199 (1963); Maron, Pierce, and Elder, *J. Colloid Sci.*, **18**, 391, 733 (1963); Maron, Pierce, and Ulevitch, *J. Colloid Sci.*, **18**, 470 (1963).
[5] Maron and Elder, *J. Colloid Sci.*, **18**, 107 (1963).

Table 21-2 shows some particle diameters obtained in this manner for several latex dispersions of butadiene-styrene ($m = 1.17$) and polystyrene ($m = 1.20$) in water.

Equations similar to (2) are obeyed also by higher minima as well as maxima,[6] and hence these can be used as well to determine D.

TABLE 21-2.   PARTICLE DIAMETERS OF LATICES OBTAINED
BY MINIMUM INTENSITY METHOD

Latex No.	m	λ_m (Å)	θ_1	D (Å)
580-G	1.20	3017	94.5°	2630
		3253	108	2600
10713	1.17	3253	66	3920
		4094	85	3980
497	1.17	3253	43	5820
		4094	54	5810
197	1.17	3253	38	6550
		4094	48	6600
597	1.17	4094	31	10060
		4336	33	10010

THE TRANSMISSION METHOD

This method is based on application of Mie theory to the total intensity of scattered light. When light of wave length λ_0 and intensity I_0 is passed through a dispersion contained in a cell of length l, the intensity of the transmitted light is reduced to I by scattering. The relation between I and I_0 is given by

$$\tau = \frac{1}{l} \ln \frac{I_0}{I} \qquad (3)$$

where τ is the *turbidity* of the dispersion. Again, if we let C be the concentration of scattering particles in grams per cc and ρ their density, then Mie theory predicts that

$$\left(\frac{2\,\rho\lambda_m}{3\,\pi}\right)\left(\frac{\tau}{C}\right)_0 = \frac{K^*}{\alpha} \qquad (4)$$

where K^* is a quantity called the *scattering coefficient* of the particles. The zero subscript on τ/C indicates that this ratio has to be plotted against C and extrapolated to $C = 0$ in order to eliminate any interactions among the particles.

All the quantities on the left-hand side of Eq. (4) are experimentally

[6] Maron, Pierce, and Elder, *J. Colloid Sci.*, **18**, 391 (1963).

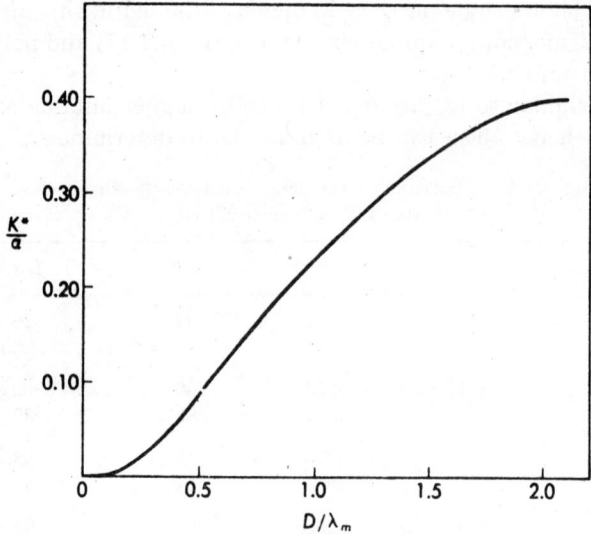

Figure 21-1.   Mie theory plot of $\dfrac{K^*}{\alpha}$ vs. $\dfrac{D}{\lambda_m}$ for $m = 1.20$.

measurable, and hence K^*/α can be found. Again, K^*/α is a function of m and α which can be calculated from Mie theory. At any constant value of m, K^*/α as a function of D/λ_m is given by a curve such as shown in Figure 21-1. Consequently, once K^*/α is found experimentally, it can be used with the plot for the appropriate m value to find D/λ_m and then D.

Table 21-3 gives some results obtained in this manner for the particle diameters of several synthetic latices.

TABLE 21-3.   PARTICLE DIAMETERS OF SYNTHETIC LATICES
OBTAINED BY TRANSMISSION METHOD[*]

	Particle diameter (Å)		
λ_0 (Å)	Latex 243 $m = 1.15$	Latex G-5 $m = 1.17$	Latex 580-G $m = 1.20$
7000	905	3660	2570
7500	920	3680	2640
8000	921	3660	2630
8500	919	3670	2640
9000	919	3710	2620
9500	931	3670	2620
10000	905	3770	2590
Average	917 ± 7	3690 ± 30	2620 ± 20

[*] Maron, Pierce, and Ulevitch, *J. Colloid Sci.*, **18**, 470 (1963).

OTHER METHODS FOR DETERMINATION OF PARTICLE SIZE

Another means of getting at the size and shape of dispersed particles is the *electron microscope*. In this instrument a beam of electrons is used instead of ordinary light to observe and photograph the colloidal particles. In Figure 21–2 is shown an electron photomicrograph of synthetic rubber latex particles[7] whose average diameter is 2300 Å. With suitable enlarge-

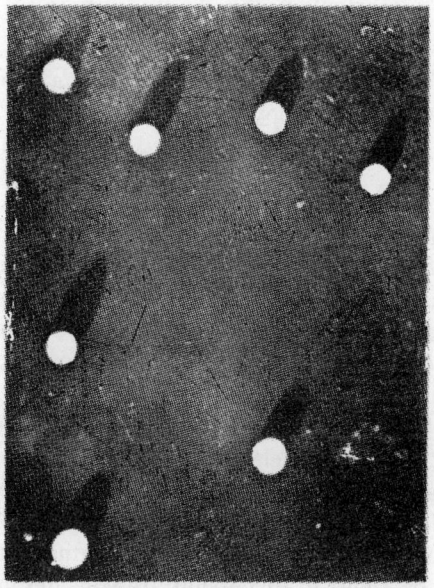

Figure 21–2.   Electron photomicrograph of synthetic rubber latex particles.

ment of such photomicrographs magnifications as high as 100,000 diameters have been obtained.

In certain systems, such as synthetic rubber latices, particle size can also be determined by adsorption methods.[8] These involve titration of the colloid with a soap solution until the surface of the particles is covered with a monolayer of soap. From the amount of soap required per unit weight of disperse phase and the area occupied by a soap molecule on the surface then is calculated the average diameter of the particles.

[7] Maron, Moore, and Powell, *J. Appl. Phys.*, **23**, 900 (1952).

[8] Maron, Elder, and Ulevitch, *J. Colloid Sci.*, **9**, 89 (1954); Maron, Elder, and Moore, *J. Colloid Sci.*, **9**, 104 (1954).

KINETIC PROPERTIES OF SOLS

The properties of sols which can be discussed conveniently under this heading are diffusion, Brownian motion, and sedimentation.

Graham observed that collodial particles diffuse much more slowly than solutes in true solution. Albert Einstein and others showed that it is possible to derive an expression for the diffusion of colloidal particles in a medium provided it is assumed that the van't Hoff equation for osmotic pressure applies, and that the colloid particles are spherical and large compared with the molecules of the medium. With these assumptions the equation for the diffusion coefficient D_f, i.e., the number of moles of colloid diffusing across unit area per unit time under a concentration gradient of 1 mole per centimeter, follows as

$$D_f = \frac{RT}{N} \left(\frac{1}{6\,\pi\eta r} \right) \tag{5}$$

where R is the gas constant in ergs mole^{-1} degree^{-1}, T the absolute temperature, N Avogadro's number, η the viscosity of the medium in poises, and r the radius of the colloidal particles in centimeters. This equation has been employed to estimate from diffusion measurements either r, when N was assumed, or N when r was known, and has been found to hold well for various types of dispersions.

Colloidal particles exhibit a ceaseless, random, and swarming motion. This kinetic activity of particles suspended in a liquid is called *Brownian motion*. Brownian motion is due to bombardment of the dispersed particles by molecules of the medium. As a result the particles acquire at equilibrium a kinetic energy equal to that of the medium at the given temperature. Since the particles in dispersion are considerably heavier than the molecules of the medium, this acquired kinetic energy manifests itself in a Brownian movement which is considerably slower than that of the molecules of the medium, and may thus be observed in a microscope.

The validity of this explanation of Brownian motion is borne out by mathematical considerations again due to Einstein. He showed that, on the basis of the assumptions mentioned in connection with diffusion, the diffusion coefficient of a colloid should be related to the *average displacement*, Δ, produced by Brownian movement in time t along the x axis. The relation between D_f and Δ is

$$D_f = \frac{\Delta^2}{2\,t} \tag{6}$$

Eliminating D_f between Eqs. (5) and (6), Δ^2 follows as

$$\Delta^2 = \frac{RT}{N} \left(\frac{t}{3\,\pi\eta r} \right)$$

This equation was used by Perrin in his classical experiments on suspensions of gamboge and mastic in water for the determination of Avogadro's number.

SEDIMENTATION OF SUSPENSOIDS

Although colloidal dispersions may be stable over long periods of time, sometimes years, they nevertheless tend to settle out slowly under the influence of gravity on prolonged standing. The rate of settling can usually be followed quite readily, for the boundary between clear medium and sol is, as a rule, quite distinct. From the observation of such rates it is possible to arrive at the dimensions of the suspended particles and their masses.

Consider a tube such as shown in Figure 21–3 containing a disperse phase of density ρ suspended in a medium of density ρ_m and viscosity η.

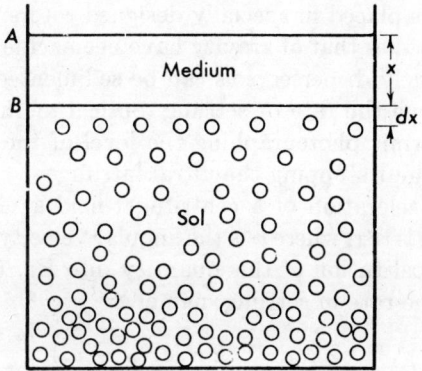

Figure 21–3.   Sedimentation under gravity.

If ρ is greater than ρ_m, the particle will settle; otherwise, the particle will be displaced upward. Any particle of radius r tending to settle under gravity will be opposed by the frictional force of the medium. Now, it was shown in Chapter 1, Eq. (96), that when equilibrium is established between these two forces a spherical particle will fall according to Stokes's law with a constant velocity v, given by

$$v = \frac{2\,r^2 g(\rho - \rho_m)}{9\,\eta} \tag{8}$$

where g is the acceleration of gravity. Since v at a point such as x from the top A is obviously dx/dt, the rate at which the height of level B changes with time, Eq. (8) may be written as

$$\frac{dx}{dt} = \frac{2\,r^2 g(\rho - \rho_m)}{9\,\eta} \tag{9}$$

Integrating this equation between the limits x_1 at t_1, and x_2 at t_2, we obtain

$$(x_2 - x_1) = \frac{2\,r^2 g(\rho - \rho_m)(t_2 - t_1)}{9\,\eta} \tag{10}$$

By knowing ρ, ρ_m, and η, and by measuring the distances x_2 and x_1 at two different times t_2 and t_1, it is possible to calculate from Eq. (10) the radius of the settling particles. Once the radius is known, the mass of a single particle follows from $m = \frac{4}{3}\,\pi r^3 \rho$, and the molar weight of the disperse phase from $M = Nm$.

Unless the suspended particles are large, sedimentation under gravity is an extremely slow process. However, it is possible to accelerate greatly the sedimentation by means of *ultracentrifuges*. An ultracentrifuge is essentially a very-high-speed centrifuge in which the centrifugal force of rotation is substituted for the force of gravity. By whirling colloidal dispersions in cells placed in specially designed rotors, accelerations as high as one million times that of gravity have been achieved. Under these conditions even finely dispersed sols can be sedimented in a relatively short time. To observe the rate of settling, optical methods have been developed which permit photographing the level of the sol at various stages of settling without stopping the ultracentrifuge.

Since the acceleration of a centrifugal field at a distance x from the axis of rotation is $\omega^2 x$, where ω is the angular velocity of rotation in radians per second, substitution of this quantity into Eq. (9) in place of g gives immediately the rate of settling, namely,

$$\frac{dx}{dt} = \frac{2\,r^2 \omega^2 x(\rho - \rho_m)}{9\,\eta} \tag{11}$$

Separating the variables, and integrating between the same limits as before, we have

$$\ln \frac{x_2}{x_1} = \frac{2\,r^2 \omega^2(\rho - \rho_m)}{9\,\eta}\,(t_2 - t_1) \tag{12}$$

Through Eq. (12) r may again be calculated from the values of x and t at two stages of the sedimentation and from a knowledge of the speed of rotation of the ultracentrifuge, ω. This procedure for obtaining r, and therefrom the molar weight of a suspensoid, is called the *sedimentation velocity* method.

An alternate procedure for obtaining these quantities is the *sedimentation equilibrium* method. If a suspensoid is whirled sufficiently long in an ultracentrifuge, a stage is reached at which the sol no longer settles. At this point an equilibrium is reached between the rate of sedimentation and the rate at which the suspended particles tend to diffuse back into the more dilute portions of a cell against the centrifugal force. By equating these two rates, it is possible to arrive at the distribution of concentration with distance in various parts of the sedimentation cell. This distribution is given by the relation

$$\ln \frac{C_2}{C_1} = \frac{M\omega^2(\rho - \rho_m)(x_2^2 - x_1^2)}{2RT\rho} \tag{13}$$

where C_1 and C_2 are the concentrations of colloid at levels x_1 and x_2, while M is the molecular weight of the colloid. By determining the concentrations C_1 and C_2 at the two levels x_1 and x_2 in the settling cell at sedimentation equilibrium, M can readily be calculated from Eq. (13).

Equation (12) is applicable to dilute dispersions of uniform spherical particles, while Eq. (13) applies to dilute dispersions of particles of any shape so long as they are all of the same size. When the particles are nonspherical, a more involved analysis of the sedimentation velocity data is required. Again, when systems are heterogeneous in particle size, it is possible by sedimentation studies to determine the particle size distribution for the systems.

Most of the sedimentation studies have been made on proteins in aqueous solution and on other substances of biological interest. Some results obtained are given in Table 21–4.

TABLE 21–4. MOLECULAR WEIGHTS OF PROTEINS
DETERMINED BY ULTRACENTRIFUGE

Protein	M (g mole^{-1})	Protein	M (g mole^{-1})
Trypsin	15,000	Urease	480,000
Egg albumin	44,000	Thyroglobulin (pig)	630,000
Insulin	46,000	Hemocyanin (octopus)	3,300,000
Serum albumin (horse)	70,000	Hemocyanin (helix)	8,900,000
Diphtheria antitoxin	92,000	Tobacco mosaic virus	31,400,000
Catalase	250,000	Rabbit virus	47,000,000

ELECTRICAL PROPERTIES OF SOLS

Colloidal dispersions possess electrical properties which are intimately associated with their ability to absorb from solution ions, molecules of medium, or both. Study of colloids has established that lyophobic sols adsorb primarily ions of electrolyte from the solutions in which they are prepared. On the other hand, the particles in lyophilic sols attract onto themselves first a layer of medium to which, depending on the conditions existing in the solution, may or may not be adsorbed ions. Although the nature of the dispersoid particles determines whether a colloid is to be lyophobe or lyophile, it is the character of the adsorbed phase on the final particle which eventually controls the stability of the sol and establishes the manner in which the particles are to behave in an electric field.

Any solid in contact with a liquid tends to develop a difference in potential across the interface between the two. Thus, when water is brought into contact with a glass surface, the latter adsorbs hydroxyl ions and becomes negatively charged with respect to the water. To counterbalance this charge, hydrogen ions are attracted to the surface to form a double layer of charges, with the negative charges on the glass and positive charges in the water immediately adjacent to these, as shown in Figure 21–4(a). Such an arrangement of charges, called a *Helmholtz double layer*,

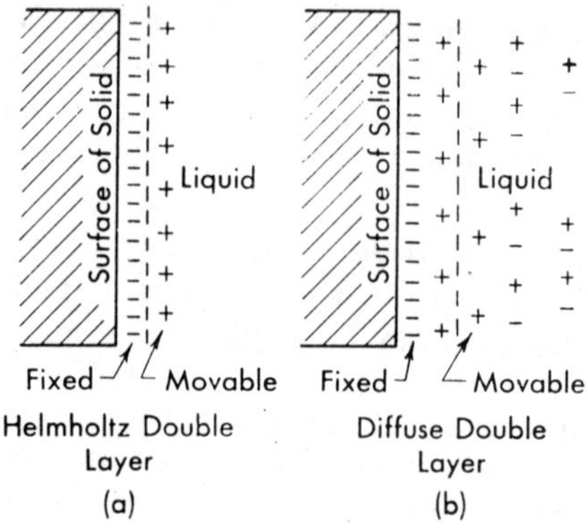

Figure 21–4. Helmholtz and diffuse double layers.

leads to a difference of electric potential between the solid and liquid.

In the original concept of the double layer the charges next to the surface were considered to be fixed, while the compensating charges in the liquid, along with the latter, were thought to be movable. However, more recent considerations indicate that the layer is more diffuse in character and extends part way into the liquid in the manner illustrated in Figure 21–4(b). Here the charges shown immediately adjacent to the surface are probably on the surface, with some of the compensating charges held in a stationary liquid layer adhering to the surface. The remaining charges are distributed through the electrolyte next to this layer in the form of a diffuse and movable atmosphere analogous to the one postulated by Debye and Hückel for electrolytes. In terms of this concept the potential drop between the surface and the liquid occurs in two parts, (a) between the surface and the stationary compensating layer, and (b) between the latter and the body of the solution. The second of these potential drops, called the *electrokinetic* or *zeta potential*, is the one involved in various nonstatic electrical properties of solid-liquid interfaces and is the one responsible for the electrical effects observed in colloids.

Since dispersions of solids in liquids produce solid-liquid interfaces, a diffuse double layer of the type described above should be established at the particle boundary, and this should lead to the appearance of a zeta potential. Further, selective adsorption of ions must result in an electric charge on the particle of sign equal and opposite to that of the surrounding medium; i.e., both the particles and medium must become electrically, though oppositely, charged. Consequently, both the particle and the medium should respond to applied electric fields and should migrate under these in opposite directions. This is actually so. When experimental conditions are adjusted to permit the migration of dispersoid but not of medium, we encounter the phenomenon of *electrophoresis*. On the other hand, when migration of dispersoid but not that of medium is prevented, we have *electroosmosis*.

ELECTROPHORESIS

The migration of electrically charged colloidal particles under an applied electric potential is called *electrophoresis*. This effect can be followed readily in the apparatus shown in Figure 21–5, called a *Burton tube*. This

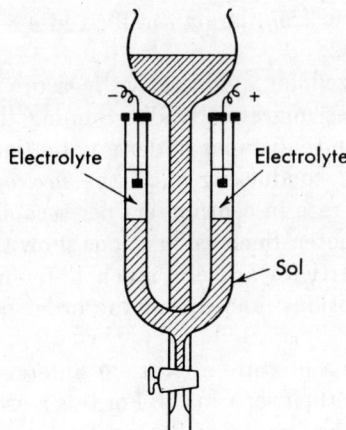

Electrolyte   Electrolyte

Sol

Figure 21–5. Burton tube for electrophoresis.

apparatus consists of a U-tube fitted with a stopcock for drainage and a funnel-shaped filling tube with stopcock (not shown) attached to the back of the U-tube. A solution of some suitable electrolyte of lower density than the sol is first placed in the tube, and the sol next introduced through the funnel so as to displace the electrolyte upward and produce sharp boundaries in the two arms. Electrodes are then inserted, as shown, and connected to a source of potential, such as a high-voltage battery. When the colloidal particles are negatively charged it is observed that the level of the sol falls gradually on the negative electrode side and rises simultaneously on the positive side, i.e., the colloidal particles move toward the positive electrode. Conversely, when the particles are positively charged, the reverse occurs. If the electrophoresis is permitted to proceed until the appropriate electrode is reached, the sol will discharge and precipitate.

By noting whether colloid particles migrate in an electric field and the direction of migration, it is readily possible to determine whether the particles in a given colloid are charged, and what is the sign of the charge. Through such means it has been established that lyophobic colloids are specific in their adsorption of either positive or negative ions. Thus sulfur, metallic sulfide, and noble metal sols are negatively charged. On the other hand, metal oxide sols, such as iron and aluminum oxide, are positively charged. This same specificity of charge is observed also with some lyophilic sols. However, with certain sols of the latter type, particularly the proteins, the sign of the charge depends on the pH of the solution. Above a certain pH value, characteristic for each sol, the particles are negatively charged, while below this pH they have a positive charge. At the pH value in question, called the *isoelectric point*, the particles are uncharged and consequently do not migrate in an electric field. As a rule the isoelectric point does not come at a definite pH but covers a pH range, being, for instance, 4.1–4.7 for casein from human milk, and 4.3–5.3 for hemoglobins from various sources.

Electrophoresis can also be utilized for quantitative measurement of the rate with which colloid particles migrate. By determining the time necessary for a sol to migrate a definite distance under a potential difference applied over a given length of conducting path, the *electrophoretic mobility* can be calculated, i.e., the rate in centimeters per second under a potential drop of 1 volt per centimeter. Such calculations show that the migration velocities of colloidal particles are not much different from those of ions under the same conditions, and are of the order of 10 to 60×10^{-5} cm^2 sec^{-1} volt^{-1}.

Since different colloidal species in a mixture migrate at different rates, electrophoresis may be employed for their separation. For this reason electrophoresis is used extensively for the fractionation and analysis of proteins, nucleic acids, polysaccharides, and other complex substances of biologic interest and activity.

ELECTROOSMOSIS

When electrophoresis of a dispersoid is prevented by some suitable means, the medium can be made to move under the influence of an applied potential. This phenomenon is referred to as *electroosmosis* and is a direct consequence of the existence of a zeta potential between the colloid particles and the medium. Figure 21–6 shows a simple apparatus for observing electroosmosis. The colloid is placed in compartment A, separated from compartments B and C by dialyzing membranes D and D'. B and C are filled with water up to the marks indicated on the sidearms. When a potential is applied across the two electrodes placed close to the membranes in B and C, the liquid level is observed to fall on one side and rise on the other owing to passage of water through D and D'. The direction of flow of the water depends on the charge of the colloid. For positively charged sols the medium is negatively charged, and hence the flow will

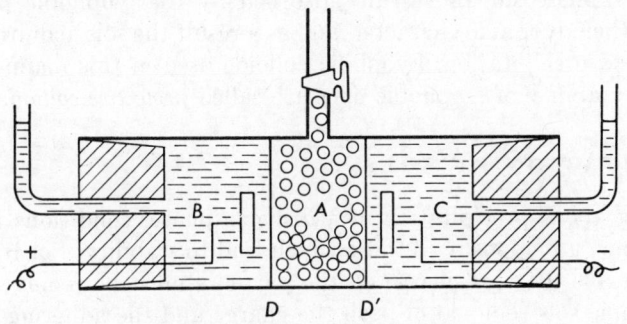

Figure 21–6.   Electroosmosis.

take place from C to B. For negatively charged colloids the reverse will be true, and the level on the C side will rise.

STABILITY OF SUSPENSOIDS

A question of considerable interest and importance is: Why are colloidal dispersions stable? The answer can be found for lyophobic colloids in their electric charge, for lyophilic colloids in both their electric charge and their extensive solvation. To bring about precipitation of a dispersion the particles must coalesce on collision into aggregates large enough to settle out. In lyophobic sols such agglomeration is prevented by the electric charges on the particles which are all of the same sign, and hence on approach they are repelled by the electrostatic forces between them. The result is that as long as these charges are present each particle is protected from every other by an electric field, and approach close enough for coalescence into larger aggregates is difficult. In lyophilic colloids the same type of electrostatic repulsion is operative, along with the added factor of solvation. The colloidal particle becomes encased in a sheath of solvent which acts as a further barrier to contact between aggregates of disperse phase.

Removal of the surface charge in lyophobic colloids would be sufficient to bring about coagulation. In lyophilic colloids, on the other hand, charge removal may decrease stability, but it does not necessarily lead to precipitation. This fact is well illustrated by proteins at their isoelectric point, where the particles are uncharged. Although in this state the protein dispersions are least stable, they do not always precipitate because they are still protected by a surrounding layer of water. However, as soon as this layer is removed by some appropriate means, coagulation takes place, and the protein settles as a precipitate.

The added stability of lyophilic colloids is frequently utilized to increase the stability of lyophobic sols by adding to them lyophilic colloidal substances. These substances are adsorbed by the lyophobic particles, impart to them lyophilic character, and as a result the sols acquire greater resistance to precipitation. Lyophilic colloids used in this manner to increase the stability of lyophobic ones are called *protective colloids*.

PRECIPITATION OF SOLS

To bring about precipitation of a disperse phase, conditions must be created which are opposite to those which lead to stability. For lyophobic sols this means that the charge on the particles must be removed, while with lyophilic sols removal of both the charge and the adhering layer of solvent is necessary.

Neutralization of electric charges on lyophobic particles may be accomplished by an applied electric field such as is used in electrophoresis. The charged particle on coming in contact with an electrode of sign opposite to itself is discharged and precipitated. A more common method of producing precipitation is by the addition of electrolytes, which, though essential in small concentrations for stability, produce in higher concentrations flocculation of the disperse phase. The amount of electrolyte required to precipitate a given sol depends on the nature of both the sol and the electrolyte added. For a given sol the precipitating power of electrolytes is determined primarily by the valence of the ions opposite in sign to that of the colloid. Table 21-5 gives some results for the precipitation of a positive and negative sol by various electrolytes. The concentrations refer to the minimum amounts of electrolyte required to produce precipitation in 2 hours. From this table it is evident that, although some specificity does exist, the precipitating action of various electrolytes depends essentially on the valence of the anions in the positive sol and on the valence of the cations in the negative sol. Further, the precipitating power of an electrolyte increases very rapidly with increase in the valence of the anion or cation, the ratios being approximately 1:40:90 for the ferric hydroxide and 1:70:500 for the arsenious sulfide.

TABLE 21-5.   PRECIPITATION OF HYDROSOLS BY ELECTROLYTES

| Ferric Hydroxide Sol (positive) | | Arsenious Sulfide Sol (negative) | |
Electrolyte	Concentration (millimoles/liter)	Electrolyte	Concentration (millimoles/liter)
NaCl	9.3	NaCl	51
KCl	9.0	KCl	50
KI	16.2	HNO_3	50
$BaCl_2$	9.6	HCl	31
K_2SO_4	0.20	$MgCl_2$	0.72
$MgSO_4$	0.22	$CaCl_2$	0.65
$K_2C_2O_4$	0.24	$BaCl_2$	0.69
$K_2Cr_2O_7$	0.19	$ZnCl_2$	0.69
$K_3Fe(CN)_6$	0.096	$AlCl_3$	0.093

Frequently a sol may be coagulated by adding to it a colloid of charge opposite to itself. In the process both sols may be partially or completely precipitated. A case in point is the precipitation of the negative arsenious sulfide sol by the positive ferric hydroxide hydrosol. Again, flocculation can occasionally be effected by boiling or freezing. In boiling, the amount of electrolyte adsorbed by the sol is reduced; and, if this reduction is sufficient to produce coagulation, the sol can be destroyed. Freezing, on the other hand, results in removal of medium. If carried far enough, there may not be enough medium left to keep the dispersoid suspended.

Flocculation in lyophilic sols may be accomplished in two ways. First, if solvents such as alcohol or acetone, which have a high affinity for water, are added to a hydrophilic colloid, the disperse phase undergoes dehydration to yield particles whose stability is due solely to their charge. A small quantity of electrolyte added to the sol in this condition produces ready flocculation. Again, precipitation can also be brought about by the use of sufficiently high concentrations of certain ions. This "salting out" of lyophilic colloids by high concentrations of electrolyte is due, apparently, to the tendency of the ions to become solvated, causing the removal of adsorbed water from the dispersed particles. The "salting out" power of various ions is given by the *Hofmeister* or *lyotropic series*, in which the ions are arranged in order of *decreasing* precipitating effectiveness. For anions this series gives

$$Citrate^{---} > Tartrate^{--} > SO_4^{--} > PO_4^{---} > Acetate^- > Cl^- >$$
$$NO_3^- > I^- > CNS^-$$

while for cations

$$Mg^{++} > Ca^{++} > Sr^{++} > Ba^{++} > Li^+ > Na^+ > K^+ > Cs^+$$

However, this order is not always valid. Depending on the nature of the colloid being precipitated, the relative positions of various anions and cations may be changed.

EMULSIONS

A colloidal dispersion of one liquid in another immiscible with it is called an *emulsion*. Emulsions can be prepared by agitating a mixture of the two liquids or, preferably, by passing the mixture through a colloid mill referred to as a homogenizer. Such emulsions prepared from the pure liquids only are generally not stable and settle out on standing. To prevent this, small quantities of substances called *emulsifying agents* or *emulsifiers* are added during preparation to stabilize the emulsions. These are usually soaps of various kinds, long-chain sulfonic acids and sulfates, or lyophilic colloids.

If we employ the term *oil* to designate any liquid immiscible with water and capable of forming an emulsion with it, we may classify emulsions into two classes, namely, (a) emulsions of oil in water, in which the disperse phase is the oil while water is the medium, and (b) emulsions of water in oil, in which the functions of the two are reversed. The type of emulsion that results on agitation of two liquids depends on the relative proportions of the two in the mixture. As a rule the one in excess will act as the "outside phase" or medium, the other as the "inner phase" or dispersoid. The exact excess required for the formation of a particular type of emulsion is hard to define. In fact, Stamm and Kraemer[9] showed that emulsions of oleic acid in water can be prepared as long as the proportion of acid in the mixture does not exceed 40 per cent. Once this percentage is exceeded, an emulsion of water in the acid results. The type of emulsion obtained depends also on the nature of the emulsifying agent used. Water-soluble alkali metal soaps and basic metal sulfates generally favor the formation of emulsions of oil in water. On the other hand, water-insoluble soaps, such as those of zinc, aluminum, iron, and the alkaline earth metals, favor the formation of emulsions of water in oil. Frequently the first type of emulsion can be converted to the second by addition of heavy metal ions. Under these conditions a water-soluble emulsifier is converted to one that is water-insoluble, and a reversal of phase is brought about.

[9] Stamm and Kraemer, *J. Phys. Chem.*, **30**, 992 (1926).

Whether a particular emulsion belongs to one type or another can be ascertained in several ways. If water is the outside phase, then any water added to the emulsion will be readily miscible with it, while oil will not. Similarly, for oil as the outside phase the miscibility will be with oil but not water. By observing the behavior on such additions under a microscope, it is readily possible to identify the nature of the emulsion. Another means of distinguishing the two depends on the fact that a small amount of electrolyte added to an emulsion will make the latter conducting if water is the outside phase but will have very little effect on the conductance if oil is the disperse medium.

The particle size of the distributed phase is usually larger in emulsions than in sols, ranging from 0.1 to more than 1 μ in diameter. Otherwise the properties of emulsions are not very different from those of lyophobic sols. They show the Tyndall effect, and Brownian motion can be observed provided the particles are not too large. Again, the globules are generally negatively charged, migrate in an electric field, and exhibit sensitivity to added electrolytes, particularly those containing multivalent positive ions.

Emulsions can be broken to yield the constituent liquids by heating, freezing, severe jarring, centrifuging, by addition of appreciable quantities of electrolyte to salt out the disperse phase, or by chemical destruction of the emulsifying agent. Centrifuging is commonly resorted to for the separation of cream from milk and the separation of water from oil. Because of the negative charge of the globules, electrolytes containing divalent and trivalent positive ions are particularly effective in bringing about a salting out of the emulsified substance. The means employed for the destruction of the emulsifying agent depend on its nature. As an example may be mentioned the precipitation of an oil in water emulsion stabilized by a sodium or potassium soap through addition of a strong acid. The acid hydrolyzes the soap to liberate the free fatty acid, and, since the latter is not a good emulsifying agent, the emulsion separates. The same result can be accomplished by careful addition of positive ions which tend to bring about a reversal of phase. If an excess is avoided, a segregation of the two phases will take place before reversal of the oil in water to a water in oil emulsion becomes possible.

GELS

Coagulation of a lyophobic or lyophilic sol yields usually a precipitate which may or may not be gelatinous. However, if the conditions are right, it is possible to obtain the disperse phase as a more or less rigid mass enclosing within it all of the liquid. The product in this form is called a gel, while the process by which it is formed is called *gelation*.

Depending on their nature, gels may be prepared generally by one of the following three methods: (a) cooling, (b) double decomposition or metathesis, or (c) change of solvents. Gels of agar-agar, gelatin, and other substances of this kind are prepared readily by cooling a not too dilute dispersion of these substances in hot water. As the sol cools the highly hydrated dispersed particles lose stability, agglomerate into larger masses, and eventually mat together to form a semirigid gel structure that entraps any free medium. The second method is illustrated by the formation of silicic acid gels on addition of an acid to a water solution of sodium silicate. The free silicic acid thus liberated is highly gelatinous due to hydration and sets more or less rapidly to a solid gel. Finally, some gels are formed on changing rather suddenly the solvent in which a substance is dissolved to one in which it is insoluble. An example of this is a gel of calcium acetate. When alcohol is added rapidly to a solution of calcium acetate in water, the salt is suddenly thrown out of solution as a colloidal dispersion which subsequently sets to a gel containing all of the liquid.

Gels may be subdivided into two kinds, *elastic*, of which agar-agar and gelatin are examples, and *nonelastic*, such as silica gel. A completely dehydrated elastic gel can be regenerated by addition of water; however, once a nonelastic gel is freed of moisture, addition of water will not bring about gelation. This difference in hydration behavior seems to lie in a difference in the structure of the dried gels. In elastic gels the fibrils composing the gel are flexible, and on hydration they can expand again to reform the gel. In nonelastic gels, on the other hand, the structure is much more rigid, and though the gel can take on some water, it cannot expand sufficiently to enclose all of the liquid. In some cases, also, heating brings about chemical changes which modify the nature of the substance. An example is the hardening of eggs on boiling, caused by the irreversible denaturation of certain proteins.

Partially dehydrated elastic gels can imbibe water when immersed in the solvent. The amount of water absorbed may be large, leading to an appreciable expansion or *swelling*. Many gels, both elastic and nonelastic, experience also a shrinkage in volume on standing with an attendant exudation of solvent. This process is called *syneresis*. Again, some gels, particularly those of hydrous oxides and gelatin, liquefy readily when shaken to form a sol which on standing reverts to a gel. The sol-gel transformation is reversible and is referred to generally as *thixotropy*. Such reversibility is observed also in gels that can be prepared by cooling, except that the liquefaction is brought about by heating, the gelation by reducing the temperature. The temperature at which the latter type of transformation occurs is usually quite definite and reproducible.

SOLUTIONS OF MACROMOLECULES

Through polymerization reactions it is possible to join by primary valence linkages a large number of relatively low molecular weight molecules into single, giant molecules. Such *macromolecules* may consist of hundreds or thousands of smaller molecules combined into the final structure in a more or less repetitive manner. Thus molecules of butadiene ($M = 54$) can be joined to yield polybutadiene of molecular weights as high as 5–6 million. High-polymer molecules of this type are generally linear, with small cross section and very great length. Furthermore, the molecules are very flexible. As a result they can take on various shapes ranging all the way from spheres when completely folded to rods on complete extension. However, the most probable configuration of such coiling chains is neither a sphere nor a rod, but a randomly coiled arrangement intermediate between the two.

High-polymeric materials such as synthetic rubbers, polystyrene, polyethylene, lucite, nylon, etc., are prepared synthetically. But many high polymers occur naturally, and these include the proteins, polysaccharides, gums and resins of various kinds, and natural rubber. Some of these high polymers are crystalline, but most are amorphous. Again, some are quite soluble in suitable solvents, while others are practically insoluble. Apparently the length of the chain is no drawback to solubility, but cross-link ing of chains to form a three-dimensional network leads to insoluble polymers or *gels.*

Polymers dissolve to form true solutions just as do solutes of lower molecular weight. The solutions are formed spontaneously and they are thermodynamically stable. Further, a polymer precipitated from a solution can be redissolved, whereas in dispersions this is not the case. However, because of the large size of the solute molecules the solutions behave as colloids, and this is the reason they are considered here.

PROPERTIES OF MACROMOLECULAR SOLUTIONS

Solutions of high polymers behave in many ways like lyophilic sols. Their physical properties are generally quite different from those of the solvents, and there may be present considerable interaction between the polymer molecules and the medium. Particularly noticeable is the relatively high viscosity of the solutions, especially at higher concentrations. Further, these solutions exhibit non-Newtonian flow at all but the lower concentrations, i.e., the viscosity coefficient of a given solution is not a constant, but depends on the shear conditions under which it is measured. Thus a solution of a polymer in a good solvent may become non-Newtonian at a concentration as low as 1–2 per cent by weight, whereas in lyophobic dispersions non-Newtonian behavior may not appear until a concentration of 25 per cent by volume is reached.

The colligative behavior of solutions of high polymers is very similar to that described for dispersions. The only property measured frequently is the osmotic pressure. Polymer solutions also scatter light and sediment in ultracentrifuges. Both of these phenomena are employed for determination of the molecular weight of the macromolecules, with sedimentation supplying a means of obtaining molecular weight distributions as well.

High-polymeric molecules in solution do not carry an electric charge and hence do not migrate in an electric field, unless the molecules themselves form ions. To do this they must contain carboxyl, sulfonic, amino, and other groups of this type. In the proteins, where both carboxyl and amino groups are present, the charge above the isoelectric point is due to the ionization of the —COOH group, while below the isoelectric point to the conversion of the amino group into an ammonium ion by acceptance of a hydrogen ion from solution. Hence the charge is negative above the isoelectric point and positive below it. Macromolecules forming ions in solution are called *polymeric or colloidal electrolytes*.

Macromolecules can be precipitated from solution by evaporation of the solvent or by addition to the solution of a solvent in which the polymer is insoluble. In the latter instance the largest molecules precipitate first, followed by the lighter molecules in order of decreasing molecular weight. Consequently, controlled addition of a nonsolvent to a solution of a polymer containing different molecular weights can be used to separate the whole polymer into fractions, each of which has a more or less narrow range of molecular weights.

DETERMINATION OF MOLECULAR WEIGHT BY OSMOMETRY

The methods employed most frequently to determine the molecular weights of polymers in solution are: (a) osmotic pressure, (b) light scattering, (c) viscosity, and (d) sedimentation. Since the procedure with the sedimentation method is very similar to that described for dispersions, only the first three will be considered here.

Thermodynamic theory of dilute high-polymer solutions predicts that at any given temperature the dependence of osmotic pressure, Π, on concentration, C, should be given by

$$\frac{\Pi}{C} = \frac{RT}{M} + AC + BC^2 + \cdots \tag{14}$$

where R is the gas constant, M the molecular weight of the solute, and

A and B are constants. Consequently, if Π/C be plotted vs. C and extrapolated to $C = 0$, the limiting value of the ordinate, $(\Pi/C)_0$, should equal

$$\left(\frac{\Pi}{C}\right)_0 = \frac{RT}{M} \tag{15}$$

and

$$M = \frac{RT}{(\Pi/C)_0} \tag{16}$$

A very common type of osmometer consists of a small glass or stainless steel chamber fitted at the top with a true-bore capillary tube, while across the open bottom is clamped a membrane. The membranes most frequently used are denitrated nitrocellulose or nonwaterproofed cellophane. The chamber is filled with the solution to be studied until the level projects into the lower end of the capillary, and the osmometer is introduced into a larger tube containing pure solvent. The level of solvent is kept just a short distance above the membrane. The entire assembly is then placed in a thermostat, and an initial reading of the solution level is made followed by a final reading when osmotic equilibrium has been established. The rise in height of the solution level is a measure of the osmotic pressure of the solution.

Figure 21–7 shows a plot of Π/C vs. C for polystyrene in toluene at 25°C. Here Π is expressed in centimeters of toluene, and the concentration in

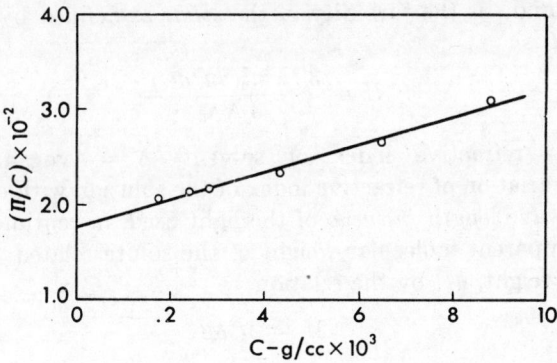

Figure 21–7. Osmotic pressures of polystyrene solutions in toluene at 25°C.

grams per cubic centimeter. From the graph $(\Pi/C)_0$ follows as 1.77×10^2. To apply Eq. (16) the pressure in $(\Pi/C)_0$ has to be converted from centimeters of toluene to atmospheres. To do this we multiply $(\Pi/C)_0$ by the density of toluene, 0.8610 at 25°C, and divide it by the weight of 76 cm of mercury at 0°C, namely, 76×13.60. For Π in atmospheres and volume in cubic centimeters $(\Pi/C)_0$ is then

$$\left(\frac{\Pi}{C}\right)_0 = \frac{1.77 \times 10^2 \,(0.8610)}{76(13.60)}$$

and, therefore,

$$M = \frac{(82.06)(298.2)(76)(13.60)}{1.77 \times 10^2 \,(0.8610)}$$

$$= 166,000 \text{ g mole}^{-1}$$

If the polymer contains molecules of varying molecular weight, as is usually the case, the value of M obtained is an average given by

$$M_n = \frac{\Sigma n_i M_i}{\Sigma n_i} \tag{17}$$

In this equation n_i is the number of molecules having molecular weight M_i, while M_n is the *number average molecular weight* of the polymer.

MOLECULAR WEIGHTS BY LIGHT SCATTERING

The determination of the molecular weights of polymers in solution by observation of the intensity of scattered light depends on the following equation due to Debye,[10] namely,

$$\frac{HC}{\tau} = \frac{1}{M'} + A'C + B'C^2 + \cdots \tag{18}$$

Here C is the concentration in grams per cubic centimeter, A' and B' are constants, and τ is the turbidity of the *solute* as defined by Eq. (3). The quantity H is

$$H = \frac{32 \,\pi^3 n_0^2 (dn/dC)^2}{3 \,N\lambda_0^4} \tag{19}$$

where $n_0 =$ refractive index of solvent, $N =$ Avogadro's number, $dn/dC =$ variation of refractive index of the solution with concentration, and $\lambda_0 =$ wave length *in vacuo* of the light used, in centimeters. Finally, M' is an apparent molecular weight of the solute related to the correct molecular weight, M, by the relation

$$M = M'\alpha\beta \tag{20}$$

The quantities α and β are correction terms necessitated by the size and nature of the solute molecules. The first, α, is called the dissymmetry correction and the second the depolarization correction.

To use these equations it is necessary to determine the turbidity at several dilute concentrations of the solution, n_0, dn/dC, and the correc-

---

[10] P. Debye, *J. Applied Phys.*, **15**, 338 (1944); *J. Phys. and Colloid Chem.*, **51**, 18 (1947).

tion factors α and β, all for the wave length of light involved. The turbidity is obtained generally from the intensity of the light scattered at
90° to the direction of the incident beam.[11] n_0 is measured with a refrac-
tometer and dn/dC with special differential refractometers. The correction α follows from measurements of the intensities of the scattered light
at angles of 45° and 135° to the primary beam[12] and β from the ratio of
the horizontal to the vertical components of the light scattered by the
solute at 90° to the direction of the incident beam.

The procedure for getting M from these data is as follows: First are
calculated the quantities HC/τ; these are then plotted against C, and the
plot extrapolated to $C = 0$. According to Eq. (18) this extrapolated value
of HC/τ is $1/M'$, and hence M' is obtained. The latter, used along with the
values of α and β in Eq. (20), gives then the correct molecular weight M.

In Figure 21–8 is shown a plot of HC/τ vs. C for polystyrene in methyl
ethyl ketone[13] at 25°C. When $C = 0$ this plot yields $1/M' = 1.19 \times 10^{-6}$,

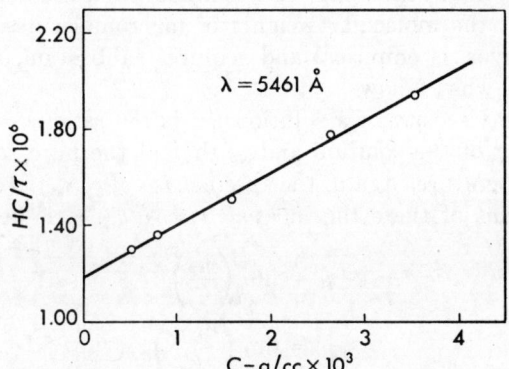

Figure 21–8. HC/τ plot for polystyrene in methyl ethyl ketone at 25°C.

and therefore, $M' = 0.840 \times 10^6$. For these solutions $\alpha = 1.25$ and $\beta = 0.961$. Hence

$$M = 0.840 \times 10^6 (1.25)(0.961)$$
$$= 1.01 \times 10^6 \text{ g mole}^{-1}$$

The average obtained by light scattering for M is different from that
given by osmotic pressure and is defined by

$$M_w = \frac{\Sigma n_i M_i^2}{\Sigma n_i M_i} \tag{21}$$

[11] See, for instance, Maron and Lou, *J. Polymer Sci.*, **14**, 29 (1954).
[12] Doty and Steiner, *J. Chem. Phys.*, **18**, 1211 (1950).
[13] Maron and Lou, *J. Polymer Sci.*, **14**, 29 (1954).

where M_w is the *weight average molecular weight*. For polymers containing a distribution of molecular weights $M_w > M_n$. Only when the molecular species are all the same, or when the distribution is very narrow, are M_n and M_w identical.

Light-scattering measurements are considerably more difficult and involved than osmotic pressure determinations. However, the light-scattering method is much more versatile. Thus, whereas osmotic pressure measurements are generally limited to molecular weights of solutes between about 20,000 and 600,000, light scattering has been used to determine molecular weights of substances ranging all the way from sucrose,[14] with $M = 342$, to polymers of M greater than 5,000,000.

MOLECULAR WEIGHTS FROM VISCOSITY MEASUREMENTS

Osmometry, light scattering, and sedimentation are absolute methods for arriving at the molecular weights of macromolecules. The viscosity method, however, is empirical and requires calibration, as will become apparent from what follows.

The relative viscosity of a solution, η_r, is the ratio $\eta_r = \eta/\eta_o$, where η is the viscosity of the solution and η_o that of the pure solvent, both at the same temperature. Again, the *specific viscosity*, η_{sp}, is given by $\eta_{sp} = \eta_r - 1$. In terms of these the *intrinsic viscosity*, $[\eta]$, is defined as either

$$[\eta] = \lim_{C \to 0} \left(\frac{\eta_{sp}}{C} \right) \tag{22}$$

$$[\eta] = \lim_{C \to 0} \left(\frac{\ln \eta_r}{C} \right) \tag{23}$$

where C is the concentration of solute, expressed generally in grams per 100 cc of solution. Both of these expressions lead to the same value for $[\eta]$. Now, the intrinsic viscosity has been shown to be related to the molecular weight M by the expression

$$[\eta] = KM^a \tag{24}$$

where at a given temperature K and a are constants for a specific polymer in a specific solvent. Consequently, once K and a are known for a polymer-solvent combination, M may be calculated from a determined value of $[\eta]$.

To obtain $[\eta]$, viscosities of several dilute solutions of a polymer in a solvent as well as η_0 are measured, and the data are then plotted as either η_{sp}/C or $(\ln \eta_r)/C$ vs. C. Extrapolation to $C = 0$ yields then the intrinsic

[14] Maron and Lou, *J. Phys. Chem.*, **59**, 232 (1955).

viscosity.[15] Such a plot for polyvinyl chloride in cyclohexanone[16] at 25°C is shown in Figure 21–9. In turn, to obtain K and a, intrinsic viscosities are determined for a series of fractionated samples of the polymer. Next the molecular weights of these fractions are ascertained by either light scattering or osmometry. Finally, log $[\eta]$ is plotted vs. log M to yield usually a straight line of slope a and intercept log K.

When the intrinsic viscosities of polymers heterogeneous in molecular weight are substituted into Eq. (24), the M obtained is an average

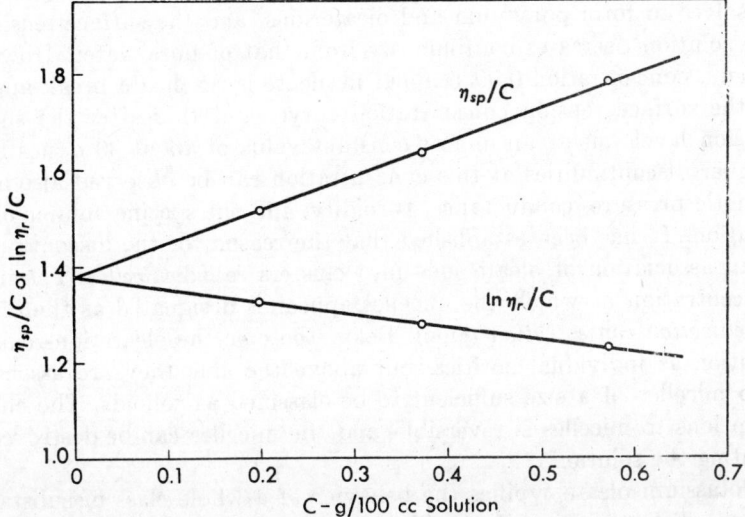

Figure 21–9. Plots of η_{sp}/C and $\ln \eta_r/C$ vs. C for polyvinyl chloride in cyclohexanone at 25°C.

dependent upon the value of a. Most observed values of a lie between 0.5 and 1.0. For $a = 1$, M is the weight average molecular weight M_w. However, for $a < 1$, M lies between M_n and M_w, but much closer to the latter.

Because of its convenience, the viscosity method is employed very extensively for characterization of polymers.

MOLECULAR WEIGHTS OF POLYMERIC ELECTROLYTES

The above methods as described are applicable only to polymers that are nonelectrolytes. When polymeric electrolytes are involved, anomalous

[15] A method for determining intrinsic viscosities from one-point measurements has been described by Maron, *J. Appl. Polymer Sci.*, **5**, 282 (1961).
[16] Mead and Fuoss, *J. Am. Chem. Soc.*, **64**, 277 (1942). In the plot the concentration units given in the paper have been changed to grams per 100 cc of solution.

effects are observed in all the methods unless suitable precautions are taken. In such cases it is necessary to work in the presence of appreciable quantities of added neutral electrolyte, and sometimes also controlled pH, in order to suppress either the effect of the charge or the extent of ionization.

ASSOCIATION COLLOIDS

When potassium oleate is added incrementally to water at 50°C, it dissolves to form potassium and oleate ions, and the surface tension of the solution decreases continuously from that of pure water. However, when a concentration 0.0035 molar in oleate is reached a break appears in the surface tension–concentration curve, and thereafter the surface tension levels off at an almost constant value of about 30 dynes cm^{-1}. Similar discontinuities at this concentration can be observed also in the osmotic pressure, conductance, turbidity, and the specific volume of the solution. It has been established that the reason for the discontinuities is an association of oleate ions into clusters called *micelles*. The initial concentration at which the micelles appear is designated as the *critical micellization concentration* (cmc). Below the cmc the oleate ions exist in solution as individual entities,˙but above the cmc they are associated into micelles of a size sufficient to be classified as colloids. The change from ions to micelles is reversible, and the micelles can be destroyed by diluting the solution.

Potassium oleate typifies the behavior of a whole class of substances known as *association colloids*. These include the soaps, higher alkyl sulfates, sulfonates, amine salts, certain dyes, long chain esters of glycerides, and polyethylene oxides. Some of these substances yield anions which micellize, like the soaps, sulfates, and sulfonates. Others, like the amine salts, yield micellizing cations. Finally, substances like the polyethylene oxides are nonionic, and in these the whole molecule undergoes micellization. These various types are referred to as anionic, cationic, and nonionic association colloids.

The micellization of an association colloid occurs at a definite concentration at each temperature. Increase in temperature raises the cmc; and, if the temperature is made high enough micellization can actually be prevented in some cases. On the other hand, presence of electrolytes has a tendency to lower the cmc.

The molecular weights of some association colloids at the cmc have been determined by light scattering. The values found range from about 10,000–30,000 g mole^{-1}. Compared with dispersed particles and polymer solutes, micelles represent, therefore, rather small aggregates.

Association colloids are of great practical utility because of the properties which they exhibit. Many anionic, cationic, and nonionic colloids are excellent emulsifying agents, detergents, and dispersion stabilizers. Some also are excellent solubilizers of many types of organic compounds in water. Finally, micellizing substances have been found to be of great importance in emulsion polymerization, where the presence of micelles has been shown to be essential for rapid initiation of the process.

REFERENCES

1. Alexander and Johnson, *Colloid Science*, Oxford University Press, New York, 1949.
2. F. W. Billmeyer, *Textbook of Polymer Science*, Interscience Publishers, Inc., New York, 1962.
3. P. Flory, *Principles of Polymer Chemistry*, Cornell University Press, Ithaca, N.Y., 1953.
4. Jirgensons and Straumanis, *Colloid Chemistry*, The Macmillan Company, New York, 1962.
5. H. R. Kruyt, *Colloid Science*, Elsevier Publishing Company, New York, 1952.
6. Mark and Tobolsky, *Physical Chemistry of High Polymeric Systems*, Interscience Publishers, Inc., New York, 1950.
7. J. W. McBain, *Colloid Science*, D. C. Heath and Company, Boston, 1950.
8. K. A. Stacey, *Light Scattering in Physical Chemistry*, Academic Press Inc., New York, 1956.
9. A. Weissberger, *Physical Methods of Organic Chemistry*, Interscience Publishers, Inc., New York, 1960, Chaps. 5, 12, 15, 16, 17, and 24.

PROBLEMS

1. The refractive index of water at 25°C is given by the equation

$$n_0^2 = 1.7521 + \frac{8.11 \times 10^5}{\lambda_0^2}$$

where λ_0 is the wave length of light in vacuo expressed in Å. Find the wave length of the light in water when λ_0 is (a) 4000 Å, and (b) 10,000 Å. *Ans.* (a) 2978 Å.

2. A colloidal dispersion ($m = 1.17$) shows a first minimum in the intensity of the vertical component of scattered light at an angle of 37° when light of $\lambda_m = 4094$ Å is used. Find the diameter of the scattering particles.

3. In a particular light scattering photometer 160° is the largest angle at which a minimum in $I_{\theta v}$ can be observed. What is the diameter of the smallest particles which can be measured in this instrument by the minimum intensity method at (a) $m = 1.20$ and (b) $m = 1.50$ when $\lambda_m = 4000$ Å? *Ans.* (a) 2620 Å.

4. Suppose that in a particular light scattering photometer 10° is the smallest angle at which a minimum in intensity can be detected. What is the diameter of the largest particles which can be measured in this instrument by the minimum intensity method with light of $\lambda_m = 4000$ Å at (a) $m = 1.20$ and (b) $m = 1.50$?

5. The average diameter of the spherical latex particles shown in Figure 21-2 is 2300 Å, while their density is 0.930 g/cc. What is the molecular weight of the latex particles?

6. A solution containing 0.2 mg/liter of a suspended material whose density is 2.2 g/cc is observed under the ultramicroscope in a field of view 0.04 mm in diameter and 0.03 mm in depth. On the average the field of view was found to contain 8.5 particles. Assuming them to be spherical, what is the diameter of the particles?   *Ans.* 9.2×10^{-6} cm.

7. A certain substance in colloidal dispersion is found to adsorb at 50°C 2.45×10^{-4} mole of sodium oleate per gram. The particles in dispersion have a density of 0.930 and are known to be spherical. If the area of the sodium oleate molecule is 28.2 Å², find (a) the surface area of the solid in Å² per gram, and (b) the diameter of the particles in Å.

Ans. Area = 4.16×10^{21} Å²; D = 1550 Å.

8. How many dispersed particles of the substance described in problem **7** will be present in 1 ml of a solution containing 10 g of disperse phase per liter of dispersion?

9. Using the data listed in Table 21-4, estimate the freezing point and osmotic pressure at 27°C of a solution containing 1 g of egg albumin per 100 g of H_2O.

10. Calculate the freezing point depression and osmotic pressure at 20°C of a solution containing 1 g of polystyrene (M = 200,000 g mole^{-1}) in 100 cc of benzene. Assume the density of the solution to be the same as that of benzene, 0.879 g/cc.

11. The mean diameter of the particles in a certain aqueous colloidal solution is 42 Å. Assuming the viscosity of the solution to be the same as that of pure water, calculate the diffusion coefficient at 25°C.

Ans. $D_f = 1.16 \times 10^{-6}$ cm² sec^{-1}.

12. What will be the average displacement along an x axis produced by Brownian motion in 1 sec on the particles described in problem **11**?

13. A portion of a certain suspension at 20°C requires 80 min to fall under the influence of gravity through a column of H_2O 1 m in height. If the density of the suspended particles is 2.0 g/cc, what is the radius of the individual particles? Assume the viscosity to be that of water.   *Ans.* 9.83×10^{-4} cm.

14. How many times that of gravity will be the acceleration of a body revolving at 10,000 rpm at a distance of 5 cm from the axis of rotation?

15. In measuring particle size in an aqueous suspension at 25°C by the sedimentation velocity method in a centrifugal field, the boundary of the suspension moves from a position 10 cm from the axis of rotation to a position 16 cm away in a period of 30.0 min when the centrifuge performs 20,000 rpm. If the density of the suspended material is 1.25 g/cc, what is the radius of the suspended particles?   *Ans.* 3.10×10^{-6} cm.

16. The molecular weight of a substance was determined by sedimentation equilibrium. After equilibrium was established it was found that the aqueous sol was 1.90 times as concentrated at a distance 10.0 cm from the axis of rotation as at a distance of 6.0 cm. The rotor was run at 6000 rpm, the temperature was 25°C, and the density of the dispersed substance was 1.20 g/cc. From these data find the molecular weight of the substance.

17. In an electrophoretic mobility experiment it is observed that a colloid moves toward the negative electrode a distance of 3.82 cm in 60 min when the potential gradient is 2.10 volts/cm. Calculate the electrophoretic mobility of the colloid.

18. Assuming the polybutadiene molecule to be linear, and its cross-sectional area to be 20 Å^2, calculate the fully extended length in Å of the molecules in the polymer when the molecular weight is 100,000 g mole^{-1}. The density of polybutadiene may be taken as 0.920 g cc^{-1}.   *Ans.* 9030 Å.

19. Using the data given in problem 18, and assuming the molecules to be spherical, calculate the diameter in Å of each polybutadiene sphere. Consider the volume occupied by the spheres to be 74% of the total volume.

20. For a certain polymer dissolved in CCl$_4$ were obtained the following osmotic pressure data at 20°C:

C (g/cc)	0.00200	0.00400	0.00600	0.00800
Δh (cm of CCl$_4$)	0.40	1.00	1.80	2.80

The density of CCl$_4$ at 20°C is 1.594 g/cc. Find the molecular weight of the polymer.   *Ans.* $M = 1.04 \times 10^5$ g mole^{-1}.

21. At a wave length of 5461 Å and a temperature of 25°C, Maron and Lou obtained the following light-scattering data for sucrose solutions in water:

C (g/cc)	0.0352	0.0614	0.106	0.163
$(HC/\tau) \times 10^3$	2.84	2.91	3.08	3.32

For sucrose solutions $\alpha = 1.00$ and $\beta = 0.935$. Determine the molecular weight of the sucrose, and compare with the expected value of 342 g mole^{-1}.

22. At 25°C and a wave length of 4358 Å, the refractive index, n, of sucrose solutions can be represented by the equation

$$n = 1.3397 + 0.1453\ C - 0.00265\ C^2$$

where C is the concentration in grams per cubic centimeter. Find the value of H in Eq. (19) for a sucrose solution of $C = 0.100$ g/cc.   *Ans.* $H = 5.74 \times 10^{-6}$.

23. At 25°C and $\lambda_0 = 4358$ Å, the following light-scattering data were obtained for a sample of polystyrene in toluene:

$C \times 10^3$ (g/cc)	0.750	1.670	3.330	5.000
$(HC/\tau) \times 10^6$	2.94	3.51	4.87	6.27

For these solutions $\alpha = 1.248$ and $\beta = 0.970$. Find the molecular weight of the polystyrene.

24. A polymer contains the following distribution of molecular weights:

f_n	0.10	0.30	0.40	0.10	0.10
$M \times 10^{-5}$ (g mole^{-1})	1	2	3	4	6

where f_n is the fraction of the total number of molecules. Find the number and weight average molecular weights of the polymer.

25. The following relative viscosities were measured for a polymer in a given solvent at 25°C:

C (g/100 cc)	0.152	0.271	0.541
η_r	1.226	1.425	1.983

Find the intrinsic viscosity of the polymer.   *Ans.* $[\eta] = 1.36$ deciliters g^{-1}.

26. For polyisobutylene in cyclohexane at 30°C $[\eta] = 2.60 \times 10^{-4} M^{0.70}$. What is the molecular weight of a polymer whose intrinsic viscosity in cyclohexane at 30°C is 2.00 deciliters g^{-1}?

27. Fractions of a polymer, when dissolved in an organic solvent, gave the following intrinsic viscosities at 25°C:

M (g mole^{-1})	34,000	61,000	130,000
$[\eta]$	1.02	1.60	2.75

Determine a and K in Eq. (24) for this system.   *Ans.* $a = 0.73; K = 5.1 \times 10^{-4}$.

Index

ATOMIC WEIGHTS—1961* (Basis: $C^{12} = 12.0000$)

Element	Symbol	Atomic No.	Atomic Weight	Element	Symbol	Atomic No.	Atomic Weight
Actinium	Ac	89	—	Mercury	Hg	80	200.59
Aluminum	Al	13	26.9815	Molybdenum	Mo	42	95.94
Americium	Am	95	—	Neodymium	Nd	60	144.24
Antimony	Sb	51	121.75	Neon	Ne	10	20.183
Argon	Ar	18	39.948	Neptunium	Np	93	—
Arsenic	As	33	74.9216	Nickel	Ni	28	58.71
Astatine	At	85	—	Niobium	Nb	41	92.906
Barium	Ba	56	137.34	Nitrogen	N	7	14.0067
Berkelium	Bk	97	—	Nobelium	No	102	—
Beryllium	Be	4	9.0122	Osmium	Os	76	190.2
Bismuth	Bi	83	208.980	Oxygen	O	8	15.9994ᵃ
Boron	B	5	10.811ᵃ	Palladium	Pd	46	106.4
Bromine	Br	35	79.909ᵇ	Phosphorus	P	15	30.9738
Cadmium	Cd	48	112.40	Platinum	Pt	78	195.09
Calcium	Ca	20	40.08	Plutonium	Pu	94	—
Californium	Cf	98	—	Polonium	Po	84	—
Carbon	C	6	12.01115ᵃ	Potassium	K	19	29.102
Cerium	Ce	58	140.12	Praseodymium	Pr	59	140.907
Cesium	Cs	55	132.905	Promethium	Pm	61	—
Chlorine	Cl	17	35.453ᵇ	Protactinium	Pa	91	—
Chromium	Cr	24	51.996ᵇ	Radium	Ra	88	—
Cobalt	Co	27	58.9332	Radon	Rn	86	—
Copper	Cu	29	63.54	Rhenium	Re	75	186.2
Curium	Cm	96	—	Rhodium	Rh	45	102.905
Dysprosium	Dy	66	162.50	Rubidium	Rb	37	85.47
Einsteinium	Es	99	—	Ruthenium	Ru	44	101.07
Erbium	Er	68	167.26	Samarium	Sm	62	150.35
Europium	Eu	63	151.96	Scandium	Sc	21	44.956
Fermium	Fm	100	—	Selenium	Se	34	78.96
Fluorine	F	9	18.9984	Silicon	Si	14	28.086ᵃ
Francium	Fr	87	—	Silver	Ag	47	107.870ᵇ
Gadolinium	Gd	64	157.25	Sodium	Na	11	22.9898
Gallium	Ga	31	69.72	Strontium	Sr	38	87.62
Germanium	Ge	32	72.59	Sulfur	S	16	32.064ᵃ
Gold	Au	79	196.967	Tantalum	Ta	73	180.948
Hafnium	Hf	72	178.49	Technetium	Tc	43	—
Helium	He	2	4.0026	Tellurium	Te	52	127.60
Holmium	Ho	67	164.930	Terbium	Tb	65	158.924
Hydrogen	H	1	1.00797ᵃ	Thallium	Tl	81	204.37
Indium	In	49	114.82	Thorium	Th	90	232.038
Iodine	I	53	126.9044	Thulium	Tm	69	168.934
Iridium	Ir	77	192.2	Tin	Sn	50	118.69
Iron	Fe	26	55.847ᵇ	Titanium	Ti	22	47.90
Krypton	Kr	36	83.80	Tungsten	W	74	183.85
Lanthanum	La	57	138.91	Uranium	U	92	238.03
Lead	Pb	82	207.19	Vanadium	V	23	50.942
Lithium	Li	3	6.939	Xenon	Xe	54	131.30
Lutetium	Lu	71	174.97	Ytterbium	Yb	70	173.04
Magnesium	Mg	12	24.312	Yttrium	Y	39	88.905
Manganese	Mn	25	54.9380	Zinc	Zn	30	65.37
Mendelevium	Md	101	—	Zirconium	Zr	40	91.22

* Because of indefinite isotopic composition, no atomic weights are given for a number of the radioactive elements.

ᵃ The atomic weight varies because of natural variations in the isotopic composition of the element. The observed ranges are: Boron, ±0.003; carbon, ±0.00005; hydrogen, ±0.00001; oxygen, ±0.0001; silicon, ±0.001; sulfur, ±0.003.

ᵇ The atomic weight is believed to have the following experimental uncertainty: Bromine, ±0.002; chlorine, ±0.001; chromium, ±0.001; iron, ±0.003; silver, ±0.003. For other elements the last digit is believed to be reliable to ±0.5.

PHYSICO-CHEMICAL CONSTANTS*

Acceleration of gravity, standard	980.665 cm sec^{-2}
Atmosphere, normal	1.013250 $\times$ 10^6 dynes cm^{-2}
Avogadro's number, N	6.02252 $\times$ 10^{23} mole^{-1}
Bohr magneton, μ_B	9.2732 $\times$ 10^{-21} erg gauss^{-1}
Bohr radius, a_0	0.529167 $\times$ 10^{-8} cm
Boltzmann constant, k	1.38054 $\times$ 10^{-16} erg degree^{-1}
Calorie, defined	4.1840 $\times$ 10^7 ergs
Electron charge, e	4.80298 $\times$ 10^{-10} esu
" "	1.60210 $\times$ 10^{-20} emu
" "	1.60210 $\times$ 10^{-19} abs. coulomb
Electron rest mass, m	9.1091 $\times$ 10^{-28} g
e/m for electron	1.75880 $\times$ 10^8 abs. coulombs g^{-1}
Electron volt, eV	1.60210 $\times$ 10^{-12} erg molecule^{-1}
" "	23,061 cal mole^{-1}
Faraday, $\mathcal{F}$	96,487 abs. coulombs equiv^{-1}
Gas Constant, R	8.3143 $\times$ 10^7 ergs mole^{-1} degree^{-1}
" "	8.3143 joules mole^{-1} degree^{-1}
" "	1.9872 cal mole^{-1} degree^{-1}
" "	0.082054 lit-atm mole^{-1} degree^{-1}
Ice Point, 0°C	273.150°K
Neutron rest mass	1.67482 $\times$ 10^{-24} g
Planck's constant, h	6.6256 $\times$ 10^{-27} erg sec
Proton rest mass	1.67252 $\times$ 10^{-24} g
Ratio of proton to electron mass	1836.1
Rydberg constant, R_H	1.0967758 $\times$ 10^5 cm^{-1}
Rydberg constant, R_∞	1.0973731 $\times$ 10^5 cm^{-1}
Triple point of H_2O, defined	273.16000°K
Speed of light in vacuo, c	2.997925 $\times$ 10^{10} cm sec^{-1}
Volume of ideal gas, standard conditions	22.4136 lit mole^{-1}
1 international ampere	0.999835 abs. ampere
1 international coulomb	0.999835 abs. coulomb
1 international watt	1.000165 abs. watts
1 international joule	1.000165 abs. joules

* Constants based on atomic weight of C^{12} = 12.0000. Compiled from consistent set of physical constants proposed by committee on fundamental constants of the National Academy of Sciences—National Research Council, and adopted by National Bureau of Standards, U.S.A. See Nat. Bur. Stds. Tech. News., 47, 175 (1963); Chem. & Eng. News, 41, 43 (1963).

LOGARITHMS

Natural Numbers	0	1	2	3	4	5	6	7	8	9	PROPORTIONAL PARTS								
											1	2	3	4	5	6	7	8	9
10	0000	0043	0086	0128	0170	0212	0253	0294	0334	0374	4	8	12	17	21	25	29	33	37
11	0414	0453	0492	0531	0569	0607	0645	0682	0719	0755	4	8	11	15	19	23	26	30	34
12	0792	0828	0864	0899	0934	0969	1004	1038	1072	1106	3	7	10	14	17	21	24	28	31
13	1139	1173	1206	1239	1271	1303	1335	1367	1399	1430	3	6	10	13	16	19	23	26	29
14	1461	1492	1523	1553	1584	1614	1644	1673	1703	1732	3	6	9	12	15	18	21	24	27
15	1761	1790	1818	1847	1875	1903	1931	1959	1987	2014	3	6	8	11	14	17	20	22	25
16	2041	2068	2095	2122	2148	2175	2201	2227	2253	2279	3	5	8	11	13	16	18	21	24
17	2304	2330	2355	2380	2405	2430	2455	2480	2504	2529	2	5	7	10	12	15	17	20	22
18	2553	2577	2601	2625	2648	2672	2695	2718	2742	2765	2	5	7	9	12	14	16	19	21
19	2788	2810	2833	2856	2878	2900	2923	2945	2967	2989	2	4	7	9	11	13	16	18	20
20	3010	3032	3054	3075	3096	3118	3139	3160	3181	3201	2	4	6	8	11	13	15	17	19
21	3222	3243	3263	3284	3304	3324	3345	3365	3385	3404	2	4	6	8	10	12	14	16	18
22	3424	3444	3464	3483	3502	3522	3541	3560	3579	3598	2	4	6	8	10	12	14	15	17
23	3617	3636	3655	3674	3692	3711	3729	3747	3766	3784	2	4	6	7	9	11	13	15	17
24	3802	3820	3838	3856	3874	3892	3909	3927	3945	3962	2	4	5	7	9	11	12	14	16
25	3979	3997	4014	4031	4048	4065	4082	4099	4116	4133	2	3	5	7	9	10	12	14	15
26	4150	4166	4183	4200	4216	4232	4249	4265	4281	4298	2	3	5	7	8	10	11	13	15
27	4314	4330	4346	4362	4378	4393	4409	4425	4440	4456	2	3	5	6	8	9	11	13	14
28	4472	4487	4502	4518	4533	4548	4564	4579	4594	4609	2	3	5	6	8	9	11	12	14
29	4624	4639	4654	4669	4683	4698	4713	4728	4742	4757	1	3	4	6	7	9	10	12	13
30	4771	4786	4800	4814	4829	4843	4857	4871	4886	4900	1	3	4	6	7	9	10	11	13
31	4914	4928	4942	4955	4969	4983	4997	5011	5024	5038	1	3	4	6	7	8	10	11	12
32	5051	5065	5079	5092	5105	5119	5132	5145	5159	5172	1	3	4	5	7	8	9	11	12
33	5185	5198	5211	5224	5237	5250	5263	5276	5289	5302	1	3	4	5	6	8	9	10	12
34	5315	5328	5340	5353	5366	5378	5391	5403	5416	5428	1	3	4	5	6	8	9	10	11
35	5441	5453	5465	5478	5490	5502	5514	5527	5539	5551	1	2	4	5	6	7	9	10	11
36	5563	5575	5587	5599	5611	5623	5635	5647	5658	5670	1	2	4	5	6	7	8	10	11
37	5682	5694	5705	5717	5729	5740	5752	5763	5775	5786	1	2	3	5	6	7	8	9	10
38	5798	5809	5821	5832	5843	5855	5866	5877	5888	5899	1	2	3	5	6	7	8	9	10
39	5911	5922	5933	5944	5955	5966	5977	5988	5999	6010	1	2	3	4	5	7	8	9	10
40	6021	6031	6042	6053	6064	6075	6085	6096	6107	6117	1	2	3	4	5	6	8	9	10
41	6128	6138	6149	6160	6170	6180	6191	6201	6212	6222	1	2	3	4	5	6	7	8	9
42	6232	6243	6253	6263	6274	6284	6294	6304	6314	6325	1	2	3	4	5	6	7	8	9
43	6335	6345	6355	6365	6375	6385	6395	6405	6415	6425	1	2	3	4	5	6	7	8	9
44	6435	6444	6454	6464	6474	6484	6493	6503	6513	6522	1	2	3	4	5	6	7	8	9
45	6532	6542	6551	6561	6571	6580	6590	6599	6609	6618	1	2	3	4	5	6	7	8	9
46	6628	6637	6646	6656	6665	6675	6684	6693	6702	6712	1	2	3	4	5	6	7	7	8
47	6721	6730	6739	6749	6758	6767	6776	6785	6794	6803	1	2	3	4	5	5	6	7	8
48	6812	6821	6830	6839	6848	6857	6866	6875	6884	6893	1	2	3	4	4	5	6	7	8
49	6902	6911	6920	6928	6937	6946	6955	6964	6972	6981	1	2	3	4	4	5	6	7	8
50	6990	6998	7007	7016	7024	7033	7042	7050	7059	7067	1	2	3	3	4	5	6	7	8
51	7076	7084	7093	7101	7110	7118	7126	7135	7143	7152	1	2	3	3	4	5	6	7	8
52	7160	7168	7177	7185	7193	7202	7210	7218	7226	7235	1	2	2	3	4	5	6	7	7
53	7243	7251	7259	7267	7275	7284	7292	7300	7308	7316	1	2	2	3	4	5	6	6	7
54	7324	7332	7340	7348	7356	7364	7372	7380	7388	7396	1	2	2	3	4	5	6	6	7

LOGARITHMS

Natural Numbers	0	1	2	3	4	5	6	7	8	9	PROPORTIONAL PARTS								
											1	2	3	4	5	6	7	8	9
55	7404	7412	7419	7427	7435	7443	7451	7459	7466	7474	1	2	2	3	4	5	5	6	7
56	7482	7490	7497	7505	7513	7520	7528	7536	7543	7551	1	2	2	3	4	5	5	6	7
57	7559	7566	7574	7582	7589	7597	7604	7612	7619	7627	1	2	2	3	4	5	5	6	7
58	7634	7642	7649	7657	7664	7672	7679	7686	7694	7701	1	1	2	3	4	4	5	6	7
59	7709	7716	7723	7731	7738	7745	7752	7760	7767	7774	1	1	2	3	4	4	5	6	7
60	7782	7789	7796	7803	7810	7818	7825	7832	7839	7846	1	1	2	3	4	4	5	6	6
61	7853	7860	7868	7875	7882	7889	7896	7903	7910	7917	1	1	2	3	4	4	5	6	6
62	7924	7931	7938	7945	7952	7959	7966	7973	7980	7987	1	1	2	3	3	4	5	6	6
63	7993	8000	8007	8014	8021	8028	8035	8041	8048	8055	1	1	2	3	3	4	5	5	6
64	8062	8069	8075	8082	8089	8096	8102	8109	8116	8122	1	1	2	3	3	4	5	5	6
65	8129	8136	8142	8149	8156	8162	8169	8176	8182	8189	1	1	2	3	3	4	5	5	6
66	8195	8202	8209	8215	8222	8228	8235	8241	8248	8254	1	1	2	3	3	4	5	5	6
67	8261	8267	8274	8280	8287	8293	8299	8306	8312	8319	1	1	2	3	3	4	5	5	6
68	8325	8331	8338	8344	8351	8357	8363	8370	8376	8382	1	1	2	3	3	4	4	5	6
69	8388	8395	8401	8407	8414	8420	8426	8432	8439	8445	1	1	2	2	3	4	4	5	6
70	8451	8457	8463	8470	8476	8482	8488	8494	8500	8506	1	1	2	2	3	4	4	5	6
71	8513	8519	8525	8531	8537	8543	8549	8555	8561	8567	1	1	2	2	3	4	4	5	5
72	8573	8579	8585	8591	8597	8603	8609	8615	8621	8627	1	1	2	2	3	4	4	5	5
73	8633	8639	8645	8651	8657	8663	8669	8675	8681	8686	1	1	2	2	3	4	4	5	5
74	8692	8698	8704	8710	8716	8722	8727	8733	8739	8745	1	1	2	2	3	4	4	5	5
75	8751	8756	8762	8768	8774	8779	8785	8791	8797	8802	1	1	2	2	3	3	4	5	5
76	8808	8814	8820	8825	8831	8837	8842	8848	8854	8859	1	1	2	2	3	3	4	5	5
77	8865	8871	8876	8882	8887	8893	8899	8904	8910	8915	1	1	2	2	3	3	4	4	5
78	8921	8927	8932	8938	8943	8949	8954	8960	8965	8971	1	1	2	2	3	3	4	4	5
79	8976	8982	8987	8993	8998	9004	9009	9015	9020	9026	1	1	2	2	3	3	4	4	5
80	9031	9036	9042	9047	9053	9058	9063	9069	9074	9079	1	1	2	2	3	3	4	4	5
81	9085	9090	9096	9101	9106	9112	9117	9122	9128	9133	1	1	2	2	3	3	4	4	5
82	9138	9143	9149	9154	9159	9165	9170	9175	9180	9186	1	1	2	2	3	3	4	4	5
83	9191	9196	9201	9206	9212	9217	9222	9227	9232	9238	1	1	2	2	3	3	4	4	5
84	9243	9248	9253	9258	9263	9269	9274	9279	9284	9289	1	1	2	2	3	3	4	4	5
85	9294	9299	9304	9309	9315	9320	9325	9330	9335	9340	1	1	2	2	3	3	4	4	5
86	9345	9350	9355	9360	9365	9370	9375	9380	9385	9390	1	1	2	2	3	3	4	4	5
87	9395	9400	9405	9410	9415	9420	9425	9430	9435	9440	0	1	1	2	2	3	3	4	4
88	9445	9450	9455	9460	9465	9469	9474	9479	9484	9489	0	1	1	2	2	3	3	4	4
89	9494	9499	9504	9509	9513	9518	9523	9528	9533	9538	0	1	1	2	2	3	3	4	4
90	9542	9547	9552	9557	9562	9566	9571	9576	9581	9586	0	1	1	2	2	3	3	4	4
91	9590	9595	9600	9605	9609	9614	9619	9624	9628	9633	0	1	1	2	2	3	3	4	4
92	9638	9643	9647	9652	9657	9661	9666	9671	9675	9680	0	1	1	2	2	3	3	4	4
93	9685	9689	9694	9699	9703	9708	9713	9717	9722	9727	0	1	1	2	2	3	3	4	4
94	9731	9736	9741	9745	9750	9754	9759	9763	9768	9773	0	1	1	2	2	3	3	4	4
95	9777	9782	9786	9791	9795	9800	9805	9809	9814	9818	0	1	1	2	2	3	3	4	4
96	9823	9827	9832	9836	9841	9845	9850	9854	9859	9863	0	1	1	2	2	3	3	4	4
97	9868	9872	9877	9881	9886	9890	9894	9899	9903	9908	0	1	1	2	2	3	3	4	4
98	9912	9917	9921	9926	9930	9934	9939	9943	9948	9952	0	1	1	2	2	3	3	4	4
99	9956	9961	9965	9969	9974	9978	9983	9987	9991	9996	0	1	1	2	2	3	3	3	4